Springer Series in Materials Science

Volume 154

For further volumes:
http://www.springer.com/series/856

The Springer Series in Materials Science covers the complete spectrum of materials physics, including fundamental principles, physical properties, materials theory and design. Recognizing the increasing importance of materials science in future device technologies, the book titles in this series reflect the state-of-the-art in understanding and controlling the structure and properties of all important classes of materials.

Aline Auroux
Editor

Calorimetry and Thermal Methods in Catalysis

Springer

Editor
Aline Auroux
Institut de Recherches sur la
Catalyse et l'Environnement de Lyon
Villeurbanne
France

ISSN 0933-033X ISSN 2196-2812 (electronic)
ISBN 978-3-642-11953-8 ISBN 978-3-642-11954-5 (eBook)
DOI 10.1007/978-3-642-11954-5
Springer Heidelberg New York Dordrecht London

Library of Congress Control Number: 2013944530

Printed on acid-free paper

Springer is part of Springer Science+Business Media (www.springer.com)

Preface

Calorimetry and other thermal methods play an increasingly important role as tools for the study of catalysts, supports, adsorbents, and for the characterization of their surfaces. This makes it particularly timely to collect in a single volume a set of texts on the fundamentals of adsorption and the bases of thermal analysis techniques such as microcalorimetry, differential scanning calorimetry, thermogravimetry, temperature-programmed desorption, temperature-programmed reduction/oxidation, inverse gas chromatography, etc. The use of many of these techniques is now fairly routine, but their application in the domain of catalysis often requires coupling them to other methods such as volumetry, gas chromatography, mass spectrometry, infrared spectroscopy, etc., in order to allow for an in-depth study of the successive stages of the life cycle of a catalyst, from its preparation to its use in the catalytic reaction and finally its regeneration. Tools that allow the measurement of the heats evolved during the various steps of a chemical reaction are particularly useful given the current emphasis on energy efficiency. However, among the various techniques mentioned above, used alone, or in couplings, each have their own strengths and weaknesses that need to be carefully discussed.

This book, based on a series of summer schools held in Lyon every year since 2007, aims to provide students, engineers, and confirmed researchers alike with an introduction to the major thermal analysis techniques used to characterize solid materials and investigate their surface reactivity, including both physical and chemical processes occurring at gas–solid or liquid–solid interfaces. The main topics covered include:

- the basic phenomena (adsorption, competitive adsorption, desorption, thermodynamics, and kinetics) involved at the solid–gas and solid–liquid interfaces;
- the main thermal analysis and calorimetry techniques used to investigate catalytic materials, alone or linked to other techniques, and their relative advantages; including the main types of calorimetric techniques (adsorption calorimetry, flow calorimetry, titration calorimetry, immersion calorimetry, etc.), as well as temperature-programmed desorption, reduction and oxidation, along with a number of examples and a discussion of experimental considerations and constraints;

- applications of these techniques, such as the study of competitive or selective adsorption processes, the characterization of acid/base sites in oxides and zeolites, the adsorption or capture of gas or liquid pollutants (CO, CO_2, VOCs, nicotine, phenol, etc.), and processes for new energies (biodiesel production, hydrogen production and storage, etc.).

We hope that this book will thus serve as a practical guide to end users about selecting and implementing the most appropriate thermal analysis techniques for solving a specific problem.

In closing, I would like to express my sincere gratitude to my colleagues who kindly agreed to write up the various lecture notes and contributions that make up this volume, as well as to all those who helped with the design, redaction, and editing of this book.

Villeurbanne, France Aline Auroux

Contents

Contributors

Aline Auroux Institut de Recherches sur la Catalyse et l'Environnement de Lyon, UMR 5256 CNRS/Université Lyon1, 2 avenue Einstein, 69626 Villeurbanne, France, e-mail: aline.auroux@ircelyon.univ-lyon1.fr

Karine Ballerat-Busserolles Institut de Chimie de Clermont-Ferrand, Clermont Université, Université Blaise Pascal, BP 10048, 63000 Clermont-Ferrand, France, e-mail: karine.ballerat@univ-bpclermont.fr

Jean-Pierre Bellat Laboratoire Interdisciplinaire Carnot de Bourgogne, Université de Bourgogne, UMR 6303 CNRS, 9 avenue A. Savary, BP 47870, 21078 Dijon, France, e-mail: jean-pierre.bellat@u-bourgogne.fr

Simona Bennici Institut de Recherches sur la Catalyse et l'Environnement de Lyon, UMR 5256 CNRS/Université Lyon1, 2 avenue Einstein, 69626 Villeurbanne, France, e-mail: simona.bennici@ircelyon.univ-lyon1.fr

Vera Bolis Dipartimento di Chimica, Università di Torino, Via Pietro Giuria 7, 10125 Turin, Italy, e-mail: vera.bolis@gmail.com

Christian Brylinski Laboratoire des Multimatériaux et Interfaces, Université Claude Bernard, Lyon 1, UMR CNRS 5615 Bât. Berthollet, 43 Bddu 11 Novembre 1918, 69622 Villeurbanne, France, e-mail: christian.brylinski@univ-lyon1.fr

Paolo Carniti Dipartimento di Chimica, Università degli Studi di Milano, Via Camillo Golgi, 19, 20133 Milan, Italy, e-mail: paolo.carniti@unimi.it

Rodica Chiriac Laboratoire des Multimatériaux et Interfaces, Université Claude Bernard, Lyon 1, UMR CNRS 5615 Bât. Berthollet, 43 Bddu 11 Novembre 1918, 69622 Villeurbanne, France, e-mail: rodica.chiriac@univ-lyon1.fr

Jean-Yves Coxam Institut de Chimie de Clermont-Ferrand, Clermont Université, Université Blaise Pascal, BP 10448, 63000 Clermont-Ferrand, France, e-mail: j-yves.coxam@univ-bpclermont.fr

Ljiljana Damjanovic Faculty of Physical Chemistry, University of Belgrade, Studentski trg 12-16, P.O. Box 47, 11158 Belgrade 118, Serbia, e-mail: ljiljana@ffh.bg.ac.rs

Eva Diaz Faculty of Chemistry, Department of Chemical Engineering and Environmental Technology, University of Oviedo, 33006 Oviedo, Spain, e-mail: diazfeva@uniovi.es

Antonella Gervasini Dipartimento di Chimica, Università degli Studi di Milano, Via Camillo Golgi 19, 20133 Milan, Italy, e-mail: antonella.gervasini@unimi.it

Pierre Leparlouer SETARAM Instrumentation, 7 Rue de l'Oratoire, 69300 Caluire-et-Cuire, France, e-mail: leparlouer@setaram.com

Salvador Ordonez Faculty of Chemistry, Department of Chemical Engineering and Environmental Technology, University of Oviedo, 33006 Oviedo, Spain, e-mail: sordonez@uniovi.es

Georgeta Postole Institut de Recherches sur la Catalyse et l'Environnement de Lyon, UMR 5256 CNRS/Université Lyon1, 2 avenue Einstein, 69626 Villeurbanne, France, e-mail: georgeta.postole@ircelyon.univ-lyon1.fr

Vesna Rakic Faculty of Agriculture, Department of Chemistry, University of Belgrade, Nemanjina 6, 11080 Zemun, Serbia, e-mail: vesna.rakic@ffh.bg.ac.rs

Dusan Stosic Institut de Recherches sur la Catalyse et l'Environnement de Lyon, UMR 5256 CNRS/Université Lyon1, 2 avenue Einstein, 69626 Villeurbanne, France, e-mail: dusan.stosic@ircelyon.univ-lyon1.fr

François Toche Laboratoire des Multimatériaux et Interfaces, Université Claude Bernard, Lyon 1, UMR CNRS 5615 Bât. Berthollet, 43 Bddu 11 Novembre 1918, 69622 Villeurbanne, France, e-mail: francois.toche@univ-lyon1.fr

Jerzy Zajac Institut Charles Gerhardt Montpellier, UMR-5253 CNRS-UM2-ENSCM-UM1, C. C. 1502 Place Eugène Bataillon, 34095 Montpellier cedex 5, France, e-mail: jerzy.zajac@univ-montp2.fr

Part I
Fundamentals and Techniques

Chapter 1
Fundamentals in Adsorption at the Solid-Gas Interface. Concepts and Thermodynamics

Vera Bolis

Abstract Some fundamental concepts about the features of a solid material surface and the adsorption at the gas-solid interface are illustrated. The basic tools dealing with the thermodynamics aspects of adsorption processes are also discussed. The stepwise adsorption microcalorimetry technique, which is a tool of greatest quantitative merit in surface chemistry studies, is described in detail through a selection of gas-solid interface systems, taken from different materials science fields. Criteria for discriminating physical and chemical adsorption are given, based on the nature of the forces involved in the process and the heat of adsorption values. The molecular interpretation of the volumetric-calorimetric data, favored by the joint use of adsorption microcalorimetry, spectroscopic and/or *ab initio* modeling techniques, is also stressed by illustrating a number of examples dealing with either physical or associative/dissociative chemical adsorption.

1.1 Introduction

Phenomena taking place at the solid-fluid interface are governed by specific and/or aspecific interactions between the atoms at the solid surface and the molecules approaching the surface from the gas (or the liquid) phase. In particular, heterogeneous catalysis is based on a sequence of steps which involve adsorption of reactants at the surface of the solid material, surface reactions and desorption of final products [1, 2]. On the other hand, the adsorption of (bio)molecules at the solid surface of biomaterials in contact with physiological fluids is recognized to be the initial step of a chain of molecular events leading to the favorable integration of the implanted material [3, 4]. The adsorption features of (probe) molecules on solid surfaces have been

V. Bolis (✉)
Dipartimento di Chimica and NIS Centre of Excellence, Università di Torino,
Via Pietro Giuria 7, 10125 Torino, Italy
e-mail: vera.bolis@gmail.com

A. Auroux (ed.), *Calorimetry and Thermal Methods in Catalysis*,
Springer Series in Materials Science 154, DOI: 10.1007/978-3-642-11954-5_1,

studied over the years by a variety of different techniques, which allowed to describe the nature of the fluid-solid interactions and give insight into the properties of the solid surface [1, 2, 5, 6]. On one hand, spectroscopic techniques (in particular, IR and Raman, UV-vis, NMR, XPS, EXAFS-XANES) are suitable methods to describe the microscopic features of the solid, i.e., the nature and structure of both surface terminations and adsorbed species formed upon contact with molecules [7–9]. On the other hand, adsorption microcalorimetry represents a tool of greatest quantitative merit in that the heat evolved when a fluid contacts the solid surface is related to the nature and energy of the adsorbed species/surface atoms interactions [10–16]. Further, the knowledge of the energetics of chemical and physical events responsible for the process as well as the assessment of the associated thermodynamic parameters contributes to a thorough understanding of phenomena taking place at both catalytic and biological interfaces [13–17]. Coupling the molar volumetric-calorimetric data with the molecular information on the nature of the interaction arising from both spectroscopic methods and *ab initio* computational results, has been proved very fruitful in characterizing in the surface acidity of materials [16, 18–29].

1.2 The Solid Surface

When a molecule (or an atom) from the gas-phase approaches a solid, it is more or less strongly attracted by the atoms exposed at the surface, according to the nature of both the molecule and the solid material.

A crystalline solid is described through the periodic infinite repetition of an elemental pattern (unit cell) [1]. A real solid is however necessarily finite: the periodic repetition of the unit cell terminates generating a surface, the structure of which depends on the cleavage of the crystal, the chemical (either ionic or covalent) nature of the solid and the origin of the surface (either chemical or mechanical). The surface atoms arrangement depends on the plane preferentially exposed during the formation of the surface, according to the preparation conditions of the real material (either in the single crystal form or as nanosized powder) [30]. If no major reconstruction processes are required in order to minimize the surface atoms energy, and if no structural/compositional defects are present, an ideal perfect homogeneous surface is obtained which can be properly represented by cutting a slab of the solid structure. Such an ideal perfect homogeneous surface is very rarely encountered, unless especially prepared for surface science studies. Real solid surfaces (mostly in the case of finely divided, nanometric sized solids) are made up of a combination of flat regions (terraces), structural defects (steps, kinks, corners, edges), point defects (vacancies of ions/atoms in the solid), as schematically illustrated in Fig. 1.1 Compositional defects may contribute to the "imperfections" of the solid surface. They include a variety of oxidation states of the atoms constituting the solid and/or a variety of heteroatoms present either as impurities, or especially introduced in order to modify the physico-chemical properties of the surface. This means that in a real solid

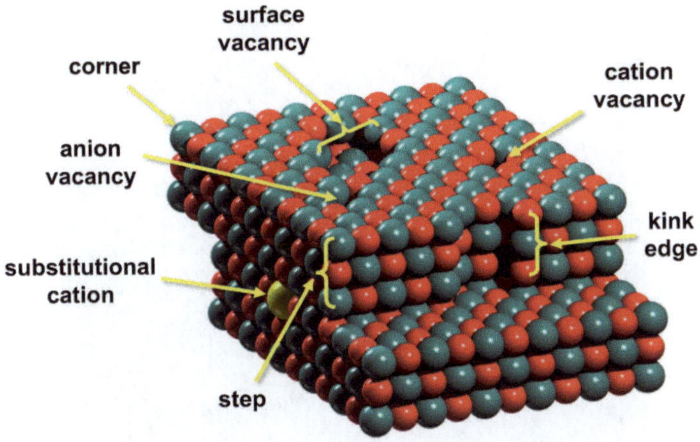

Fig. 1.1 Cartoon of a piece of realistic MgO nanocrystal, which exhibits both structural (steps, kinks edges and corners) and point (anionic and cationic vacancies) defects along the flat regions (terraces). The presence of a compositional defect (substitutional cation) is also outlined. By courtesy of Prof. Piero Ugliengo, University of Torino

a heterogeneous distribution of surface sites potentially active in catalytic reactions and/or in interface processes is generally expected.

Another type of heterogeneous solid surface is represented by an active material dispersed *ad hoc* over the surface of another solid (the support) [31].

In recent years, with the advent of high resolution electron microscopes it has become possible to image atomic details in nanocrystals [9]. Some of the above mentioned structural defects can be imaged by the high resolution transmission electron microscopy (HR-TEM), as illustrated in Fig. 1.2 for monoclinic ZrO_2 nanocrystals, which terminate with structural defects as steps, kinks edges and corners [19].

Owing to their intrinsic coordinative and/or valence unsaturation, species making up such defects act as surface highly reactive sites capable of taking up molecules from the environment.

1.2.1 Porous Materials

Finely divided solids possess not only a geometrical surface, as defined by the different planes exposed by the solid, but also an internal surface due to the primary particles aggregation, which generates pores of different size according to both the nature of the solid and origin of the surface. Pores are classified on the basis of their width *w*, which represents either the diameter of a cylindrical pore, or the distance between the sides of a slit-shaped pore [32]. The smallest pores, characterized by a width $w < 20$ Å (2 nm) are defined *micropores*; the intermediate pores, characterized by a width comprised in the 20 Å $\leq w \leq$ 500 Å (2 and 50 nm) range are classified as

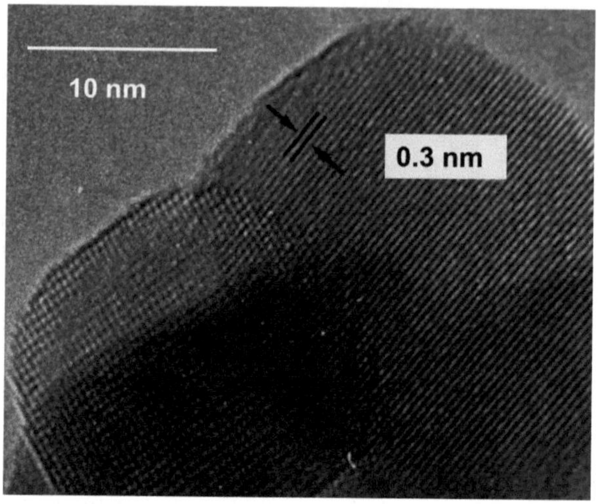

Fig. 1.2 High resolution transmission electron microscopy (HR-TEM) of monoclinic ZrO2 nanocrystals. Adapted from Ref. [19], Fig.2a

mesopores, whereas the largest pores, characterized by a width $w > 500$ Å (50 nm) as *macropores* [31, 32].

Some materials, like charcoal and silico-alumina, have irregular pores with widely variable diameters in a normal shape. Conversely, other materials such as zeolites and clay minerals are entirely micro- or meso-porous, respectively. In these cases, the porosity does not arise from the primary particles aggregation but is an intrinsic structural feature of the solid material [31, 33, 34].

Zeolites are either natural or synthetic crystalline alumino-silicates, the structure of which is based upon a three dimensional polymeric framework, with nanosized cages and channels [1, 2, 33–35]. The basic building block of such materials, of general formula $M^{n+}_{x/n}\left[(AlO_2)_x\,(SiO_2)_y\right]^{x-}\cdot zH_2O$, is the $[TO_4]$ unit with T = Si, Al. This unit is a tetrahedron centered on one T atom bound to four O atoms located at the corners; each O atom is in turn shared between two T atoms. These tetrahedral units join each another through T–O–T linkages in a variety of open-structure frameworks characterized by (interconnected) channels and voids which are occupied by cations and water molecules. The presence of charge-balancing (extra-framework) cations is required in order to compensate the negative charge of the tetrahedral $[AlO_4]^-$ units in which Al is in isomorphous substitution of Si atoms. The density of charge-balancing cations depends upon the $\frac{Si}{Al}$ ratio, which span in the 1 to ∞ range (for $\frac{Si}{Al} \rightarrow \infty$ the so-called all-silica zeolites are obtained) [36–38]. The nature and distribution of extra-framework cations, which are intrinsically mobile and can be exchanged by other cations (including the acidic proton) give specific chemical properties to the material. On the other hand, the presence of nanosized cages and channels within the crystalline structure of zeolites gives to these materials unique molecular sieve

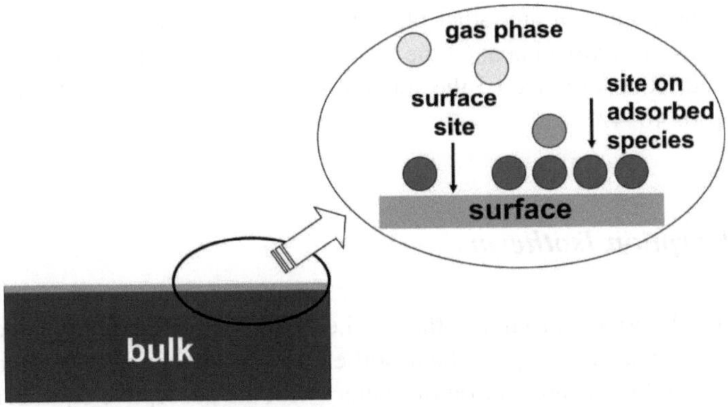

Fig. 1.3 Schematic picture of adsorption process at the surface of a solid material

and shape selectivity properties, of greatest interest in catalysis and in gas separation processes [1, 2, 33, 34, 38, 39].

In all cases, porous materials exhibit high surface areas, which maximize the extension of the interface region.

1.3 Adsorption Processes

The surface atoms of a solid, which are coordinatively unsaturated with respect to the bulk atoms, become saturated thanks to the interaction with molecules of the environment. Adsorption is the process whereby molecules from the gas (or liquid) phase are taken up by a solid surface; it is distinguished from absorption which refers to molecules entering into the lattice (bulk) of the solid material. The adsorptive is the material in the gas phase capable of being adsorbed, whereas the adsorbate is the material actually adsorbed by the solid. The solid, which exposes the surface sites responsible for the process is called the adsorbent. In Fig. 1.3 the adsorption process at the surface of a solid material is schematically illustrated.

Adsorption is governed by either physical or chemical forces. In the former case the adsorption is named physical adsorption (*physisorption*) whereas in the latter case chemical adsorption (*chemisorption*). Details on the nature of these forces will be dealt in Sect. 1.6, through the description of a selection of examples.

When a solid is exposed in a closed space to a gas at pressure p, the weight of the solid typically increases and the pressure of the gas decreases: the gas is adsorbed by the solid. After a time, the pressure p does not change any more and correspondingly the weight ceases to increase any further: a dynamic equilibrium is reached. The amount of gas adsorbed is experimentally determined: (i) by gravimetry (the increase in weight of the solid is monitored by a spring balance); (ii) by volumetry (the fall in the gas pressure is monitored by manometers/transducer gauges); (iii) by monitoring the change of any other physical parameter related to the adsorption of matter, such as the evolved heat (if the heat of adsorption is known and constant) or the integrated IR absorbance (if the specific molar absorbance of adsorbed species is known) [20, 40–42].

1.3.1 Adsorption Isotherms

Adsorption is described through isotherms, i.e., through the functions connecting the amount of adsorbate taken up by the adsorbent (or the change of any other physical parameter related to the adsorption of matter) with the adsorptive equilibrium pressure p, the temperature T and all other parameters being constant. Below the critical temperature the pressure is properly normalized to the saturation vapor pressure p°, and the adsorbed amounts are so referred to the dimensionless relative pressure, $\frac{p}{p^\circ}$.

Adsorption isotherms are currently classified in five classes (I - V) according to the Brunauer, Deming, Deming, Teller (BDDT) original classification, [43] which is often referred to as the Brunauer, Emmet, and Teller (BET), [44] or simply to as the Brunauer [45] classification. An extra type of isotherm (the stepped Type VI isotherm, which is relatively rare) is also reported. Type IV and V isotherms typically exhibit a hysteresis loop, which is characteristic of porous systems, involving capillary condensation [32].

The fractional coverage θ of the adsorbate, at a given equilibrium pressure p, is defined as the ratio of N_S surface sites occupied by the adsorbate over the total available adsorption sites N, i.e. the total number of substrate surface sites which are active towards the given adsorptive. The first layer of adsorbed phase is due to either *chemisorption* or *physisorption*, or both, according to the nature of the forces governing the adsorbate/adsorbent interactions (*vide infra* Sect. 1.6). Conversely, the second layer is originated by physical forces, similar to the forces that lead to the non-ideal behavior of gases and eventually to the condensation to the liquid. Subsequent layers are expected to approach a liquid-like phase.

When the number of N_S occupied sites matches the number of total available sites N, the adsorbate monolayer is complete ($\theta = 1$). In Fig. 1.4 the formation of subsequent layers of adsorbate at the surface of a solid sample is schematically illustrated.

The amount of gas taken up by a solid surface depends upon the solid and the gas nature, the pressure p of the gas and the temperature T. The uptake being proportional to the mass m and the surface area A of the sample, adsorbed amounts (often expressed as mass or volume *STP* of the gas) are properly normalized either to the unit mass or to the unit surface area. Here, in view of describing the process at molecular details, the adsorbed amounts n_{ads} are properly expressed as adsorbate moles (or molecules) per either unit mass or unit surface area of the adsorbent.

As an example, in Fig. 1.5 the equilibrium data for CO adsorbed at $T = 303$ K on Na− and K−MFI zeolites, are reported as volumetric (Fig. 1.5a) and calorimetric (Fig. 1.5b) isotherms. In volumetric isotherms the adsorbed amounts (n_{ads}), in calori-

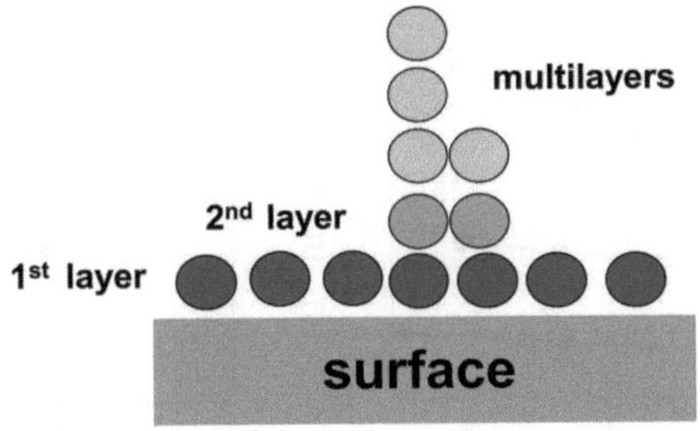

Fig. 1.4 Schematic illustration of the first and second layer of adsorption; the multi-layers adsorption approaches a liquid-like phase

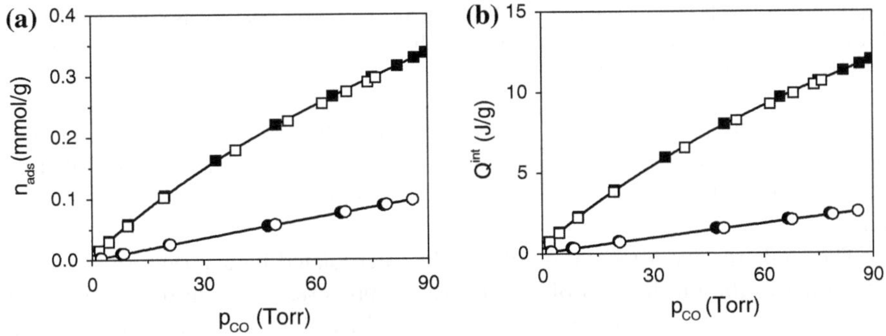

Fig. 1.5 Adsorption of CO at $T = 303$ K on Na–MFI (*square*) and K–MFI (*circle*) zeolites outgassed at $T = 673$ K. **a** Volumetric isotherms (adsorbed amounts vs. equilibrium pressure). **b** Calorimetric isotherms (evolved heats vs. equilibrium pressure). *Solid symbols* first run, *open symbols* second run of adsorption. Experimental points interpolated by the Langmuir equation (*vide infra*)

metric isotherms the integral heat (Q^{int}) evolved during the process are plotted as a function of the CO equilibrium pressure (p_{CO}), expressed in Torr (1 Torr = 133.3 Pa).

In this particular case, the first (1st) and second (2nd) run isotherms are virtually coincident indicating that CO adsorption was entirely reversible upon evacuation of the CO equilibrium pressure. For the experimental and samples details *vide infra* Sect. 1.4. It is here only recalled that the 2nd run isotherms were performed after the overnight outgassing of the reversible adsorbed phase. The isotherms experimental points were interpolated by the Langmuir model equation (*vide infra*).

As an example of the influence of the adsorption temperature, the equilibrium data for water (H_2O) adsorbed at $T = 303$, 353 and 423 K on a H-BEA zeolite specimen (outgassed at $T = 873$ K) are illustrated in Fig. 1.6. In Fig. 1.6a the three volumetric isotherms are reported (for experimental and samples details *vide infra* Sect. 1.4): as far as the equilibrium pressure p_{H2O} increases, the adsorbed amounts also increase more or less steeply according to the adsorption temperature. In Fig. 1.6b the amounts adsorbed at a constant equilibrium pressure ($p_{H2O} = 6$ Torr) are plotted against the adsorption temperature giving rise to an adsorption isobar (n_{ads} vs. T_{ads}). Note that

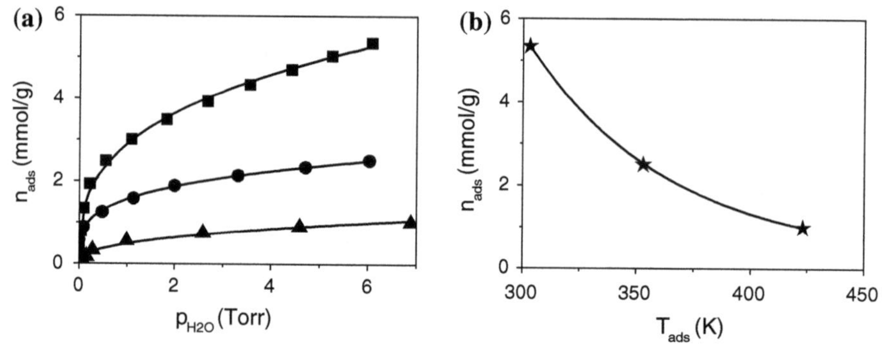

Fig. 1.6 **a** 1st run adsorption volumetric isotherms of H_2O on H-BEA zeolite at $T = 303\,K$ (*square*), $T = 353\,K$ (*circle*) and at $T = 423\,K$ (*triangle*). Experimental points interpolated by the Freundlich equation (*vide infra*). **b** Adsorption isobar (n_{ads} vs. T_{ads}) at constant $p_{H2O} = 6$ Torr; best fitting by exponential decay

in this case the adsorption process was depressed by the increasing temperature, according to the enhanced mobility of the adsorptive molecules. In all isotherms, the experimental points were interpolated by using the Freundlich isotherm equation (*vide infra*).

The isotherm experimental points can be interpolated by a variety of equations according to the mechanism of the adsorption process, which in turn depends on the nature of the gas/solid interaction. Among the different equations proposed to describe quantitatively the isotherms, the only one based on a physical model is the Langmuir equation [1, 2, 30, 32, 46].

The Langmuir model assumes a dynamic equilibrium at constant T between the gas (at pressure p) and the adsorbed layer, and requires a number of well-defined conditions: (i) the adsorption, which is limited to a monolayer, takes place at a surface consisting of a distribution of energetically equivalent, non-interacting sites; (ii) the ability of a molecule to bind at a site is independent of whether or not a nearby site is occupied (absence of lateral interactions); (iii) once adsorbed, the molecules are localized in that the activation barrier hindering migration to an adjacent site is supposed to be much larger than kT; (iv) the enthalpy of adsorption $\Delta_a H$ (per site) is constant with θ.

The Langmuir adsorption isotherm is derived from a kinetic mechanism. Let us assume for sake of simplicity that the molecule M is adsorbed molecularly (i.e. without rupture/formation of chemical bonds) from the gas at a surface site S. The fractional monolayer coverage of the sites occupied by adsorbate molecules is $\theta = \frac{N_S}{N}$. The rate of adsorption is given by the Eq. 1.1:

$$\text{adsorption rate} = k_a p\,(1 - \theta) \tag{1.1}$$

k_a being the rate constant for the adsorption and $(1 - \theta)$ the fractional monolayer coverage of sites not occupied yet by the adsorbate molecules.

The rate of desorption, k_d being the rate constant for desorption, is given by the Eq. 1.2:

$$\text{desorption rate} = k_d\theta \tag{1.2}$$

When the dynamic equilibrium is reached (adsorption rate = desorption rate) the Eq. 1.3 is obtained:

$$k_a p\,(1-\theta) = k_d\theta \tag{1.3}$$

Equation 1.4 represents the Langmuir equation:

$$\frac{\theta}{1-\theta} = \mathrm{K}p \tag{1.4}$$

Note that the constant K is obtained by the ratio of the rate constant for adsorption over the rate constant for desorption $\left(\frac{k_a}{k_d}\right)$.

The Langmuir equation is often written as reported by the Eq. 1.5

$$\theta = \frac{V}{V_{mon}} = \frac{Kp}{(1+Kp)} \tag{1.5}$$

The term V represents the adsorbate volume and V_{mon} the monolayer volume, i.e. the volume of adsorbate required to complete the monolayer.

At very low pressure the equation reduces to a linear dependence of the coverage upon the equilibrium pressure ($\theta = hp$). Conversely, at high pressure the equation reduces to the case of coverage approaching the monolayer ($\theta \approx 1$).

The monolayer coverage (V_{mon}) is hardly determined experimentally with accuracy. So, for practical purposes the Langmuir equation is suitably transformed in the so-called reciprocal linear form, as illustrated by the Eq. 1.6:

$$\left(\frac{1}{V}\right) = \frac{1}{KV_{mon}}\left(\frac{1}{p}\right) + \frac{1}{V_{mon}} \tag{1.6}$$

For isotherms obeying the Langmuir model, the reciprocal volume $\left(\frac{1}{V}\right)$ against reciprocal pressure $\left(\frac{1}{p}\right)$ plot is linear (Langmuir-type isotherms). Conversely, if the experimental data plot is not linear, Langmuir equation does not hold in describing the given adsorption process.

The monolayer capacity is obtained from the intercept $i = \frac{1}{V_{mon}}$ of the straight line. Once determined V_{mon}, the equilibrium constant K is obtained by the slope $s = \frac{1}{KV_{mon}}$ of the plot.

The monolayer volume and the equilibrium constant are typical of the adsorbent/adsorbate pairs at a given temperature. In particular, the value of K is bound to the strength of the adsorbent-adsorbate interaction: high values of K indicate large strength, low values little strength.

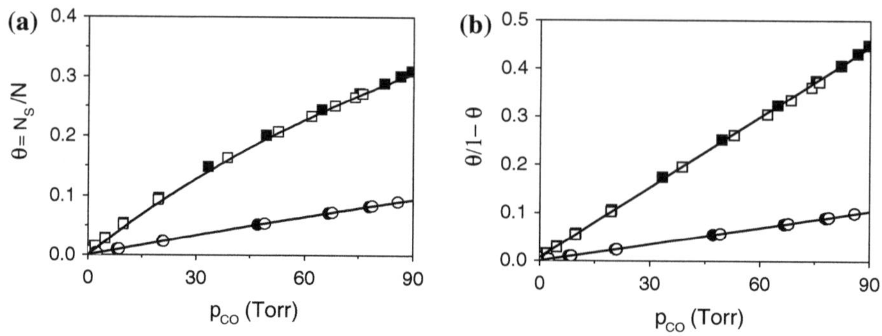

Fig. 1.7 Adsorption of CO at $T = 303\,\text{K}$ on dehydrated zeolites Na–MFI (*square*) and K–MFI (*circle*). *Solid symbols* 1st run, *open symbols* 2nd run of adsorption. **a** θ_p versus p_{CO} plot. **b** $\frac{\theta}{1-\theta}$ *versus* p_{CO} plot

In the following, the very simple case of CO adsorbed at $T = 303\,\text{K}$ on dehydrated Na− and K−MFI will be discussed (*vide supra* in Fig. 1.5 the experimental volumetric and calorimetric isotherms). The number of CO molecules adsorbed per gram of zeolite at p_{CO} represents the number of occupied sites (N_S), whereas the number of charge-balancing cations exposed per gram of zeolite represents the total available sites (N).

In Fig. 1.7, the coverage $\theta = \frac{N_S}{N}$ (Fig. 1.7a) and the $\frac{\theta}{1-\theta}$ quantity (Fig. 1.7b) are plotted against p_{CO}. The slope of the $\frac{\theta}{1-\theta}$ versus p_{CO} plot is the Langmuir constant K (see Eq. 1.4).

CO is a soft Lewis base which is easily polarized by the electrostatic field generated by the extra-framework alkaline-metal cations located in the MFI zeolite nanocavities. As a consequence, it is reversibly taken up by the surface when put in contact with the activated zeolite [23].

The equilibrium constant K for Na–MFI ($4.88 \pm 0.02\ 10^{-3}\ \text{Torr}^{-1}$) is larger than for K–MFI ($1.15 \pm 0.02\ 10^{-3}\ \text{Torr}^{-1}$), in agreement with the different polarizing power of the cations. In fact, the local electric field generated by the coordinatively unsaturated (*cus*) cations depends on the charge/ionic radius ratio, which is larger for Na^+ than for K^+, the ionic radius of the former being 0.97 Å and that of the latter 1.33 Å [47]. Note also that, according to the charge/ionic radius ratio, the maximum coverage attained at $p_{CO} = 90$ Torr was larger for Na–MFI ($\theta \approx 0.3$) than for K–MFI ($\theta \approx 0.1$).

The standard free energy $\Delta_a G^\circ$ for CO adsorption at the two alkaline-metal sites is obtained by the Langmuir equilibrium constant K by employing the Eq. 1.7:

$$\Delta_a G^\circ = -RT \ln K \tag{1.7}$$

In both cases the adsorption process in standard conditions is endoergonic, being $\Delta_a G^\circ = +13.4\,\text{kJ mol}^{-1}$ for Na−MFI and $+17.0\,\text{kJ mol}^{-1}$ for K−MFI. The endoergonic character of the process is witnessed by the non-spontaneity of the adsorption unless a CO pressure is applied. In fact, by evacuating the CO pressure, the electrostatic $Na^+ \cdots CO$ and $K^+ \cdots CO$ adspecies are completely destroyed, as confirmed by the overlap of the 1st and 2nd run of adsorption (see Figs. 1.5 and 1.7).

From $\Delta_a G^\circ$ the standard entropy of adsorption $\Delta_a S^\circ$ is obtained, if $\Delta_a H^\circ$ is known. The CO adsorption enthalpy change was measured calorimetrically during the same experiments in which the adsorbed amounts were measured (*vide infra*, Sect. 1.4.2.3). We will come back to this point and to the evaluation of the $\Delta_a S^\circ$ in the section devoted to the entropy of adsorption (*vide infra*, Sect. 1.5.3).

Deviations from the Langmuir model are often observed in real systems. The Langmuir model assumptions listed above are indeed very limitative and severe: (i) the solid surface is rarely uniform: there are always "imperfections" at the surface, (ii) the mechanism of adsorption is not the same for the first molecules as for the last to adsorb. When two or more kind of sites characterized by different adsorption energies are present at the surface (as stated in point i), and when lateral interactions among adsorbed species occur (as stated in point ii), the equivalence/independence of adsorption sites assumption fails. The most energetic sites are expected to be occupied first, and the adsorption enthalpy $\Delta_a H$ (per site) instead of keeping a constant, coverage-independent value, exhibits a declining trend as far as the coverage θ increases.

Further, the adsorbed molecules are not necessarily inert, and on the top of the monolayer other molecules may adsorb and multi-layers build up: this is properly described by the Brunauer, Emmet and Teller (BET) model [2, 30, 32].

Freundlich and Temkin isotherms, which refer to the case of the adsorption at surfaces characterized by a heterogeneous distribution of active sites, will be discussed briefly [30].

Freundlich isotherm is mathematically expressed by the Eq. 1.8:

$$V_{ads} = kp^{1/n} \tag{1.8}$$

This is a purely empirical formula, where the term V_{ads} represents the adsorbed amount, p the adsorptive pressure, whereas k and n are suitable empirical constants for a given adsorbent-adsorbate pair at temperature T. The adsorbed amount are normalized either to the mass of the adsorbent or to the exposed surface area. As an example see. Fig. 1.6a, where the adsorption volumetric isotherms of H_2O on H–BEA zeolite are reported: the experimental points were interpolated by the Freundlich isotherm equation. The Freundlich isotherm assumes that the adsorption enthalpy $\Delta_a H$ (per site) varies exponentially with increasing equilibrium pressure. In fact, the experimental points in the correspondent heat of adsorption versus coverage plot were properly interpolated by an exponential fitting, as illustrated in Fig. 1.15 (*vide infra* Sect. 1.4.2.3).

Temkin isotherm is mathematically expressed by the Eq. 1.9:

$$V_{ads} = k_1 \ln (k_2 p) \tag{1.9}$$

Temkin equation too is a purely empirical formula, where V_{ads} represents the adsorbed amount and p the adsorptive pressure; k_1 and k_2 are suitable empirical constants for a given adsorbent-adsorbate pair at temperature T. Also in this case, the adsorbed amount are normalized either to the mass of the adsorbent or to the exposed surface area.

The Temkin isotherm assumes that the adsorption enthalpy $\Delta_a H$ (per site) decreases linearly upon increasing coverage. Examples of heats of adsorption decreasing linearly with coverage are reported in the literature, as for instance in the case of NH_3 adsorbed on hydroxylated silica, either crystalline, [48] or amorphous, [49] as well as in the case of CH_3OH adsorption on silica-based materials [26].

Further, it is worth noticing that at sufficiently low pressure all adsorption isotherms are linear and may be regarded as obeying the Henry's law, which is reported in Eq. 1.10:

$$V_{ads} = h\,p \tag{1.10}$$

The Henry constant h is typical of the individual adsorbate-adsorbent pair, and is obtained by the slope of the straight line representing the isotherm at low coverage.

The isotherms classification, which is of high merit in terms of generality, deals with ideal cases which in practical work are rarely encountered. In fact, most often the adsorption process over the whole interval of pressure is described by an experimental isotherm which does not fit into the classification. Nonetheless, each of the equations described above may be used over restricted ranges of equilibrium pressure, so allowing to describe the experimental isotherm through the combination of individual components to the process. In such a way the surface properties of the solid, and the thermodynamics features of processes taking place at the interface can be quantitatively described [30].

As an example, it is here mentioned that the adsorption NH_3 on a highly dehydrated silica specimen was satisfactorily described by the combination of the Langmuir and Henry isotherms. The former accounted for H–bonding interactions on isolated silanols (Si-OH), whereas the latter accounted for the aspecific adsorption on dehydrated patches of the surface, dominated by dispersion forces interactions [28].

1.4 Adsorption Microcalorimetry

The measurement of the heat of adsorption by a suitable calorimeter is the most reliable method for evaluating the strength of adsorption (either physical or chemical). Tian-Calvet heat-flow microcalorimeters are an example of high sensitivity apparatus which are suitably adapted to the study of gas-solid interactions when connected to sensitive volumetric systems [10–14, 50–55]. Volumetric-calorimetric data reported in the following were measured by means of either a C-80 or MS standard heat-flow microcalorimeter (both by Setaram, F), connected to a high vacuum (residual pressure

$p \leq 10^{-5}$ Torr) gas-volumetric glass apparatus. During the same experiment, both integral heats evolved and adsorbed amounts were measured for small increments of the adsorptive, from the gas or vapor phase. Two identical calorimetric vessels, one containing the sample under investigation, the other (usually empty) serving as reference element were connected in opposition. Thanks to the differential construction of the apparatus, all parasitic phenomena (i.e., all thermal effects other than the one due to the interaction of the gas with the solid surface) were successfully compensated. C-80 microcalorimeters allow the heats of adsorption to be measured at constant T in the room temperature -573 K range, whereas MS standard microcalorimeters in the room temperature -473 K range. The adsorptive pressure in the measurements were monitored by either a Varian Ceramicell or a Baratron MKS transducer gauge $(0 - 100$ Torr).

A well-established stepwise procedure was followed [16, 23, 25, 56]. Small successive doses of the adsorptive were admitted and left in contact with the adsorbent until the thermal equilibrium was attained. The 1st run of adsorption performed on the activated sample (pretreated in high vacuum conditions and/or in controlled atmosphere) will be hereafter referred to as ads. I. At any individual dose of gas introduced in the system, the evolved heat ΔQ^{int} was measured within the calorimetric cells, while the adsorbed amount Δn_{ads} was measured by volumetry. Ads. I was followed by a desorption run (des. I), performed by simple evacuation of the cell. In such a way the reversibly adsorbed phase was desorbed and either the pristine surface was restored, in case of an entirely reversible adsorption, or the pristine surface was not recovered, in case of a (partially) irreversible adsorption. Ads. II was subsequently performed in order to assess which fraction (if any) of the pristine surface sites was irreversibly occupied by the adsorbed phase (in the adopted conditions). By subtracting the ads. II curve from the ads. I one, the adsorbed fraction not removed by evacuation is evaluated. The ads. II component will be hereafter referred to as the reversible adsorbed phase, whereas the (ads. I - ads. II) component will be referred to as the irreversible phase (in the adopted conditions). Subsequent runs of adsorption (ads. III, IV etc.) are performed in some cases, if the irreversible modification of the surface is expected/suspected not to be extinguished during the ads. I [21, 23, 26]. Adsorption measurements are usually performed at least twice on a virgin portion of the same batch of the material, activated in the same conditions, to check the experiments reproducibility. The routinely run protocol of adsorption-desorption-adsorption cycles is schematically illustrated in Fig. 1.8.

1.4.1 Materials

Before illustrating an instructive selection of experimental data obtained by the method described above, it is worth doing to report a brief description of the investigated materials. Some data have been already published (as will be reported), other are original.

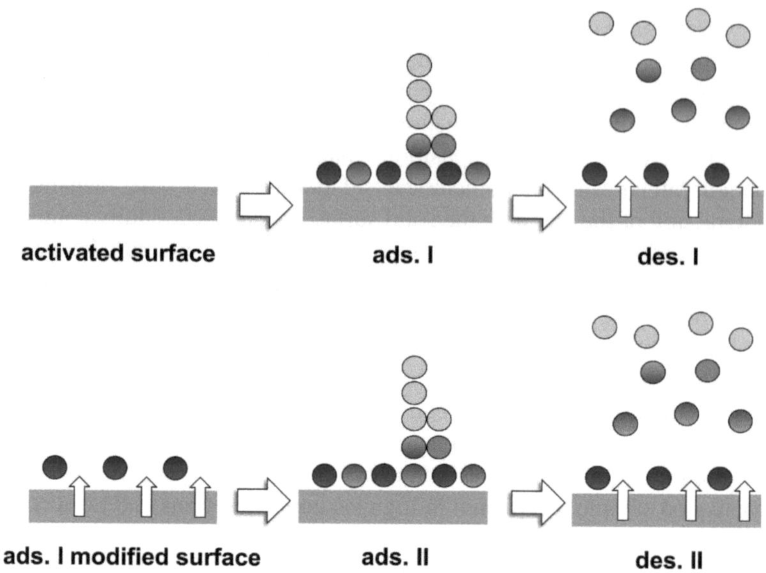

Fig. 1.8 Schematic illustration of the adsorption-desorption-adsorption cycle routinely run in order to collect the equilibrium data (evolved heats and adsorbed amounts) which are measured within the calorimetric cells at increasing equilibrium pressure

1.4.1.1 Solid Materials

H-BEA (H-BETA): a proton exchanged BEA zeolite specimen, characterized by a three-dimensional network of pores consisting of three families of 12-ring interconnected channels [57]. The specimen here illustrated was characterized by a silica-to-alumina ratio $\frac{SiO_2}{Al_2O_3} = 4.9$, corresponding to a distribution of Al atoms per unit cell $\frac{Al}{uc} = 5.9$ (see Ref. [25]. The acidic strength of such material is related to the presence of both Brønsted and Lewis acidic sites. Such latter kind of sites gives H–BEA zeolites unique catalytic properties [58–60].

BEA (BETA): an all-silica BEA zeolite specimen, characterized by the same three-dimensional network as H–BEA, but virtually free of Al species in that the silica-to-alumina ratio was $\frac{SiO_2}{Al_2O_3} = 255$, corresponding to a negligible distribution of Al atoms per unit cell $\frac{Al}{uc} \ll 0.1$ (see Ref. [25, 61]).

Prior to the adsorption experiments, H–BEA sample was outgassed for 2 h at $T = 873$ K, a temperature which ensured a maximum surface dehydration, still compatible with the stability of the structure, and yielding the maximum density of Brønsted and Lewis acidic sites. For the all-silica specimen, which was less hydrophilic than the proton-exchanged counterpart, a 2h-outgassing at $T = 673$ K was sufficient to get rid of all adsorbed water, so yielding the maximum density of Si–OH hydroxyl nests.

H−MFI (H−ZSM5): a proton exchanged MFI zeolite specimen, characterized by a three-dimensional network of pores consisting of sinusoidal and intersecting straight 10-ring channels [35]. The specimen here illustrated was characterized by a silica-to-alumina ratio $\frac{SiO_2}{Al_2O_3} = 7.5$, corresponding to a distribution of Al atoms per unit cell $\frac{Al}{uc} = 6.0$, very close to that of the H-BEA specimen illustrated above [25].

The acidic strength of H–MFI materials is related to the presence of Brønsted acidic sites whereas, opposite to H-BEA zeolites, Lewis acidic sites represent only a minor feature.

MFI–Silicalite: a Na- and Al-free defective all-silica MFI specimen, characterized by the same three-dimensional network as H–MFI, but with an extremely large silica-to-alumina ratio $\frac{SiO_2}{Al_2O_3} \rightarrow \infty$. See Ref. [25] for details.

The all-silica zeolites, both BEA and MFI, are in general characterized by a variable amount of internal defects consisting of hydroxyl nests made up of H–bonding interacting Si-OH species, located within the zeolite nanopores. A virtually perfect (i.e. defect-free) MFI–Silicalite was also investigated for comparison purposes. See Ref. [24] and references therein for details on both defective and perfect all-silica specimens.

Prior to the adsorption experiments, both H–MFI and MFI samples were outgassed for 2 h at $T = 673$ K in order to achieve the maximum dehydration of the surface compatible with the stability of the structure and yielding the maximum density of Brønsted acidic sites in H–MFI and of polar Si-OH hydroxyl nests in MFI–Silicalite.

Me(I)−MFI (Me(I)−ZSM5): cation-exchanged zeolites (MFI) with Me(I) = Cu^+ or Ag^+ (both belonging to the *group 11* of transition metals), or Me(I) = Na^+ or K^+ (both belonging to the *group 1* of alkaline-metals) as extra-framework species. The samples were prepared starting from the same NH_4–MFI precursor (characterized by a silica-to-alumina ratio $\frac{SiO_2}{Al_2O_3} = 7$) either by conventional wet exchange (Ag(I)−, Na− and K−MFI), or by direct CuCl gas phase exchange (Cu(I)−MFI). In all cases a nearly total exchange of the parent material extra-framework cations was achieved, as confirmed by IR spectroscopy: one Me(I) cation for every framework Al atom was present in all examined materials. For samples details see Ref. [21, 23, 62]

TiO_2 : (a) a crystallographic pure anatase obtained by a sulphate preparation and thoroughly freed from sulphate impurities, following the preparation route described in Ref. [63]; (b) the same crystallographic pure anatase but still carrying sulphate surface impurities (4–5 % SO_4 by weight) [64]. Prior to the adsorption experiments all samples were outgassed at $T = 673$ K for 2 h, in order to dehydrate the surface and yield the maximum density of Lewis acidic sites (i.e. *cus* Ti^{4+} cations). After outgassing, the samples were contacted at the same temperature with ≈100 Torr of O_2 for 10 min to ensure stoichiometry.

A Ca-modified silica (8% mol CaO) was obtained by adding dosed amounts of an aqueous solution of $Ca(NO_3)_2{\cdot}4H_2O$ to a dry amorphous nonporous silica (Aerosil 200 from Degussa, Frankfurt A.M., D) using the incipient-wetness impregnation technique described in Refs. [26, 65] Prior to the adsorption experiments the sample was outgassed at $T = 423$ K for 2 h, in order to get rid of physically adsorbed water but without inducing any appreciable surface dehydroxylation. The choice of a vacuum activation temperature only slightly higher than room temperature was determined by the need to study the surface properties of still highly hydrated samples (i.e. of solids taken under conditions not too far from those experienced by biomaterials in contact with the biological medium).

1.4.1.2 Molecular Probes

CO specpure from either Matheson or Praxair was used as a molecular probe in order to assess the Lewis acidic properties of coordinatively unsaturated (*cus*) cations either located in the dehydrated zeolite nanocavities as charge-balancing cations, or exposed at the dehydrated surface of oxidic materials. CO is capable of interacting with the *cus* cations leading to the formation of adducts of different stability according to the chemical nature of the cation. Weak electrostatic adducts are formed on alkaline metal cations, σ-coordinated species of intermediate stability on non d/d^0 metal cations, whereas high-stability carbonyl-like species originated by a σ-coordination $+\pi$-back donation of d electrons are formed on d block metal cations.

H_2O_{vap} was used as a molecular probe to assess the hydrophilic and/or hydrophobic features of protonic (H−BEA) and all-silica (BEA) zeolites. Water (from Millipore) was distilled several times in vacuo and rendered gas-free by several 'freeze-pump-thaw' cycles. The vapor pressure of H_2O at $T = 303\,\mathrm{K}$ is 31.8 Torr, and the standard molar enthalpy of liquefaction (i.e. the latent heat of liquefaction, q_L) is $-\Delta_L H^\circ = 44\,\mathrm{kJ\,mol^{-1}}$.

NH_3 gas (from Praxair) was used as molecular probe of moderate basic strength ($PA = 854\,\mathrm{kJ\,mol^{-1}}$), [66]. in order to characterize the acidic strength of Brønsted (and Lewis, if any) acidic sites in protonic and all-silica zeolites.

CH_3OH vapor, obtained by distilling in vacuo liquid methanol (Sigma-Aldrich), was rendered gas-free by several "freeze-pump-thaw" cycles. The vapor pressure of CH_3OH at $T = 303\,\mathrm{K}$ is 164 Torr, and the standard molar enthalpy of liquefaction (i.e. the latent heat of liquefaction, q_L) is $-\Delta_L H^\circ = 38\,\mathrm{kJ\,mol^{-1}}$.

1.4.2 Equilibrium Data

1.4.2.1 Volumetric and Calorimetric Isotherms

Adsorbed amounts and integral heat evolved will be suitably reported as a function of the increasing equilibrium pressure, i.e. as volumetric and calorimetric isotherms, respectively. Adsorbed amounts $n_{ads} = \sum \Delta n_{ads}$ were obtained by adding the individual doses amounts, Δn_{ads}, and will be reported either as mol per unit mass ($\mathrm{mol\,g^{-1}}$) or per unit surface area ($\mathrm{mol\,m^{-2}}$), or as molecules per square nanometer. In zeolites, in order to compare from a structural point of view the affinity of different zeolites towards the given adsorptive, the adsorbed amounts will be more suitably

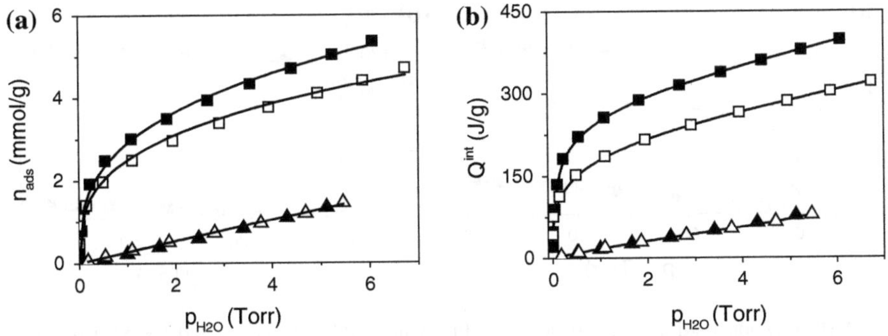

Fig. 1.9 Adsorption of H_2O_{vap} adsorbed at $T = 303$ K on proton-exchanged (H-BEA, square) and all-silica (BEA, up triangle) zeolites pre-outgassed at $T = 873$ and 673 K, respectively. **a** Volumetric isotherms. **b** Calorimetric isotherms. *Solid symbols* ads. I; *open symbols* ads. II. Adapted from Ref. [25] Fig.4

reported as molecules per unit cell (uc) or per Al atom. Integral heats $Q^{\text{int}} = \sum \Delta Q^{\text{int}}$ were obtained by adding the individual doses evolved heats, ΔQ^{int}, and will be reported per gram (J g^{-1}) or per unit surface area (J m^{-2}) of the adsorbent. Integral heats plotted as a function of the adsorbed amounts will be referred to as the integral heat curve: Q^{int} *versus* n_{ads}.

A selection of adsorption isotherms obtained for a variety of materials and probe molecules will be illustrated. Note that all the adsorption measurements reported in the following were performed at $T = 303$ K.

H_2O_{vap} *adsorbed on* H–BEA *and all-silica BEA zeolites*. In Fig. 1.9, ads. I and ads. II volumetric (section a) and calorimetric (section b) isotherms of H_2O_{vap} adsorbed on proton-exchanged (H–BEA) and all-silica (BEA) zeolites are reported. In H–BEA an irreversible adsorption component was revealed by the non-coincidence of the ads. I and ads. II isotherms (both volumetric and calorimetric). Conversely, in the all-silica case the process was entirely reversible upon evacuation of the vapor phase, as witnessed by the coincidence of ads. I and ads. II isotherms. The isotherms experimental points were interpolated by the Freundlich equation.

H–BEA exhibited, as expected, a much higher affinity towards H_2O than the Al–free systems, owing to the presence of $Si(OH)^+Al^-$ species, characteristic of proton-exchanged zeolites and acting as Brønsted acidic sites. Such species are located within the zeolite nanocavities and are able to adsorb guest molecules by strong H–bonding interactions, often leading to the formation of protonated species [25, 67–73]. In addition, in H–BEA zeolites structural defects acting as Lewis acidic sites (i.e. strong electron acceptors) are often present [58, 60, 74–76]. It is still under debate whether such species consist of framework trigonal Al (III) atoms, [73, 77, 78] or of extra-framework Al (III) species (EFAL) located within the pores [75, 76]. Anyway, both Lewis and Brønsted acidic sites are responsible for the formation of water complexes which are stable upon room temperature evacuation.

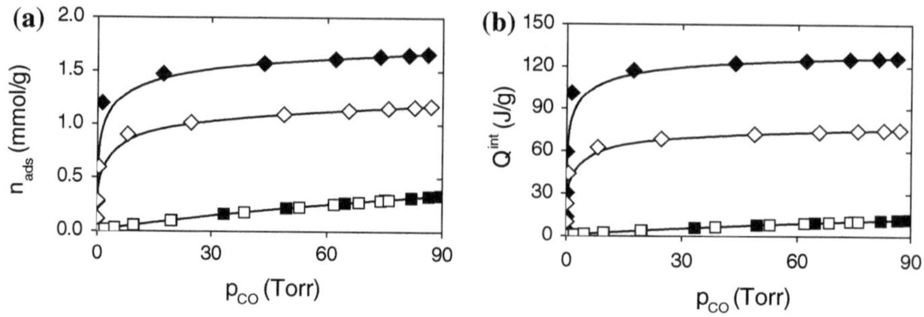

Fig. 1.10 CO adsorbed at $T = 303\,K$ on zeolites Cu(I)–MFI (*diamond*) and Na–MFI (square): volumetric (**a**) and calorimetric (**b**) isotherms. Both samples were pre-outgassed at $T = 673\,K$. *Solid symbol* ads. I; *open symbols* ads. II. Volumetric isotherms: adapted from Ref. [23] Fig. 3a

The affinity towards water of the all-silica counterpart was lower than that of the proton exchanged zeolite, as expected, but it was not negligible. The reported isotherms indicated that hydrophilic sites, responsible for weak and reversible water H–bonding adducts, are developed in zeolites even in the absence of framework Al atoms. Structural defects generating polar species consisting of Si–OH nests (which are characterized by a weak Brønsted acidic strength),[25, 61] are always present in Al-free zeolites, unless especially prepared in order to obtain hydrophobic, inert materials, as claimed by Flanigen et al.[36]. See also Ref. [24].

CO *adsorbed on* Me(I)-*exchanged* MFI *zeolites*. In Fig. 1.10 the ads. I and ads. II volumetric (section a) and calorimetric (section b) isotherms of CO adsorbed on Cu(I)– and Na–MFI are reported. In Fig. 1.11 ads. I and ads. II volumetric (section a) and calorimetric isotherms (section b) of CO adsorbed on Ag(I)– and K–MFI are reported. Note that the ordinate scale of the isotherms plots for Ag(I)– and K–MFI is twice as large as the Cu(I)– and Na–MFI scale, owing to the much lower adsorption capacity of the Ag(I)– and K–MFI zeolites with respect to the Cu(I)– and Na–MFI ones.

Cu(I)– and Ag(I)–MFI ads. II isotherms (both volumetric and calorimetric) lie below the ads. I correspondent isotherms, indicating the presence of irreversible phenomena. The irreversible adsorption component was quantified by taking the (ads. I – ads. II) difference in the volumetric isotherms at $p_{CO} = 90$ Torr. It was $\approx$30 % of total uptake (ads. I) for copper- and $\approx$20 % for silver-exchanged zeolites.

Conversely, in both Na– and K–MFI cases the coincidence of ads. I and ads. II isotherms confirmed the reversibility of CO adsorption. The much lower adsorption capacity witnessed by the *group 1* metal cations volumetric isotherms with respect to the two *group 11* ones is striking. The calorimetric isotherms too confirmed the much lower affinity of the former with respect to the latter. The isotherms experimental points of the *group 1* metals exchanged zeolites were interpolated by the Langmuir equation (*vide supra*, Sect. 1.3.1), whereas in the case of the *group 11* metals isotherms the curves were drawn by employing a B-Spline function, just as an aid to the eye. The two d-block metals isotherms dramatically deviate from the

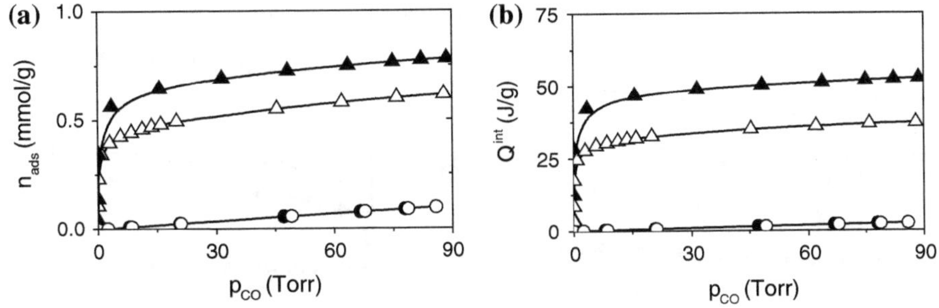

Fig. 1.11 CO adsorbed at $T = 303\,\mathrm{K}$ on zeolites Ag(I)–MFI (*triangle*) and K–MFI (*circle*): volumetric (**a**) and calorimetric (**b**) isotherms. K–MFI was pre-outgassed at $T = 673\,\mathrm{K}$, Ag(I)–MFI at $T = 400\,\mathrm{K}$. *Solid symbol* ads. I; *open symbols* ads. II. Volumetric isotherms: adapted from Ref. [23], Fig. 3b

Langmuir behavior, in that in this latter case the adsorption of CO was driven by forces other than the simple electrostatic polarization, and concerns a heterogeneous distribution of active sites. In fact, Cu(I)− and Ag(I)−cations hosted in the zeolite nanocavities, besides the electrostatic polarization of CO molecule, interacted chemically with the molecule. Stable carbonyl-like species were formed at the d-block metal cations sites, through a σ-coordination of the C− end lone pair plus a partial π-back-donation of d electrons. The stoichiometry of the heterogeneous di-carbonyl $\left[\mathrm{Cu}\,(\mathrm{CO})_2\right]^+$ and mono-carbonyl $\left[\mathrm{Ag}\,(\mathrm{CO})\right]^+$ complexes was obtained from the quantitative data reported in Figs. 1.10 and 1.11. These results, confirmed also by IR spectroscopic data (see Ref. [23] and references therein for details), are in good agreement with the stoichiometry reported for the correspondent Cu(I) and Ag(I) complexes formed in homogeneous conditions [79, 80].

The differences between the two d-block metal cations can be explained on one hand from an electrostatic point of view, since the charge density of Cu(I) is much larger than that of Ag(I) cations (rCu(I) $= 0.96$ Å and rAg(I) $= 1.26$ Å) [47]. On the other hand, the overlap of the metal cations and CO orbitals in the carbonyl bond is expected to be larger for Cu(I) than for Ag(I). The adsorption of CO on Na^+ and K^+ cations hosted in the same zeolite framework allowed to roughly single out the electrostatic contribution to the two d-block metal cations/CO interaction. Na^+ and K^+ cations possess indeed a charge/radius ratio very close to that of Cu(I) and Ag(I), respectively (0.97 Å for Na^+ and 1.33 Å for K^+) [47].

NH_3 adsorbed on H−MFI and all-silica MFI zeolites. Adsorption of NH_3 has been widely used to assess the acidic strength of both Brønsted and Lewis sites at solid surfaces [68, 70, 71]. In Fig. 1.12, the ads. I and ads. II volumetric (section a) and calorimetric (section b) isotherms of NH_3 adsorbed on: (i) the proton-exchanged H−MFI zeolite, (ii) one defective MFI–Silicalite (MFI−def) and (iii) the perfect (defect-free) all-silica MFI–Silicalite (MFI−perf), are reported. In H−MFI, the adsorption was only partially reversible in agreement with the proton-transfer from the Brønsted acidic site $\mathrm{Si(OH)^+Al^-}$ to NH_3, as reported in Ref. [81]. Conversely,

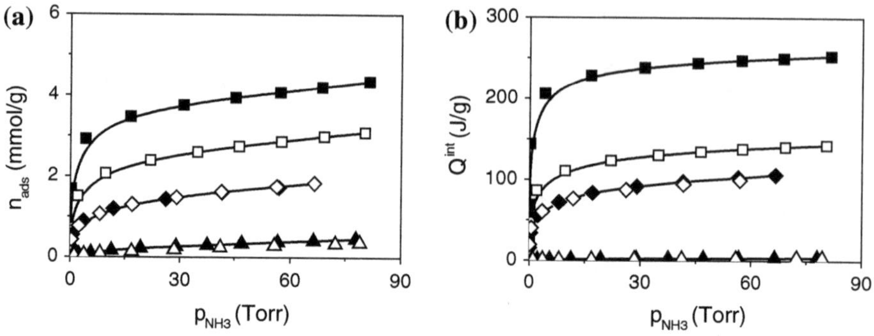

Fig. 1.12 Adsorption of NH_3 at $T = 303\,K$ on proton-exchanged H–MFI zeolite (*square*), defective MFI–Silicalite (MFI−def, diamond), and perfect (defect-free) MFI–Silicalite (MFI−perf, up triangle). **a** Volumetric isotherms. **b** Calorimetric isotherms. H−MFI zeolite was pre-outgassed at $T = 873\,K$, MFI–Silicalite samples at $T = 673\,K$. *Solid symbols* ads. I, *open symbols* ads. II

in both defective and perfect all-silica zeolites the process was entirely reversible upon evacuation of the gas phase. In this latter case NH_3 interacted only *via* hydrogen bond with Si−OH nests [24]. In the defect-free MFI–Silicalite, which exposes only unreactive siloxane bridges, the interaction was aspecific in that governed by dispersion forces due to the nanopores walls (*confinement effect*) [82–84].

1.4.2.2 Integral Heat of Adsorption

Integral heats normalized to the adsorbed amounts are referred to as the integral molar heat of adsorption at the given equilibrium pressure p: $(q_{\mathrm{mol}})_p = \left(\frac{Q^{\mathrm{int}}}{n_{ads}}\right)_p$ expressed in kJ mol^{-1}. The $(q_{\mathrm{mol}})_p$ quantity is an intrinsically average value, as it refers to the thermal response of the surface as a whole, and is comprehensive of all thermal contributions from the variety of interactions the gas molecules have experienced from the beginning of the process up to the chosen equilibrium pressure p.

By plotting the integral heats evolved against the adsorbed amounts the so-called integral heats curve is obtained (*vide infra* as an example the insets of Fig. 1.14). In the Langmuir-like adsorption characterized by a uniform distribution of equivalent, non-interacting sites the heat of adsorption is constant upon increasing coverage: the integral heats curve is thus a straight line through the origin, the slope of which gives the differential heat of adsorption (q^{diff}). In cases other than this particular one, q^{diff} is obtained by differentiating the non-linear $Q^{\mathrm{int}} = f\,(n_{ads})$ function.

The interest in dealing with differential heats stems on the fact that the differential quantities are more adequate than the average ones in describing the evolution with the increasing coverage of the adsorbate/surface sites energy of interaction.

1.4.2.3 Differential Heat of Adsorption

Differential heats of adsorption represent a reasonable measure of the energy of interaction of a (probe) molecule with the individual sites, at any adsorbate coverage. The magnitude of the heat evolved during adsorption, which depends on the nature of the adsorbate/surface sites bonding, varies upon increasing coverage as a consequence of the presence of either a heterogeneous distribution of surface sites, or lateral interactions among adsorbed species (*vide infra* Sect. 1.5.1). The shape of the q^{diff} *versus* n_{ads} plots depends on, and actually describes, the surface heterogeneity.

Differential heats of adsorption are properly defined as $q^{diff} = \frac{\delta Q^{\mathrm{int}}}{\delta n_{ads}}$, i.e., the derivative of the $Q^{\mathrm{int}} = f\,(n_{ads})$ function which best fits the Q^{int} *versus* n_{ads} equilibrium data.

An alternative route for evaluating q^{diff} is however first discussed here. The method is based on the use of the partial molar heats $\frac{\Delta Q^{\mathrm{int}}}{\Delta n_{ads}}$ quantities,[10, 12, 85] i.e. the ratio of the integral heat evolved over the correspondent amount adsorbed for the individual incremental doses of the adsorptive. The $\frac{\Delta Q^{\mathrm{int}}}{\Delta n_{ads}}$ quantity, expressed in kJ mol^{-1} is still an integral molar heat (and thus average in nature) but it refers to the thermal response of very small regions of the surface, provided that the individual doses were prepared as small as possible. Note that the limit of $\frac{\Delta Q^{\mathrm{int}}}{\Delta n_{ads}}$ quantity for Δn_{ads} approaching zero is the true differential heat, as illustrated by the Eq. 1.11:

$$\lim_{\Delta n_{ads} \to 0} \frac{\Delta Q^{\mathrm{int}}}{\Delta n_{ads}} = q^{diff} \tag{1.11}$$

By plotting, in the form of an histogram, $\frac{\Delta Q^{\mathrm{int}}}{\Delta n_{ads}}$ values as a function of the adsorbed amounts n_{ads}, the evolution of the heat of adsorption upon increasing coverage is properly described. By taking the middle point of each partial molar heat $\frac{\Delta Q^{\mathrm{int}}}{\Delta n_{ads}}$ block, the mean heat values correspondent to small portions of the surface are obtained, which represent a reasonable measure of the differential heat, as reported by several authors [10, 14, 56, 85–88].

The q^{diff} experimental points versus the n_{ads} adsorbed amounts are interpolated by functions which best fit the experimental points (*vide infra* in Fig. 1.13a the case of H_2O_{vap} adsorbed on zeolites).

For most purposes, it is convenient to define the zero-coverage differential heat of adsorption q_0, which corresponds to the energy of interaction of the molecular probe with the most energetic sites, expected to be active in the earliest stages of the adsorption process. The q_0 value is estimated by extrapolating the q^{diff} *versus* n_{ads} plot to vanishing coverage. The extrapolated quantities of experimental origin can be properly and often fruitfully compared to the computed energy of interaction of a probe molecule with an individual model site, as obtained through *ab initio* calculations [24–26, 29, 73, 89].

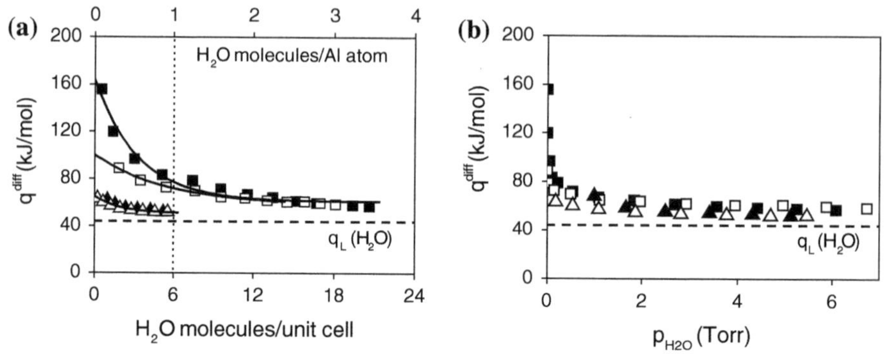

Fig. 1.13 **a** q^{diff} *versus* n_{ads} of H_2O_{vap} at $T = 303$ K on proton-exchanged (H−BEA, square) and all-silica (BEA, up triangle) zeolites, pre-outgassed at either $T = 873$ K (H−BEA) or $T = 673$ K (BEA). **b** q^{diff} *versus* p_{H2O}. *Solid symbols* ads. I; *open symbols* ads. II. Adapted from Ref. [25],Fig. 5

H_2O_{vap} adsorbed on H − BEA and BEA zeolites. In Fig. 1.13 the differential heats of adsorption of H_2O_{vap} on H−BEA and BEA zeolites are reported as a function of water adsorbed amounts (Fig. 1.13a) or water equilibrium pressure (Fig. 1.13b).

Figure 1.13a curves are the exponential functions which best fitted the partial molar heats experimental points. The q^{diff} values were plotted against the number of H_2O molecules adsorbed per unit cell, so facilitating the interpretation of the results in terms of structural features. As already outlined in describing the correspondent volumetric isotherms (*vide supra* Fig 1.9), in H−BEA case water adsorption was only partially reversible upon evacuation of p_{H_2O}. The ads. I and ads. II q^{diff} *versus* n_{ads} curves are well distinguished in the early stage of the process, namely up to the adsorption of ≈1H_2O molecule per Al, i.e. per acidic (either Brønsted or Lewis) site. Afterwards, the two curves merge and remain constant at a heat value larger than the latent heat of liquefaction of water ($q_L = 44$ kJ mol^{-1}). The ads. I and ads. II curves for the all-silica BEA zeolite are virtually coincident and rapidly approach q_L.

In all cases, the q^{diff} *versus* n_{ads} curves are typical of heterogeneous surfaces. In H−BEA several interactions, of different strength, take place on sites of different nature located either within the zeolite nanocavities or at the external surface.

All interactions taking place simultaneously contributed to the calorimetrically measured heat, and consequently it is hard to single out the energetics of the individual contributions to the interaction. At the Brønsted Si(OH)$^+$Al$^-$ sites H_2O molecules are either strongly H−bonded or protonated, whereas at the Lewis acidic sites, i.e. the *cus* framework Al (III) cations, H_2O molecules are oxygen-down coordinated [25]. H_2O molecules interacted also *via* H−bonding with Si−OH nests located within the zeolite nanocavities and, more weakly, with (isolated) Si−OH species exposed at the external surface. Aspecific interactions generated by *confinement effect*,[82–84] also contributed to the overall measured heat.

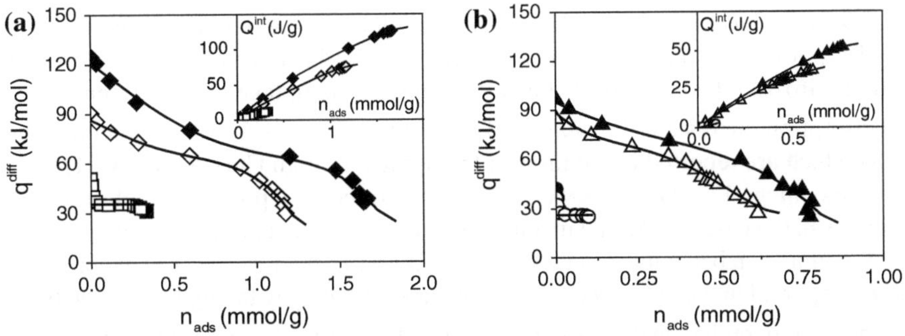

Fig. 1.14 Differential heats of adsorption versus CO uptake ($T = 303$ K). **a** zeolites Cu(I)–MFI (*diamond*) and Na–MFI (*square*). **b** zeolites Ag(I)–MFI (*triangle*) and K–MFI (*circle*). Insets: interpolated integral heats of adsorption curves Q^{int} *versus* n_{ads}. Zeolites Cu(I)−, Na− and K−MFI were pre-outgassed at $T = 673$ K, Ag(I)–MFI at $T = 400$ K. *Solid symbols:* ads. I; *open symbols* ads. II. Agreement between the experimental points (partial molar heats) and the derivative of the integral heat curves is quite good (see the text for details). Adapted from Ref. [23], Fig.4

The heats of adsorption started from a quite high zero-coverage value ($q_0 \approx 160$ kJ mol^{-1}), which is compatible with a chemisorption process, either the protonation of H_2O at the Brønsted acidic site or the strong oxygen-down coordination at the Lewis acidic sites. According to the *ab initio* modeling results, indicating that the H_2O/Lewis site energy of interaction is comprised in the 160-109 kJ mol^{-1} range (depending on the local coordination of Al (III) atom),[25] the zero-coverage heat of adsorption for H−BEA could be assigned to the H_2O/Lewis complex formation, which dominated the early stage of the adsorption. At increasing coverage the heat values decreased exponentially but remained well above the latent heat of liquefaction of water even after the adsorption of ≈4 molecules per Al atom. For one H_2O molecule adsorbed per Al atom, on average, the heat values were comprised in the $160 < q^{diff} < 80$ kJ mol^{-1} range, whereas for the second-to-fourth H_2O adsorbed molecules in the $80 < q^{diff} < 60$ kJ mol^{-1} range.

In the all-silica BEA specimen the zero-coverage heats of adsorption were much lower than for H–BEA ($q_0 \approx 70$ *vs.* 160 kJ mol^{-1}, respectively) and both ads. I and ads. II curves dropped very fast down to an almost constant value, only slightly higher than $q_L = 44$ kJ mol^{-1}. This result indicates that, despite the absence of specific Brønsted/Lewis acidic sites, the Si−OH nests manifest a H−bonding capacity of medium strength giving to the all-silica zeolite a substantial non-hydrophobic character.

Figure 1.14b plot indicates that for adsorption leading to the same residual pressure in different runs and/or on different samples, q^{diff} values virtually coincide. Differences between the protonic H−BEA and the all-silica BEA zeolites (and between ads. I and ads. II for H−BEA) are evident at a residual pressure close to zero, at which the irreversible adsorption took place only on H−BEA.

Two examples of q^{diff} obtained by differentiating the $Q^{int} = f(n_{ads})$ functions best fitting the Q^{int} *versus* n_{ads} experimental points will be discussed in the following.

CO adsorbed on Me(I)−exchanged MFI zeolites. In Fig. 1.14 the differential heat of adsorption of CO on Cu(I)− and Na−MFI (Fig. 1.14a) and on Ag(I)− and K−MFI (Fig. 1.14b) zeolites are reported as a function of the uptake. The differential heat of adsorption curves are the analytical derivative $q^{diff} = \frac{\delta Q^{\text{int}}}{\delta n_{ads}}$ of the integral heats plots which are reported in the inset of the figures. In both Cu(I)− and Ag(I)−MFI cases the integral heats curves were reasonably fitted by a polynomial of order five, whereas in the case of alkaline metal cations by a linear equation (in agreement with the Langmuirian behavior of such systems, *vide supra* Sect. 1.3.1). The experimental points reported in the q^{diff} *versus* n_{ads} plot were obtained by taking the middle points of the partial molar heats $\frac{\Delta Q^{\text{int}}}{\Delta n_{ads}}$ histogram. A reasonably good agreement exists between the two methods over the whole examined adsorption range. From both Fig. 1.14 plots it is clearly evident that at least a fraction of CO adspecies formed at the Cu(I) and Ag(I) sites are stable complexes.

The ads. I q_0 was estimated as high as ≈120 and ≈100 kJ mol^{-1} for Cu(I) and Ag(I) sites, respectively. Such values are compatible with the bond energies typical of a chemisorption process, i.e. the formation of stable carbonyl-like species. The value for the early formation of the reversible carbonyl-like species (ads. II) was lower (≈90 kJ mol^{-1}) than for ads. I and very close for the two systems. As far as the coverage increased, the heat decreased to values typical of labile species and eventually fell down to values as low as ≈35 kJ mol^{-1} for Cu(I)−MFI and ≈25 kJ mol^{-1}for Ag(I)−MFI. These latter values were ascribed to the aspecific interaction of CO with the zeolite nanopores walls. It is worth noting that the heat of adsorption associated to this interaction was much larger than the latent heat of liquefaction of CO ($q_(L) = 6$ kJ mol^{-1}).

In agreement with the Langmuir-like behavior of the Na− and K–MFI isotherms (*vide supra* Sect. 1.3.1), their Q^{int} *versus* n_{ads} curves were reasonably fitted by linear equations. A constant value for the differential heat was obtained: q^{diff} ≈35 kJ mol^{-1} for Na–MFI and ≈28 kJ mol^{-1} for K–MFI. The linear fit of the integral heat curves seemed the most realistic, in spite of the fact that in both cases at very low and at high coverage the middle points of the experimental histogram deviated from the constant value. At both low and high coverage, however, the heat values cannot be assigned to processes involving a specific $Me^+ \cdots CO$ interaction and must be disregarded in evaluating the heat of formation of the adducts. In fact, the low-coverage heterogeneity was due to the presence of a few defective centers (1–2 % of the total active sites) interacting with CO more strongly than the alkaline metal cations. Conversely, the high-coverage low heat values correspond to the aspecific interaction with the zeolite nanopores walls, similarly to what observed for Cu(I)− and Ag(I)–MFI as explained above.

NH_3 adsorbed on H−MFI and all-silica MFI zeolites. In Fig. 1.15 integral (Fig. 1.15a) and differential (Fig. 1.15b) heats of the reversible adsorption of NH_3 on a variety of defective MFI–Silicalite (Sil−A, Sil−B, Sil−C) and on a perfect (i.e. defect-free) MFI–Silicalite (Sil−D) are illustrated as a function of NH_3 uptake. Note that Sil−A and Sil−D are the same specimens as the ones named in Fig. 1.12 MFI−def and MFI−perf, respectively. See Ref.[24] for the experimental details.

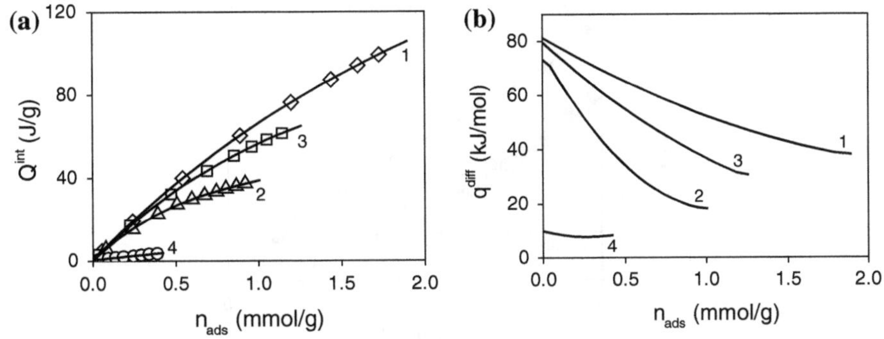

Fig. 1.15 Reversible adsorption (ads. II) of NH_3 at $T = 303$ K on defective MFI–Silicalite: Sil−A(*diamond*, 1), Sil-B(*up triangle*, 2), Sil−C(*square*, 3) and on perfect (i.e. defect-free) MFI–Silicalite: Sil−D(*circle*, 4). **a** Integral heats of adsorption versus NH_3 uptake. Curves interpolating the experimental points (1–4) are polynomials of order 3. **b** Differential heats of adsorption versus NH_3 uptake. Differential heats were obtained by differentiating the section (a) polynomial functions. All samples were outgassed at $T = 673$ K. Adapted from Ref. [24] Fig. 4. Note that Sil−A and Sil−D were the same specimens as the ones named in Fig. 1.12 MFI-def and MFI-perf, respectively

The curves interpolating the integral heats plot experimental points (Fig. 1.15a) were in all cases polynomials of order 3. Differential heats reported in Fig. 1.15b were obtained by differentiating the integral heats polynomial functions. The $q^{diff} = \frac{\delta Q^{int}}{\delta n_{ads}}$ curves were all typical of heterogeneous surfaces, in that the heat values decreased upon increasing coverage, but for the perfect Silicalite (Sil−D). The defective MFI–Silicalite heat curves (Sil−A, Sil−B and Sil−C) decreased from the initial values $q_0 \approx 80$ kJ mol^{-1} down to $q \approx 20$ kJ mol^{-1}, a value close to the latent heat of liquefaction of NH_3, $q_L = 21$ kJ mol^{-1}. The evolution of q^{diff} with increasing coverage was found to vary according to the different population and/or geometrical arrangement of the polar sites (Si−OH nests) in the defective MFI–Silicalite specimens [24, 90]. Conversely, the heat of adsorption on defects-free MFI–Silicalite (Sil−D) was found to be virtually coverage-independent and assessed to a very low value ($q \approx 10$ kJ mol^{-1}), according to the lack of polar sites capable of specifically interacting with NH_3. In this latter case, the adsorption was bound to the *confinement effect* due to dispersion forces. The constant heat value, lower than latent heat of liquefaction of NH_3, could be reasonably taken as a measure of the aspecific interaction of ammonia with the siloxane surface, which takes place also at the flat surface of amorphous non-porous silica, as documented in Ref. [28].

In Fig. 1.16 the differential heat of NH_3 adsorption on one of the all-silica defective specimens discussed above (Sil−A) will be compared with the correspondent heat of adsorption on the Brønsted acidic H−MFI zeolite. Note that Sil−A will be hereafter named MFI−def, in agreement with the nomenclature of the correspondent volumetric-calorimetric isotherms illustrated in Fig. 1.12.

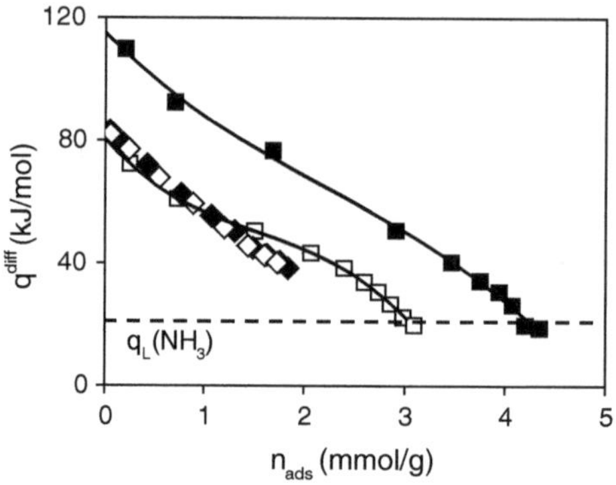

Fig. 1.16 Differential heat of adsorption versus NH_3uptake at $T = 303$ K on proton-exchanged H−MFI (*square*) and defective all-silica MFI−def (*diamond*) zeolites. The samples were pre-outgassed at either $T = 873$ K (H−MFI) or $T = 673$ K (MFI−def). *Solid symbols* ads. I; *open symbols* ads. II. *Solid lines*: polynomial best fitting of H−MFI ads. I and ads. II partial molar heats experimental points .

The larger affinity towards water of the Brønsted acidic H−MFI zeolite with respect to the all-silica MFI−def is well evident, as already pointed out by the Fig. 1.12 volumetric-calorimetric isotherms. The zero-coverage heat of adsorption on the proton-exchanged zeolite ($q_0 \approx 120$ kJ mol^{-1}) was compatible with the proton transfer from the $Si(OH)^+Al^-$ Brønsted acidic site to the NH_3 molecule, in agreement with the standard enthalpy of NH_3 adsorption $\Delta_a H^\circ = -128 \pm 5$ kJ mol^{-1} measured as isosteric heat (*vide infra* Sect. 1.5.2) reported in Ref. [81]. Afterwards, as far as the coverage increased the H–MFI heat values decreased similarly to what observed for the all-silica MFI−def specimen, indicating that a heterogeneous distribution of surface sites was present at the surface of H−MFI as well. The contribution of the aspecific interaction with the nanopores walls (*confinement effect*) characterized by a very low heat of adsorption ($q \approx 10$ kJ mol^{-1}, as measured for Sil−D/MFI−perf, see Fig. 1.15b) was most likely one of the causes for the progressive decrease of the heat of adsorption.

The H−MFI ads. II heat curve was initially virtually coincident with that correspondent to the reversible adsorption of NH_3 on the all-silica MFI−def: $q_0 \approx 80$ kJ mol^{-1}. The heat values for H−MFI ads. II and for MFI−def ads. I = ads. II followed lying in a common curve up to the coverage of ≈ 1.5 mmol g^{-1}. Afterwards, the trend of the H−MFI ads. II curve changed and reached a coverage much larger than that of MFI−def.

In conclusion, the most accurate method for evaluating the differential heat depends on the specific features of the investigated systems, i.e. on the mechanism of the surface sites filling. It is worth mentioning that if the volumetric/calorimetric isotherms are properly fitted by suitable equations and if the mechanism of filling the sites is known, the q^{diff} values are obtained by processing the isotherms, as it was done in Ref. [28], dealing with the adsorption of NH_3 on a highly dehydrated amorphous silica specimen. In that case the assumption of two independent sites was suggested/confirmed by IR spectroscopy and *ab initio* modeling: both volumetric and calorimetric isotherms were successfully simulated by the superposition of two local isotherms, one of which was Langmuir-like.

1.5 Thermodynamics of Adsorption

The adsorption of a gas at a solid surface is exothermic. This is required by the thermodynamic condition for a spontaneous process, illustrated by the Eq. 1.12:

$$\Delta_a G = \Delta_a H - T\,\Delta_a S < 0 \tag{1.12}$$

In fact, adsorption being necessarily accompanied by a decrease in entropy ($\Delta_a S < 0$) in that the degrees of freedom of the molecules in the adsorbed state are lower than in the gaseous state (*vide infra* Sect. 1.5.3 for details), it turns out that the $\Delta_a H$ term, i.e. the enthalpy change accompanying adsorption, must be negative [1, 13, 30].

Heat is not a state function and the value of the heat of adsorption depends on both the experimental conditions and the employed method of measurement. As a consequence, any physical interpretation of the experimentally determined heats of adsorption requires an accurate thermodynamic definition.

Only a few fundamental concepts will be summarized and discussed here, in that a detailed description of the thermodynamics of adsorption is out of the scopes of this Chapter. The interested reader is addressed to the exhaustive review on this subject published in 1992 by Cardona-Martinez and Dumesic [13].

Thermodynamics requires a precise and accurate use of quantities; the convention suggested by IUPAC for a proper thermodynamics nomenclature is here summarized. For any extensive quantity X we define: (i) the mean molar quantity $x = \frac{X}{n}$, which is the quantity normalized to the amount of matter expressed as moles n, and is indicated by the lower case; ii) $\frac{dX}{dn}$, the correspondent differential quantity, which is defined as the derivative of the quantity X with respect to the amount of matter expressed as moles n, and is indicated by the bar lower case: $\overline{x}$.

The quantities of interest are: (i) n, moles of adsorbate; (ii) m, mass of adsorbent; (iii) V, volume; (iv) p, pressure; (v) T, absolute temperature; (vi) R, molar ideal gas constant; (vii) A, surface area of the adsorbent; (viii) Q heat; (ix) U, internal energy; (x) H, enthalpy; (xi) S, entropy and (xii) G, Gibbs free energy. Superscripts refer to: differential quantities (d); experimentally measured quantities (exp); integral quantities (int); gas phase (g), adsorbed phase (s) and solid adsorbent (sol) quantities; standard state quantities (°). Subscript (a) refers to adsorption phenomena (e.g. $\Delta_a H$) [13, 91].

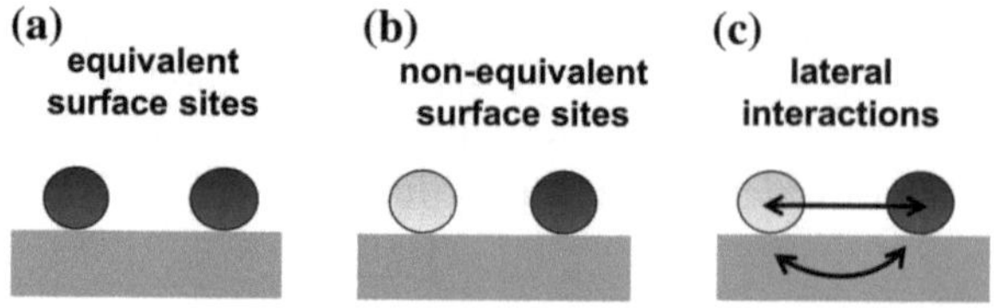

Fig. 1.17 Schematic picture of a real solid surface. Adsorption on: **a** homogeneous distribution of surface sites; **b** heterogeneous distribution of surface sites; **c** homogeneous distribution of surface sites in presence of lateral interaction among adsorbed species, either through the space or through the solid

1.5.1 Heat of Adsorption from Direct Calorimetric Methods

Heats of adsorption derived from direct calorimetric methods are based on the measurement of the heat evolved when a known amount of gas is allowed to adsorb onto a "clean" surface. A "clean" surface is a solid surface kept in high vacuum conditions after having been activated either in vacuo in order to eliminate (either totally or partially) the surface contaminants, or in controlled atmosphere/conditions.

The simplest way envisaged to perform this experiment is to measure the temperature rise in the solid. This very simple and intuitive way is of limited interest because it does not give any information about the kinetics of the heat release [10]. Further, it is worth recalling that not only the magnitude of the heat evolved during adsorption but also its variation upon increasing coverage may reveal useful information concerning the type of adsorbate/surface sites bonding, and its evolution according to the surface heterogeneity. For these reasons the equipment described above (a heat-flow Calvet-type microcalorimeter connected to a high-vacuum volumetric apparatus) is suitable for characterizing in detail the surface chemical features of a solid material. As pointed out in Sect. 1.2 (see Figs. 1.1 and 1.2), the surface of a real solid material is in general characterized by a structural and/or a chemical heterogeneity of the sites, owing to the presence of either structural defects and/or (hetero)atoms in different oxidation state. Another kind of surface heterogeneity, originated by the presence of lateral interactions among adsorbed species, is the so-called induced heterogeneity. The adsorption on either a homogeneous or a heterogeneous distribution of surface sites are schematically illustrated in Fig. 1.17a and b, respectively. Lateral interactions among adsorbed species, either through space or through solid, are envisaged in Fig. 1.17c.

In the following, the thermodynamic features of the direct calorimetric methods for measuring the heat of adsorption will be discussed. The integral heat Q^{int} is the heat evolved when n^s moles are adsorbed at constant T in a *closed gas-solid* system. Because no volume work is done, according to the First Law, the change in internal energy during the adsorption of n^s moles:

$$n^s \left(u^s - u^g\right) = n^s \Delta_a U \tag{1.13}$$

is given by the integral heat Q^{int}:

$$Q^{\mathrm{int}} = n^s\left(u^s - u^g\right) = n^s \Delta_a U \tag{1.14}$$

and the molar integral heat of adsorption is the molar change in internal energy:

$$q^{\mathrm{int}} = \frac{Q^{\mathrm{int}}}{n^s} = u^s - u^g = \Delta_a U \tag{1.15}$$

Differential heat of adsorption q^d is related to the integral heat according to the Eq. 1.16:

$$q^d = \left[\frac{\delta Q^{\mathrm{int}}}{\delta n^s}\right]_{T,A} = \Delta_a \bar{u} \tag{1.16}$$

In a *gas-solid open* system, as the one described above (i.e., a Tian-Calvet type microcalorimeter connected to a high-vacuum gas-volumetric apparatus) the system will exchange with the environment not only heat but also work and matter. Work is due to the reversible, isothermal transfer of matter to both gas ($V^g dp$) and adsorbed ($RT\ dn^s$) phase. So, the isothermal heat measured in the calorimeter at constant T is defined by the Eq. 1.17:

$$Q^{\mathrm{exp}} = n^s\left(u^s - u^g\right) - RT\ n^s - V^g\ p \tag{1.17}$$

and is related to the molar change in internal energy by the Eq. 1.18:

$$\frac{(Q^{\mathrm{exp}} + RT\ n^s + V^g\ p)}{n^s} = u^s - u^g = \Delta_a u \tag{1.18}$$

It turns out that the two Q^{exp} and Q^{int} quantities differ by the amount of work exchanged, as illustrated by the Eq. 1.19:

$$Q^{\mathrm{int}} = Q^{\mathrm{exp}} + RT\ n^s + V^g\ p \tag{1.19}$$

The differential heat of adsorption q^d, obtained by differentiating the experimental integral heats Q^{int} measured in isothermal conditions, is expressed by the Eq. 1.20:

$$q^d = \left[\frac{\delta Q^{\mathrm{int}}}{\delta n^s}\right]_{T,A} = \left[\frac{\delta Q^{\mathrm{exp}}}{\delta n^s}\right]_{T,A} + RT + V^g\left[\frac{\delta p}{\delta n^s}\right]_{T,A} \tag{1.20}$$

The last term in Eq. 1.20 represents either the compression work during adsorption or the expansion work during desorption, and is null when two calorimetric cells (one with the sample, the other as reference element) are connected differentially, as in the case of the equipment here described. Equation 1.20 so simplifies in Eq. 1.21:

$$q^{d} = \left[\frac{\delta Q^{\text{int}}}{\delta n^{s}}\right]_{T,A} = \left[\frac{\delta Q^{\text{exp}}}{\delta n^{s}}\right]_{T,A} + RT \tag{1.21}$$

which is the differential heat of adsorption at constant T and surface area A.

By combining Eqs. 1.21 and 1.18 it turns out that the change in internal energy is related to the experimental heat measured in the way illustrated by the Eq. 1.22:

$$\Delta_{a}U = \left[\frac{\delta Q^{\text{int}}}{\delta n^{s}}\right]_{T,A} = \left[\frac{\delta Q^{\text{exp}}}{\delta n^{s}}\right]_{T,A} + RT \tag{1.22}$$

For processes at constant pressure, we are in general more interested to the enthalpy than to the internal energy change. By considering that the condensed phases depend very little upon pressure p and that $H^{s} \approx U^{s}$, it turns out that the differential form of the enthalpy change is defined by the Eq. 1.23:

$$\Delta_{a}\bar{h} = \Delta_{a}\bar{u} - RT = \left[\frac{\delta Q^{\text{exp}}}{\delta n^{s}}\right]_{T,A} \tag{1.23}$$

From Eq. 1.23 it turns out that the experimental heat measured in a *gas-solid open* system, operating in a differential assembly of calorimetric cells, represents the enthalpy change associated to the adsorption. This result applies to adsorption processes performed in a *gas-solid open* system through the admission of the adsorptive on the solid material kept isothermally within a heat-flow microcalorimeter consisting of two cells in opposition.

The IUPAC nomenclature adopted in this section is not strictly followed by most Authors, as witnessed by the quoted literature and will not be followed here in describing the reported examples.

1.5.2 Heat of Adsorption from Indirect Non-Calorimetric Methods

Another classical method for determining the heat of adsorption is based on the application of the Clausius-Clapeyron equation:

$$q^{\text{st}} = RT^{2}\left[\frac{\delta \ln p}{\delta T}\right]_{ns,A} = \Delta_{a}H \tag{1.24}$$

to a series of adsorption isotherms obtained at (at least) two different temperatures. No calorimetric measurements are needed: the heat of adsorption is evaluated by using only the equilibrium data from the isotherms. The isostere of adsorption$(p\ \textit{versus}\ T)_{\theta}$ is obtained by plotting the equilibrium pressure as a function of the adsorption temperature, at constant coverage θ. The so called isosteric heat of adsorption q^{st} is determined by introducing in Eq. 1.25 (which is obtained by integrating Eq. 1.24) the

p_1 and p_2 values at T_1 and T_2, for a given constant coverage θ:

$$\ln\left(\frac{p_1}{p_2}\right) = \frac{q^{st}}{R}\left[\left(\frac{1}{T_2}\right) - \left(\frac{1}{T_1}\right)\right] \tag{1.25}$$

It is worth recalling that severe restrictions are imposed in order to obtain accurate isosteric heats data: (i) the adsorption is reversible; (ii) the gas partial molar volume is much greater than the adsorbate volume; (iii) the gas behaves ideally; (iv) the surface state does not vary during measurements; (iv) the heat of adsorption does not vary with T. For this last reason T_1 and T_2 should be as close as possible.

Dunne et al. [54] employed the Clapeyron equation for determining the isosteric heat of a variety of molecules adsorbed on Silicalite. The obtained data were compared by the same authors to the heat of adsorption measured calorimetrically by means of a home-made volumetric-calorimetric apparatus. The agreement between the two methods was fairly good.

The isotherms required to evaluate the isosteric heat may be obtained by IR spectroscopy measurements. In Ref. [92] it was demonstrated that the variable-temperature FTIR spectroscopy, with the simultaneous measurement of temperature and equilibrium pressure, may be a suitable method for obtaining the isosteric heat for an ideal Langmuir-like adsorption process. Optical isotherms of N_2 adsorption on H–ZSM5 were determined (in the 104 –183 K temperature range) from the change upon adsorption of the integrated intensity of the ν_{OH} stretching band at 3616 cm^{-1}, typical of the Brønsted acidic site $Si(OH)^+Al^-$. The adsorption enthalpy so evaluated was $\Delta_a H° = -19.7 \pm 0.5$ kJ mol^{-1}, in good agreement with the calorimetric heats ($q = 19$ kJ mol^{-1}) measured at $T = 195$ K for N_2 on H–ZSM5 by Savitz et al.[93]. By the same method the standard enthalpy of adsorption of NH_3 on a H–ZSM5 zeolite was evaluated as $\Delta_a H° = -128 \pm 5$ kJ mol^{-1}, [81] which is in reasonably good agreement with the zero-coverage calorimetric heat of adsorption reported in Fig. 1.16 ($q_0 \approx 120$ kJ mol^{-1}).

1.5.3 Entropy of Adsorption

In order to illustrate the role played by entropy S in adsorption processes, the entropy change $\Delta_a S$ accompanying the adsorption in a very simple case was computed, by using the standard statistical mechanics rigid rotor/harmonic oscillator formula[94]. In Fig. 1.18 the adsorption of an Ar atom at the surface of an apolar solid is schematically illustrated. The Ar atom approaches the solid surface from the gas phase: the translation entropy of the solid, which is fixed in the space, is taken as zero, whereas the free Ar atoms, before interacting with the solid surface, possess a translational entropy S_t which amounts to 150 and 170 J mol^{-1} K^{-1} at $T = 100$ and 298 K, respectively, at $p_{Ar} = 100$ Torr.

The adsorbed atom, which is not allowed to translate freely anymore, starts vibrating. Taking as reasonable vibrational frequency $\approx$100 cm^{-1}, the adsorbed atoms

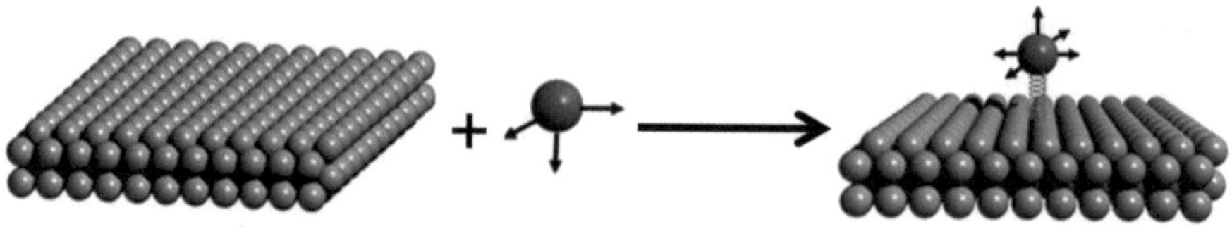

Fig. 1.18 Cartoon representing the adsorption of an Ar atom approaching a solid (apolar) surface from the gas phase

vibrational entropy is $S_v = 18$ and $43\,\mathrm{J\,mol^{-1}\,K^{-1}}$ at $p_{Ar} = 100$ Torr and at $T = 100$ and 298 K, respectively. As expected, the adsorbed atoms entropy is much lower than the free atoms entropy, and the entropy change $\Delta_a S = (S_v - S_t)$ is in all cases negative: $\Delta_a S° = -132\,\mathrm{J\,mol^{-1}\,K^{-1}}$ at $T = 100\,\mathrm{K}$ and $\Delta_a S° = -127\,\mathrm{J\,mol^{-1}\,K^{-1}}$ at $T = 298\,\mathrm{K}$.

This means that in a spontaneous process, which requires a negative free energy change (see Eq. 1.12), the enthalpy of adsorption must be negative in order to compensate the loss of entropy. In other words, the process must be exothermic of an amount of heat evolved at least as high as the decrease of the $T\Delta_a S$ term.

The adsorbate can be held at the surface more or less strongly and so the negative entropy change (with respect to the free molecules in the gas phase) will be more or less pronounced, according to the nature/structure of the adsorbate and to the nature and strength of the surface atoms/molecules bonds.

From a molar point of view, the term entropy of adsorption covers a great number of different functions and it is required to specify whether the function considered is a derivative or an integral, and if it refers to an equilibrium state (p, T) or to a standard state ($p°$, T) [95, 96].

The integral molar entropy of adsorption is the difference between the entropy of an adsorbed mole and the entropy of the adsorptive in the ideal gas state, at given p and T. It is a mean integral quantity taken over the whole amount adsorbed and it is characteristic of a given state of equilibrium. This is distinguished by the standard integral molar entropy of adsorption, which is the entropy of one adsorbed mole with respect to the entropy of the adsorptive in the ideal gas state at the same T, but under standard pressure.

It is worth recalling that the entropy of adsorption may be obtained from calorimetric experiments only if the heat exchange is reversible. A formula for evaluating the standard adsorption entropy $\Delta_a S°$ from a reversible adsorption volumetric-calorimetric data was proposed by Garrone et al. [91] and applied to a selection of quasi-ideal systems, [97] consisting of CO adsorbed on non d/d^0 metal oxides, at the surface of which *cus* cations acting as Lewis acidic sites were exposed. An isothermal microcalorimeter with a discontinuous (stepwise) introduction of the adsorptive, as the one described here, was fruitfully employed.

The formation of surface Lewis complexes through a plain σ-coordination being fully reversible upon evacuation of p_{CO}, this process entirely fulfilled the requirements imposed by the method. Equation 1.26 was employed:

$$\Delta_a S = \frac{q}{T} + R - RT\ \ln \frac{p_{1/2}}{p^\circ} \tag{1.26}$$

In the right-hand side of the formula, the enthalpic and the ergonic contributions to the standard entropy of adsorption are recognized. The enthalpic term represented by the Eq. 1.27:

$$\Delta_a H = \frac{q}{T} + R \tag{1.27}$$

is obtained from the calorimetric data, whereas the ergonic term represented by the Eq. 1.28:

$$\Delta_a G = -RT\ \ln \frac{p_{1/2}}{p^\circ} \tag{1.28}$$

is obtained from the adsorption isotherms. The term $p_{1/2}$ is the equilibrium pressure at half-coverage.

The method is here applied to a very simple but instructive case: CO adsorbed *via* electrostatic polarization,[23] on Na–MFI and K–MFI pre-outgassed at $T = 673$ K. The adsorption represents in both cases an ideal process, which is characterized (within the experimental error) by a Langmuir-like behavior, as evidenced by the adsorption isotherms (*vide supra* Fig. 1.7 in Sect. 1.3.1). The calorimetric heat of adsorption was ca. 35 and 28 kJ mol^{-1} for Na–MFI and K–MFI, respectively (*vide supra* Sect. 1.4.2.3, Fig. 1.14a, b). The half-coverage equilibrium pressure (obtained by the adsorption isotherms) were $p_{1/2} = 200$ Torr for Na$-$MFI and 850 Torr for K$-$MFI. In both cases the reference state for the gas phase was $p^\circ = 1$ Torr, as done in previous work (see Ref. [18, 97]). The obtained standard adsorption entropy was $\Delta_a S^\circ = -151$ J mol^{-1} K^{-1} for Na–MFI and -140 J mol^{-1} K^{-1} for K$-$MFI.

Note that: (i) the loss of entropy for CO adsorbed at a polar surface through electrostatic forces was slightly larger than for the aspecific interaction of Ar atoms adsorbed at an apolar surface discussed above; (ii) the loss of entropy for CO adsorbed on Na$-$MFI was larger than on K$-$MFI, in agreement with the larger strength of adsorption of CO on Na^+ than on K^+ sites.

In the following, the evaluation of the adsorption entropy change for the slightly more complex case of CO σ-coordinated (at $T = 303$ K) on a variety of TiO_2 – anatase specimens (pre-outgassed at $T = 673$ K) will be illustrated. At the TiO_2 dehydrated surface, CO was adsorbed giving rise to two adspecies, as witnessed by two distinct IR bands located at $\nu_{CO} = 2188$ and 2206 cm^{-1}, as reported in Ref. [18] As illustrated schematically in Fig. 1.19 the two adspecies were formed on two different Lewis acidic sites made up of structurally different *cus* Ti^{4+} cations. They were named species A and B, and their spectroscopic and energetic features are summarized in the figure.

Species A ($\nu_{CO} = 2188$ cm^{-1}) were formed at the 5-coord Ti^{4+} cations typically exposed at the flat planes of anatase nanocrystals, whereas species B ($\nu_{CO} = 2206$ cm^{-1}) were formed at the 4-coord Ti^{4+} cations, which are exposed at the steps, corners and kinks of the flat planes [98]. As extensively illustrated in Ref. [18] species – A ν_{CO} moved from 2188 down to 2184 cm^{-1}as far as the

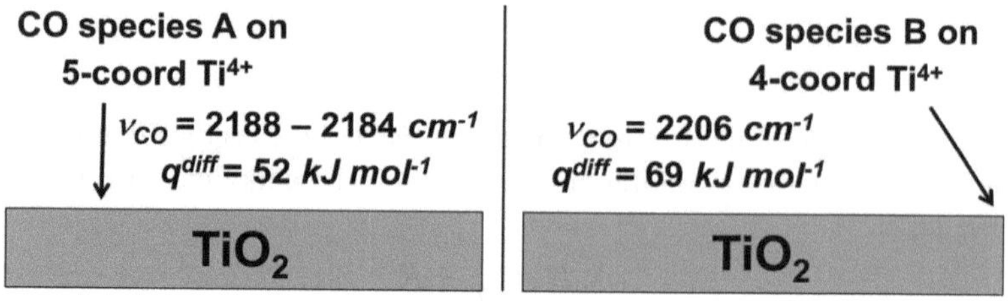

Fig. 1.19 Spectroscopic and energetic features of species A and B formed upon adsorption of CO at $T = 303$ K on TiO_2–anatase pre–outgassed at $T = 673$ K. *Left side* Species A formed on 5-coord Ti^{4+} cations exposed at the flat planes of anatase nanocrystals. *Right side* Species B formed on 4-coord Ti^{4+} cations, exposed at steps, corners and kinks of the flat planes

coverage increased, suggesting the presence of surface heterogeneity among CO adspecies. Conversely, species-B ν_{CO} remained constant at 2206 cm^{-1} at any coverage examined, suggesting that such species were formed at structurally equivalent and non-interacting sites. The heat of adsorption values were found to be coverage-independent for both species A and B: $q^{diff} \approx 52$ kJ mol^{-1} and ≈ 69 kJ mol^{-1}, respectively.

According to the constancy of both ν_{CO} spectral position and heat of adsorption, species–B adsorption was considered Langmuirian and the Eq. 1.26 could be applied straightforward. The standard molar entropy of adsorption was so evaluated from the calorimetric data ($q^{diff} \approx 69$ kJ mol^{-1}) and the Langmuir adsorption isotherm ($p_{1/2} \approx 7$ Torr), and turned out to be $\Delta_a S° = -237$ J mol^{-1} K^{-1}. Note that this value is larger than both $\Delta_a S° = -151$ J mol^{-1} K^{-1} and $\Delta_a S° = -140$ J mol^{-1} K^{-1} obtained for the weak physical adsorption of CO on Na– and K–MFI (*vide supra*). The larger loss of entropy in the TiO_2 case with respect to the alkaline-metal exchanged zeolites was due to the stronger $Ti^{4+} \leftarrow CO$ bond, responsible for the associative chemisorption.

Conversely, as far as species A are concerned, the adsorption was found to deviate from the ideal Langmuirian behavior, as witnessed by the variation of ν_{CO} with coverage. The apparent contradiction with the constancy of the heat of adsorption was interpreted as due to the quite scarce population of sites A active towards CO at $T = 303$ K, which prevented an accurate determination of the evolution of the heat of adsorption with coverage [18].

In any case, Langmuir model did not hold in describing species–A thermodynamic features and Eq. 1.26 employed for species B did not apply. So, the entropy of CO adsorption at A sites was obtained by considering that, in the very early stage of adsorption, the coverage θ increased linearly with the p_{CO} [18]. By employing the Eq. 1.29, based on the Henry-like isotherm:

$$\Delta_a S = \frac{q}{T} + R + R \ln h \tag{1.29}$$

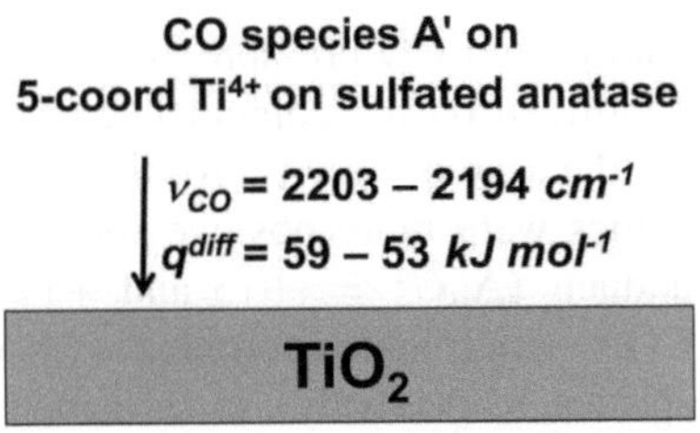

Fig. 1.20 Spectroscopic and energetic features of species formed on 5-coord cations exposed at the flat planes of dehydrated sulphated TiO_2–anatase nanocrystals

the low-coverage standard molar entropy of adsorption was evaluated from the calorimetric data ($q^{diff} \approx 52\,\text{kJ mol}^{-1}$) and the Henry adsorption isotherm (h = 0.027 Torr^{-1}). It turned out to be $\Delta_a S° = -194\ \text{J mol}^{-1}\ \text{K}^{-1}$. This datum, which is lower than $\Delta_a S° = -237\,\text{J mol}^{-1}\ \text{K}^{-1}$ obtained for species B, is in agreement with the less stronger interaction of CO with sites A with respect to sites B, which implied a lower loss of degree of freedom of the CO molecule.

The heterogeneity features of the 5-coord $Ti^{4+} \leftarrow$ CO complex were investigated in more detail on the dehydrated sulphated–TiO_2 anatase, which exhibited a population of 5-coord $Ti^{4+} \leftarrow$ CO adducts larger than that of the unsulphated specimen [18]. No B – *type* CO adspecies were formed, as witnessed by the presence of a single ν_{CO} band, initially located at $2203\,\text{cm}^{-1}$, as reported in Ref. [18]. The spectroscopic and energetic features of the single CO adspecies formed at the *sulphated* – TiO_2 surface, which will be named species A′, are summarized and schematically illustrated in Fig. 1.20. The presence at the surface of strong electron-withdrawing sulphate moieties caused the CO oscillators frequency to increase for sulphated anatase with respect to the unsulphated specimen ($2203\,\text{cm}^{-1}$ vs. $2188\,\text{cm}^{-1}$, respectively). The same occurred for the heat of adsorption which was found to increase from $q^{diff} \approx 52\,\text{kJ mol}^{-1}$ (unsulphated anatase) up to $\approx 59\,\text{kJ mol}^{-1}$ (sulphated specimen). Note also that both CO stretching frequency ($\nu_{CO} = 2203 - 2194\,\text{cm}^{-1}$) and heat of adsorption ($59 - 53\,\text{kJ mol}^{-1}$) varied upon increasing coverage, clearly indicating the presence of induced heterogeneity among the 5-coord $Ti^{4+} \leftarrow$ CO adspecies. This datum confirmed the ability of TiO_2 to transmit electronic effects [18, 22].

The adsorption of species A′ being non-Langmuirian, Eq. 1.26 could not be applied and the low-coverage entropy change for species A′ was evaluated by employing the formula based on the Henry isotherm, with $h = 0.073\ \text{Torr}^{-1}$. A value of $\Delta_a S° = -208\,\text{J mol}^{-1}\ \text{K}^{-1}$ was obtained, which is larger than that obtained for species A ($\Delta_a S° = -194\,\text{J mol}^{-1}\ \text{K}^{-1}$) and lower than that obtained for species B ($\Delta_a S° = -237\,\text{J mol}^{-1}\ \text{K}^{-1}$). According to the fact that the CO entropy loss is lower when the molecule is less tightly bound, the sequence of entropy values for species A, A′ and B was in agreement with the sequence of the enthalpy values ($q^{diff} \approx 52$, 59 and $69\,\text{kJ mol}^{-1}$, respectively).

$\Delta_a G°$ values, obtained by combining the $\Delta_a S°$ and $\Delta_a H°$ terms, were positive in all cases reported. The endoergonic character of the process indicated the non-spontaneity of the adsorption unless a CO pressure is applied: by evacuating the equilibrium CO pressure, $Ti^{4+} \leftarrow CO$ complexes were completely destroyed, as confirmed by the coincidence of the 1st and 2nd run of adsorption (not reported for the sake of brevity; for details see Ref. [18]). It is worth of noticing that $\Delta_a G^\circ$ values were more positive for the weak electrostatic $Na^+ \cdots CO$ and $K^+ \cdots CO$ adducts ($\Delta_a G^\circ = +13$ and $+17\,kJ\ mol^{-1}$, respectively), than for the $Ti^{4+} \leftarrow$ CO σ−coordinated complexes ($\Delta_a G^\circ = +9$, $+7$ and $+6\,kJ\ mol^{-1}$ for A, A′ and B species, respectively), in agreement with the increased strength of the coordinative chemisorption with respect to the plain polarization of the CO molecule.

In the reported examples, an isothermal microcalorimeter with a discontinuous (stepwise) introduction of the adsorptive was fruitfully employed. An alternative, suitable way to collect the experimental data required for determining the enthalpy and entropy changes accompanying the adsorption process is to follow the procedure implying the slow-and-constant adsorptive introduction which was extensively described in Ref. [99].

The adsorption entropy of N_2 and of NH_3 on a H−MFI zeolite was evaluated starting from the correspondent optical Langmuir-like isotherms (*vide supra* Sect. 1.5.2), as reported for N_2 in Ref. [92] and for NH_3 in Ref. [81]. The entropy change was negative for both adsorptives, but much more negative for the latter ($\Delta_a S^\circ = -184 \pm 10\,J\ mol^{-1}\ K^{-1}$) than for the former adsorbate ($\Delta_a S^\circ = -125 \pm 5\,J\ mol^{-1}\ K^{-1}$), as expected on the basis of the larger adsorption strength of NH_3 than of N_2.

1.6 Adsorption of a Single Component: Physisorption Versus Chemisorption

Forces governing adsorption depend on the nature of both adsorbent and adsorptive. They are distinguished in: (i) van der Waals/London-dispersion forces, similar to the forces leading to the non-ideal behavior of gases and eventually to the formation of a liquid phase; (ii) chemical (covalent) forces leading to the formation of a true chemical bond between the adsorbate and surface atoms. In the former case the process is called physical adsorption (*physisorption*), in the latter case chemical adsorption (*chemisorption*), [2, 32]; (iii) H–bonding interactions are variably classified as responsible for either physical or chemical adsorption. In fact, H–bonding interactions originate surface complexes of variable stability according to the nature of the involved species [100].

The total potential energy of the interacting moieties is quantitatively described by the Lennard-Jones potential, [1, 2, 30, 32] which includes attractive and repulsive energy terms.

Physisorption is intrinsically weak and is characterized by heats of adsorption relatively small, close to the enthalpy of liquefaction of the adsorptive, typically comprised in the 5−45 $kJ\ mol^{-1}$ range; it is in general favored by temperatures close to the boiling point of the adsorptive. As a consequence of the non-covalent nature of the interaction, adsorbate and adsorptive are chemically equivalent moieties.

Chemisorption, consisting of a chemical reaction confined to the solid surface, does involve rearrangement of electrons of both adsorptive molecules and surface atoms, yielding new surface terminations. Adsorptive and adsorbate being chemically different species in dissociative chemisorption, spectroscopic and/or *ab initio* modeling methods are required to assess the nature of surface species formed upon contact of the adsorptive with the reactive surface atoms [25, 26, 29]. Further, *chemisorption* is structure-sensitive in that the features of the process depend on the solid crystal structure (see for instance anatase vs. rutile, [56, 101] and amorphous silica vs. crystalline quartz, [15, 85, 102]) and on the crystal faces exposed by the solid material [103].

Chemisorption heats are generally larger than *physisorption* heats, and are typically comprised in the 80–400 kJ mol^{-1} range. Dissociative chemisorption is distinguished from associative chemisorption, which involves the coordination of the molecules as such [16, 23, 26, 29]. In this latter case, the enthalpy change is not in general as large as for dissociation: it can be even lower than $\approx$80 kJ mol^{-1}, as for instance for the case of CO adsorbed on *cus* non d/d^0 metal cations described above. The enthalpy change for adsorption driven by H–bonding interactions may cover a large range of values, in that the correspondent interaction energy depends in a complex way on the chemical nature of the participating atoms, as well as on the distances and angles between them. For instance, the zero-coverage enthalpy change for H_2O adsorption on hydroxylated silica, which is characterized by two H–bonds per adsorbed molecule on adjacent Si–OH terminations, is $-(\Delta_a H)_0 \approx$60 kJ mol^{-1} for amorphous silica and $\approx$80 kJ mol^{-1} for crystalline silica [15, 49]. This datum does suggest that (at least at low coverage) H–bonding interactions exhibit a certain degree of specificity, similarly to *chemisorption*. The energy of H–bonding interactions at the water/all-silica zeolites interface is even larger than for plain silica as indicated by a zero-coverage adsorption enthalpy of $\approx$100 kJ mol^{-1} [25].

Conversely, the enthalpy change for H_2O adsorption on dehydroxylated silica, which is characterized by only one H–bond per adsorbed molecule on isolated Si–OH terminations, is quite low: $-\Delta_a H < 44$ kJ mol^{-1}, and virtually surface-coverage independent [15, 56, 85].

Also CH_3OH adsorption on hydroxylated silica is structure-sensitive: on crystalline silica the H–bonding adsorption enthalpy at vanishing coverage is $-(\Delta_a H)_0$ $\approx$90 kJ mol^{-1},[15] whereas on amorphous silica is only $\approx$70 kJ mol^{-1} [26].

NH_3 is adsorbed on all-silica zeolites and on hydroxylated amorphous silica with a zero-coverage adsorption enthalpy as high as $\approx$80 kJ mol^{-1},[24] whereas on dehydroxylated silica $-(\Delta_a H)_0$ is only $\approx$60 kJ mol^{-1} [28].

In all cases, at increasing equilibrium pressure the adsorption approaches the adsorptive liquid phase, and the adsorption enthalpy approaches the adsorptive latent enthalpy of liquefaction $-\Delta_L H$, as typically occurs for physical adsorption at $\frac{p}{p^\circ} \to 1$ [28, 56, 85, 104].

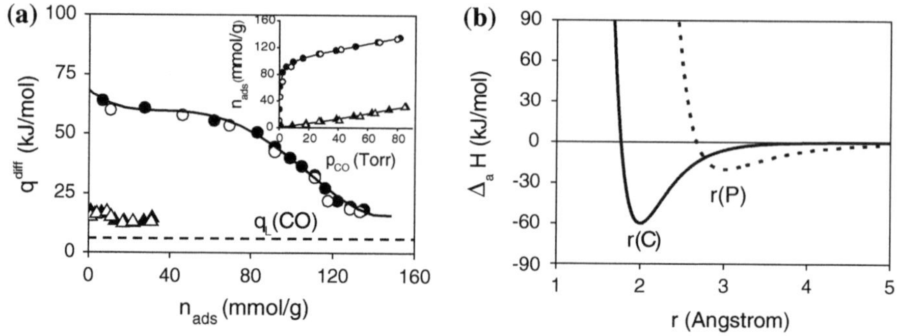

Fig. 1.21 **a** Differential heats of adsorption of CO, at $T = 303$ K, on proton-exchanged zeolite H–BEA zeolite (*circle*) and on defective MFI–Silicalite (*up triangle*) as a function of the increasing coverage. Inset: volumetric isotherms. H–BEA zeolite was pre-outgassed at $T = 873$ K, MFI–Silicalite at $T = 673$ K. *Solid symbols* ads. I, *open symbols* ads. II. *Dashed line* Latent heat of liquefaction of CO, $q_L = 6\,\text{kJ mol}^{-1}$. **b** One dimension potential energy diagram for physical adsorption (*dotted line*) and associative chemical adsorption (*solid line*). See text for details

In the following, the adsorption of CO on a H–BEA zeolite specimen, characterized by both Brønsted and Lewis acidic sites (consisting these latter of structural defects in zeolite framework), will be discussed.

The room temperature adsorption of CO on protonic zeolites allowed Lewis acidic sites to be calorimetrically discriminated from the Brønsted acidic ones: Al(III) $\leftarrow$ CO adducts of relative stability were formed at the *cus* Al(III) Lewis sites, whereas only very weak H–bonding adducts were formed at the Brønsted $Si(OH)^{+}Al^{-}$ acidic sites [73].

In Fig. 1.21a, the differential heats of adsorption of CO on H–BEA zeolite and on MFI–Silicalite are reported as a function of the adsorbed amounts. Volumetric isotherms are illustrated in the figure inset. In both cases the adsorption was fully reversible upon evacuation of the CO pressure, as typical of both physical and weak, associative chemical adsorption. For H–BEA a constant heat plateau at $q \approx 60\,\text{kJ mol}^{-1}$ was measured. This value is typical of a specific interaction of CO with coordinative unsaturated Al(III) atoms, as it was confirmed by combining adsorption microcalorimetry and molecular modeling [73, 74, 78, 89] Note that the heat value was close to the heat of adsorption of CO at *cus* Al(III) sites on transition catalytic alumina, a typical Lewis acidic oxide [55, 73]. Once saturated the Al(III) defects, the heat of adsorption started decreasing down to values typical of the H–bonding interaction of CO with the Brønsted acidic sites ($-\Delta_a H \approx 30\,\text{kJ mol}^{-1}$, as reported by Savitz et al. [93]) and with polar defects, either confined in the zeolite nanopores or at the external surface.

Conversely, the heat of adsorption of CO at polar sites located in the all-silica nanopores (Si–OH nests) was much lower ($q \approx 17\,\text{kJ mol}^{-1}$) than on H–BEA, and virtually constant over the whole coverage examined in agreement with the formation of very weak H–bonding adducts, as described in Refs. [73, 105]. The adsorption capacity of the all-silica zeolite was dramatically lower than that of the proton-exchanged zeolite, as witnessed by the correspondent isotherms reported in the inset

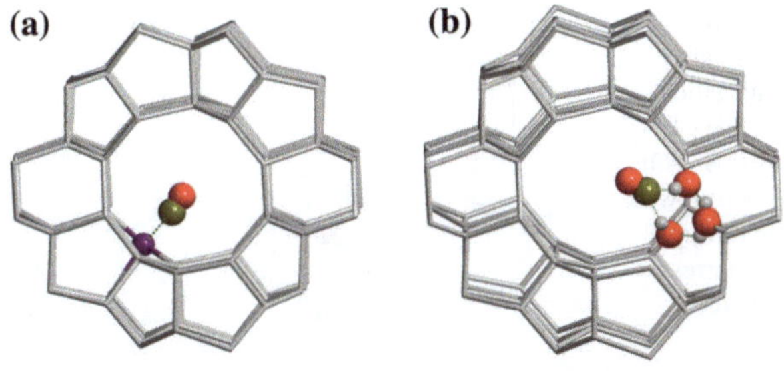

Fig. 1.22 Cartoon representing the CO adducts formed within a zeolite nanopore. **a** CO σ-coordinated on trigonal Al(III) atom (Lewis acidic site). **b** CO H−bonded at polar sites Si−OH (nests). Courtesy of Prof. Piero Ugliengo, University of Torino

of Fig. 1.21. In Fig. 1.21b a schematic one-dimension potential energy diagram is illustrated, including on one hand the CO physical adsorption on MFI–Silicalite and on the other hand the associative chemical adsorption on H−BEA. The classical Lennard-Jones formula,[2, 30, 32, 106] was used. Note that both physical and chemical adsorption of CO, which occur in different zeolites in the present example, are reported in the same plot just for comparison purposes. The right-side curve represents *physisorption*. The Si−OH nests in MFI–Silicalite are the P sites at which CO is adsorbed *via* weak H−bonding interaction and r(P) is the equilibrium separation between the two interacting moieties (in this case set to ≈3 Å). The enthalpy change $-\Delta_a H$(P) is set at ≈17 kJ mol^{-1}, according to the heat of adsorption reported in Fig. 1.21a. The left-side curve represents the enthalpy diagram for chemisorption. C are the associative-chemisorption sites at which CO is σ−coordinated, i.e. the *cus* Al(III) cations acting as Lewis acidic sites. r(C) is the equilibrium separation between the two interacting moieties (in this case set to ≈2.5Å); $-\Delta_a H$(C) is the negative enthalpy change (≈60 kJ mol^{-1}) according to the heat of adsorption plateau reported in Fig. 1.21a.

In Fig. 1.22 a cartoon illustrating the formation in a zeolite nanopore of the Al(III) ← CO complex (Fig. 1.22a) at the Lewis acidic site, and of the Si−OH···CO H-bonding adduct (Fig. 1.22b) is reported.

The same kind of one-dimension potential energy diagram (not reported for the sake of brevity) can be applied to compare NH_3 interacting *via* H−bonding with polar sites (Si−OH nests) and NH_3 interacting *via* dispersion forces with the nanopore walls of an all-silica zeolite (*confinement effect*). In Fig. 1.23a the former interaction is illustrated, in Fig. 1.23b the latter one. The H−bonding interaction of NH_3 with Si−OH nests described in Fig. 1.23a cartoon is quite strong, as witnessed by the relatively high zero-coverage heat values measured for defective MFI–Silicalite (q_0 ≈80 kJ mol^{-1}, see the heat plots in Figs. 1.15b and 1.16). The process could be reasonably classified as associative chemisorption. Conversely, the aspecific interaction of NH_3 with the nanopore walls (Fig. 1.15b) is much weaker, as inferred by the adsorption heat measured for NH_3 adsorbed on defects-free MFI–Silicalite

Fig. 1.23 Cartoon representing the NH_3 adducts formed within an all-silica zeolite nanopore. **a** NH_3 H–bonded at polar sites (Si–OH nests). **b** NH_3 interacting with the nanopore walls through dispersion forces. Courtesy of Prof. Piero Ugliengo, University of Torino

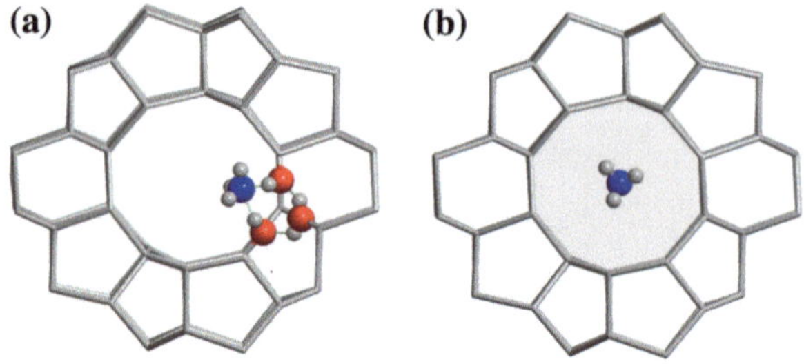

($q \approx 10\,\mathrm{kJ\,mol^{-1}}$, see Fig. 1.15b), which remained constant with increasing coverage, in that insensitive to possible surface heterogeneity.

As an example of associative/dissociative chemisorption, the interaction of CH_3OH with a Ca-modified amorphous silica specimen will be discussed. This latter material was adopted as a model system to investigate the role of Ca species in the development of bioactivity in sol-gel glasses (see Ref. [26] and references therein). The interaction involved both physical and chemical adsorption. The former process consisted of H–bonding interactions (the only ones occurring at the surface of the parent plain silica specimen, data not reported for the sake of brevity), which originated rather labile adducts on silica polar terminations and were eliminated by pumping off at the adsorption temperature. *Chemisorption* involved both associative and dissociative processes. The associative adsorption at the surface *cus* Ca^{2+} cations (acting as Lewis acidic sites), yielded coordinated CH_3OH species, a fraction of which was eliminated by evacuation of the methanol pressure at the adsorption temperature (*weak associative chemisorption*). Another fraction was eliminated only by outgassing at temperatures as high as $T = 423$ K, according to the relatively high energy of interaction, $q_0 \approx 100\,\mathrm{kJ\,mol^{-1}}$, measured in the early stage of the adsorption (*strong associative chemisorption*). A further fraction was found to resist outgassing up to $T > 573$ K, and was assigned to a dissociative chemical adsorption. Dissociative adsorption consisted of a chemical reaction at Ca sites, yielding Si – OCH_3 and Ca–OH as new surface terminations. This process was slow, slightly less exothermic than associative *chemisorption* and was found to depend on both CH_3OH pressure and contact time, as illustrated in Fig. 1.24, in which the volumetric isotherms (Fig. 1.24a) and the q^{diff} *versus* n_{ads} plots (Fig. 1.24b) are reported for a first (ads. IA and ads. IIA) and a second set of measurements (ads. IB and ads. IIB). A large difference between ads. IA and ads. IIA curves is well evident in both volumetric isotherms and heat of adsorption plots, indicating the presence of an irreversible component due to a chemical adsorption. The standard ads-des-ads protocol was followed until a fourth adsorption run without getting evidence that irreversible adsorption was extinguished. Data for the third (ads. IIIA) and the fourth (ads. IVA) runs, see Ref. [26], are not reported for the sake of brevity. The reported results indicated that the process originating irreversibly adsorbed species, consisting of both strong associative and dissociative *chemisorption*, as demonstrated by IR spectroscopy and *ab initio* modeling, [26] was not accomplished within the second adsorption (ads. IIA), as typically observed, but

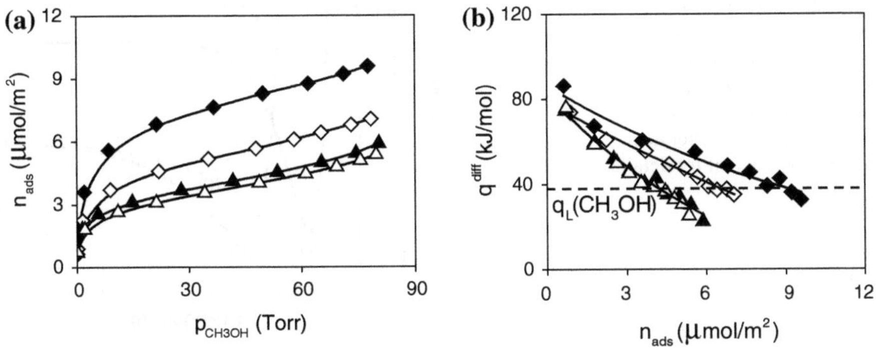

Fig. 1.24 Adsorption (at $T = 303$ K) of CH_3OH vapor on Ca–modified silica outgassed at $T = 423$ K. **a** Volumetric isotherms. **b** q^{diff} *versus* n_{ads} plots. Ads. IA: *diamond, solid symbols*; ads. IIA: diamond, *open symbols*. Ads. IB (run after evacuating overnight the sample previously kept 3 days in contact with ≈80 Torr of CH_3OH): *triangle, solid symbols*; ads. IIB: *triangle, open symbols*. Adapted from Ref. [26] Fig. 12

continued in all subsequent adsorption steps. The irreversible adsorption was found to represent a progressively decreasing component of the overall uptake, and for the fourth adsorption run became almost negligible. After the fourth adsorption the sample (still in the calorimeter at $T = 303$ K) was kept in contact with CH_3OH pressure ($p \approx 80$ Torr) for 3 days, and then it was evacuated overnight at the same temperature. A new set of ads-des-ads runs were then performed (ads. IB and ads. IIB). The presence of a new measurable irreversible component was detected, as witnessed by the non-coincidence of ads. IB and ads. IIB isotherms reported in Fig. 1.24a. This datum confirmed that the prolonged contact of the surface with a large CH_3OH pressure allowed (some of the) coordinated species to react, yielding an increased amount of Si–OCH_3 and Ca–OH surface species. The q^{diff} *versus* n_{ads} plots reported in Fig. 1.24b were typical of highly heterogeneous surfaces, in that heat values progressively decreased with increasing coverage. The zero-coverage heat values for the first set of data were $q_0 \approx 100\,\mathrm{kJ\,mol^{-1}}$ for ads. IA and $q_0 \approx 80\,\mathrm{kJ\,mol^{-1}}$ for ads. IIA. The ads. IA curve crossed the heat of liquefaction of methanol line ($q_L = 38\,\mathrm{kJ\,mol^{-1}}$) at a CH_3OH coverage of ≈9.0 μmol m^{-2}, corresponding to ≈5.4 molecules nm^{-2}, whereas the ads. IIA curve at a CH_3OH coverage of ≈6.5 μmol m^{-2}, corresponding to ≈3.9 molecules nm^{-2}. The ads. IB and IIB curves, even starting from a q_0 value close to the of ads. IIA, decreased together much more sharply than that of ads. IA and ads. IIA and approached steeply values lower than the heat of liquefaction of methanol, already at a CH_3OH coverage of ≈4.2 μmol m^{-2}, corresponding to ≈2.5 molecules m^{-2}. IR spectroscopy data, also combined with *ab initio* calculations, confirmed a precursor-mediated kinetics, in which coordinated CH_3OH was a precursor for the methoxylation reaction [26].

In Fig. 1.25 the one-dimension potential enthalpy diagram for physical adsorption (dotted line) and for associative (dot-dashed line) and dissociative (solid line) chemical adsorption schematically illustrates the features of the interactions described

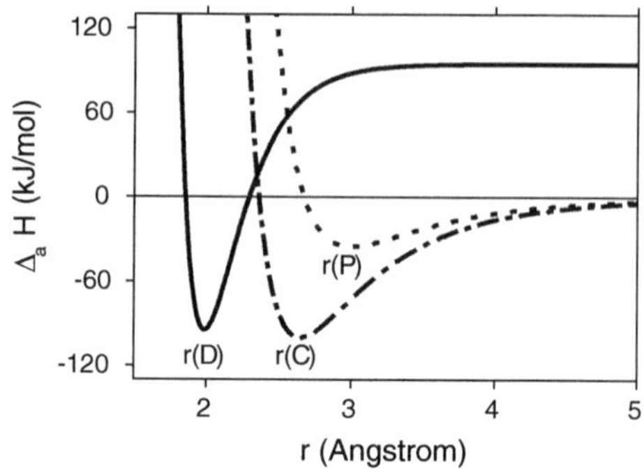

Fig. 1.25 One-dimension potential energy diagram for physical adsorption (P, *dotted line*), and for associative (C, *dashed line*) and dissociative (D, *solid line*) chemical adsorption. See text for details

above. The classical Lennard-Jones formula,[106] was used for *physisorption* and coordinative *chemisorption*, whereas a Morse expression was used for dissociative *chemisorption* [107].

The right-side curve represents physisorption. P indicates the physical adsorption site exposed at the silica matrix surface (polar terminations Si−OH) at which CH_3OH is weakly H−bonded; *r (P)* is the equilibrium separation between the two interacting moieties (set to ≈3Å) and $-\Delta_a H\ (P) \approx 35\,\mathrm{kJ\ mol^{-1}}$ is the enthalpy of adsorption. The left-side curves represent the enthalpy diagram for *chemisorption*. Curve C represents the coordination of CH_3OH at Ca sites (associative *chemisorption*), whereas curve D represents the methoxylation reaction of coordinated methanol (dissociative *chemisorption*). The equilibrium separation r(C) between the two interacting moieties is set to ≈2.5Å, whereas the equilibrium separation r(D) in dissociative *chemisorption* is set to ≈2.0Å, a lower value than r(C). The enthalpy of adsorption $-\Delta_a H$(C) at r(C) was set to $\approx 100\,\mathrm{kJ\ mol^{-1}}$, whereas $-\Delta_a H$(D) at r(D) was set to $\approx 95\,\mathrm{kJ\ mol^{-1}}$, according to the obtained experimental calorimetric and computed *ab initio* values [26]. By the inspection of the enthalpy profile illustrated in Fig. 1.25, it is clearly evident that the presence of molecularly adsorbed precursor C facilitates the rupture and formation of new chemical bonds at methoxylation site D, in that the energy barrier hindering the dissociation of methanol is dramatically lowered starting from the coordinated state instead of from the gas phase. It was calculated that coordination of CH_3OH at Ca sites is thermodynamically favored with respect to the methoxylation reaction, in agreement with the experimental results indicating that, before the prolonged contact with methanol pressure, only a fraction of the total uptake remained irreversibly bound to the surface.

1.7 Conclusions

In this preliminary Chapter it has been illustrated that adsorption microcalorimetry is very fruitfully employed in describing quantitatively the processes occurring at the gas-solid interface. The population of the surface sites active towards suitably chosen probe molecules is evaluated through the volumetric adsorption isotherms; the surface sites heterogeneity (either structural, chemical and/or induced) is described by monitoring the evolution of the heat of adsorption with increasing coverage.

By combining the adsorption microcalorimetry results with spectroscopic and/or *ab initio* modeling methods, the knowledge of the chemical nature of the pristine surface terminations and of the adsorbed species allows to interpret at molecular level the intrinsically molar volumetric-calorimetric data. We will come back to develop this point in more details in Chap. 15.

Further, it is worth noting that much of the concepts and methods developed for characterizing the surface chemistry of solid materials of interest in heterogeneous catalysis may be used for a better understanding of the structure and properties of solid materials of interest in other fields of the material science as for instance the materials for biomedical applications (bioactive glasses, biosensors, materials for controlled drug release).

Acknowledgments Prof. Bice Fubini (University of Torino) teachings in the early stage of my work in adsorption microcalorimetry are greatly acknowledged. Further, I would like to acknowledge the contribution of Dr Claudia Busco (zeolites work, University Piemonte Orientale "A. Avogadro") and Dr Valentina Aina (Ca-modified silica work, University of Torino) in collecting the volumetric-calorimetric data published in the quoted references and employed to describe the use of adsorption microcalorimetry in surface chemistry studies. Without their enthusiasm and helpful contribution much of this work would not have been carried out. Mr Raffaele Disa (Disa Raffaele e F.lli s.a.s - Milano, I) is also greatly acknowledged for the endless, patient help in building up and maintaining the high-vacuum volumetric line.

References

1. J.M. Thomas, W.J. Thomas, *Principles and Practice of Heterogeneous Catalysis* (VCH, Weinheim (Germany), 1997)
2. R.A. Van Santen, P.W.N.M. van Leeuwen, J.A. Moulijn, B.A. Averill, *Catalysis an Integrated Approach. Studies in Surface Science and Catalysis*, vol 123 (Elsevier, Amsterdam, 1999)
3. L.L. Hench, J. Wilson, *Introduction to Bioceramics* (World Scientific, Singapore, 1993)
4. D.F. Williams, J. Black, P.J. Doherty, *Advances in Biomaterials: Biomaterial-Tissue Interfaces*, vol. 10 (Elsevier Science, Amsterdam, 1992)
5. B. Kasemo, Biological surface science. Curr. Opin. Solid State Mater. Sci. **3**(5), 451–459 (1998). doi:10.1016/S1359-0286(98)80006-5
6. B. Kasemo, Biological surface science. Surf. Sci. **500**(1–3), 656–677 (2002). doi:10.1016/S0039-6028(01)01809-X
7. R. Schlogl, in *Handbook of Heterogeneous Catalysis*, vol. 8, 2nd edn., ed. by G. Ertl, H. Knozinger, F. Schuth, J. Weitkamp (Wiley-VCH Verlag, Weinheim, 2008)
8. G. Ertl, in *Encyclopedia of Catalysis*, vol. 1, ed. by J.T. Horvath (John Wiley & Sons, Hoboken, NJ, 2003), pp. 329–352
9. A. Zecchina, D. Scarano, S. Bordiga, G. Spoto, C. Lamberti, Surface structures of oxides and halides and their relationships to catalytic properties. Adv. Catal. **46**, 265–397 (2001). doi:10.1016/S0360-0564(02)46024-5
10. Gravelle PC, in *Heat-Flow Microcalorimetry and Its Application to Heterogeneous Catalysis*, ed. by D.D. Eley HP, Paul BW. Advances in Catalysis, vol 22 (Academic Press, 1972). pp. 191–263

11. V. Bolis, G. Della Gatta, B. Fubini, E. Giamello, L. Stradella, G. Venturello, Identification of surface sites on potentially catalytic solids by adsorption calorimetry. Gazz Chim Ital **112**, 83–89 (1982)
12. B. Fubini, Adsorption Calorimetry in Surface-Chemistry. Thermochim. Acta **135**, 19–29 (1988)
13. N. Cardona-Martinez, J.A. Dumesic, Applications Of adsorption microcalorimetry to the study of heterogeneous catalysis. Adv. Catal. **38**, 149–244 (1992). doi:10.1016/s0360-0564(08)60007-3
14. A. Auroux, *Physical Techniques for Solid Materials* (Plenum Press, New York, 1994)
15. V. Bolis, A. Cavenago, B. Fubini, Surface heterogeneity on hydrophilic and hydrophobic silicas: Water and alcohols as probes for H-bonding and dispersion forces. Langmuir **13**(5), 895–902 (1997). doi:10.1021/la951006i
16. V. Bolis, S. Bordiga, C. Lamberti, A. Zecchina, A. Carati, F. Rivetti, G. Spano, G. Petrini, Heterogeneity of framework Ti(IV) in Ti-silicalite as revealed by the adsorption of NH_3. Combined calorimetric and spectroscopic study. Langmuir **15**(18), 5753–5764 (1999). doi:10.1021/la981420t
17. V.A. Basiuk in *Encyclopedia of Surface and Colloid Science*, ed. by S. Ponisseril. Adsorption of Biomolecules at Silica, (Marcel Dekker, Inc., New York, 2002), p. 277
18. V. Bolis, B. Fubini, E. Garrone, C. Morterra, Thermodynamic and vibrational characterization of CO adsorption on variously pretreated anatase. J. Chem. Soc. Faraday Trans. I **85**, 1383–1395 (1989). doi:10.1039/f19898501383
19. C. Morterra, V. Bolis, B. Fubini, L. Orio, T.B. Williams, A Ftir and Hrem study of some morphological and adsorptive properties of Monoclinic ZrO_2 microcrystals. Surf. Sci. **251**, 540–545 (1991). doi:10.1016/0039-6028(91)91051-X
20. V. Bolis, G. Magnacca, C. Morterra, Surface properties of catalytic aluminas modified by alkaline-earth metal cations: a microcalorimetric and IR-spectroscopic study. Res. Chem. Intermed. **25**(1), 25–56 (1999). doi:10.1163/156856799X00374
21. V. Bolis, S. Maggiorini, L. Meda, F. D'Acapito, G.T. Palomino, S. Bordiga, C. Lamberti, X-ray photoelectron spectroscopy and x-ray absorption near edge structure study of copper sites hosted at the internal surface of ZSM-5 zeolite: A comparison with quantitative and energetic data on the CO and NH_3 adsorption. J. Chem. Phys. **113**(20), 9248–9261 (2000). doi:10.1063/I.1319318
22. E. Garrone, V. Bolis, B. Fubini, C. Morterra, Thermodynamic and spectroscopic characterization of heterogeneity among adsorption Sites—CO on anatase at ambient-temperature. Langmuir **5**(4), 892–899 (1989). doi:10.1021/la00088a002
23. V. Bolis, A. Barbaglia, S. Bordiga, C. Lamberti, A. Zecchina, Heterogeneous nonclassical carbonyls stabilized in Cu(I)- and Ag(I)-ZSM-5 zeolites: Thermodynamic and spectroscopic features. J. Phys. Chem. B **108**(28), 9970–9983 (2004). doi:10.1021/Jp049613e
24. V. Bolis, C. Busco, S. Bordiga, P. Ugliengo, C. Lamberti, A. Zecchina, Calorimetric and IR spectroscopic study of the interaction of NH_3 with variously prepared defective silicalites—Comparison with ab initio computational data. Appl. Surf. Sci. **196**(1–4), 56–70 (2002). doi:10.1016/S0169-4332(02)00046-6
25. V. Bolis, C. Busco, P. Ugliengo, Thermodynamic study of water adsorption in high-silica zeolites. J. Phys. Chem. B **110**(30), 14849–14859 (2006). doi:10.1021/Jp061078q
26. V. Bolis, C. Busco, V. Aina, C. Morterra, P. Ugliengo, Surface properties of silica-based biomaterials: Ca species at the surface of amorphous silica as model sites. J. Phys. Chem. C **112**(43), 16879–16892 (2008). doi:10.1021/Jp805206z
27. M. Corno, A. Rimola, V. Bolis, P. Ugliengo, Hydroxyapatite as a key biomaterial: quantum-mechanical simulation of its surfaces in interaction with biomolecules. Phys. Chem. Chem. Phys. **12**(24), 6309–6329 (2010). doi:10.1039/C002146f
28. M. Armandi, V. Bolis, B. Bonelli, C.O. Arean, P. Ugliengo, E. Garrone, Silanol-Related and Unspecific Adsorption of Molecular Ammonia on Highly Dehydrated Silica. J. Phys. Chem. C **115**(47), 23344–23353 (2011). doi:10.1021/Jp206301c

29. V. Bolis, C. Busco, G. Martra, L. Bertinetti, Y. Sakhno, P. Ugliengo, F. Chiatti, M. Corno, N. Roveri, Coordination chemistry of Ca sites at the surface of nanosized hydroxyapatite: interaction with H_2O and CO. Philos. Trans. R. Soc. A: Math. Phys. Eng. Sci. **370**(1963), 1313–1336 (2012). doi:10.1098/rsta.2011.0273
30. G. Attard, C. Barnes (1998) *Surfaces. Oxford Chemistry Primers N. 59* (Oxford Science Publications, Oxford, 1998)
31. I.M. Campbell, *Catalysis at Surfaces* (Chapman and Hall, London and New York, 1988)
32. S.J. Gregg, K.S.W. Sing, *Adsorption, Surface Area and Porosity* (Academic Press, London, 1982)
33. J.A. Rabo, *Zeolites Chemistry and Catalysis*, vol. 171 (ACS Monograph American Chemical Society, Washington, D.C., 1976)
34. D.W. Breck, *Zeolites Molecular Sieves*, vol. 4 (John Wiley, New York, 1974)
35. W.M. Meier, D.H. Olson *Atlas of Zeolites Structure Types*, 2nd edn. (Butterworths, London, 1993)
36. E.M. Flanigen, J.M. Bennett, R.W. Grose, J.P. Cohen, R.L. Patton, R.M. Kirchner, J.V. Smith, Silicalite, a new hydrophobic crystalline silica molecular sieve. Nature **271**(5645), 512–516 (1978)
37. W.O. Haag, R.M. Lago, P.B. Weisz, The active site of acidic aluminosilicate catalysts. Nature **309**(5969), 589–591 (1984). doi:10.1038/309589a0
38. A. Corma, From microporous to mesoporous molecular sieve materials and their use in catalysis. Chem Rev **97**(6), 2373–2419 (1997). doi:10.1021/cr960406n
39. A. Corma, State of the art and future challenges of zeolites as catalysts. J. Catal. **216**(1–2), 298–312 (2003). doi:10.1016/s0021-9517(02)00132-x
40. E. Garrone, B. Fubini, B. Bonelli, B. Onida, C.O. Arean, Thermodynamics of CO adsorption on the zeolite Na-ZSM-5—a combined microcalorimetric and FTIR spectroscopic study. Phys. Chem. Chem. Phys. **1**(4), 513–518 (1999). doi:10.1039/a806973e
41. C. Morterra, V. Bolis, E. Fisicaro, Thc hydrated layer and the adsorption of CO at the surface of TiO_2 (Anatase). Colloids Surf. **41**(1–2), 177–188 (1989). doi:10.1016/0166-6622(89)80051-4
42. C. Morterra, G. Magnacca, V. Bolis, On the critical use of molar absorption coefficients for adsorbed species: the methanol/silica system. Catal. Today **70**(1–3), 43–58 (2001). doi:10.1016/S0920-5861(01)00406-0
43. S. Brunauer, L.S. Deming, W.E. Deming, E. Teller, On a theory of the van der Waals adsorption of gases. J. Am. Chem. Soc. **62**(7), 1723–1732 (1940). doi:10.1021/ja01864a025
44. S. Brunauer, P.H. Emmett, E. Teller, Adsorption of gases in multimolecular layers. J. Am. Chem. Soc. **60**(2), 309–319 (1938). doi:10.1021/ja01269a023
45. S. Brunauer, *The Adsorption of Gases and Vapours* (Oxford University Press, Oxford, 1945)
46. I. Langmuir, The adsorption of gases on plane surfaces of glass, mica and platinum. J. Am. Chem. Soc. **40**(9), 1361–1403 (1918). doi:10.1021/ja02242a004
47. R. Shannon, Revised effective ionic radii and systematic studies of interatomic distances in halides and chalcogenides. Acta Crystallogr. A **32**(5), 751–767 (1976). doi:10.1107/S0567739476001551
48. B. Fubini, V. Bolis, A. Cavenago, P. Ugliengo, Ammonia and water as probes for the surface reactivity of covalent solids—cristobalite and silicon-carbide. J. Chem. Soc., Faraday Trans. **88**(3), 277–290 (1992). doi:10.1039/ft9928800277
49. B. Fubini, V. Bolis, A. Cavenago, E. Garrone, P. Ugliengo, Structural and induced heterogeneity at the surface of some SiO_2 polymorphs from the enthalpy of adsorption of various molecules. Langmuir **9**(10), 2712–2720 (1993). doi:10.1021/la00034a034
50. P.C. Gravelle, Application of adsorption calorimetry to the study of heterogeneous catalysis reactions. Thermochim. Acta **96**(2), 365–376 (1985). doi:10.1016/0040-6031(85)80075-7
51. G. Della Gatta, Direct determination of adsorption heats. Thermochim. Acta **96**(2), 349–363 (1985). doi:10.1016/0040-6031(85)80074-5
52. A. Auroux, Acidity characterization by microcalorimetry and relationship with reactivity. Top. Catal. **4**(1–2), 71–89 (1997). doi:10.1023/A:1019127919907

53. B. Fubini, Rev. Gen. Thermique **18**, 297 (1979)
54. J.A. Dunne, M. Rao, S. Sircar, R.J. Gorte, A.L. Myers, Calorimetric heats of adsorption and adsorption isotherms.2. O_2, N_2, Ar, CO_2, CH_4, C_2H_6, and SF_6 on NaX, H-ZSM-5, and Na-ZSM-5 zeolites. Langmuir **12**(24), 5896–5904 (1996)
55. V. Bolis, G. Cerrato, G. Magnacca, C. Morterra, Surface acidity of metal oxides. Combined microcalorimetric and IR-spectroscopic studies of variously dehydrated systems. Thermochim. Acta **312**(1–2), 63–77 (1998). doi:10.1016/S0040-6031(97)00440-1
56. Fubini B, Bolis V, Bailes M, Stone FS (1989) The reactivity of oxides with water vapor. Solid State Ionics **32–33**, (Part 1 (0))258–272. doi:10.1016/0167-2738(89)90230-0
57. J.M. Newsam, M.M.J. Treacy, W.T. Koetsier, C.B.D. Gruyter, Structural characterization of zeolite beta. Proc. R. Soc. Lond. Math. Phys. Sci. **420**(1859), 375–405 (1988). doi:10.1098/rspa.1988.0131
58. M.M. Huang, A. Auroux, S. Kaliaguine, Crystallinity dependence of acid site distribution in HA, HX and HY Zeolites. Microporous Mater. **5**(1–2), 17–27 (1995). doi:10.1016/0927-6513(95)00028-8
59. A.L. Blumenfeld, J.J. Fripiat, Acid sites topology in aluminas and zeolites from high-resolution solid-state NMR. Top. Catal. **4**(1–2), 119–129 (1997). doi:10.1023/A:1019119718089
60. C. Otero Arean, G. Turnes Palomino, E. Escalona Platero, M. Penarroya Mentruit, Zeolite-supported metal carbonyls: sensitive probes for infrared spectroscopic characterization of the zeolite surface. J. Chem. Soc., Dalton Trans. **5**, 873–880 (1997). doi:10.1039/A604775K
61. M. Bregolato, V. Bolis, C. Busco, P. Ugliengo, S. Bordiga, F. Cavani, N. Ballarini, L. Maselli, S. Passeri, I. Rossetti, L. Forni, Methylation of phenol over high-silica beta zeolite: effect of zeolite acidity and crystal size on catalyst behaviour. J. Catal. **245**(2), 285–300 (2007). doi:10.1016/j.jcat.2006.10.024
62. V. Bolis, S. Bordiga, G.T. Palomino, A. Zecchina, C. Lamberti, Calorimetric and spectroscopic study of the coordinative unsaturation of copper(I) and silver(I) cations in ZSM-5 zeolite - Room temperature adsorption of NH3. Thermochim. Acta **379**(1–2), 131–145 (2001). doi:10.1016/S0040-6031(01)00612-8
63. C. Morterra, A. Chiorino, G. Ghiotti, E. Fisicaro, Spectroscopic study of anatase properties. Part 5—Surface modifications caused by K_2O addition. J. Chem. Soc., Faraday Trans. 1: Phys. Chem. Condensed Phases **78**(9), 2649–2659 (1982). doi:10.1039/F19827802649
64. C. Morterra, G. Ghiotti, E. Garrone, E. Fisicaro, Spectroscopic study of anatase properties. Part 3—Surface acidity. J. Chem. Soc., Faraday Trans. 1: Phys. Chem. Condensed Phases **76**, 2102–2113 (1980). doi:10.1039/F19807602102
65. M. Cerruti, G. Magnacca, V. Bolis, C. Morterra, Characterization of sol-gel bioglasses with the use of simple model systems: a surface-chemistry approach. J. Mater. Chem. **13**(6), 1279–1286 (2003). doi:10.1039/B300961k
66. E.P.L. Hunter, S.G. Lias, Evaluated gas phase basicities and proton affinities of molecules: an update. J. Phys. Chem. Ref. Data **27**(3), 413–656 (1998)
67. A. Auroux, Microcalorimetry methods to study the acidity and reactivity of zeolites, pillared clays and mesoporous materials. Top. Catal. **19**(3–4), 205–213 (2002). doi:10.1023/A:1015367708955
68. W.E. Farneth, R.J. Gorte, Methods for characterizing zeolite acidity. Chem. Rev. **95**(3), 615–635 (1995). doi:10.1021/cr00035a007
69. A. Zecchina, Otero Arean C (1996) Diatomic molecular probes for mid-IR studies of zeolites. Chem. Soc. Rev. 25 (3):187–197. doi:10.1039/cs9962500187.
70. A. Zecchina, C. Lamberti, S. Bordiga, Surface acidity and basicity: general concepts. Catal. Today **41**(1–3), 169–177 (1998). doi:10.1016/S0920-5861(98)00047-9
71. R.J. Gorte, What we do know about the acidity of solid acids? Catal. Lett. **62**, 1–13 (1999)
72. C. Otero Arean, E. Garrone, Trends in infrared spectroscopy of zeolites. Trends Inorg. Chem. **7**, 119–133 (2001)

73. V. Bolis, M. Broyer, A. Barbaglia, C. Busco, G.M. Foddanu, P. Ugliengo, Van der Waals interactions on acidic centres localized in zeolites nanocavities: a calorimetric and computer modeling study. J. Mol. Catal. A: Chem. **204**, 561–569 (2003). doi:10.1016/S1381-1169(03)00339-X
74. C. Busco, A. Barbaglia, M. Broyer, V. Bolis, G.M. Foddanu, P. Ugliengo, Characterisation of Lewis and Bronsted acidic sites in H-MFI and H-BEA zeolites: a thermodynamic and ab initio study. Thermochim. Acta **418**(1–2), 3–9 (2004). doi:10.1016/j.tca.2003.11.050
75. A. Poppl, T. Rudolf, D. Michel, A pulsed electron nuclear double resonance study of the Lewis acid site nitric oxide complex in zeolite H-ZSM-5. J. Am. Chem. Soc. **120**(19), 4879–4880 (1998). doi:10.1021/ja9741685
76. R.D. Shannon, K.H. Gardner, R.H. Staley, G. Bergeret, P. Gallezot, A. Auroux, The nature of the nonframework aluminum species formed during the dehydroxylation of H-Y. J. Phys. Chem. **89**(22), 4778–4788 (1985). doi:10.1021/j100268a025
77. P.A. Jacobs, H.K. Beyer, Evidence for the nature of true Lewis sites in faujasite-type zeolites. J. Phys. Chem. **83**(9), 1174–1177 (1979). doi:10.1021/j100472a013
78. C. Busco, V. Bolis, P. Ugliengo, Masked Lewis sites in proton-exchanged zeolites: a computational and microcalorimetric investigation. J. Phys. Chem. C **111**(15), 5561–5567 (2007). doi:10.1021/Jp0705471
79. H. Willner, F. Aubke, Homoleptic metal carbonyl cations of the electron-rich metals: their generation in superacid media together with their spectroscopic and structural characterization. Angewandte Chemie-Int Ed. Engl. **36**(22), 2403–2425 (1997). doi:10.1002/anie.199724021
80. A.J. Lupinetti, S.H. Strauss, G. Frenking, Nonclassical metal carbonyls. Prog. Inorg. Chem. **49**, 1–112 (2001). doi:10.1002/9780470166512.ch1
81. M. Armandi, B. Bonelli, I. Bottero, C.O. Arean, E. Garrone, Thermodynamic features of the reaction of ammonia with the acidic proton of H-ZSM-5 as studied by variable-temperature IR Spectroscopy. J. Phys. Chem. C **114**(14), 6658–6662 (2010). doi:10.1021/Jp100799k
82. E.G. Derouane, C.D. Chang, Confinement effects in the adsorption of simple bases by zeolites. Microporous Mesoporous Mater. **35–6**, 425–433 (2000)
83. L. Yang, K. Trafford, O. Kresnawahjuesa, J. Sepa, R.J. Gorte, D. White, An examination of confinement effects in high-silica zeolites. Russ. J. Phys. Chem. B **105**(10), 1935–1942 (2001). doi:10.1021/jp002964i
84. R.J. Gorte, D. White, Measuring sorption effects at zeolite acid sites: pursuing ideas from W.O. Haag. Microporous Mesoporous Mater. **35–36**, 447–455 (2000)
85. V. Bolis, B. Fubini, L. Marchese, G. Martra, D. Costa, Hydrophilic and hydrophobic sites on dehydrated crystalline and amorphous silicas. J. Chem. Soc., Faraday Trans. **87**(3), 497–505 (1991). doi:10.1039/ft9918700497
86. V. Bolis, C. Morterra, M. Volante, L. Orio, B. Fubini, Development and suppression of surface-acidity on monoclinic zirconia—a spectroscopic and calorimetric investigation. Langmuir **6**(3), 695–701 (1990). doi:10.1021/la00093a028
87. V. Bolis, B. Fubini, E. Garrone, E. Giamello, C. Morterra, in *Studies in Surface Science and Catalysis: Structure and Reactivity of Surfaces* , vol. 48, ed. by C. Morterra, A. Zecchina, G. Costa (Elsevier Sci. Publ. B.V., Amsterdam, 1989), pp. 159–166
88. V. Solinas, I. Ferino, Microcalorimetric characterisation of acid-basic catalysts. Catal. Today **41**(1–3), 179–189 (1998). doi:10.1016/S0920-5861(98)00048-0
89. V. Bolis, A. Barbaglia, M. Broyer, C. Busco, B. Civalleri, P. Ugliengo, Entrapping molecules in zeolites nanocavities: a thermodynamic and ab-initio study. Orig. Life Evol. Biosph. **34**(1–2), 69–77 (2004). doi:10.1023/B:ORIG.0000009829.11244.d1
90. S. Bordiga, I. Roggero, P. Ugliengo, A. Zecchina, V. Bolis, G. Artioli, R. Buzzoni, G. Marra, F. Rivetti, G. Spano, C. Lamberti, Characterisation of defective silicalites. J. Chem. Soc., Dalton Trans. **21**, 3921–3929 (2000). doi:10.1039/B004794p
91. E. Garrone, F. Rouquerol, B. Fubini, G. Della Gatta, Entropy of adsorption and related thermodynamic functions in an open isothermal system. J Chim Phys **76**, 528–532 (1979)

92. C. Otero Arean, O.V. Manoilova, G.T. Palomino, M.R. Delgado, A.A. Tsyganenko, B. Bonelli, E. Garrone, Variable-temperature infrared spectroscopy: an access to adsorption thermodynamics of weakly interacting systems. Phys. Chem. Chem. Phys. **4**(23), 5713–5715 (2002). doi:10.1039/B209299a
93. S. Savitz, A.L. Myers, R.J. Gorte, Calorimetric investigation of CO and N_2 for characterization of acidity in zeolite H-MFI. J. Phys. Chem. B **103**(18), 3687–3690 (1999). doi:10.1021/jp990157h
94. D. McQuarrie, *Statistical Mechanics* (Harper and Row, New York, 1986)
95. D.H. Everett, Thermodynamics of adsorption. Part I—General considerations. Trans. Faraday Soc. **46**, 453–459 (1950)
96. T.L. Hill, Statistical mechanics of adsorption. V. Thermodynamics and heat of adsorption. J. Chem. Phys. **17**(6), 520–535 (1949). doi:10.1063/1.1747314
97. E. Garrone, G. Ghiotti, E. Giamello, B. Fubini, Entropy of adsorption by microcalorimetry. Part 1—Quasi-ideal chemisorption of CO onto various oxidic systems. J. Chem. Soc., Faraday Trans. 1: Phys. Chem. Condensed Phases **77**(11), 2613–2620 (1981)
98. G. Busca, H. Saussey, O. Saur, J.C. Lavalley, V. Lorenzelli, Ft-Ir Characterization of the surface-acidity of different titanium-dioxide anatase preparations. Appl. Catal. **14**(1–3), 245–260 (1985). doi:10.1016/S0166-9834(00)84358-4
99. F. Rouquerol, J. Rouquerol, C. Letoquart, Use of isothermal microcalorimetry data for the determination of integral molar entropies of adsorption at the gas–solid interface by a quasi-equilibrium procedure. Thermochim. Acta **39**(2), 151–158 (1980). doi:10.1016/0040-6031(80)80008-6
100. H. Knozinger, *The Hydrogen Bond* (North Holland, Amsterdam, 1976)
101. U. Diebold, The surface science of titanium dioxide. Surf. Sci. Rep. **48**(5–8), 53–229 (2003)
102. V. Bolis, B. Fubini, E. Giamello, Effect of form on the surface-chemistry of finely divided solids. Mater. Chem. Phys. **29**(1–4), 153–164 (1991). doi:10.1016/0254-0584(91)90012-J
103. L. Kieu, P. Boyd, H. Idriss, Trends within the adsorption energy of alcohols over rutile TiO_2(110) and (011) clusters. J. Mol. Catal. A: Chem. **188**(1–2), 153–161 (2002). doi:10.1016/S1381-1169(02)00210-8
104. A.C. Zettlemoyer, F.T. Micale, K. Klier, *Water in dispersed Systems*, vol. 5 (Plenum, New York, 1975)
105. S. Savitz, A.L. Myers, R.J. Gorte, A calorimetric investigation of CO, N_2, and O_2 in alkali-exchanged MFI. Microporous Mesoporous Mater. **37**(1–2), 33–40 (2000). doi:10.1016/S1387-1811(99)00190-0
106. J.E. Jones, On the determination of molecular fields. II. From the equation of state of a gas. Proc. R. Soc. Lond. Ser. A **106**(738), 463–477 (1924). doi:10.1098/rspa.1924.0082
107. P.M. Morse, Diatomic molecules according to the wave mechanics. II. Vibrational levels. Physical Review **34**(1), 57–64 (1929). doi:10.1098/rspa.1924.0082

Chapter 2
Thermal Analysis and Calorimetry Techniques for Catalytic Investigations

Pierre Le Parlouër

Abstract The use of thermal analysis and calorimetry techniques is quite an old and known field of applications for the catalytic investigations and many publications have been published on the various topics including analysis of catalysts, investigation of the processes during the preparation of catalysts, desactivation of catalysts and interaction of reactants or catalytic poisons with the catalysts. Differential thermal analysis, calorimetry and thermogravimetry are also used to characterize the catalysts, especially in the field of gas–solid and gas–liquid interactions. Since the last years, many technical improvements have appeared in the design and the use of thermal analyzers and calorimeters, particularly for the characterization of catalysts. This chapter gives a detailed overview of the uptodate thermal techniques covering various techniques including Differential Thermal Analysis (DTA), Differential Scanning Calorimetry (DSC), the calorimetric techniques (including Isothermal Calorimetry, Titration Calorimetry), Thermogravimetric Analysis (TGA), the combined techniques (including TG-DTA and TG-DSC), the Evolved Gas Analysis (including TG-MS, TG-FTIR). Some examples of applications are given to illustrate the catalyst characterizations.

2.1 Introduction

The use of thermal analysis and calorimetry techniques is quite an old and known field of applications for the catalytic investigations and many publications have been published on the various topics. In 1977, Habersberger [19] has published a short review of the possibilities of applications of thermal analysis to the investigation of catalysts including analysis of catalysts, investigation of the processes during the preparation of catalysts, desactivation of catalysts and interaction of reactants

P. Le Parlouër (✉)
Setaram, 7 Rue de l'Oratoire, 69300 Caluire, France
e-mail: leparlouer@setaram.com

A. Auroux (ed.), *Calorimetry and Thermal Methods in Catalysis*,
Springer Series in Materials Science 154, DOI: 10.1007/978-3-642-11954-5_2,

or catalytic poisons with the catalysts. In 1994, Auroux [4, 8] has described the different thermal techniques including differential thermal analysis, calorimetry and thermogravimetry that are used to characterize the catalysts, especially in the field of gas–solid and gas–liquid interactions. More recently in 2003, Pawelec and Fierro [45] have prepared a chapter called "Applications of thermal analysis in the preparation of catalysts and in catalysis" for the "Handbook of Thermal Analysis and Calorimetry". Since the last years, many technical improvements have appeared in the design and the use of thermal analyzers and calorimeters, particularly for the characterization of catalysts. Following the summer school of calorimetry named "Calorimetry and thermal methods in catalysis" organized in Lyon (France) since 2007, this chapter will give a detailed overview of the uptodate thermal techniques presented during this meeting with some examples of applications for catalyst characterization.

2.2 Thermal Analysis and Calorimetry: Techniques and Applications

Thermal Analysis (TA) means the analysis of a change in a sample property which is related to an imposed temperature alteration and calorimetry means the measurement of heat [24]. The corresponding property of the sample is measured according to time and temperature with the main following techniques:

- Temperature difference >>> Differential Thermal Analysis (DTA).
- Heat flux variation >>> Differential Scanning Calorimeter (DSC).
- Heat variation >>> Calorimetry.
- Mass change >>> Thermogravimetry (TG).
- Length or volume change >>> Thermomechanical Analysis (TMA).

There is no one method of measurement in thermal analysis and calorimetry that can be used for all materials or to cover all the possible thermal properties and applications. A measurement method has to be selected depending on the following criteria:

- the parameter to be measured (temperature, heat, mass or length),
- sample type (solid or liquid),
- sample size,
- sample reactivity (which affects choice of crucible, protective gas),
- range of temperature,
- atmosphere around the sample (inert, oxidative, reducing),
- operation under pressure,
- operation in a corrosive gas,
- operation in humid atmosphere,
- sensitivity of detection in changes of mass, heat and length,
- whether a combined technique is necessary, ...

According to the different thermal techniques and the selection of the experimental parameters, a large selection of applications is available for the characterization of catalysts and related materials, and also the simulation of the catalytic processes. The Table 2.1 provides a brief overview of the main applications that have been developed.

2.3 The Differential Thermal Analysis Technique

2.3.1 Principle

ICTAC (International Confederation for Thermal Analysis and Calorimetry) gives the following definition for the DTA technique: "A technique where the temperature difference between a sample and a reference material is measured while they are subjected to the same temperature variation (heated or cooled) in a controlled atmosphere" [25]

The Differential Thermal Analysis technique is based on a differential mounting of thermocouples in the sample (S) crucible and reference (R) crucible (Fig. 2.1). The device is located in a block heated (or cooled) at a controlled temperature.

The electric signal (emf) measured at the ends of the thermocouple is proportional to the difference of temperature, ΔT, between the sample and reference sides.

The temperature of the furnace (T_p) is programmed with a linear heating rate (Fig. 2.2). The reference temperature (T_r), measured in an inert material (that means a material that does not undergo a thermal transition for the given temperature range

Table 2.1 Techniques and applications in thermal analysis

Physical property	Technique	Application
Temperature	Differential thermal analysis (DTA)	Phase change, dehydration, desorption, decomposition, reduction, oxidation, ...
Heat flow	Differential scanning calorimetry (DSC)	Phase change, adsorption, absorption, hydrogenation, dehydration, desorption, decomposition, dehydrogenation, reduction, oxidation, catalysis, heat capacity, kinetics, ...
Heat	Calorimetry	Reaction, adsorption, absorption, hydration, mixing, formation, catalysis, thermodynamics, heat capacity, kinetics, ...
Mass	Thermogravimetric analysis (TGA)	Adsorption, hydrogenation, dehydration, desorption, decomposition, dehydrogenation, oxidation, kinetics, ...
Length, volume	Thermomechanical analysis (TMA)	Phase change, expansion, sintering, ...

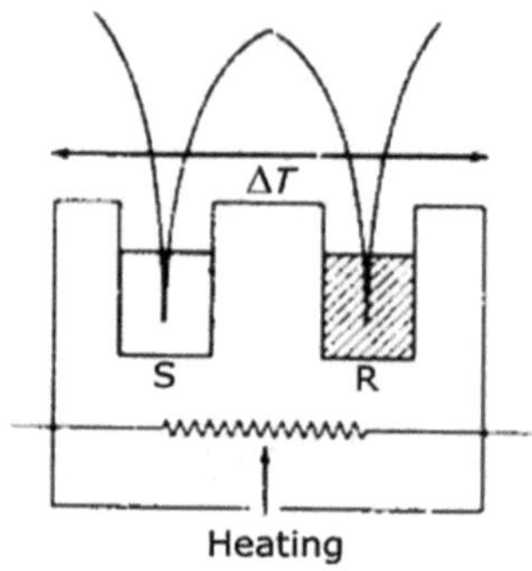

Fig. 2.1 DTA principle

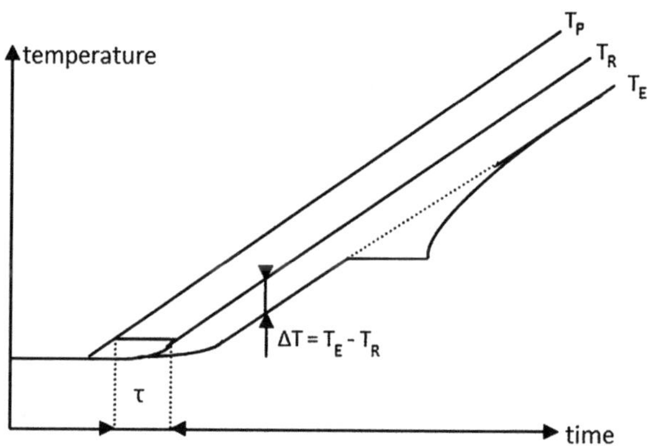

Fig. 2.2 DTA principle—recording of the different temperatures

of investigation) follows this heating profile with a small delay due to the thermal gradient in the crucible.

The sample temperature (T_e), measured in the sample material under investigation, follows in the same way the heating profile until a thermal transition in the sample is reached. In the case of melting, the temperature of the sample remains constant during melting (this is the melting plateau). When melting is complete the sample temperature changes to then follow the heating profile.

By measuring the difference between the sample temperature and the reference temperature, the thermogram (Fig. 2.3) is obtained, describing the melting peak.

From this curve, a first information is obtained: the temperature of melting of the material. In a case of a pure sample, this temperature corresponds to the onset temperature of the melting peak.

Attention: Such a principle described on Fig. 2.1 does not experimentally exist. In fact there is a risk of interaction between the thermocouple and the material that will affect the measure. In the experimental DTA device, a crucible has to be used as an interface between the thermocouple and the sample.

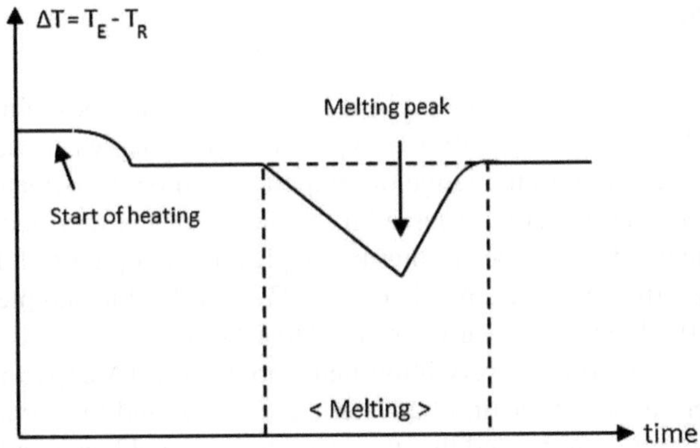

Fig. 2.3 Melting thermogram

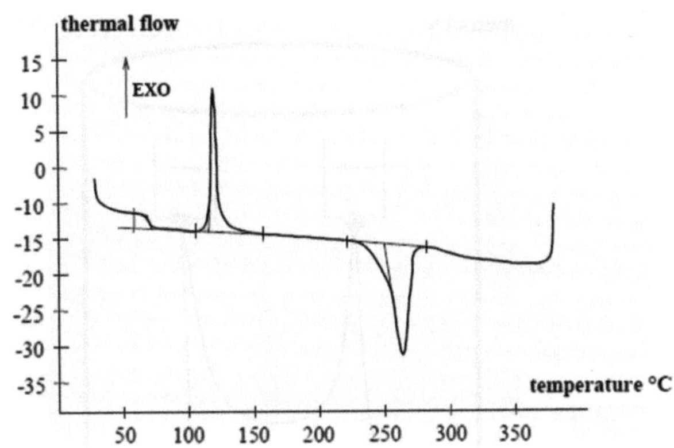

Fig. 2.4 The DTA thermogram

The DTA technique also gives another important information on the type of transition, either endothermic or exothermic effect. The endothermic tranformation produces a decrease of the temperature that means that ΔT is negative. For an exothermic transformation, a positive ΔT is recorded. The DTA thermogram (Fig. 2.4) allows making a clear identification of the different types of transformations according to the side of the peaks.

Among the different types of endothermic and exothermic effects, the following list can be given when catalysts are under investigation:

- Endothermic: melting, evaporation, sublimation, dehydration, dehydroxylation, desorption, and pyrolysis.
- Exothermic: crystallization, adsorption, oxidation, combustion, hydrogenation, and decomposition.

A variation of the base line may also be detected on the DTA curve (the first transition on Fig. 2.4). This is the signature of a glass transition, expressing that the material contains an amorphous phase.

2.3.2 Detectors

A standard DTA detector is shown in Fig. 2.5. Two different types of metals are used to build the DTA arrangement. Different types of thermocouples are used to produce DTA detectors according to the range of temperature to be investigated (Table 2.2). The sensitivity of the detector varies with the nature of the thermocouple. Higher sensitivity is obtained with types of thermocouples in the top part of Table 2.2.

A crucible is adjusted on each side of the DTA rod for the sample and the inert material. The crucible volume varies from 20 to 100 μl.
The choice of the crucible is a very important step in the DTA experimentation. It is needed that there is no interaction between the crucible and the sample during the test. In the same time it is needed that the crucible is a good heat conductor in order

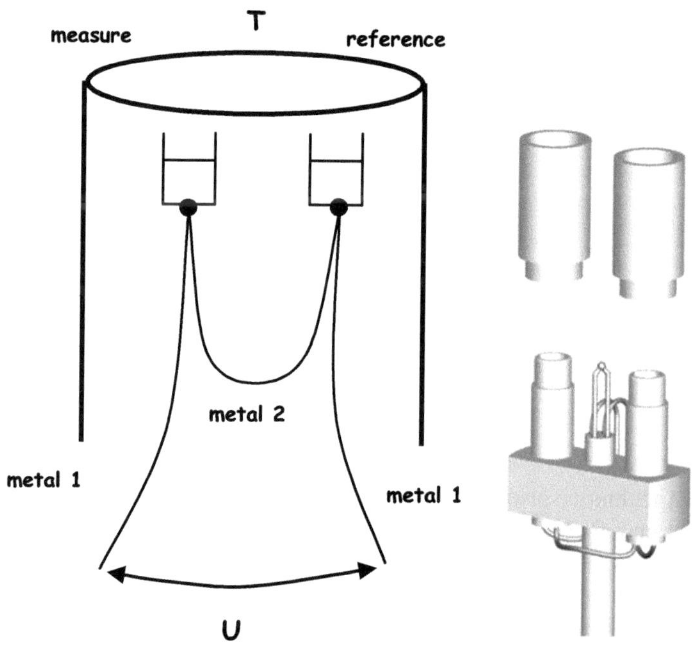

Fig. 2.5 Principle and cross section of a DTA detector

Table 2.2 Different types of thermocouples used for DTA detectors

Type	Composition	Temperature range
Platinel	55Pd 31Pt 14Au/65Au 35Pd	−150 °C → 1000 °C
E	Ni Cr/Cu Ni = Chromel/Constantan	−50 °C → 800 °C (500 °C under air)
S	Pt Rh10 % / Pt	1600 °C (1000 °C under H_2)
B	Pt Rh30 % / Pt Rh6 %	1750 °C (1000 °C under H_2)
W5	W Re26 % / W Re5 %	2400 °C / inert atmosphere

to ensure a good heat transfer between the sample and the thermocouple. Different materials are used to produce adequate crucibles:

- Aluminium (up to 500 °C),
- Alumina (especially required for metallic samples),
- Platinum (especially required for inorganic samples),
- Graphite and tungsten (for temperature above 1750 °C).

According to experimentation and corrosion problems, other types of materials are available such as magnesia, zirconia, boron nitride, ...

2.3.3 *Operation*

In most experimentations, the DTA crucible is used open. A cover is needed especially in the two following situations:

- The vapour has to be retained above the sample, in the case of dehydration or decomposition. For such an experiment a cover with a pin hole is used.
- There is a risk of overflowing of the sample when decomposing. In such a case it is recommended to close the crucible with a lid to avoid the destruction of the DTA rod.

As the crucible is open, the control of the gas above the sample is a very important parameter. Different types of gas atmosphere are available according to the experiment to be run:

- Inert atmosphere (nitrogen, argon, helium) to protect the sample from any oxidation.
- Oxidative atmosphere (air, oxygen) to be used for oxidation and combustion investigations.
- Reactive gas (hydrogen, carbon monoxide, ammonia, ...) for adsorption and absorption reactions.
- Water vapour for hydration reactions.

2.3.4 *Applications*

The applications of the DTA technique are quite wide for the characterization of the catalysts and related products, but they are mainly oriented on the determination of their structural properties. Here below is a non exhaustive list of DTA applications:

- Melting and crystallization,
- Phase transitions (glass transition, order–disorder, ...),
- Preparation of catalyst: dehydration and dehydroxylation,
- Rapid screening of potential catalysts,
- Evaluation of the effects of various pretreatments,

- Adsorption, desorption,
- Decomposition,
- Oxydation, reduction,
- Regeneration of catalyst, ...

The DTA applications applied to catalysts and more generally to inorganic materials have been extensively reviewed in various books [23, 33, 44, 45, 48, 61].

2.4 The Differential Scanning Calorimetry Technique

2.4.1 Principle

Differential Scanning Calorimetry is defined as followed by ICTAC:
"A technique in which the heat flux (thermal power) to (or from) a sample is measured versus time or temperature while the temperature of the sample is programmed, in a controlled atmosphere. The difference of heat flux between a crucible containing the sample and a reference crucible (empty or not) is measured." [25].

In a heat flux DSC, the specimen surroundings (generally the furnace and detector) are at constant temperature (isothermal mode) or at a variable temperature (scanning temperature mode). A defined exchange of heat takes place between the specimen and its surroundings (Fig. 2.6). The amount of heat flux (heat flow rate) is determined

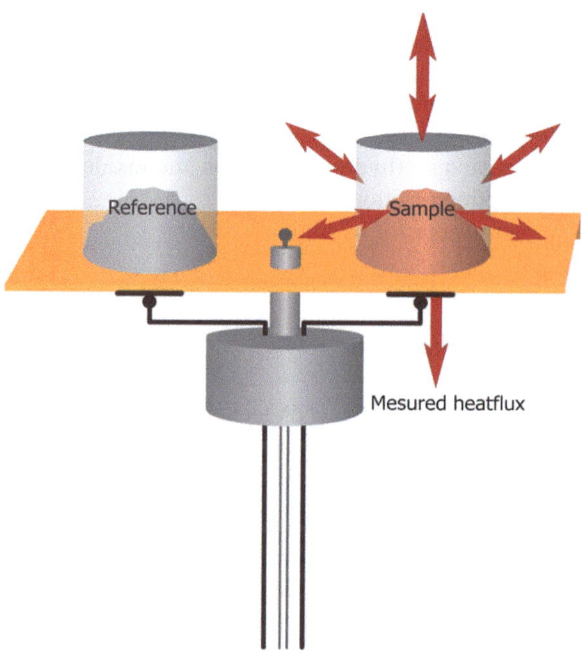

Fig. 2.6 Heat flux DSC principle

on the basis of the temperature difference along a thermal resistance between the specimen and its surroundings.

The heat flux for a given sample at a temperature T_s is equivalent to:

$$dq_s/dt = -dh/dt + C_s dT_s/dt \tag{2.1}$$

where:

- dh/dt: Heat flux produced by the transformation of the sample or the reaction.
- C_s: Heat capacity of the sample, including the container.

The heat flux dq_s/dt is exchanged with the thermostatic block at a temperature T_p through a thermal resistance R according to the following relation;

$$dq_s/dt = (T_p - T_s)/R. \tag{2.2}$$

From the relation (2.1), it is seen that the thermal contribution due to the heat capacity of the sample and container is significant and will provide a major disturbance at the introduction of the container in the calorimeter. From the relation (2.2), it is also evident that any temperature perturbation of the thermostatic block will affect the calorimetric measurement.

In order to solve these different problems, a symmetrical design is used. An identical crucible with an inert material (the reference container can also be empty) is placed on the detector at the same $T_p$temperature.

The difference of heat flux is measured between the two sides:

$$dq/dt = dq_s/dt - dq_r/dt = -dh/dt + C_s dT_s/dt - C_r dT_r/dt \tag{2.3}$$

where:

- C_r: Heat capacity of the reference, including the container.
- T_r: Temperature of the reference.

The relation (2.2) becomes:

$$dq/dt = (T_r - T_s)/R$$

or by derivation

$$Rd^2q/dt^2 = dT_r/dt - dT_s/dt \tag{2.4}$$

By combination of relations (2.3) and (2.4), the characteristic equation for the calorimetric measurement is obtained:

$$dh/dt = -dq/dt + (C_s - C_r)dT_p/dt - RC_s d^2q/dt^2. \tag{2.5}$$

If dh/dt corresponds to an absorbed thermal power due to an endothermic transformation or reaction, the dh/dt value is positive.

If dh/dt corresponds to a released thermal power due to an exothermic transformation or reaction, the dh/dt value is negative.
dh/dt (heat flow rate) is related to the kinetics of the transformation. The shape of the DSC peak gives a first indication on the rate of reaction.

When dh/dt is null (no transformation), the Eq. (2.5) allows to access the heat capacity of the sample. DSC provides a direct and accurate measurement of the specific heat of any type of material.

If the DSC test is run isothermally, the parameter dT_P/dt is null. In the case of a small perturbation of the temperature T_P of the thermostatic block, the corresponding thermal effect will be minimized if the C_s and C_r heat capacities are similar.
The last term R $C_s d^2q/dt^2$ (called also thermal lag) that affects the measurement mostly depends on the thermal resistance or also the time of response of the DSC on one side, and the heat capacity of the sample and container on the other side. For long period of time ($t >> RC_s$) it will be negligible.

As seen before with DTA (Fig. 2.4), the DSC curve will show endothermic and/or exothermic effects according to the transformations and the reactions.
If DTA is more considered as a qualitative technique, the sophisticated DSC detectors define a quantitative and accurate determination of the heat effects.

From the DSC curve different important information are obtained (Fig. 2.7):

- the temperature of the transformation (generally measured at the top of the peak),
- the heat related to the transformation, and
- the rate of the transformation.

In order to get accurate measurements of temperature of transition, the DSC has to be calibrated using reference materials [20]. The experimental procedures for such a correction are described in different standards, especially ISO standard [26].

As the DSC signal is equivalent to a thermal power (expressed in mW), the integration of the DSC peak between the time t_i (start of the peak) and t_f (end of the

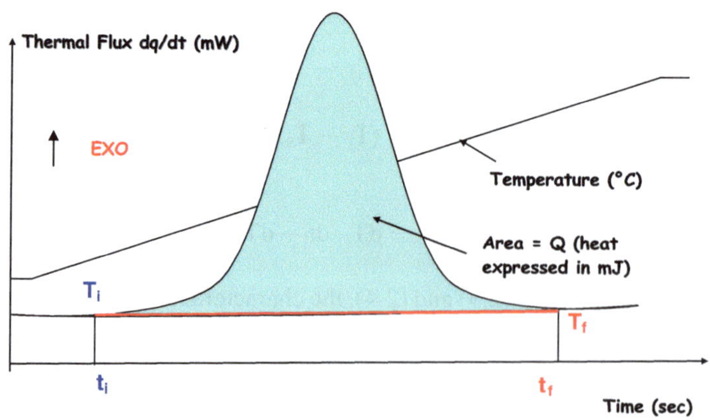

Fig. 2.7 DSC curve

peak) corresponds to the area under the peak, delimited by the line drawn between the two points.
Before getting the corresponding heat value, a calibration of the DSC detector is also needed. The raw DSC signal is an electrical signal provided by the thermocouples, expressed in microvolt. To convert this signal in microwatt, the use of reference materials with known heat of melting is required [20].

However another calibration technique is available when the Calvet type DSC is used. The principle is to apply a known amount of power in a dedicated calibration vessel. To reach this target, a resistance is embedded in the crucible. A known current I is delivered and the corresponding tension U measured, providing the power $P = UI$ that is applied. The corresponding Joule effect provides a DSC exothermic deviation in microvolt (Fig. 2.8). Such an electrical calibration is very interesting as it can apply at any temperature, even at constant temperature. This will be more detailed in paragraph 5 for the calorimetric techniques.

The other important information associated to the DSC peak is related to the kinetics of the transformation. As the DSC signal corresponds to a heat flow rate, the shape of the peak directly informs on the rate of the transformation. For example a sharp peak indicates a high rate of reaction. At the top of the peak, dh/dt_{max} (Fig. 2.7), the rate of the reaction reaches the maximum value. This information will be used in many models for the determination of the kinetic parameters of a transformation.

2.4.2 Detectors

Different types of DSC techniques are available and are described in Table 2.3.

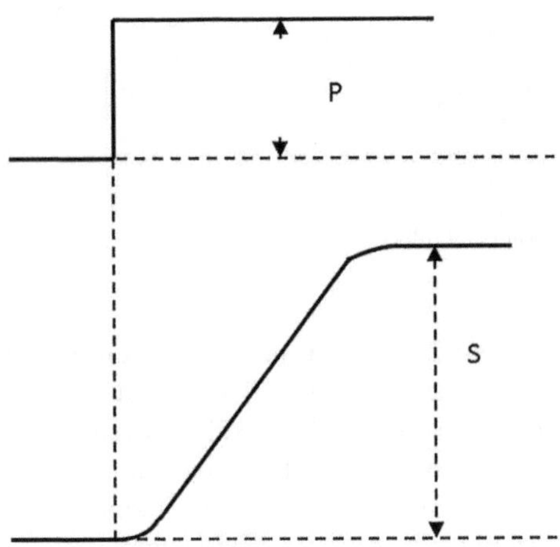

Fig. 2.8 Principle of the DSC electrical calibration using the Joule effect

Table 2.3 Different types of DSC techniques

Property	Technique	Description
Heat flow rate	Heat flux DSC	A technique in which the difference in heat flow rate to a test sample and to a reference sample is analysed while they are subjected to the same temperature variation (heated or cooled). The detector is a plate-DSC detector.
Heat flow rate	Power compensated DSC	A technique in which the temperature difference between a test specimen and a reference specimen occurring through subjecting both specimens to the same controlled temperature program is compensated by appropriately adjusting the difference of heating power to the test and reference specimen. The differential heating power is recorded against temperature or time.
Heat flow rate	Calvet DSC	A technique in which the difference of heat flow rate to the sample and to the reference sample is analysed while they are subjected to a temperature variation (heated or cooled). The detector is a cylindrical DSC detector.

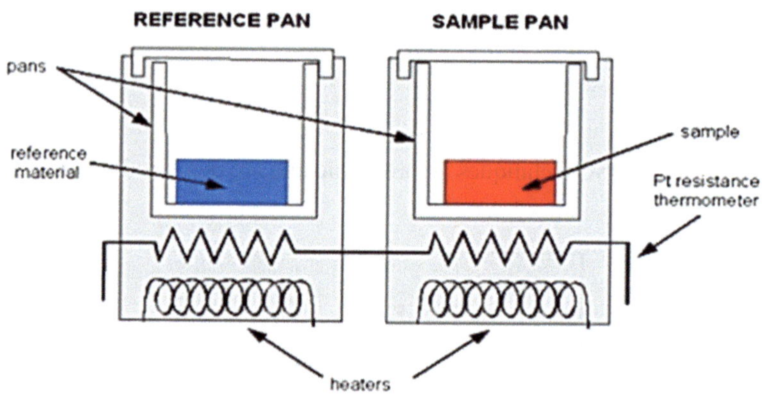

Fig. 2.9 DSC power compensation principle

For the heat flux plate DSC (Fig. 2.6) and the power compensation DSC (Fig. 2.9), the common characteristic is that they both use a flat-shaped sensor.

The principle of the power compensated DSC [37] is different from the heat flux DSC described previously. The heaters beneath the pans aim to minimise the difference in temperature between a specimen and an inert material. When a transition occurs in the specimen, the reference heater will aim to compensate for this and keep the reference pan at a similar temperature. Individual platinum resistance thermometers measure the temperature of each pan, and the power required to maintain this state is recorded. Endothermic and exothermic transitions are recorded depending if power is added or subtracted.

Even if the measurement principles are quite different, the heat transfer from (or to) the sample is about the same. The sample, contained in a metallic crucible, is placed and centered on the plate acting as a flat-shaped sensor. A reference crucible (empty or containing an inert material) is placed on the other plate.
The basics of the plate DSC is that the heat exchange between the sample and the detector is done through the bottom of the crucible, corresponding to a two dimensional detection. In fact only a part of this heat transfer is measured, as a significant part is dissipated through the walls and the cover of the crucible (Fig. 2.6). If the efficiency ratio is considered, that means the heat flux measured by the sensor to the total heat flux produced by the thermal event, a simulation using a thermal modeling software shows that only around half of the heat flux is dissipated through the plate [13, 31]. The Fig. 2.10 clearly shows that the efficiency rapidly decreases with the temperature and the thickness for the plate. The efficiency is also affected by the amount of the sample under investigation. This is why it is recommended to work with small amounts of material (about 5–10 mg) when using a plate DSC, in order to minimize the heat losses. The thermal conductivities of the crucible and the gas used in the experimental chambers also are very important parameters to be considered in the efficiency of the heat exchange. For example a very heat conductive gas (helium) will favour the heat transfer between the crucible and the detector, but in the same time increase the heat losses. So the calibration of a plate-type DSC is very critical and has to be run with the experimental conditions that are selected for the sample test.
The Calvet type DSC offers another technical option [38].
The detection concept is based on a three dimensional fluxmeter sensor. The fluxmeter element consists in a ring of several thermocouples in series (Fig. 2.11). The corresponding thermopile of high thermal conductivity surrounds the experimental space within the calorimetric block. The radial arrangement of the thermopiles guarantees an almost complete integration of the heat. This is verified by the calculation of the efficiency ratio that indicates that an average value of 94 ± 1 % of heat is transmitted through the sensor on the full range of temperature of the Calvet type DSC (Fig. 2.12). Another very important point is that the sensitivity of the DSC is no more affected by the type of crucible, the type of purge gas and flow rate.

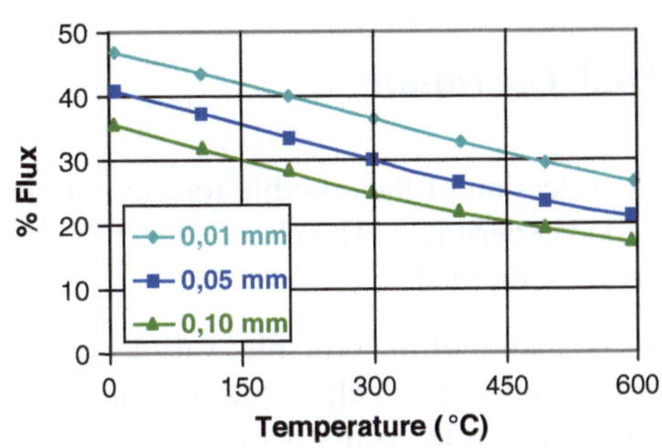

Fig. 2.10 Efficiency ratio of a flat-shaped DSC as a function of the sensor plate thickness

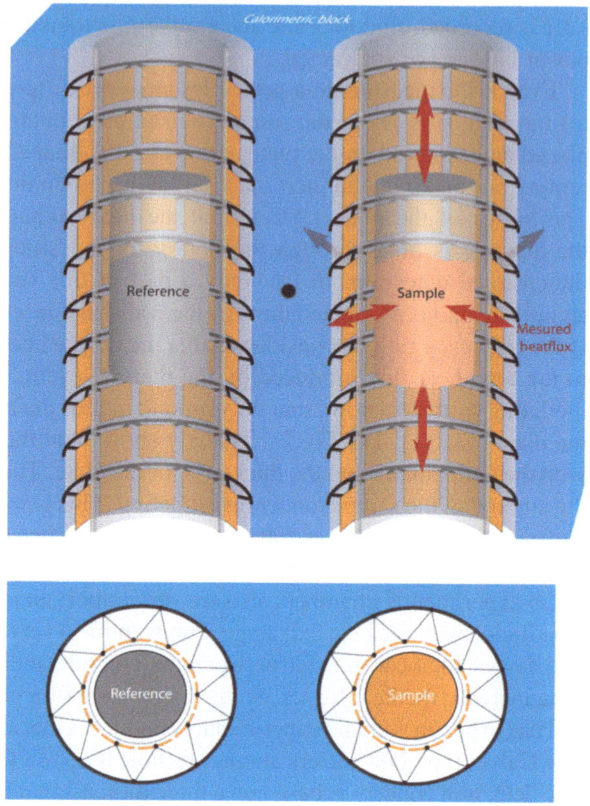

Fig. 2.11 Schematic of the Calvet principle

One of the main advantages of the Calvet-type DSC is that larger amounts of sample can be investigated. The open tube detection allows the adaptation of different types of experimental crucibles, especially with the possibility of introduction of various types of gas under normal or high pressure. This design specificity is of a great interest for the applications on catalysts, and more generally for all the adsorption investigations.

2.4.3 Operation

The selection of the crucible for the sample is one of the most important task to get a good experiment. The choice depends on a series of parameters that has to be very clearly identified:

- The nature of the crucible: various materials are available (aluminium, platinum, gold, stainless steel, ...) according to the temperature range. It is important that no interaction occurs between the sample and the crucible during the test.

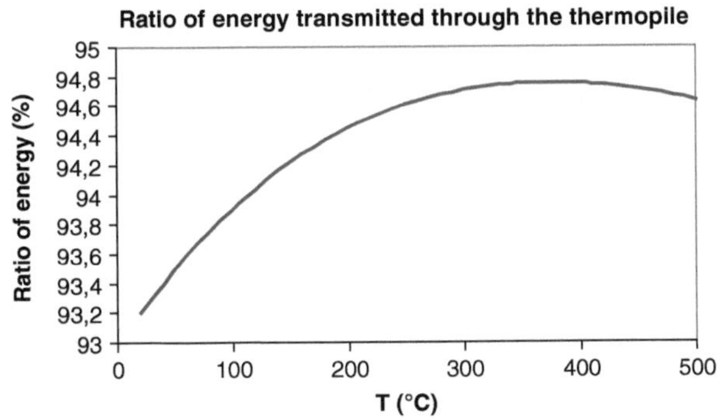

Fig. 2.12 Efficiency ratio of a Calvet type calorimeter versus temperature

- The way of using the crucible:
  - closed with a lid: this is the most common use for a DSC crucible. It is very convenient for any type of transition or transformation as soon as the vapour pressure of the sample remains low during the test
  - closed with a pierced lid: the hole is to favour the escape of vapour from the crucible, for example during a dehydration
  - open: this situation is needed when a gas interaction with the sample is investigated, for example an adsorption
  - tightly closed: dedicated crucibles are available to keep the vapour inside the crucible during a reaction. As this will produce an increase of the vapour pressure, a tight closing of the crucible is strictly needed.

Two other types of crucibles, that can be only used with a Calvet type DSC are also described as they find specific applications for catalytic investigations.

The first type of crucible is a quartz reactor that is designed for the investigation of gas adsorption on powders (Fig. 2.13).

The quartz tube is introduced in the Calvet type DSC (set in the vertical position). A fritted glass substrate is located in the middle of the tube to receive the powdered sample in order to be surrounded by the calorimetric detector. Tight connections are adjusted at both ends of the tubes for the gas inlet and outlet.

By using a gas injection loop, pulses of known volume of reactive gas are introduced on the sample [4, 9, 16, 46, 50]. The DSC signal gives the corresponding heat of adsorption. In order to know if the total volume of gas is adsorbed on the sample, a GC detector is set on-line with the DSC. The outlet gas is analyzed and the corresponding amount of non adsorbed gas measured.

This DSC-GC coupling allows correlating the heat of adsorption to the amount of gas adsorbed on the sample.

The use of a quartz tube makes also possible to use any type of gas (even corrosive gas) for such an investigation.

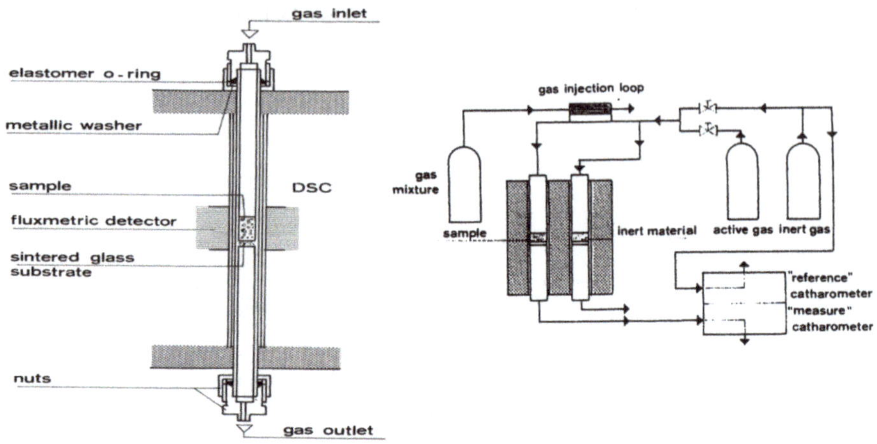

Fig. 2.13 Quartz reactor for adsorption studies and coupling with the gas introduction and analysis device

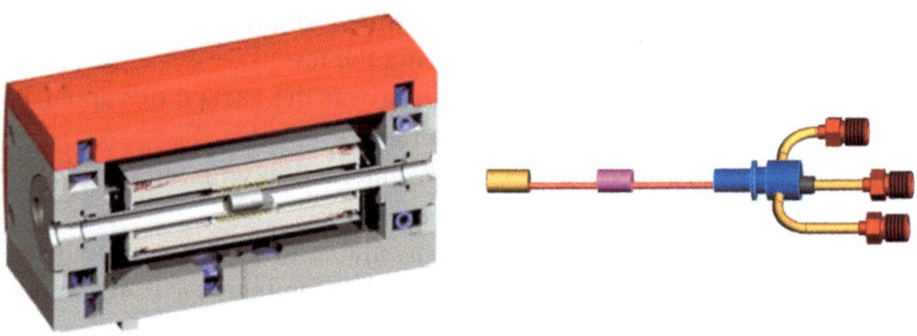

Fig. 2.14 Controlled high pressure crucible with the Calvet type DSC

The second interesting type of crucible is dedicated to work under a controlled gas pressure. Today more and more works are done on the adsorption of hydrogen to form metal hydrides used for the storage of hydrogen. Such investigations have to be run under a controlled pressure of hydrogen and a given temperature, or range of temperature.

After introducing the sample, the crucible (made of incoloy) is tightly closed with a screwable stopper, then introduced in the tube of the Calvet type DSC (Fig. 2.14). A pressure of gas is applied through an external high pressure panel.
Using such an experimental design, only the small volume containing the sample is under pressure. The DSC detector itself remains under normal pressure.
This is very convenient for safety reasons especially when hydrogen is used, but also for the DSC calibration as it is not affected by the gas pressure. A current pressure of 400 bar is available on such a vessel up to 600 °C.
Two different modes are used for the investigation of absorption and decomposition of metal hydrides:

- Operation in temperature scanning mode with a constant gas pressure.
- Operation at constant temperature with a variable pressure.

2.4.4 Applications

The DSC applications for catalytic investigations are very similar to the ones described previously for the DTA technique. However the DSC measurement gives more possibilities as described in Sect. 4.3, especially for adsorption investigations under normal [57] or high pressure [51]
The DSC technique also offers more flexibility for coupling with other techniques (thermogravimetry, gas analysis).
DSC can also be used for the Thermal Programmed Desorption (TPD) determination coupled with a gas analyzer [11, 34]

Among the applications that are specific to the DSC technique, the determination of heat capacity is one of the most important. Different methods can apply for this measurement and are described below.

2.4.4.1 Heat Capacity Determination

Heat capacity plays an important role in thermal processes in any type of industry. Heat loads and processing times, and industrial equipment sizes are influenced by the heat capacity of the material. Combined with thermal conductivity and thermal diffusivity, heat capacity data are needed for the modelisation of the thermal processes. Heat capacity varies with temperature, composition and also water content. As the material can be under the solid or the liquid form, different ways of measuring heat capacity using the calorimetric techniques have been developed.
Heat capacity is thermodynamically defined as the ratio of a small amount of heat δQ added to the substance, to the corresponding small increase in its temperature dT:

$$C = \delta Q/dT$$

For processes at constant pressure, the heat capacity is expressed as:

$$C_p = (\delta H/\delta T)_p$$

Though DSC is a very well adapted technique to measure heat capacity [52], only one procedure has been essentially developed using a continuous heating mode for solid samples. In this chapter, another procedure is described using a step heating mode.

c_p Determination in the Temperature Scanning Mode

If there is no transformation for the considered temperature range, the calorimetric signal for a given mass of sample heated at a constant heating rate dT/dt is relative to the following relation for the sample side:

$$(dq/dt)_s = (m_s c_{p(s)} + m_{cs} c_{p(cs)})dT/dt \tag{2.6}$$

where:

- m_s and m_{cs} respectively sample mass and vessel mass (including the cover),
- $c_{p(s)}$ and $c_{p(cs)}$ respectively specific heat capacity of sample and its vessel.

On the reference side, an empty vessel is used giving the corresponding signal:

$$(dq/dt)_r = m_{cr} c_{p(cr)} dT/dt \tag{2.7}$$

where:

- m_{cr} respectively reference vessel mass,
- $c_{p(cr)}$ respectively specific heat capacity of reference vessel (equal to $c_{p(cs)}$ as the nature of the vessel is identical).

The differential calorimetric signal is given by the following relation:

$$(dq/dt)_s = (m_s c_{p(s)} + m_{cs} c_{p(cs)} - m_{cr} c_{p(cr)})dT/dt \tag{2.8}$$

In order to get rid of the thermal effect of both vessels, the same test (called blank test) is run with identical empty containers. From a practical point of view, the reference vessel is not removed from the calorimeter. The corresponding relation is obtained:

$$(dq/dt)_b = (m_{cs} c_{p(cs)} - m_{cr} c_{p(cr)})dT/dt \tag{2.9}$$

By subtracting the two calorimetric traces, the specific heat capacity of the sample is extracted (Fig. 2.15):

$$C_{p(s)} = (1/m_s)[(dq/dt)_s - (dq/dt)_b]/dT/dt. \tag{2.10}$$

As seen before in the calibration paragraph, the calibration using the Joule effect technique allows converting any calorimetric signal in mW without the need of standard reference materials. That means that in relation (2.10) all the parameters (sample mass, calorimetric signals, heating rate) are accurately known to determine the specific heat capacity of the sample $C_{p(s)}$ (expressed in $J.g^{-1}.°C^{-1}$) at a given temperature. The variation of $C_{p(s)}$ versus temperature can be determined.
With the DSC technique, a third test is needed using a standard reference material (sapphire) that has a known specific heat capacity.

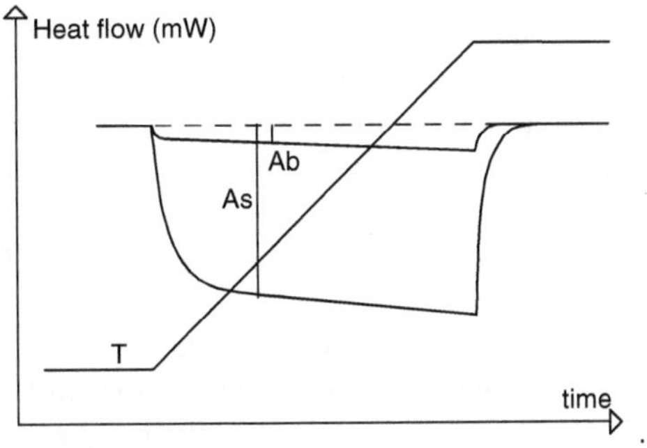

Fig. 2.15 c_p determination in the temperature scanning mode

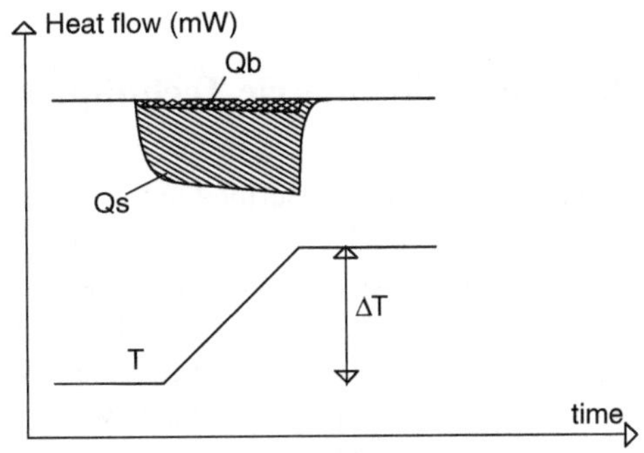

Fig. 2.16 c_p determination in the step heating mode

c_p Determination in the Temperature Step Mode

The previous described technique is very easy to use, but shows some drawback as soon as the accuracy of the c_p determination is concerned. With the temperature scanning mode, the sample is continuously heated and is never at the thermal equilibrium. However c_p is a thermodynamical parameter, defined at the thermal equilibrium. The temperature step mode has been developed to answer this limitation. A step of temperature is applied to the sample and the thermal equilibrium (characterized by the signal return to the baseline) is waited after each step (Fig. 2.16).

If the relation (2.8) is integrated between time t_o (beginning of the step) and time t_n (return to the baseline):

$$\int (dq/dt)_s dt = \int [(m_s c_{p(s)} + m_{cs} c_{p(cs)} - m_{cr} c_{p(cr)}) dT/dt] dt. \quad (2.11)$$

the corresponding equation is obtained:

$$[Q]_{to}^{tn} = (m_s \bar{c}_{p(s)} + m_{cs} \bar{c}_{p(cs)} - m_{cr} \bar{c}_{p(cr)}) \Delta T \quad (2.12)$$

where $\overline{c_p}$ corresponds to the mean c_p value between the two temperatures defining the step of temperature.

Q is obtained by integrating the corresponding surface defined by the calorimetric signal between t_o and t_n.
If again the signal corresponding to the blank test is subtracted when an identical step of temperature is applied, the final relation giving the mean c_p of the sample is obtained:

$$C_{p(s)} = (1/m_s)[Q_s - Q_b]/\Delta T.$$

In that case, the c_p value is obtained by a difference of surface. That means that the result is not depending on fluctuations of the baseline between the different tests that can be seen with the previous technique.

2.5 The Calorimetric Techniques

The DSC technique, described in the previous paragraph, has a lot of advantages but also some drawbacks as soon as it is needed to work on larger amounts of sample, to investigate gas or liquid interactions, to simulate mixing or reactions between two or more components, to work with higher sensitivity.

A calorimeter is mainly characterized by a measurement chamber surrounded by a detector (thermocouples, resistance wires, thermistors, thermopiles) to integrate the heat flux exchanged by the sample contained in an adapted vessel. The measurement chamber is insulated in a surrounding heat sink, made of a high thermal conductivity material.

The main improvement with calorimetry is that it becomes possible to increase the size of the experimental vessel, and consequently the size of the sample without affecting the accuracy of the calorimetric measurement. According to this fundamental property, calorimeters of different sizes have been produced to adapt various types of applications. The calorimeter offers an experimental space in which different types of vessels are designed to especially make possible the investigations of interactions between solid and liquid materials.

2.5.1 Calorimetric Principles

Many different types of calorimeters are commercially available and most of the calorimetric principles are summarized in Table 2.4:

The main calorimetric modes can be resumed using Fig. 2.17,
where:

- A is the sample contained in a vessel,
- B is the thermostated jacket (liquid or metallic),
- T_i is the temperature of the sample, and
- T_j is the temperature of the jacket.

Table 2.4 Different calorimetric principles

Property	Technique	Description
Heat	Isothermal calorimetry	In an isothermal calorimeter, a defined exchange of heat occurs between the sample and the surroundings at a given environment temperature. The heat flow rate is determined on the basis of the temperature difference along a thermal resistance between the sample and its surroundings.
Heat	Isoperibolic calorimetry	In an isoperibolic calorimeter, the temperature of the surroundings remains constant, while the temperature of the sample can differ from the surrounding temperature.
Heat	Adiabatic calorimetry	An adiabatic calorimeter is used to study chemical reactions and is one in which there is no net heat gain or loss during the chemical reaction.
Heat	Calvet Calorimetry	A Calvet calorimeter is a heat exchanging calorimeter with a cylindrical type detector working in isothermal and scanning modes.
Heat	Titration calorimetry	In a titration calorimeter, a small volume of solution A is injected into the calorimetric vessel containing the solution B. The corresponding heat of interaction is measured.
Heat	Flow calorimetry	In a flow calorimeter, the sample is a flowing liquid.
Heat	Solution calorimetry	In a solution calorimeter the mixing of the samples is performed in a static (batch) mode.
Heat	Reaction calorimetry	A reaction calorimeter is designed for the investigation of reactions between liquids or solids. The calorimetric technique can be isothermal, isoperibolic or adiabatic.
Heat	Bomb calorimetry	A bomb calorimeter is designed for measuring heat of combustion. The sample is contained in a heavy-walled metal container containing oxygen under pressure, followed by ignition to obtain the heat of combustion of the sample.

In the isothermal mode, the sample temperature T_i has to remain constant. To reach this target, the temperature of the jacket T_j is permanently adjusted. In this situation, a heat flux is measured between A and B and relates to any transformation or reaction that occurs in A.

In the adiabatic mode, the temperature of the sample T_i and the thermostated jacket T_j are kept at the same temperature in order to prevent any heat transfer between A and B. In that case, the temperature of the sample increases if the occurring reaction in A is exothermic or decreases if the reaction is endothermic.

The isoperibolic principle is an intermediate mode where the temperature of the thermostated jacket T_j is maintained at a constant temperature. The sample

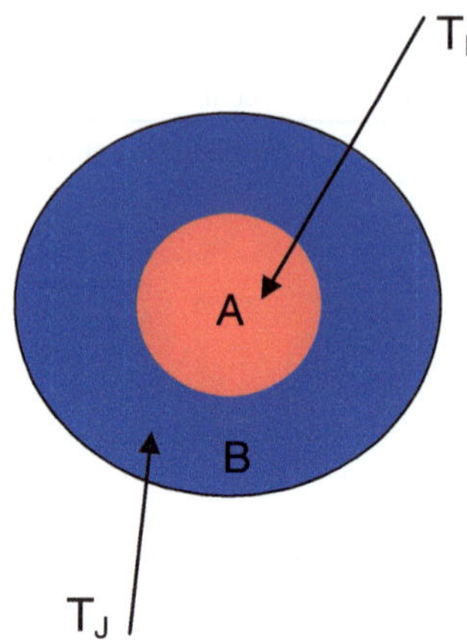

Fig. 2.17 Sample and thermostated jacket in calorimetric principles

temperature T_i varies according to the transformation in A. The difference of temperature is measured and is related to the heat generated by the transformation.

According to the catalyst to be investigated or the catalytic process to be simulated, the calorimetric mode, and accordingly the type of calorimeter, has to be carefully selected.

2.5.2 *Isothermal Calorimetry*

The Calvet type calorimeter is one of the most commonly used for the catalytic investigations. Different types of experimental vessels have been developed to fulfill the different experimental requirements.

2.5.2.1 Calvet Principle

The Calvet principle has already been described in Sect. 2.4.2.

To understand the direct correlation between the electrical signal and the heat flux, it is needed to consider that a power W is fully dissipated in a calibration vessel surrounded by a fluxmeter composed of crowns of thermocouples (Fig. 2.11). An elementary power w_i is dissipated through each thermocouple producing an elementary variation of temperature ΔT_i between the internal and external weldings (Fig. 2.18):

$$w_i = \delta_i \Delta T_i \tag{2.13}$$

where δ_i is the conductance of the thermocouple.

The corresponding variation of temperature generates an elementary electromotice force (emf) according to the Oersted law:

Fig. 2.18 Joule effect calibration principle

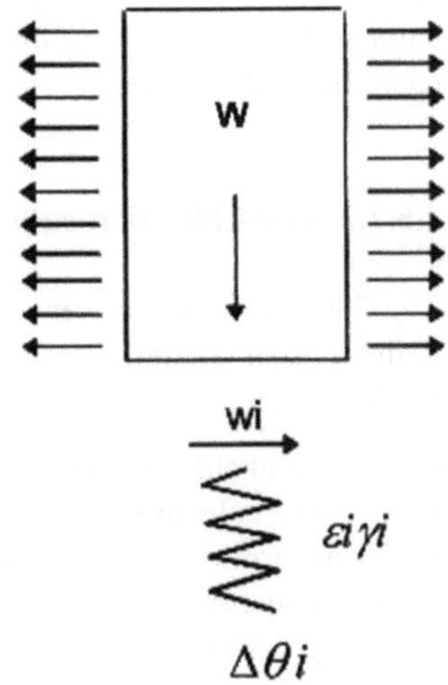

$$e_i = \varepsilon_i \Delta T_i \tag{2.14}$$

where ε_i is the thermoelectric constant of the thermocouple.

By combining the relations (2.13) and (2.14), and considering that all the thermocouples are in series:

$$E = \Sigma e_i = \Sigma(\varepsilon_i/\delta_i)_{wi} \tag{2.15}$$

As all the thermocouples are identical and made from the same materials, the relation (2.15) can be expressed as:

$$E = (\varepsilon/\delta)\Sigma w_i \text{ or } E = (\varepsilon/\delta)W \tag{2.16}$$

The relation (2.16) shows that the power dissipated in the vessel is directly correlated to the heat flux. The term ε/δ corresponds to the calibration factor of the calorimeter. This relation provides the correlation between the electrical output of the calorimetric detector and the corresponding heat flux exchanged with the sample.

The main advantages of this type of calibration:

- it is an absolute calibration;
- it is not needed to use metallic reference materials;
- the calibration can be performed at a constant temperature, in the heating mode and in the cooling mode;
- it can apply to any experimental vessel volume; and
- it provides a very accurate calibration of the calorimetric detector.

According to the Calvet principle, many different calorimeters have been designed with various temperature ranges, from small to large size volumes, with a large variety of sensitivity. In the next paragraph is more precisely described one Calvet calorimeter, produced by the SETARAM company, and that is used worldwide for catalytic investigations.

2.5.2.2 The C80 Microcalorimeter

The C80 microcalorimeter (produced by the SETARAM company) is a versatile tool in the field of catalytic investigations as it works according to different calorimetric modes:

- Isothermal calorimetry.
- Scanning calorimetry.
- Gas adsorption calorimetry.
- Liquid adsorption calorimetry.
- Mixing calorimetry.
- Percolation calorimetry.
- Reaction calorimetry.
- Pressure calorimetry.

In a Calvet calorimeter, the most important part is the thermoelectric element that provides the performances of the calorimetric detector.

The microcalorimeter is built around a high metallic conductive block with two cavities containing the thermopiles. Each thermopile is made of crowns of thermocouples, mounted on a metallic tube, and defines the experimental zone (Fig. 2.19). The two thermopiles (measure and reference) are inserted in a thermostated heating block that fixes the temperature of the calorimeter (Fig. 2.20). The block itself is surrounded by the heating element and arranged in an insulated chamber. The detectors define the experimental zone, in which the vessels are very tightly introduced. The top of the calorimeter is removable in order to make possible the fixation of dedicated vessels (mixing) inside the calorimeter and also for the arrangements of temperature prestabilisation features for fluid samples before introduction in the calorimetric zone.

The microcalorimeter is characterized by a large experimental volume ($15\,cm^3$) that has allowed the design of a large variety of experimental vessels according to investigations to be carried out.

The main characteristic of the calorimeter is to be fitted on a rotating mechanism. A special mixing vessel has been designed to be used in such experimental conditions.

The microcalorimeter offers a large choice of experimental vessels according to the applications to be run:

- Isothermal and scanning calorimetry.
  - The batch standard vessel is designed for investigating any type of transformation at a constant temperature, when heating or cooling large volume of samples in the solid or liquid form. It is also dedicated to the determination of heat capacity.
- Pressure calorimetry

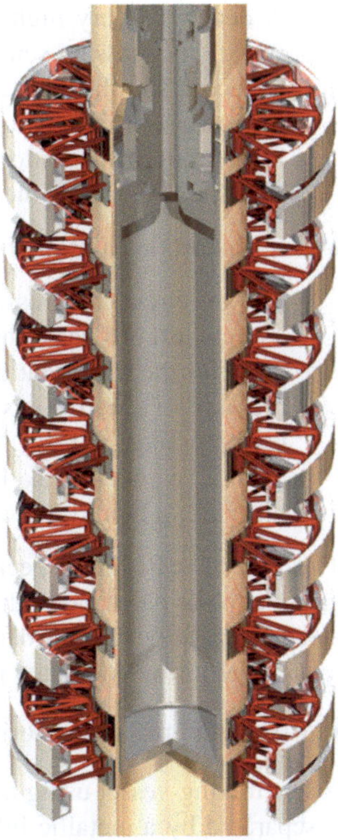

Fig. 2.19 Cross-section of the C80 thermopile

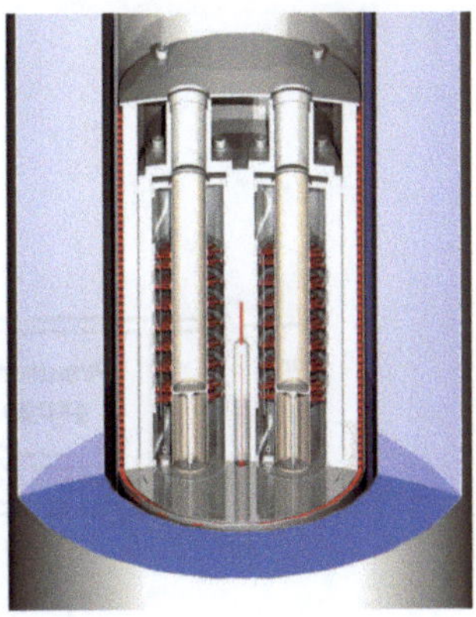

Fig. 2.20 Cross section of the calorimeter

– The batch high pressure vessel is designed for the simulation of reaction and decomposition under pressure in a closed vessel or under controlled pressure (max : 100 bar). It is used to define the safety conditions of reaction operations but also investigations under supercrital gas conditions (e.g. adsorption of CO_2 on zeolites).

A dedicated very high pressure vessel (max 350 bar) is connected to a high sensitivity pressure sensor through a special capillary tube. This vessel allows to record simultaneously the heat and pressure released by the sample during the reaction as a function of time or temperature [32].
For studies of adsorption at very high pressures (e.g. formation of gas hydrates), specific vessels are able to operate up to 1000 bar.

- Gas adsorption calorimetry.
  - The gas circulation vessel is fitted with two coaxial tubes and is used to produce a circulation of gas (inert or active) around the sample. It is especially convenient for the investigation of adsorption/absorption on a catalyst under normal pressure of reactive gas such as hydrogen, ammonia, CO, ... [18, 22, 35] or high pressure [15].
  Such a vessel can also be fitted with a relative humidity generator in order to introduce a humid gas with a known rate of humidity on a solid sample.

This type of vessel has especially been used by Auroux [5, 14] to adapt a volumetric gas line to the calorimeter that is described on Fig. 2.21. Using such a calorimetric and volumetric device, the determination of the surface acidity and basicity of various types of zeolites and related materials was performed [6].

- Mixing and reaction calorimetry.

- The mixing vessel using the rotating mechanism is divided in two chambers and separated by a metallic lid (Fig. 2.22). The two materials are placed, one in the

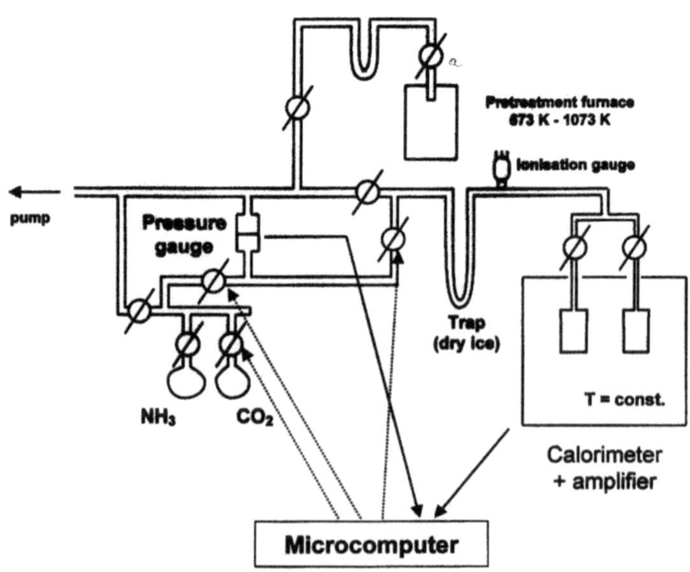

Fig. 2.21 Calorimetric et volumetric devices coupling (from Ref. [14])

Fig. 2.22 Mixing vessel (by reversing)

lower chamber (i.e. powder) and the other in the upper chamber (i.e. liquid). The mixing of the two components is obtained by rotating the calorimeter, the metallic lid acting as a stirrer. This mixing vessel is designed for investigating any liquid–liquid mixing (dilution, neutralization, reaction, ...), solid–liquid mixing (dissolution, hydration, wetting, reaction, ...)

– The membrane mixing vessel is especially dedicated to the mixing of viscous samples and for applications when the rotation of the calorimeter cannot be used. In such a vessel, the separation between both chambers is performed with a very thin membrane (metal or PTFE) (Fig. 2.23). The vessel is fitted with a metallic rod that is operated from outside the calorimeter. In this situation, the mixing of both components is obtained by pushing the rod in order to break the membrane. The rod is also used as a stirrer during the test. The applications are similar to the ones previously described.

- Sorption calorimetry.
  - The ampoule sorption vessel is designed for slow dissolution process and for wetting operation [43, 56]. The sample is introduced inside a special glass ampoule the upper part of which is connected to a vacuum pump and the bottom part of which is very thin. After outgassing during 30 min, the ampoule is sealed with a torch and the upper part of the ampoule is cut and withdrawn. The vacuum operation allows desorbing the surface of the solid sample (especially powder) and makes easier the interaction. The sealed ampoule is introduced in the vessel

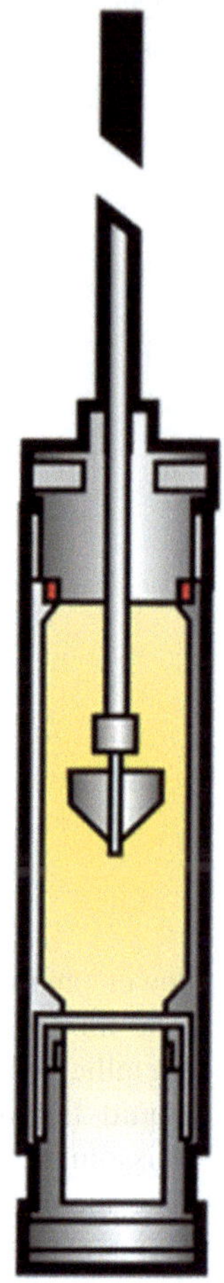

Fig. 2.23 Mixing vessel (metallic membrane)

containing the solution. By breaking the ampoule using a piston rod, the solid and liquid samples are put into contact

- Percolation calorimetry.
  - The percolation vessel is used for adsorption of liquid on a powder (especially catalyst) [41]. However when a liquid is flowed through a powder, the first interaction that occurs is wetting of the powder followed by adsorption. In order to make distinct these two thermal interactions, a special calorimetric design has been developed with the percolation vessel. The powder is introduced in a metallic cylinder on a sintered metallic section. A first carrier liquid is flowed through the powder to get the wetting (Fig. 2.24). Then the pump is switched to the liquid solute for the adsorption phase. According to the reaction, it will be possible to be back with the first carrier liquid in order to proceed

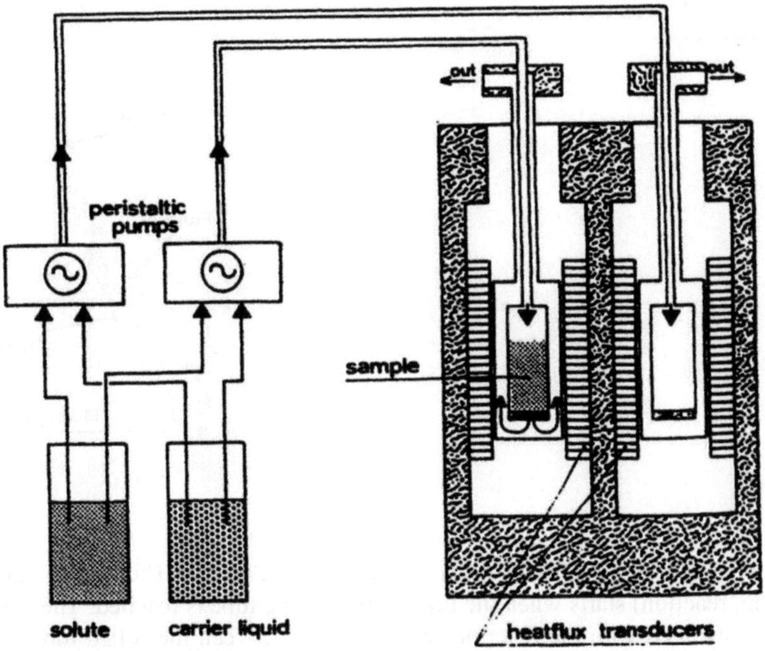

Fig. 2.24 Experimental set-up for percolation calorimetry

to the desorption phase. This calorimetric vessel is very convenient for the investigations of surface reactivity, catalyst contamination, chemisorption under normal and high pressure.

- Flow mixing calorimetry.
  - The use of the chemical sorbents is today one of the most popular absorption technique for the CO_2 capture in postcombustion techniques. In such an industrial process, the amine solution is introduced at the top of an absorption tower while the exhausted fume containing carbon dioxide is introduced at the bottom. As an intimate contact is reached in the absorption tower, the amine solution chemically absorbs the carbon dioxide from the gaseous stream. Such a process especially requires two types of thermodynamic parameters: gas solubility and enthalpy of absorption. The enthalpy of absorption, according to the amount of absorbed gas and the corresponding heat capacities of solutions, define the temperatures of the fluids when they exit the absorption columns. Flow mixing calorimetry is the ideal technique for measuring such enthalpies of absorption [2, 28]. In order to work under pressure, a dedicated high pressure mixing vessel is adapted to be used on the Setaram C80 calorimeter. The mixing vessel is made of a stainless steel tube in a helicoidal shape into a cylindrical container (Fig. 2.25). The length of the tube in closed thermal contact with the cylinder is about 240 cm. The fluids (CO_2 and amine solution) are introduced at the bottom

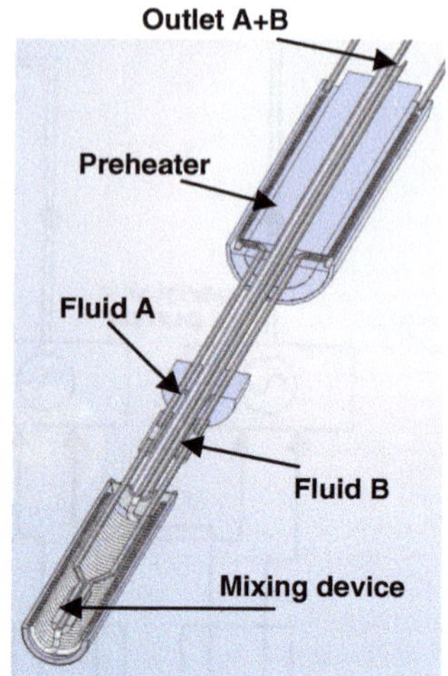

Fig. 2.25 The high pressure flow mixing vessel

part of the vessel in two vertical and concentrical tubes. The mixing (dissolution, reaction) starts when the thinner part of the tube is reached. The heat that is associated with the reaction, is exchanged between the rolled tube and the calorimetric block through the wall of the vessel in an isothermal mode.
The flow mixing vessel operates from room temperature to 200 °C and for a range of fluid pressure from 0.1 to 20 MPa. The fluid flowrates vary from 50 to 1500 $\mu \cdot min^{-1}$, that allow to cover a wide range of mixture composition.

2.5.3 *Isothermal Titration Calorimetry*

The different mixing vessels that are described in the previous paragraph are designed for the interaction between two components and are not well adapted for titration studies. In the Isothermal Titration Calorimetry, aliquots of sample B is added in a volume of sample A, located in the calorimetric vessel and maintained under a constant stirring (Fig. 2.26).

Another sample C can also be added according to the reaction to be simulated. To allow such experimentations, the C80 calorimetric block is modified to adapt a magnetic stirring system at its bottom part. This leaves more space to adapt the different liquid and gas introduction lines. A heating cover has also to be used for the thermostatisation of the fluids at the temperature of the calorimeter. The liquids are introduced through a motor-driven syringe pump that allow a continuous or a step injection of defined volumes.

Such a calorimetric technique is especially designed to follow the liquid adsorption of organic compounds on catalysts or zeolites [12, 21, 49]. The adsorption of n-butylamine (in decane) on a zeolite (in a solution of decane) illustrates the calori-

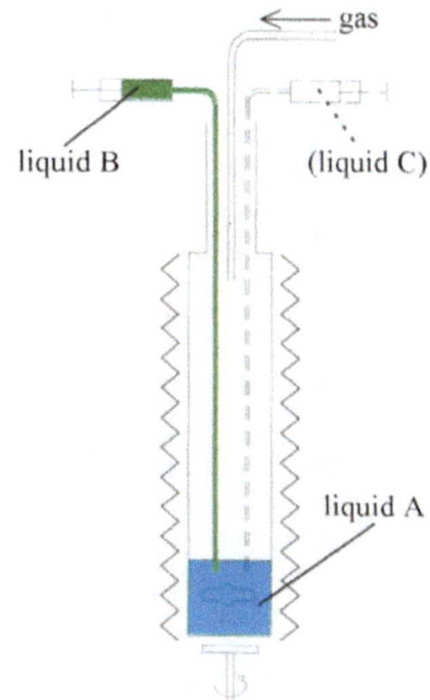

Fig. 2.26 Principle of the Titrys calorimeter

metric titration mode. After activation of the zeolite at 400 °C, the calorimetric test is run at 40 °C. Aliquots of 0.25 ml of adsorbant are added until the saturation of the zeolite is reached (Fig. 2.27).

Such a vessel is adapted to follow an organic reaction when several compounds have to be added during the process.
It is also used to simulate the injection of very small amounts of water to produce a hydrolysis reaction (hydrolysis of ammonia borane for the production of hydrogen).

2.6 The Thermogravimetric Technique

2.6.1 Principle

Thermogravimetric Analysis (TGA) or Thermogravimetry (TG) is defined as followed by ICTAC: "A technique in which the mass of the sample is recorded versus time or temperature while the temperature of the sample is programmed, in a controlled atmosphere. The instrument is called a thermogravimetric analyzer (TGA) or a thermobalance." [17].

Two different types of transformations can occur:

- Transformation with a mass loss: dehydration, dehydroxylation, evaporation, decomposition, desorption, pyrolysis, . . .
- Transformation with a mass gain: adsorption, hydration, reaction, . . .

The Fig. 2.28 shows a classical mass loss with the determination of the derivative curve (DTG).

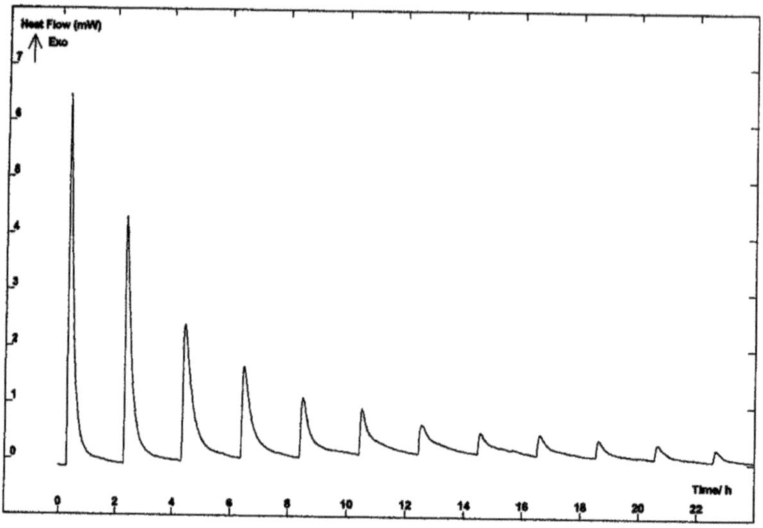

Fig. 2.27 Adsorption of n-butylamine (in decane) on a zeolite at 40 °C

Fig. 2.28 A TG mass loss curve with the derivative curve (DTG)

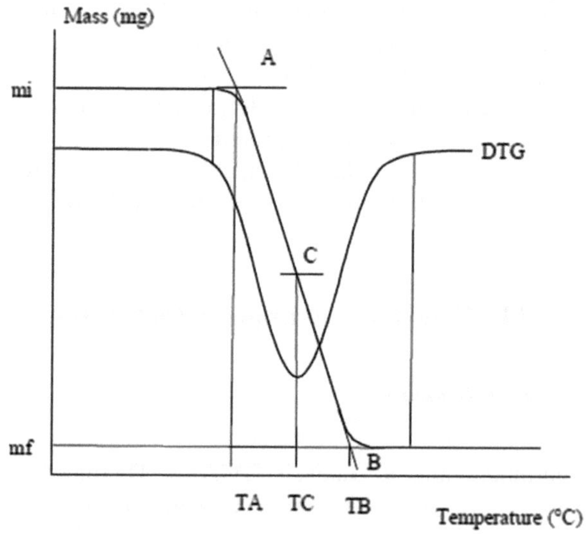

If m_i is the initial mass and m_f the final mass, the mass loss is equal to $(m_i - m_f)$ or $(m_i - m_f)/m_i$ in percentage.

It is interesting to notice that the DTG peak is very similar to the DTA/DSC peak (see also paragraph on couplings). The integration of the DTG peak also gives access to the variation of mass.

The DTG peak is related to the kinetics of the transformation, as the DTG signal corresponds to a mass variation rate. The shape of the peak directly informs on the rate of the transformation.

2.6.2 Detectors

Different types of TGA techniques are available and are described in Table 2.5.

A thermobalance is built around a highly sensitive balance module, a furnace and a controlled atmosphere cabinet. Different types of balance detectors are available but the most commonly used is based on the principle of the null position balance. An example is given in Fig. 2.29.

The balance is made of an articulated beam, suspended on a torsion wire. A rod is fixed in the middle of a beam and is coming with a window in the upper part. Four magnets are mounted in the middle part of the rod. Each of them is plunging in a solenoid fixed on the mechanical frame. On the same frame, a device of optical detection delivers a light that is fully crossing the slot of the window. The light signal is measured by means of a photodiode.
When there is a mass variation inducing a movement of the balance beam, the light is partially hidden. A compensation current is sent through the solenoids in order to have the beam back to the null position. With such a detection principle, the mass variation is proportional to the compensation current. It is positive or negative depending if it is a mass loss or a mass gain.

Based on this principle, different types of thermobalances are available: top loading balance, bottom loading balance, horizontal balance.

Table 2.5 Different types of TGA techniques

Property	Technique	Description
Mass	Mono TGA	A technique in which the change of sample mass is analysed while the sample is subjected to a temperature variation. A mono TGA has only one furnace, for the sample.
Mass	Symmetrical TGA	A technique in which the change of sample mass is analysed while the sample is subjected to a temperature variation. A symmetrical TGA has two furnaces, one for the sample and one for the reference material.
Mass	High pressure TGA	A high pressure TGA is a TGA (mono or symmetrical) working with the test sample under pressure.
Mass	Corrosive TGA	A corrosive TGA is a TGA (mono or symmetrical) working with the test sample in a corrosive atmosphere.

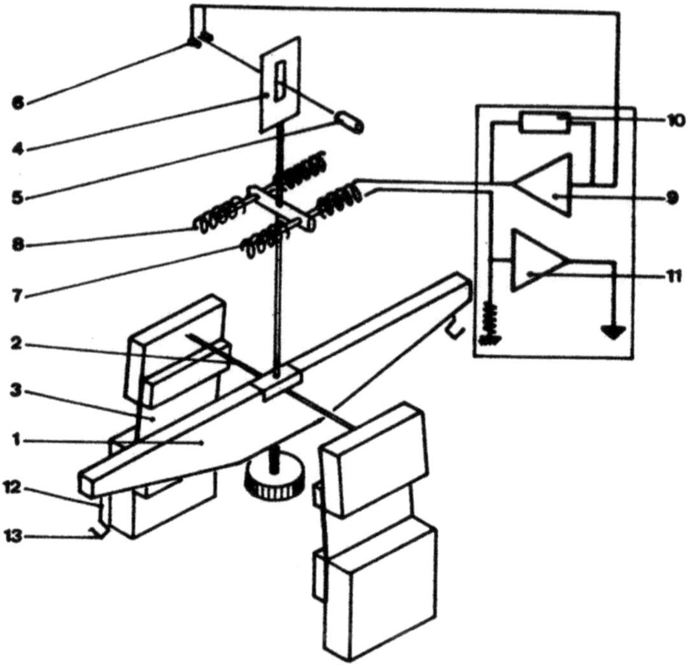

Fig. 2.29 Principle of a null position balance (SETARAM balance)

The nature of the furnace heating element depends on the range of temperature for the investigations. The most used materials are metallic (nickelchrome, kanthal, platinum, tungsten) but also SiC and graphite.

As for the DTA experimentation, the choice of the crucible for the TGA test has to be done very carefully. The main point is to select a material that will not react with the sample, especially at high temperature;

- Aluminum (up to 500 °C),
- Silica (especially used below 1000 °C as it is very inert and easy to clean),
- Alumina (especially required for metallic samples),
- Platinum (especially required for inorganic samples),
- Graphite and tungsten (for temperature above 1750 °C).

Other types of materials are available such as magnesia, zirconia, boron nitride, according to the sample under investigation.

2.6.3 Operation

2.6.3.1 Selection of the Atmosphere

As the crucible is open for the TGA test, the atmosphere around the sample has to be carefully controlled, especially in the case of catalysts. Different gas options are available to work in a thermobalance:

- Operation in inert gas:
  - This is the most common situation where it is needed to protect the sample from oxidation during the heating process. A vacuum purge is recommended prior to the introduction of the inert gas in order to empty the balance, the furnace and all the gas lines.
    If it is needed for desorbing the sample prior to the heating, a primary vacuum or even a secondary vacuum pumping is to be used for this operation before the introduction of the inert gas.
    Most of the investigations on catalysts need a good initial desorption of their surfaces before starting the adsorption investigations.
- Operation in reactive gas:
  - During catalytic investigations, different types of reactive gases are used and it is needed to know the risk of interactions with the different metallic and ceramic parts of the thermobalance. Here is the situation for the most common reactive gases:
    Air, oxygen for oxidation studies up to 1750 °C without specific problem.
    Reactive gas such as butane for investigations of catalytic reactions [36].
    Hydrogen: with such a gas two main concerns have to be considered: the concentration and the corrosion.
    A non-explosive mixture of less than 4 % H_2 in an inert gas is recommended. For higher concentrations, a safety device is required on the thermobalance. The use of H_2 requires that no trace of air is present in the thermobalance to avoid an explosion.
    Above 1000 °C, H_2 becomes a poison for platinum that is becoming brittle. The platinum–rhodium temperature control thermocouple has to be replaced to work above this temperature. Tungsten rhenium thermocouple is one option.
- Operation in corrosive gas:
  - Adsorption or reaction under corrosive gas, such as CO, NH_3 but also halogens (chlorine, fluorine), can be investigated using the TGA technique but caution has to be taken to avoid the contact between the corrosive gas and any metallic part of the thermobalance (especially the thermocouple).
    Dedicated experimental set-up need to be used to guarantee such a protection, as described on Fig. 2.30. A silica tube, with a restriction in the lower part, is introduced in the furnace. This tube shape allows having the thermocouple out

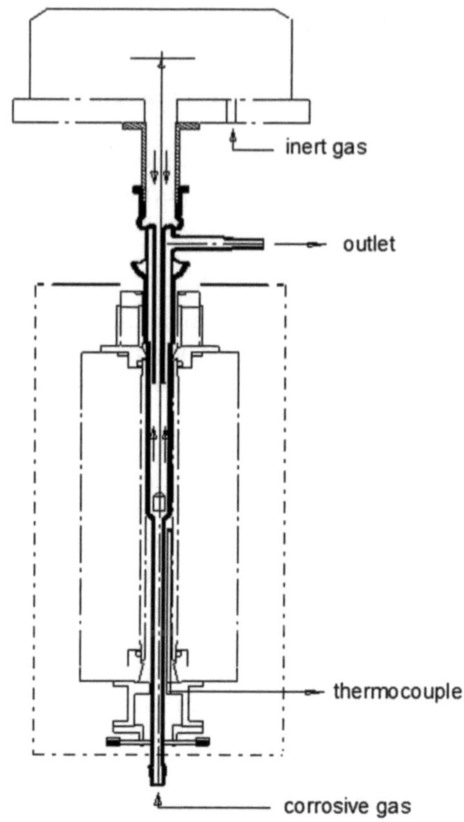

Fig. 2.30 TGA corrosive set-up

of the corrosive gas stream. The protection of the balance has to be carried out with an inert gas. The crucible and the suspension also are in silica. With such an adaptation it is possible to work with TGA with any type of corrosive gas, even at high concentration up to 1000 °C.

- Operation in humid atmosphere:
  - Adsorption of water vapour at a given pressure or at a relative humidity percentage is an important test to characterize the capabilibity of solid sorbents to fix water. It is also known that water vapour will enhance the adsorption of CO_2 on solid sorbents.

    To perform such a measurement a relative humidity generator has to be connected to the TGA through a thermostated gas line (Fig. 2.31).

    A dry gas is splitted in two lines, one remaining dry and the other one saturated with water. The two gas streams are mixed in a chamber. According to the desired relative humidity, the respective flowrates are adjusted for the gas mixture. The humid gas resulting from this operation is then transferred to the TGA via a heated transfer line.

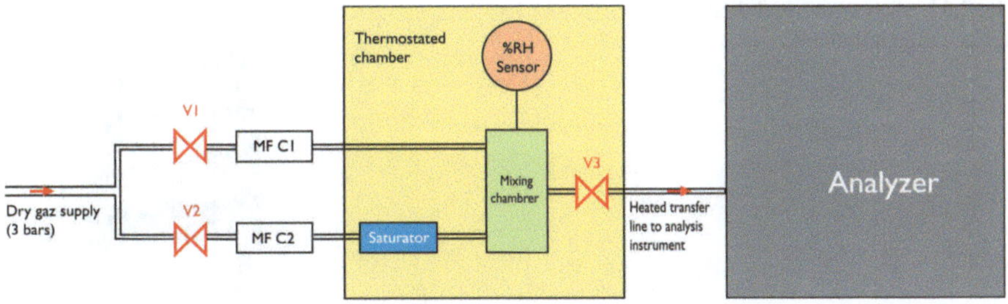

Fig. 2.31 Principle of the Wetsys relative humidity generator (Setaram)

2.6.3.2 Buoyancy Effect

The buoyancy effect is most of the time a problem that is misunderstood in the TGA measurement or often hidden by the numerical treatment of the TGA curves.
Buoyancy is the upward force exerted on an object when it is immersed, partially or fully, in a fluid (liquid or gas). Its value is equal to the weight of the fluid displaced by the object. In the case of the TGA experiment with a given crucible (the object), this results in an apparent increase of the mass when the sample is heated. The effect is observed on all conventional balances. In fact the term buoyancy includes different parameters: the true buoyancy as described before, the convection currents, the gas flow drag effects, the gas velocity effects, the thermomolecular forces, the thermal effects on the balance mechanism.
The buoyancy, according to these differents factors, is especially significant at low temperature and will decrease at high temperature. When a mass variation has to be accurately measured at low temperature (for example water content), a correction of the buoyancy has to be performed. The common way is to run a blank test with an empty crucible with the same experimental conditions. However this numerical correction remains dependent on the reproducibility of such a blank curve and the correction is affected with a certain uncertainty.
Another way to solve the problem is to use a symmetrical thermobalance.
In such a configuration (Fig. 2.32), the crucibles containing respectively the sample and the inert reference material are hung on each side of the balance. One or two furnaces are used according to the symmetrical model.
If the gas flowrates are adjusted on both sides, the buoyancy effect coming from the sample and the reference, as it is identical, is compensated.

2.6.4 Applications

As described in the previous paragraphs, the TGA technique provides a wide range of applications for the investigation of catalysts and related material:

- Gas adsorption and desorption.

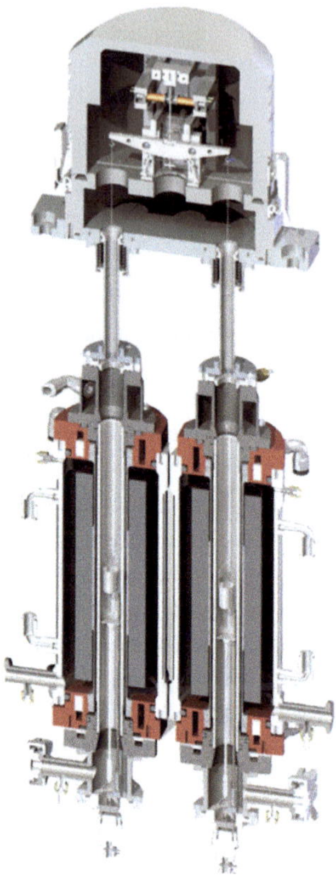

Fig. 2.32 A symmetrical dual furnace thermobalance

- Decomposition.
- Dehydration and dehydroxylation.
- Oxidation.
- Preparation of catalysts.
- Regeneration of catalysts.
- Investigation under various reactive gas (H_2, CO, ...).
- Investigation under relative humidity.
- Investigation under corrosive gas.

In a more general way, thermogravimetry, with simple and short experiments allow preliminary screenings of catalysts where multiple variables are being considered. In just one experiment, the capacity of an adsorbent can be evaluated over an entire temperature range. It is also possible to collect qualitative information about the initial adsorption rates. Thermogravimetry, which has been applied to the preparation and characterization of adsorbents, has also proved to be a useful technique for preliminary adsorption capacity assessment. This is especially the case for the CO_2 capture performance of the sorbents and their thermal stability. For example, the evaluation of aminated solid sorbents for the CO_2 capture performance was evaluated [47].

Thermogravimetry is a very useful tool to evaluate the performances of catalytic filters used for the removal of soot from diesel exhaust gas [42].
TGA is also very widely used to investigate the NO_X-carbon reactions to control the emission of NO_X during the combustion process [60].

More recently, thermogravimetry has been applied to the evaluation of the chemical-looping combustion (CLC) process. It is considered as an energy-efficient method for the capture of carbon dioxide from combustion and provides an advantage of no energy loss for CO_2 separation without NO_X formation. This process consists of oxidation and reduction reactors where metal oxides particles are circulating through these two reactors [29, 54, 58]. The TGA set up for such an application is described on Fig. 2.33.

2.7 The Simultaneous Techniques

The TGA technique provides information on the mass variations that occurs within the sample, but most of the time, if the composition of the sample is unknown, the interpretation of the data are not easy. This is why the TGA technique is mostly used combined with other techniques such as DTA or DSC [59].

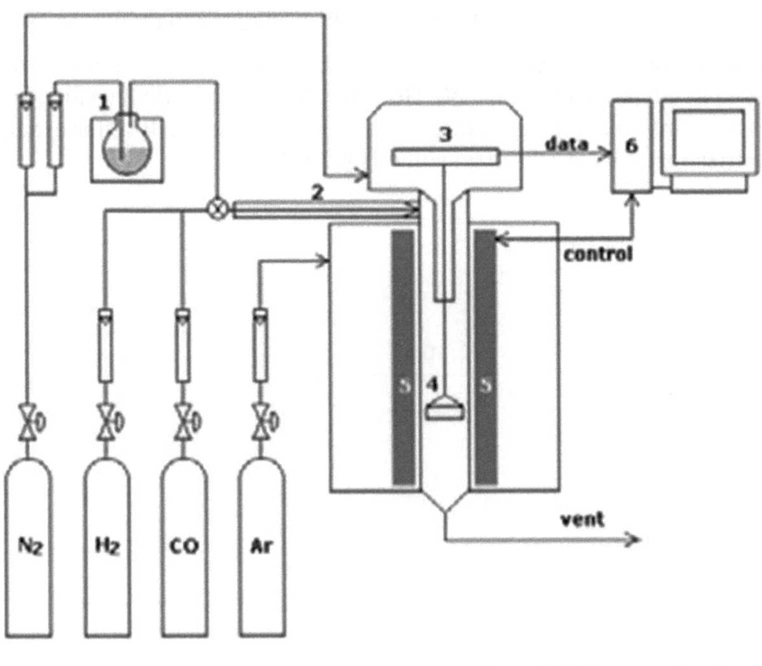

Fig. 2.33 Schematic diagram of a thermal gravimetric analyzer (TGA) system for the reactivity test; **1** water reservoir and bubbler, **2** gas pre-heater, **3** electric balance, **4** sample basket, **5** heater, **6** personal computer (from Ref. [29])

The TG-DTA or TG-DSC combination allows obtaining the simultaneous TGA and DTA or DSC measurements on the same sample (Fig. 2.34). The heat effects are correlated with the mass variations. The DTA or DSC signal gives information on heat effects not correlated with mass variation (melting, phase transitions, ...)
The main interest of such a simultaneous measurement is that only one sample is used. This is especially very important for the characterization of catalysts as sample size (mass, surface, porosity, and composition), the gas flowrate and composition are critical parameters for an accurate determination.
Another important factor of the combination is that any endo or exo effect is related to the mass fraction that has been adsorbing or reacting.
The derivative DTG curve is similar to the DTA or DSC curve for the transformation with mass variation. Both curves are used for the kinetic interpretation of the sample transformation.

As an example of the interest of such a coupling technique, the investigation of the hydrogenation of a poisoned catalyst is given (Fig. 2.35).
A sample of dead catalyst is heated under pure hydrogen using a protected DTA rod. The combination of both techniques allows to understand the different processes that are occurring during the regeneration reaction:

- The first mass moss is associated with a small endothermic effect corresponding to the degradation of the organic compounds adsorbed on the catalyst.
- The second mass loss gives rise to a significant exothermic effect, due to the reduction of sulphur contained in the poisoned catalyst and producing H_2S.

The applications of the TG-DTA (or DSC) technique are very similar to the ones developed in the previous paragraph. However the DTA (or DSC) technique allows to give more information on the thermal properties of the catalyst, for example the characterization of materials after synthesis [1, 3, 27].

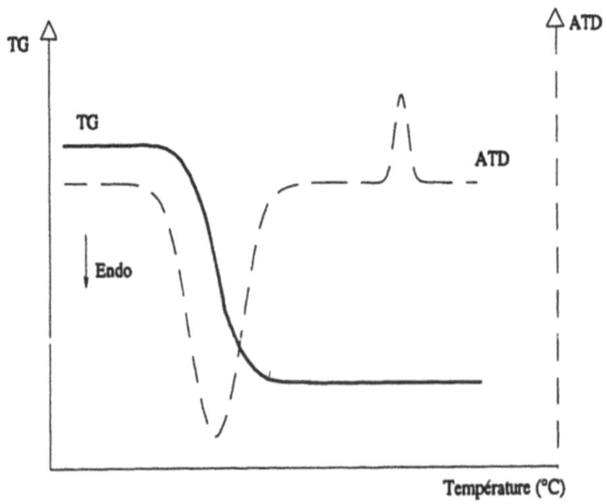

Fig. 2.34 The TG-DTA (or DSC) curve

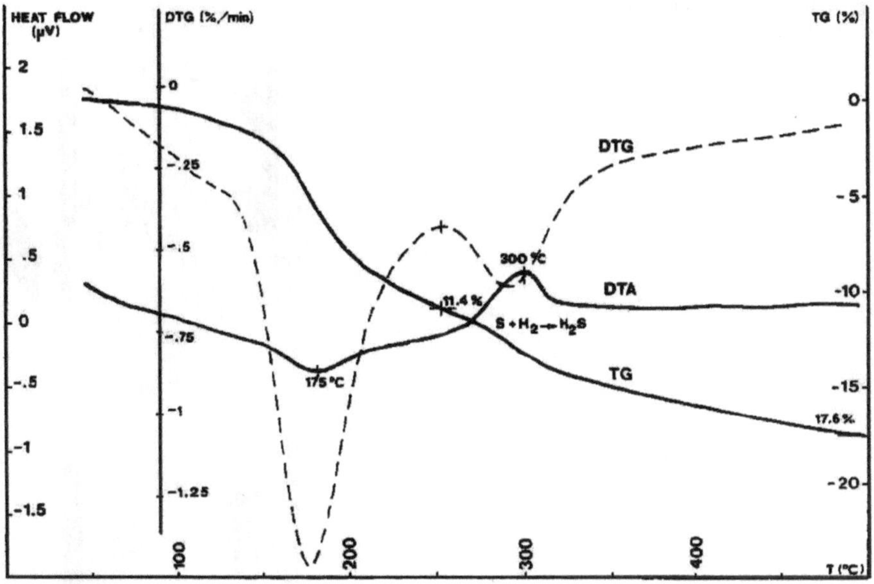

Fig. 2.35 TG, DTG and DTA curves of the hydrogenation of a poisoned catalyst

In most TG-DTA or DSC instruments, the DTA or DSC probe is attached to the balance. Such a combination results in a reduction of the performances of the TGA and the DTA or DSC detectors compared to the characteristics when they are used alone.

The Sensys TG-DSC is an exception in the field as the TGA and the DSC detectors are not mechanically attached (Fig. 2.36).

The Sensys TG-DSC is based on the Calvet type DSC (Fig. 2.36) used in the vertical position [30]. On top of the DSC, is adjusted a symmetrical balance corresponding to the principle described on Fig. 2.29. The crucibles containing the sample and the inert material are hung on each side of the balance and introduced in the calorimetric zone of the DSC without touching the walls. In such a situation, the crucibles are fully surrounded by the fluxmeters, providing an accurate DSC determination. In the same time, the symmetrical balance allows a compensation of the buoyancy effect resulting on a very high sensitive TG determination.

This technique is very powerful for the investigation of gas interaction on solid or liquid sample, such as adsorption, absorption, desorption, catalytic reaction under various types of gas (inert, oxidative, reducing, corrosive, humid) [55], the synthesis and catalytic behaviour of catalysts [53].

An example of the use of such a TG-DSC combination is given on the adsorption and desorption of CO_2 on a catalyst (Fig. 2.37). After preparation of the sample under helium, a mixture of helium (90 %) and CO_2 (10 %) is introduced on the catalyst. The exothermic effect is related to the amount of CO_2 adsorbed on the catalyst. When

Fig. 2.36 Cross section of the Sensys TG-DSC

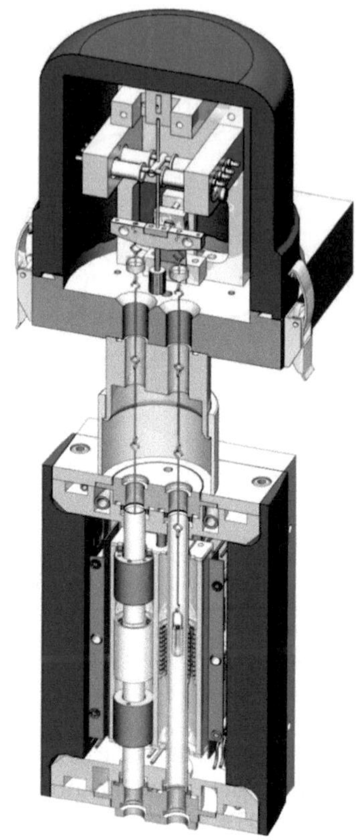

switching back to helium, it is noticed that desorption of the catalyst is not complete at this temperature.

The TG-DSC technique was also used for the investigation of amminoborane for the production of hydrogen [7]

2.8 Evolved Gas Analysis

Evolved Gas Analysis (EGA) is defined as followed by ICTAC:
"A technique of determining the nature and amount of volatile product or products formed during thermal analysis." [40].

As indicated in the previous chapter, the TGA technique only informs on the variation of sample mass, but hardly on the mechanisms linked to the decomposition or the reaction. The idea behind the EGA coupling is to analyze the vapours emitted by the sample during the test [39]. Three main gas coupling techniques are used for such an investigation as described in Table 2.6:

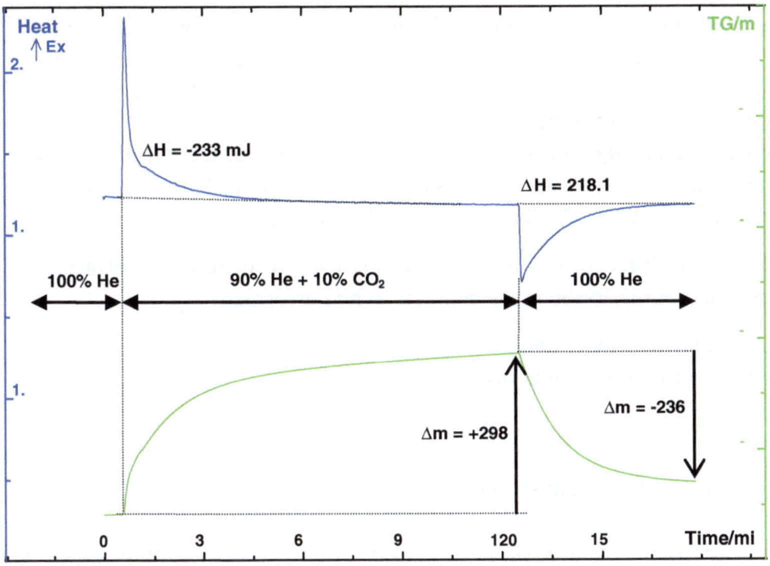

Fig. 2.37 Adsorption/desorption of CO_2 on a catalyst at 40 °C (Sensys TG-DSC)

Table 2.6 The main EGA techniques

Mass and MS spectrum	TGA-MS	In TGA-MS, the TG is connected online with a mass spectroscopy detector to determine the nature and amount of volatile product emitted by the sample during a thermogravimetric experiment
Mass and FTIR spectrum	TGA-FTIR	TGA-FTIR is a thermogravimetric analyzer (TGA) coupled online with a Fourier-Transform Infrared (FTIR) spectrometer to determine the nature and amount of volatile product emitted by the sample during a thermogravimetric experiment
Mass and GC chromatogram	TGA-GC	In TGA-GC, the TGA is connected online with a gas chromatograph(GC) detector to determine the nature and amount of volatile product emitted by the sample during a thermogravimetric experiment

2.8.1 TG-MS Coupling

During a long period, the main difficulty in the TG-MS coupling has been to define the right interface that will solve the following technical coupling problem:

- Pressure in the balance is equal to 1 bar, and about 10^{-6} bar in the mass spectrometer. The pressure has to be reduced by 6 decades during the gas transfer.
- Gas sampling has to be representative of the gas emitted by the sample.
- Emitted gas has not to be degradated in the transfer line.
- Condensation of the emitted gas have to be prevented in the transfer line.
- Time of gas transfer has to be short (about 100 ms) in order to have a simultaneous information between the TGA and the MS curves.

To solve these different technical requirements, two different interfaces are today available, according to the mass analysis to be performed:

- the capillary interface, and
- the skimmer or supersonic interface.

2.8.1.1 The Capillary Interface

The transfer line is made of a very thin silica capillary contained in a heated jacket to prevent the condensation, especially at the outlet of the furnace (Fig. 2.38). A fraction of the gas emitted by the sample is picked as closed as possible from the

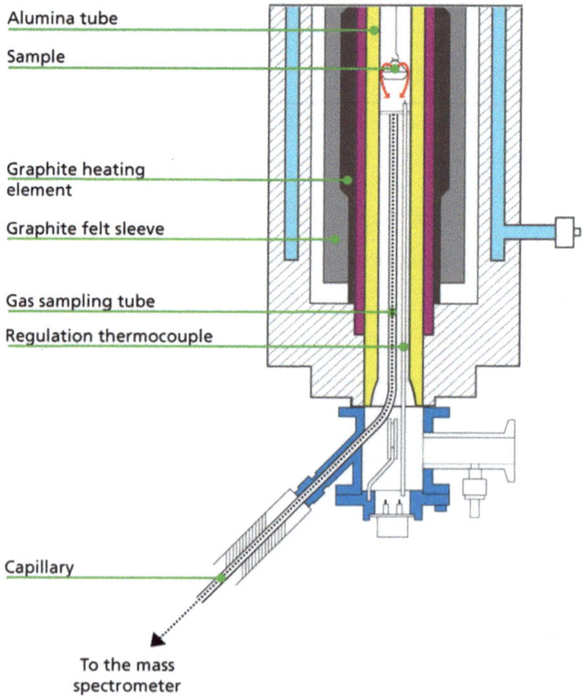

Fig. 2.38 The heated capillary interface (Setaram)

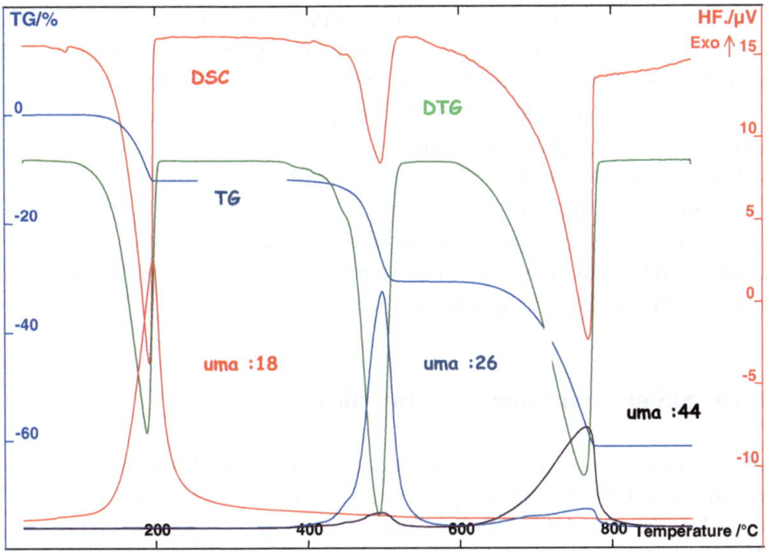

Fig. 2.39 TG, DTG, DSC and MS curves of the calcium oxalate decomposition. TG and FTIR curves of the calcium oxalate decomposition

TGA crucible to avoid gas dilution. The gas is then transferred through the capillary towards the MS detector within 100 ms.

The MS detector will provide a spectrum for each injection. According to the sample under study in the TGA, two situations can occur:

- The nature of the sample is known and it is possible to predict the types of gas that will be emitted by the sample during the TGA experiment. For such a test, characteristic mass numbers (amu), corresponding to the emitted gas, are selected and their variation is recorded versus time to be correlated with the TGA and DTG curves.
- The nature of the sample is unknown. In that case, it is needed to investigate the different MS spectra, especially at temperatures (times) corresponding to the different mass variations. When the significant mass numbers (amu) are identified, their variation can be drawn as described previously.

An example of the use of the heated capillary is given with the investigation of calcium oxalate CaC_2O_4, H_2O (Fig. 2.39)

The different uma traces show very clearly and respectively the different transformations that are expected and seen on the TG and DSC curves:

- the decomposition of the hydrate (H_20^+);
- the decomposition of calcium oxalate in calcium carbonate (CO^+); and
- the decomposition of the calcium carbonate in calcium oxide (CO_2^+).

However the MS curves also allow to detect some CO_2 emission during the decomposition of the oxalate, and some CO during the decarbonatation.

It indicates the presence of some traces of air in the furnace that reacts with CO. Such a reaction is slightly seen on the DSC curve (small exothermic deviation at the end of the second endothermic peak) but it is not detectable on the TGA curve. This example, even based on a very common sample decomposition, shows the contribution of the MS coupling to the TG-DSC technique.

When coupled to MS, the thermogravimetric equipment can be used as a Temperature Programmed Oxidation (TPO) or/and a Temperature Programmed Desorption (TPD) technique to follow catalytic oxidation processes [62].
The TG-MS technique is also used to run Temperature Programmed Reduction (TPR) experiments in reducing atmosphere [10].

2.8.1.2 The Skimmer or Supersonic Interface

The capillary interface is easy to install but its use remains limited to a certain range of mass numbers. For high mass species (above 150 amu), condensation occurs in the transfer line. To avoid this problem, the skimmer, also called supersonic interface has to be selected.

Using a mass spectrometer in molecular mode requires to work under very low pressures, in order to avoid any saturation of the quadrupole and to make the production of the electron beam easier. Using the supersonic system, it is necessary to have an important difference of pressure between the thermobalance and the mass spectrometer, in order to obtain a free molecular beam.

A gas at a stable temperature and at atmospheric pressure is in a disturbed molecular state. The rates of the molecules have an isotropic distribution in the space, varying as the temperature square root varies. If a chamber under atmospheric pressure and a chamber under low pressure are connected together through a small hole, it is possible to obtain a thermal effusion or a supersonic beam. The isentropic pressure reduction enables the molecules to reach some supersonic speeds in a few micro seconds. Therefore, intermolecular recombinations are not possible and a molecular beam in only one direction is obtained. Pressure reduction releases the impact between the molecules.

On a technical point of view, the supersonic system is adjusted on the thermobalance furnace, in order to have the MS detector straight to the sample container (Fig. 2.40).
Gas emitted by the sample flows through a taking hole at the top of a first chamber working under primary vacuum. A second chamber, working under secondary vacuum, is located inside the first chamber. The secondary chamber receives a cone with a hole at the top. The chamber is in line with the quadrupole detector of the gas analysis system.

The Supersonic or skimmer system is more costly and more difficult to use but it gives many advantages on the capillary transfer line: no condensation in the transfer line, no collision between the molecules and with the transfer line wall, possibility to detect high mass species (above 500 amu)

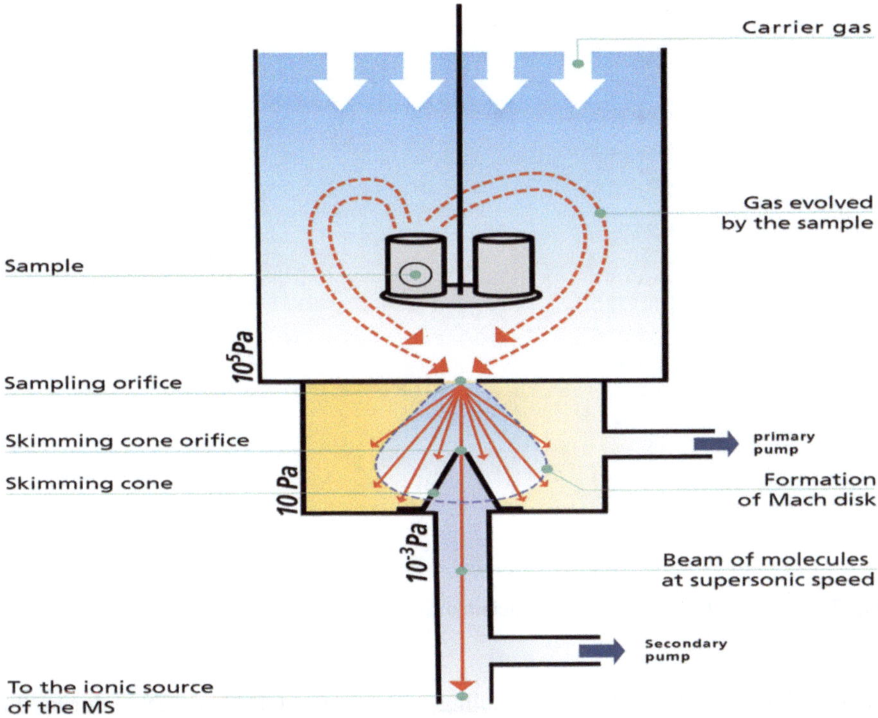

Fig. 2.40 The supersonic interface (Setaram)

2.8.2 TG-FTIR Coupling

The TG-FTIR coupling is easier to carry out as both analyzers are working at the same normal pressure. A heated transfer line has to be installed between the outlet of the thermobalance and the inlet of the FTIR spectrometer.
As in the TG-MS coupling, the time of transfer is very short. Caution has to be taken to avoid any condensation in the transfer line.
The infra red cell that is mostly KBr made of, has to be adapted (use of ZnSe cell) in order to avoid degradation with water vapour.

The same investigation of calcium oxalate, described previously with the TG-MS coupling, is used to see the difference between the two coupling techniques (Fig. 2.41).

The characteristic wavelengths of the expected emitted gas are recorded together with the three mass losses corresponding to the different steps of decomposition of the calcium oxalate:

- H_2O (3734 cm^{-1}) for the decomposition of the hydrate.
- CO (2068 cm^{-1}) for the decomposition of the calcium oxalate in calcium carbonate.
- CO_2 (2359 cm^{-1}) for the decomposition of the calcium carbonate in calcium oxide.

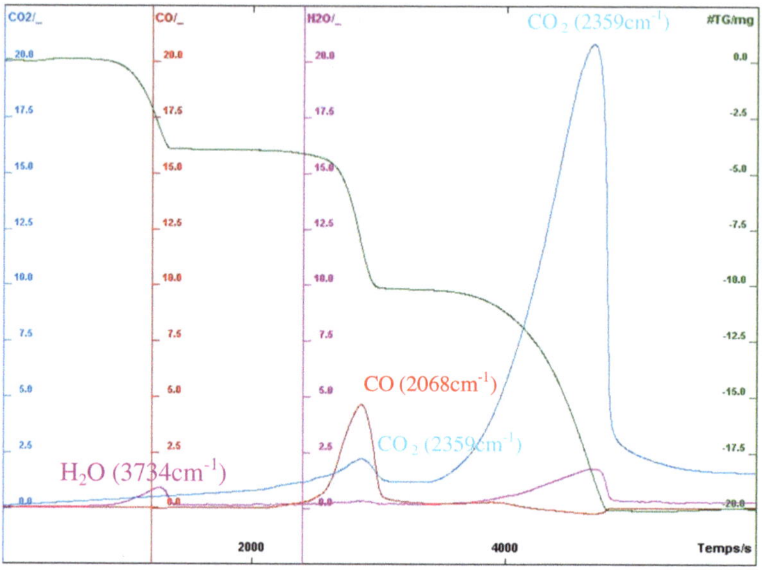

Fig. 2.41 TG and FTIR curves of the calcium oxalate decomposition

As seen before with the TG-MS coupling, a fraction of CO_2 is detected during the second loss, and a fraction of CO during the third mass loss, indicating the presence of some trace of air in the furnace during the experiment.

On such an example, very similar results are obtained with both EGA techniques.

2.9 Conclusion

The thermal analysis and calorimetric techniques provide a very large variety of possibilities of experimentations, of combinations with other analytic techniques that make unlimited the number of applications, especially in the field of characterization of catalysts and evaluation of catalytic processes.

The different techniques, together with the corresponding detectors and the experimental vessels, can be adapted and adjusted to the investigations of new materials or new processes. This is the case for example with the research on hydrogen storage on metal hydrides, all the works related to the capture and sequestration of CO_2 with the evaluation of solid and liquid sorbents, the development of the Chemical Looping Combustion with improved catalysts...

With this efficient faculty of adaptation to new research fields, the thermal analysis and calorimetric techniques are promised to a brilliant future

References

1. M.A. Aramendia, V. Borau, C. Jiménez, A. Marinas, J.M. Marinas, J.A. Navio, J.R. Ruiz, F.J. Urbano, Synthesis and textural-structural characterization of magnesia, magnesia–titania and magnesia–zirconia catalysts. Colloids Surf. A: Physicochem. Eng. Aspects **234**, 17–25 (2004)
2. H. Arcis, L. Rodier, J.Y. Coxam, Enthalpy of solution of CO_2 in aqueous solutions of 2-amino-2-methyl-1-propanol. J. Chem. Thermodyn. **39**, 878–887 (2007)
3. A. Arnold, M. Hunger, J. Weitkamp, Dry-gel synthesis of zeolites [Al]EU-1 and [Ga]EU-1. Microporous Mesoporous Mater. **67**, 205–213 (2004)
4. A. Auroux, Thermal methods: calorimetry, differential thermal analysis and calorimetry (Chap. 22), in *Catalyst Characterization: Physical Techniques for Solid Materials*, ed. by B. Imelik, J.C. Vedrine (Plenum Press, New York, 1994)
5. A. Auroux, Acidity characterization by microcalorimetry and relationship with reactivity. Top. Catal. **4**, 71–89 (1997)
6. A. Auroux, *Acidity and Basicity: Determination by Adsorption Microcalorimetry Molecular Sieves* (Springer, Berlin, 2006)
7. J. Baumann, F. Baitalow, G. Wolf, Thermal decomposition of polymeric aminoborane $(H_2BNH_2)\times$ under hydrogen release. Thermochim. Acta **430**, 9–14 (2005)
8. S. Bennici, A. Auroux, Thermal analysis and calorimetry methods, in *Metal Oxide Catalysis*, vol. 1, ed. by S.D. Jackson, J.S.J. Hargreaves (Wiley, Weinheim, 2009)
9. J.J.P. Biermann, P.P. Coelen, H. Den Daas, F.J.J.G. Janssen, The use of a heat-flow differential scanning calorimeter as a plug-flow fixed bed reactor in heterogeneous catalysis. Thermochim. Acta **144**, 329–337 (1989)
10. V.Y. Bychkov, Y.P. Tyulenin, V.N. Korchak, E.L. Aptekar, Study of nickel catalyst in oscillating regime of methane oxidation by means of gravimetry and mass-spectrometry. Appl. Catal. A: Gen. **304**, 21–29 (2006)
11. K. Chakarova, K. Hadjiivanov, G. Atanasova, K. Tenchev, Effect of preparation technique on the properties of platinum in NaY zeolite: a study by FTIR spectroscopy of adsorbed CO. J. Mol. Catal. A: Chem. **264**, 270–279 (2007)
12. F. Dan, M.H. Hamedi, J.P.E. Grolier, New developments and applications in titration calorimetry and reaction calorimetry. J. Thermal Anal. Calorim. **85**, 531–540 (2006)
13. J.L. Daudon, Heat flux devices and methods for optimum specific heat measurements, in *Fourteenth European Conference on Thermophysical Properties*, Lyon, France, 1996
14. L. Dussault, J.C. Dupin, E. Dumitriu, A. Auroux, C. Guimon, Microcalorimetry TPR and XPS studies of acid–base properties of NiCuMgAl mixed oxides using LDHs as precursors. Thermochim. Acta **434**, 93–99 (2005)
15. A.F.P. Ferreira, M.C. Mittelmeijer-Hazeleger, A. Bliek, Adsorption and differential heats of adsorption of normal and iso-butane on zeolite MFI. Microporous Mesoporous Mater. **91**, 47–52 (2006)
16. S.P. Felix, C. Savill-Jowitt, D.R. Brown, Base adsorption calorimetry for characterising surface acidity: a comparison between pulse flow and conventional "static" techniques. Thermochim. Acta **433**, 59–65 (2005)
17. P. Gallagher, Thermogravimetry and thermomagnetometry, in *Handbook of Thermal Analysis and Calorimetry*, vol. 1, Principles and Practice, ed. by M.E. Brown (Elsevier, The Netherlands, 1998)
18. A. Gervasini, P. Carniti, J. Kerainen, L. Niinisto, A. Auroux, Surface characteristics and activity in selective oxidation of o-xylene of supported V_2O_5 catalysts prepared by standard impregnation and atomic layer deposition. Catal. Today **96**, 187–194 (2004)
19. K. Habersberger, Application of thermal analysis to the investigation of catalysts. J. Thermal Anal. **12**, 55–58 (1977)
20. P.J. Haines, M. Reading, F.W. Wilburn, Differential thermal analysis and differential scanning calorimetry, in *Handbook of Thermal Analysis and Calorimetry*, vol. 1, Principles and Practice, ed. by M.E. Brown (Elsevier, The Netherlands, 1998)

21. M. Hart, G. Fuller, D.R. Brown, J.A. Dale, S. Plant, Sulfonated poly(styrene-co-divinylbenzene) ion-exchange resins: acidities and catalytic activities in aqueous reactions. J. Mol. Catal. A: Chem. **182–183**, 439–445 (2002)
22. M.P. Hart, D.R. Brown, Surface acidities and catalytic activities of acid-activated clays. J. Mol. Catal. A: Chem. **212**, 315–321 (2004)
23. T. Hatakeyama, Z. Liu, *Handbook of Thermal Analysis* (Wiley, New York, 1998)
24. W. Hemmiger, S.M. Sarge, Definitions, nomenclature, terms and litterature, in *Handbook of Thermal Analysis and Calorimetry*, vol. 1, Principles and Practice, ed. by M.E. Brown (Elsevier, The Netherlands, 1998)
25. International Confederation for Thermal Analysis, in *For Better Thermal Analysis and Calorimetry*, 3rd edn., ed. by J.O. Hill (RSC Publishing, Cambridge, 1991). http://www.ictac.org
26. ISO 11357-1, *Plastics: Differential Scanning Calorimetry—Part1: General principles*, (Universidade De Aveiro, Aveiro, 1997)
27. D.S. Kim, J.S. Chang, J.S. Hwang, S.E. Park, J.M. Kim, Synthesis of zeolite beta in fluoride media under microwave irradiation. Microporous Mesoporous Mater. **68**, 77–82 (2004)
28. D. Koschel, J.Y. Coxam, L. Rodier, V. Majer, Enthalpy and solubility data of CO_2 in water and $NaCl_{(aq)}$ at conditions of interest for geological sequestration. Fluid Phase Equilib. **247**, 107–120 (2006)
29. A. Lambert, C. Delquié, I. Clémençon, E. Comte, V. Lefebvre, J. Rousseau, B. Durand, Synthesis and characterization of bimetallic Fe/Mn oxides for chemical looping combustion. Energy Procedia **1**, 375–381 (2009)
30. P. Le Parlouër, Simultaneous TG-DSC: a new technique for thermal analysis. Thermochim. Acta **121**, 307–322 (1987)
31. P. Le Parlouër, C. Mathonat, Sensys: an innovative concept for the Calvet DSC111 and TG-DSC111, in *33rd NATAS Proceedings*, Universal City, USA (2005), p. 44
32. X.R. Li, H. Koseki, Hazard evaluation of self-decomposition materials by the combination of pressure and heat flux measurements. Thermochim. Acta **423**, 77–82 (2004)
33. R.C. Mackenzie, *Differential Thermal Analysis*, vols. 1, 2 (Academic, London, 1972)
34. E. Manova, T. Tsoncheva, C. Estournès, D. Paneva, K. Tenchev, I. Mitov, L. Petrov, Nanosized iron and iron-cobalt spinel oxides as catalysts for methanol decomposition. Appl. Catal. A: Gen. **300**, 170–180 (2006)
35. A. Maroto-Valiente, I. Rodrıguez-Ramos, A. Guerrero-Ruiz, Surface study of rhodium nanoparticles supported on alumina. Catal. Today **93–95**, 567–574 (2004)
36. V. Martin, J.M.M. Millet, J.C. Volta, Thermogravimetric analysis of vanadium phosphorus oxide catalysts doped with cobalt and iron. J. Thermal Anal. **53**, 111–121 (1998)
37. J.L. McNaughton, C.T. Mortimer, *Physical Chemistry Series*, vol. 2 (Butterworths, London, 1975)
38. J. Mercier, J. Thermal Anal. **14**, 161 (1978)
39. L. Meublat, P. Le Parlouër, Couplage thermogravimétrie: analyse de gaz. Spectra 2000 **19**(161), 59–62 (1991)
40. J. Mullens, EGA: evolved gas analysis, in *Handbook of Thermal Analysis and Calorimetry*, vol. 1, Principles and Practice, ed. by M.E. Brown (Elsevier, The Netherlands, 1998)
41. F.T.T. Ng, A. Rahman, T. Ohasi, M. Jiang, A study of the adsorption of thiophenic sulphur compounds using flow calorimetry. Appl. Catal. B: Environ. **56**, 127–136 (2005)
42. Y. Nguyen Huu Nhon, H.M. Magan, C. Petit, Catalytic diesel particulate filter: evaluation of parameters for laboratory studies. Appl. Catal. B: Environ. **49**, 127–133 (2004)
43. F. Omota, A.C. Dimian, A. Bliek, Partially hydrophobized silica supported Pd catalyst for hydrogenation reactions in aqueous media. Appl. Catal. A: Gen. **294**, 121–130 (2005)
44. J. Paulik, F. Paulik, in *Comprehensive Analytical Chemistry*, vol. 12, ed. by G. Svehla (Elsevier, Amsterdam, 1981)
45. B. Pawelec, J.L.G. Fierro, Applications of thermal analysis in the preparation of catalysts and in catalysis, in *Handbook of Thermal Analysis and Calorimetry*, vol. 2, Applications to Inorganic and Miscellaneous Materials, ed. by M.E. Brown, P.K. Gallagher (Elsevier, The Netherlands, 2003)

46. L. Pinard, J. Mijoin, P. Magnoux, M. Guisnet, Dichloromethane transformation over bifunctional PtFAU catalysts. Influence of the acidobasicity of the zeolite. C. R. Chim. **8**, 457–463 (2005)
47. M.G. Plaza, C. Pevida, B. Arias, J. Fermoso, A. Arenillas, F. Rubiera, J.J. Pis, Application of thermogravimetric analysis to the evaluation of aminated solid sorbents for CO_2 capture. J. Thermal Anal. Calorim. **92**(2), 601–606 (2008)
48. M.I. Pope, M.D. Judd, *Differential Thermal Analysis: A Guide to the Technique and its Applications* (Heyden, London, 1977)
49. V. Quaschning, A. Auroux, J. Deutsch, H. Lieske, E. Kemnitz, Microcalorimetric and catalytic studies on sulfated zirconia catalysts of different preparations. J. Catal. **203**, 426–433 (2001)
50. V. Rakic, V. Dondur, R. Hercigonja, Thermal effects of the interaction of carbon monoxide with zeolites. Thermochim. Acta **379**, 77–84 (2001)
51. M. Resan, M.D. Hampton, J.K. Lomness, D.K. Slattery, Effects of various catalysts on hydrogen release and uptake characteristics of $LiAlH_4$. Int. J. Hydrogen Energy **30**, 1413–1416 (2005)
52. M.J. Richardson, E.L. Charsley, Calibration and standardisation in DSC, in *Handbook of Thermal Analysis and Calorimetry*, vol. 1, Principles and Practice, ed. by M.E. Brown (Elsevier, The Netherlands, 1998)
53. R. Ruiz, C. Pesquera, F. González, C. Blanco, Synthesis and catalytic behaviour of heterogenized rhodium catalysts on modified clays. Appl. Catal. A: Gen. **257**, 165–175 (2004)
54. L. Shen, M. Zheng, J. Xiao, R. Xiao, A mechanistic investigation of a calcium-based oxygen carrier for chemical looping combustion. Combust. Flame **154**, 489–506 (2008)
55. S. Siffert, L. Gaillard, B.L. Su, Alkylation of benzene by propene on a series of beta zeolites: toward a better understanding of the mechanisms. J. Mol. Catal. A: Chem. **153**, 267–279 (2000)
56. J. Silvestre-Albero, C. Gomez de Salazar, Characterization of microporous solids by immersion calorimetry. Colloids Surf. A: Physicochem. Eng. Aspects **187–188**, 151–165 (2001)
57. P.F. Siril, A.D. Davison, J.K. Randhawa, D.R. Brown, Acid strengths and catalytic activities of sulfonic acid on polymeric and silica supports. J. Mol. Catal. A: Chem. **267**, 72–78 (2007)
58. S.R. Son, S.D. Kim, Semi-continuous operation of chemical looping combustion with metal oxides supported on bentonite in an annular fluidized bed reactor, in *Proceedings of the Asian Conference on Fluidized Bed and Three Phase Reactors* (2006)
59. J. van Humbeeck, Simultaneous thermal analysis, in *Handbook of Thermal Analysis and Calorimetry*, vol. 1, Principles and Practice, ed. by M.E. Brown (Elsevier, The Netherlands, 1998)
60. S. Wang, V. Slovak, B.S. Haynes, Kinetic studies of graphon and coal-char reaction with NO and O_2: direct non-linear regression from TG curves. Fuel Process. Technol. **86**, 651–660 (2005)
61. W.W. Wendlandt, *Thermal Analysis*, vol. 19, 3rd edn. (Wiley, New York, 1986)
62. J. Zhao, Z. Liu, D. Sun, TPO-TPD study of an activated carbon-supported copper catalyst-sorbent used for catalytic dry oxidation of phenol. J. Catal. **227**, 297–303 (2004)

Chapter 3
Couplings

Dušan Stošić and Aline Auroux

Abstract Basic principles of calorimetry coupled with other techniques are introduced. These methods are used in heterogeneous catalysis for characterization of acidic, basic and red-ox properties of solid catalysts. Estimation of these features is achieved by monitoring the interaction of various probe molecules with the surface of such materials. Overview of gas phase, as well as liquid phase techniques is given. Special attention is devoted to coupled calorimetry–volumetry method. Furthermore, the influence of different experimental parameters on the results of these techniques is discussed, since it is known that they can significantly influence the evaluation of catalytic properties of investigated materials.

3.1 Introduction

Catalytic cycle in heterogeneous catalysis can be divided in five consecutive steps; diffusion of reactants to surface, adsorption of reactants, surface reaction processes, desorption of products and diffusion of reactants from the surface [1]. A fundamental understanding of the great diversity of chemistry occurring at the surface requires familiarity with molecule–surface interaction. The energetics of chemical events occurring on the surface play an important role in determining the catalytic properties of the surface, since different surfaces will cause a substance to react in different ways. It is known from the literature that oxidation of methanol can lead to different reaction products, and different selectivities depending on surface properties (structural and chemical) of catalyst used in this reaction [2, 3]. Study of adsorption of probe and reactive gas molecules onto surface of catalysts lead to the better understanding of the

D. Stošić (✉) · A. Auroux
Institut de Recherches sur la Catalyse et l'Environnement de Lyon,
UMR 5256 CNRS/Université Lyon 1,
2 avenue Albert Einstein, 69626 Villeurbanne Cedex, France
e-mail: dusan.stosic@ircelyon.univ-lyon1.fr; aline.auroux@ircelyon.univ-lyon1.fr

A. Auroux (ed.), *Calorimetry and Thermal Methods in Catalysis*,
Springer Series in Materials Science 154, DOI: 10.1007/978-3-642-11954-5_3,

nature of gas-solid interactions and give valuable information about the properties of the adsorbent surface and thus is of primary importance in catalysis.

Adsorption is an exothermic process and when a reactive molecule interacts with the surface of the solid heat is evolved. This heat is related to the energy of the bonds formed between the adsorbed species and the adsorbent and hence to the nature of the bonds and to the chemical reactivity of the surface. From a number of techniques used to study this interaction only a few provide information about the strength of chemisorption itself. The determination of the differential heats evolved, by a suitable microcalorimeter, when known amounts of gas probe molecules are adsorbed on catalytic surface is the most suitable and accurate method which allows the determination of the number, strength and energy distribution of the adsorption sites [4–12]. The effective use of this technique in heterogeneous catalysis depends on choice of probe molecule, since the nature of the probe determines which surface property is going to be investigated.

The definition of various heats of adsorption/desorption and their relationship with thermodynamic quantities directly obtainable by calorimetry are given in Table 3.1.

3.2 Coupled Calorimetry–Volumetry

Adsorption microcalorimetry is a powerful technique for the energetic characterization of solid surfaces, as well as for the thermodynamic description of solid–gas interfacial phenomena.

To access the energetic heterogeneity of surface, small doses of probe gas (typically $<10\,\mu\text{mol g}^{-1}$) have to be added successively on the solid to saturate active sites progressively. Corresponding heats range from 100 to 1000 mJ, and require few hours for being evolved. Since the development of commercial instrumentation able

Table 3.1 Definition of various heats of adsorption/desorption

Name and symbol	Unit	Definition	Measurement method
Integral heat of adsorption: q_{int}	J	Quantity of heat evolved when n_{a} moles are adsorbed at constant temperature on an adsorbent initially *in vacuo*, without a change in the volume of the cell: $Q_{\text{int}} = n_{\text{a}}(U_{\text{a}} - U_{\text{g}})$	Isotherm calorimetry
Molar integral heat $Q_{\text{int}} = \frac{q_{\text{int}}}{n^a}$	J mol^{-1}	Integral heat per mole adsorbed	Isotherm calorimetry
Differential heat of adsorption: q_{diff}	J mol^{-1}	Defined from integral heat by: $Q_{\text{diff}} = \left(\frac{\partial q_{\text{int}}}{\partial n^a}\right)_{T,A}$	Isotherm calorimetry
Isosteric heat: q_{st}	J mol^{-1}	Heats of adsorption calculated from adsorption isotherms: $q_{\text{st}} = -RT^2\left(\frac{\partial \ln P}{\partial T}\right)_{n^a}$	Adsorption isotherm

to measure quantitatively such low heats and adsorbed amounts, microcalorimetry has gained importance as one of the most reliable methods for the study of gas solid interactions [5, 7, 8].

A calorimeter that is suitable for measurement of heats of adsorption is required to have the following characteristics:

(1) High sensitivity, with sufficient precision, to follow the changes of differential and integral heats with coverage, allowing the use of small doses of probe molecules.
(2) High thermal stability, to have a well-stabilized base line, which is necessary to study the slow process of adsorption.
(3) Large interval of utilization temperature. This is important for studying the interactions of different reactants with surface.
(4) Good accessibility of the calorimeter proper allowing the connections with volumetric apparatus. Relatively large dimensions as well as suitable geometry of the calorimetric cells are also required for same purposes.

Most commonly used are heat-flow microcalorimeters of the Tian-Calvet type [5, 8]. The detailed theory and operation of this calorimeter can be found elsewhere [11]. The apparatus is composed of an experimental vessel, where the studied system is located, which is placed into a calorimetric block (Fig. 3.1). The temperature of the block, which functions as heat sink, is controlled very precisely. The heat generated in the system flows to the heat sink and is accurately measured by means of detector. This is made of a large numbers of identical thermocouples (a thermopile) that surrounds the vessel and connected to the block (Fig. 3.2) in such a way that vessel and block temperature are always close to each other. A signal is generated by the detector that is proportional to the heat transfer per unit time. Undesired signals due to the external temperature fluctuations in the calorimetric block are minimized by connecting in opposition two heat flow detectors from two identical vessels, one of which is used to perform the experiment, the other being used as a reference. Heat related to the introduction of the probe and other parasitic phenomena are thus compensated.

The heat generated by the process inside the vessel is transferred completely through the thermocouples, provided that radiation, conduction along the wall of the vessel, conduction and convection through the fluid phase are minimized by the design of the apparatus. A voltage signal, *E*, is generated by the thermocouples and recorded over the time of the thermal event, whose end is indicated by the returning of the thermopile output signal to the base line. The total heat evolved during the event is given by the area under the thermogram:

$$Q_{\text{exp.}} = K \int_{\text{exp.}} E dt \tag{3.1}$$

where K is the instrument constant [11].

This highly sensitive heat flow calorimeter linked to sensitive volumetric system makes it possible to study gas–solid interactions and catalytic reactions. The

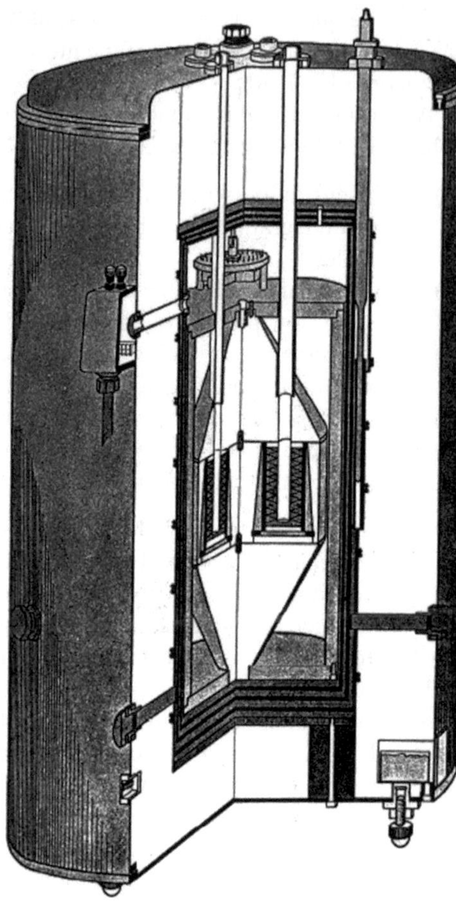

Fig. 3.1 Calvet microcalorimeter

apparatus used for this kind of measurement is presented in Fig. 3.3. The volumetric determination of the adsorbed amount of gas is performed in a constant volume vessel linked to a vacuum pump. The apparatus consists of two parts: the measuring element equipped with a capacitance manometer and a cell section, which includes the cells placed in the calorimeter (a sample cell in which the adsorbent solid is placed, and an empty reference cell). The volume of this vessel is determined by the expansion of a known quantity of gas, contained in the measuring part of the assembly, into the previously evacuated cell section. This calibration must be made with the same gas and at the same temperature as the proposed study is going to be done.

In a typical experimental procedure the sample is first outgassed at the desired pretreatment temperature and under high vacuum in the calorimetric cell. After cooling to the adsorption temperature and when the thermal equilibrium of the system is reached, successive doses of gaseous probe molecules are brought into contact with catalyst sample. The heat flow signal and the corresponding pressure change are

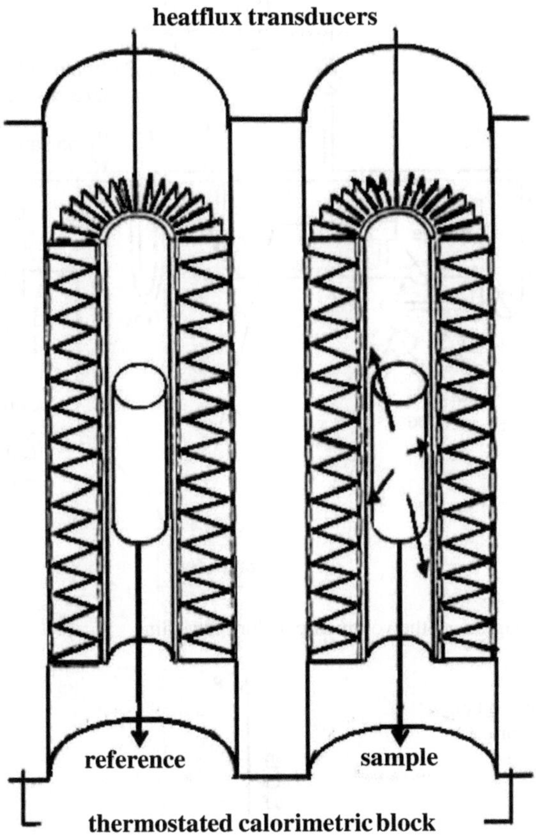

Fig. 3.2 Thermostated calorimetric block

monitored and recorded until equilibrium is reached after admission of each dose. The volume of doses is minimized in order to detect any change in adsorption heats evolved with increasing the coverage of surface active sites. The adsorption is considered complete when for a significant increase in pressure there is no detectable heat evolution or gas adsorption. The adsorption temperature is maintained constant during the experiment. The irreversibly chemisorbed amount (V_{irr}) can be evaluated from the difference between the primary adsorption isotherm (adsorbed amount of gas as function of equilibrium pressure over sample) and a secondary isotherm obtained after desorption under vacuum and re-adsorption of the gaseous probe at the same temperature. By subtraction of re-adsorption isotherm from the first adsorption isotherm amount of strong sites can be roughly estimated. The data obtained directly from adsorption calorimetry measurements are represented in Fig. 3.4, where both

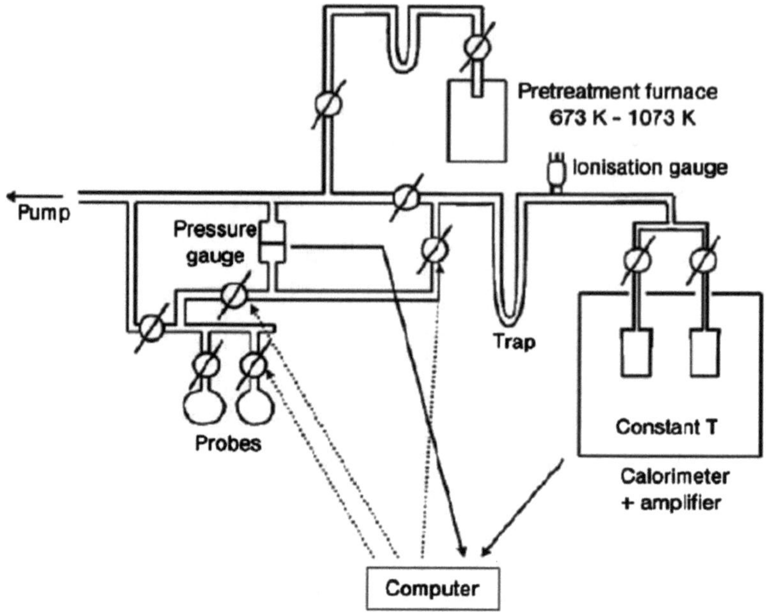

Fig. 3.3 Schematic diagram of the volumetric–calorimetric line

Fig. 3.4 Recorded data, pressure and heat flow signal, as a function of time for each dose of probe molecules

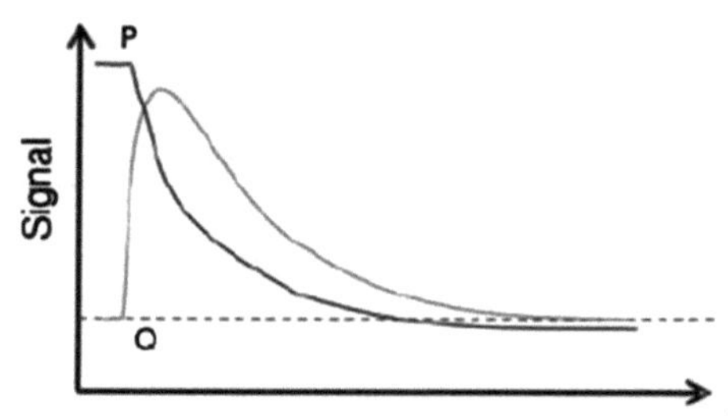

the calorimetric signal and the evolution of pressure are plotted as a function of time for each dose added.

From accumulation of these raw data until the full coverage is attained we can get information about the sample properties that can be summarized on various plots (Fig. 3.5):

(a) The volumetric isotherms ($n_a = f(P)$) for a cycle of adsorption (I), desorption by pumping at the same temperature and then re-adsorption (II). The irreversibly adsorbed volume (V_{irr}), which characterizes the strong sites of the catalyst, can then be calculated as the difference between the adsorption volume and the re-adsorption volume at a given equilibrium pressure.
(b) The corresponding calorimetric isotherms ($Q_{int} = f(P)$).

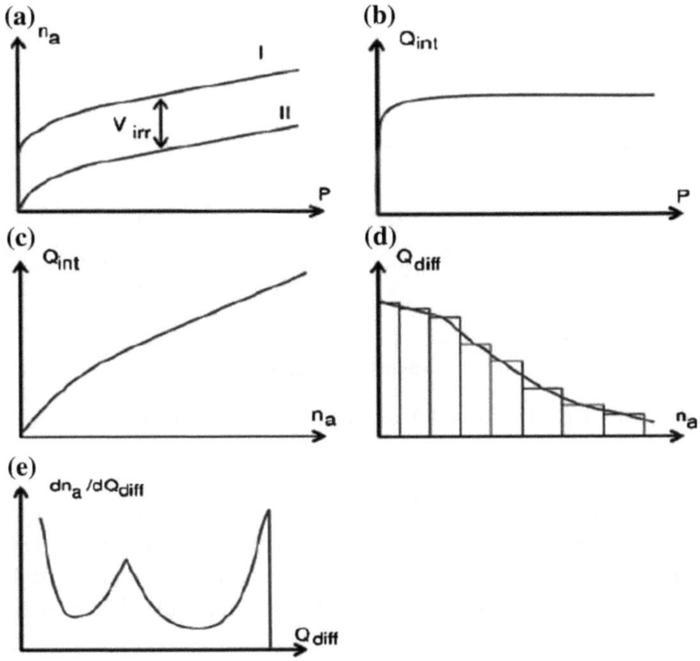

Fig. 3.5 Data obtained from volumetric–calorimetric adsorption experiments

(c) The integral heats (Q_{int}) as a function of the adsorbed quantities (n_{a}). This representation leads to the detection of coverage ranges with constant heat of adsorption, those for which the evolved heat is a linear function of coverage.

(d) The differential heat, $q_{\text{diff}} = Q_{\text{int}}/\partial n_{\text{a}}$ (molar heat of each dose of adsorbate) as a function of n_{a}. The ratio of the amount of heat evolved for each increment to the number of moles adsorbed in the same period is equal to the average value of the differential enthalpy of adsorption in the interval of the adsorbed quantity. The curve of differential heat as a function of surface coverage is traditionally represented as histograms. For simplification the steps of histogram are often replaced by a continuous curve connecting the centres of steps. Differential heats of chemisorption usually decrease with increasing the surface coverage. The way in which the profile of differential heat changes with increasing surface coverage depends on nature of adsorbate and adsorbent. Sharp heat falls at low surface coverage are in general regarded to surface heterogeneity.

(e) The energy distribution spectra in which $-\mathrm{d}n_{\text{a}}/\mathrm{d}q_{\text{diff}}$ is expressed as a function of q_{diff}. The area under this curve represents the number of molecules that are adsorbed with a given evolved heat. This kind of representation is not as accurate as the previous one.

Although it is not possible to determine the nature of adsorbed species, or even to distinguish between different kinds of adsorbed species from the calorimetric

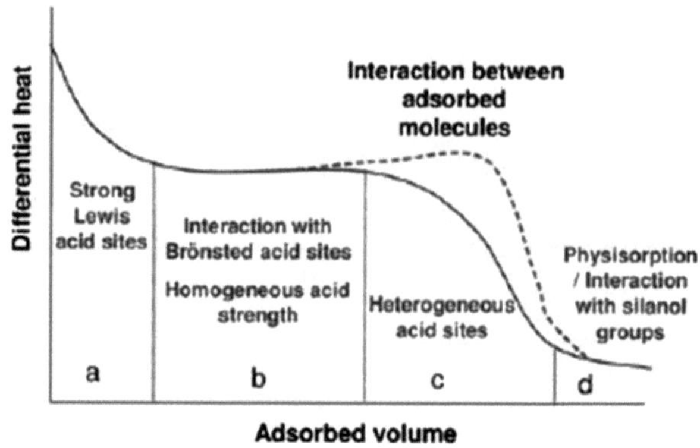

Fig. 3.6 Regions in a typical curve of differential heats of adsorption versus adsorbed amount

data, the variation of the differential heats of adsorption with coverage quite clearly shows energy distribution of surface active sites with respect to a given adsorbate and their varying reactivity on given adsorbents. A general representation of differential heats of adsorption as a function of uptake of probe molecule (Fig. 3.6) presents the following features:

(a) An initial region of high heats of adsorption, representing adsorption on the strongest sites, which are usually thought to be of Lewis type. After this region the curve drops abruptly.
(b) One or more regions of intermediate strength sites. Regions of constant heat in this region are attributed to sets of sites of constant strength. This is taken as an indication of discrete inhomogeneity. These features, when adsorption of base molecule on the surface of zeolite is in question, are attributed to Brönsted acid sites.
(c) A region where heats decrease more or less steeply depending on the heterogeneity of the sites. In this area a bump on the curve can sometimes be observed, which indicates lateral interactions between adsorbed molecules.
(d) At a high coverage heat of adsorption approaches a constant value characteristic of hydrogen bonding between the molecule and the sample or physisorption of the molecules. This constant value is dependent on the nature of the probe molecule.

All regions (a, b, c, d) can be observed for zeolite samples presenting both Lewis and Brönsted acid sites, as probed by ammonia adsorption. For oxides presenting only Lewis acid sites, the regions a, c and d are observed.

This kind of representation of calorimetric data can be used to assess the uniformity–non-uniformity of the surface of the adsorbent with respect to energy, the energy of the lateral (adsorbate–adsorbate) interactions, and the structural or substructural (textural) changes that the adsorbent often undergoes as a result of interaction with the adsorbed substance [13].

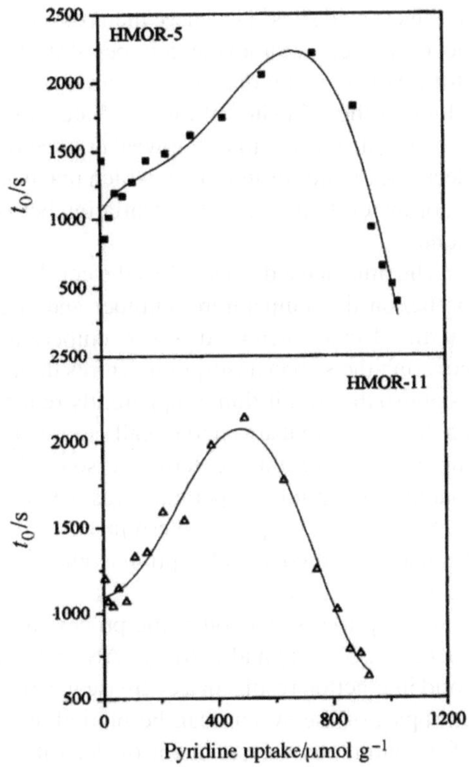

Fig. 3.7 Thermokinetic parameter as a function of pyridine uptake on mordenite samples (from [16])

Study of the thermokinetics of the heat evolution combined with volumetric and calorimetric data can provide a better insight into the adsorption mechanism and into the location of adsorbing sites on the catalyst surface [14–16]. It has been shown that interpretation of the shapes of individual thermograms facilitates identification of simultaneous processes with different kinetics as well as the variation of kinetics with coverage.

The kinetics of heat release during adsorption can be monitored by the change in thermokinetic parameter τ [14]. The calorimetric signal decreases exponentially with the time of adsorption after the maximum of each adsorption peak. This can be expressed in the form $D = D_m e^{-t/\tau}$, where D and D_m are the deviation at time t and maximum deviation of calorimetric signal, respectively. The thermokinetic parameter in this expression can be computed as the inverse of the slope of the logarithm of the evolved heat curve during the return to equilibrium. Variations of the thermokinetic parameter, determined in this way, can then be plotted versus the amount of adsorbed probe.

When ammonia or pyridine were used as probe molecules to study the acidity of zeolites at a given temperature [14, 16], the time needed to establish equilibrium after the addition of doses at first increases with increasing coverage (Fig. 3.7). This is due to the fast adsorption rate because the molecules are bonded irreversibly to the strongest acid sites. The curve then passes through a maximum as the adsorption rate decreases because a smaller number of strong sites is available and the molecules must diffuse over greater distances on the surface. This coverage should correspond to the filling of the acid sites of intermediate strength. Finally only the exchange between probe molecules among the weaker sites occurs. This is a fast process resulting in decrease of equilibrium time, which reaches a value close to the time constant of the calorimeter. In this way differentiation between strong and weak adsorptions can be done.

The time needed to establish the equilibrium depends on the quantity of adsorbed probe, on the temperature, number and strength of surface active sites and on the inertia of calorimeter. At lower temperatures, a slower adsorption is observed in covering the strong adsorption centres than at higher temperatures. The long time to establish the equilibrium is apparently related to redistribution of the adsorbed probe on the centres that are energetically more favourable [17]. When the time to establish thermal equilibrium is determined solely by inertia of calorimeter, one can be sure that the adsorption temperature was well chosen.

Entropy of adsorption, determined from adsorption equilibrium constants obtained from adsorption data (adsorption isotherms and differential heats of adsorption) can be also estimated.

Adsorption is an exothermic process and takes place with decrease in both free energy ($\Delta G < 0$) and entropy ($\Delta S < 0$). With respect to the adsorbate, the gas–solid interaction results in a decrease in entropy of the system. The differential molar entropies of the system can be plotted as a function of the coverage. The interest of a continuous measurement of derivative entropy of adsorption stems from the fact that any variation of its magnitude provides information on: (i) a variation of the adsorbate-adsorbent interactions as a function of coverage (it allows the heterogeneities of the adsorbing surface to be detected), (ii) a collective modification of the state of the interaction of the molecules already adsorbed, (iii) or a combination of both [17, 18].

3.3 Coupled Calorimetry–Gravimetry

Gravimetry can be associated with adsorption calorimetry either in vacuum or in gas flow system. When vacuum system is used, methodology of experiment is similar as used in calorimetry–volumetry technique. Desired amount of the sample is placed on microbalance which is surrounded by Tian-Calvet thermopile. Adsorbed amount of gas is determined by measuring the change in the sample mass. The heat flow signal and the corresponding mass change are monitored and recorded until equilibrium is reached after admission of each dose. Results obtained by this technique are comparable to that obtained when volumetric system is used to determine the amount of adsorbed gas. Volumetric measurements are to be preferred to gravimetric measurements for these determinations because it is very difficult to ensure a good

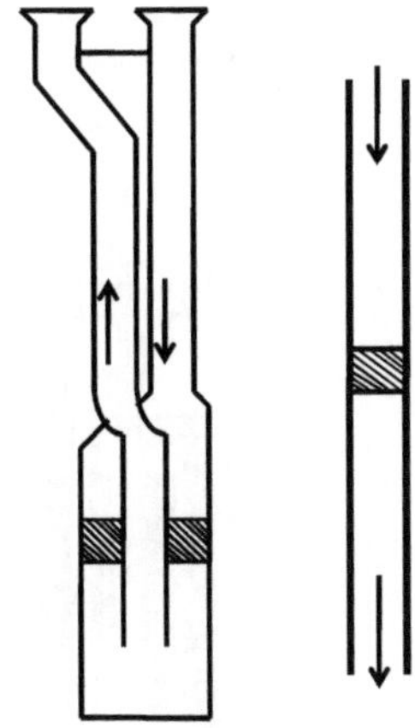

Fig. 3.8 Gas flow cells

and reproducible thermal contact between a sample of adsorbent, hanging from a balance beam, and the inner cell of a heat-flow calorimeter [11].

Flow adsorption calorimetry involves the use of carrier gas passing continuously through the adsorption cell. The catalyst is placed on a glass frit in a gas circulation cell in the calorimeter. Examples of gas flow cells are represented in Fig. 3.8. In order to determine the amounts of adsorbed gas, flow calorimetry is used in combination with thermogravimetry (TG). When this system is used in pulse mode it allows the measurement of heats of adsorption of a gaseous reactant on a solid or interaction heats between a gaseous reactant and pre-adsorbed species. When used as a flow reactor, it allows the kinetic study of catalytic reactions as well as the study of the activation or the aging of the catalyst. This is also a suitable system to perform calorimetric temperature programmed reduction (TPR), temperature programmed oxidation (TPO) or temperature programmed desorption (TPD) experiments. In addition to calorimetry, temperature programmed desorption (TPD) of adsorbed probe molecules can in principle also be used to estimate heats of adsorption [19].

3.4 Temperature Programmed Desorption Technique

The temperature programmed desorption of probe molecules is a commonly used method for measuring basic or acidic properties of catalyst surfaces. In this procedure the sample is subjected to temperature programmed heating and pre-adsorbed gas is progressively desorbed. The rate of desorption increases with temperature increase, goes through a maximum, and finally goes back to zero as the surface is cleared from adsorbate. TG-DSC-MS is experimental setup (Fig. 3.9) that can be used for this kind of measurements. During this experiment endothermic heat signal, temperature of desorption and concentration of desorbed molecules is measured. Desorption energy distribution of probe molecules that is available from these data, can be related to the

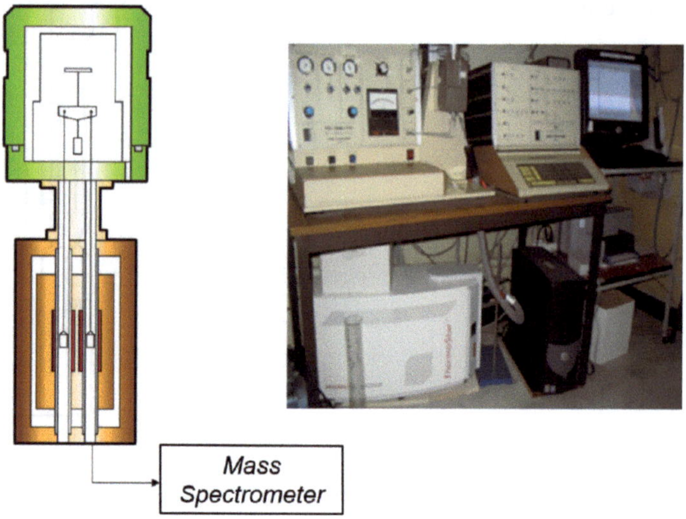

Fig. 3.9 Schematic representation and a photograph of a TG-DSC-MS apparatus

acid strength distribution of surface active sites by hypothesis that desorption energy is equal to the energy of acid re-protonation.

Despite the fact that TPD is widely used because of its simplicity the main drawbacks of this technique have to be pointed out.

- TPD gives an average value of acid strength rather than a distribution.
- Results obtained by this technique are strongly affected by mass or heat transfer and diffusion.
- TPD enables only to distinguish the strength of active sites, but not their types.
- Desorption on weak sites is hindered by adsorbates on strong sites.
- During desorption, a readsorption may occur.
- Peaks overlap, so TPD can be used only to roughly distinguish the various acid site strengths.

TPD of reactive amines is a very useful technique for measuring Brönsted site densities in zeolites [20–23]. Alkylammonium ions, formed by protonation of amines at Brönsted sites react in a very narrow temperature range in TPD through a reaction similar to Hoffman elimination reaction. Figure 3.10 shows a result for TPD-TGA-MS of isopropylamine from an H-MFI sample having a bulk Si/Al ratio of 26 [24]. The TGA results are useful since one can determine whether there are residual species on the sample after desorption and one can quantify the amounts which desorb without having to calibrate the mass spectrometer.

From MS results the nature of acid sites can be determined. Upon heating in vacuum two stages of weight loss and two stages of product evolution can be observed. The molecules associated with low-temperature desorption can probably be assigned to a number of different types of species, including molecules associated with Lewis

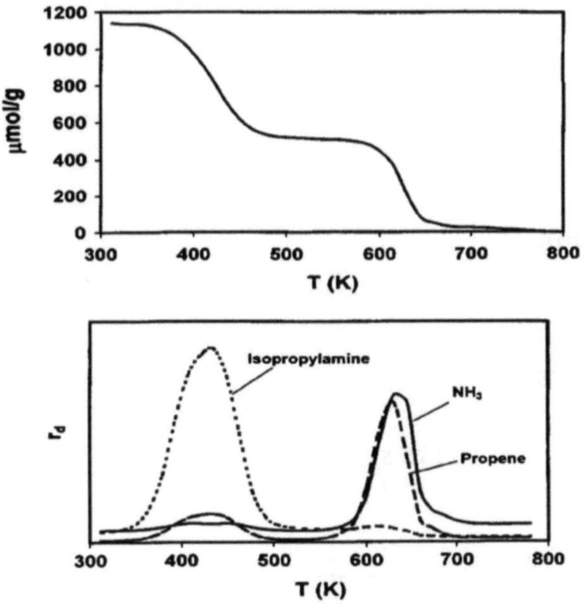

Fig. 3.10 TPD-TGA-MS results for isopropyl amine in H-MFI sample. The desorption of propylene and ammonia above 550 K corresponds to decomposition of the amine at the Brönsted acid sites (from [24])

acid sites, molecules associated with hydroxyl defects, and molecules which are hydrogen bonded to protonated amines at the Brönsted sites. At high temperature molecules desorb as propylene and ammonia. This desorption feature can be associated with protonated molecules at Brönsted sites. Fact that the measured site density is independent of the choice of alkylamine gives further confirmation that TPD-TGA-MS results provide a true evaluation measure of the Brönsted site density [25].

3.5 Temperature Programmed Reduction

From calorimetric TPR-TPO experiments heats of reduction and reoxidation of cationic or metal species can be determined. Data obtained from these experiments may provide kinetic data of theoretical significance as well as an insight into the mechanism of the reduction processes.

TPR profiles of copper exchanged ETS-10, obtained by differential scanning calorimetry (DSC) and carried out with a hydrogen stream in helium, are presented in Fig. 3.11 [26]. All the reduction profiles show complex structure, in which a stepwise reduction of CuO to Cu through Cu_2O intermediates occurred with overlapping of the two peaks. By analysis of this profiles authors gained useful information con-

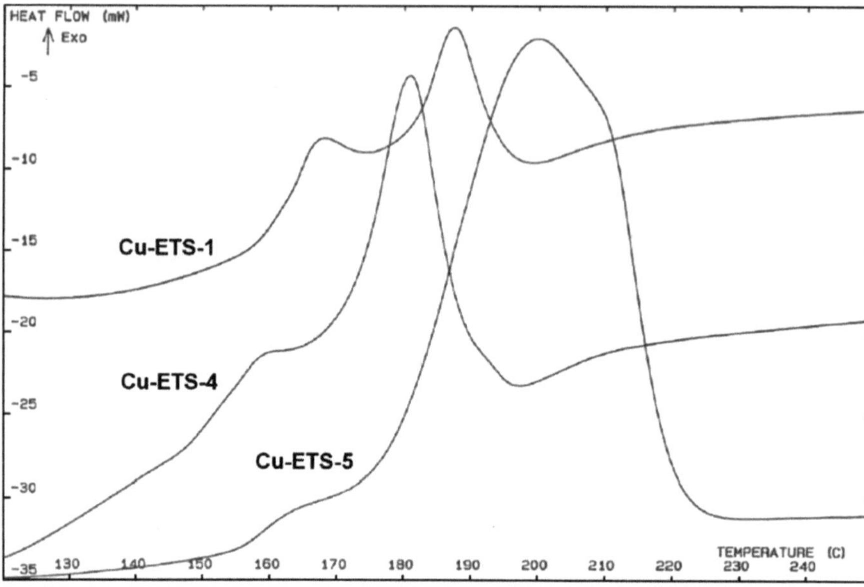

Fig. 3.11 DSC curves of reduction by hydrogen as a function of temperature for Cu-ETS samples (from [26])

cerning the identification of supported species and the determination of intermediate oxidation states of metals during reduction.

In another study in order to compare reducibility power of hydrogen, ethane, and ethylene authors performed the reduction of fresh vanadium pentoxide by each of them (diluted in helium) in a TG-DSC apparatus [27]. Figure 3.12 shows the heats of reduction which are exothermic and the derivative of the thermogravimetric curves which are negative if associated with a weight loss occurring during the reduction. According to the weight loss, a similar reduction extent was recorded for all three cases. Heat flow measurements have shown that reducibility levels of V_2O_5 by ethane and hydrogen are similar, whereas ethylene is much more active that ethane over that oxide.

3.6 Calorimetry–Gas Chromatography/Mass Spectrometry

This coupling can be used for characterization of acidic or basic character of solid catalysts, and for study of solid–gas reactions. Systems used for these measurements are usually modified DSC instruments connected with GC/MS by a heated capillary tube [28–32]. In a typical experiment small pulses of probe gas are injected at regular intervals in to the carrier gas stream from a gas sampling valve. The gas flows are regulated by mass flow controllers. The concentration of ammonia down-stream of the sample is monitored with the GC/MS and heat evolution with the calorimeter.

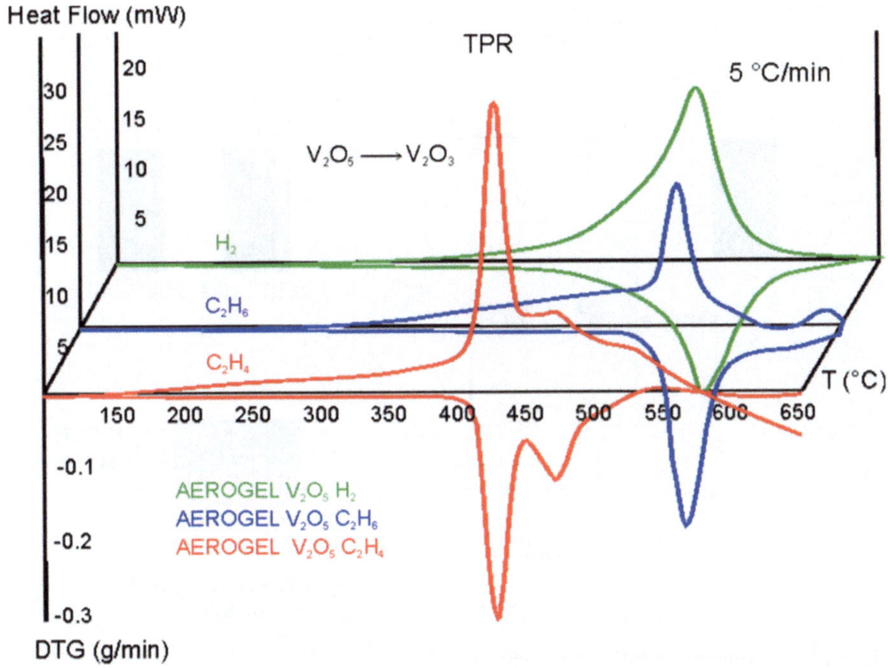

Fig. 3.12 Heats of reduction of V_2O_5 and associated derivatives of the thermogravimetric curves as a function of temperature, for various reducing agents (from [27])

When gas–solid reactions are under investigation this system allows simultaneous determination of heats of reaction and identification of products. Scheme of the apparatus is shown in Fig. 3.13.

3.7 Calorimetry-Syringe Pump-UV-Vis Fluorescence Spectrometry

Isothermal titration calorimetry is a technique that can directly measure liquid–solid or liquid–liquid interactions. This method has been used for determination of acid–base properties of solid catalysts in an inert phase (n-decane), for determination of heats of reaction in liquid phase (alkylation, anisole + benzoic anhydride on H-Beta zeolite), and for determination of heats of adsorption of different pollutants on an adsorbent [33]. It is also recognized as the only technique that can directly measure the binding energies of biological processes such as protein–ligand binding and protein–protein binding [34].

Instrument that is usually employed in this technique is differential heat flow calorimeter of Tian-Calvet type equipped by a stirring system. A programmable

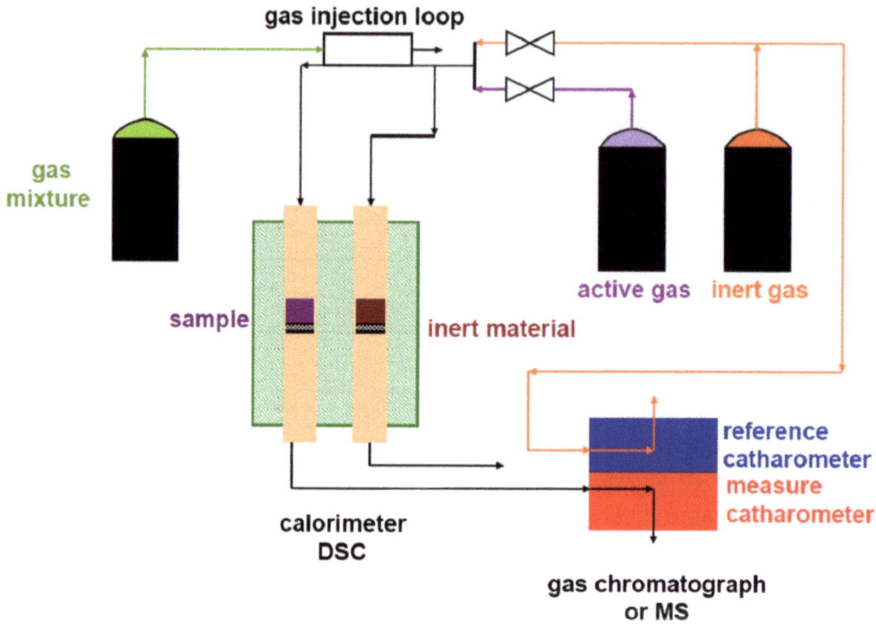

Fig. 3.13 Calorimetry-GC/MS experimental device

syringe pump is linked to the calorimeter by capillary tubes. Using this syringe pump, successive pulses of known amounts of a solution of a probe molecule can be sent on the sample which is maintained at constant temperature during the experiment. UV-Vis spectrometer is connected to the calorimetric cell by an optic fiber, and it is used to monitor the changes in concentration of the probe molecule. From equilibrium concentration after each dose of solution, adsorption isotherms can be constructed. Scheme of the instrument is presented in Fig. 3.14.

This method has been used for characterization of acidity of solid catalysts, or to investigate adsorptive properties of solid materials by measuring the adsorption heats evolved upon injection of dilute solutions of chosen molecules onto the solid itself in slurry of appropriate solvent. The output of typical microcalorimetric experiment is shown in Fig. 3.15. Each pulse injection of dosed amount of solution of probe molecule gives as a result the exothermic signal that formed a specific peak related to the heat of adsorption. This heat derives from two different contributions: the exothermic enthalpy of adsorption and the endothermic enthalpy of displacement of adsorption; while the enthalpy effects originating from dilution and mixing phenomena can be neglected due to the differential design of the calorimeter and the preheating of the probe solution. The amount of adsorbate in solution is measured simultaneously with a UV-Vis spectrometer, leading to an adsorption isotherm which is measured over the range of base addition used in calorimetric titration. Study of evolution of UV-Vis signal with time (Fig. 3.16) during adsorption of each dose added gives a better insight into the adsorption mechanism.

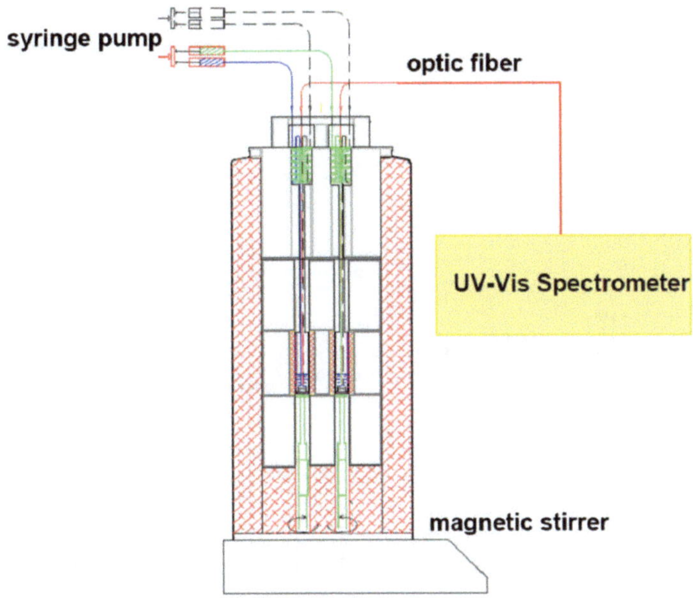

Fig. 3.14 Schematic representation of a TITRYS calorimeter connected to UV-Vis spectrometer

Combination of data obtained from calorimetric measurements and adsorption isotherms permits a determination of the number of active sites, the equilibrium constants, and the enthalpy of a binding to each type of the site [35–39].

For microcalorimetric determination of the strength of solid acids in solution, pyridine, n-butylamine and aniline are the usual probe molecules that are used. The measurements are done in a non-interacting hydrocarbon solvent (e.g. cyclohexane) whose molecular mass is close to that of donor in order to cancel out contribution from a dispersion component to the measured enthalpy [36]. Throughout the experiment the molar enthalpy of neutralization, ΔH°_{neut}, is measured. If ΔH°_{neut} values are going to be used as measurements of relative acid strength, some assumptions have to be made. This is because the measured enthalpy of neutralization is made up from contributions from several sources: proton loss from the acid and proton gain by the base, changes in solvation as the acid and base are converted to the conjugate base and acid, plus any enthalpy changes associated with formation of ion pairs that form or break up during the neutralization process. In order for the differences in measured ΔH°_{neut} values to reflect the differences in the acid strength of acid solids requires that all contributions to ΔH°_{neut} other than the proton affinity of the acid are constant. This can be taken as reasonable assumption only when closely related solid acids are compared [36, 40–42]. Another point is that titration calorimetric measurements of this kind are usually performed at temperatures close to room temperature so the process of adsorption is more likely to be kinetically controlled. In this way adsorption could occur simultaneously on different types of active sites and average

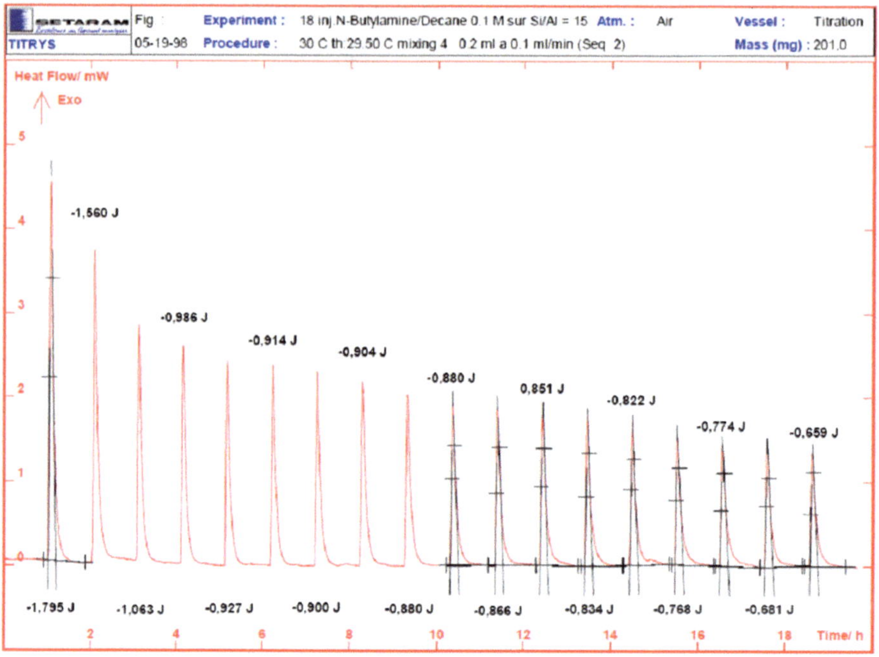

Fig. 3.15 The output of typical microcalorimetric experiment for successive doses

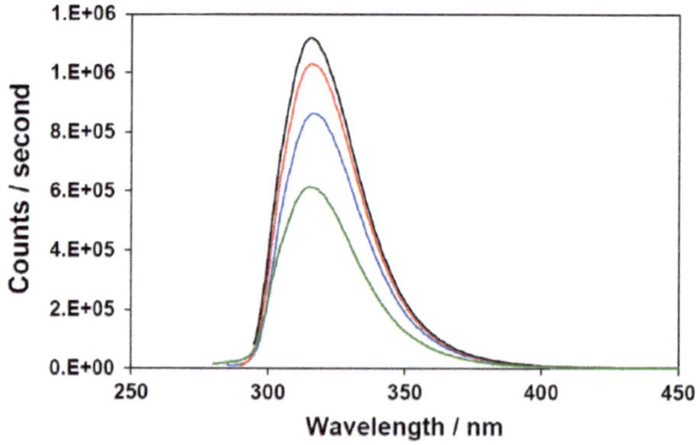

Fig. 3.16 UV-Vis signal evolution as a function of time for one injection

value of ΔH°_{neut} is obtained for each dose of probe molecule. Because of this, real strength distribution of acid sites is not always available from these measurements.

Useful feature of this method is possibility to change solvent and investigate the influence of solvent on acid strength. Niobium oxide and niobium phosphate were investigated by this technique, using aniline and 2-phenyl-ethylamine as probe molecules, in order to understand the effective acidity of these solids measured in various solvents (decane, cyclohexane, toluene, methanol, and isopropanol) of very different polarities and proticities. By using this technique it is possible to discriminate the acid site strength distribution more accurately than with conventional gas-solid phase titration with ammonia [43]. In another study acid strength of sulfonic acids supported on polystyrene and silica solids were determined in water and cyclohexane. It was shown how the relative acid strength of these very similar types of supported acids depends markedly on the nature of solvent. The relative acidities of these solid materials are reversed when going from water to cyclohexane. This leads to a conclusion that if these measuremenrs are to be used for predicting catalytic activity, it is essential that they are made in solvents similar to those in which the catalysts are ultimately to be used [44].

3.8 Limitations of Technique

Results obtained by adsorption microcalorimetry are not sufficient to determine the nature of adsorbed species, or even to distinguish between different kinds of adsorbed species. When examination of catalyst surface is done, that possesses both Lewis and Brönsted acidity, by adsorption of base probe molecule such as ammonia, it is difficult to discriminate strong Lewis from strong Brönsted acid sites solely by adsorption microcalorimetry. This is due to the fact that differential heats of NH_3 interaction with strong Lewis and Brönsted acid sites are relatively close to each other. For this reason complementary informations from suitable IR, MAS NMR and XPS investigations are necessary to identify these sites [45]. However, because of the complex nature of the active site strength distribution it is not possible to make a detailed correlation between sites of different nature and their strength.

Initial values representing very small concentrations of the strongest active sites can be easily missed in the measurement if the gas doses are not small enough.

If kinetics of adsorption is determined by diffusion limitations, determined heats of adsorption may reflect an average heat of the various strong sites, rather than a specific effect. In this case adsorption on active but less-accessible sites may occur only after better-exposed but less-active sites have interacted. Differential heat profiles determined under diffusion restrictions may reflect less surface heterogeneity than actually exists on surface of the adsorbent [11]. The sample bed thickness has also to be considered because the molecular mobility in the sample bed is a limitation to rapid data collection [7].

Finding correlations of surface properties of catalyst (in the case of adsorption calorimetry: acid, basic and redox properties) with its catalytic behaviour is sometimes difficult. The lack of correlation is due to the fact that by microcalorimetry measurements the total amount of acid or basic sites is obtained and only a fraction of these active sites may actually be involved during the catalytic reaction [46]. There could be surface sites that are either too weak to activate the reactants or are too strong, leading to strongly held species which block and deactivate these sites or causing excessive fragmentation of reactant or products. It has been recognized that the catalytic properties of Brönsted acid sites (proton donors) can be different from Lewis acid sites (electron pairs acceptors). For example, it has been suggested that Brönsted acids are much more active than Lewis acids for skeletal transformations of hydrocarbons [47], and as said before it is impossible to differentiate them solely by adsorption microcalorimetry technique.

3.9 Influence of the Adsorption Temperature on the Acid/Base Determination

In order that adsorption microcalorimetry can give an accurate representation of active site strength distribution the adsorbed probe molecule must be equilibrated among all sites on the catalyst surface within the time frame of the experiment. If external or internal mass-transfer limitation exists the adsorption on surface sites is controlled by kinetics instead of thermodynamics [5]. In this case the profile of differential heats as function of coverage would not accurately represent energy spectrum of the surface active sites and the adsorbent surface would appear to be more homogeneous. For these reasons, it is important to study the effect of the adsorption temperature and verify that molecules possess sufficient thermal energy to obtain the thermodynamically stable site occupation.

When the process of adsorption was studied at various temperatures two different kinds of temperature dependence were observed [48, 49]. One type was described as a change of the profile of differential heat of adsorption with increasing coverage of pyridine on HY zeolite and silica–alumina as temperature was raised. The adsorption of pyridine was performed at 423 and 473 K. For lower coverage, the profile of integral heats obtained for the adsorption at 473 K, lies above that obtained at 423 K. This difference is decreasing as the surface coverage is increasing, and at high coverage the situation is opposite. The difference between the differential heat curves with temperature suggests that there is difference in selectivity when adsorption is performed at higher temperatures compared to adsorption at lower temperatures. Adsorption occurred on stronger acid sites in preference to weaker sites at higher temperature, whereas random adsorption occurred simultaneously on both weak and strong sites at low temperatures. The authors [48, 49] proved this hypothesis by monitoring the change in the intensity of the hydroxyl infrared adsorption bands on a HY zeolite at different temperatures for the progressive titration of acid sites with pyridine. This behaviour proves the importance of high surface mobility of probe molecules on the catalysts surface for equilibration on active sites.

The second type of behaviour was a slight decrease in the differential heats of ammonia adsorption on HY above 473 K and NaY above 313 K with an increase of temperature. This decrease was attributed to the fact that the temperature dependence of the heat of adsorption was thermodynamically defined by the difference in molar heat capacity between adsorbed state and gaseous state.

Unusual behaviour was observed for adsorption of ammonia on protonated ZSM zeolite at 416 K [50]. Differential heat curve passed through a maximum at relatively low coverage. This behaviour was explained by conjunction of three independent phenomena: immobile adsorption, mass-transfer limitations and preferential location of the most energetic acid sites in the internal pores of zeolite structure. The maximum was eliminated by heating of samples between doses to increase the surface mobility of the preadsorbed ammonia and allow it to migrate and adsorb on the most acidic sites. In this way most accessible sites are made available for new doses of ammonia at lower temperature. After this the heat curve showed a slightly higher value than the initial heat for the conventional method of adsorption, and it did not show any maximum.

From the discussed experimental results it is clear that results obtained by adsorption microcalorimetry technique are strongly dependent on the temperature at which the experiments are performed. For samples which possess strong sites, increasing the adsorption temperature yields a better description of the thermodynamic site strength distribution. For samples which possess weaker sites adsorption at higher temperatures produces a large decline in the adsorption capacity. In any case the choice of adsorption temperature is constrained and involves compromise between using high temperatures to reduce equilibrium times, and using lower temperatures to achieve high coverage of the active sites at reasonable pressures.

3.10 Probe Molecules

The key for effective utilisation of microcalorimetry in heterogeneous catalysis is the judicious choice of gas-phase molecules for study. Although total number of surface active sites and potentially active centres can be estimated by this technique, the obtained values are strongly dependent on the nature and size of the probe molecule. As a general rule, no matter which surface property is examined (acidic, basic or redox); probe molecule should be stable with temperature and with time. Furthermore, as discussed before, the adsorbed probe should be sufficiently mobile to equilibrate on active sites at temperature of the experiment.

3.10.1 Probing Surface Acidic Properties

Though it is very reliable, the calorimetric measurement of surface acidity of solid catalysts depends on the choice of the basic probe molecule used to neutralize the acid sites. Ammonia ($pK_a = 9.24$, proton affinity in gas phase = $857.7\,kJmol^{-1}$) and pyridine ($pK_a = 5.19$, proton affinity in gas phase = $922.2\,kJmol^{-1}$) are the favoured probe molecules to probe overall surface acidity of catalysts, because both Lewis and Brönsted acid sites retain these molecules.

Ammonia is among the smallest strongly basic molecules and its diffusion is little affected by the porous structure, if at all. Because of this, it is the most commonly used probe molecule for testing surface acid properties. It is adsorbed as an ammonium ion, and the corresponding heat of adsorption depends both on the proton mobility and on the affinity of ammonia for the proton.

Heats of adsorption of strong base molecules, are related to intrinsic acidity of the site and to the interaction energy between the deprotonated surface of the sample and the protonated base [51]. From this it is a logical question how strongly do different bases interact with the Brönsted and Lewis sites, and which base interacts more strongly. A study of surface acidity by adsorption of various basic probe molecules (ammonia, pyrrole, dimethyl ether and acetonitrile) on dealuminated Y type zeolites proved that obtained acid site distribution strongly depends on the basicity of probe molecule [52]. Ammonia proved to be a reliable probe molecule when Brönsted acid sites were being investigated. It helped to unveil two site populations whose properties varied with Si/Al ratio. These two populations differ by local environment of the associated AlO_4 tetrahedra. Ammonia, however, failed to reveal inhomogeneity of one particular acid population. Dimethyl ether appeared as a too weak base, whereas pyrrole, showing no steric hindrance in this case, appeared as a rather amphoteric probe which helped visualize the basicity difference between the parent and dealuminated materials. Acetonitrile proved to be a reliable probe to monitor quantitatively and qualitatively Lewis acidity.

In another study with various base probe molecules experiments have been performed on ferrierite sample, covering a wide range of basic strength, in the order acetonitrile > dimethyl ether > water > pyrrole > ammonia [53]. It has been shown that the strength distribution depends on the basicity of the probe molecule. Moreover the total number of neutralized sites appeared constant irrespective of the nature of the probe, except for pyrrole, and nearly equal to the number of protons (Fig. 3.17).

Using a series of amines as probe molecules, Parrilo et al. found a good correlation between heats of adsorption and gas phase proton affinities, but not with the proton transfer energies of those bases in aqueous phase [54–56]. These results indicate that the proton transfer dominates the interaction between the adsorbate and the acid sites. However, in a theoretical study published by Teraishi the fact that the heat of ammonia adsorption depends both on the proton affinity and the ammonium ion affinity was underlined [57]. With regard to the catalytic reaction in zeolites, the activity depends not only on the proton affinity but also on the stability of the cationic intermediate in the zeolite. The heat of ammonia adsorption, which includes the later effect, is thus in disagreement with proton affinity and provides a different measure of acidity, which is better suited to evaluate the acid strength of the zeolite in relation with its catalytic activity.

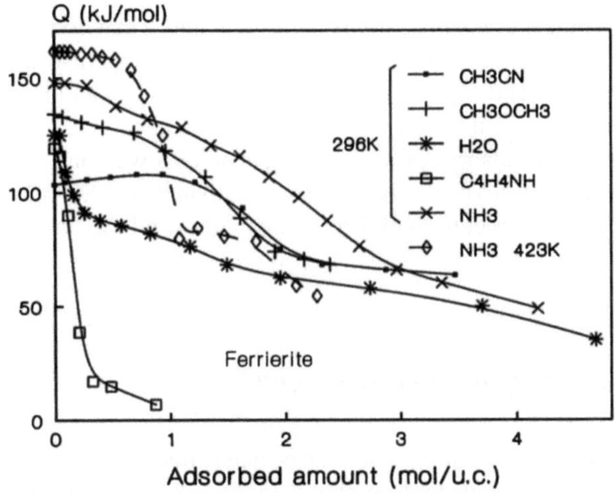

Fig. 3.17 Differential heats of adsorption of various probe molecules on a ferrierite sample versus the probe uptake (from [53])

3.10.2 Probing Surface Basic Properties

The number of acidic probe molecules that are able to cover wide range of strength is rather small. Moreover the difficulty in evaluating surface basic properties stems from the fact that these molecules may interact simultaneously with cations (such as Na^+). The ideal probe molecule should be specific to basic sites, it should distinguish interaction with oxide ion and hydroxyls and it should not give rise to chemical reactions.

Carbon dioxide ($pK_a = 6.37$), owing to its acidic character, is commonly used to characterize the basicity of solids. A large number of species of this molecule on solid surfaces can be formed; on basic hydroxyl groups there is a formation of hydrogen carbonate species, on basic oxygen ions different kinds of carbonate species can be formed. IR evidence concerning the formation of carbonate-like species of different configurations has been reported for metal oxides, which accounts for the heterogeneity of the surface revealed by microcalorimetric measurements [58]. Localization of the carbonate species formed either on surface (in the form of unidentate, bidentate of bridged species) or in the bulk (polydentate species) is essential for the results obtained by microcalorimetry. The heat evolved from the CO_2 interaction can result not only from its adsorption on basic surface sites, but also from its reaction with the bulk. Moreover, if basic character of zeolites is investigated, CO_2 can interact with extra framework cation.

Sulphur dioxide ($pK_a = 1.89$) is another molecule commonly used to investigate basicity of solid catalysts. SO_2 adsorption on the surface of metal oxides is complex. Several types of species can be formed according to hydroxylation state and the acidity or basicity of surface [58]. Interaction of SO_2 with basic O^{2-} leads to formation of sulfite type species. The sulfite ion is pyramidal and it can coordinate to metal in monodentate, bidentate or bridged form. Sulfur dioxide can also interact with surface hydroxyl groups and form hydrogen sulfite species. It can also be coordinated to Lewis acid sites where it forms weakly bonded species.

The number of CO_2 adsorption sites is always much lower than that found for SO_2 suggesting that adsorption of CO_2 is more specific than that of SO_2. The latter, having a very strong acidic character, therefore probes almost all the basic sites regardless of their strength.

3.10.3 Probing Surface Redox Properties

The heats of reduction of oxide samples can be determined by studying the adsorption of hydrogen, CO and various hydrocarbons on the fully oxidized catalysts [27, 59]. The extent of reduction of the catalyst surface can be evaluated in particular using H_2. The measurement of hydrocarbon (e.g. propene, propane, acrolein, etc.) adsorption heats is complicated by the subsequent reaction of the adsorbed species or by incomplete desorption of the products [60].

In the case of CO reduction, the catalyst–oxygen bond energy has to be calculated by subtracting the heat of formation of CO_2.

However, it is known that, in the absence of processes other than plain surface coordination, CO acts as a weak Lewis base and can interact with the strongest surface Lewis acid sites. NO can also be employed either as a probe to identify Lewis acid sites or as a reducing agent. However, NO may disproportionate into N_2O and oxygen and it is also very likely to form nitrosyl complexes in the presence of transition metal ions [60].

The heats of oxidation of the reduced oxides can be further measured using O_2 adsorption. Large variations of the reoxidation heat can be sometimes observed when any further oxidation is limited by the diffusion of oxygen into the reduced portion of the particle [61].

References

1. L. Arnaut, S. Formosinho, H. Burrows, *Chemical Kinetics: From Molecular Structure to Chemical Reactivity* (Elsevier, Amsterdam, 2007)
2. J.M. Tatibouet et al., Methanol oxidation as a catalytic surface probe. Appl. Catal. A Gen. **148**, 213–252 (1997)
3. M. Baldani, I.E. Wachs et al., Methanol: a "smart" chemical probe molecule. Catal. Lett. **75**, 137–149 (2001)
4. W.E. Farneth, J. Gorte et al., Methods for characterizing zeolite acidity. Chem. Rev. **95**, 615–635 (1995)
5. N. Cardona-Martinez, J.A. Dumesic, Applications of adsorption microcalorlmetry to the study of heterogeneous catalysis, in *Advances in Catalysis*, vol. 42, ed. by D.D. Eley, H. Pines, P.B. Weisz (Academic, San Diego, 1992), pp. 149–244

6. P.J. Andersen, H.H. Kung, Characterization of catalysts with microcalorimetry, in *Catalysis*, vol. 11, ed. by J.J. Spivey, S.K. Agarwal (Royal Society of Chemistry, Cambridge, 1994), pp. 441–466
7. A. Auroux et al., Acidity characterization by microcalorimetry and relationship with reactivity. Top. Catal. **4**, 71–89 (1997)
8. V. Solinas, I. Ferino et al., Microcalorimetric characterisation of acid–basic catalysts. Catal. Today **41**, 179–189 (1998)
9. B.E. Spiewak, J.A. Dumesic et al., Microcalorimetric measurements of differential heats of adsorption on reactive catalyst surfaces. Thermochim. Acta **290**, 43–53 (1996)
10. A. Auroux, Thermal methods: calorimetry, differential thermal analysis, and thermogravimetry, in *Catalyst Characterization: Physical Techniques for Solid Materials*, ed. by B. Imelik, J.C. Védrine (Plenum Press, New York, 1994), pp. 611–635
11. P.C. Gravelle et al., Heat-flow microcalorimetry and its application to heterogeneous catalysis, in *Advances in Catalysis*, vol. 22, ed. by D.D. Eley, H. Pines, P.B. Weisz (Academic, San Diego, 1972), pp. 191–263
12. A. Auroux et al., Acidity and basicity: determination by adsorption microcalorimetry, in *Molecular Sieves—Science and Technology: Acidity and Basicity*, vol. 6, ed. by H. Karge, J. Weitkamp (Springer, Heidelberg, 2008), pp. 45–152
13. Y.I. Tarasevich et al., Interaction of water and other polar substances with various sorbents according to calorimetric and chromatographic data. Theor. Exp. Chem. **37**, 197–214 (2001)
14. A. Auroux, M. Huang, S. Kaliaguine et al., Decrystallization process of HNaY zeolites upon mechanical milling: a microcalorimetric and thermokinetic study. Langmuir **12**, 4803–4807 (1996)
15. N. Cardona-Martinez, J.A. Dumesic et al., Acid strength of silica–alumina and silica studied by microcalorimetric measurements of pyridine adsorption. J. Catal. **125**, 427–444 (1990)
16. I. Ferino, R. Monaci, E. Rombi, V. Solinas et al., Microcalorimetric investigation of mordenite and Y zeolites for 1-methylnaphthalene isomerisation. J. Chem. Soc. Faraday Trans. **95**, 2647–2652 (1998)
17. G.I. Kapustin, T.R. Brueva, A.L. Klyacho, A.M. Rubinshtein et al., Analysis of distribution of adsorbate molecules within zeolite crystals by thermokinetics. Kinet. Catal. **22**, 183–195 (1981)
18. A. Gervasini, A. Auroux et al., Thermodynamics of adsorbed molecules for a new acid–base topochemistry of alumina. J. Phys. Chem. **97**, 2628–2639 (1993)
19. Lj Damjanović, A. Auroux et al., Determination of acid/base properties by temperature programmed desorption (TPD) and adsorption calorimetry, in *Zeolite Chemistry and Catalysis: An Integrated Approach and Tutorial*, ed. by A. Chester, E. Derouane (Springer, Berlin, 2009), pp. 107–167
20. T.J. Gricus Kofke, R.J. Gorte, W.E. Farneth et al., Stoichiometric adsorption complexes in H-ZSM-5. J. Catal. **114**, 34–45 (1988)
21. T.J. Gricus Kofke, R.J. Gorte, G.T. Kokotailo, W.E. Farneth et al., Stoichiometric adsorption complexes in H-ZSM-5, H-ZSM-12, and H-mordenite zeolites. J. Catal. **115**, 265–272 (1989)
22. T.J. Gricus Kofke, G.T. Kokotailo, R.J. Gorte et al., Stoichiometric adsorption complexes in [B]- and [Fe]-ZSM-5 zeolites. J. Catal. **116**, 252–262 (1989)
23. J. Tittensor, R.J. Gorte, D. Chapman et al., Isopropylamine adsorption for the characterization of acid sites in silica-alumina catalysts. J. Catal. **138**, 714–720 (1992)
24. R.J. Gorte et al., What do we know about the acidity of solid acids? Catal. Lett. **62**, 1–13 (1999)
25. C. Pereira, R.J. Gorte et al., Method for distinguishing Brönsted-acid sites in mixtures of H-ZSM-5 H-Y and silica–alumina. Appl. Catal. A **90**, 145–157 (1992)
26. A. Gervasini, C. Picciau, A. Auroux et al., Characterization of copper-exchanged ZSM-5 and ETS-10 catalysts with low and high degrees of exchange. Micropor. Mesopor. Mater. **35–36**, 457–469 (2000)
27. J. Le Bars, J.C. Védrine, A. Auroux, B. Pommier, G.M. Pajonk et al., Calorimetric study of vanadium pentoxide catalysts used in the reaction of ethane oxidative dehydrogenation. J. Phys. Chem. **96**, 2217–2221 (1992)

28. P.F. Siril, A.D. Davison, J.K. Randhava, D.R. Brown et al., Calorimetric study of vanadium pentoxide catalysts used in the reaction of ethane oxidative dehydrogenation. J. Mol. Catal. A: Chem. **267**, 72–78 (2007)
29. P.R. Westmoreland, T. Inguilizian, K. Rotem et al., Flammability kinetics from TGA/DSC/GCMS, microcalorimetry and computational quantum chemistry. Thermochim. Acta **367–368**, 401–405 (2001)
30. A.J. Groszek et al., Flow adsorption microcalorimetry. Thermochim. Acta **312**, 133–143 (1998)
31. P.F. Siril, D.R. Brown et al., Acid site accessibility in sulfonated polystyrene acid catalysts: calorimetric study of NH_3 adsorption from flowing gas stream. J. Mol. Catal. A: Chem. **252**, 125–131 (2006)
32. S.P. Felix, C.S. Jowitt, D.R. Brown et al., Base adsorption calorimetry for characterising surface acidity: a comparison between pulse flow and conventional "static" techniques. Thermochim. Acta **433**, 59–65 (2005)
33. V. Rakić, Lj Damjanović, V. Rac, D. Stošić, V. Dondur, A. Auroux et al., The adsorption of nicotine from aqueous solutions on different zeolite structures. Water Res. **44**, 2047–2057 (2010)
34. J.A. Thomson, J.E. Ladbury, Isothermal titration calorimetry, in *Biocalorimetry 2 Applications of Calorimetry in the Biological Sciences*, ed. by J.E. Ladbury, M.L. Doyle (Wiley-VCH, Weinheim, 2004), pp. 35–58
35. Y.Y. Lim, R.S. Drago, M.W. Babich, N. Wong, P.E. Doan et al., Thermodynamic studies of donor binding to heterogeneous catalysts. J. Am. Chem. Soc. **109**, 169–179 (1987)
36. C.W. Chronister, R.S. Drago et al., Determination of hydrogen-bonding acid sites on silica using the Cal-Ad method. J. Am. Chem. Soc. **115**, 4793–4798 (1993)
37. R.S. Drago, S.C. Dias, M. Torrealba, L. de Lima et al., Calorimetric and spectroscopic investigation of the acidity of HZSM-5. J. Am. Chem. Soc. **119**, 4444–4452 (1997)
38. N. Kob, R.S. Drago, V. Young et al., Preparation, characterization, and acidity of a silica gel/tungsten oxide solid acid. Inorg. Chem. **36**, 5127–5131 (1997)
39. R.S. Drago, C.S. Dias, J.M. McGilvray, A.L.M.L. Mateus et al., Acidity and hydrophobicity of TS-1. J. Phys. Chem. B **102**, 1508–1514 (1998)
40. R.S. Drago, N. Kob et al., Acidity and reactivity of sulfated zirconia and metal-doped sulfated zirconia. J. Phys. Chem. B **101**, 3360–3364 (1997)
41. E.M. Arnet, R.A. Haaksma, B. Chawla, M.H. Healy et al., Thermochemical comparisons of homogeneous and heterogeneous acids and bases. 1: sulfonic acid solutions and resins as prototype Brönsted acids. J. Am. Chem. Soc. **108**, 4888–4896 (1986)
42. E.M. Arnet, T. Absan, K. Amarnath et al., Thermochemical comparisons of solid and homogeneous acids and bases: pyridine and polyvinylpyridine as prototype bases. J. Am. Chem. Soc. **113**, 6858–6861 (1991)
43. P. Carniti, A. Gervasini, S. Biella, A. Auroux et al., Intrinsic and effective acidity study of niobic acid and niobium phosphate by a multitechnique approach. Chem. Mater. **17**, 6128–6136 (2005)
44. S. Koujout, D.R. Brown et al., Calorimetric base adsorption and neutralisation studies of supported sulfonic acids. Thermochim. Acta **434**, 158–164 (2005)
45. C. Guimon, H. Martinez, *Recent Research Developments in Catalysis*, vol. 2 (Research Signpost, Kerala, 2003)
46. H. Knözinger, Specific poisoning and characterization of catalytically active oxide surfaces, in *Advances in Catalysis*, vol. 25, ed. by D.D. Eley, H. Pines, P.B. Weisz (Academic, San Diego, 1976), pp. 184–271
47. H.A. Benesi, B.H. Winquist, Surface acidity of solid catalysts, in *Advances in Catalysis*, vol. 27, ed. by D.D. Eley, H. Pines, P.B. Weisz (Academic, San Diego, 1997), pp. 97–182
48. Y. Mitani, K. Tsutsumi, H. Takahashi et al., Direct measurement of the interaction energy between solids and gases X. acidic properties of hydroxyl sites of H-Y zeolite determined by high-temperature calorimetry. Bull. Chem. Soc. Jpn. **56**, 1917–1920 (1983)
49. K. Tsutsumi, Y. Mitani, H. Takahashi et al., Direct measurement of the interaction energy between solids and gases IX. heats of adsorption of ammonia and pyridine on several solids at high temperature. Bull. Chem. Soc. Jpn. **56**, 1912–1916 (1983)

50. J.C. Védrine, A. Auroux, G. Coudurier, Combined physical techniques in the characterization of zeolite ZSM-5 and ZSM-11 acidity and basicity, in *Catalytic Materials: Relationship Between Structure and Reactivity*, vol. 248, ASC Symposium Series, ed. by T.W. Whyte Jr, R.A. Dalla Betta, E.G. Derouane, R.T.K. Baker (American Chemical Society, Washington, 1984), pp. 253–273
51. Lj Damjanović, A. Auroux, Heterogeneous catalysis on solids, in *Handbook of Thermal Analysis and Calorimetry*, vol. 5, ed. by M.E. Brown, P.K. Gallagher (Elsevier, Amsterdam, 2008), pp. 387–438
52. A. Auroux, Z.C. Chi, N. Echoufi, Y. Ben Taarit, Calorimetric investigation of the acidity of dealuminated Y-type zeolites using various basic probes, in *Zeolites as Catalysts, Sorbents and Detergent Builders*, vol. 46, ed. by H.G. Karge, J. Weitkamp (Elsevier, Amsterdam, 1989), pp. 377–387
53. A. Auroux, New probes for an accurate calorimetric determination of the acidity of zeolites, in *Innovation in Zeolite Materials Science*, vol. 37, Studies in Surface Science and Catalysis, ed. by P.J. Grobet, W.J. Mortier, G. Schulz-Ekloff (Elsevier, Amsterdam, 1988), pp. 385–391
54. D.J. Parrillo, R.J. Gorte, W.E. Farneth et al., A calorimetric study of simple bases in H-ZSM-5: a comparison with gas-phase and solution-phase acidities. J. Am. Chem. Soc. **115**, 12441–12445 (1993)
55. D.J. Parrillo, A. Biaglow, R.J. Gorte, D. White, Quantification of acidity in H-ZSM-5, in *Zeolites and Related Microporous Materials: State of the Art*, vol. 34, Studies in Surface Science and Catalysis, ed. by J. Weitkamp, H.G. Karge, H. Pfeifer, W. Hölderich (Elsevier, Amsterdam, 1994), pp. 701–708
56. D.J. Parrillo, R.J. Gorte et al., Characterization of stoichiometric adsorption complexes in H-ZSM-5 using microcalorimetry. Catal. Lett. **16**, 17–25 (1992)
57. K. Teraishi et al., Effect of Si to A1 substitution at next-nearest neighbor sites on the acid strength: ab initio calculation of the proton affinity and the heat of ammonia adsorption. Micropor. Mater. **5**, 233–244 (1995)
58. J.C. Lavalley et al., Infrared spectrometric studies of the surface basicity of metal oxides and zeolites using adsorbed probe molecules. Catal. Today **27**, 377–401 (1996)
59. M. Cabrejas Manchado, J.M. Guil, A. Perez Masia, A. Ruiz Paniego, J.M. Trejo Menayo et al., Adsorption of H_2, O_2, CO, and CO_2 on a y-alumina: volumetric and calorimetric studies. Langmuir **10**, 685–691 (1994)
60. F. Witzel, H.G. Karge, A. Gutze et al., in *Proceedings from the 9th International Zeolite Conference*, Montreal, Canada, 5–10 July 1992, vol. 2, ed. by R. von Ballamos, J.B. Higgins, M.M.J. Treacy (Butterworth-Heineman, Boston, 1993), p. 283
61. L. Yang, O. Kresnawahjuesa, R.J. Gorte, A calorimetric study of oxygen-storage in Pd/ceria and Pd/ceria–zirconia catalysts. Catal. Lett. **72**, 33–37 (2001)

Chapter 4
Temperature-Programmed Desorption (TPD) Methods

Vesna Rakić and Ljiljana Damjanović

Abstract This chapter presents the fundamentals, the experimental setups and the applications of temperature-programmed desorption (TPD), method used to investigate the events that take place at the surface of solid material while its temperature is changed in a controlled manner. At the beginning, fundamental principles of adsorption and desorption phenomena, as well as the data concerning first experimental setups are given. Further, important information related to the construction of nowadays used equipment and the organization of common experiments are underlined. The significance of data directly obtained from temperature-programmed experiment—TPD profile, which are the area under it and the position of peak maximum, are highlighted. Particular attention is given to the results that can be derived from these data—characterization of active sites that can be found on the surface of solid material and determination of kinetic and thermodynamic parameters of desorption process. In this regard, the influence of important experimental parameters on derived values is explained. Besides, the distinctions between TPD experiments performed in ultra-high vacuum and in the flow systems (differences in experimental setups and in the derivation of kinetic and thermodynamic parameters) are explained. Also, the modification of temperature-programmed techniques, known as temperature-programmed oxidation and temperature-programmed reduction are shortly explained and compared with temperature-programmed desorption method. In the end, a brief comparison of the TPD and adsorption calorimetry, two most widely used techniques for the study of acid/base properties of catalysts, is given.

V. Rakić (✉)
University of Belgrade, Nemanjina 6, 11080 Zemun, Serbia
e-mail: vesna.rakic@ffh.bg.ac.rs

L. Damjanović
University of Belgrade, Studentski trg 12-16, 11000 Belgrade, Serbia
e-mail: ljiljana@ffh.bg.ac.rs

A. Auroux (ed.), *Calorimetry and Thermal Methods in Catalysis*,
Springer Series in Materials Science 154, DOI: 10.1007/978-3-642-11954-5_4,

4.1 Introduction

Temperature-programmed desorption (TPD) belongs to the group of techniques in which an event that takes place at the surface of solid substance is monitored, while the temperature of investigated sample is changed with a temperature program $\beta(t) = dT/dt$. When the technique is applied to a system in which the adsorption process is (at least in part) irreversible and surface reaction takes place, this technique is often known as temperature-programmed reaction spectroscopy (TPRS). Although most often investigated surface process is desorption, reactions such as reduction, oxidation or sulfidation can be monitored using the techniques named temperature-programmed reduction (TPR), temperature-programmed oxidation (TPO) or temperature-programmed sulfidation (TPS), respectively. Usually, temperature T is a linear function of the time t; in that case the heating rate β is constant value.

In general, temperature-programmed methods are applicable for the investigation of both porous materials (such as real catalysts) and well-defined surfaces of single-crystalline samples. In addition, their application is experimentally simple and inexpensive, what explains their wide application in several scientific domains.

Temperature-programmed desorption originates from so-called flash desorption, which was originally developed in early fifties of twentieth century to quantitatively investigate the kinetics of molecular desorption from well-defined single crystal surfaces in high vacuum [1]. Flash desorption involved the adsorption of a known gas on the sample (in the form of a ribbon or wire, rigorously cleaned previously) while kept in vacuum. Subsequently, desorption was provoked by heating while the pressure in the system was recorded: as the temperature increased, certain previously absorbed species had enough energy to desorb from the surface and would be detected as a rise in pressure. The resulting pressure-time curve was referred to as a "desorption spectrum" [1, 2]. In flash desorption, temperature was raised very quickly (from 1 to 1000 Ks^{-1}). Two heating schedules were applied: a linear variation of sample temperature with time: $T = T_0 + \beta t$ and its reciprocal variation $1/T = 1/T_0 - \alpha t$ (T_0 is starting temperature, α and β are constants) [1]. From the beginning, the possibility to extract the heat of adsorption from the obtained results was demonstrated.

Flash desorption was mainly applied for the investigations of low-surface area substances such as metals. The method was later adapted for the investigations of high-surface area materials, under carrier gas and ambient pressure. In the domain of catalysis, the pioneering work in this direction was done by Amnenomiya and Cvetanović [3], who studied the catalysts' active sites, and needed the methodology that enabled conditions similar to those ordinarily used in catalytic reaction. Their apparatus allowed the pre-treatment of solid material by heating and evacuation, its exposure to molecular species of interest (at low temperature) and subsequent programmed desorption performed by heating in a controlled manner with the possibility to detect the desorbed gas in the carrier. This equipment consisted of two important parts: a temperature programming controller and a thermistor type thermal conductivity cell for measuring the rate of desorption of pre-adsorbed molecules from the surface. In this way, it was enabled to increase the temperature in a controlled manner, slowly (about 1 to tenth degrees K per minute), in a manner similar to those applied in real catalytic reactions.[1] An additional important feature of their modified technique was that it permitted simultaneous study of a chemisorption process and the surface reaction which accompanied it. Applying the "temperature-programmed desorption" (the name they used for this modification of flash desorption), Amnenomiya and Cvetanović discovered the existence of two different active sites for the adsorption of ethylene on alumina [4]. Importantly, they employed the method to calculate the values of energies for ethylene desorption.

Since that time, the method is widely developed; experimental setups are improved and adjusted to many different purposes (e.g. for the investigations of oxidation and reduction reactions). Today, two main types of equipment are available: those operating under ultrahigh vacuum and so-called "flow" systems. Well-defined surfaces of single-crystalline samples are investigated in a continuously pumped ultra-high vacuum (UHV) chamber (this technique is often referred to as thermal desorption spectroscopy—TDS [5]). The equipment that is constructed to allow adsorption–desorption in the gas flow are most often used for the investigation of porous materials (catalysts, for example). Vacuum setups are customarily used for surface science studies, but they can be also useful for the characterization of porous materials.

Generally, TPD can be described as the measurement of the rate of desorption of adsorbed molecules, as a function of temperature. Therefore, this method can be useful in the extraction of very important information. It can be used in the identification and characterization of sites active in adsorption and catalytic reactions, in the study of adsorption states, binding energies, surface concentration and desorption kinetics. To summarize, this method is nowadays very important and often applied for the characterization of materials used as catalysts. In this domain, two main areas of applications are: the characterisation of acid/base properties of solid materials, what is essential for understanding their reactivity[2]; and the determination of kinetic and thermodynamic parameters of desorption processes or decomposition reactions.

In the following sections, the basic concepts of temperature-programmed methods (primarily temperature-programmed desorption, but also temperature-programmed reactions) are outlined. At the beginning, fundamental principles of adsorption and desorption—their thermodynamic and kinetic aspects, are presented. Furthermore, the descriptions of experimental setups, the data that can be obtained from the experiments and their interpretation are given. The possibilities to extract the adsorption energies and kinetic parameters from experimental results are discussed. Finally, the examples of possible applications and the comparison of results obtained by TPD with those obtained from adsorption calorimetry, are presented.

---

[1] Significant difference in heating rates makes main distinction between "flash desorption", where the heating rate is very high (the desired temperature is reached in seconds) and temperature-programmed desorption (where the sample is heated in minutes or even hours).

[2] TPD is perhaps the most often used for estimation of acid/base properties of solid catalysts.

4.2 Adsorption–Desorption; Fundamental Principles

Adsorption is defined as the enrichment of gas or liquid (adsorbate) at the surface of a solid material (adsorbent); or as the increase in the density of the fluid in the vicinity of an interface. Adsorption takes place at the active sites—specific points on the solids' surface that possess affinity toward the particles coming from the gas or liquid phase; it happens when an attractive interaction between a particle from adsorbate phase and a surface is strong enough to overcome the disordering effect of thermal motion [6, 7].

In discussing the fundamentals of adsorption it is usual to distinguish between physisorption, which takes place when the attractive interactions are essentially the result of weak intermolecular forces; and chemisorption, which involves the overlapping between the molecular orbitals of the adsorbed particle and the surface atoms (i.e. electron transfer which leads to the formation of chemical—covalent bond between adsorbate molecule and the active site at the surface). Although this distinction is very useful, it has to be pointed out that there are many intermediate cases, and that it is not always possible to categorize a particular event at the surface as physi- or chemisorption. However, there are some general characteristics which distinguish these two possible types of adsorption. Chemisorption is highly specific, while quite contrary, physisorption is non-specific; chemisorption may involve dissociation, it is possible over a wide range of temperature and a monolayer is formed; while physisorption takes place only at relatively low temperatures, multilayer can be formed, dissociation of adsorbed species does not happen. However, the most important distinction is the amount of heat that is associated to either one of these two general types of adsorption: physisorption is characterized by low heat of adsorption (below approximately 50 kJ mol^{-1}); while chemisorption is characterized by high heat of adsorption, typically exceeding 50 kJ mol^{-1}. Hence, as a result of molecular chemisorption, the weakening of intramolecular bonds inside the adsorbed molecule can happen, which may lead to its dissociation [6–8]. The differences in potential energies curves that present processes of physisorption and chemisorption (dissociative and non-dissociative) are presented in Fig. 4.1.[3]

In addition, chemisorption is often an activated process, which means that the formation of a chemisorptive bond requires overcoming the activation barrier (Fig. 4.1b); it may be slow and irreversible. By contrast, physisorption is rapid, non-activated and reversible process.

---

[3] Dissociative chemisorption of a diatomic molecule can also happen through the dissociation in a gas phase and a creation of two gas phase atoms; these two atomic species can be then adsorbed on the surface (this way is almost always non-activated). If the curves describing molecular and atomic adsorption intersect at or below the zero potential energy line, then the precursor physisorbed molecule can experience non-activated dissociation, followed by chemisorption (Fig. 4.1a). In contrast, if the energetic for these two pathways are such that the intersection occurs above the zero energy plane, then chemisorption will be activated with activation energy, E_{ad}, as indicated in Fig. 4.1b.

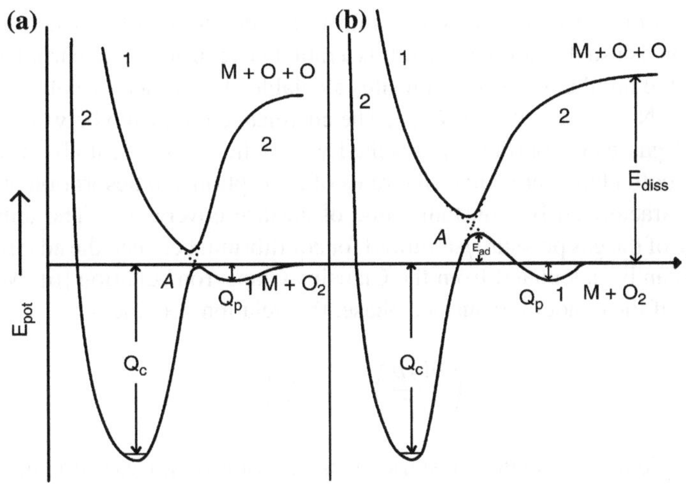

Fig. 4.1 Potential energy curves for (*1*) physical and (*2*) chemical adsorption: (**a**) non-activated (**b**) activated. E_{pot} - potential energy, Q_c - heats of chemisorption, Q_p - heats of physisorption, E_{ad} - energy of activation for desorption, E_{diss} - dissociation energy for the diatomic molecule. The sum: $\Delta E_{des} = E_{ad} + Q_c$ is the the heat of hemisorption, in the activated processes [8]

4.2.1 Thermodynamic View

Both physisorption and chemisorption are exothermic processes, what can be concluded from a simple thermodynamic consideration. The adsorbed molecule has at most two degrees of translational freedom on the surface; in addition, its rotational freedom must always be less than that of the gas phase molecule. In total the adsorbed molecule possesses less degree of freedom than the same molecule in the gas phase. Consequently, the entropy change of adsorption $\Delta S_{ads} = S_{ads} - S_{gas}$ is obligatory negative. However, the adsorption is spontaneous process, which means that the free energy change ($\Delta G = \Delta H - T\Delta S$) must be negative. Negative value of entropy change ΔS_{ads} means that the second term in previous relation is positive ($-T\Delta S_{ads}$), what requires the value of ΔH necessarily negative. Hence, adsorption is always exothermic process [6].

The heats of adsorption provide a direct measure of the strength of the bonding between adsorbate and the active site at the surface of solid substance. Therefore, it is of importance to estimate these values, particularly in the domain of catalysis where the strength of active sites determines the mechanism and the yield of certain process. One possible way to determine the heat of adsorption is to apply calorimetry, experimental technique which provides the heats of adsorption as a function of the adsorbed amount ($-\Delta H = f(n_a)$, where n_a is the adsorbed amount and $-\Delta H$ is, in that case, so-called differential heat) [9]. The heats of adsorption can be derived also from the variation of adsorption with temperature. In that case, Clausius-Clapeyron equation and the data from isosteric measurements are used (in that way, so-called isosteric enthalpy of adsorption can be obtained).

The extent of surface coverage (or simply surface coverage), reached as a result of adsorption, is usually denoted as θ. It is a ratio between adsorbed particles number (N_{ads}) and the number of adsorption sites available at a surface (usually denoted as active sites $-N_{surf}$): $\theta = N_{ads}/N_{surf}$. The chemical equilibrium between adsorbed species and gas phase particles is reached when chemical potentials of adsorbate particles in both phases are equal (the rates of adsorption and desorption are equal); and it is characterized by constant value of surface coverage θ. The temperature dependence of the gas pressure p required for equilibrium between the adsorption and desorption can be calculated from the Clausius-Clapeyron equation [6]. Neglecting the volume of the condensed surface phase, this relation becomes:

$$\left(\frac{\partial \ln p}{\partial T}\right)_\theta = \frac{q_{isost}}{RT^2} \tag{4.1}$$

where q_{isost} denotes so-called isosteric heats of adsorption. Evidently, these values can be calculated from the temperature dependence of the adsorption isotherms, i.e. from the isosteres, for each average temperature:

$$q_{isost} = \left(\frac{\partial H}{\partial \theta}\right)_T = -RT^2\left(\frac{\partial(\ln P)}{\partial T}\right)_{n_r} \tag{4.2}$$

where T is the absolute temperature, R the gas constant and n_r the number of reversibly adsorbed molecules.[4]

Third possibility to obtain the adsorption heats is to extract them from the data acquired from temperature-programmed desorption experiments. This possibility will be exposed in detail later, in the Sect. 4.5.2. However, for that purpose, it is obligatory to know some basic postulates about the kinetics of adsorption and desorption; what is given in the following section.

4.2.2 Kinetics of Adsorption and Desorption

In the case of gas-phase adsorbate, the surface coverage θ is dependent on the gas pressure. Adsorption isotherms relay the surface coverage and the gas pressure (at constant temperature); the most known equation of this type is Langmuir adsorption isotherm. It is based on the following assumptions [6]:

– one adsorbed particle interacts with one active site at the surface (once adsorbed, it is immobile on the surface);

[4] Instead to define equilibrium by constant surface coverage, it is possible to keep constant pressure at the surface; in that case the equilibrium heat of adsorption q_{eq} is incorporated in Clausius-Clapeyron equation.

- the surface of adsorbent is saturated when all adsorption sites are covered with the particles of adsorbate, *i.e.* the adsorption goes on until monolayer of adsorbate particles is spread over the solid (then, θ equals 1);
- there are no interactions between the adsorbed particles.

The rates of adsorption (r_{ads}) and desorption (r_{des}) are proportional to the numbers of empty or occupied active sites, θ or $(1 - \theta)$, respectively:

$$\begin{aligned} r_{ads}(\theta) &= \mathrm{A_n} p\,(1-\theta)^x \\ r_{des}\,(\theta) &= \mathrm{B_n}\theta^x \end{aligned} \tag{4.3}$$

where r_{ads} and r_{des} are the rates of adsorption and desorption, respectively, A_n and B_n are constants, while x is the kinetic order of surface event (x can be 1 or 2, although adsorption and desorption are usually considered as the first order events). In the equilibrium, the rates of adsorption and desorption are equal: $|r_{ads}| = |r_{des}|$, therefore, surface coverage θ can be expressed as:

$$\theta_n = \frac{(b_n p)^{1/x}}{1 + (b_n p)^{1/x}} \tag{4.4}$$

(where $b_n(T) = A_n/B_n = \mathrm{const}$, $x = 1, 2$).

If adsorption is a reversible process (i.e. backward process—desorption, passes through exactly the same states), the rates of both processes can be described using the same equation:

$$r = -\frac{d\theta}{dt} = k \cdot \theta^{\mathrm{X}} \tag{4.5}$$

However, in contrast to adsorption which may or may not be activated process, desorption is always activated, with a minimum activation energy denoted as activation energy for desorption (ΔE^a_{des}). The rate constant for desorption can be expressed by Arrhenius equation:

$$k_{des} = \nu\,(\theta) \cdot \exp\left(-\frac{\Delta E^a_{des}}{RT}\right) \tag{4.6}$$

where $\nu\,(\theta)$ is pre-exponential factor, which is in general dependent on surface coverage θ. Now, the rate of desorption becomes:

$$r_{des} = -\frac{d\theta}{dt} = \nu(\theta) \cdot \exp\left(-\frac{\Delta E^a_{des}}{RT}\right) \cdot \theta^x \tag{4.7}$$

The relation (4.7) is customarily used to describe the rate of desorption, and is known as Polanyi-Wigner equation.

Desorption is often explained assuming the existence of transition state:

$$(A)_{ads} \rightarrow (A)^{\#}_{ads} \rightarrow A_{gas} + ()_{ads} \tag{4.8}$$

where $(A)_{ads}$ is molecule of specific adsorbate adsorbed on the active site at the surface $()_{ads}$, A_{gas} is the gas-phase molecule and $(A)^{\#}_{ads}$ is the transition complex. The equilibrium between the adsorbate and transition state is defined as:

$$K^{\#} = \frac{(A)_{ads}}{(A)^{\#}_{ads}} \tag{4.9}$$

Equilibrium constant for overall desorption process $(A)_{ads} \rightarrow A_{gas} + ()_{ads}$, (K) can be considered as equal to $K^{\#}$. The reaction rate constant is related with the equilibrium constant $K^{\#}$ (hence, with equilibrium constant K), as:

$$k_{des} = \frac{kT}{h} K^{\#} = \frac{kT}{h} K \tag{4.10}$$

From the other side, thermodynamic functions are related with the equilibrium constant through the known relation:

$$\Delta G = \Delta H - T\Delta S = -RT \ln K \tag{4.11}$$

The rate of desorption is customarily extracted from the experiments of temperature-programmed desorption. Since the rate of desorption is related to the equilibrium constant through the relations (4.5) and (4.10); it is evident that the rate constant for desorption can be expressed as:

$$k_{des} = \frac{kT}{h} K = \frac{kT}{h} \exp\left(-\frac{\Delta G_{des}}{RT}\right) = \frac{kT}{h} \cdot \exp\left(\frac{\Delta S_{des}}{R}\right) \cdot \left(-\frac{\Delta H_{des}}{RT}\right) \tag{4.12}$$

Obviously, thermodynamic quantities (ΔH_{des}, ΔS_{des} and ΔG_{des}) for the activated process such as desorption can be extracted from the data obtained from desorption experiments. Importantly, since the adsorption is spontaneous process (it does not need the energy of activation), the heat of adsorption equals, in general, the activation energy for desorption. Hence, desorption experiments provide also thermodynamic parameters of adsorption.

4.3 Experimental Setups

The experimental setups for temperature-programmed desorption have evolved with time. As it has been already stated, there are numerous experimental designs that allow the application of this method under the conditions that are the same or very similar to those applied in real catalytic reaction (or any other surface event). Nevertheless, the various equipment used for these experiments, although different, is all constructed to allow two main steps that are common for all thermal desorption methods:

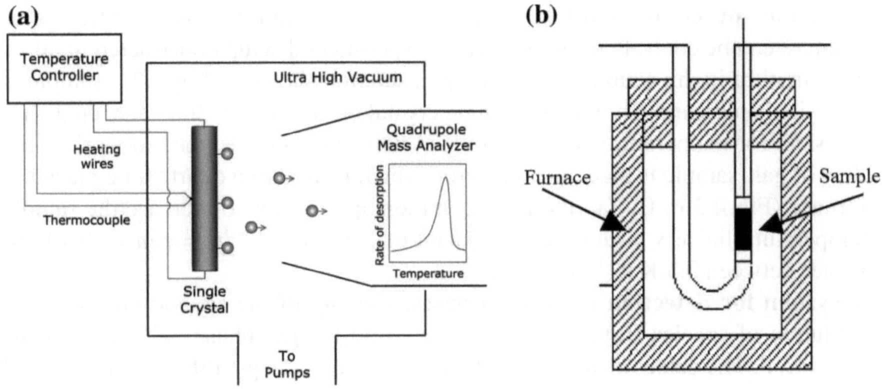

Fig. 4.2 (**a**) Experimental setup for temperature-programmed desorption performed in vacuum. The sample (monocrystal or a thin layer of powder) is exposed at the sample holder, which is connected to the system for temperature control. The pumping speed has to be high enough to allow the monitoring of desorbed species by mass spectrometer. Inset presents the output of one TPD experiment—the TPD profile [5]. (**b**) In the flow systems, sample is placed in the sample holder, inside the furnace [10]

1. **The admission of desired gas to the sample.**
2. **The heating of the sample in a programmed way.**

In TPD experiment, gaseous molecules (atoms) of interest are adsorbed at the surface, at constant temperature. The adsorption is very often performed at ambient temperature, but can be sometimes done at sub-ambient or at elevated temperature. In the modifications of technique such as TPO or TPR, gaseous species are consumed while temperature is increased in a programmed manner. In the case of TPD procedure, desorption of adsorbate is monitored while increasing the solid sample temperature in a controlled fashion; while in the case of TPR/TPO, the consumption of active gas is monitored during temperature increase, as explained later in more details (see Sect. 4.4).

From previously stated, it follows that temperature-programmed experiments can be performed under ultrahigh vacuum or in the flow of gas. Still, whatever is the experimental design, three main parts of equipment are always necessary to perform this kind of investigations:

- **The system for the controlled admission of (different) gases.** The adsorption is commonly performed as isothermal process. Nowadays, it is possible to construct the equipment which enables the adsorption of desired and precisely known amount of adsorbate. In the past, small polar molecules (NH_3, CO, CO_2, SO_2, H_2O) have been usual adsorbates in the TPD studies. More recently, larger molecules (such as hydrocarbons) and non-polar molecules (such as Ar_2 or N_2) have been applied as adsorbates. Usually, adsorbates are denoted as "probes" or "probe species". Most often used are the probes customarily applied to titrate acidic or basic surface sites (NH_3, pyridine, CO, CO_2, and SO_2).

- **"Reactor" or sample holder**, placed in a heated area (or furnace) where temperature can be controlled. In the case of experimental setup constructed to allow TPD in ultra-high vacuum (schematic presentation shown in Fig. 4.2a) sample is deposited on a sample holder as a single crystal or monolayer, and connected with a system for temperature control. In the systems that are designed to work in the flow of gas, sample is placed in a reactor, which is usually a quartz tube placed in furnace (Fig. 4.2b). By far, the most common approach is to increase the sample temperature linearly with time at constant rates ($\beta = dT/dt = \text{const}$) that have values between 0.5 K s^{-1} and 25 K s^{-1}.
- **A system for detection of evolved gases.** Heating of the sample provokes the evolution of species from the surface back into the gas phase, which has to be monitored. The detectors used for detection and possible quantification of evolved gases are: thermal conductivity detector (catharometer), flame ionization detector, conductometric titration and mass spectrometer.

- *Catharometer* serves for measuring the difference in thermal conductivity between reference gas and the gas that flows through the sample (it is used only in flow systems). This kind of detectors is often used in the variations of temperature-programmed (TP) methods, known as TPO or TPR.
- *Flame ionization detector* is in specific use for the detection of organic effluents: effluent enters in a flame obtained by combustion of hydrogen and air, then, the ions that are formed are trapped by two electrodes (with a potential difference between them). As a result, electric current appears and can be detected.
- *Conductometric titration* is applied if it is possible to entrap the evolved gas in an aqueous solution; then, the change in conductivity can be detected.
- In modern implementations of temperature-programmed techniques the detector of choice is a small, *quadrupole mass spectrometer (QMS)*. The application of mass spectrometer enables simultaneous acquisition of single or multiple masses desorbed during heating. In fact, the application of this kind of detector enabled the distinction between different species desorbed in the same time from the surface.

Nowadays, two main techniques that are most often used for detection of effluents are mass spectrometry and thermal conductivity; the whole process is most often controlled by computer.

4.4 The Design of Temperature-Programmed Experiment; Obtained Data

The equipment used for TPD experiments have to be designed in a way which allows the performance of certain steps that may be necessary in particular experiment. Firstly, sample is placed in the sample holder (reactor) and pre-treated in the appropriate way (in vacuum or in the flow of desired gas, at desired temperature); the pre-treatment procedure depends on the characteristics of investigated material and

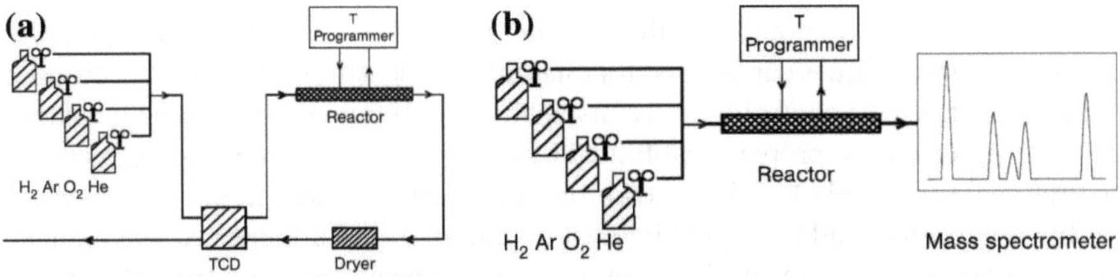

Fig. 4.3 Experimental setups for temperature-programmed desorption, reduction and oxidation. (**a**) The reactor is placed inside the furnace which is connected with temperature programmer. Detection of evolved gas(es) is performed by monitoring the variations in thermal conductivity of gas mixture. (**b**) The TPD apparatus equipped with mass spectrometer as a detector [5]

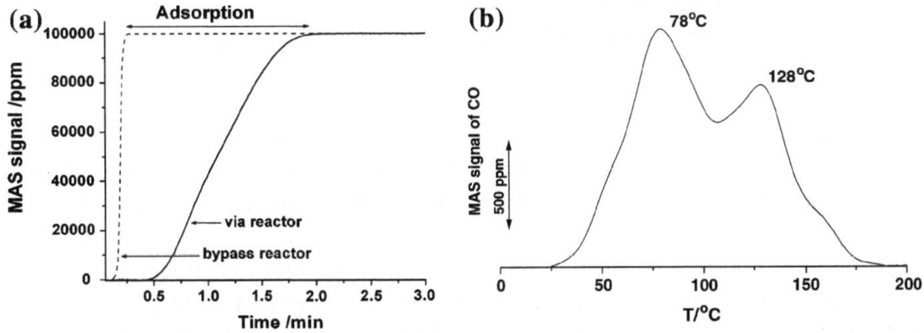

Fig. 4.4 (**a**) Time dependence of adsorbate concentration upon exposing the solid sample. (**b**) An example of TPD profile, drown as a signal of detector versus temperature. Common TPD profile is a complex shaped curve. Figure presents the interaction (adsorption and desorption) of CO with CoY zeolite [11]

the purpose of TPD experiment. Afterwards, the sample is exposed to the adsorbate. Usually, the adsorption is performed isothermally, at appropriate temperature (frequently at 300 K, but also at temperatures higher then this one, or even at sub-ambient temperatures). Subsequently, physisorbed part of adsorbed gas is removed from the surface, either by the evacuation, or by inert gas flow. The residual chemisorbed adsorbate is desorbed by heating the sample in a controlled manner, preferably in a way to give a linear temperature ramp; the analysis of the evolved gas (gases) is performed to establish its identity and the amount, using the appropriate detection system. The whole procedure is performed *in situ*.

Schematic presentation of one experimental setup that enables realisation of all mentioned steps in vacuum is shown in Fig. 4.2, while Fig. 4.3 shows two typical constructions designed for the experiments in the flow of appropriate gases.

The data obtained from one temperature-programmed experiment are presented as the variation of a detector signal intensity (presented at y axis) as a function of time (or temperature, presented at x axis). Consequently, as detector signal is proportional to the concentration of the species desorbed from the surface, y axis values are proportional to the rate of desorption (r_{des}).

It is important to point out that in the case of flow systems, temperature-programmed techniques can be used for quantitative measurements. For that purpose, the gas flow has to be controlled and constant in time. If this condition is achieved, the detector signal can be properly calibrated by using dilute gas streams of known concentrations: the signal of such a known mixture is passed through the empty reactor (or bypass reactor) and the signal intensity is monitored. Furthermore, the intensity of signal obtained as a result of the same gaseous species evolved from the sample can be considered as proportional to the value obtained for the known mixture. Once calibrated in that way, the intensity of detector signal can be given in concentration units (Fig. 4.4), and the measurement of evolved gas concentration becomes possible. The calibration of a detector signal enables the precise determination of both adsorbed and desorbed amounts.

The precise amount of adsorbed gas can be obtained by passing it through "bypass" reactor, and subsequently, through the sample (Fig. 4.4a). If the detector signal has been previously properly calibrated, the monitoring of two signals (bypass and through the sample) provides the true adsorption amount of respective gas at a given temperature, which can be derived from the surface in between these two signals. Precise amount of desorbed gas can be obtained from the surface under the TPD profile (Fig. 4.4b).

It is very important to keep in mind that many different chemical species can evaporate from the sample in the same temperature region (particularly in the case of real catalytic systems). In the old versions of TPD setups, the overall desorbed amount would be recorded as a rise in the pressure. It is especially important to point out that the incorporation of mass spectrometer as a detector enabled the discrimination of different products, desorbed in the same time (in the same temperature region).

Figure 4.5 shows one example: evidently, the evolution of ammonia and hydrocarbons happens in the same temperature region (from ~370 °C up to ~440 °C), during the thermal decomposition of 1,2-diaminopropane entrapped inside a zincophosphate structure.

The insight in the typical TPD profiles presented in Figs. 4.4 and 4.5 reveals what information can be obtained from one TPD experiment. There are two main classes of data that we can "read" from the desorption profile:

1. **The area under the TPD profile**, which is proportional to the amount of adsorbate originally adsorbed, in other words, to the surface coverage θ. Under particular circumstances, *id est*, if the limitations such are diffusion and/or readsorption can be neglected (see later, in the Sect. 4.5.2.2), ***temperature-programmed technique can be employed as an excellent tool for determination of surface coverage***.
2. **The position of the peak maximum** (along the temperature scale, T_{max}), which is related to the activation energy for desorption. Generally, the higher the value of temperature of peak maximum (T_{max}) is, desorption is more difficult, which is an indication of stronger interaction between the adsorbate species and the active site on the surface. **As it has been explained previously, the adsorption is spontaneous process, therefore, the heat of adsorption equals, in general, the activation energy for desorption.** Hence, the value of T_{max} is related to the

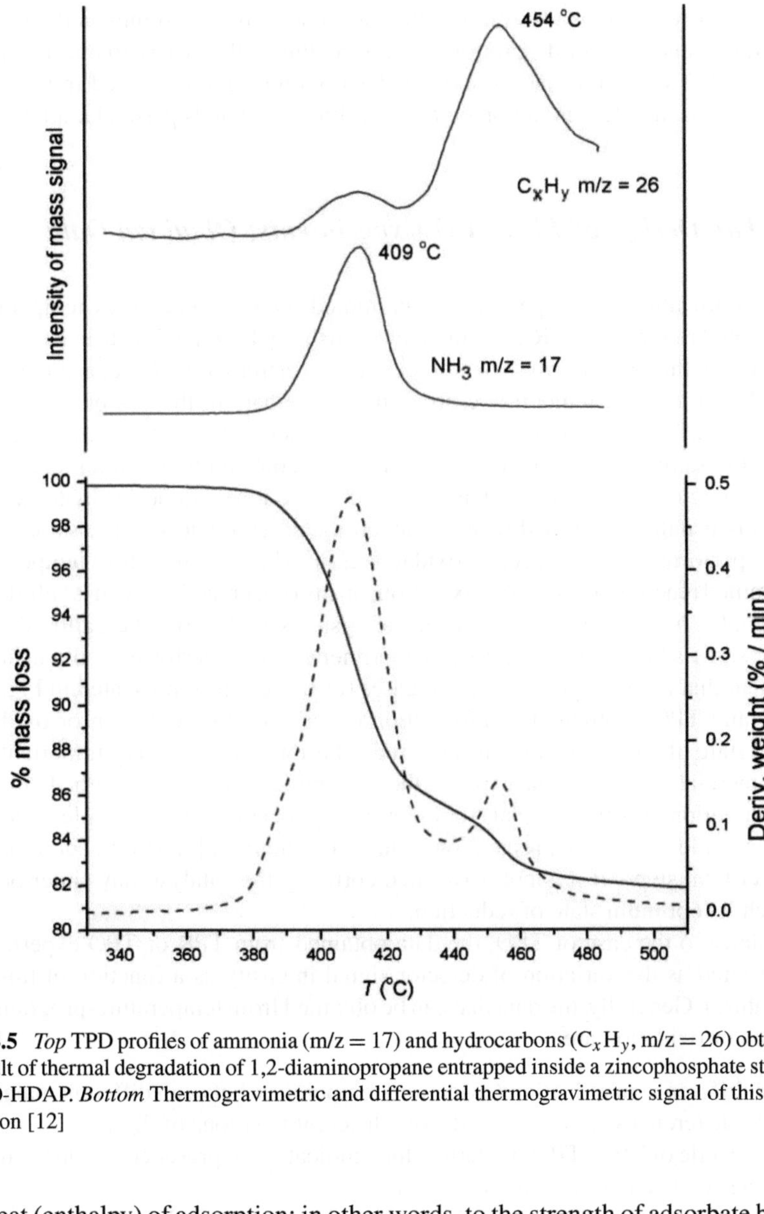

Fig. 4.5 *Top* TPD profiles of ammonia (m/z = 17) and hydrocarbons (C_xH_y, m/z = 26) obtained as a result of thermal degradation of 1,2-diaminopropane entrapped inside a zincophosphate structure-ZnPO-HDAP. *Bottom* Thermogravimetric and differential thermogravimetric signal of this decomposition [12]

heat (enthalpy) of adsorption; in other words, to the strength of adsorbate binding to the surface. The methods for deriving the activation energy for desorption from the position of T_{max} will be discussed latter. However, it is very important to note here that a simple TPD profile that possesses only one peak (and one T_{max}, like a desorption profile of mass 17, presented in Fig. 4.5) is not a common case that can be found for the majority of investigated systems. In fact, very often there is more than one binding state for the adsorbate molecules on a surface, which express significantly different adsorption enthalpies (consequently, significantly different activation energies for desorption); therefore, this will give rise to multiple peaks in the TPD spectrum (one example is desorption profile presented in Fig. 4.4b). In that case, the determination of T_{max} positions demands particular attention.

4.4.1 The Design of TPR/TPO Experiments; Obtained Data

In the modification of temperature-programmed methods known as temperature-programmed reduction (TPR), a reductive gas (usually H_2 mixed with an inert gas) is consumed by the sample, while the temperature is increased with a constant heating rate β. In the case of temperature-programmed oxidation, the gas or gas mixture which can perform oxidation (O_2 mixed with an inert gas, air, N_2O, etc.) is passed through the sample. The reductive or oxidant gas consumption is monitored either by mass spectrometer or by catharometer. If reduction is done by hydrogen, the analysis is usually performed using a catharometer as a detector; the same stands for TPO performed with oxygen as oxidative gas.[5] The experiments of temperature-programmed reactions (reduction, oxidation, or any other reaction, such is sulfidation, for example) have to be performed in the systems that allow the calibrations of detector signals [14]. Therefore, these experiments can be performed only using the equipment that allows flow of different gases (as those setups presented in Fig. 4.3).

TPR and TPO profiles give information concerning the reduction or oxidation (red-ox) state of the solid which is analyzed. These features are very important data for commercial catalysts; what explains the vast application of TPR and/or TPO in the characterization of solid materials for industrial applications. For example, reduction is an inevitable step in the preparation of metallic catalysts [5, 15]. In addition, it is often a critical step—if it is not performed correctly the catalyst may sinter or may not reach its optimum state of reduction.

Similarly to the case of TPD, the data obtained from TPR or TPO experiments are presented as the variation of detector signal intensity as a function of time (or temperature). Generally, the data that can be obtained from temperature-programmed reduction or oxidation are:

1. The difference in reduction (oxidation) temperature between different materials; these differences are recognized from different positions of T_{max};
2. The profile of TPR (TPO) pattern, which indicates the presence of one (or more) species that can be reduced (oxidized);
3. The area under a TPR or TPO curve represents the total hydrogen (oxygen) consumption, and is commonly expressed in moles of reductive (oxidative) gas consumed per mole of metal atoms. Hence, the calculation of exact amounts of those species which were reduced (oxidized) at the surface is possible.

---

[5] In those cases, the consumption of either reductive or oxidative gas by the catalyst is derived from the change in thermal conductivity of the gas mixture.

Table 4.1 The organization of temperature-programmed techniques

Temperature-programmed desorption	Temperature-programmed reduction or oxidation
Sample is pre-treated if necessary (by heating, or by flushing isothermally, in the desired atmosphere).	Sample is pre-treated if necessary (by heating, or by flushing isothermally, in the desired atmosphere)
The sample is exposed to the adsorbate, isothermally.	The sample is purged by inactive gas, isothermally
Desorption of physisorbed part of a gas by inert gas flow (isothermally, at the appropriate temperature).	The exposure of a sample to the reductive or oxidative gas, in the linear heating regime. The detector monitors the consumption of this active gas. Subsequently, the sample is cooled and purged in the inert atmosphere, in order to remove the traces of active gas
Heating of a sample in the appropriate atmosphere, with the analysis of the evolved gases.	If necessary, the sample is exposed to oxidative (if the sample was previously reduced) or reductive gas (if the sample was previously oxidized). The detector monitors the consumption of this active gas

4. The activation energy of the reduction can be estimated from the temperature T_{max} at which the reduction rate is maximal by using appropriate equations.

The organization of one TPR or TPO experiment is somehow different in comparison with that one applied for temperature-programmed desorption. For example, the sample has to be purged with inactive gas, before exposure to active (reductive or oxidative) gas.[6] These differences can be seen in Table4.1, which presents the organisation of both TPD and TPR/TPO experiments.

4.5 The Interpretation of Results Obtained from Temperature-Programmed Desorption Experiments

Temperature-programmed desorption technique offers very useful and important methodology which can be applied for the characterization of materials used as catalysts. There are two main fields of applications:

1. **The characterisation of active sites of solid materials**. It is of outmost importance to determine and understand the acid/base character of solid catalysts, because these features are essential for their reactivity [16–20]. There are several groups of techniques developed and particularly adapted for the investigation of acidity/basicity of solid catalysts; most of these methods are based on

[6] If catharometer is used as detector, it is very important to remove traces of water or any other impurities from the gas flows, because they would affect the thermal conductivity measurements.

the adsorption of gas-phase probe molecules, which are chosen on the basis of their reactivity, molecular shape and size [21–26, 13]. Among the other techniques, TPD is of particular importance because its experimental conditions can be organized in the same (or very similar) way as the conditions of real catalytic reaction. The investigation of acid/base character of solids is perhaps the most common application of TPD. For that purpose, many different chemical species can be used as adsorbates (probes). In addition, the strength and the population of specific active sites can be estimated, using the appropriate probes and applying appropriate experimental conditions.

2. **The determination of kinetic and thermodynamic parameters of desorption processes, decomposition or other reactions**. The interpretation of experimentally obtained data and derivation of kinetic and thermodynamic parameters from TPD results depends on the type of TP experiment: specific experimental conditions have influence on the overall TP profile and on the position of T_{max} obtained either in ultra-high vacuum or in the flow system.

The details that explain more closely how the data obtained from TPD experiment can be used to get the information concerning the characterization of active sites, kinetic and thermodynamic parameters, are given in the following sections.

4.5.1 The Application of Temperature-Programmed Desorption in Active Sites Characterisation

The characterization of active sites of solid catalysts includes the determination of active sites nature, the estimation of their density (or population, i.e. the number of active sites per unit of mass or per unit of surface area), their strength and strength distribution. Active sites can be acidic, basic and, in certain cases, amphoteric. All mentioned characteristics are very important for catalysts functionality; therefore, many experimental techniques are invented and adapted for their investigation. Among others, mainly spectroscopic methods (like NMR, IR, XPS, XRF...), temperature-programmed desorption is particularly important because it can be useful in the characterization of all mentioned features.

The strategy which is employed in order to get the above mentioned features of catalysts' active sites is the adsorption of appropriate gas phase probe, under the specific experimental conditions (that are chosen in a way to be similar to those applied in the particular catalytic reaction), followed by subsequent desorption, monitored with appropriate detector. One experiment, in which the characterisation of acid/base properties of solid material is performed, is designed as follows:

- The sample is pre-treated *in situ*, in desired atmosphere (or in vacuum), at appropriate temperature, and during the appropriate time. Usually, the purpose of pretreatment is to remove water (eventually, some impurities) and/or to perform degasification;

- The probe gas is admitted and adsorbed on the solid surface up to some specific surface coverage or up to the saturation;
- Desorption process is performed.

According to the Lowry-Brönsted theory, a Brönsted acid is a proton donor, while a Brönsted base is a proton acceptor. In Lewis' concept, acid acts as electron-pair acceptor, while base is electron donor (such as molecules possessing electron lone pairs). Hence, a Lewis base is in practice equivalent to a Brønsted base. However, the concepts of acidity are markedly different [27].

In the case of solid catalysts, any atomic (ionic) group at the surface that can donate a proton is a Brönsted acid; while any place where one empty electron orbital exists is Lewis acid. For example, in the case of zeolites, Brönsted acid site is a part of microporous aluminosilicate framework—a bridging [≡ Si · · ·(OH) · ·· Al ≡] configuration which is able to donate a proton to an acceptor; while Lewis acid site is either tri-coordinated Al atom or charge-balancing cation Me^{z+} which are able to accept the electron pair. Accordingly to the same theories, any place at the solid surface which can accept proton is a Brönsted base; while any place which can donate electron(s) is a Lewis basic site. For example, in the case of MeO_x (metallic oxides), the oxygen ions (O^{z-}) behave as Brønsted bases (because they are proton acceptors); while cations at thc surface possess Lewis acidity (they are electron acceptors) [27, 28].

The probe molecules that are used to investigate surface acidity should be chosen accordingly to their ability to accept proton from the surface active site, or to donate electron pair to the solid surface. The molecules that fulfil these demands are, for example, ammonia, pyridine, or hydrocarbons. Similarly, the probe molecules that can be used to "trace" the basic site of solid catalysts must be able either to donate a proton or to accept electron(s). Importantly, many species (that even do not contain hydrogen in their formula, which is a demand according to Lowry-Brönsted theory) can function as Lewis acid, accepting electron pair. Hence, the molecules that could be chosen to investigate surface basicity are, for example, dioxides of carbon or sulphur.

However, acidity or basicity of a gas-phase adsorbate is not a sole criterion for its choice as a probe molecule. Firstly, the strength of an acidic or basic probe should be distinguished accordingly to its acid- or base-dissociation constant (K_a or K_b). In addition, very important feature of probe molecule is its radius. If there is a need to locate all active sites in the structure of microporous solid material, the radius of probe molecule has to be smaller then the diameter of pore(s) opening(s). In other words, probe molecules have to be of appropriate size, so the entrance in the micropores of the solid and the access of adsorbate to each active site become possible. For example, ammonia, which is frequently used to reveal the acidic property of solids, is selected as a probe due to its basicity and due to the size of the molecule. Its molecule is smaller than the diameter of the pores in the zeolites' structures, and also in many other solids. The other probe often used for investigation of solids' acidity is pyridine; however, the application of other chemical species is also possible.

As it has been already stated, the value of T_{max} is the indication of the strength of the interaction adsorbate—active site. The stronger the active site is, the stronger the

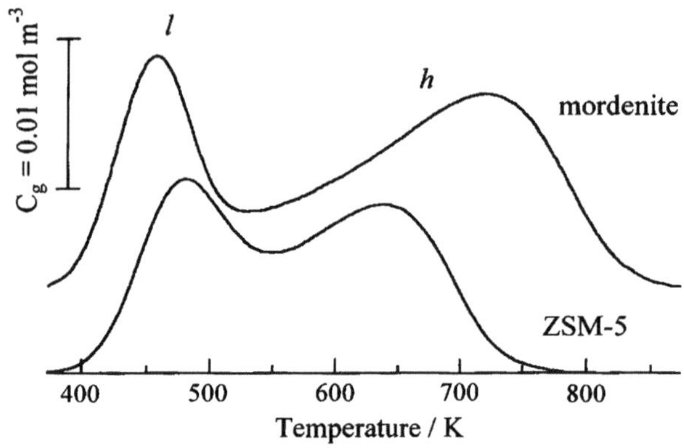

Fig. 4.6 TPD profiles of ammonia obtained from H-mordenite ($SiO_2/Al_2O_3 = 15.0$) and H-ZSM-5 ($SiO_2/Al_2O_3 = 23.8$) [54]

interaction with probe molecule is, which causes more difficult desorption: higher T_{max} indicates that desorption is more difficult. The energy that have to be consumed for desorption is related with the bond energy between surface active site and adsorbate; hence, the position of peak maximum provides information on the strength of this bond. Solid materials possess active sites of different strength, i.e. they express energetic heterogeneity. The origin of active sites strengths heterogeneity is usually the consequence of the solids' structure, or it can be result of different topologies and chemical environments of active sites.

In the case of energetically heterogeneous surface, TPD curves are generally complex-shaped profiles. Figure 4.6 presents two typical cases. Sometimes, desorption profile is composed of well resolved peaks, like upper TPD curve in the Fig. 4.6, where two desorption peaks are denoted as low (*l*) and high (*h*), accordingly to the temperature region of appearance. More often, desorption of probe molecules takes place simultaneously from different sites, what gives more or less pronounced overlapping of the peaks (bottom TPD profile in Fig. 4.6, TPD profiles already presented in the Figs. 4.4 and 4.5).

It is important to notice the influence of adsorption temperature (T_{ads}) on the shape of TPD profile. Desorption takes part consequently to adsorption as a result of thermal motion which kinetic energy is high enough to break the bond between the adsorbate and weak active sites. If adsorption is performed at high temperature, TPD profiles are either single-peak shaped, or overlapping of peaks is less pronounced. By contrast, low temperature of adsorption allows the bonding of adsorbate with all active sites present in the investigated structure. In that case, complex-shaped TPD profile is obviously obtained, in the case of heterogeneous solid surface. The adsorption temperature is apparently very important experimental condition: its influence on the shape of TPD profile (hence, on the conclusions that can be derived from TPD experiment) is illustrated by the example shown in Fig. 4.7. Evidently, the lower the T_{ads} value is, the more complex TPD profile is obtained, and *vice versa*. Of course,

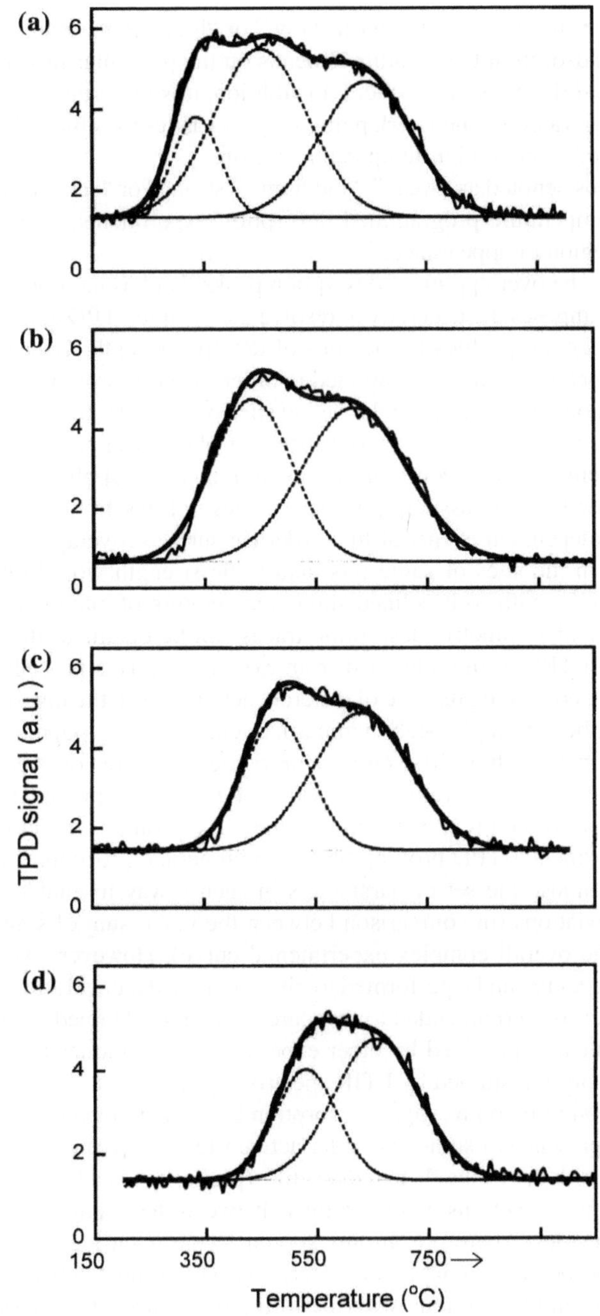

Fig. 4.7 TPD signals obtained after pyridine was adsorbed at Y-type zeolite. The applied T_{ads}: (**a**) 150 °C (**b**) 200 °C (**c**) 250°C (**d**) 300 °C [29]

it has to be kept in mind that the terms "low" or "high temperature" are relative—the choice of adsorption temperature depends on the particular investigated system (on the pair adsorbate—solid surface). In addition, it is important to notice that, the higher the T_{ads} value is, the overlapping of peaks is less visible, while desorption profile is shifted more to high-temperature region.

Usually, sites denoted as "weak", "medium", "strong" or "very strong" are recognized from temperature-programmed desorption experiments, accordingly to the temperature region of appearance.

Apparently, the overlapping of desorption peaks that has its origin in the surface heterogeneity imposes the necessity to resolve the complex TPD curve and to assign particular desorption profiles to the sites of definite strength. Generally, there are two possible methods that can be applied in order to get information concerning the presence and population of particular active sites on the surface of solid material.

Firstly, if a complex desorption profile has been obtained as result of probe adsorption, the mathematical procedure of deconvolution can be applied. Usually, deconvolution is based on the assumption that the desorptions from different sites are parallel and independent events of first order (in surface coverage) [31, 33]. Then, desorption from the sites of same type and same strength would give symmetric desorption profile, with well-defined single temperature of maximum. In that way, certain number of symmetric desorption traces can be obtained, their sum should give the overall TPD profile obtained from experiment. This procedure enables to get information about the presence of different active sites in the investigated system (from the number of single-peak symmetrical curves), their population (from the percent with which each of these curves contribute in the area of overall desorption profile) and their strength (from the positions of T_{max}). The procedure of deconvolution can be performed relatively simply: the T_{max} positions could be recognized from the experimental TPD profile, the whole numerical procedure should be performed as to choose the set of parameters in such a way to enable minimization of standard deviations (in comparison between the linear sum of single desorption profiles and the overall complex experimental curve). However, even though this numerical procedure can be performed to give a unique deconvolution of the experimental curves, it is recommended to compare the results obtained by deconvolution with the information provided by other experimental techniques (for example, the adsorption-desorption studied by FTIR spectroscopy).

Another possibility to investigate desorption from a heterogeneous surface and to recognize the presence of some particular active sites is to perform step-wise filling of the surface with the probe. When the active (probe) gas is admitted to the solids' surface, the first interactions would be those between the strongest active sites and the probe molecules. Therefore, in this approach, the main idea is to admit small quantities of probe gas to enable the adsorption on the most active (the strongest) sites separately, and to continue with the filling of surface, step by step. This task can be fulfilled in two ways:

– The usage of experimental setup which enables the admission of controlled amounts of a gas-phase probe;

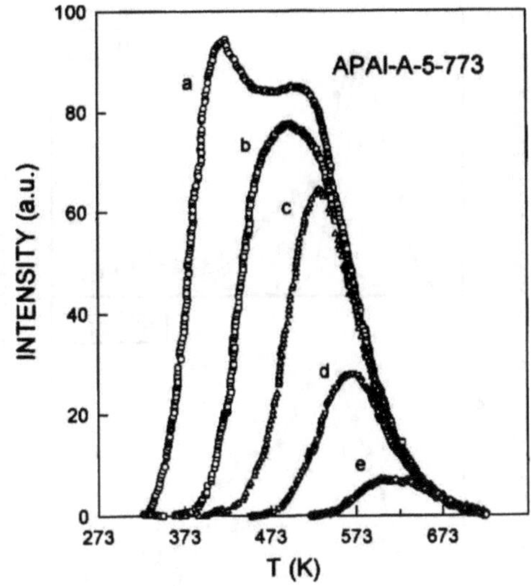

Fig. 4.8 TPD signals, obtained after pyridine was adsorbed on $AlPO_4 - Al_2O_3$. The step-wise filling of the surface is achieved by applying different T_{ads}: (*a*) 50 °C (*b*) 100 °C (*c*) 150 °C (*d*) 200 °*C* (*e*) 300 °C [30]

– The variation of adsorption temperature in order to start the adsorption at the highest possible temperature.

It is evident from the example presented in Fig. 4.7 that when the lower the temperature of adsorption is applied more complex TPD profiles are obtained. By contrast, the higher the adsorption temperature is, less complicated TPD profile is obtained. Hence, if high enough temperature is applied, the interaction with one single type of energetically homogeneous centres should be expected; what should give a symmetric, single-peak desorption curve. Figure 4.8 presents more obvious example.

Evidently, step-wise filling of $AlPO_4 - Al_2O_3$ surface with pyridine has enabled more information about the population and strength distribution of active sites of this solid catalyst. For example, the adsorption of pyridine performed at high temperature (300 °C) enabled to reveal the existence of some very strong active sites, which population is low.

This presentation of active sites characterisation by TPD has been started with the statement that the applied strategy in investigation of active sites' characteristics is the adsorption of appropriate gas phase probe, under the specific experimental conditions. Evidently, the first criterion that has to be applied in order to choose the gas-phase probe is its acidity/basicity. Through this text, the importance of gas-phase nature, the size of its molecules and the temperature of adsorption, has been considered. At this place, the importance of the nature of gas-phase probe will be underlined. Figure 4.9 presents the example in which the investigation of same solids has been performed using two different probes: HY and dealuminated HY zeolites have been investigated using ammonia and pyridine, respectively. It is possible to see from Fig. 4.9a that the adsorption of ammonia revealed the existence of three or four different types of acid centres (in the case of HY and dealuminated HY zeolite, respectively). However, the same solids possess only two different types of sites

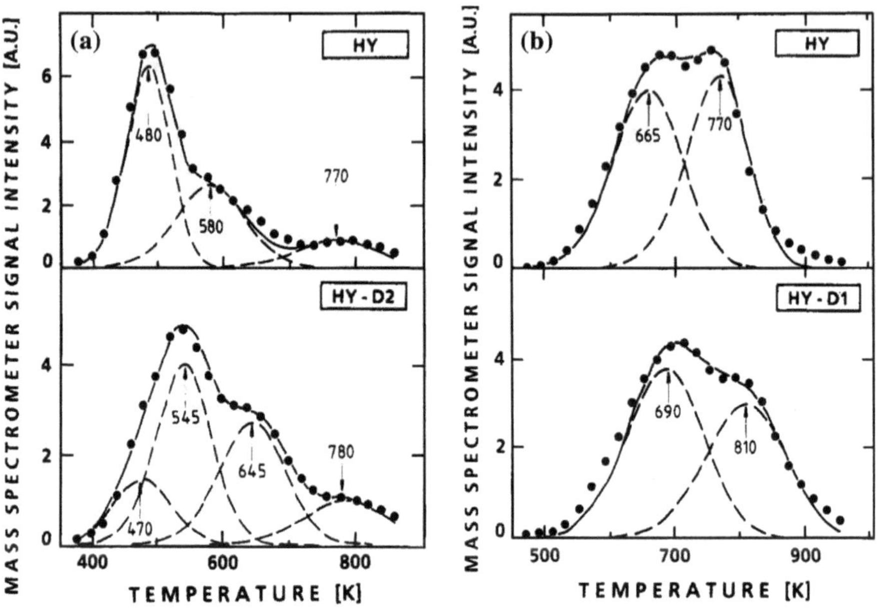

Fig. 4.9 (**a**) TPD profiles of ammonia obtained from HY and dealuminated HY zeolites. (**b**) TPD profiles of pyridine obtained from HY and dealuminated HY zeolites. Overall TPD profile is presented by *spotted line*; *dashed lines*, obtained by deconvolution, present desorption from the acid sites of different strength [31]

for pyridine adsorption, Fig. 4.9b. Apparently, pyridine molecule could not reach all active sites in microporous zeolitic structure. Evidently, not only the basicity of gas-phase molecule is important for its application as a probe, the diameter of its molecule seems to be decisive in the example presented in Fig. 4.9.

It should be emphasized that the choice of a probe molecule should be done by taking into account all relevant parameters, and having in mind the features of solid material at which surface this probe should be adsorbed. In fact, the solid surface and the gas which is chosen as a probe for the characterization of its active sites should be considered as a pair. Very often, the separate adsorption-desorption experiments of more then one gas-phase probe is necessary in order to obtain reliable information concerning all active sites for particular solid material. The adsorption-desorption of more than one probe molecule should complete the picture about the catalysts' active sites, particularly in the case of complex systems, where different types of active sites and energetic heterogeneity could be expected.

It is worth noting that the improvement of equipment available for temperature-programmed desorption, made from the first experiments in the field until nowadays, has enabled the application of many gases as probes. In fact, the possibility to perform adsorption at low temperatures (even sub-ambient) and the improvements in the detection systems allowed to introduce the gases which molecules are poorly polar (such as CO), or even non-polar (such are inert gases and saturated hydrocarbons).

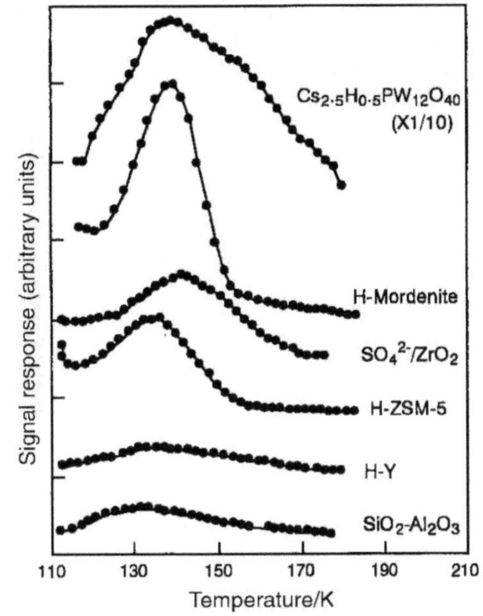

Fig. 4.10 TPD profiles of argon, obtained from different solid acids. $T_{ads} = 113$ K [32]

The interactions of these probes with the active sites are often realised through dispersion forces. Applied as probes, these kinds of molecules enable the recognition and "titration" of very weak active sites. Being on the level of dispersion forces, the interactions of inert gases with the active sites can be considered as specific, in some cases. In Fig. 4.10, the application of argon as probe gas is shown.

The application of same probe for the characterization of different solids revealed important differences in the strength and strength distributions of acid sites. It can be seen from the results presented in Fig. 4.10 that T_{max} positions and the shapes of desorption profiles differ for all investigated systems; while the strongest sites and the most pronounced heterogeneity are found for two specific solids (mordenite-type zeolite and Cs-salt of heteropolyacid). It is important to note that if some less specific probe would be used (such are ammonia or pyridine) all investigated solids should express significant acidity and heterogeneity. In that case, the differences among these catalysts would not be noticed.

The adsorption of non-polar gases offers the possibility to find an appropriate probe which would allow desorption in the desired range of temperature. Figure 4.11 presents one example of hydrocarbons' application as probe gases. Single crystals possess energetically homogeneous surface for adsorption-desorption of these molecules. Figure 4.11 shows that, in the case of adsorption on Rh(111), progressively higher peak desorption temperature is noticed with increasing molecular weight of adsorbate (saturated hydrocarbon).

At the end of this section, it can be concluded that in the domain of active sites characterisation, there is a large body of methods and techniques that are developed

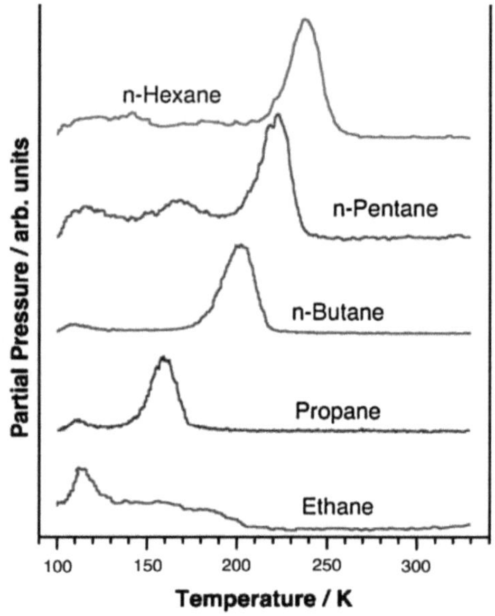

Fig. 4.11 TPD profiles of different saturated hydrocarbons obtained from single-crystal Rh (111) surface [34]

for the determination of nature, population, strength and strength distribution of active sites. Among them, temperature-programmed desorption, a well-established and simple technique, continues to be a very useful and often applied in active sites characterisation. It is important to keep in mind that, although TPD of appropriate probe can be successfully used for distinction between the sites of different strength, it does not enable to distinguish their nature. From the interaction with the probe gas it can be concluded whether some active sites are acidic or basic; but, it can not be concluded if they belong to Brönsted or Lewis type. In order to get precise data on the active sites' nature, the application of other techniques is needed.

Despite the evident richness of data derived from TPD experiment, there are several significant limitations of this technique. A short summary of data that can be obtained from TPD experiments and the limitations of technique are given in Table 4.2.

4.5.2 The Application of TPD in the Determination of Kinetic and Thermodynamic Parameters of Desorption Processes

Temperature-programmed desorption is by far the most often used technique for determination of kinetic parameters on both model and real systems. From these experiments, kinetic and thermodynamic information can be extracted under the conditions of variable temperature. In the following section, the procedures of evaluation of these important parameters will be presented.

Table 4.2 Experimental data that can be obtained from TPD experiments; data that can be derived and the limitations of technique

The obtained data	Evaluated data	Comments/limitations
Surface under the TPD profile, which is proportional to the amount of desorbed (i.e. adsorbed) gas	Surface coverage (θ)	The possibility to determine the surface coverage is one of the major advantages of TPD. *However, for that purpose, the quantification of detector signal is necessary*
The values presented on *y* axis are proportional to the rate of desorption (r_{des}). The position of the peak maximum (T_{max}), which is related to the activation energy for desorption. T_{max} is a temperature where r_{des} is maximal	The activation energy for desorption, hence, the enthalpy of adsorption. Pre-exponential factor for desorption	The strength and population of sites active for adsorption (surface reactions) can be evaluated. Often, complex desorption profiles are obtained; additional procedures of deconvolution or adsorption under different experimental conditions is necessary. *It is not possible to distinguish the type of active sites*

Most often, the estimation of kinetic parameters is based on the assumption of independent desorption that takes part from different active sites as the first order event. In the interpretation of data obtained from TPD experiments, it is also assumed that desorption is the sole surface event. However, in reality the readsorption and diffusion of probe molecules take part, as events consecutive to desorption. These effects are particularly prominent in the case of microporous solids, and in the experiments performed in the flow systems. In those cases, the results obtained in TPD experiment can be often misinterpreted. The readsorption and diffusion can be avoided by adjusting some important experimental parameters. First of all, high heating rate would not favour these processes. However, the choice of very high heating rate is not recommended, particularly in modern implementations of TPD; where the investigation of solid materials used in real catalytic systems is the most common application.

If readsorption is important part of surface event, its influence on temperature-programmed desorption is that TPD profile broadens towards higher desorption temperatures; the same stands for diffusion limitations. Therefore, the task to neglect or minimize the effects of these processes is imposed; and it is relatively easy if TPD experiments are performed in UHV setups. The pumping speed should be sufficiently high to prevent readsorption of the desorbed species back onto the surface. However, in the case of flow systems there are many experimental conditions that have to be adjusted; even though, the interpretation of data obtained from TPD experiment is not simple—in the estimation of kinetic parameters, those experimental conditions have to be taken into consideration. Therefore, the derivation of kinetic and thermodynamic parameters from the results obtained in UHV and in the flow system will be discussed separately.

4.5.2.1 The Interpretation of Data Obtained in Ultra-High Vacuum Systems

If the pumping speed is high enough, readsorption may be ignored and the rate of desorption, defined as the change in adsorbate coverage per unit of time, is given by Eq. (4.7). In a TPD experiment temperature (T) is usually increased linearly with time from some initial temperature T_0, with the heating rate β:

$$T = T_0 + \beta \cdot t \quad , \quad dT = \beta \cdot dt \tag{4.13}$$

where all symbols have the same meaning as previously stated (see Sect. 4.2.2).

The intensity of the desorption signal, $I(T)$, which is proportional to the rate at which the surface concentration of adsorbed species is decreasing, *i.e.* to the rate of desorption, can be expressed by combining Eqs. (4.7) and (4.13):

$$I(T) \propto -\frac{d\theta}{dT} = \frac{\nu(\theta)\,\theta^x}{\beta} \exp\left(\frac{-\Delta E^a_{des}}{RT}\right) \tag{4.14}$$

Molecular adsorption and desorption are often the first order events ($x = 1$). The maximum desorption signal will occur when the first derivative of signal intensity with temperature equals 0 ($dI/dT = 0$):

$$\frac{d}{dT}\left[\frac{\nu(\theta)\,\theta}{\beta} \exp\left(\frac{-\Delta E^a_{des}}{RT}\right)\right] = 0 \tag{4.15}$$

Since surface coverage changes with temperature, *i.e.* $\theta = f(T)$, this derivative is:

$$\frac{\nu(\theta)\,\theta}{\beta} \cdot \frac{E^a_{des}}{RT^2} \exp\left(\frac{-E^a_{des}}{RT}\right) + \frac{\nu(\theta)}{\beta} \exp\left(\frac{-E^a_{des}}{RT}\right) \cdot \frac{d\theta}{dT} = 0 \tag{4.16}$$

The derivative of surface coverage with temperature ($d\theta/dT$) can be substituted from Eq. (4.14). In that way, Eq. (4.16) is transformed to:

$$\frac{\nu(\theta)\,\theta}{\beta}\left[\frac{E^a_{des}}{RT^2} - \frac{\nu(\theta)}{\beta} \exp\left(\frac{-E^a_{des}}{RT}\right)\right] \exp\left(\frac{-\Delta E^a_{des}}{RT}\right) = 0 \tag{4.17}$$

The solution of Eq. (4.17) can be obtained by setting the expression in square brackets to be equal to zero; from where the relation between the temperature at which the desorption maximum (T_{max}) appears and E^a_{des} is obtained as:

$$\frac{E^a_{des}}{RT^2_{\max}} = \frac{\nu(\theta)}{\beta} \exp\left(\frac{-E^a_{des}}{RT_{\max}}\right) \text{ and: } E^a_{des} = RT_M \ln\left(\frac{RT^2_{\max}\,\nu}{E^a_{des}\beta}\right) \tag{4.18}$$

These equations give the relations between the temperature T_{max} at which the desorption maximum appears and the activation energy for desorption. Hence, a simple approach to obtain the value E^a_{des} should be to analyze the TPD curve in order to get easily accessible parameter such as the temperature T_{max}. Unfortunately, the differential equation in (4.7) and (4.14) can not be solved analytically, so the value E^a_{des} can not be obtained simply by substituting T_{max} value in Eq. (4.18). Therefore, the derivation of kinetic parameters can be rather complicated task, in particular because the kinetic parameters usually depend on surface coverage. However, we can note that several facts can be stated from each temperature-programmed experiment:

- as the activation energy for desorption increases peak temperature T_{max} increases;
- the peak temperature is not dependent upon, and consequently, does not change with the initial coverage, $\theta_{t=0}$;
- after the desorption maximum, the shape of the desorption peak tend to be asymmetric, with the signal which decreases rapidly.

Consequently, the values of T_{max} are evident from the experimental result. The procedure that can be applied to derive kinetic parameters is to solve Eq. (4.18) iteratively, applying a suitable choice for pre-exponential factor $\nu(\theta)$ (for chemisorption this value is typically $10^{13}\ \mathrm{s}^{-1}$). The procedure is to read T_{max} from the measurement, to insert an estimated value for E^a_{des} in the right-hand side of Eq. (4.18) and to calculate the resulting E^a_{des} value. The obtained value has to be fed back into Eq. (4.18) to yield an improved value. The iterations should be done until the difference between two subsequent iterations becomes negligible [5].

In the case of second-order desorption, a similar, although more complicated expression exists for second-order desorption kinetics. In this case, the maximum desorption signal will occur when the second derivative of surface coverage is equal to zero:

$$\frac{d^2\theta}{dt^2} = \frac{E^a_{des}\beta}{RT^2_{\max}}\frac{d\theta}{dt} - \nu \exp\left(\frac{-E^a_{des}}{RT_{\max}}\right) 2\frac{d\theta}{dt}\theta = 0 \tag{4.19}$$

From Eq. (4.19) the relation analogue to (4.18) can be derived:

$$E^a_{des} = RT_{\max} \ln\left(\frac{RT^2_{\max}\nu}{E^a_{des}\beta 2\theta}\right) \tag{4.20}$$

Again, iterations are necessary in order to estimate the desorption energy. The insight in the Eq. (4.20) reveals evidence that θ has to be known (or estimated) at the point where T_{max} is reached.

Apart from this approach which implies the evidence of T_{max}, there is another which includes the value of peak width in the analysis. Also, many authors rely on the application of other, even more simplified methods that enable the calculation of kinetic parameters. Particularly popular among surface scientists are the Redhead's and Kissinger's methods.

From all previously stated, it can be inferred that the starting point for extraction of kinetic parameters from thermodesorpion profiles is desorption rate equation proposed by Polanyi and Wigner (Eqs. (4.7) and (4.14)) [6]. However, it has to be kept in mind that the term θ^n is just one particular case of one general function $f(\alpha)$, where α denotes the reacted (desorbed) fraction (the degree of surface event) and $f(\alpha)$ is the reaction kinetic model. Therefore, generalized form of Eq. (4.14) can be written as:

$$\frac{d\theta}{dT} = f(\alpha)\frac{\nu(\theta)}{\beta}\exp\left(\frac{E_{des}^{a}}{RT}\right) \tag{4.21}$$

The methods that are derived for the calculation of kinetic parameters from TPD profiles can bee divided in two big groups, shortly presented by following text.

(i) **Integral methods** are based on the temperature of desorption rate maximum (T_{max}) and/or peak half-widths. These methods assume that pre-exponential factor, reaction order and activation energy are coverage independent values. The most known is Redhead method [1], where Eq. (4.7) is solved in order to find the temperature at which desorption rate expresses its maximum. For the first-order desorption ($x = 1$), the relation between the temperature of peak maximum (T_{max}), activation energy, heating rate and pre-exponential factor is:

$$\frac{E}{RT_{\max}^{2}} = \frac{\nu_1}{\beta}\exp\left(-\frac{E_{des}^{a}}{RT_{\max}}\right) \tag{4.22}$$

where ν_1 is pre-exponential factor for the first-order desorption. The relation between activation energy and T_{max} is almost linear; therefore for ν/β values which are between 10^8 and 10^{13} °, Eq. (4.22) can be written as:

$$\frac{E}{RT_{\max}^{2}} = \ln\frac{\nu_1 T_{\max}}{\beta} - 3.46 \tag{4.23}$$

The activation energy can be determined by varying heating rate β and plotting $ln(T_{max})$ values against $ln\beta$, without assuming the value of rate constant.
For the second-order desorption ($x = 2$), the relation analogue to (4.23) is:

$$\frac{E_{des}^{a}}{RT_{\max}^{2}} = \frac{2\theta_{\max}\nu_2}{\beta}\exp\left(-\frac{E_{des}^{a}}{RT_{\max}}\right) \tag{4.24}$$

where ν_2 is pre-exponential factor for second-order desorption, θ_{max} is the adsorbate coverage at $T = T_{max}$, and it is assumed that $\theta_{max} = \theta_0/2$, with θ_0 being the initial surface coverage.
Equation is approximately correct for the first-order desorption and for values of ν/β between 10^8 and 10^{13} °. It is very often applied to determine E_{des} from a single TPD spectrum.

(ii) **Differential methods** are based on the assumption that at the temperature of maximum, the second derivative of desorption rate (Eqs. (4.7), (4.14) or (4.21)) is equal to zero. The most known is Kissinger method [40]. If reaction rate is expressed by (4.21), the second derivative is:

$$\frac{d^2\alpha}{dt^2} = \left(\frac{E^a_{des}\beta}{RT^2_{\max}} + \nu f'(\alpha_{\max}) \exp\left(-\frac{E^a_{des}}{RT_{\max}}\right)\right)\left(\frac{d\alpha}{dt}\right)_{\max} = 0 \qquad (4.25)$$

where α_{max} and $(d\alpha/dt)_{max}$ are reacted fraction and reaction rate at the maximum; while heating rate β should be constant. From (4.25) it follows:

$$\frac{E\beta}{RT^2_{\max}} = -\nu f'(\alpha_{\max}) \exp\left(-\frac{E^a_{des}}{RT_{\max}}\right) \qquad (4.26)$$

Equation (4.26) can be rearranged after taking logarithms:

$$\ln\left(\frac{\beta}{T^2_{\max}}\right) = \left(-\frac{\nu R}{E^a_{des}} f'(\alpha_{\max})\right) - \frac{E^a_{des}}{RT_{\max}} \qquad (4.27)$$

Evidently, for the first order reaction, $f'(\alpha) = -1$, and (4.27) becomes:

$$\ln\left(\frac{\beta}{T^2_{\max}}\right) = \ln\frac{\nu R}{E^a_{des}} - \frac{E^a_{des}}{RT_{\max}} \qquad (4.28)$$

The procedure of extracting the activation energy for desorption is to analyse a set of TPD profiles measured with different constant heating rates β, and to plot graphs of left hand side of (4.28) versus $1/T_{max}$, what should lead to a straight line whose slope gives the activation energy, independently of the value of reacted fraction, α_{max}, at this point.

Evidently, for application of either integral or differential methods, the values of T_{max} have to be detectable. In case of poorly resolved TP profiles, their application would not provide reliable kinetic parameters. In those cases, either deconvolution of complex desorption profiles should be performed or desorption would be done under different experimental conditions, so the resolving of simple desorption profiles becomes possible.

4.5.2.2 The Interpretation of Data Obtained in the Flow Systems

Previously presented procedures for evaluation of kinetic parameters would give reliable values only if desorption is a lone surface process that takes part as a result of temperature increase. In the case of UHV systems, where samples are usually spread in a thin layer, diffusion takes part in a very limited extent; while readsorption can be avoided using sufficiently high pumping speed. However, if experiment is organized in a flow of a gas, in one usual physical situation, desorption and readsorption are occurring simultaneously with diffusion. In these systems, the construction of sample holders (Fig. 4.2b) does not allow neglecting mass transfer and readsorption limitations. These effects are particularly significant in the case of porous samples.

Therefore, the interpretation of results obtained in the systems designed to allow temperature-programmed experiments in the flow of gas requires consideration of all parameters that may induce mass transfer and readsorption limitations. The parameters that have to be considered are related to the gas (carrier or adsorbate), the geometry of furnace and sample holder, and the features of the sample.

The nature of both carrier and probe gas should be important, as well as their purities and flow rate. High gas flow can provoke desorption of weakly bound species. High amount of desorbed species that arrive in carrier gas can change its purity, so the sensitivity of experiment can be reduced. The consequences of low flow are: diffusion and readsorption effects become more probable, the time between desorption and detection is longer. In addition, appropriate gas flow has to be chosen to avoid concentration gradients within the catalyst particles and along the length of the bed. Hence, a compromise between low and high flow must be found.

The characteristics of furnace and sample holder that may influence the desorption profile are bed length, diameter and porosity, while the characteristics of the sample that could be important are its weight, particle size radius, sample density, particle porosity and number of active sites. In order to avoid temperature gradients, the reactor can not be of big size; hence, mass of the sample is limited by its size and geometry. Diameters of sample particles are another important factor - small particles decrease the possibility of intra-granular diffusion.

Some additional parameters such are temperature range, heating rate, the temperature detection and monitoring, and distance between sample holder and detector may influence the shape of temperature-programmed profile. Distance between reactor and detector has to be the smallest possible, so the answer of detector is instantaneous. Thermocouple has to be precise enough to enable the time of furnace response appropriate, so the temperature rise is absolutely linear. In addition, it has to be kept in mind that temperature of desorbed gas (which has to be analysed) can change during the experiment, what can cause a so-called "apparent" concentration. This effect can be minimized if high flow of gas, low mass of the sample and the equipment situated in the constant temperature area, are employed.

Evidently, in order to calculate reliable kinetic parameters, the temperature of peak maximum T_{max} has to be exactly the temperature at which the rate of desorption is maximal. However, additional surface events (diffusion and readsorption) can influence the TPD profile. The coupling of readsorption and mass transfer effects with desorption can shift the peak of TPD curve significantly.

Therefore, it is necessary to select suitable operating conditions that enable to avoid effects that could have influence on temperature-programmed profile. The ways how to find experimental conditions required to obtain reliable activation energies have been discussed in the literature. The recommendations that help to find appropriate sets of experimental parameters for experiments of temperature-programmed desorption [41–43] or other tempetrature-programmed techniques can be found [14, 44]. Once limitations that arrive from diffusion and readsorption are minimized, simplified procedures can be applied to evaluate kinetic parameters.

In the case of desorption which takes part in the flow of gas, material balance can be obtained from the assumption of equilibrium which is reached in a time t, between the adsorbed and the gas-phase molecules. In the absence of diffusion, this equilibrium can be expressed by following relations:

$$-C_{sm}\frac{d\theta}{dt} = C_{sm}k_{des}\theta - k_{ads}C_g(1-\theta) \tag{4.29}$$

where C_{sm} is concentration of adsorbed molecules (in mol kg^{-1}, for $\theta = 1$) and C_g is concentration of adsorbate molecules in the gas phase (in mol dm^{-3}); while k_{des} and k_{ads} are rate constants of desorption and adsorption. If temperature increase is linear, with constant heating rate (β), previous relation is transformed:

$$-C_{sm}\beta\frac{d\theta}{dT} = C_{sm}k_{des}\theta - k_{ads}C_g(1-\theta) \tag{4.30}$$

If the gas flow (F, dm^{3}s^{-1}) is constant in time and the mass of adsorbent is known (W, kg), the same equilibrium can be expressed as:

$$FC_g = C_{sm}Wk_{des}\theta - Wk_{ads}C_g(1-\theta) \tag{4.31}$$

From Eq. (4.31), the concentration of adsorbate molecules in the gas phase can be obtained from:

$$C_g = \frac{C_{sm}Wk_{des}\theta}{F + Wk_{ads}(1-\theta)} \tag{4.32}$$

The same value can be related with the rate of desorption ($d\theta/dT$) through the equation obtained from (4.30) and (4.31):

$$C_g = -\frac{C_{sm}W\beta}{F}\frac{d\theta}{dT} \tag{4.33}$$

Readsorption is negligible if the gas flow is high enough ($F >> Wk_{\text{ads}}(1-\theta)$). In that case the concentration of adsorbate molecules in the gas phase can be obtained from:

$$C_g = \frac{C_{sm}Wk_{des}\theta}{F} \tag{4.34}$$

At the temperature of peak maximum (T_{max}), the concentration of adsorbate molecules in gas phase reaches its maximum, so $dC_g/dT = 0$. Since, the rate of desorption is expressed by Eq. (4.7), for the first order desorption which takes part without readsorption and diffusion limitations, the value of rate constant reached at T_{max}, $(k_{des})_m$, is related with T_{max} value as:

$$(k_d)_m = \beta\frac{E^a_{des}}{RT^2_{\max}} \tag{4.35}$$

The relation between T_{max} and activation energy for desorption is given by equation identical to (4.27):

$$\ln\left(\frac{T_{\max}^2}{\beta}\right) = \ln\frac{E_{des}^a}{\nu R} + \frac{E_{des}^a}{RT_{\max}} \tag{4.36}$$

It can be seen that for constant heating rate β, the value T_{max} is independent on surface coverage θ. The activation energy for desorption can be obtained from the slope of a plot: $\ln\left(\frac{T_m^2}{\beta}\right) = f(1/T_{\max})$.

If a flow of carrier gas is not high enough ($F << Wk_{ads}(1-\theta)$) Eq. (4.32) is transformed to:

$$C_g = \frac{C_{sm}k_{des}\theta}{k_{ads}(1-\theta)} = C_{sm}K\frac{\theta}{1-\theta} \tag{4.37}$$

where $K = \exp\left(\frac{\Delta S_{des}}{R}\right)\cdot\left(-\frac{\Delta H_{des}}{RT}\right)$. Again, having in mind that at the temperature of peak maximum dC_g/dT is equal to zero, the relation between T_{max} and the heat of adsorption can be obtained:

$$\ln\left(\frac{T_{\max}^2}{\beta}\right) = \ln\left(\frac{(1-\theta_{\max})^2 W\Delta H}{RF\exp\left(\frac{\Delta S_{des}}{R}\right)}\right) + \frac{\Delta H}{RT_{\max}} \tag{4.38}$$

Evidently, the heat of adsorption can be obtained from the slope of the same plot as previously. Similarly, it can be shown that for desorption which takes part as a second-order surface event, the activation energy for desorption (in the case of negligible readsorption) or the heat of adsorption (in the case of significant readsorption) could be obtained from the slope of the same plot. Hopefully, even in the cases when desorption is significantly affected by side effects such are readsorption or diffusion, kinetic parameters can be obtained using relatively simple procedures.

After everything that has been said about temperature-programmed methods, few examples will be considered further in the text. The two experimental techniques most commonly used for the study of acid/base properties of porous solid materials are temperature-programmed desorption (TPD) and adsorption calorimetry. Application of these techniques for characterisation of several different classes of materials will be presented as well as comparison of data obtained by both techniques.

4.6 The Examples of TPD Application; the Comparison with Data Obtained by Adsorption Calorimetry

As discussed in details previously, in TPD experiments, temperature increases linearly and the concentration of desorbed gas is recorded as a function of temperature, whereas calorimetry involves the adsorption of gases onto the sample's surface while it is kept at a constant temperature and a heat-flow detector emits a signal proportional to the amount of heat transferred per unit time. The peak maxima temperatures in the TPD spectra are influenced by the active site strength, the number of active sites, the structure of investigated material and the heating rate [45, 46]. In particular, adsorption microcalorimetry gives access to the number, strength, and strength distribution of the acid sites in a single experiment [21]. This information is of outmost importance for design of catalysts with high activity and selectivity.

Every micro or mesoporous material can be investigated by these two techniques. Among many, few examples will be presented for the most often investigated catalysts; such are zeolite, oxides and metals.

4.6.1 Zeolites

Zeolites are known to be important catalysts for a number of industrially important reactions. A question of basic interest, which provides opportunity for development of catalyst with suitable and tailored characteristics, is to determine the correlation between number, strength and strength distribution of active sites and the promotion of catalytic activity. Therefore, the investigation of acid sites, both Lewis and Brönsted type, is very important subject. Properties of zeolites as catalysts will depend on many factors: the adsorption or desorption temperature of the probe, pretreatment of the sample, proton exchange level, influence of coking as well as Si/Al ratio and dealumination and influence of exchanged cations [47].

4.6.1.1 Influence of the Si/Al Ratio and Dealumination

The Si/Al ratio plays a significant role, since the aluminum atom is directly related to the acidic site. Dealumination processes can promote porous structure modifications, which may improve some interesting properties of zeolites, like thermal and hydrothermal stability, acidity, catalytic activity, resistance to aging and low coking rate, and matter transfer. However, a severe dealumination may also cause a loss of crystallinity [47].

Different dealumination processes have been proposed, namely steaming and acid treatments, as well as reactions with $SiCl_4$ or SiF_6^{2-} [47]. The removal of aluminum from zeolite crystals leads to products with high framework Si/Al ratios. Some of the aluminum atoms are released from the framework and form non-framework aluminum-containing species. The non-framework aluminum species can be eliminated by treatment with diluted hydrochloric acid. Dealumination generally brings a decrease in the acid site concentration. However, the extent of the indicated decrease varies with the kind of base probe, and a significant change was observed by Mitani et al. [48] in the ratio of acid site concentrations when titrated with pyridine instead of ammonia. An important increase of the initial heat values and of the site strength heterogeneity was observed for samples presenting many extra-framework aluminum species. Samples subjected to a moderate dealumination and nearly total removal of the extra-framework aluminum displayed a homogeneous acid strength [49].

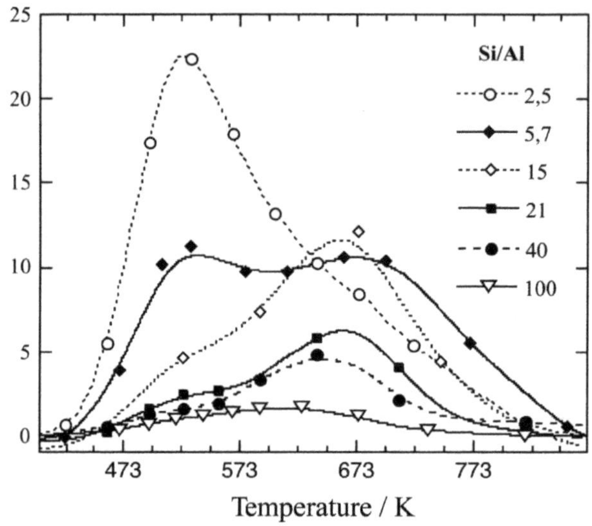

Fig. 4.12 Influence of dealumination on NH_3-TPD profiles of HY zeolites [14]

The effect of dealumination on NH_3-TPD profiles of HY zeolites is shown in Fig. 4.12 [14]. Evidently *l*-peak became smaller upon dealumination, while the *h*-peak increased up to a maximum for Si/Al = 15 and then decreased upon further dealumination.

Figure 4.13 shows the effect of varying the Si/Al ratio of a MFI sample on its TPD profile. As shown at the figure, both the *l*- and *h*-peaks became smaller upon dealumination. A curve-fitting analysis led to the determination of average adsorption heats that were almost constant for all investigated MFI samples, ca. 130 kJ mol^{-1} [35].

The acidity of H-ZSM-5 zeolites synthesized with different Al contents has been characterized by microcalorimetric measurements of the differential heats of adsorption of ammonia [50]. The strength of the strongest acid sites increased with the Al content to a maximum for Si/Al = 17.5 and then decreased notably. The total acidity increased regularly with Al content. The importance of selecting appropriate Si/Al ratios for specific catalytic applications is therefore obvious.

Figure 4.14 represents the differential heats of adsorption on dealuminated H-ZSM-5 zeolite samples at 393 K versus the adsorbed amount of ammonia [36]. It appears that the adsorption heat decreases gradually with the amount of adsorbed ammonia and exhibits various steps, attributed to populations of sites of different strengths. A heterogeneous distribution of sites is clearly evidenced by the shape of the curves and can be attributed to the presence of extra-framework aluminium spices. The samples adsorb ammonia with initial differential heats that vary between 160 and 177 kJ mol^{-1}. Karge et al. [51] have reported that, in first approximation, the sites evolving $Q_{diff} > 150\,kJ\,mol^{-1}$ can be assigned to strong Lewis sites, while the sites with $Q_{diff.} = 150 - 100\,kJ\,mol^{-1}$ typically correspond to strong Brönsted

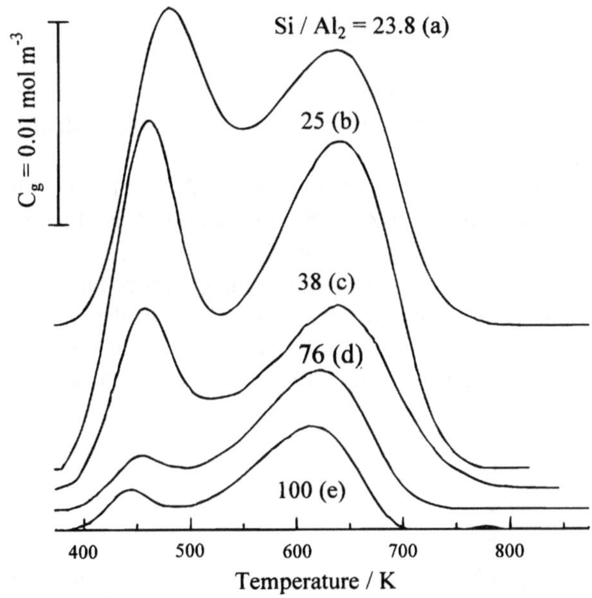

Fig. 4.13 NH3-TPD profiles observed on H-MFI with Si/Al ratios: (*a*) 11.9 (*b*) 12.5 (*c*) 19 (*d*) 38 (*e*) 50 [35]

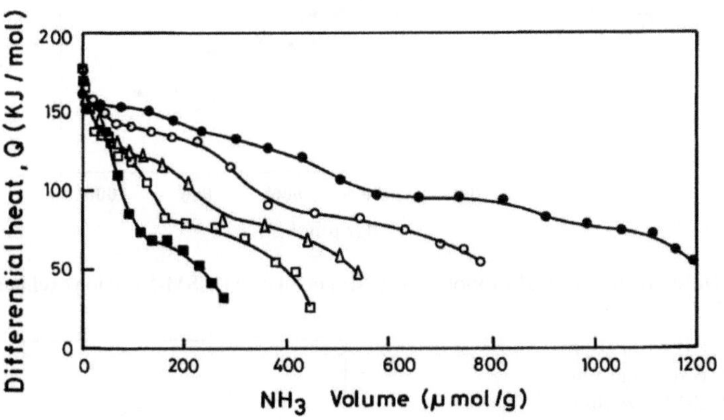

Fig. 4.14 Differential heats of ammonia adsorption at 393 K on H-ZSM-5 zeolites with Si/Al ratios: (•) 14 (○) 25 (Δ) 37.5 (□) 50 (■) 75 [36]

acid sites. This assumption is supported by a comparison of the numbers of acidic sites obtained from infrared and microcalorimetric measurements. The initial heat of adsorption for the zeolite with Si/ Al = 75 is lower than that of the other zeolites except for Si/Al = 14, and the decrease in heat with coverage is steeper. The heat evolved falls abruptly from an initial value of 150 kJ mol^{-1} to 70 kJ mol^{-1} at around 50 % coverage.

4.6.1.2 Influence of Substitution by Other Cations

The nature of the exchanged cation is one of the key points that determine acidity in zeolites. It is very important to use an acidic probe able to distinguish the alkali cations from the basic sites.

Differential heats of ammonia adsorption and NH_3-TPD profiles of HZSM-5 zeolite as well as FeZSM-5, Cu-ZSM-5 and MnZSM-5 zeolites are presented in Figs. 4.15 and 4.16. As can be seen in Fig. 4.15, the overall acidity of investigated samples was not significantly modified by ion-exchange procedure. However, changes in the Q_{diff} versus NH_3 uptake profiles, particularly in their middle parts (140 – 65 kJ mol^{-1}), indicate that the distribution of strength of acid sites was affected by ion exchange. Ion exchange with Cu and Mn resulted in enhanced heterogeneity of

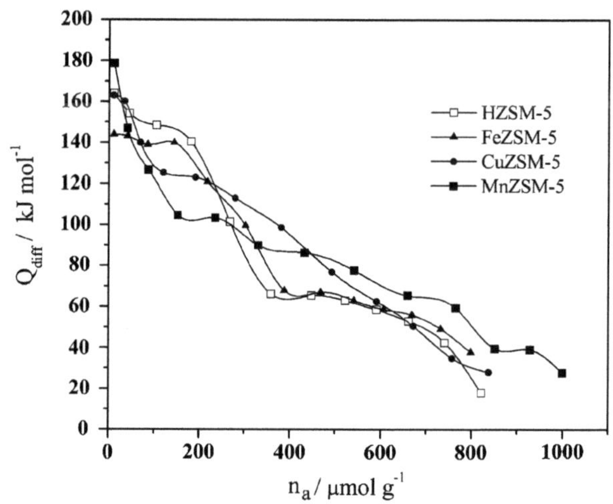

Fig. 4.15 Differential heat of adsorption versus NH_3 uptake on HZSM-5 and ion exchanged ZSM-5 zeolites [37]

Fig. 4.16 NH_3 – TPD profiles for HZSM-5 and ion exchanged ZSM-5 zeolites [37]

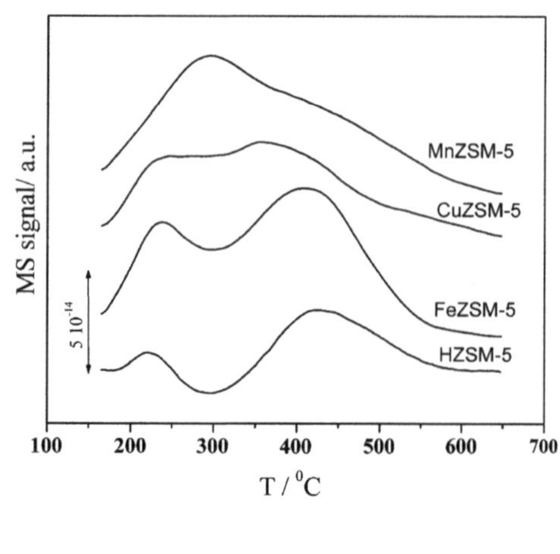

the acid site strength, as confirmed by the NH_3-TPD profiles, which present poorly separated desorption peaks and even a single very broad peak for sample MnZSM-5 (Fig. 4.16) [37].

This example clearly shows that TPD and adsorption microcalorimetry are complementary techniques, and combination of these two techniques provides very good characterisation of active sites number, strength and distribution on particular solid material.

4.6.2 Metal Oxides

Metal oxides, either bulk, doped, supported or mixed, are widely used as catalysts in chemical industry. Catalytic behavior of these materials, in terms of activity and selectivity, is related to their acid/base properties.

Metal oxide surfaces react with gases or solutions; they can be used as active phases or as supports for catalysts. The behavior of metal oxide surfaces is controlled by: (i) coordination—sites of low coordination are in general more reactive than sites of high coordination; (ii) acid/base properties—clean and anhydrous metal oxide surfaces present two different types of active sites, cations and anions (acid/base pairs) which determine reactivity towards gas-phase adsorbates; (iii) the redox mechanism—when the oxide deviates from the stoichiometry due to the presence of defects such as vacancies or adatoms, the oxidation state of surface atoms varies [52].

Molecular and dissociative adsorption can be understood as acid/base processes. Molecules adsorbing without dissociation always bind to one or several metal cations. Ammonia and pyridine are the most commonly used probes for determining the acid site strength of oxides.

In the case of supported metal oxide catalysts, the role of the support is to disperse the active phase and to create new active surface species by host (active phase) – guest (support) interaction. The dispersion of the active phase plays a fundamental role, and very often a maximum of strength of the active sites is observed when the monolayer coverage is reached. The pure oxides, such as Al_2O_3, ZrO_2, TiO_2 (the most frequently used catalyst supports) carry both basic and acidic Lewis sites on their surface; depending on the probe molecules used (CO_2 or NH_3), they can exhibit either acidic or basic character. Excess negative or positive charges can be induced, and therefore acidity (Brönsted or Lewis) or basicity can be generated by mixing oxides. Modifying the surface with a minor anionic, cationic or metallic component enhances or decreases the acidic or basic strength of the sites. For example, the incorporation of chloride, fluoride or sulfate ions increases the acidity of carrier oxides (Al_2O_3, ZrO_2, TiO_2), while alkali cations enhance the basic strength of alumina or silica [53].

Silica-aluminas (amorphous aluminosilicates) are widely used as catalyst supports due to their high acidity and surface area. The behavior of silica-alumina surfaces is similar to that of zeolites concerning the initial differential heats of ammonia and

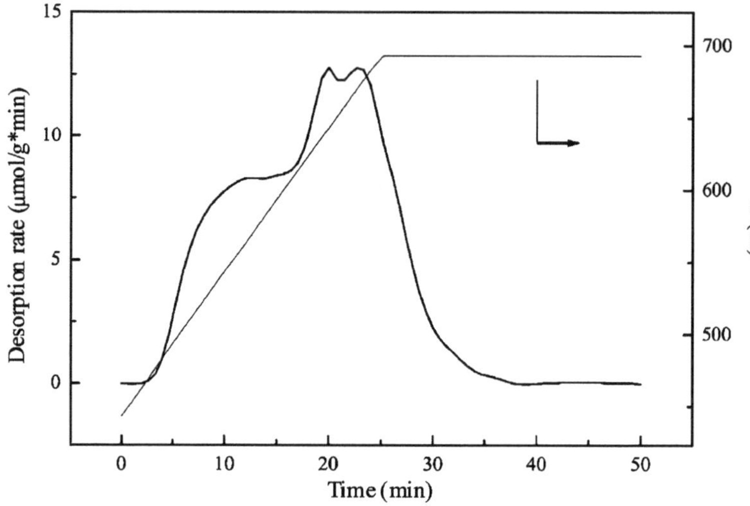

Fig. 4.17 TPD curve for γ-alumina, using pyridine as probe molecule [38]

pyridine, but the total number of acidic sites varies with the preparation method and the Si/Al ratio.

Auroux et al. [38] have studied acid properties of silica, alumina and multilayers of silica on alumina (SA) and alumina on silica (AS), obtained by grafting. The surface acidity of the pure oxides and samples obtained by grafting, SA and AS, of both Lewis and Brönsted type, has been investigated by TPD and microcalorimetry, using pyridine as probe molecule.

Figure 4.17 shows the TPD curve of pure alumina; a semi-quantitative estimation of the number of acid site can be derived from the TPD plots such as this one. Three classes of acid sites have been considered, depending on the desorption temperature: weak, ($T_D < 523$ K), medium ($523 < T_D < 673$ K) and strong, ($T_D > 673$ K) acid sites.

Calorimetric results are presented in Fig. 4.18 where the differential heat of adsorption (Q_{diff}) of pyridine is plotted versus coverage. From the calorimetric data, the number of sites and their distribution, according to the adsorption energies, can be determined. Analyzing calorimetric profile of differential heats *vs.* surface coverage, following sites active for adsorption can be distinguished: weak acid sites, ($90 \leq Q_{\text{diff}} < 120 kJmol^{-1}$), medium acid sites, m ($120 \leq Q_{diff} < 150 kJmol^{-1}$) and strong acid sites, s ($Q_{\text{diff}} \geq 150 kJmol^{-1}$).

From Fig. 4.18 it can be seen that calorimetric curve for pure alumina shows three type of sites with different strength, in accordance with TPD result. Also, it can be seen that the grafted mixed oxides SA and AS have acidic properties different from those of the pure alumina and silica supports used as starting materials, *i.e.* properties of catalysts depend on preparation procedure.

It can be noticed that some differences exist between results obtained from calorimetric measurements and those revealed from TPD results, with regard to the sites

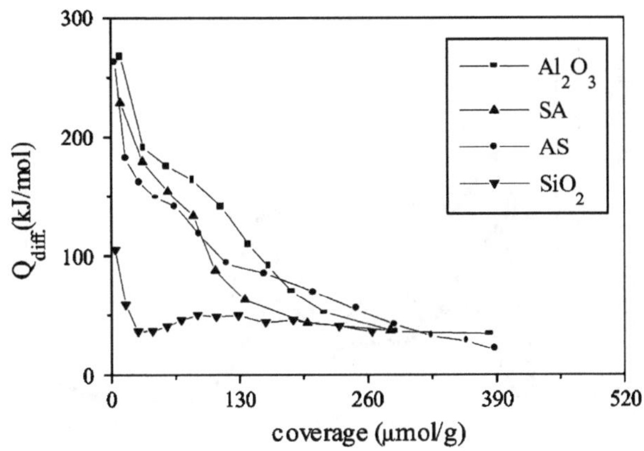

Fig. 4.18 Differential heats of pyridine adsorption as function of the coverage degree [38]

strength distribution: the percent of medium acid sites, derived from TPD results is always remarkably higher than one found using microcalomitery. This is due to the fact that strong acid sites having $Q_{diff} > 180\,\mathrm{kJ\,mol^{-1}}$ might not completely release the organic base at 693 K and therefore are not disclosed by the TPD technique. This implies importance of knowing limitations of techniques used for characterizations as well as operative conditions adopted during experiments [38].

4.6.3 Metals

TPD and microcalorimetric methods provide an effective means for measuring the strengths of adsorbate-surface interactions, not only on clean metal surfaces, but also on metal surfaces that have been exposed to reaction conditions. Many recent studies use H_2, CO, O_2 and hydrocarbons as probe molecules since they are involved in numerous commercial catalytic processes.

Bimetallic catalysts have been the subject of great interest for a long time because of their exceptional properties compared to the monometallic catalysts, yet the reason behind their improved activity is still a question of debate and they are subject of many recent studies. Tanskale et al. [39] have studied the promoting effect of Pt and Pd in bimetallic Ni–Pt and Ni–Pd catalysts supported on alumina nanofibre (Alnf) for the liquid phase reforming of sorbitol to produce hydrogen. Fig. 4.19 shows TPD profiles for CO desorption for several monometallic and bimetallic catalysts dispersed on alumina nanofibre.

Results obtained by temperature-programmed desorption suggested that in the case of bimetallic catalysts there was a reduction in the number of strong CO-adsorption sites. This finding allows conclusion that the alloying effect of these systems leads to the lowering of the CO heat of adsorption. This finding was confirmed by direct measurement of differential heat of CO chemisorption in the microcalorimetry experiment (see Fig. 4.20).

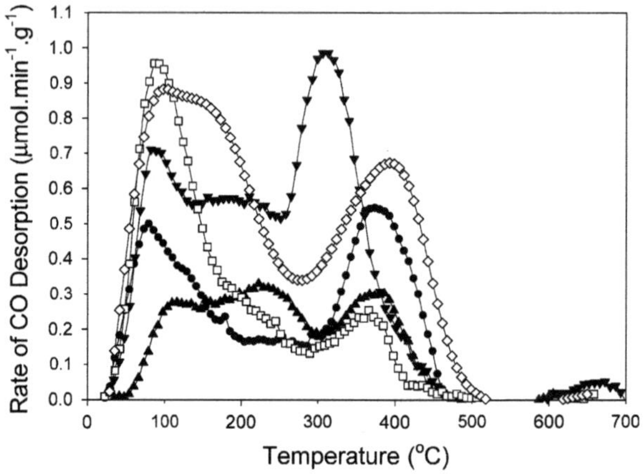

Fig. 4.19 Results from CO-TPD of monometallic and bimetallic catalysts over alumina nanofibre support; rate of CO desorption. Pt/Alnf (▼), Ni/Alnf (•), Pd/Alnf (▲), Ni/Pd/Alnf (□), Ni-Pd/Alnf(◇) [39]

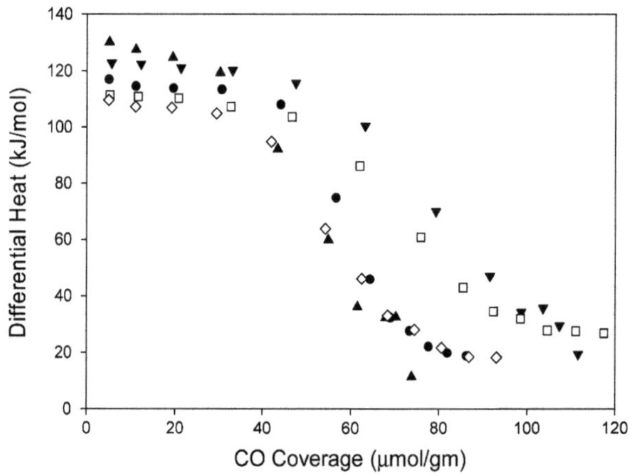

Fig. 4.20 Differential heats of CO adsorption as a function of coverage, adsorption temperature 3 °C. Pt/Alnf (▼), Ni/Alnf (•), Pd/Alnf (▲), Ni/Pd/Alnf (□), Ni-Pd/Alnf (◇) [39]

The differential heat of adsorption for Ni–Pt/Alnf was reduced to 111.28 kJ mol^{-1}, which was 11.45 and 5.51 kJ mol^{-1} lower than Pt/Alnf and Ni/Alnf, respectively. A similar result was obtained for the Ni–Pd bimetallic catalyst. This was interesting because the atomic ratio of Pt:Ni and Pd:Ni was only 1:33 and 1:18, hence even small amount of surface concentration of solute in the alloy catalysts is sufficient to significantly alter the properties of constituent monometallic catalysts.

The differential heat profiles are characterised by a plateau of nearly constant heat of adsorption at low coverage (15–20 μmol g m^{-1}) followed by an abrupt decrease as the surface saturation limit is reached. In the low coverage plateau region the adsorption of CO can be considered to be strongly bonded and at high coverage the differential heat of CO adsorption represents an average heat from the various adsorption sites on the surface of a given catalyst particle. Therefore, one can compare only the initial heat of adsorption when the CO chemisorption may be considered to be equilibrated. The initial heat of CO adsorption on Pd/Alnf was found to be the highest (130.09 kJ mol^{-1}) followed by Pt/Alnf (122.73 kJ mol^{-1}) and Ni/ALnf (116.79 kJ mol^{-1}). The initial heat of CO adsorption on Ni–Pt/Alnf was measured as 111.28 kJ mol^{-1}, which is lower than both Pt/Alnf and Ni/Alnf. Similarly, the initial differential heat of Ni–Pd/Alnf (109.60 kJ mol^{-1}) was lower than both the corresponding monometallic catalysts. This indicates that the addition of noble metals, even in small fractions, to the Ni catalyst has a significant but weaker promoting effect on the adsorption of CO. Finding that the CO differential heat of adsorption is lowered in the bimetallic catalysts is substantial because with reduction of the CO binding strength poisoning of the active metal sites can be avoided which will increase its catalytic activity [39].

4.7 Conclusion

This chapter has presented the fundamentals, the experimental setups and the applications of temperature-programmed desorption (TPD). TPD is widely utilized for characterization of active sites present on the surface of solid materials and determination of kinetic and thermodynamic parameters of desorption processes. It is a powerful technique, even if it does not provide direct information about molecular nature of adsorbed species. Given its relative simplicity and low cost, this technique will continue to find more applications in the future.

Thermal method which is complementary to TPD, adsorption calorimetry, provides tools necessary for measuring the energy of an adsorption system as a function of coverage, allowing precise determination of number of surface active sites as well as their relative populations and strengths. In general, adsorption calorimetry provides a much better description of the surface active site strength distribution than TPD. Combination of these two methods allows detailed characterization of various materials, in particular catalysts, and will surely find more application in different research areas such as hydrogen production and storage or environmental chemistry.

References

1. P.A. Redhead, Thermal desorption of gases. Vacuum **12**, 203–211 (1962). doi:10.1016/0042-207X(62)90978-8
2. G. Ehrlich, Modern Methods in surface kinetics: flash desorption, field emission microscopy, and ultrahigh vacuum techniques. Adv. Catal. **14**, 255–427 (1963). doi:10.1016/S0360-0564(08)60341-7

3. R.J. Cvetanović, Y. Amenomiya, Application of temperature-programmed dessorption technique to catalyst studies. Adv. Catal. **17**, 103–149 (1967). doi:10.1016/S0360-0564(08)60686-0
4. R.J. Cvetanović, Y. Amenomiya, A temperature programmed desorption technique for investigation of practical catalysts. Catal. Rev. **6**, 21–48 (1972). doi:10.1080/01614947208078690
5. I. Chorkendorff, J.W. Niemantsverdriet, *Concepts of Modern Catalysis and Kinetics* (Wiley-VCH Verlag GmbH & Co. KGaA, Weinheim, 2003)
6. F. Rouquerol, J. Rouquerol, K. Sing, *Adsorption by Powders and Porous Solids, Principles Methodology and Applications* (Academic Press, San Diego, 1999)
7. D.M. Ruthven, *Principles of Adsorption and Adsorption Processes* (Wiley, New York, 1984)
8. M.A. Vannice, *Kinetics of Catalytic Reactions* (Springer Science Business Media, Inc., New York, 2005)
9. Lj. Damjanović, A. Auroux, Determination of acid/base properties by temperature-programmed desorption (TPD) and adsorption calorimetry, in *Zeolite Chemistry and Catalysis: An Integrated Approach and Tutorial*, ed. by A.W. Chester, E.G. Derouane (Springer, Heidelberg, 2009)
10. J.M. Kanervo, T.J. Keskitalo, R.I. Slioor, A.O.I. Krause, Temperature-programmed desorption as a tool to extract quantitative kinetic or energetic information for porous catalysts. J. Catal. **238**, 382–393 (2006). doi:10.1016/j.jcat.2005.12.026
11. V. Rakić, V. Dondur, U. Mioč, D. Jovanović, Microcalorimetry in the identification and characterization of the most reactive active sites of heterogeneous catalysts. Top Catal. **19**, 241–247 (2002). doi:10.1023/A:1015328526702
12. Dj. Stojaković, N. Rajić, V. Rakić, N. Zabukovec Logar, V. Kaučić, Structure and thermal behavior of the layered zincophosphate $[NH_3-CH_2-CH(NH_3)-CH_3](ZnPO_4)_2$. Inorg. Chim. Acta. **362**, 1991–1995 (2009). doi:10.1016/j.ica.2008.09.020
13. B. Brunner, Solid state NMR–a powerful tool for the investigation of surface hydroxyl groups in zeolites and their interactions with adsorbed probe molecules. J. Mol. Struct. **355**, 61–85 (1995). doi:10.1016/0022-2860(95)08867-U
14. D. Delahay, Méthodes en température programmée, in *Les matériaux micro et mésoporeux, Charactérisation*. ed. by F. Thibault-Starzyk, Groupe français des zéolithes, (EDP sciences, Les Ulis, France, 2004)
15. S. Bennici, A. Auroux, Thermal analysis and calorimetric methods, in *Metal Oxide Catalysis*, eds. by S.D. Jackson, J.S.J. Hargreaves (Wiley, New York, 2009)
16. G.A. Somorjai, Modern surface science and surface technologies: an introduction. Chem. Rev. **96**, 1223–1236 (1996). doi:10.1021/cr950234e
17. W.E. Farneth, R.J. Gorte, Methods for analyzing zeolite acidity. Chem. Rev. **95**, 615–635 (1995). doi:10.1021/cr00035a007
18. A. Corma, Inorganic solid acids and their use in acid-catalyzed hydrocarbon reactions. Chem. Rev. **95**, 559–614 (1995). doi:10.1021/cr00035a006
19. J. Weitkamp, U. Weis, E. Ernst, New aspects and trends in zeolite catalysis, in *Catalysis by Microporous Materials, Studies in Surface Science and Catalysis*, vol. 94, eds. by H.K. Beyer, H.G. Karge, I. Kiricsi, J.B. Nagy (Elsevier, Amsterdam), p. 363
20. V. Solinas, I. Ferino, Microcalorimetric characterisation of acid-base catalysts. Catal. Today. **41**, 179–189 (1998). doi:10.1016/S0920-5861(98)00048-0
21. A. Auroux, Microcalorimetry methods to study the acidity and reactivity of zeolites, pillared clays and mesoporous materials. Top Catal. **19**, 205–213 (2002). doi:10.1023/A: 1015367708955
22. A. Auroux, Innovation in zeolite materials science, in *Proceedings of the International Sumposium on Studies in Surface Science and Catalysis*, Nieuwpoort, vol. 37 (Elsevier, Amsterdam, 1988), 13–17 September 1987, eds. by P.J. Grobet, W.J. Mortier, E.F. Vansant, G.G Schulz-Eklo, p. 385
23. A. Zecchina, S. Bordiga, G. Spoto, L. Marchese, G. Pterini, G. Leofanti, M. Padovan, Silicalite characterization. 2. IR spectroscopy of the interaction of carbon monoxide with internal and external hydroxyl groups. J. Phys. Chem. **96**, 4991–4997 (1992). doi:10.1021/j100191a048

24. M.A. Makarova, K.M. Al-Gefaili, J. Dwyer, Brönsted acid strength in US-Y: FTIR study of CO adsorption. J. Chem. Soc. Faraday Trans. **90**, 383–386 (1994). doi:10.1039/FT9949000383
25. Y. Kuroda, T. Mori, Y. Yoshikawa, S. Kittaba, R. Kumashiro, M. Nagao, What are the important factors determining the state of copper ion on various supports? Analysis using spectroscopic methods and adsorption calorimetry. Phys. Chem. Chem. Phys. **1**, 3807–3816 (1999). doi:10.1039/A904754I
26. E. Garrone, B. Fubini, B. Bonelli, B. Onida, C.O. Arean, Thermodynamics of CO adsorption on the zeolite Na-ZSM-5 A combined microcalorimetric and FTIR spectroscopic study. Phys. Chem. Chem. Phys. **1**, 513–518 (1999). doi:10.1039/A806973E
27. J. Sauer, Acidic sites in heterogeneous catalysis: structure, properties and activity. J. Mol. Catal. **54**, 312–323 (1989). doi:10.1016/0304-5102(89)80149-X
28. P.A. Jacobs, *Carboniogenic Activity of Zeolites* (Elsevier Scientific Publishing Company, Amsterdam, 1977)
29. E. Selli, L. Forni, Comparison between the surface acidity of solid catalysts determined by TPD and FTIR analysis of pre-adsorbed pyridine. Microporous Mesoporous Mater. **31**, 129–140 (1999). doi:10.1016/S1387-1811(99)00063-3
30. J.M. Campelo, A. Garcia, D. Luna, J.M. Marinas, A.A. Romero, Characterization of acidity in $AlPO_4$-Al_2O_3 (5–15 wt% Al_2O_3) catalysts using pyridine temperature-programmed desorption. Thermochim. Acta. **265**, 103–110 (1995). doi:10.1016/0040-6031(95)02379-G
31. H.G. Karge, V. Dondur, J. Weitkamp, Investigation of the distribution of acidity strength in zeolites by temperature-programmed desorption of probe molecules. 2. Dealuminated Y-type zeolites. J. Phys. Chem. **95**, 283–288 (1991). doi:10.1021/j100154a053
32. H. Matsuhashi, K. Arata, Temperature-programmed desorption of argon for evaluation of surface acidity of solid superacids. Chem. Commun. **387–388**, (2000). doi:10.1039/A909844E
33. H.G. Karge, V. Dondur, Investigation of the distribution of acidity in zeolites by temperature-programmed desorption of probe molecules. I. Dealuminate mordenites. J. Phys. Chem. **94**, 765–772 (1990). doi:10.1021/j100365a047
34. J. Wilson, H. Guo, R. Morales, E. Podgornov, I. Lee, F. Zaera, Kinetic measurements of hydrocarbon conversion reactions on model metal surfaces. Phys. Chem. Chem. Phys. **9**, 3830–3852 (2007). doi:10.1039/B702652H
35. N. Katada, H. Igi, J.-H. Kim, M. Niwa, Determination of the acidic properties of zeolite by theoretical analysis of temperature-programmed desorption of ammonia based on adsorption equilibrium. J. Phys. Chem. B. **101**, 5969–5977 (1997). doi:10.1021/jp9639152
36. S. Narayanan, A. Sultana, Q.T. Le, A. Auroux, A comparative and multitechnical approach to the acid character of templated and non-templated ZSM-5 zeolites. Appl. Catal A-Gen. **168**, 373–384 (1998). doi:10.1016/S0926-860X(97)00368-2
37. V. Rac, V. Rakić, S. Gajinov, V. Dondur, A. Auroux, Room-temperature interaction of n-hexane with ZSM-5 zeolites. Microcalorimetric and temperature-programmed desorption studies. J. Therm. Anal. Cal. **84**, 239–245 (2006). doi:10.1007/s10973-005-7164-z
38. A. Auroux, R. Monaci, E. Rombi, V. Solinas, A. Sorrentino, E. Santacesaria, Acid sites investigation of simple and mixed oxides by TPD and microcalorimetric techniques. Thermochim. Acta. **379**, 227–231 (2001). doi:10.1016/S0040-6031(01)00620-7
39. A. Tanksale, J.N. Beltramini, J.A. Dumesic, G.Q. Lu, Effect of Pt and Pd promoter on Ni supported catalysts–A TPR/TPO/TPD and microcalorimetry study. J. Catal. **258**, 366–377 (2008). doi:10.1016/j.jcat.2008.06.024
40. K. Kissinger, Reaction kinetics in differential thermal analysis. Anal. Chem. **29**, 1702–1706 (1957). doi:10.1021/ac60131a045
41. R.J. Gorte, Design parameters for temperature programmed desorption from porous catalysts. J. Catal. **75**, 164–174 (1982). doi:10.1016/0021-9517(82)90131-2
42. R.A. Demmin, R.J. Gorte, Design parameters for temperature-programmed desorption from a packed bed. J. Catal. **90**, 32–39 (1984). doi:10.1016/0021-9517(84)90081-2
43. R.J. Gorte, Temperature-programmed desorption for the characterization of oxide catalysts. Catal. Today **28**, 405–414 (1996). doi:10.1016/S0920-5861(96)00249-0

44. D.A.M. Monti, A. Baiker, Temperature-programmed reduction. Parametric sensitivity and estimation of kinetic parameters. J. Catal. **83**, 323–335 (1983). doi:10.1016/0021-9517(83)90058-1
45. J. Shen, A. Auroux, The determination of acidity in fluid cracking catalysts (FCCs) from adsorption microcalorimetry of probe molecules, in *Proceedings of International Symposium on Fluid Catalytic Cracking VI, Preparation and Characterization of Catalysts*, New York, 7–11 September 2003, eds. by M. Occelli, p. 35. Studies in Surface Science and Catalysis, vol. 149, pp. 35–70 (2004). doi:10.1016/S0167-2991(04)80756-0
46. A. Corma, From microporous to mesoporous molecular sieve materials and their use in catalysis. Chem. Rev. **97**, 2373–2420 (1997). doi:10.1021/cr960406n
47. A. Auroux, Acidity and basicity, in *Molecular Sieves- Science and Technology*, vol. 6. (Springer, Berlin, 2008), p. 45
48. Y. Mitani, K. Tsutsumi, H. Takahashi, Direct measurement of the interaction energy between solids and gases. XI. Calorimetric measurements of acidities of aluminium deficient H-Y zeolites. Bull. Chem. Soc. Japan. **56**, 1921–1923 (1983). doi:10.1246/bcsj.56.1921
49. A. Auroux, Y. Ben Taarit, Calorimetric investigation of the effect of dealumination on the acidity of zeolites. Thermochim. Acta. **122**, 63–70 (1987). doi:10.1016/0040-6031(87)80105-3
50. A. Auroux, P.C. Gravelle, J.C. Védrine, M. Rekas, in *Proceedings of the 5th International Conference on Zeolite*, Naples, 2–6 June 1980, ed. by L.V.C. Rees, L.V. Heyden, p. 433
51. H.G. Karge, L.C. Jozefowicz, A comparative study of the acidity of various zeolites using the differential heats of ammonia adsorption as measured by high-vacuum microcalorimetry, in *Proceedings of the 10th International Conference on Zeolites and Related Microporous Materials: State of the Art 1994*, Garmisch-Partenkirchen, Elsevier, Amsterdam, 17–22 July 1994, eds. by J. Weitkamp, H.G. Karge, H. Pfeifer, Hö.W. lderich, p. 685. Studies in Surface Science and Catalysis, vol. 84, pp. 685–692 (1994). doi:10.1016/S0167-2991(08)64174-9
52. M. Calatayud, A. Markovits, C. Minot, Electron-count control on adsorption upon reducible and irreducible clean metal-oxide surfaces. Catal. Today **89**, 269–278 (2004). doi:10.1016/j.cattod.2003.12.015
53. Lj. Damjanović, A. Auroux, Heterogeneous catalysis on solids, in *The Handbook of Thermal Analysis and Calorimetry. Further advances, techniques and applications*, vol. 5, ed. by M. Brown, P. Gallagher (Elsevier, Amsterdam, 2008)
54. M. Niwa, N. Katada, Measurements of acidic property of zeolites by temperature programmed desorption of ammonia. Catal. Surv. Japan. **1**, 215–226 (1997). doi:10.1023/A:1019033115091

Chapter 5
Temperature Programmed Reduction/Oxidation (TPR/TPO) Methods

Antonella Gervasini

Abstract The redox properties of the metal oxides impart them peculiar catalytic activity which is exploited in reactions of oxidation and reduction of high applicative importance. It is possible to measure the extent of oxidation/reduction of given metal oxide by thermal methods which are become very popular: TPR and TPO analyses. By successive experiments of reduction and oxidation (TPR-TPO cycles) it is possible to control the reversible redox ability of a given oxide in view of its use as catalyst. The two methods are here presented with explanation on some possibility of exploitation of kinetic study to derive quantitative information on the reduction/oxidation of the oxide. Examples of selected metal oxides with well-established redox properties which have been used in catalytic processes are shown.

5.1 Redox Properties of Metal Oxides and Catalytic Implications

The important properties of metal oxides that are of certain interest in different fields of applied science and technology are connected with magnetic, electrical, dielectric, optical, acid-base, and redox behavior. In particular, transition metal oxides with redox properties are of interest for applications in catalysis [1]. Metal oxides and in particular transition metal oxides can possess more than one stable oxidation state making them possible to catalyze reactions which necessitate electron exchanges between reactant(s) and surface active sites (e.g., oxidation, dehydrogenation, etc.). Conversely, other kinds of metal oxides are known to be almost complete irreducible. Oxides of copper, nickel, cobalt, molybdenum, are only few examples of reducible

A. Gervasini (✉)
Dipartimento di Chimica,
Università degli Studi di Milano,
Via Camillo Golgi, Milan, Italy
e-mail: antonella.gervasini@unimi.it

A. Auroux (ed.), *Calorimetry and Thermal Methods in Catalysis*,
Springer Series in Materials Science 154, DOI: 10.1007/978-3-642-11954-5_5,

oxides and oxides of gallium or niobium are examples of *quasi*-irreducible oxides (for Ga_2O_3, any reduction is not observed up to 1000 °C and for Nb_2O_5, less than 50 % of Nb(V) can be reduced to Nb(IV) at temperature higher than 900 °C).

The reducing properties of metal oxides can be quantitatively established evaluating the reaction between the metal oxide (MO) and hydrogen to form metal (M) and water vapor:

$$MO_{(s)} + H_{2(g)} \rightarrow M_{(s)} + H_2O_{(g)} \quad (5.1)$$

The standard free energy change for the reduction (ΔG°) is negative for a number of oxides, thus indicating that for these oxides the reactions are thermodynamically feasible (Fig. 5.1). However, it may still possible for the reduction to proceed even when ΔG° is moderately positive. The following equation:

$$\Delta G = \Delta G^\circ + RT \ln (P_{H2O}/P_{H2}) \quad (5.2)$$

where P_{H2O} and P_{H2} denote partial pressures of water and hydrogen, respectively, has to be considered. If the experimental method of performing the reductions is that the water vapor is constantly swept from the reaction zone as it is formed, then P_{H2O} is lowered and it is possible that the term RT ln (P_{H2O}/P_{H2}) could be sufficiently negative to nullify the positive ΔG° value. This is likely to occur at high temperatures where it is possible to observe reduction phenomena for the oxides of vanadium, tin, chromium, which have positive ΔG° values in the range 40–100 kJ/mol (Fig. 5.1).

During some reactions catalyzed by metal oxides, they undergo reduction and reoxidation simultaneously by loss and gain of surface lattice oxygen to and from the gas phase. This phenomenon is called *redox catalysis.* The redox property as well as the acidic and basic feature is the most important properties of metal oxide catalysts. Some simple metal oxides, like V_2O_5, CoO_2, NiO, MnO_2, CeO_2, MgO and some mixed metal oxides have redox properties and they found application as they are or supported on ceramic carriers as oxidation catalysts.

The use of oxides and mixed metal oxides as heterogeneous catalysts in selective oxidation reactions, in particular, is of great industrial importance. In selective oxidation reactions, organic feeds are converted to useful products by oxygen insertion without any change of the number of carbons. The major processes include allylic oxidation (i.e., selective oxidation of olefins at the allylic position) to give aldehydes, nitriles, and acids; aromatic oxidation to acids and anhydrides, epoxidation to olefins, methanol oxidation to formaldehyde, etc. [3]. The desired selective oxidation reactions are, of course, thermodynamically favorable and the oxide catalysts are necessary to overcome kinetic limitations. Moreover, they have to select the desired oxidized product limiting the formation of the more thermodynamically favored deep oxidation products, like CO_2, H_2O, and HCN (Table 5.1 [3]).

The research and study of selective oxidation oxide catalysts with developed redox properties began more than five decades ago with the concept that the lattice oxygen of a reducible metal oxide could serve as a more versatile and useful oxidizing agent for hydrocarbons than would gaseous molecular oxygen. The role of molecular oxygen is only to replenish lattice oxygen vacancies and therefore an oxidation-

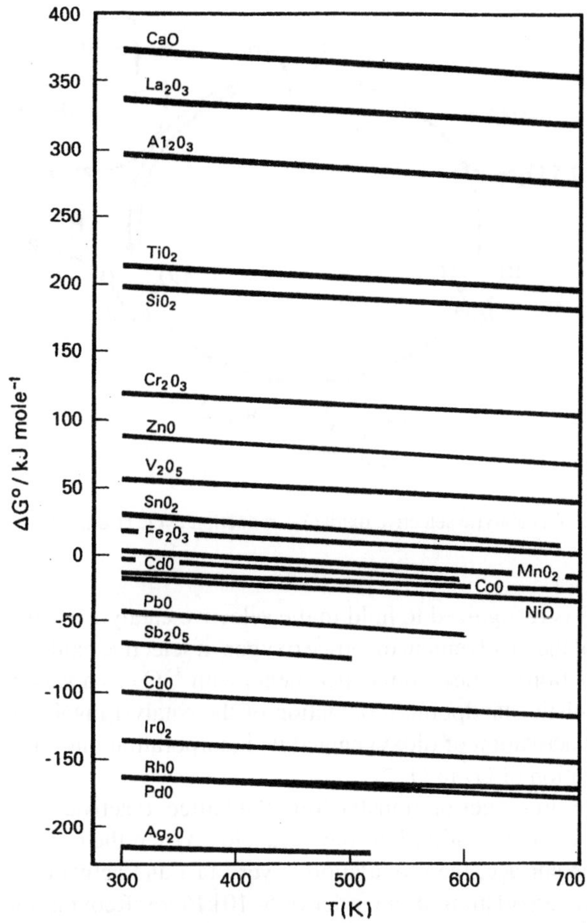

Fig. 5.1 Standard free energy change as a function of temperature (from Ref. [2])

Table 5.1 Thermodynamics of several selective and total oxidation reactions

	Reactions	$\Delta G^{\circ}_{427^{\circ}C}$(kcal/mol)
(A)	$C_3H_6 + O_2 \rightarrow CH_2 = CHCHO + H_2O$	− 80.82
(B)	$C_3H_6 + \frac{3}{2}O_2 \rightarrow CH_2 = CHCOOH + H_2O$	−131.42
(C)	$C_3H_6 + 3O_2 \rightarrow 3CO + 3H_2O$	−304.95
(D)	$C_3H_6 + \frac{9}{2}O_2 \rightarrow 3CO_2 + 3H_2O$	−463.86
(E)	$C_3H_6 + NH_3 + \frac{3}{2}O_2 \rightarrow CH_2 = CHCN + 3H_2O$	−136.09
(F)	$C_3H_6 + \frac{3}{2}NH_3 + \frac{3}{2}O_2 \rightarrow \frac{3}{2}CH_3CN + 3H_2O$	−142.31
(G)	$C_3H_6 + 3NH_3 + 3O_2 \rightarrow 3HCN + 6H_2O$	−273.48

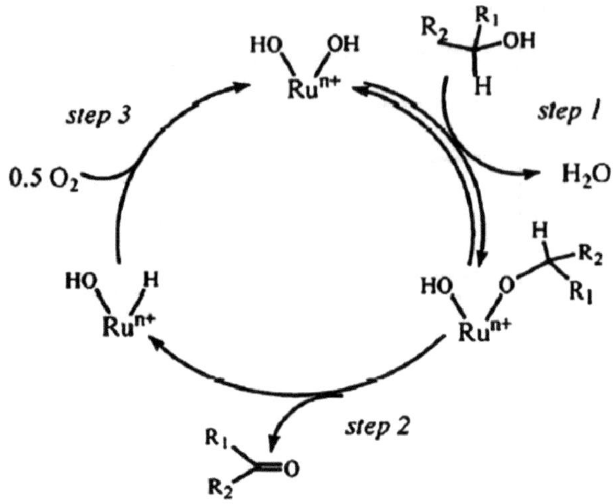

Fig. 5.2 Scheme of a catalytic selective oxidation-reduction cycle (from Ref. [3])

reductant cycle is recognized to hold in the selective catalytic oxidations (Fig. 5.2). Confirmation of the involvement of lattice oxygen in selective oxidation reaction (i.e., propylene oxidation) comes from experiments with $^{18}O_2$ as a vapor phase oxidant. From a mechanistic standpoint, reoxidation of the catalyst involves two processes: adsorption and activation of dioxygen and its incorporation into the solid vacancies (i.e., transformation of O_2 to O^{2-}).

This step requires electron transfer from the lattice to activate and dissociate the dioxygen before incorporation into the vacancies. When the oxygen vacancies are located near the surface, the reoxidation is very fast and activation energy for this surface vacancy reoxidation of the order of 5–10 kJ/mol. Reoxidation of subsurface vacancies is much slower and it is limited by the ability of the catalyst to transport the oxygen from the surface O_2-chemisorption sites to the subsurface vacancies. For example, γ-bismuth molybdate with its layer structure results in low-energy pathway for diffusion of the oxygen anions and reoxidation proceeds with an activation energy of only 35–38 kJ/mol. In more closed packed structures, like α-bismuth molybdate, rapid diffusion cannot occur and the reoxidation rate is diffusion limited with activation energy as high as 105–110 kJ/mol. Moreover, the presence of a redox couple in a catalyst can create a lower energy pathway for diffusion by promoting electron and oxygen transfer between the surface and the bulk. Incorporation of iron (Fe^{3+}/Fe^{2+} redox couple) into the α-bismuth molybdate structure, single $Bi_3FeMo_2O_{12}$ phase is formed, decreases the activation energy for the reoxidation process to 29–35 kJ/mol.

The real breakthrough in the appreciation of the idea of lattice oxygen depletion and subsequent replenishment governing the selective oxidation reactions took place after the appearance of a series of papers by Mars and Van Krevelen (MvK), in 1954 [4] describing some applications of their idea in the kinetics (Fig. 5.3). However,

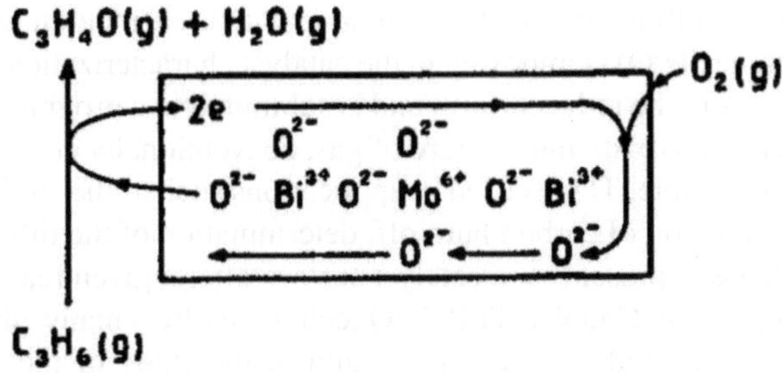

Fig. 5.3 Conversion of propylene into acroleine on bismuth molybdate following the Mars and Van Krevelen mechanism (from Ref. [4])

all oxidations are redox reactions, even when the prevailing mechanism can not be called a MVK mechanism

A property of key importance for an efficient selective oxidation catalyst is its tendency to donate the structural oxygens. If reduction of the catalyst is too facile, it may be well active but it ceases to be selective. Although the reoxidation rate is usually not rate determining in the overall redox cycle, the catalyst ability to replenish its reservoir of lattice oxygen is of fundamental importance in order to sustain the catalyzed reaction. While reoxidation at higher temperatures is generally fast for all the selective oxidation catalysts, it becomes more difficult at low temperatures, where selectivities to the usual products are highest.

Examples of reactions of industrial and environmental importance that run according MvK mechanism are alcohol selective oxidations, VOC total oxidations, selective catalytic reductions of NO_X by ammonia or hydrocarbons, and more recently photocatalytic oxidation of hydrocarbons [5] etc. Revisitation of MvK mechanism for various reactions appeared in Ref. [6] For each one of these reactions, a metal oxide-based catalyst with well-developed redox properties has been optimized.

The redox property of any metal oxide catalyst can be characterized by using the techniques of temperature-programmed reduction/temperature programmed oxidation (TPR/TPO). TPR is a relatively new technique used for the chemical characterization of solid materials. Since the first publications in this area (1975–1985), many papers have appeared in the literature showing TPR applicability in a variety of scientific fields, most particularly in characterization of catalysts. In 1982 the first review on the subject of TPR appeared in the literature [2]. TPR technique is based on the reducibility of species in/on solids by a gas, in general by hydrogen, at the same time that the temperature of the system is changed in a predetermined way (in general, linear temperature increase). The decrease in reducing gas concentration in the effluent gas with respect to its initial concentration allows monitoring the reduction progress. The obtained results enable obtaining information not only of a purely analytical nature but also, and more importantly, about the condition of species present in/on solids. The technique is relatively simple in concept and application and as a result, TPR and related techniques are becoming commonly used among catalysis scientists.

More recently, combination of TPR technique with the temperature programmed oxidation technique (TPO) is imposing in the catalyst characterization studies [7]. In TPO, the catalyst is in the reduced form and is submitted to a programmed temperature increase, while an oxidizing mixture of gas, conventionally containing oxygen, is flowed over the sample. TPO is used in applications such as the study of the kinetics of coking, evaluation of carbon burn off, determination of the different forms of carbonaceous deposits present on a catalyst surface after a given reaction involving organic compounds, etc. Coupled TPR/TPO technique offers many advantages. The two techniques can provide useful information in the study of the reactivity and redox behavior of catalysts. The combination of the two reactions of reduction and oxidation represents a real titration of any reducing/oxidant phase by quantification of hydrogen/oxygen consumption.

5.2 Temperature-Programmed Reduction/Oxidation Technique

The developed technique foresees an increase of the sample temperature at a uniform rate ($T_i = T_0 + \beta t$, where β is the rate of temperature increase) measuring and recording the rate at which the sample (or the active component in the sample) reduces/oxidizes under a reducing/oxidizing atmosphere of known concentration. The following reactions hold:

$$\text{TPR:}\ M_xO_{y(s)} + H_{2(g)} \rightarrow M_{(s)} + H_2O$$
$$\text{TPO:}\ xM_{(s)} + y/2O_2 \rightarrow M_xO_{y(s)}$$

A typical TPR/TPO apparatus can be assembled following some general principles. The apparatus essentially consists of three parts (i) the gas line for pre-treatments and analysis; (ii) the reactor electrically controlled; (iii) the detector for quantitative evaluation of gas-consumption. At present, all the experimental apparatus used operate at ambient (in flowing gas) or sub-ambient (under vacuum) pressure.

Typically, TCD detectors are mounted (even if many other different types are reported in the literature: detection by pressure changes of the reacting gas, detection by weight change of the solid, etc.), they produced on output signal that is proportional to the concentration of hydrogen or oxygen in the carrier gas. A TC-detector works by measuring the difference in thermal conductivity between the pure carrier gas and the mixture of carrier and reactant gas. For this reason, it is most sensitive when the thermal conductivities of the two gases in mixture are very different. This is why argon or nitrogen are preferentially used as carrier gases in TPR and helium in TPO experiments (thermal conductivities of Ar, N_2, and He of 45.46×10^{-6}, 65.71×10^{-6}, and 376.07×10^{-6} cal/(cm · s · K), respectively and thermal conductivities of H_2 and O_2 of 471.11×10^{-6} and 68.19×10^{-6} cal/(cms K), respectively). Moreover, the change in thermal conductivity is proportional to the mole fraction of reactant gas in the mixture only at low concentrations of reactant gas. To maintain a high detector sensitivity the concentration of reactant gas in the mixture has to be limited in the 1–10 % range.

When setting a TPR/TPO apparatus (Fig. 5.4), some practical considerations have to be followed. Particular care has to be taken concerning the determination of time/temperature of the reactive events. If the period between the reduction/oxidation of the sample and the detection of hydrogen/oxygen concentration is significantly long, then the measured temperatures (T_{meas}) of the reactive events require correction according to:

$$T_{corr} = T_{meas} - (\beta \cdot V/F) \quad (5.3)$$

where T_{corr} is the corrected temperature, V is the volume between the reactor and detector, and F is the used flow rate. If such corrections are large, the precision of the temperature measurement is reduced and it is advised maintaining high flow rates during experiment. The flow rate must also be compatible with the geometry of the reactor such that the gas flow maintains uniform velocity distribution conditions. Moreover, the flow rate has also to be in relation to the amount of reducible/oxidized sample, as detailed here below.

During the TPR/TPO experiments, several products as H_2O, CO, or CO_2 can be formed. It is important to remove all the undesired gas molecules that can interfere in the signal output. Besides a correct pre-treatment procedure of the sample, use of suitable traps to stop secondary products are necessary and they have to be present in the apparatus configuration.

5.2.1 General Operative Procedure

The procedure to collect TPR/TPO data is also comprehensive of the sample pre-treatment. Several types of procedures can be chosen in relation with the sample nature and type of information required. Generally, before starting a TPR analysis, the sample should be in its oxidized form. In this case, the pretreatment consists in oxidizing the sample in flowing air or pure oxygen and then purging the product formed (like water and carbon residues) by flowing an inert gas. The two procedures have to be carried out at given temperature to assure the effectiveness of the actions. In the case of TPO analysis, the sample has to be previously reduced; in this case too, the reducing procedure has to be carried out at given temperature.

At the beginning of a TPR or TPO experiment, reducing/oxidizing gas is made to flow over a fixed amount of solid at a temperature low enough to prevent reaction. The starting temperature of the experiment has then to be below the expected reduction/oxidation temperature of the phase under study. In this condition, a perfect flat baseline is observed because any hydrogen/oxygen is consumed from the sample (the two branches of TCD are at same concentration of active gas component). Next, the temperature of the sample is increased at a constant rate (β, °C/min) and, as reduction/oxidation begins, hydrogen/oxygen is consumed from the gas mixture (conventionally, H_2/Ar and O_2/He mixtures), which is detected by the detector (TCD).

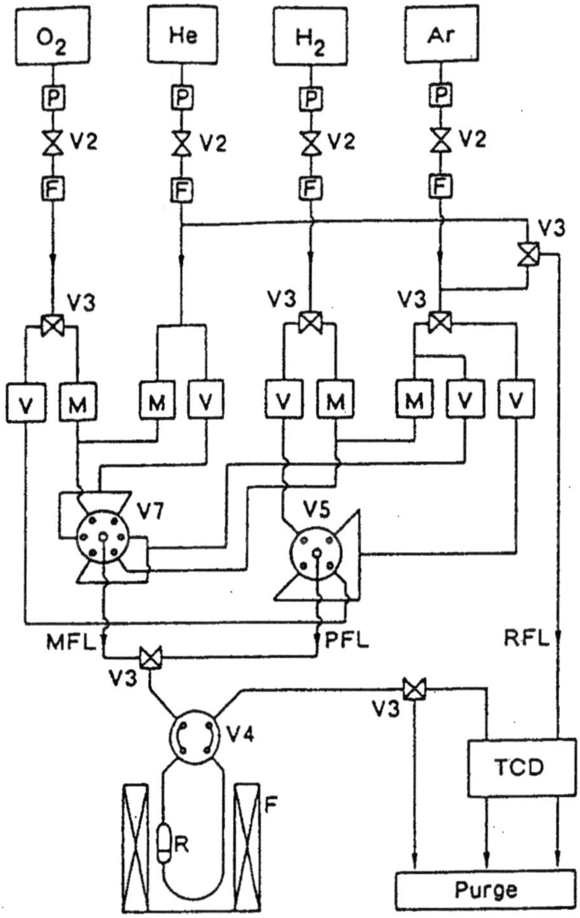

Fig. 5.4 Scheme of a typical TPR/TPO apparatus. Items: *P* pressure controller; *V2* shutoff valve; *F* purification stage; *M* mass flow controller; *V* metering valve; *V3, V5, V7* 3-way, 5-way, and 7-way switching valves, respectively; *V4* crossover valve; *F* temperature controlled furnace; *R* reactor; *PFL* pretreatment flow line; *RFLK* and *MFL* reference and measure flow line of TCD

The TCD signal rises proportionally to the rate of hydrogen/oxygen consumption. When reduction/oxidation ceases, no more hydrogen/oxygen is consumed and the TCD signal returns to its baseline. A peak is obtained (Fig. 5.5) which maximum position represents the maximum rate of hydrogen/oxygen consumed (i.e., temperature of maximum rate of reduction/oxidation, T_{max}). Several peaks ($T_{max,i}$) may be detected over the course of the temperature ramp if different reduction/oxidation steps may occur and each one requires given thermal energy value. The amplitude of each peak is proportional to the reaction rate. On the basis of suitable calibration tests, the amount of consumed hydrogen/oxygen can be obtained from the graphical integration of each peak. Calibration tests are readily achieved by injecting into the

Fig. 5.5 Example of typical TPR and TPO profiles

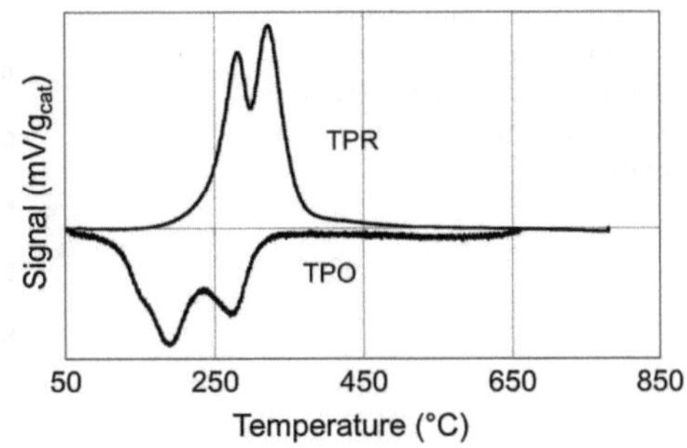

TPR apparatus known amounts of hydrogen, for example, usually several different amounts. The linear calibration curve obtained (detector signal vs. moles of H_2) is used for the quantitative calculation of the extent of reduction (or oxidation) of the actual sample under study.

5.2.2 Analytical Parameters

The choice of the analytical parameters for a TPR/TPO experiment, in particular sample volume/mass, temperature increasing rate, and flow concentration and flow rate, is fundamental to obtain significant reaction profiles.

Theory predicts that the T_{max} values should be independent of the mass of the sample [8, 9]. Minimal changes in T_{max} were experimentally observed from many authors. The volume (and mass) of the sample is best kept to a minimum as this diminishes problems due to back pressure, temperature differences between sensor and sample, and no uniform gas concentration within the solid. The smaller amount of sample used, it is easier to handle the whole analysis, in particular, less water is formed thus avoiding problems in the water removal system. The main effect of changing the sample mass was that resolution of two separate chemical events obtained with, e.g., 50-mg sample was lost when the sample mass was increased to, e.g., 400 mg with higher value of T_{max} for the composite resultant peak. This behavior is due to the temperature difference between the temperature sensor and the sample that becomes marked for large mass of sample. Therefore, care must be taken in comparing results when sample sizes are markedly different. For example, the two well known rate processes for the reduction of Cu^{2+} to Cu^0 through Cu^+ intermediate, are well distinguished if TPR is performed with low sample mass and low hydrogen concentration [9, 10].

Sample particle size has to be controlled. Particles that are too small can result in the creation of back pressure and consequent problems in maintaining constant flow along the sample, while too large particles can result in intraparticle diffusion problems, causing distorted peaks. Mass transport processes, both inter- and intraparticles, are characterized by low-activation energies and they can thus alter

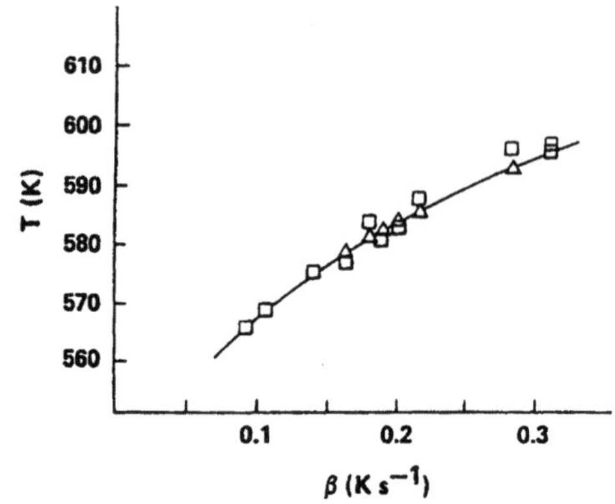

Fig. 5.6 Sensitivity of the temperature of the maximum reduction rate for the reduction process of NiO with β variation

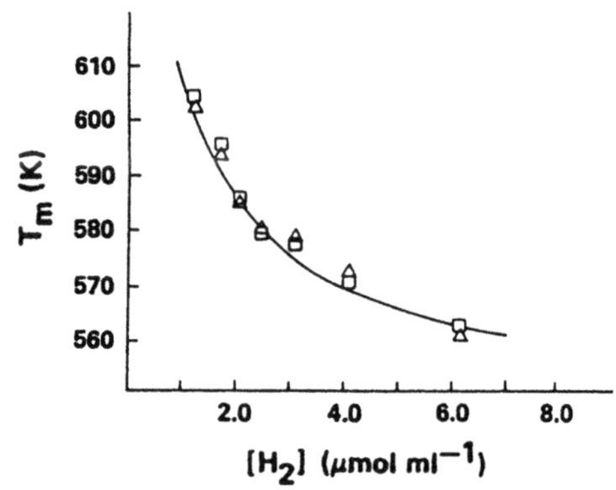

Fig. 5.7 Sensitivity of the temperature of the maximum reduction rate for the reduction process of NiO with H_2 concentration

the shape of reduction profiles. It is common practice to experimentally check for mass transport limitations by looking for changes in TPR/TPO profiles when particle size is varied. Heat transfer limitations can be also considered to be present during reduction/oxidation processes. To avoid or highly limited them, sample dilution with quartz or inert material is commonly performed, this procedure does not alter the TPR/TPO profile.

Concerning the rate of temperature increase, there are advantages in having the heating rate as fast as all the other factors allow. This results in the obtainment of better defined-peaks. Moreover, variation of heating rate (β) leads to variation of T_{max}, this behavior is most commonly used to obtain activation energies for the reactive process. When the change in T_{max} is due to kinetic parameters, other parameters remaining constant, the sensitivity of T_{max} to changes in heating rate is established. For example, systematic study of hydrogen reduction of NiO (simple one-step reduction, from Ni^{2+} to Ni(0), first-order in hydrogen and NiO), showed that an increase in heating rate from 5 to 20 °C/min produced a shift in T_{max} of 30 °C (Fig. 5.6). Exploiting the observed T_{max} with β, activation energy of about 134 kJ/mol was calculated for the NiO reduction process [8].

Concentration of hydrogen/oxygen in TPR/TPO experiment affects the sensitivity of the detector and the T_{max} values vary. For example, an increase on H_2 concentration from 3 % (1.23 μmol/cm^3) to 15 % (6.15 μmol/cm^3) produced a decrease in T_{max} from 330 to 290 °C for the reduction process of NiO [8] (Fig. 5.7). This indicates that the sensitivity is inversely proportional to the H_2 concentration in the reactant mixture.

The variation of the total flow rate influences the results obtained in a TPR/TPO experiment, too. In the case of hydrogen reduction of Cu-exchanged zeolite, an increase in total flow rate from 10 to 20 cm^3/min (with 4 % H_2 concentration) lowered the value of T_{max} by 15−20 °C [9]. Recalling the flow reactor theory, it is expected that an increase in total flow rate for a reactant consumed by a first-order process results in a lowering of the degree of conversion and then an increase of reactant concentration in the reactor. The observed increase of H_2 concentration in the reactor and decrease of T_{max} value of the reduction event with total flow rate are in agreement with theory.

5.2.3 Selection of Operating Parameters

To obtain meaningful TPR/TPO profiles, certain restrictions on the choice of combinations of operating parameters are self-evident. Of particular importance in the case of a TPR experiment, is that the hydrogen feed rate should be equal to the maximum reduction rate, otherwise distortion of the reduction profile will occur. Certain combinations of total flow rate, reactant concentration, sample volume (mass), and heating rate are allowed while other ones are not.

The possibility to carry out a TPR experiment (but the same holds for TPO experiment) with quantitative evaluation of results is based on the simplifying assumption that the mean hydrogen concentration between the reactor inlet and outlet is the driving force for the reduction; that is, the mean hydrogen concentration into the reactor. This holds only if low fractional conversion of hydrogen occurs all along the experiment. On the other hand, the difference of hydrogen concentration between reactor inlet and outlet has to be well detectable, that is, the change in hydrogen concentration has to be large compared to the statistical noise of the detector baseline. Two criteria have to be fulfilled to meet the two requirements above exposed: (i) the amount of hydrogen consumed at the peak maximum should not exceed 2/3 of the hydrogen fed to the reactor and (ii) the minimum conversion at the peak maximum should be 10 %. The two criteria can be met with sets of operating variables. A characteristic number, K (usually expressed in s), has been defined [8] to facilitate the selection of appropriate operating variables:

$$K = S^o/(V^*C^o) \tag{5.4}$$

where V* is the total flow rate of the reducing gas (cm^3/s); S^o, the initial amount of reducible species in the sample (μmol); C^o is the initial concentration of the reducing gas (μmol/cm^3). The values of K have to be comprised between 55 s and 140 s, with possibility to match S^o, V*, and C^o parameters to obtain the appropriate K value. For values of K below 55 s, the sensitivity of the analysis becomes too low, while for values exceeding 140 s, the amount of reducing gas consumed is too large, so violating the assumption of a linear concentration profile. Concerning the physical meaning of the K parameter, it represents the *sensitivity* parameter. Once optimized K, a second characteristic number can be determined, P (usually expressed in °C), that is defined as:

$$P = K \cdot \beta \tag{5.5}$$

P represents the shape and *resolution* parameter and it has to be comprised between 20 and 50 °C. With the K and P numbers in the correct range, accurate and reliable results from the TPR/TPO analysis and above all comparable data are obtained. The sensitivity of the analysis becomes lower for small heating rates with an inverse proportionality to the hydrogen concentration in the feed. A typical example is in the TPR analysis of CuO: changing the rate of heating and of total flow, two different reaction profiles result. It is possible or not to distinguish the two steps of the reduction process identified by two reducing peaks, Cu(II) → Cu(I) → Cu(0), or to obtain one only peak comprehensive of the total hydrogen consumption involved in the two steps. In the second case, the advantage is to calculate more easily the total amount of the consumed reactive gas. In general, when the sample contains only one component it is useful to perform the analysis with a low temperature rate (b) to observe the mechanism of the reaction process.

5.3 Kinetics and Reduction Mechanisms

The aim of the analysis of the TPR/TPO data is to derive kinetic parameters relating to the reduction/oxidation process. It is common to observe the reduced/oxidized fraction (α) as a function of time for various temperatures and pressures/concentrations of reducing/oxidant gas. The reduction/oxidation of a sample with redox property or of part of it (in the case of supported phase system) is a bulk phenomenon and its degree of reduction/oxidation is interpreted in terms of mechanism by which the reaction occurs. *Nucleation* model or *contracting sphere* model are the most successful utilized kinetic models [2].

In synthesis, at constant temperature, the rate of the reaction: $MO_x + H_2 \rightarrow$ reduced products, may be expressed as:

$$rate = -d[MO_x]/dt \text{ or } -d[H_2]/dt = k[H_2]^P[MO_x]^Q \tag{5.6}$$

with P and Q, the reaction orders for the MO_x and H_2 species, respectively.

If the rate constant k is expressed with its dependence on temperature by the classical Arrhenius equation (with A, the pre-exponential factor and E_a, the activation energy) and the linear dependence of temperature on time is introduced by the linear heating rate ($\beta = dT/dt$), we can write the following expression:

$$-\beta\, d[MO_x]/dT = A \exp(-E_a/RT)[H_2]^P[MO_x]^Q \tag{5.7}$$

If the H_2 consumed is little (in accord with the criteria above cited in Sect. 5.2.3, the reaction rate can be considered independent of H_2 concentration and Eq. (5.7) can be rewritten as:

$$-\beta\, d[MO_x]/dT = A \exp(-E_a/RT)[MO_x]^Q \tag{5.8}$$

we can now introduce the reduced fraction at given time t (α_t), instead of the residual concentration of sample to be reduced, and being: $\beta\, d(1-\alpha)/dT = \beta\, d(\alpha)/dT$, we can write:

$$\beta\, d(\alpha)/dT = A \exp(-E_a/RT)(1-\alpha)^Q \tag{5.9}$$

integration of Eq. (5.9) gives:

$$\int_0^{\alpha} d(\alpha)/(1-\alpha)^Q = A/\beta \int_0^{T} \exp(-E_a/RT)dT \tag{5.10}$$

or

$$\int_0^{\alpha} d(\alpha)/(1-\alpha)^Q = A \int_0^{t} \exp(-E_a/R(T_o + \beta\, t)dt \tag{5.11}$$

with T_o representing the starting temperature of the linear heating rate.

The problem is now to define the correct expression for the variation of α versus t to solve Eq. (5.11) that is to describe the reduction process by a suitable model.

For the derivation of the kinetic parameters of reaction, it is more generally useful to utilize equations able to directly interpret the raw TPR data. Most experimental TPR systems monitor the rate of change of reducing gas concentration as a function of temperature, producing reduction profiles with peaks corresponding to maxima for the reaction rates. Methods originally developed for other temperature-programmed techniques are employed, like thermogravimetry analysis [11–16]. In these approaches, multiple reaction analyses have to be carried out with several different constant heating rates (by varying β), from which the shift of peak maximum position (T_{max}) with β is exploited. The proposed equations converge on linear expressions like that here below recalled, from which both E_a and A can be obtained from the slope and intercept of the relevant line:

$$\ln(T_{max}^2[H_2]/\beta) = E_a/R(1/T_{max}) + \ln(E_a/RA) \tag{5.12}$$

normally, first order kinetics with respect to the reducing sample and hydrogen is assumed and mean hydrogen concentration at the temperature of the maximum reaction rate is considered.

5.3.1 Nucleation Model

When a reducing sample (e.g., oxide) and hydrogen come into contact, the reaction starts and after some time, t_1 (induction time), the first nuclei of the reduced product form. Reduction removes oxygen ions from the lattice and when the concentration of vacancies reached a certain critical value they are annihilated by rearrangement of the lattice with formation of metal nuclei. Oxygen ions may be removed by inward diffusion of hydrogen to the metal/metal-oxide interface or outwards diffusion of oxygen ions from the metal oxide to the metal gas interface. The reaction interface increases more and more rapidly thanks to two processes: (i) the growth of the nuclei already formed and (ii) the appearance of new ones. At a given time of reduction, the metal nuclei have grown at the surface of the oxide grains to such an extent that they begin to make contact with each other. Starting from this moment, a decrease of the reaction interface is observed because of the overlapping of the metal nuclei and the steady consumption of the oxide grains. These processes continue until the complete reduction of the sample; during the final stage in which the reaction interface is shrinking as the reduced layer grows, the process is equivalent to the *contraction sphere* model. The nucleation mechanism results in the S-shaped α against t plot and in the maximum in the $d\alpha/dt$ against α, as illustrated in Fig. 5.8 (left). The same plots are observed for autocatalytic reductions; for example, hydrogen can be dissociated and activated by the reducing oxide; this phenomenon is observed on oxide promoted by metal dopants (doped-CuO and NiO samples).

The mathematical treatment that describes the nucleation mechanism of reduction can be found in Ref. [2]. The final integral equation rate, Eq. 5.13), represents the rising part of the α-vs-t curve before the inflection point (Fig. 5.8 left) and shows that the rate of nucleation is proportional to $(\text{time})^q$:

$$\alpha = \frac{1 C_1 C_2}{V_{\text{final}}\ {}^{p+q+1}} (t - t_1)^{q+p+1} \tag{5.13}$$

with C_1 and C_2, proportionality constants; p, the dimension of nuclei; V, the final volume of nuclei; and q, an integer number representing the order of nucleation rate.

5.3.2 Contracting Sphere Model

In many cases, the reaction interface between the oxidized and freshly reduced sample decreases continuously from the beginning of the reaction when it has its maximum value. This can be interpreted in terms of a very rapid nucleation resulting in the total coverage of the reducing grain with a thin layer of the reduced product in the first instant of the reaction. The reaction interface then decreases as the substrate grain is continuously consumed in the course of the reaction. Figure 5.8 (right) shows the contracting sphere model with its decreasing reaction interface resulting in continuously decreasing rate ($d\alpha/dt$ vs. t).

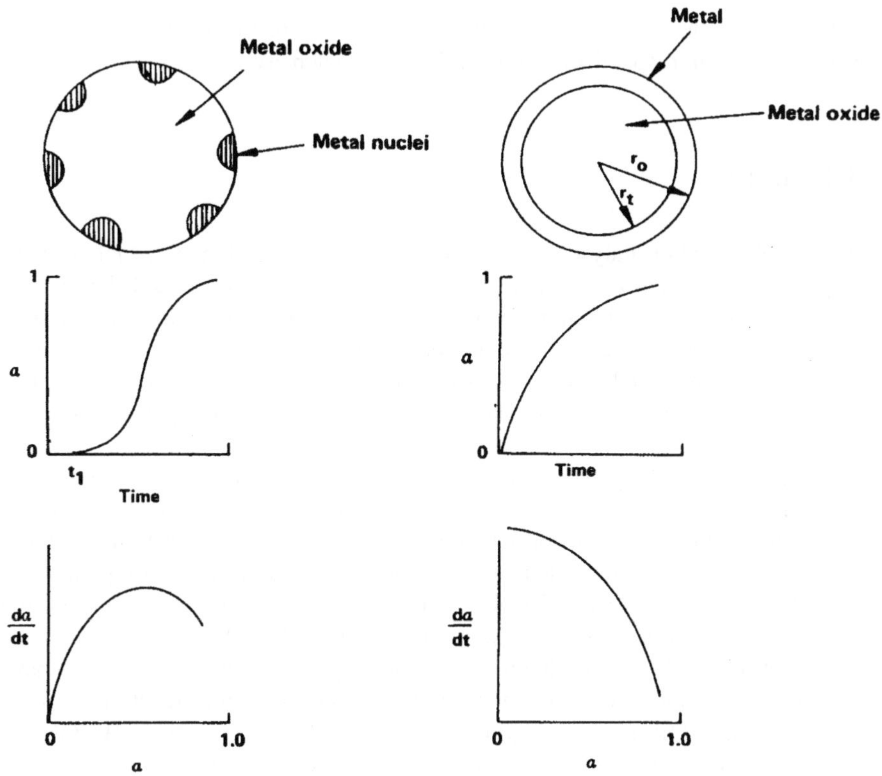

Fig. 5.8 Reduction process by the nucleation mechanism (*left*) and the contracting sphere mechanism (*right*)

The distinction between the nucleation and contracting sphere model is somewhat artificial; in fact the contracting sphere model starts with very rapid nucleation and the nucleation mechanism finishes by an essentially contracting sphere model. The real distinction is between a reaction interface, and reaction rate, that is increasing in the early stages of the reactive process (nucleation model) and a reaction interface that is contracting throughout the reaction (contracting sphere model).

The mathematical treatment that describes the contracting sphere mechanism of reduction can be found in Ref. [2]. The rate of the reaction is controlled by both diffusion from the gas stream to the sample particle and through the reduced layer freshly formed and by chemical reaction at the interface. The final integral equation rate, Eq. (5.14) gives the degree of reduction α as a function of time. The rate constant, k_p, may be higher or lower than the constant rate of hydrogen diffusion, k_d, with the chemical or hydrogen diffusion stage controlling the reaction kinetics, respectively.

$$\frac{k_p}{r_0 d_0}(c_0 - C_{eq})t = [1 - (1-\alpha)^{1/3}] + \frac{r_0 k_p}{k_d}[\frac{1}{2} - \frac{\alpha}{3} - (1-\alpha)^{\frac{2}{3}}] \qquad (5.14)$$

with C_o and C_{eq}, the hydrogen concentration when reaction rate is zero, r_o, and when it is at equilibrium, respectively; and d_o, the sample density.

5.4 Examples

Typical TPR and TPO analyses of catalysts (bulk, supported, doped, reducing/oxidizing phases) are aimed at evaluating the position of T_{max} peaks and the amount of consumed reducing/oxidizing gas which provides information on (i) easiness to reduction/oxidation; (ii) degree of reduction/oxidation attained; (iii) average oxidation state of given (metal) phases after a given reducing/oxidizing treatment; (iv) dispersion and aggregation state of supported (reducing/oxidizing) phases; *v*) ability to complete a redox cycle.

In the preparation of supported metal catalysts, salt and molecular complex precursors of the metal active components are used; procedures of air-calcination or H_2-reduction are then needed for activating the catalyst for the reaction. Platinum is one among the most important metals widely used as catalytic component for a host of chemical processes. For the preparation of Pt-containing catalysts, the highest Pt-dispersion over suitable supports is highly desirable to avoid the consumption of the costly metal. Therefore, a wide range of platinum compounds have been studied to determine their reduction conditions to obtain the Pt-phase, e.g., $PtCl_4$, H_2PtCl_6, $Pt(NH_3)_2(NO_2)_2$, $Pt(NH_3)_4(OH)_2$, $[Pt(NH_3)_4]Cl_2$, $H_2[Pt(OH)_6]$, ecc. Some Pt-compounds require a four-electron process for reduction ($PtCl_4$, H_2 $PtCl_6$, and $H_2[Pt(OH)_6]$) while the reduction of some others require a two-electron process. The TPR profiles of some dried Pt-samples are shown in Fig. 5.9. It is interesting to note that the Cl-containing salts/complexes are significantly more difficult to reduce than the OH- containing compounds (e.g., in Fig. 5.9 compares the TPR profiles A with B and C; and the D with the E profiles).

In the catalyst preparation, not only the choice of the active phase precursor is crucial, the method of catalyst preparation is decisive, too, for obtaining good dispersion of the active phase. Active phase can be deposited on supports by impregnation, ion-exchange, adsorption, etc. Once selected the nature of support and active phase, the observed differences in dispersion should only be due to the method of preparation. Dispersed iron oxide catalysts (FeO_x) have received much attention because their potentiality for many applications in environmental catalysis (N_2O decomposition and reduction) and in fine chemical industry (Friedel-Crafts, isomerisations, etc.). For most applications, high dispersion of the metal centres is desirable to enhance the activity-selectivity pattern of the catalysts.

Iron oxide is a well reducible phase, the comparison of reduction profiles of supported Fe-catalysts may give information on the dispersion state of iron. In Fig. 5.10, two TPR spectra of two dispersed FeO_x catalysts on zirconia prepared by conventional impregnation and adsorption methods are shown [17]. Reduction of pure hematite (Fe_2O_3) by hydrogen is a complex event that can proceed in different steps

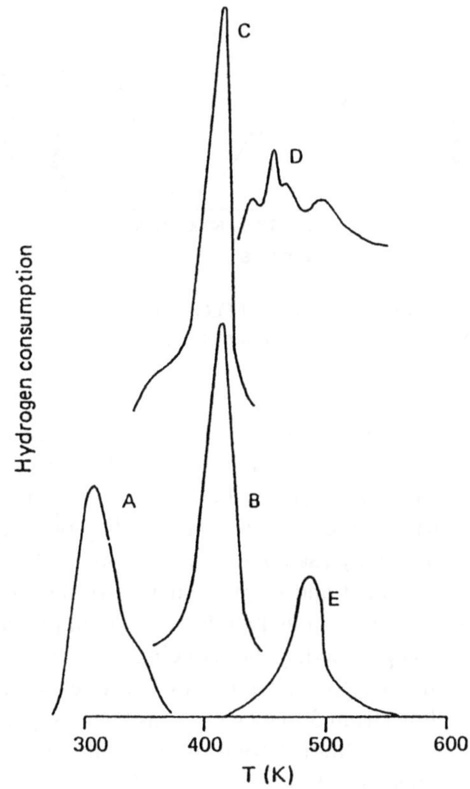

Fig. 5.9 TPR profiles for some Pt salts used in catalyst preparation (from Ref. [2])

via intermediate oxides (i.e., magnetite, Fe_3O_4, and wüstite, FeO_x). The possible reduction reactions, involving hematite, are written here below.

(i) $Fe_2O_3 + 3H_2 \rightarrow 2Fe + 3H_2O$
(ii) $3Fe_2O_3 + H_2 \rightarrow 2Fe_3O_4 + H_2O$
(iii) $3Fe_2O_3 + 3H_2 \rightarrow 6FeO + 3H_2O$
(iv) $Fe_3O_4 + 4H_2 \rightarrow 3Fe + 4H_2O$
(v) $Fe_3O_4 + H_2 \rightarrow 3FeO + H_2O$
(vi) $(1 - x)Fe_3O_4 + (1 - 4x)H_2 \rightarrow 3Fe_(1 - x)O + (1 - 4x)H_2O$
(vii) $FeO + H_2 \rightarrow Fe + H_2O$
(viii) $Fe_(1 - x)O + H_2 \rightarrow (1 - x)Fe + H_2O$

Generally, the first step in the hematite reduction is the formation of Fe_3O_4. Accordingly, observation of a TPR peak at low temperature (the exact position diverges to a large extent dependent on impurities) corresponds to the Fe_3O_4 formation. The Fe_3O_4 reduction profile gave one symmetrical peak with T_{max} at around 550 °C, while the reduction of FeO was observed at very higher temperature (> 750 °C).

A very different and complex situation occurs when supported iron phases are concerned. The TPR profile of the FeO_x/ZrO_2 catalysts prepared by impregnation presents a convoluted curve (Fig. 5.10, left). The complexity of the TPR profile reveals

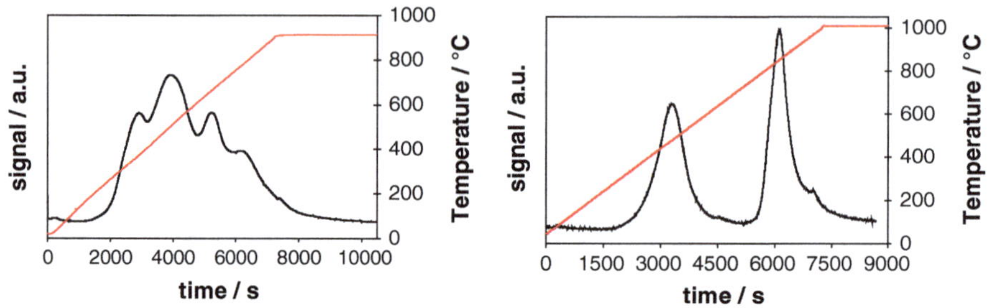

Fig. 5.10 TPR profiles of FeO_x/ZrO_2 (with 5 wt% of Fe) prepared by the impregnation (*left*) and by adsorption equilibrium method (*right*)

the high heterogeneity of the iron surface in terms of nature of the oxide species, dimension of aggregates, and oxide-support interaction. A completely different situation emerges when FeO_x/ZrO_2 was prepared by adsorption method (Fig. 5.10, right). The spectrum is dominated by two well defined and intense peaks positioned at low ($T_{max,1}$at *ca*.430 °C) and high ($T_{max,2}$at *ca*.850 °C) temperature. The two well defined and sharp reduction peaks suggest that iron oxide aggregates are of smaller dimension than those of the impregnated catalyst. The nature of the support can deeply modify the reducing path of the supported Fe_2O_3 phase. Examples can be found in Ref. [18], in which a series of dispersed Fe-catalysts with increasing Fe loading (from 3 to 17 wt.%) has been prepared over a mesoporous silica as support. From a qualitative point of view, all the reducing profiles present the same feature independent of the Fe-content (Fig. 5.11): a main reduction peak with well defined maximum at ca. 400 °C dominates the spectra, other weakly resolved maxima with lower intensity are between 700 and 800 °C. The samples at progressively higher Fe loading consume increasing amount of H_2 without any important modification of the shape and position of the reducing peaks. Concerning the quantitative point of view, the experimental H_2-consumptions are lower than those that can be calculated assuming total reduction of the F_2O_3, assumed as the starting species. It has been determined that the silica matrix had a strong inhibiting effect on the complete Fe(3+) reduction to Fe(0) due to fayalitic phase formation (Fe_2SiO_4). Only the reducible iron oxide aggregates took part in the reaction [18], the high dispersion of the Fe-oxide species imparts acidic- properties of Lewis type to the surface.

The reducibility of any metal oxide can be improved or got worse by doping the oxide. Detailed studies were made on the reducibility of copper oxide. Hurst et al. [2] report an exhaustive study on the CuO phase doped with 2 mol% of a large series of metal ions (Cr, Mn, Fe, Co, Ni, Ru, Rh, Pd, Ag, and Ir, Pt, and Au). The doping with the first-row transition metal ions gives TPR profiles with the same shape as that obtained from CuO with a shift to lower temperatures (293 °C $< T_{max} <$ 313 °C) compared with T_{max} of 333 °C for CuO under the same experimental conditions. The improving effect of these ions on the reducing property of CuO likely arises from nonspecific modification of the CuO lattice, resulting in the increase of copper nucleation sites. The TPR profiles of Pd- and Ru-doped CuO are distinguished from

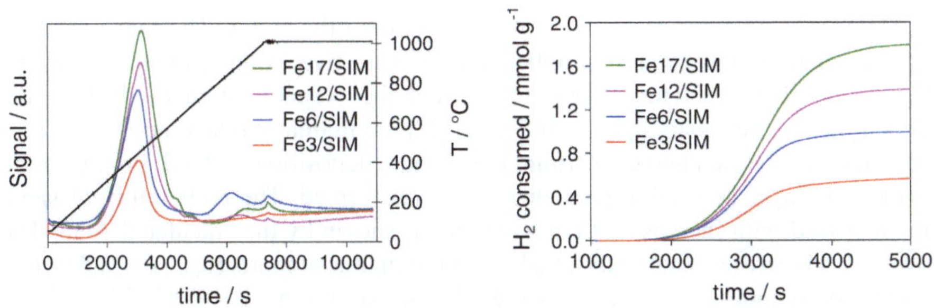

Fig. 5.11 TPR profiles of FeO_x catalysts dispersed on a mesoporous silica (SIM) with different amount of Fe_2O_3, from 3 to 17 wt.% (*left*) and H_2 consumed in the reduction (*right*) (from Ref. [18])

all the others in that reduction is complete at very low temperatures (150–230 °C) and the TPR profiles consist of two peaks which indicate a low-temperature process followed by a more intense high-temperature reduction. Preferential reduction of Pd and Ru ions to form Pd and Ru nucleation sites occurred followed by an improvement of the CuO reducibility. The ion-doping effect of Ag, Au, Rh, Ir, and Pt on CuO is more complex; CuO reduction is completed at about the same temperature as that observed for CuO with a more complex shape of the TPR profile.

The knowledge of the average oxidation state (AOS, $n_{ox,av}$) attained for a given oxide under defined reducing conditions (nature of the reducing agent and temperature, in particular) is of importance for several applications. TPR offers a simple way to compute AOS for any given reducible metal species as an alternative to expensive analytical analyses (i.e., XPS). The simple starting idea is to compare the theoretical amount of electrons needed for the reduction of one mole of species (Ne°_{theor}) with the amount of effective electrons exchanged during reduction (Ne°_{exptl}), obtained from the TPR experiment (derived from the experimental knowledge of H_2 consumed: one mole of H_2 gives 2 electrons during the reduction process). For a mole of reducing species, $n_{ox,av}$ can be calculated as:

$$n_{ox,av} = (Ne^{\circ}_{theor} - Ne^{\circ}_{exptl})/SF \quad (5.15)$$

where SF indicates the stoichiometric factor, number of reducible atoms in the species; for example, SF is 2 for V_2O_5.

In alternative, AOS can be calculated from the balanced redox reaction; here below written for a trivalent metal species, M^{3+}, which can be reduced to bivalent or univalent or zerovalent metal species:

$$M^{3+} + (y-2)/2H_2 = M^{(5-y)+} + (y-2)H^+ \quad (5.16)$$

where y indicates the experimental amount of hydrogen consumed, obtained from TPR analysis. Once obtained y, the final average oxidation state ($n_{ox,av}$) of M can be calculated on the basis of Eq. (5.16).

Many examples can be found in the literature on this point. The reducibility of vanadia catalysts has catalytic implications for selective oxidation reactions where they found real use. The support nature and the preparation method affect the reducibility of the vanadia phase. In Ref. [19]; pure titania or bilayered titania/silica supports were chosen and concerning the vanadia deposition method, impregnation and atomic layer deposition procedures were performed. The reducibility of vanadia improved with increasing titania loading as shown by the calculated AOS. The lowest AOS were associated to vanadia on pure titania supports ($n_{ox,av} = 3.5$) while vanadia on titania-silica supports achieved at maximum $n_{ox,av}$ of 3.7–3.8. AOS of vanadium after reduction was independent of the preparation method.

When supported metal oxide phases are concerned, the kind of surface species that can be present at the surface depends on the support nature, being active a strong or weak metal-support interaction. CuO is an important catalytic phase, easy to disperse on acid supports that can interact with it with strong metal-support bond; this interaction can influence the redox properties of CuO. Modified silicas with amount of alumina (SA), titania (ST), and zirconia (SZ) in 12–14 wt.% concentration were used to support CuO (8–9 wt.%) [20]; a commercial silica-alumina support was comparatively studied (SAG). On the different supports, the CuO redox properties were controlled by combining TPR and successive TPO experiments.

As depicted in Fig. 5.12, the profiles for the two silico–aluminas (SA and SAG) were well different between each other, in terms of T_{max} values and line widths. For Cu/SAG, two well distinguished peaks of similar intensity and for Cu/SA, four peaks with very different line widths were otained. A set of three peaks described

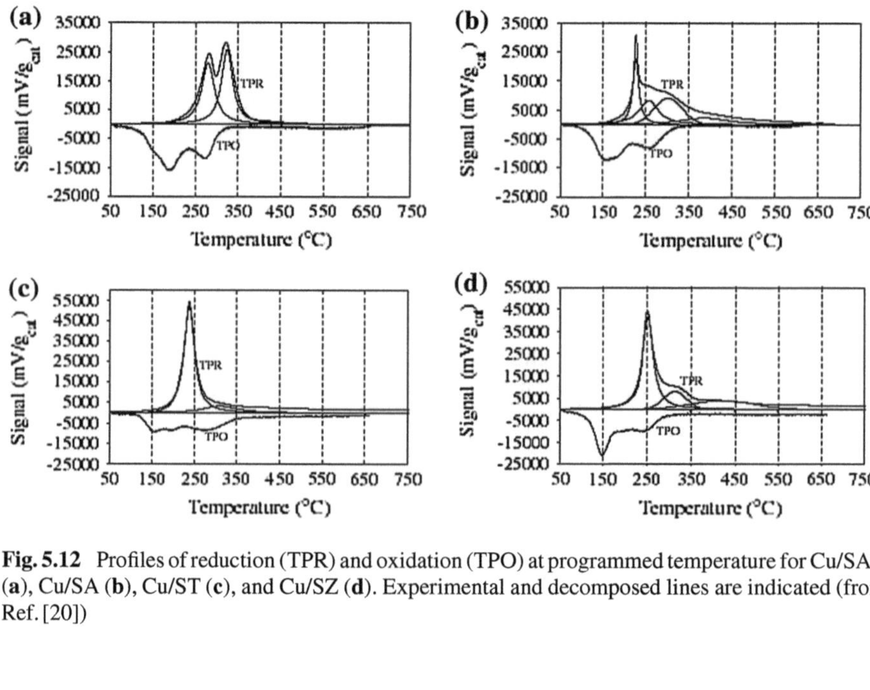

Fig. 5.12 Profiles of reduction (TPR) and oxidation (TPO) at programmed temperature for Cu/SAG (**a**), Cu/SA (**b**), Cu/ST (**c**), and Cu/SZ (**d**). Experimental and decomposed lines are indicated (from Ref. [20])

the TPR profile of Cu/SZ, while quite an unique peak was obtained for Cu/ST. The reduction of the copper phase on the supports of high acidity (SA and SZ) showed low defined and very broadened TPR peaks at very high temperatures, (380–425 °C). The temperature and feature of these peaks suggest that they could be assigned to copper species in interaction with the support. These compounds were more stable to reduction than the dispersed CuO species.

The reoxidation of the reduced copper phase was accomplished by temperature programmed oxydation (TPO). All the studied catalysts could be almost completely reoxidized to CuO, as indicated by the values of oxidation-percent calculated (>60 %). Quite the same TPO profiles were observed on all the catalysts with two oxidation steps: at .first, oxidation of surface copper species and at higher temperature, bulk oxidation. The effective redox properties of CuO allow it to find application in several reaction processes demanding cyclic reduction and reoxidation process with the reactants (i.e., selective catalytic reduction of NO_x with hydrocarbons as reducing agent, [20–22]).

References

1. J. Hu, L. Chen, R. Richards, in *Metal Oxide Catalysis*, vol. 2, ed. by S.D. Jackson, J.S.J. Hargreaves (WILEY-VCH Verlag GmbH & Co. KGaA, Weinheim, 2009)
2. N.W. Hurst, S.J. Gentry, A. Jones, B.D. McNicol, Catal.-Rev.-Sci. Eng. **24**, 233 (1982)
3. R.K. Grasselli, J.D. Burrington, Adv. Catal. **30**, 133 (1981)
4. P. Mars, D.W. van Krevelen, Chem. Eng. Sci. Spec. Suppl. **3**, 41 (1954)
5. A.R. Almeida, J.A. Moulijn, G. Mul, J. Phys. Chem. C. **115**, 1330 (2011)
6. C. Doornkamp, V. Ponec, J. Mol. Catal. A **162**, 19 (2000)
7. B. Jouguet, A. Gervasini, A. Auroux, Chem. Eng. Technol. **18**, 243 (1995)
8. D.A.M. Monti, A. Baiker, J. Catal. **83**, 323 (1983)
9. S.J. Gentry, N.W. Hurst, A. Jones, J. Chem. Soc. Faraday Trans. I, **75**, 1688 (1979)
10. P.A. Jacobs, J.P. Linart, H. Nijs, J.B. Uytterhoeven, J. Chem. Soc. Faraday Trans. I, **73**, 1745 (1977)
11. R.J. Cvetanovic, Y. Amenomiya, Catal. Rev. **16**, 21 (1972)
12. V. Rakic, V. Dondur, D. Misljenovic, J. Therm. Alnal. **38**, 879 (1992)
13. J.H. Chan, S.T. Balke, Polym. Degrad. Stabil. **57**, 135 (1997)
14. P. Carniti, A. Gervasini, Thermochim. Acta **379**, 51 (2001)
15. H. Tanaka, Thermochim. Acta **267**, 29 (1995)
16. F. Eigenmann, M. Maciejewski, A. Baiker, Thermochim. Acta **359**, 131 (2000)
17. C. Messi, P. Carniti, A. Gervasini, J. Thermal Anal. **91**, 93 (2008)
18. A. Gervasini, C. Messi, P. Carniti, A. Ponti, N. Ravasio, F. Zaccheria, J. Catal. **262**, 224 (2009)
19. J. Keranen, P. Carniti, A. Gervasini, E. Iiskola, A. Auroux, L. Niinistö, Catal. Today **91–92**, 67 (2004)
20. S. Bennici, P. Carniti, A. Gervasini, Cat. Lett. **98**, 187 (2004)
21. A. Gervasini, M. Manzoli, G. Martra, A. Ponti, N. Ravasio, L. Sordelli, F. Zaccheria, J. Phys. Chem. B **110**, 7851 (2006)
22. S. Bennici, A. Gervasini, Appl. Catal. B **62**, 336 (2006)

Chapter 6
Calorimetry at the Solid–Liquid Interface

Jerzy Jozef Zajac

Abstract Broad principles of Solid-Liquid calorimetry together with some illustrative examples of its use in the field of catalysis are presented here. The first use is related to the determination of surface properties of catalysts, adsorbents and solid materials in contact with liquids. In particular, it is shown how to evaluate the capacity of a given solid to establish different types of interaction with its liquid environment or to calculate its specific surface area accessible to liquids. The second use includes the measurement of the heat effects accompanying catalytic reactions and the related interfacial phenomena at Solid-Liquid and Liquid-Liquid interfaces. Examples of competitive ion adsorption from dilute aqueous solutions, as well as the formation of surfactant aggregates either in aqueous solution or at the Solid-Liquid interface are considered in view of potential applications in Environmental Remediation and Micellar Catalysis.

6.1 Introduction

*Micro*calorimetry, also *nano*calorimetry to follow the recent trends in thermal instrumentation and analysis, is a measuring technique that can be used to study interfacial phenomena occurring at the Solid-Liquid interface. Immersion of a solid in a pure liquid or a solution, wetting of a solid initially in contact with a gas or vapour by a liquid, adhesion between two condensed phases upon their "molecular" contact are examples of exothermic phenomena which are accompanied by significant heat evolvement. Competitive adsorption from solution is an important exception to the exothermicity of interfacial phenomena. This is because certain components of the

J. J. Zajac (✉)
Institut Charles Gerhardt Montpellier, UMR-5253 CNRS-UM2 Equipe Agrégats, Interface, et Matériaux pour l'Energie (AIME), Université Montpellier 2, C.C. 1502 Place Eugène Bataillon, 34095 Montpellier Cedex 5, France
e-mail: jerzy.zajac@univ-montp2.fr

A. Auroux (ed.), *Calorimetry and Thermal Methods in Catalysis*, Springer Series in Materials Science 154, DOI: 10.1007/978-3-642-11954-5_6, solution may compete against each other to adsorb at the interface. Adsorption from solution is thus considered as an exchange process and formally split into several adsorption and desorption steps. If one of the components is to be preferentially accumulated at the interface, the transfer of its molecules to the interface must be accompanied by the transfer of an equivalent amount of molecules of another component in the reverse direction, i.e., from the interface to the interior of the solution. The "displacement" is a frequent term for this process. The overall effect of such a displacement may be endothermic in numerous systems, thereby giving rise to an entropy-driven phenomenon. The complexity of displacement process is the main reason why the van't Hoff procedure for heat determination based on the measurement of the temperature dependence of adsorption isotherms frequently leads to unreliable values. In consequence, direct measurement of the thermal effect in these systems by calorimetry is strongly recommended.

Frequently the heat values for the displacement are relatively small, and thereby difficult to be detected. With the recent progress in ultra-sensitive heat flow measurements and the use of a wide variety of accessories to control the experimental conditions, new commercially available calorimeters offer maximum sensitivity, flexibility, and productivity. Nowadays it becomes possible to study the competitive adsorption phenomena with increased sensitivity and lower detection limits than previously possible.

In the field of catalysis, calorimetry may be used in two manners. The first use is related to the determination of surface properties of catalysts, adsorbents and solid materials in contact with liquids. In particular, it is possible to evaluate the capacity of a given solid to establish different types of interaction with its liquid environment or to calculate its specific surface area accessible to liquids. The second use of calorimetry includes the measurement of the heat effects accompanying catalytic reactions and the related interfacial phenomena at Solid-Liquid and Liquid-Liquid interfaces. In the present chapter, this group of calorimetry applications will be illustrated by following the examples of competitive ion adsorption from dilute aqueous solutions and the formation of surfactant aggregates either in aqueous solution or at the Solid-Liquid interface.

The present chapter does not pretend to be an exhaustive record of Solid-Liquid calorimetry applications in Surface Science and Technology. It should be rather regarded as an introductory course with some illustrative examples. It is important to realise that the individual author's experience in the field has been the principal criterion for selection of specific instruments and their uses, without any intention of neglecting other contributions. The presentation of calorimetry methods will be restricted only to interfacial systems composed of a pure liquid or a dilute binary, at the most, solution in contact with a solid which does not dissolve in the liquid phase. This formalism may be still employed in the case of solutions which are not strictly binary but may be viewed as such (e.g., solutions containing ionizable solutes, background electrolytes or other additives that may be lumped together as constituting a mean solvent or a mean solute).

6.2 Thermodynamic Treatment of the Solid–Liquid Interface and the Related Interfacial Phenomena

This opening paragraph reviews some basic ideas and methods relating to interfacial phenomena at the Solid-Liquid interface. Usually, the subject is characterised by two main approaches: the presentation of these phenomena in terms of thermodynamics and their molecular interpretation. The detailed treatment of such general concepts and relationships can be found in numerous standard texts [1–5] and the interested reader should consult these texts. Here only a very brief review is provided on which to base the entire text, in particular the language that will be widely used in further discussion.

6.2.1 Surface Excess Functions and Surface Phase Model

Contrary to the bulk liquid phase which is homogeneous in three directions in space, has a characteristic composition, and is also autonomous (i.e., its extensive properties depend only on the intensive variables characterising this phase such as the temperature T, the pressure P, and the chemical potentials of the solvent μ_1 and the solute μ_2), the formal thermodynamic description of a Solid-Liquid interface presents a serious difficulty. In the interfacial region, the density ω of any extensive quantity Ω changes continuously throughout the thickness (Fig. 6.1a).

For real solids, even the two-dimensional homogeneity of the interface is very difficult to attain, because the solid boundary is heterogeneous both in a physical and in a chemical sense (surface heterogeneity and roughness) [6, 7]. In order to overcome this difficulty, the only possible way is to introduce the so-called *excess* thermodynamic functions [1, 6, 8].

The widely used definitions of excess functions are based on the Gibbs model of the system, in which a flat interface is regarded as a mathematical dividing plane (the *Gibbs dividing surface*—GDS) [1, 6]. The two phases α and β in contact are assumed to remain homogeneous up to the GDS (Fig. 6.1b depicts such a model). Since the Gibbs model provides a complete description of the heterogeneous system in physico-chemical equilibrium, the formal thermodynamic study of interfacial phenomena is commonly based on this approach. Compared to bulk phases, the thermodynamic expressions for the interface contain additional terms relating to *interfacial tension* and *adsorption* of chemical species. This model has the advantage of leading rapidly to the principal thermodynamic relationships between the interfacial quantities (e.g., the so-called *Gibbs adsorption equation*) [1, 6].

Nevertheless, the Gibbs formalism has some serious drawbacks. Firstly, the interfacial properties (i.e., the Gibbs excess functions) adopt different values depending on the position of the GDS and, consequently, have no direct experimental significance. For dilute solutions, this imperfection may be removed through introducing the so-called *relative interfacial quantities*. Secondly, the Gibbs surface has zero

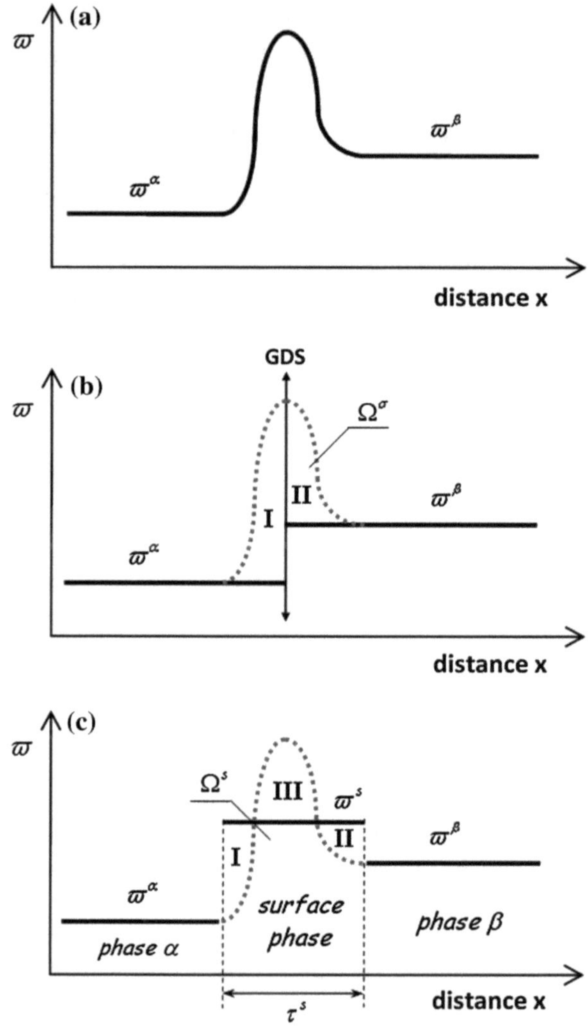

Fig. 6.1 A hypothetical profile of the density ϖ of some extensive property Ω (e.g., number of moles, internal energy, free energy, enthalpy, Gibbs energy, entropy) in the heterogeneous system as a function of the distance x perpendicular to the planar interface. (**a**) Real system: values of ϖ^{α} and ϖ^{β} are determined at such a distance from the interfacial region that the two phases have their bulk properties, (**b**) Gibbs model of the interface: value of the interfacial excess Ω^{σ} is given by a sum of areas I and II [1, 6]; (**c**) Surface phase model of the interface: interfacial property Ω^{s} is defined such that areas I, II and III compensate for one another [6, 8]

thickness and volume and this is at variance with the obvious physical picture of an interface. Furthermore, the physical meaning of the Gibbs excess functions is difficult to translate into molecular terms.

The alternative *surface phase* model popularized by Guggenheim is conceptually simpler since the interfacial region is approximated by a thin, homogeneous layer having an arbitrary thickness [6, 8]. This is illustrated in Fig. 6.1c where the two phases α and β remain homogeneous up to the imaginary planes which constitute the boundaries of the surface phase. This representation may be directly used for the purpose of constructing molecular models of the interface. The related interfacial properties have a real physical significance and are compatible with the experimentally measured quantities. For systems containing a *thermodynamically inert* solid in contact with a pure liquid or a dilute solution, one of the separating planes is chosen to coincide with the surface of the solid. Therefore, the area A of the interface (i.e., its cross-sectional area) is identified with the surface area of the solid phase accessible to the liquid. In practice, the main challenge is always to evaluate correctly the thickness τ^s of the interfacial region.

In the Guggenheim convention [8], the value of any *extensive property of the surface phase*, per unit area of the interface, may be expressed as

$$\Omega^s = \tau^s \cdot \varpi^s = \frac{1}{A}\left[\Omega - \left(\varpi^\alpha V^\alpha + \varpi^\beta V^\beta\right)\right] \tag{6.1}$$

where Ω is the total extensive property of the whole system; ϖ^α, ϖ^β and ϖ^s are the densities of Ω, respectively, in the two bulk phases and in the surface phase of thickness τ^s and surface area A; V^α and V^β represent the volume of the bulk phases α and β. The interfacial enthalpy H^s, interfacial Gibbs energy G^s, and interfacial entropy S^s are defined in such a manner.

It is satisfactory to define the *interfacial tension* γ as the work required to create isothermally and reversibly a unit area of an interface [1, 8]: γ_{SG} (Solid-Gas interface), γ_{SL}(Solid-Liquid interface), and γ_{LG}(Liquid-Gas interface). Since γ is referred to as an energy per unit area in this formulation, the privileged SI unit is J m^{-2}. Nevertheless, interfacial tensions reported in J m^{-2} and N m^{-1} have the same numerical value. Usually more convenient is the submultiple mJ m^{-2} or mN m^{-1} (numerically equivalent to the previously used c.g.s. units). Conceptually, the interfacial tension can be also seen as a new excess quantity that is attributed to the surface phase (or to the GDS) for the Guggenheim (or Gibbs) model to be *thermodynamically* equivalent to the real system [1, 8]. In heterogeneous systems where adsorption does not occur, like those containing the interface between one-component liquid and gas phases, interfacial tension is numerically equal to interfacial Gibbs energy G^s (per unit area of the interface). Otherwise, adsorption takes place with a change in interfacial tension.

From a mechanical standpoint, the interface between a pure liquid and its own equilibrium vapour (or air, *as adsorption is to be neglected here*) behaves as a membrane of infinitesimal thickness stretched uniformly and isotropically by a force exerted tangential to it. Rapid relaxation towards equilibrium is the hallmark of liquid surfaces: when the viscosity of the liquid is not too high, the freshly formed area has enough time to relax completely and the equilibrium interfacial tension will attain the same value in all surface parts. It is important to realise that, owing to non-equilibrated cohesive forces operating in the interfacial region, the liquid squeezes itself together until it has the locally lowest surface area possible. Therefore, γ_{LG} is regarded as a force (per unit length of surface edge) which opposes any attempt to increase the surface area. The terms *surface* and *interfacial tensions* are used interchangeably for γ_{LG}. The surface tension of most liquids (against equilibrium vapour or air) near room temperature ranges between 10 and 80 mJ m^{-2}, and decreases in a nearly linear fashion as the temperature rises. The description of the common experimental methods with comments on their suitability may be found in Refs. [4, 9].

In pure water, the collective action of intermolecular hydrogen bonds together with classical Van der Waals forces make water molecules stay close to one another (one water molecule is capable of forming four hydrogen bonds since it can accept two and donate two hydrogen atoms). When pure water is in contact with air, the great surface tension tends to minimize the area of hydrophobic-hydrophilic contact: the experimental value of γ_{LG} at 298.15 K is equal to 71.99 ± 0.05 mJ m^{-2} [10]. In the temperature range 273.15–323.15 K, the effect of temperature on the surface tension of water against air is given by [11, 12]:

$$\gamma_{LG} = 75.716 - 0.1416 \cdot (T - 273.15) + \\ - 0.25054 \times 10^{-3} \cdot (T - 273.15)^2 \tag{6.2}$$

Figure 6.2 illustrates the temperature dependence of the surface tension and the numerical procedure leading to the estimate of the surface enthalpy H^s_{LG} of water at room temperature.

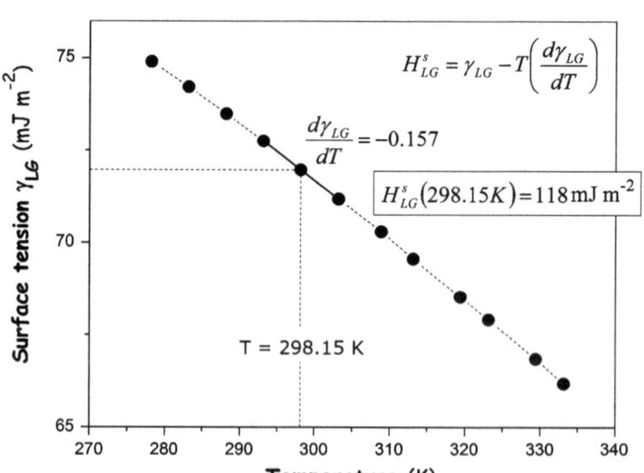

Fig. 6.2 Temperature dependence of the surface tension γ_{LG} of water in contact with air [13] and the numerical determination of the surface enthalpy H^s_{LG} at 298.15 K. The surface entropy S^s_{LG} at 298.15 K is equal to 0.157 mJ m^{-2} K^{-1}

The value of γ_{LG}, which is still equal to 58.9 mN m^{-1} at T = 373.15 K, approaches zero in the vicinity of the critical temperature T = 647.4 K, where there is no longer an interface between the liquid and the vapour.

In the case of solids, the term *surface tension* is used only to designate the interfacial tension γ_{S0} operating at the boundary between the solid phase and the surrounding vacuum. The determination of γ_{S0} for solid surfaces by stretching the surface area against the surface stress is not possible, since the latter is not equal to γ_{S0}. Solids do not deform reversibly and are capable of retaining their non-equilibrium shapes for a long time [4]. The determination of γ_{S0} is sometimes possible from the calculation of reversible work of cleaving a crystal, i.e., by creating fresh surface having the same properties as the original. Several known examples are given in Table 6.1. These results are, however, subject to considerable uncertainty, because the cleavage technique is not entirely reversible [14]. Furthermore, the different crystallographic faces have somewhat different surface tensions owing to the differences in packing density of the atoms.

6.2.2 Adhesion and Cohesion

Adhesion between a pure liquid and a solid may be described in terms of the interfacial and surface tensions [18]. Consider the reversible process of splitting a unit area of the Solid-Liquid interface in such a way as to create a unit area of the Solid-Vacuum interface and a unit area of the Liquid-Gas interface (as shown in Fig. 6.3a).

Adhesion between the two phases is defined as the reversed process and the *Gibbs energy of adhesion* is given, at fixed P and T, by the Dupré equation [4, 19–21]

$$\Delta_{adh} G_{SL} = \gamma_{SL} - \gamma_{S0} - \gamma_{LG}, \textit{per unit area of the interface} \qquad (6.3)$$

where γ_{SL} is the interfacial tension between both phases; γ_{S0} and γ_{LG} are the individual surface tensions of the solid against vacuum and the liquid against its equilibrium vapour (or air). For a single solid or liquid phase an analogous procedure (as shown in Fig. 6.3b) yields the *Gibbs energy of cohesion* (per unit area of the interface)

$$\Delta_{coh} G_S = -2\gamma_{S0} \quad \text{or} \quad \Delta_{coh} G_L = -2\gamma_{LG} \qquad (6.4)$$

Table 6.1 Surface tensions of several solid crystals against vacuum, as obtained from the work of cleavage, $w_{cliv} = 2 \cdot \gamma_{S0}$ [14–17]

Solid	Cleavage plane	γ_{S0} (mJ m^{-2})
Mica	–	4500
MgO	(100)	1200
CaF_2	(111)	450
LiF	(100)	340
$CaCO_3$	(001)	230
NaCl	(100)	110

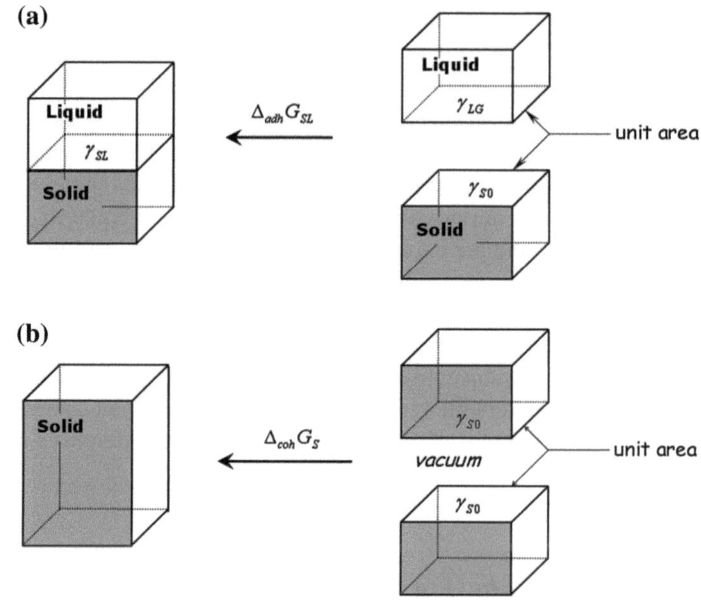

Fig. 6.3 Schematic illustration of the reversible process of (**a**) adhesion and (**b**) cohesion

Combination of Eqs. (6.3) and (6.4) yields

$$\gamma_{SL} = \Delta_{adh}G_{SL} - \frac{1}{2}\left(\Delta_{coh}G_S + \Delta_{coh}G_L\right) \quad (6.5)$$

In a sense, the right-hand side of Eq. (6.5) may be considered as a generalised Gibbs energy of mixing. If the two phases mix spontaneously in all proportions, the Gibbs energy will decrease during such a process, thereby rendering γ_{SL} negative. In thermodynamic terms this means that there is no stable interface. When the phases are immiscible, separation is spontaneous and the interfacial tension becomes positive. In that case, the interface is stable.

In broad outline, the *cohesion* of molecules (atoms, ions) to form the bulk phase of matter is due to long-range physical interactions (mainly of the van der Waals type) and short-range chemical forces (giving rise to covalent, ionic, metal, or hydrogen bonds). Amongst these various types of interactions encountered more or less frequently in interfacial phenomena involving liquid and solid phases, the London (dispersion) forces and the Lewis acid-base ones are really crucial in the construction of a thermodynamic treatment of interfaces [19, 21–23]. Solid and liquid substances may be classified according to their capacity of forming Lewis acid-base bonding. Materials that can be both Lewis acids (electron acceptors) and Lewis bases (electron donors) are termed *bipolar*. The *monopolar acidic* or *monopolar basic* substance can act exclusively as a Lewis acid or a Lewis base (the other property is negligible).

Inert materials, capable of neither acid nor base interactions, are called *apolar*. It should be noted here that this nomenclature [22, 23] has nothing in common with polarity of molecules per se, as measured by their respective dipole moments.

Dispersion forces are universal because they attract all molecules together, regardless of their specific chemical nature. The potential energy of dispersion attraction between two isolated molecules decays with the sixth power of the separation distance. Based on the so-called *Hamaker theory* (i.e., the method of pair-wise summation of intermolecular forces) or the more modern Lifshitz macroscopic treatment of strictly additive London forces, it is possible to develop the so-called Lifshitz-Van der Waals expression for the *macroscopic interactions* between macroscopic-in-size objects (i.e., macrobodies) [19, 21]. Such an expression strongly depends on the shapes of the interacting macrobodies as well as on the separation distance (non-retarded or retarded interaction). For two portions of the same phase of infinite extent bounded by parallel flat surfaces, at a distance h apart, the potential energy of macroscopic attraction is:

$$U(h) = -\frac{A_{11}}{12\pi \cdot h^2} \tag{6.6}$$

where A_{11} is the so-called *Hamaker constant* which depends on the chemical nature of the molecules (atoms, ions) constituting the phase under consideration and the number of molecules per unit volume in two interacting bodies. The more gradual fall-off of the potential energy (6.6) with distance compared to the molecule-molecule interaction indicates that macroscopic attractions are of a more long-range type and they are expected to make a significant contribution to the total energy of attraction even at longer distances.

The values of A_{11} for the various substances interacting across vacuum or across a medium can be found in Refs. [24, 25]. It should be noted that Hamaker constants for interaction across a medium are usually much lower in comparison with the related values under vacuum (e.g., in the case of two macroscopic bodies of quartz at short distances apart, A_{11}equals 41.3×10^{-20} J for interaction across vacuum and 1.3×10^{-20} J for interaction across water [25]).

The macroscopic effect of cohesion due to dispersion forces is usually calculated from the Lifshitz-Van der Waals expression (6.6), providing that the separation distance h is known. Israelachvili [21, 22] has proposed a universal value of 0.165 nm to describe the effective spacing h between molecular planes in all liquids with molecules interacting solely through dispersion forces. In this case, the Gibbs energy of cohesion may be evaluated as

$$\Delta_{coh} G_L \approx 9.74 \times 10^{17} A_{11}\, in\ Jm^{-2}, \quad \textit{for apolar liquids} \tag{6.7}$$

where A_{11} is the Hamaker constant for the liquid substance. In general, considerable theoretical and experimental evidence is consistent with the postulate that the Keesom-Debye contribution to the Gibbs energy of cohesion is very small and has no significant importance between macroscopic bodies in the condensed systems [21, 26]. As a result, the Lifshitz-Van der Waals (LW) component $\Delta_{coh} G^{LW}$ of the

Gibbs energy of cohesion for any liquid or solid material is commonly identified with the dispersion contribution.

According to Eq. 6.4, the apolar component of the surface tension of a solid against vacuum or a liquid against its equilibrium vapour (or air) becomes:

$$\gamma_{SO}^{LW} = -\frac{1}{2}\Delta_{coh}G_S^{LW} \quad \text{or} \quad \gamma_{LG}^{LW} = -\frac{1}{2}\Delta_{coh}G_L^{LW} \tag{6.8}$$

Lewis acid-base interaction between molecules (atoms, ions) differs from a classical covalent bond in that only one of the partners supplies the pair of electrons [27]. Electron pair donors (EPD) are molecules which donate the lone pair of non-bonding electrons (n-EPD), the electron pair of a σ-bond (σ-EPD), or the pair of π-electrons (π-EPD). Electron pair acceptor (EPA) molecules may use a vacant valence orbital (n-EPA), a nonbonding σ-orbital (σ-EPA), or a π-bond system with electron-withdrawing substituents (π-EPA). The combinations between all the above donor and acceptor types result in nine types of EPD-EPA complexes, with the bond strength ranging from high values for n-EPD/n-EPA associations to very week π-EPD/π-EPA interactions between neutral molecules. Formally, Lewis acid-base interaction includes a hydrogen bond which is usually situated at the lower end of the chemical bond range.

It has long been an operational premise that the Gibbs energy of cohesion for any *non-metallic* liquid or solid phase can be split into two contributions: an *apolar* one $\Delta_{coh}G^{LW}$ originating chiefly from dispersion forces and a *polar* one $\Delta_{coh}G^{AB}$ arising from the Lewis acid-base interaction between the constituent molecules (atoms, ions). Several semi-empirical methods have been proposed in the literature, e.g., [18, 26, 28–30], to determine both contributions. The results can be collated for different liquids and solids and subsequently used to predict the behaviour of new interfaces. Nevertheless, it is important to understand the approximate nature of this approach and to consider the resulting conclusions with caution.

In consequence, the total surface tension of a given *non-metallic* material against vacuum or its own equilibrium vapour is expressed by the sum of the Lifshitz-van der Waals (LW) and Lewis acid-base (AB) contributions [19, 22]

$$\gamma_{SO} = \gamma_{SO}^{LW} + \gamma_{SO}^{AB} = -\frac{1}{2}\left(\Delta_{coh}G_S^{LW} + \Delta_{coh}G_S^{AB}\right) \tag{6.9a}$$

$$\gamma_{LG} = \gamma_{LG}^{LW} + \gamma_{LG}^{AB} = -\frac{1}{2}\left(\Delta_{coh}G_L^{LW} + \Delta_{coh}G_L^{AB}\right) \tag{6.9b}$$

The change of Gibbs energy during *adhesion* between two phases is the macroscopic outcome of interactions between the microscopic constituents of the different phases. According to the empirical *Berthelot principle* [21], the London energy of attraction between two dissimilar macrobodies is a geometric mean of the mutual interactions between similar objects, so that

$$A_{12} \approx \sqrt{A_{11}A_{12}} \tag{6.10}$$

where A_{12} is the Hamaker constant referring to macroscopic interaction between two different phases.

To obtain the apolar (LW) contribution to the Gibbs energy of adhesion $\Delta_{adh} G_{SL}^{LW}$ between a solid and a liquid, it is assumed that expressions analogous to that given by Eq. 6.7 are still valid. The Berthelot principle Eq. 6.10 may be therefore used to evaluate $\Delta_{adh} G_{SL}^{LW}$. The combining rule for this component of $\Delta_{adh} G_{SL}$ is given by the Good-Girifalco-Fowkes relation [18, 31]:

$$\Delta_{adh} G_{SL}^{LW} = \sqrt{\Delta_{coh} G_S^{LW} \cdot \Delta_{coh} G_L^{LW}} = -2\sqrt{\gamma_{S0}^{LW} \cdot \gamma_{LG}^{LW}} \tag{6.11}$$

If one of the condensed phases is apolar, dispersion forces are the only important type of interaction operating across the interface and $\Delta_{adh} G_{SL} = \Delta_{adh} G_{SL}^{LW}$. The apolar (LW) surface tension component of any solid can be thus determined by the measurement of the Gibbs energy of adhesion between this material and an apolar probe substance. The latter may be a liquid alkane, methylene iodide, or α-bromonaphthalene, for which the surface tension $\gamma_{LG} = \gamma_{LG}^{LW}$ has already been measured.

The two-condensed-phase analog of Eq. 6.9a, 6.9b is:

$$\Delta_{adh} G_{SL} = \Delta_{adh} G_{SL}^{LW} + \Delta_{adh} G_{SL}^{AB} \tag{6.12}$$

The polar (AB) contributions to the surface tension and Gibbs energy of adhesion are sometimes expressed in terms of Van Oss-Chaudhury-Good parameters [18, 19, 22, 32]

$$\gamma^{AB} = 2\sqrt{\gamma^A \cdot \gamma^B} \tag{6.13a}$$

$$\Delta_{adh} G_{SL}^{AB} = -2\sqrt{\gamma_{S0}^A \cdot \gamma_{LG}^B} - 2\sqrt{\gamma_{S0}^A \cdot \gamma_{LG}^B} \tag{6.13b}$$

where γ^A and γ^B are the Lewis acid and Lewis base parameters of surface tension γ, respectively. This approximation leads to the following interesting conclusions: the polar (AB) surface tension component of a pure substance is equal to zero, if the substance is monopolar or apolar. There is no acid-base interaction across the interface, i.e., $\Delta_{adh} G_{SL}^{AB} = 0$, if one of the components is apolar or if the two components are monopolar in the same sense, i.e., both being monofunctional acids or both monofunctional bases.

The Lewis acid and base contributions to the surface tension of solids can be derived from measurements of the Gibbs energy of adhesion between the material and a polar probe liquid. At least two different polar liquids (e.g., water and appropriate monofunctional liquid) must be used as probes, provided that their surface components γ_{LG}^{LW}, γ_{LG}^A, γ_{LG}^B are known.

6.2.3 Wetting in Solid–Liquid Systems

Wetting includes the spreading of a pure liquid over the surface of a solid, displacing the gas (or vapour) initially in contact with that surface [18, 33]. Hence the phenomenon involves three interfaces, namely Solid-Gas, Solid-Liquid, and Liquid-Gas ones. The *spreading coefficient* W_S is defined as [4, 33]

$$W_S = \gamma_{SG} - (\gamma_{SL} + \gamma_{LG}) \tag{6.14}$$

where γ_{SG}, γ_{SL} and γ_{LG} are the appropriate interfacial tensions for the three interfaces at equilibrium. When a portion of the liquid is placed on a uniform, perfectly flat, and non-deformable solid surface and the two phases are allowed to come to equilibrium with the surrounding gas phase, one of the two events may happen:

1. When W_S is positive or zero, the liquid wets the solid material, i.e., spreads out spontaneously over its surface, providing there is enough liquid to eliminate a unit area of the Solid-Gas interface while exposing a corresponding amount of the Solid-Liquid and Liquid-Gas interfaces.
2. When W_S is negative, the liquid remains as a drop having, at equilibrium, a definite angle of contact with the solid surface (the liquid does not wet the solid). This case is illustrated in Fig. 6.4.

The equilibrium *contact angle* Θ between the liquid and the solid phases is determined by the following balance of interfacial tensions [4, 33, 34]:

$$\cos\Theta = \frac{\gamma_{SG} - \gamma_{SL}}{\gamma_{LG}} \tag{6.15}$$

known as Young's equation. This relation applies for contact angles Θ less than, equal to, or greater than 90°. In the limiting case where $\Theta = 0$, the liquid wets out the solid.

The classical form of Young's equation, which describes the equilibrium balance of forces meeting at the three-phase contact line in the plane of the solid surface (see Fig. 6.4), is one of the most controversial expressions in Surface Science and there is

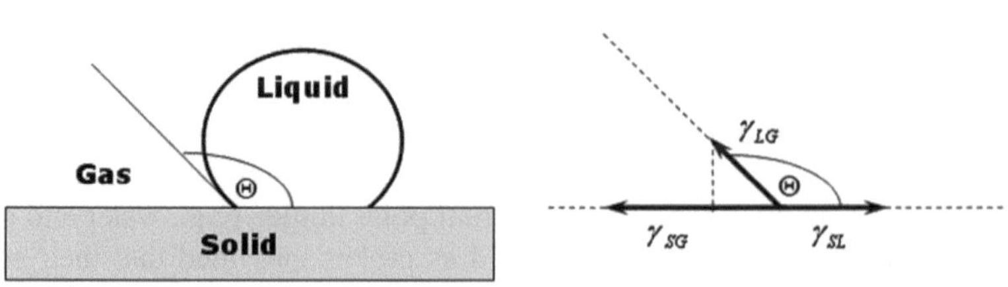

Fig. 6.4 A drop of a non-spreading liquid on a flat solid surface together with the traditional representation of the vectorial equilibrium between respective interfacial tensions viewed as forces acting along the perimeter of the drop

a long list of objections to it (e.g., see Refs. [4, 33, 34] for details). In spite of that, it is still very widely encountered in the literature. To derive it, one must assume an ideal solid: chemically homogeneous, thermodynamically inert (e.g., it cannot swell under the action of the liquid neither can dissolve in the liquid), and flat at an atomic scale. In practice, appreciable hysteresis of the contact angle is observed in real systems (chemical heterogeneity and roughness of solid surfaces), depending on whether the liquid is advancing or receding across the solid surface [33]. Advancing contact angles are larger than receding angles, and the difference may be sometimes as much as 20°–30°. A very detailed critical discussion of the various methods for measuring contact angles can be found in Ref. [34]. The contact angles of powdered solids (e.g., clay minerals) are technically important but are difficult to measure. In the case of numerous fine-grained minerals, which do not occur as large, perfect single crystals with well-developed faces, the contact angles are determined indirectly by column and thin layer wicking [35, 36].

If the liquid is volatile, the gas phase will contain its vapour. In consequence, even though both the fluid phases are nominally pure components, there is in general finite adsorption at the Solid-Gas interface. The equilibrium value of γ_{SG} for this interface will be, therefore, lower than its pure-component value γ_{S0} by an experimentally determinable quantity, which is called the *two-dimensional* or *surface pressure* π_{SG} [4, 5]. For the one-component gas phase under the conditions of sufficiently low pressures in contact with an inert solid adsorbent, the value of π_{SG} can be evaluated using the adsorption isotherm for the vapour of the liquid on the solid surface:

$$\pi_{SG} = \gamma_{S0} - \gamma_{SG}\left(p^{*}\right) = RT \int_{0}^{p*} \Gamma^{s} d \ln p \quad \mathrm{T, P = const} \tag{6.16}$$

where R is the molar gas constant; Γ^s denotes the number of moles of gas adsorbed per unit area of the solid adsorbent related to the equilibrium bulk pressure p in the bulk gas phase, at constant temperature T and pressure P; p^* is the equilibrium pressure at which the actual adsorbed film has been formed on the solid surface: the integration of the adsorption isotherm $\Gamma^s = \Gamma^s(p)$ is carried out over a p-range from 0 to p^*. On applying Eq. 6.16 to the solid surface saturated with the vapour, i.e., when the latter forms an equilibrium, physically adsorbed film on the available adsorbent surface at p equal to the saturation vapour pressure p_{sat}, the surface concentration Γ^s and the surface pressure π_{SG} are found to reach their (positive) maximum values, Γ^s_m and π^m_{SG}, respectively.

In the case of solids having relatively small values of surface tension γ_{S0} against vacuum (usually less than 100 mN m^{-1} [2]), the effect of gas adsorption is thought to be of little importance. Polymers and many other solid organic compounds are usually given as examples of this category of substrates. Solid materials for which π^m_{SG} is small for any adsorbate are named the *low-energy solids*. For *high-energy solids* (e.g., mineral oxides, metal sulphides, inorganic salts), the decrease in surface tension due to adsorption is significant [30, 37, 38].

Equations 6.3, 6.15, and 6.16 may be combined to give the Gibbs energy of adhesion between the solid and liquid phases

$$\Delta_{adh} G_{SL} = -\gamma_{LG} \left(1 + \cos \Theta\right) - \pi_{SG}^{m} \quad (6.17)$$

This is a very useful relation, in which γ_{LG}, $\cos \Theta$, and π_{SG}^{m} can be measured and calculated quite easily and accurately. Therefore, the operating procedures for the adhesion experiment are usually based on Eq. 6.17.

6.2.4 Hydrophobic and Hydrophilic Substances

Commonly the distinction between hydrophobic and hydrophilic substances is based on the analysis of interactions between their molecules and water as a solvent. A more precise classification of liquid and solid substances as hydrophobic and hydrophilic may be constructed basing on the apolar (LW) and polar (AB) components of their surface tensions. This three-parameter approach is of great importance for the understanding of surface behaviour [22, 32, 39, 40]. A non-metallic substance is *hydrophobic* if it interacts with water by exhibiting only LW character. It has very little (or none at all) Lewis acid or Lewis base character. Typical substances at the hydrophobic end have low γ^{LW} surface parameters and their γ^{A} and γ^{B} components are equal to zero. *Hydrophilic* substances have non-zero γ^{LW} components of the surface tension and at least one of their γ^{A} and γ^{B} parameters is significant.

The values of surface tension components that have been derived from the appropriate measurements of the Gibbs energy of adhesion for numerous liquids and solids are listed in Table 6.2. The determination of a set of γ_{LG}^{A} and γ_{LG}^{B} values for probe liquids is based on the choice of the first reference liquid. For this purpose, Van Oss et al. [39] assumed that $\gamma_{LG}^{A} = \gamma_{LG}^{B}$ for water. In consequence, all of acid-base parameters in Table 6.2 are relative to those of water.

6.3 Calorimetry Applied to Evaluate Surface Properties of Solids

The determination of the apolar and polar components of the surface tension of liquids and solids may be a powerful tool for classification of various substances with respect to their hydrophobic-hydrophilic character. In the case of solids, this macroscopic approach provides important information about the potential "global" behaviour of their surfaces against a given environment, however, giving no direct indication in regard with the heterogeneity of the solid surface. In consequence, it cannot replace methods based on the adsorption of probe molecules from the gas phase, but it does complement them by providing a different level of surface scanning.

Table 6.2 Surface tension parameters (in mJ m^{-2}) of some liquids and polymers [23, 32, 39]

Substance	γ^{LW}	γ^{A}	γ^{B}	γ_{SO}/γ_{LG}
Water	21.8	25.5	25.5	72.8
Liquids				
n-Heptane	20.1	0	0	20.1
n-Decane	23.8	0	0	23.8
Chloroform	27.15	3.8	0	27.15
n-Hexadecane	27.5	0	0	27.5
α-Bromonaphthalene	44.4	≈0	≈0	44.4
Methylene iodide	50.8	≈0	≈0	50.8
Ethylene glycol	29.0	1.92	47.0	48.0
Formamide	39.0	2.28	39.6	58.0
Glycerol	34.0	3.92	57.4	64.0
Polymers				
Poly(methylmethacrylate), cast film	39–43	≈0	9.5–22.4	39–43
Poly(vinylchloride)	43	0.04	3.5	43.72
Poly(oxyethylene):				
PEG 6000	45	≈0	66	45
Cellulose acetate	35	0.3	22.7	40.2
Cellulose nitrate	45	0	16	45
Poly(styrene)	42	0	1.1	42

In practice, the experimental procedure is quite long and fastidious referring to the successive determinations of the Gibbs energy of adhesion $\Delta_{adh}G_{SL}$ between a given solid and an apolar or polar liquid, which requires, in accordance with Eq. 6.17, measurement of the surface tension γ_{LG}, contact angle Θ, and vapour adsorption isotherm (to calculate π^{m}_{SG}) for each solid-liquid couple. To make matters worse, the very precise measurement of Θ is possible only for atomically smooth surfaces. Finally, the additive approximation expressed by Eqs. 6.9a, 6.9b and 6.12 is better suited to calculation of the enthalpy term than to that of the free energy, since the interactions may have both mechanical and entropic contributions [41].

The determination of the related surface enthalpy terms H^{s}_{S0} and H^{s}_{LG} from direct calorimetry measurements provides an alternative way to evaluate the hydrophobic-hydrophilic character of a solid surface [38, 42–44], since the following deconvolution procedures may be proposed for the surface enthalpy in analogy with those holding for the Gibbs energy (Eqs. 6.8–6.13a, 6.13b):

- surface enthalpy of a non-metallic solid against vacuum:

$$H^{s}_{S0} = H^{LW}_{S0} + H^{AB}_{S0} = H^{LW}_{S0} + 2\sqrt{H^{A}_{S0} \cdot H^{B}_{S0}} \tag{6.18a}$$

- surface enthalpy of a non-metallic liquid against its equilibrium vapour (or air):

$$H_{LG}^{s} = H_{LG}^{LW} + H_{LG}^{AB} = H_{LG}^{LW} + 2\sqrt{H_{LG}^{A} \cdot H_{LG}^{B}} \tag{6.18b}$$

- Solid-Liquid interfacial enthalpy (*Berthelot principle*):

$$H_{SL}^{s} = H_{SL}^{LW} + H_{SL}^{AB} = H_{S0}^{s} + H_{LG}^{s} + \\ - 2\left(\sqrt{H_{S0}^{LW} \cdot H_{LG}^{LW}} + \sqrt{H_{S0}^{A} \cdot H_{LG}^{B}} + \sqrt{H_{S0}^{B} \cdot H_{LG}^{A}}\right) \tag{6.18c}$$

where H_{S0}^{LW}, H_{LG}^{LW} and H_{S0}^{AB}, H_{LG}^{AB} are, respectively, the apolar (LW) and polar (AB) contributions to the appropriate surface enthalpy; H_{S0}^{A}, H_{LG}^{A} denote the Lewis acid and H_{S0}^{A}, H_{LG}^{A} the Lewis base components of the surface enthalpy ; the H_{SL}^{LW}, H_{SL}^{AB} components refer to the Solid-Liquid interface.

It should be always remembered that the global treatment of surface hydrophobicity-hydrophilicity based on surface and interfacial enthalpies does not include the entropy effects.

For liquids and solids, specific orientation and conformation of *unsymmetrical* molecules (ions) in the interfacial regions result not only in the maximization of their interaction energy, but also yield entropy effects that cannot be neglected. For example, molecular dynamics calculations of the intermolecular potential function points to a predominant orientation of the water dipoles at the Liquid-Gas interface [45]. Other examples are an icelike structuring of water molecules in the vicinity of crystalline solid surfaces [46] and a specific orientation of the alcohol molecules in the interface between a liquid *n*-alkanol and water [47].

Immersional and wetting calorimetry is an important method for studying interactions at the Solid-Liquid interface [48–50], especially in the case of finely divided and porous solids where the direct measurements of contact angle are hardly possible.

6.3.1 *Enthalpy Changes in the Thermodynamic Cycle of Immersion-Adsorption–Wetting*

Consider a simple experiment in which a clean solid surface (free of adsorbed liquid and vapour impurities) is immersed in an excess of pure liquid (Path 1 in Fig. 6.5). If thermal effects arising from absorption, solubility, and swelling of a solid may be eliminated, the whole enthalpy change on *immersion* is ascribed only to the interface. Sometimes the immersion of a solid in a liquid is accompanied by the formation of an electrical double layer. For mineral oxide-water systems [51, 52], the double-layer effects (i.e., generation of surface charge by protonation or deprotonation of some surface hydroxyl groups, and adsorption of counterions in the Stern or/and diffuse layers) are clearly secondary in comparison with the basic wetting (this contribution is 10–15 % of the total heat effect, at the most).

The total enthalpy change, at constant P and T, is then written as

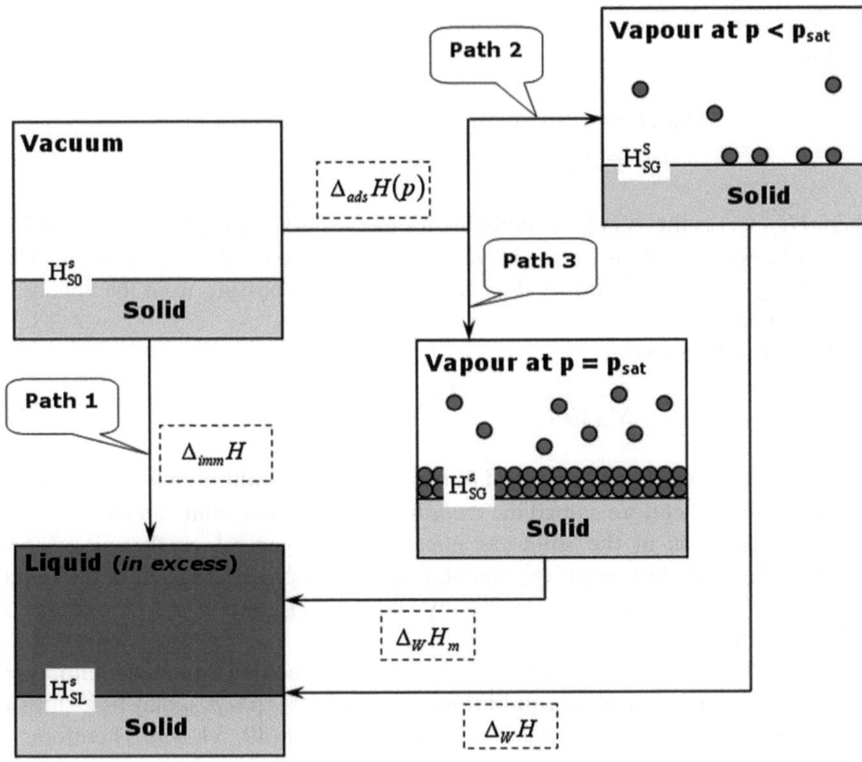

Fig. 6.5 Schematic representation of the difference between immersion (clean solid surface) and immersional wetting (solid surface pre-covered with vapour); the excess of liquid is high enough for the enthalpy change in the bulk liquid phase during adsorption or immersion to be neglected

$$\Delta_{imm} H = A_{SL} \cdot \Delta_{imm} H^* = A_{SL} \left(H^s_{SL} - H^s_{S0} \right) \tag{6.19}$$

where $\Delta_{imm} H$ is called the *enthalpy of immersion*; A_{SL} is the area of the Solid-Liquid interface (often identified with the surface area of a non-microporous solid); H^s_{SL} and H^s_{S0} are the interfacial enthalpies per unit area for the Solid-Liquid and Solid-Vacuum interfaces, respectively.

The experiment of immersion is sometimes performed under completely different conditions. The solid may be first put in equilibrium with the vapour of the immersional liquid at a given equilibrium pressure p (Paths 2 or 3 in Fig. 6.5). The adsorbed gas may be at submonolayer, monolayer or multilayer coverage, depending chiefly on the value of p, but also on the nature of liquid and solid. When the solid is subsequently immersed in the liquid, the measured enthalpy change, called the *enthalpy of immersional wetting*, $\Delta_W H$, will be different from $\Delta_{imm} H$. The various stages of the immersion-adsorption-wetting cycle are shown in Fig. 6.5.

In the case of ideal wetting (this means that the contact angle among the solid, immersional liquid, and vapour of this immersional liquid $\Theta = 0$), the final state is always the same, irrespective of the path followed. Therefore, one can write for Path 2 in Fig. 6.5:

$$\Delta_W H = A_{SL} \cdot \Delta_W H^* = A_{SL}\left[H^s_{SL} - H^s_{SG}(p)\right] \\ = \Delta_{imm} H - \Delta_{ads} H(p) \tag{6.20}$$

where $H^s_{SG}(p)$ is the interfacial enthalpy for the Solid-Gas interface at equilibrium pressure p. The area of the Solid-Gas interface is taken to be identical with A_{SL} (the accessibility of the solid surface does not change when passing from the vapour to the liquid phase).

The enthalpy change

$$\Delta_{ads} H(p) = A_{SL}\left[H^s_{SG}(p) - H^\sigma_{S0}\right] \tag{6.21}$$

refers to the formation of an adsorbed film onto solid in equilibrium with the current gas phase and is therefore named the *enthalpy of adsorption* from vapour.

Since adsorption at the solid-gas interface is in general exothermic, $\Delta_W H$ increases (becomes less negative) monotonously with precoverage from the initial value $\Delta_{imm} H$ at p = 0 to a limiting steady value $\Delta_W H_m$ at p = p_{sat}. Examples of such curves are presented in Fig. 6.6.

As far as the enthalpy is concerned, the interface between the saturated (multilayer) film on the solid surface and the equilibrium vapour phase at p = p_{sat} may be identified with the interface between the liquid and its own vapour [49, 54, 55]. Therefore,

$$\Delta_W H_m = A_{SL}\left[H^s_{SL} - H^s_{SG}(p_{sat})\right] = -A_{SL} H^s_{LG} \tag{6.22}$$

where H^s_{LG} is the surface enthalpy of the immersional liquid.

6.3.2 Immersional and Wetting Calorimetry Experiments

Measurements of the enthalpy changes accompanying the immersion and wetting phenomena may be performed with a Tian-Calvet type differential calorimeter [49, 54]. The experimental procedure includes several intermediate stages which may take much time, especially when the wetting enthalpy is measured as a function of the surface coverage by the vapour of the immersional liquid.

The first stage corresponds to the sample preparation during which a solid sample of a given mass is placed in a bulb made of high temperature glass and closed by a brittle tail (Fig. 6.7a). The subsequent sample evacuation and pre-coverage steps are performed with the solid enclosed in this bulb, as represented schematically in Fig. 6.7. The bulb fixed to the end of a glass tube is placed in the outgassing rig where the sample is evacuated at a high temperature to remove all adsorbed impurities from its surface. Then the tube is sealed off a few cm from the bulb end, fixed to the end

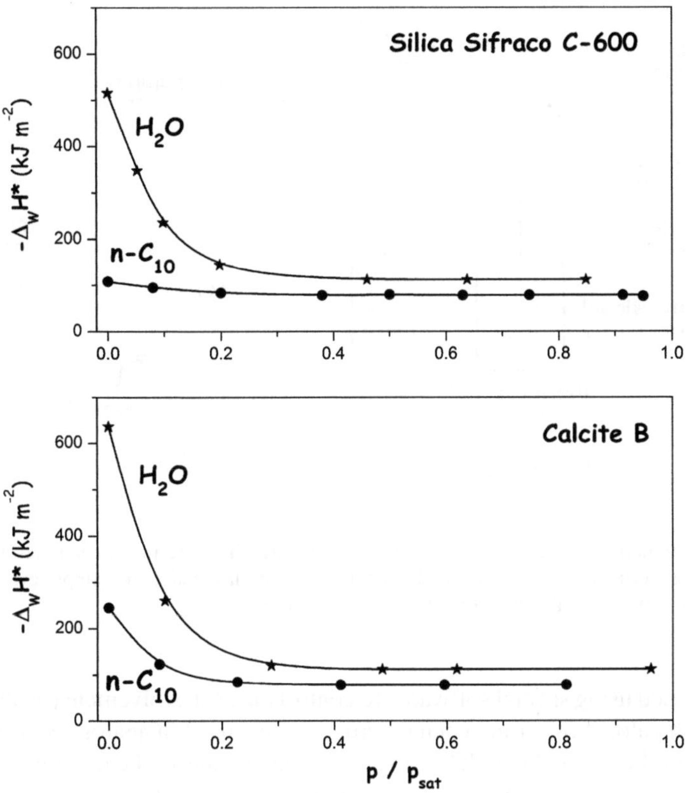

Fig. 6.6 Enthalpy of immersional wetting per unit surface area $\Delta_W H^*$ (taken with the opposite sign) as a function of the surface pre-coverage for two powdered solid samples in two immersional liquids: water (H_20) and *n*-decane (n-C_{10}) [53]

of glass rod and transferred to the calorimeter. Prior to wetting experiment, the solid sample is outgassed in a glass bulb and then brought into contact with the vapour of the immersional liquid at a given pressure and at constant temperature T_M. The equilibrium pressure of the pre-coverage step is controlled by the temperature T_b of the thermostated bath (T_b must be lower than ambient temperature and T_M).

The Tian-Calvet type calorimeter system contains a massive calorimetric bloc which acts as a heat sink (its temperature is constant) and a removable calorimetric cell made of stainless steel and designed to fit in a cylindrical hole inside the calorimetric bloc. The instrumental signal is obtained by measuring the heat flux between the cell and the bloc. The temperature signal is derived from a sensor in the bloc: two symmetrical thermal flux meters, each constructed by a series of thermocouples surrounding a cylindrical hole for the measurement cell. The electric signal delivered by the difference in output voltage of the two flux meters is proportional to the temperature difference θ between the bloc and the cell. The electric signal is directly fed into a computer; the digitized signal is recorded on the computer hard disk and

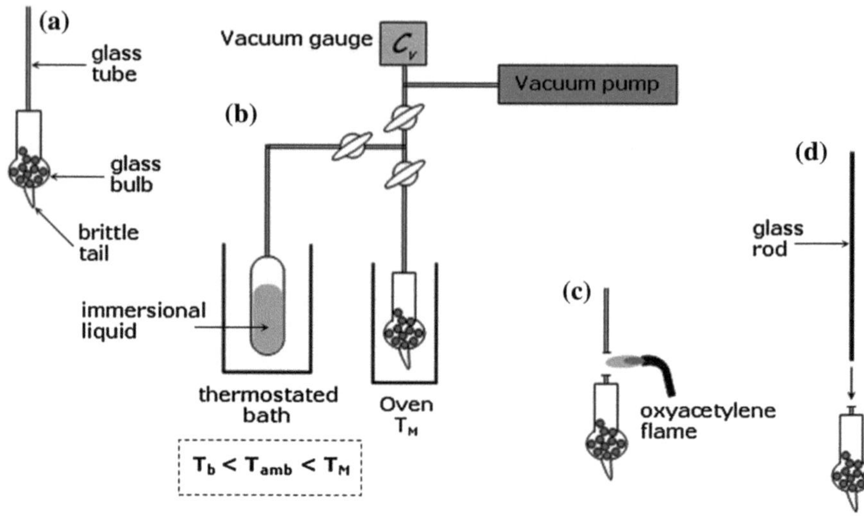

Fig. 6.7 Schematic representation of the sample preparation stage in immersional and wetting calorimetry experiments: (**a**) solid sample enclosed in the glass bulb, (**b**) sample evacuation and pre-coverage, (**c**) sealing the end of the tube, (**d**) assembling

then processed using special software. To control the heat evolvement (or absorption) during each calorimetric run, a pen recorder may be used in analog recording of the signal; here, the vertical pen deflection Δl perpendicular to the direction of feed of the recording chart is proportional to the temperature difference θ. The scheme of the calorimetric cell is shown in Fig. 6.8, together with a trace showing a representative thermal profile for immersion.

According to the theoretical equation of Tian [56–59] for a conduction calorimeter working under ideal conditions (e.g., the thermal delay between the temperature change θ in the calorimetric cell and the response of the sensor is to be neglected), the total heat effect occurring in the calorimetric cell during the time of experiment t_{exp} is given by the following expression:

$$Q_{\text{exp}} = \int_0^{t_{\text{exp}}} P(t)dt = \frac{\lambda}{g} \int_0^{t_{\text{exp}}} \Delta dt + \frac{\Lambda}{g} \int_{\Delta_1}^{\Delta_2} d\Delta \tag{6.23}$$

where *P(t)* is the heat generation (absorption) rate at time t, λ is the thermal conductivity of conducting surface separating the calorimetric cell and the calorimetric bloc, Λ is the heat capacity constant of the calorimetric cell with its content, Δ is the voltage signal produced by the sensor at time t, and g is the proportional constant in the relation $\Delta = g \cdot \theta$; Δ_1 and Δ_2 denote the initial (t = 0) and final (t = t_{exp}) sensor indications, respectively. The first term on the right hand side of Eq. 6.23 represents the heat exchange between the bloc and the cell *monitored by the sensor during the*

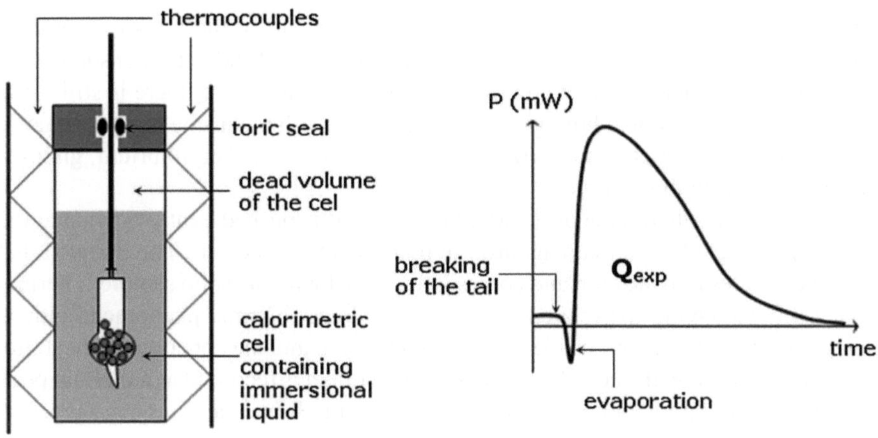

Fig. 6.8 Scheme of the calorimetric cell in a Tian-Calvet type differential calorimeter and a trace showing a representative thermal profile for immersion. Q_{exp} is the overall thermal effect recorded

experiment, whereas the second term corresponds to the temperature rise (decrease) in the calorimetric cell. When the duration of the experiment t_{exp} is chosen such that $\Delta_1 = \Delta_2$ (i.e., the sensor signal returns to the baseline), this second term is always equal to zero and the total heat effect Q_{exp} is determined by integrating the sensor record $\Delta = \Delta(\mathrm{t})$ from Δ_1 to Δ_2:

$$Q_{\mathrm{exp}} = \frac{\lambda}{g} \int_0^{t_{\mathrm{exp}}} \Delta dt = K \int_0^{t_{\mathrm{exp}}} \Delta dt \tag{6.24}$$

The calorimeter constant $K = \frac{\lambda}{g}$ is evaluated during the *calibration run*. Calibration of the area under each thermal peak in the *thermogram* $\Delta = \Delta(t)$ is carried out by dissipating a known amount of energy in the cell (heating through Joule effect). For this purpose, a special calibration resistor is placed in a glass bulb and introduced into the calorimetric cell under the conditions of real experiment. The operator decides both the power dissipated in the resistor $I^2 \cdot R$(I—current flowing through the resistor and R—its resistance) and the duration of the calibration step t_{cal}. The integration of the resulting thermal peak allows the calibration constant K to be calculated as follows:

$$K = \frac{A_{cal}}{I^2 \cdot R \cdot t_{cal}} \tag{6.25}$$

where A_{cal} is the area under the calibration peak.

The standard operating procedure for heat measurement is as follows [49, 54]. The glass bulb containing the solid sample after the sample evacuation (and pre-coverage) stage is introduced into the calorimetric cell, which has been previously filled with the immersional liquid (usually about 15cc). A large part of the glass rod remains outside when the cell is closed. Complete gas-tightnest of the cell is ensured by a special toric seal placed around the rod. At a given temperature, there is still some vapour of the immersional liquid occupying the dead volume inside the calorimetric cell. Then the calorimeter is left overnight to come to thermal equilibrium, giving a steady baseline on the recorder.

After attaining thermal equilibrium, the glass rod is pushed gently down and the glass tail of the bulb is broken against the bottom side of the cell. The immersional liquid penetrates into the bulb and comes into contact with the solid sample. Thermal effects accompanying the related exothermic and endothermic phenomena induce changes in the temperature inside the calorimetric cell and the concomitant heat flux between the bloc and the cell. The global thermal effect is recorded as a thermal peak A_{exp} which can be integrated and transformed to the heat quantity Q_{exp} (Fig. 6.8) making use of the calibration constant K:

$$Q_{\text{exp}} = K \cdot A_{\text{exp}} \tag{6.26}$$

To extract the net enthalpy of immersion $\Delta_{imm}H$ or net enthalpy of wetting $\Delta_W H$, several correction terms have to be subtracted from Q_{exp}. These correction terms are related to (i) breaking the tail of the glass bulb inside the calorimetric cell, (ii) changes in the dead volume of the calorimetric cell and the bulb during the experiment, (iii) evaporation of the immersional liquid and condensation of its vapour, (iv) decrease in the adsorbent mass during the evacuation step. They may be determined in a blank test recording. Nevertheless, only two of them were found to give a noticeable contribution to the total heat effect [43, 54].

When a given volume of the immersional liquid enters the bulb, the liquid level in the cell is lowered and there is some evaporation of the liquid to equilibrate the vapour pressure in the dead volume. This phenomenon yields an endothermic effect which is proportional to the dead volume V_0 of the bulb:

$$Q_{cor} = -V_0 \cdot \Delta_{vap}h \tag{6.27}$$

where $\Delta_{vap}h$ is the evaporation enthalpy per unit volume measured in a blank run with an empty bulb (without a solid sample). The values of $\Delta_{vap}h$ obtained for some immersional liquids are given in Table 6.3.

The second correction term of great importance concerns the loss of the adsorbent mass during the evacuation step [43]. Prior to immersion and wetting measurements, the solid sample is subjected to thermal treatment in the glass bulb under vacuum. The total mass of the sample certainly decreases, depending on the amount of liquid and vapour impurities pre-adsorbed on its surface. The present experimental procedure does not allow the actual mass to be determined precisely because the bulb containing the sample cannot be weighed at the end of the outgassing process. In consequence, the measured enthalpy values may be underestimated, especially if the weight loss is significant (e.g., hydrophilic solids with a high specific surface area may contain much water vapour adsorbed on the surface). The most reliable way of quantifying

Table 6.3 Enthalpies of evaporation per unit volume $\Delta_{vap}h$ for selected immersional liquids, as measured in a blank calorimety run with an empty bulb [60]

Immersional liquid	Δ_{vap} h(mJ cm^{-3})
n-heptane	126
Isooctane	59
Chloroforme	254
Benzene	249
Water	78
Formamide	24
l-butanol	160

the mass decrease is to carry out the outgassing procedure separately under the same conditions as those used in the calorimetry experiment (the outgassed sample may be weighed and the mass difference calculated).

Finally, the enthalpy change upon immersion of a solid in a given liquid per unit surface area of the solid is calculated as follows:

$$\Delta_{imm}H^* = \frac{\Delta_{imm}H}{A_{SL}} = -\frac{1}{m_S \cdot S}\left(Q_{\text{exp}} - V_0 \cdot \Delta_{vap}h\right), \text{ in mJ m}^{-2} \qquad (6.28)$$

where m_S and S are the mass and the specific surface area of the "dry" solid sample, respectively.

6.3.3 Hydrophilic-Hydrophobic Series and Harkins-Jura Method

Using approximation Eq. 6.18c to express the interfacial enthalpy H_{SL}^s in terms of the apolar (LW), Lewis acid (A) and Lewis base (B) components of the surface enthalpy for the solid and immersional liquid, the final explicit form for the enthalpy of immersion is as follows [42–44]:

$$\Delta_{imm}H = H_{LG}^s - 2 \cdot \left(\sqrt{H_{S0}^{LW} \cdot H_{LG}^{LW}} + \sqrt{H_{S0}^{A} \cdot H_{LG}^{B}} + \sqrt{H_{S0}^{B} \cdot H_{LG}^{A}}\right) \qquad (6.29)$$

Provided that the surface enthalpy components for the liquids, i.e., H_{LG}^{LW}, H_{LG}^{A}, H_{LG}^{B} are known, the apolar (H_{S0}^{LW}), Lewis acid (H_{S0}^{A}), and Lewis base (H_{S0}^{B}) contributions to the surface enthalpy of solids can be derived from measurements of the enthalpy of immersion. At least one apolar liquid and two polar liquids must be used. The values of H_{LG}^{LW}, H_{LG}^{A}, H_{LG}^{B} for selected immersional liquids, i.e., apolar *n*-heptane, bifunctional water, and monofunctional basic formamide, are given in Table 6.4.

The examples of the use of apolar *n*-heptane, bifunctional water, and monofunctional basic formamide to study the surface hydrophobic-hydrophilic character of several solids are shown in Figs. 6.9, 6.10, and 6.11 [38, 43, 44, 60]. *n*-heptane

Table 6.4 Surface enthalpy components for *n*-heptane, water and formamide [43]

Liquid	H_{LG}^{LW} mJ m^{-2}	H_{LG}^{A} mJ m^{-2}	H_{LG}^{B} mJ m^{-2}	H_{LG}^{s} mJ m^{-2}
n-heptane	54.5	0	0	54.5
Water	35.0	41.5	41.5	118.0
Formamide	55.6	3.25	56.4	82.6

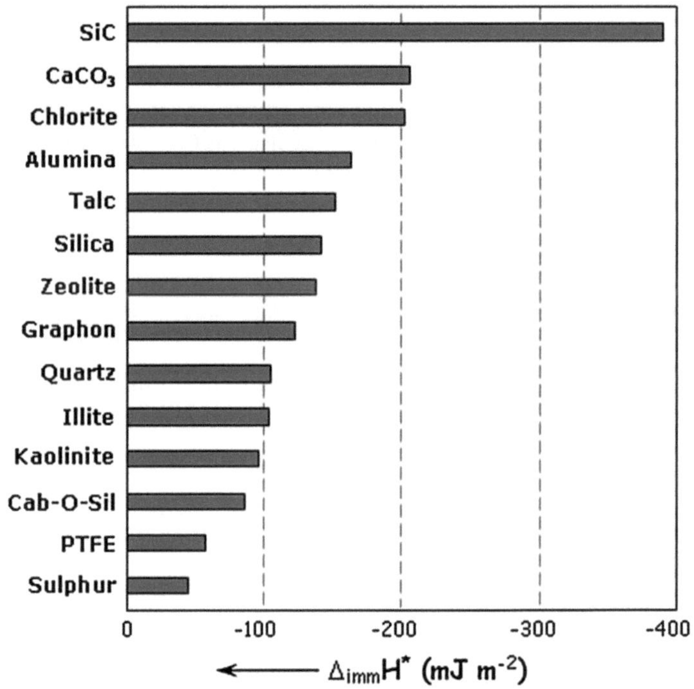

Fig. 6.9 Enthalpy of immersion per unit surface area of the solid $\Delta_{imm}H^*$ for several solid materials in *n*-heptane

and formamide (Merck HPLC grade materials with purity exceeding 99 %), were additionally dried with 3A zeolite molecular sieves. *This purification procedure to remove even traces of water from organic solvents is of great importance and strongly recommended for all immersion experiments.* Water was deionised and purified with a Millipore Super Q System.

The immersion experiment was repeated three times for each solid sample and the average value taken. The amount of immersional liquid penetrating into the glass bulb upon immersion was determined by weighing the bulb after every run (to evaluate the correction term 6.27). Reproducibility of the calorimetry measurement in regard to the enthalpy of immersion expressed by weight of the solid was within 3 %. To obtain the enthalpy of immersion per unit surface area of the solid $\Delta_{imm}H^*$,

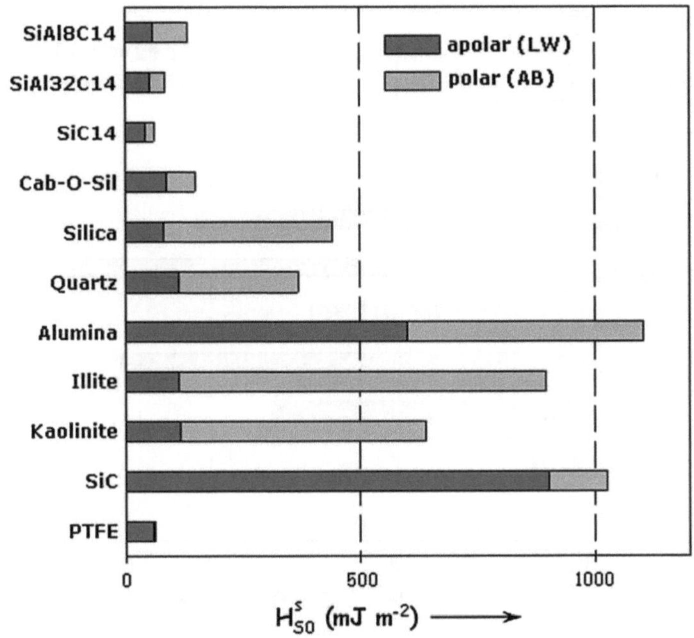

Fig. 6.10 Apolar (LW) and polar (AB) contributions to the surface enthalpy for a series of solids, as derived from measurements of the enthalpy of immersion $\Delta_{imm}H^*$ of each solid in *n*-heptane, water and formamide

the experimental enthalpy values were divided by the corresponding BET specific surface areas (the adsorption model of Brunauer, Emmet, and Teller applied to the experimental results of gaseous nitrogen adsorption at 77 K taking a cross sectional area of 0.162 nm^2 per N_2 molecule), i.e., $S = S_{BET}$. Finally, each solid was separately outgassed under the same conditions as those used in the immersion experiment with the purpose of quantifying the mass of dried sample. Then the immersion data were corrected for the mass loss during outgassing (following Eq. 6.28). In consequence of all these additional steps, the uncertainty in the enthalpy determination increased, but was, on average, better than 5 %.

Figure 6.9 shows the experimental values of $\Delta_{imm}H^*$ for one series of solids immersed in *n*-heptane. Here the interfacial phenomenon involves only van der Waals interactions between the immersional liquid and the solid, irrespective of the actual *surface hydrophobic-hydrophilic balance* (SHB) of the latter. Based on the enthalpy results obtained, it is possible to classify these solids with respect to hydrophobic character of their surface.

According to the criteria of hydrophobicity given in Sect. 6.2.4, the most hydrophobic substance has the lowest H_{SO}^{LW} (more precisely γ_{SO}^{LW}) surface parameter. Therefore, its enthalpy of immersion in *n*-heptane per unit area of the solid should have the smallest value. Among the solids presented in Fig. 6.9, only sulphur and

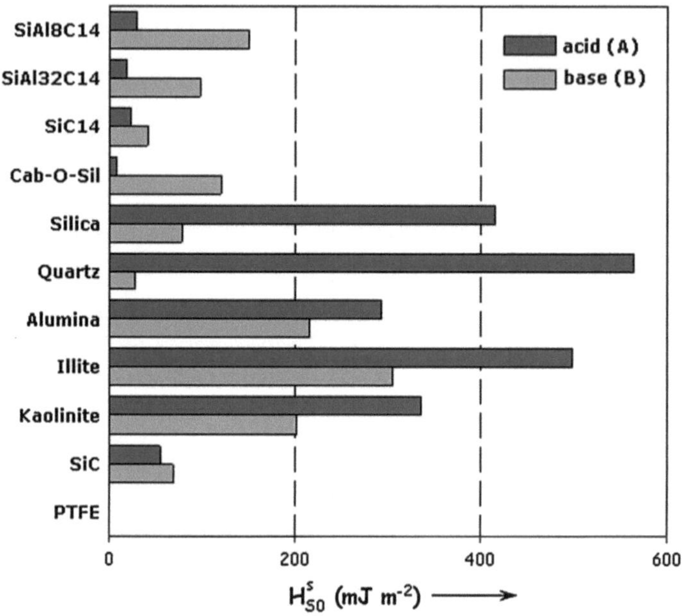

Fig. 6.11 PLewis acid (**A**) and Lewis base (**B**) contributions to the surface enthalpy for a series of solids, as derived from measurements of the enthalpy of immersion $\Delta_{imm} H^*$ of each solid in *n*-heptane, water and formamide

polytetrafluoroethylene (PTFE) appear clearly at the hydrophobic end. Such mineral oxides as clay minerals with a lamellar structure (kaolinite, illite, talc, chlorite), precipitated amorphous SiO_2 (silica), crystalline SiO_2 (quartz), amorphous Al_2O_3 (alumina), or crystalline aluminosilicates (zeolite) are known to be more or less hydrophilic (commonly, the presence of numerous functional groups with a polar character in the surface of these materials is advanced as a typical argument). The comparison of the $\Delta_{imm} H^*$ values reported in Fig. 6.9 indicates that the overall intensity of Lifshitz-Van der Waals interactions between apolar *n*-heptane and a unit area of the solid surface is not the same for various solids. This means that all microscopically unsaturated structures which may be encountered in solid surfaces differ also in "apolar character" and, consequently, the apolar (LW) component to the surface enthalpy H^{LW}_{S0} is a solid-dependent parameter. Surprisingly, non-porous graphitized carbon black (Graphon), regarded as a weakly hydrophilic solid, yield $\Delta_{imm} H^*$ greater than those of kaolinite, illite, or quartz which certainly possess more polar groups per unit surface area. Of course, it may be argued that graphitic basal planes in the surface of Graphon act as electron pair donors, thereby participating in π-EPD/π-EPA or π-complexation interactions with the foreign molecules. Nevertheless, it is more reasonable to evaluate both the apolar (LW) and the polar (AB) components to the surface enthalpy H^s_{S0} in order to compare the differences in the SHB among various solids.

It is worth noting that any arrangement of solid surfaces by hydrophobic character, as well as any ordering or ranking of solids in regard to their SHB may be used only for comparative purposes and do not give the "absolute ranking position" of a given solid within the class of hydrophobic and hydrophilic substances.

Figure 6.10 shows examples of several hydrophobic and hydrophilic solids, together with their H_{SO}^{LW} and H_{SO}^{AB} parameters obtained on the basis of Eq. 6.29. SiC14, SiAl32C14, and SiAl8C14 are mesoporous silica-based materials of the MCM-41 type prepared by a surfactant-assisted synthesis and doping with aluminium (e.g., abbreviation "SiAl32" refers to a molar Si-to-Al ratio of 32). Only PTFE is a clearly hydrophobic substance since it has low H_{SO}^{LW} and zero H_{SO}^{AB} component. Amorphous fumed SiO_2 (Cab-O-Sil) and mesoporous MCM-41 silica or aluminosilicates can be reckoned among low-energy solids possessing surfaces with a weakly hydrophilic character. For high-energy solids with surface enthalpy H_{SO}^{s} greater than 500 mJ m^{-2}, the ratio between H_{SO}^{AB} and H_{SO}^{LW} varies from 7 (illite) to 0.2 (SiC); alumina is characterized by an intermediate SHB value of 0.8.

Figure 6.11 illustrates the Lewis acid-base character of the solid surfaces presented in Fig. 6.10. The Lewis acid H_{SO}^{A} and Lewis base H_{SO}^{B} components are related to the polar (AB) contribution to the surface enthalpy of each solid: $H_{SO}^{AB} = 2\sqrt{H_{SO}^{A} \cdot H_{SO}^{B}}$. Crystalline SiO_2 (quartz) and, to a smaller extent, amorphous precipitated SiO_2 (silica) provide strongly acidic surfaces, whereas the surfaces of MCM-41 mesoporous silicas and Cab-O-Sil have a predominantly basic character.

The *Harkins-Jura method* for estimating the specific surface area S_{HJ} of a solid available to a given immersional liquid is based on Eq. 6.22 [20, 54, 55]. In practice, water is by far the most frequently used liquid because of the small size of its molecules (the van der Waals diameter of a water molecule is about 0.28 nm). The specific surface area of a hydrophilic solid is thus calculated as follows:

$$S_{HJ} = -\frac{\Delta_W H_m}{m_S \cdot H_{LG}^{s}} = -\frac{\Delta_W h_m}{H_{LG}^{s}} = -\frac{\Delta_W h_m}{0.118} \tag{6.30}$$

where $\Delta_W h_m$, expressed in J g^{-1}, denotes the minimum value of the specific enthalpy of immersional wetting, i.e., wetting enthalpy per unit mass of the solid sample $\Delta_W h_m = \frac{\Delta_W H_m}{m_S}$. This enthalpy value is determined from the plot of $\Delta_W h$ against the relative pressure $\frac{p}{p_{sat}}$ at which surface pre-coverage with water vapour has been carried out (Fig. 6.12).

The application of the above procedure is limited only to non-microporous materials, for which the solid surface area available to water is equal to the area of the interface between the adsorbed water film on the surface and the equilibrium water vapour phase. It is always necessary to check the condition of the ideal wetting. Water does not spread on hydrophobic surfaces, like the surface of the Graphon, and the value of $\Delta_W H$ decreases (becomes more negative) as the surface adsorbs more water vapour [20]. This proves that the empty surface has less affinity for water than has the surface of the adsorbed water. Equations 6.22 and 6.30 cannot be applied to such a system.

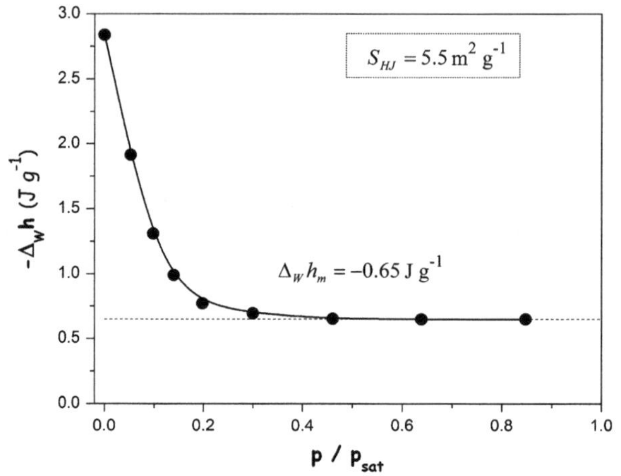

Fig. 6.12 Specific enthalpy of immersional wetting $\Delta_W h$ (taken with the opposite sign) in water for silica SIFRACO C-600 pre-covered with water vapour as a function of the relative vapour pressure [60]. The resulting HJ specific surface area is also given ($S_{HJ} = 5.5\ m^2g^{-1}$)

6.4 Enthalpy Changes Accompanying Competitive Adsorption from Dilute Solution

The *competitive aspect* of the phenomenon, nowadays commonly accepted, shows up most clearly in adsorption from concentrated solutions (e.g., mixtures of completely miscible liquids) [6]. In dilute solution, only a very detailed analysis of the adsorption system may reveal that the phenomenon is indeed competitive. In this context, calorimetry has proven very useful in studying the competitivity of adsorption from dilute solution onto solids.

The *driving force* of competitive adsorption from binary solution onto solids is a macroscopic outcome of intermolecular forces belonging to the following categories: (i) interactions of the solute and the solvent with the solid surface, (ii) interactions among the solution components in the interfacial region, and (iii) solute—solvent, solvent—solvent, and solute—solute interactions in the bulk solution. In numerous adsorption systems, a considerable enhancement of the amount adsorbed is often the result of *co-operative effects* involving various interaction types. The most spectacular examples are related to the adsorption of amphiphilic substances (e.g., surfactants and polymers) that exhibit reduced water solubility [61, 62]. The general behaviour of amphiphilic molecules in aqueous solution reflects the opposing tendencies of the hydrophobic portion of the molecule to escape from the aqueous environment while the hydrophilic moiety tends to remain immersed in the water. Depending on the surface hydrophilic-hydrophobic character of the solid particules immersed in the solution, such molecules can be compelled to accumulate in an oriented fashion at the Solid-Solution interface or to associate between themselves within the solution (i.e., *micellization*). Aggregation of surfactant units both at the Solid-Solution interface and in the bulk solution is mainly governed by *hydrophobic bonding*, which refers to the apparent attraction between hydrophobic moieties being much stronger in water than it is in non-aqueous media [26, 61, 63]. Hydrophobic bonding is a typical example of the so-called thermodynamic interactions having both mechanical and entropic contributions.

Adsorption experiment may be carried out in two different ways: as (1) *immersion* of a dry adsorbent in a binary solution of a given composition or (2) *displacement* of the solvent from the solid-solvent interface by the adsorbing solute supplied in a stock solution [6, 64, 65]. Additionally, the effects of displacement are determined either in a batch displacement or in a flow displacement experiment. Depending on the experimental procedure applied, the accompanying changes of macroscopic properties differ to a great extent. The amount adsorbed and enthalpy change on adsorption can be precisely defined in operational terms by considering models of the immersion and displacement experiments.

6.4.1 Thermal Properties of Dilute Solutions

A dilute solution typically means a solution containing no more than about 10^{-2} mol l^{-1} of solute. For the solvent, a convenient reference state is its own pure liquid state at 1 bar. However, a more useful *reference state for the solute*, including aqueous solutions of electrolytes and non-electrolytes, is that of *infinite dilution in the solvent* [66, 67].

The general form of the chemical potential of a real solute in a binary solution of non-electrolytes is therefore [66, 67]

$$\mu_2 = \mu_2^* + RT \ln\left(\frac{m_2}{1} \cdot f_2\right), \; f_2 \rightarrow 1 \text{ as } m_2 \rightarrow 0 \tag{6.31a}$$

$$\mu_2 = \mu_2^* + RT \ln\left(\frac{c_2}{1} \cdot f_2\right), \; f_2 \rightarrow 1 \text{ as } c_2 \rightarrow 0 \tag{6.31b}$$

where μ_2^* is the standard chemical potential of the solute; m_2 is the molality of this component and c_2 the corresponding molar concentration. The unity on the right-hand side is written to remind the reader that the molality (the molar concentration) is expressed in mol kg^{-1} (mol L^{-1}), whereas the activity is dimensionless. The employment of concentration units is often discouraged because the concentration of a given solution varies with temperature, while the molality is independent of temperature. The latter also allows the composition determination to be achieved with greater precision, since any solution may be prepared by weighing the solute and solvent, or the stock solution and solvent (dilution of the stock solution).

The interactions between ions contained in an ionic solution are so strong that the solution approaches ideality in the sense of obeying Henry's law only at very low values of total ion concentration, usually less than 10^{-3} mol kg^{-1} [68, 69]. In precise considerations, ion activities must be used. In the case of strong electrolytes in aqueous solution, the total Gibbs energy (and any other extensive property) of the solute is the sum of the partial molar Gibbs energies for the individual ions produced by the solute. Suppose that a strong electrolyte M_pX_q dissociates into ions according to the following reaction

$$M_pX_q = p \cdot M^{z+} + q \cdot X^{z-} \quad , \quad p \cdot z_+ = q \cdot z_- \tag{6.32}$$

where p and q are the number of ions of M and X type, respectively; z_+ and z_- represent the valencies of the respective ions (positive number for cations and negative number for anions). The chemical potential of the solute j in a real solution may be written as [68, 69]

$$\mu_j = \mu_j^* + (p+q)RT \ln m_\pm + (p+q)RT \ln f_\pm \tag{6.33a}$$

where the mean ionic molality $m_\pm$ and the mean ionic activity coefficient $f_\pm$ are defined as follows:

$$m_\pm = \left[(m_+)^p \, (m_-)^q\right]^{\frac{1}{p+q}} \quad \text{and} \quad f_\pm = \left[(f_+)^p \, (f_-)^q\right]^{\frac{1}{p+q}} \tag{6.33b}$$

Now both types of ion share equal responsibility for the non-ideality.

Strong and long-range Coulombic forces acting between ions are primarily responsible for the departures from ideality (the activity coefficients are lowered) and dominate all other contributions. The effect has been evaluated in the Debye-Hückel theory and there exist several equations, which are useful in estimating the mean activity coefficient [68, 69]. The latter is related to the *ionic strength* of the solution:

$$I = \frac{1}{2}\sum_j z_j^2 \cdot m_j \tag{6.34}$$

where the sum extends over all ions present in the solution; m_j is the molality of the jth ion. For example, the value of $f_\pm$ can be calculated from the Debye-Hückel limiting law

$$\log f_\pm = -A\,|z_+ \cdot z_-| \cdot \sqrt{\frac{I}{1}} \tag{6.35a}$$

where A is the dimensionless constant characteristic of the solvent at the specified temperature and pressure ($A = 0.5085$ for an aqueous solution at 298.15 K [68, 69]).

The approximation (6.35a) is in good agreement with the experiment at very low molalities: less than 0.01–0.001 mol kg^{-1}, depending on charge type (the range of validity becomes narrower when divalent or multivalent ions are present in the solution). Nevertheless, when the departures from the experimental results are large, the activity coefficient may be estimated from the extended Debye-Hückel law

$$\log f_{\pm} = -\frac{A\,|z_+ \cdot z_-| \cdot \sqrt{\frac{I}{1}}}{1 + B \cdot \sqrt{\frac{I}{1}}} \tag{6.35b}$$

where B is an adjustable empirical parameter (which involves numerical factors, the dielectric constant, the temperature and the mean distance of nearest approach of the ions [68, 69]).

In thermodynamics of solutions, the partial molal quantity and the apparent molal quantity for a component are usually defined [66]. In the properties of enthalpy, one deals with quantities which cannot be measured in an absolute sense. It is thus necessary to use a reference state from which to make the evaluations. Since the state of infinite dilution of the solute in the solvent is taken as reference, the specification of the composition of the solution by means of the molality m_2 of the solute is consequently most convenient for the calculation of enthalpy.

If the enthalpy H of a two-component solution is expressed as a function of the composition, at constant pressure and temperature, the *partial molal enthalpy* $\overline{h}_2$ for the solute is defined as the rate of change of the enthalpy H with change in the number of moles of this component, with the number of moles of the solvent being held constant; that is

$$\overline{h}_2 = \left(\frac{\partial H}{\partial n_2}\right)_{T,P,n_1} \tag{6.36}$$

The basic partial molal equation for the total enthalpy of the binary solution is

$$H = \sum_{j=1}^{2} n_j \cdot \overline{h}_j \quad , \mathrm{T,P} = \mathrm{const} \tag{6.37}$$

where n_j is the number of moles of the jth component in the solution.

The partial molal enthalpy of the solute in a solution may not always be conveniently measured experimentally with the required precision. Usually, the apparent molal property for the solute is most used in connection with dilute solutions. For a solution of two components, the *apparent molal enthalpy* of the solute Φ_H is

$$\Phi_H = \frac{\Delta H}{n_2} \quad , \mathrm{T,P} = \mathrm{const} \tag{6.38}$$

whereΔH is the enthalpy change produced when the solution is formed by adding n_2 moles of solute to n_1 moles of pure solvent.

The total enthalpy of the solution may be conveniently expressed in terms of the apparent molal enthalpy of the solute as

$$H = \frac{M_1}{10^3} m_2 \Phi_H + M_1 h_1^* \tag{6.39}$$

where M_1 is the total mass of the solvent in the solution and h_1^* is the specific enthalpy (per unit mass) of the pure solvent.

In the infinitely dilute solution, the value of the molal enthalpy for the solute is identical with the corresponding value of the partial molal enthalpy, namely

$$\Phi_H(\infty) = \overline{h}_2(\infty) \quad \textit{at infinite dilution} \tag{6.40}$$

For any other solution, the partial molal enthalpy of the solute referred to the state of infinite dilution may be calculated if the apparent molal enthalpy is known as a function of the composition. In such a case,

$$\overline{h}_2 - \overline{h}_2(\infty) = \Phi_H - \Phi_H(\infty) + m_2 \frac{d[\Phi_H - \Phi_H(\infty)]}{dm_2} \tag{6.41}$$

It frequently happens that the apparent molal enthalpy is expressible as an appropriate function of the composition in order to obtain a curve which does not depart greatly from a simple function and which has not too great a curvature over the given range. For aqueous solutions of strong electrolytes, the value of Φ_H is sometimes linear with the square root of the molality, over a given range of molality [68]. Converting Eq. 6.41 to the slope of Φ_H against $\sqrt{m_2}$, one obtains

$$\overline{h}_2 - \overline{h}_2(\infty) = \Phi_H - \Phi_H(\infty) + \frac{1}{2}\sqrt{m_2}\frac{d[\Phi_H - \Phi_H(\infty)]}{d\left(\sqrt{m_2}\right)} \tag{6.42}$$

The difference $\Phi_H - \Phi_H(\infty)$, called the *relative apparent molal enthalpy*, is the quantity most readily evaluated from calorimetric measurements of enthalpies of dilution. For the following reaction of dilution:

$$\begin{aligned} &\left(\frac{\mathrm{M}_1^0}{10^3}\mathrm{m}_2^0 \times \text{solute} + \mathrm{M}_1^0 \times \text{solvent}\right)(\text{liq}) + \left(\infty - \mathrm{M}_1^0\right) \times \text{solvent(liq)} \\ &\rightarrow \left(\frac{\mathrm{M}_1^0}{10^3}\mathrm{m}_2^0 \times \text{solute} + \infty \times \text{solvent}\right)(\text{liq}) \end{aligned} \tag{6.43}$$

the change in the enthalpy is equal to

$$\Delta_{dil} H\left(m_2^0 \rightarrow \infty\right) = -\frac{M_1^0}{10^3} m_2^0 [\Phi_H - \Phi_H(\infty)] \tag{6.44}$$

where M_1^0 denotes the initial mass of the solvent and m_2^0 the initial molality of the solute. Values of the relative apparent molal enthalpy may be obtained by measuring enthalpy changes $\Delta_{dil} H$ for the dilution of a given stock solution of molality m_2^0 to a series of different molalities m_2,

$$\Delta_{dil} H \left(m_2^0 \rightarrow m_2\right) = -\frac{M_1^0}{10^3} m_2^0 \left[\Phi_H \left(m_2^0\right) - \Phi_H \left(m_2\right)\right] \tag{6.45}$$

and suitably extrapolating the curve $\Delta_{dil} H = \Delta_{dil} H\,(m_2)$ to infinite dilution.

6.4.2 Macroscopic Description of Competitive Adsorption

Compared to interfacial phenomena occurring at the boundary between a solid and a pure liquid, Coulombic forces constitute an additional type of interaction that should be taken into account when analysing the adsorption phenomenon at an electrified interface from solution containing free charges (ions and electrons) and associated charges (dipolar molecules and polarised atoms). In such systems, there is a variation in the charge density across the interfacial region and an *electric double layer* forms [70–72] (the Gouy-Chapman-Stern-Grahame model of the EDL is shown in Fig. 6.13).

Transfer of the individual ionic species between the interface and the solution leads to a thermodynamic equilibrium at which *the interface is electrically neutral as a whole* (and so is the equilibrium bulk solution).

Based on isothermal reversible work done in transferring one mole of ions of the solute 2 from infinity (*in vacuum*) to a given part in the interfacial region which has a non-zero average charge and where the electrostatic potential is Ψ, it is possible to define the so-called *electrochemical potential* of component 2 [70]

$$\mu_2 = \mu_2 + z_2 \cdot \mathsf{F} \cdot \Psi \quad , \mathrm{T}, \mathrm{P} = \mathrm{const} \tag{6.46}$$

where μ_2 is the chemical potential of the solute, z_2 is the valency of an ion of the solute, and F is the Faraday constant (F is a product of the Avogadro number L and the elementary charge e). If the transferred species carries no net charge, i.e., $z_2 = 0$, the electrochemical potential $\overline{\mu}_2$ becomes simply the chemical potential μ_2. The new potential $\overline{\mu}_2$ is now the quantity which must have *the same value everywhere in the system for thermodynamic equilibrium to be established* at constant temperature and pressure [70].

According to the surface phase model (cf., Sect. 6.2.1), when the concentrations of the components of the binary solution β in the solid phase α are to be neglected, i.e., $c_j^\alpha = 0$ (j =1,2), the *interfacial concentration* Γ_j^s of component j is given by [6, 8]

$$\Gamma_j^s \equiv \frac{n_j^s}{A} = \tau^s \cdot c_j^s = \frac{1}{A}\left(n_j - c_j^\beta V^\beta\right), \quad (\mathrm{j} = 1, 2) \tag{6.47}$$

where c_j^β and c_j^s are the concentrations (in moles per unit volume of solution) of the jth component in the bulk solution β and in the surface phase of thickness τ^s and surface area A; V^β represents the volume of the bulk phase β in the surface phase model. The subscripts 1 and 2 denote the solvent and the solute, respectively; this notation will be used throughout the present chapter.

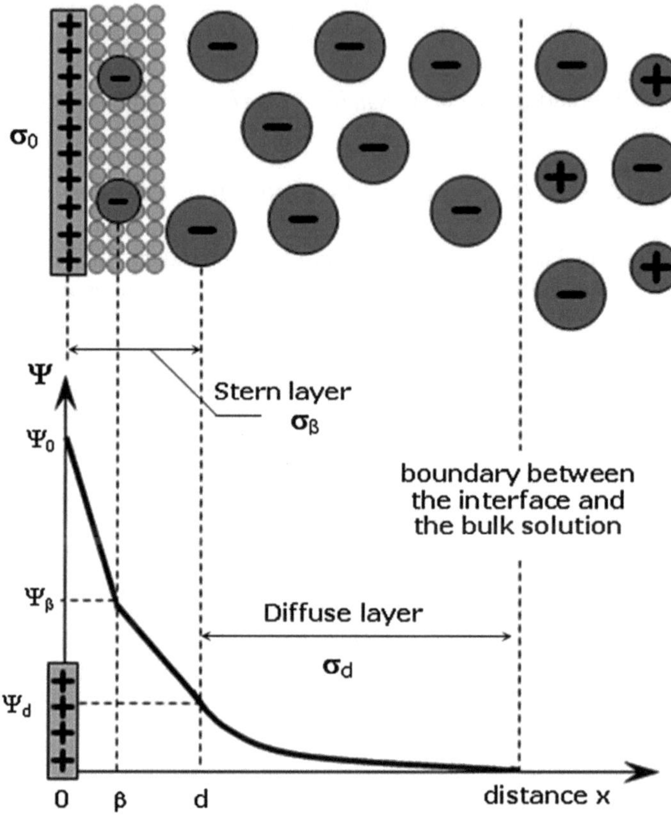

Fig. 6.13 Schematic representation of the structure of the electric double layer according to the Gouy-Chapman-Stern-Grahame model and variations of the interfacial electric potential Ψ across the interfacial region. By convention, the potential Ψ of any part of the electrified interface is defined with respect to the solution at infinite distance where the average charge is zero. The distance x is measured in reference to the position of a plane containing *solid surface charge* σ_0. The solution part of the EDL with an excess of counter-ions is composed of two distinct regions: a thin *Stern layer* (next to the surface charge plane) including specifically adsorbed (Coulombic and other non-electrostatic forces) counter-ions which may be partially dehydrated and an extended *diffuse layer* where the hydrated counter-ions are distributed non-uniformly according to the combined action of electrostatic forces and random thermal motion. Electroneutrality of the charged interface requires that $\sigma_0 + \sigma_\beta + \sigma_d = 0$; σ_β and σ_d represent the charge densities of the Stern and diffuse layers, respectively [70]

The case of $\Gamma_j^s > 0$ defines the preferential *adsorption* of component j at the Solid-Liquid interface. In a given adsorption system, the interfacial concentration Γ_j^s is a function of both the actual composition of the bulk solution (i.e., c_1^β and c_2^β) and the relative affinities of the components for the interface (chiefly, for the solid surface). Contrary to the case of adsorption at the Solid-Gas interface, the interfacial region is always completely filled with molecules of solvent and solute. Further consideration will be based on the *preferential adsorption* of solute against solvent, taking the interface between the solid and the pure solvent as a starting point for thermodynamic treatment of competitive adsorption. In such a case, the adsorption equilibrium can be represented schematically as follows:

$$\mathrm{r} \cdot \mathrm{n}_2^{\beta} + \mathrm{n}_1^{\mathrm{s}} \rightleftarrows \mathrm{r} \cdot \mathrm{n}_2^{\mathrm{s}} + \mathrm{n}_1^{\beta} \tag{6.48}$$

where r is the *molar ratio of displacement* introduced to account for unequal molecular sizes of both the components. It is worth noting that the same formalism may be applied in the treatment of adsorption at the interface between a binary liquid solution and a gas phase.

When the pure liquid phase is replaced by a binary liquid solution β, the interfacial tensions γ_{SL} and γ_{LG} will be lowered owing to the preferential adsorption of the solute at the Solid-Liquid and Liquid-Gas interface. The relationship between the extent of adsorption and the resulting interfacial (surface) tension change (T, P = const) is given by the following equation developed by Guggenheim [6, 8]

$$\begin{aligned} d\gamma &= -\left[\Gamma_2^s - \Gamma_1^s\left(\frac{c_2^{\beta}}{c_1^{\beta}}\right)\right] d\mu_2 = -\left[\Gamma_2^s - \Gamma_1^s\left(\frac{n_2^{\beta}}{n_1^{\beta}}\right)\right] d\mu_2 \\ &= -\left[\Gamma_2^s - \left(\frac{M_1^s}{10^3 A}\right) \cdot m_j^{\beta}\right] d\mu_2 \end{aligned} \tag{6.49}$$

where μ_2 is the chemical potential of solute in the bulk solution β: at equilibrium, μ_2 is the same throughout the system; when expressing the solution composition in terms of the molality m_2^{β} (in moles per kilogram of solvent), M_1^s is the mass of solvent in the surface phase. It is worth noting that Eq. 6.49 is formally analogous to the Gibbs adsorption isotherm for a binary solution [6, 8].

6.4.3 Competitive Adsorption Measurements

The amount of solute molecules preferentially adsorbed from dilute solution onto a given solid can be measured in a separate adsorption experiment, independently of the calorimetry measurement. In the case of the titration calorimetry procedure, this is even the only possibility to determine the amount adsorbed after each injection step and subsequently calculate the differential molar enthalpy of adsorption. The main difficulty here, contributing to a significant uncertainty of the experimental result, is related to *the necessity of reproducing strictly the same experimental conditions in both types of experiment* (i.e., the same solid surface-to-solution volume ratio, evolution of the pH and ionic strength in the equilibrium bulk solution, charging behaviour of the solid surface, etc.).

The quantity of adsorption is usually measured by means of the *solution depletion method* [6] in glass stoppered tubes or flasks (Fig. 6.14). A known mass of the solid

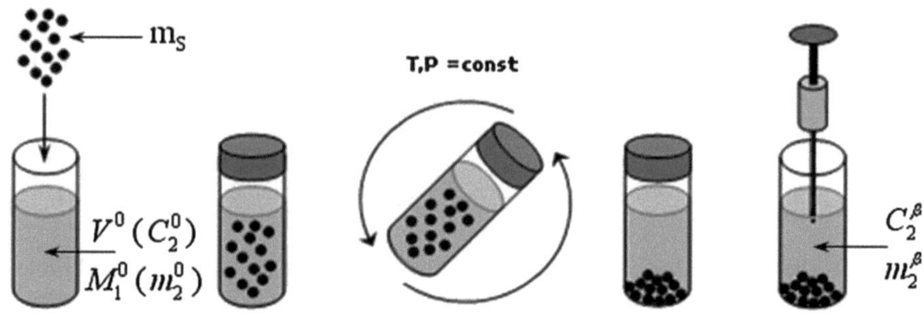

Fig. 6.14 Schematic representation of the solution depletion technique. The solid particles are separated from the supernatant solution by centrifugation or filtration. The way of calculating the quantity of adsorption depends on whether the solution composition is expressed in terms of molality m_2^β or the molar concentration C_2^β

sample and a certain amount of dilute binary solution of given composition are put into each tube or flask, and the mixture is shaken at a constant temperature (e.g., in a thermostated box) for a period of time necessary to attain the adsorption equilibrium. Then the solid particles are separated from the supernatant by centrifugation or filtration (special attention should be paid during fitration to avoid the retention of the solute on the filtration membrane). After centrifugation when the solid particles have been precipitated at the bottom of the tube or flask, a sample of the supernatant may be collected by using a special syringe with a long needle. The equilibrium composition of the supernatant is determined by referring to the appropriate analytical technique (UV spectroscopy, refractometry, total carbon analysis, etc).

The amount n_2^s of solute adsorbed at the Solid-Liquid interface is calculated by means of the following formulas:

$$\frac{n_2^s}{m_S} = \frac{V^0}{m_S}\left(C_2^0 - C_2^\beta\right) \quad \text{or} \quad \frac{n_2^s}{m_S} = \frac{M_1^0}{10^3 \cdot m_S}\left(m_2^0 - m_2^\beta\right) \tag{6.50}$$

where m_S is the mass of the solid sample in a given tube or flask; C_2^0 and m_2^0 are respectively the molarity and the molality of the initial solution (before adsorption); C_2^β and m_2^β are respectively the molarity and the molality of the supernatant solution β (after the attainment of adsorption equilibrium); V^0 denotes the volume of the solution put initially into the tube or flask and M_1^0 is the total mass of the solvent in the heterogeneous system. When the specific surface area of the solid S_{BET} or S_{HJ} is known, it is even possible to calculate the interfacial concentration Γ_2^s of the solute. It should be realised that expressions 6.50 hold only for dilute binary solutions containing a solute that is preferentially adsorbed at the Solid-Liquid interface.

For a given adsorption system, the amount of solute adsorbed at equilibrium depends on the temperature T, the pressure P, and the composition of the equilibrium solution phase β. The experimental results of adsorption measurements are usually reported in the form of *individual adsorption isotherms* showing the quantity $\frac{n_2^s}{m_S}$ or

Γ_2^s as a function of the solution composition:

$$\frac{n_2^s}{m_S} = \frac{n_2^s}{m_S}\left(C_2^\beta\right) \quad \text{or} \quad \frac{n_2^s}{m_S} = \frac{n_2^s}{m_S}\left(m_2^\beta\right), \text{T, P} = \text{const} \tag{6.51a}$$

$$\Gamma_2^s = \Gamma_2^s\left(C_2^\beta\right) \quad \text{or} \quad \Gamma_2^s = \Gamma_2^s\left(m_2^\beta\right), \text{T, P} = \text{const} \tag{6.51b}$$

Figure 6.15 shows three selected shapes of adsorption isotherms at the Solid-Liquid interface for various charged and uncharged solutes exhibiting a limited solubility in the solvent employed. In general, typical adsorption curves present the amount adsorbed as a smooth, monotonically increasing function of the solute content in the equilibrium bulk solution. Many adsorption isotherms end up in an upper composition range at a constant amount adsorbed (a *plateau* of the isotherm).

The appearance of an adsorption plateau region at equilibrium concentrations (molalities) approaching the solubility limit indicates that the phenomenon involves only single solute species that are individually dissolved in the solvent. When the adsorption plateau is observed at lower concentrations (molalities), it is usually argued that surface sites of a given type have been saturated by the adsorbing solute species.

The experimental adsorption isotherms may be plotted on different scales, thereby allowing a more detailed analysis of the subsequent adsorption stages to be made. When the adsorption of homologous substances onto the same solid is due only to their different solubilities in the solvent, the use of a *reduced concentration (molality)*

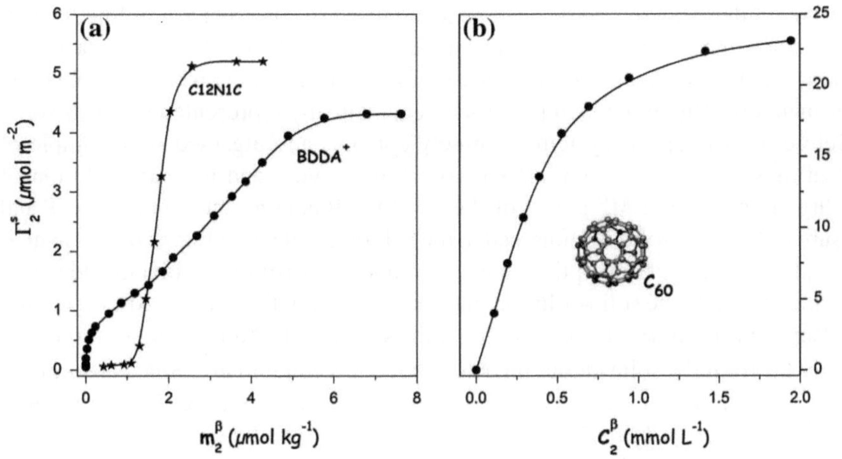

Fig. 6.15 Adsorption isotherms of (**a**) zwitterionic (C12N1C), and cationic ($BDDA^+$) surfactants from aqueous solutions onto non-porous negatively charged silica at 298 K [73, 74], (**b**) buckminsterfullerene C_{60} from its tollune solutions onto mesoporous activated carbon at 293 K [75]; C12N1C and $BDDA^+$ denote (dodecyldimethylammonio)ethanoate and benzyldimethyldodecylammonium cation, respectively

scale to represent the corresponding isotherms results in a single adsorption curve. Any changes in the adsorption mechanism with increasing bulk concentration (molality) show up most clearly in *double logarithmic* (log-log) plots. The transitions between such adsorption regions are obviously more gradual in reality than they seem to be on a log-log scale. The plateau adsorption region markedly manifests itself only on a *double linear* (lin-lin) scale. In the *linear-logarithmic* (lin-log) plot, it is difficult to observe the effects of surface heterogeneity and co-operative adsorption at low bulk concentrations (molalities) but the top parts of the isotherm become clearer.

By analogy to Eq. 6.16, the surface pressure per unit surface area of the Solid–Liquid interface may be determined as follows:

$$\pi_{SL}\left(C_2^*\right) = \gamma_{SL1} - \gamma_{SL}\left(C_2^*\right) = \frac{RT}{m_S \cdot S} \int\limits_{C_2=0}^{C_2^*} n_2^s d\ln C_2 \qquad (6.52)$$

where γ_{SL1} is the interfacial tension of the solid in equilibrium with the pure solvent; S is the specific surface area of the solid (i.e., $S = S_{BET}$ or $S = S_{HJ}$); C_2^* denotes the equilibrium bulk concentration taken into consideration.

6.4.4 Immersion in Dilute Solutions

The solution depletion procedure can be generalised to construct a simplified model of immersion experiment [64, 65]. This model is very useful to define the measurable enthalpy quantities, which does not necessarily mean that they can be easily obtained experimentally. The immersion process is schematically represented in Fig. 6.16.

Before immersion the system separately contains an outgassed solid sample of a given mass m_S and a given surface area S in vacuum and a binary solution of molality m_2^0 consisting M_1^0 grams of the solvent, all at constant temperature T and pressure P. Then the solid is immersed in the solution under conditions of constant T, P. The phenomenon of adsorption induces an uneven partition of solvent and solute molecules between the solid-solution interface and the bulk solution. After attaining the adsorption equilibrium, the molality of the solute in the bulk of the solution (sufficiently far from the solid-liquid interface) becomes equal to m_2^β. Since absorption, dissolution and swelling effects are not taken into consideration, the adsorbent is considered as *thermodynamically inert*, i.e., its mass, specific surface area and bulk phase properties do not change during adsorption. Furthermore, the adsorption of the solute at the Solution-Gas interface and the relative surface enthalpy for this interface are omitted for convenience: changes of both effects during immersion in an excess of dilute solution are negligible small.

The balance of enthalpy in the initial (init) and final (fin) states of the model system can be written as follows (cf. Sect. 6.4.1):

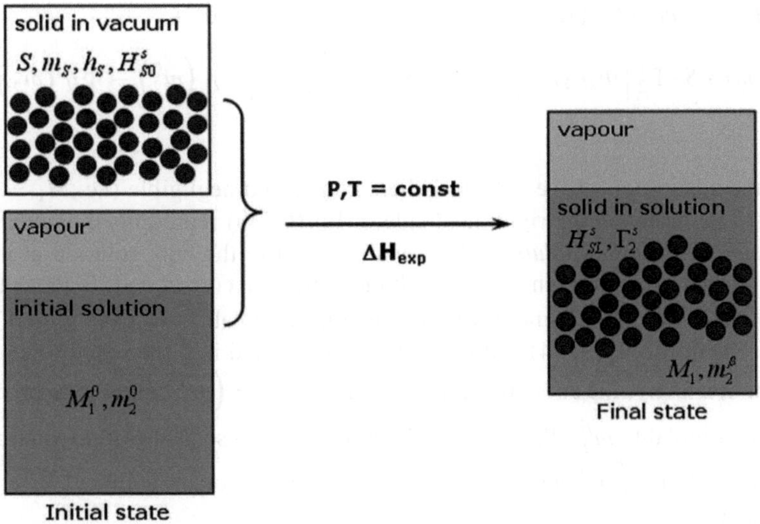

Fig. 6.16 Schematic representation of the immersion experiment (explanation of the symbols in the text). The adsorption phenomenon is assumed not to affect the enthalpy of the bulk of solid, $m_S \cdot h_S$; changes in the mass of the solvent are to be neglected when immersion is carried out in a sufficient excess of dilute solution containing a preferentially adsorbed solute (i.e., $M_1 = M_1^0$)

$$H^{init} = m_S \cdot S \cdot H^s_{S0} + m_S \cdot h_S + M_1^0 \cdot h_1^* + \frac{M_1^0}{10^3} m_2^0 \cdot \Phi_H \left(m_2^0\right) \qquad (6.53a)$$

$$H^{fin} = m_S \cdot S \cdot H^s_{SL} + m_S \cdot h_S + \frac{M_1^0}{10^3} m_2^\beta \cdot \Phi_H \left(m_2^\beta\right) + M_1^0 \cdot h_1^* \qquad (6.53b)$$

where H^s_{S0} and H^s_{SL} are the interfacial enthalpies for solid in vacuum and for Solid-Liquid interface in equilibrium with a bulk solution of molality m_2^β, respectively; h_S is the specific (per unit mass) enthalpy of the bulk of the solid phase; $\Phi_H\,(m_2)$ is the value of the apparent molal enthalpy of the solute in a binary solution corresponding to the molality m_2; h_1^* is the specific enthalpy (per unit mass) of the pure liquid solvent. Equation (6.53b) has been developed based on the hypothesis that changes in the mass of the solvent can be neglected during immersion in excess solution.

The total thermal effect of the immersion experiment is:

$$\Delta H_{\text{exp}} = H^{fin} - H^{init} = m_S \cdot S \cdot \left(H^s_{SL} - H^s_{S0}\right) + \frac{M_1^0}{10^3} \left[m_2^\beta \cdot \Phi_H \left(m_2^\beta\right) - m_2^0 \cdot \Phi_H \left(m_2^0\right)\right] \qquad (6.54a)$$

Taking into account the reference state for the solute defined in Sect. 6.4.1 and using Eqs. (6.40), (6.47), and (6.50), this expression may be transformed as follows:

$$\Delta H_{\text{exp}} = m_S \cdot S \cdot \left(H^s_{SL} - H^s_{S0}\right) - m_S \cdot S \cdot \Gamma^s_2 \cdot \overline{h}_2(\infty) + \\ - m_S \cdot S \cdot \Gamma^s_2 \left[\Phi_H\left(m^\beta_2\right) - \Phi_H(\infty)\right] - \frac{M^0_1}{10^3} m^0_2 \left[\Phi_H\left(m^0_2\right) - \Phi_H\left(m^\beta_2\right)\right] \tag{6.54b}$$

Since the adsorption of the solvent is here considered negligible (i.e.,$M_1 = M^0_1$), the first two terms on the right hand side of Eq. (6.54b) represents the *enthalpy of immersion in a binary solution (1+2)* with respect to the bulk solution at infinite dilution taken as the reference state [64]; the third term corresponds to the enthalpy of dilution of $m_S \cdot S \cdot \Gamma^s_2$ moles of the solute from molality m^β_2 to infinite dilution in the solvent (cf., Eq. 6.44) and it is often incorporated into the previous enthalpy contribution to give the enthalpy of immersion $\Delta_{imm} H_{12}\left(m^\beta_2\right)$ of the solid in excess solution of molality m^β_2 [48, 76, 77]. The last term represents the dilution effects in the bulk solution: $\frac{M^0_1}{10^3} m^0_2$ moles of the solute are diluted from molality m^0_1 to molality m^β_2. Finally,

$$\Delta H_{\text{exp}} = \Delta_{imm} H_{12}\left(m^\beta_2\right) + \Delta_{dil} H\left(m^0_2 \rightarrow m^\beta_2\right) \tag{6.55}$$

When the dilution term is evaluated independently in appropriate dilution experiment, the enthalpy of immersion $\Delta_{imm} H_{12}\left(m^\beta_2\right)$ can be determined experimentally by means of the same calorimetry equipment as that used to measure the enthalpy of immersion in a pure liquid (Sect. 6.3.2). However, great difficulties may be encountered when evaluating the usual correction terms in case of solutes which are volatile or surface-active (the composition of the vapour occupying the dead volume of the bulb is unknown). This method is also tedious. For systems containting electrified interfaces, the effects of EDL formation additionally contribute to the complexity of the adsorption phenomenon. Here the displacement experiment yields the enthalpy data easier to interpret.

6.4.5 Model of Flow Calorimetry Experiment

Liquid-flow microcalorimetry is a reliable method to measure simultaneously the enthalpy changes and amounts of adsorption under dynamic conditions. Calorimetry experiments may be carried out in two different ways by following a *pulse* or *saturation operating mode* [64, 78–83]. In the pulse mode, small aliquots of a stock solution at a known concentration are injected into the carrier liquid (pure solvent) flowing through the adsorbent bed placed inside the calorimetric cell. In this case, the calorimetric system contains an additional loop injection facility (a manual injection valve with appropriate injection loops). The interpretation of the enthalpy data obtained is straightforward only when the whole amount of the solute injected is irreversibly adsorbed on the solid surface.

The saturation mode (*continuous-flow method*) is more frequently used. Here changes in enthalpy and amount adsorbed of the solute correspond to the formation of a Solid-Liquid interface being in thermal and material equilibrium with the percolating stock solution of a given composition. Repeated adsorption and desorption cycles with the liquid phase in contact with the solid surface for a time required to reach equilibrium can be used to assess reversibility of the phenomenon, and quantify the reversible and irreversible adsorption components [79, 80]. In addition, the same equipment allows probing for some active sites in the solid surface.

The physical meaning of thermodynamic quantities measured in the flow calorimetry experiment, may be discussed based on a simplified model of the system (Fig. 6.17). The model system is composed of three parts: (i) a reservoir R containing a given volume of the stock solution of molality m_2^0, (ii) a cell C with the solid sample of mass m_S, in contact with the solvent or the stock solution, (iii) a trap T for the effluent [65, 78].

Initially, the outgassed solid sample is immersed in M_{1C}^0 grams of pure solvent and the total quantity of stock solution in the reservoir is given by the mass of solvent M_{1R}^0; the trap is empty. Then the flow of stock solution from the reservoir is directed to the cell under constant liquid-flow conditions. The temperature and pressure are assumed to be uniform throughout the system and there is neither loss of energy nor loss of matter between the reservoir, the cell and the trap. As the stock solution

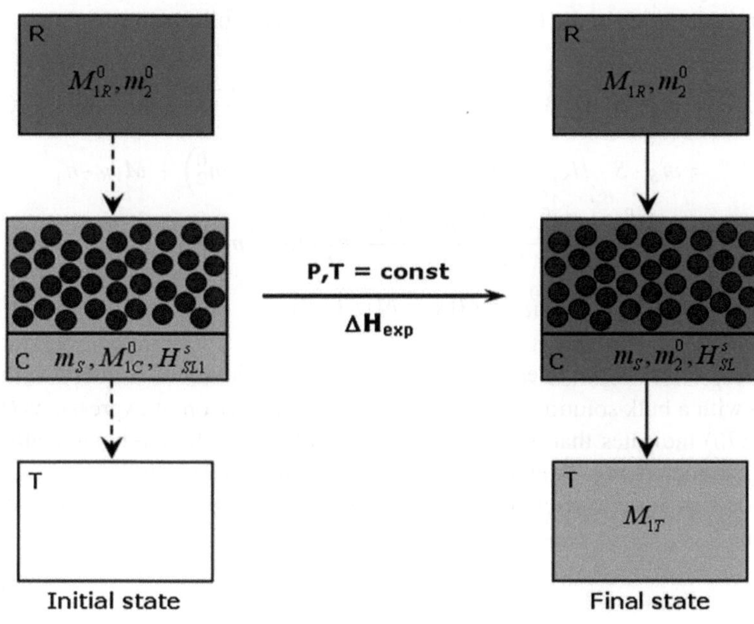

Fig. 6.17 Schematic representation of the flow calorimetry experiment (explanation of the symbols in the text). The model system is composed of three parts: *R* Reservoir, *C* Calorimetric cell, *T* Trap

flows through the cell with the adsorbent, the boundary between the solution and the solvent remains sharp, and the effects of adsorption and desorption change its position according to the *ideal chromatographic behaviour*. In practice, this means that the change of liquid composition should be sufficiently small for heat of mixing contribution to be neglected. The final state is achieved just before the front of the stock solution reaches the trap. At this point, the trap is filled with M_{1T} grams of pure solvent and the equilibrium molality of the solute within the cell (far from the solid surface) is equal to m_2^0. The composition of the stock solution in the reservoir does not change but its amount decreases; the final mass of the solvent is M_{1R}.

For the initial state, the amount of the solute n_2^{init} in the model system and the total enthalpy of the system H^{init} are given by (cf. Sect. 6.4.1):

$$n_2^{init} = \frac{M_{1R}^0}{10^3} m_2^0 \tag{6.56a}$$

$$\begin{aligned} H^{init} &= m_S \cdot S \cdot H_{SL1}^s + m_S \cdot h_S + M_{1C}^0 \cdot h_1^* \\ &+ \frac{M_{1R}^0}{10^3} m_2^0 \cdot \Phi_H\left(m_2^0\right) + M_{1R}^0 \cdot h_1^* \end{aligned} \tag{6.56b}$$

where H_{SL1}^s is the interfacial enthalpy for the solid-solvent interface; other quantities have been introduced previously (cf., Eqs. 6.53a, 6.53b).

Applying the law of mass conservation to the solvent in the system, the corresponding quantities in the final state may be written as follows:

$$n_2^{fin} = m_S \cdot S \cdot \Gamma_2^s + \frac{M_{1R}}{10^3} m_2^0 + \frac{M_{1R}^0 + M_{1C}^0 - M_{1R} - M_{1T}}{10^3} m_2^0 \tag{6.57a}$$

$$\begin{aligned} H^{fin} &= m_S \cdot S \cdot H_{SL}^s + m_S \cdot h_S + \frac{M_{1R}}{10^3} m_2^0 \cdot \Phi_H\left(m_2^0\right) + M_{1R} \cdot h_1^* \\ &+ \frac{M_{1R}^0 + M_{1C}^0 - M_{1R} - M_{1T}}{10^3} m_2^0 \cdot \Phi_H\left(m_2^0\right) \\ &+ \left(M_{1R}^0 + M_{1C}^0 - M_{1R} - M_{1T}\right) \cdot h_1^* \end{aligned} \tag{6.57b}$$

where H_{SL}^s refers to the interfacial enthalpy for the solid-solution interface in equilibrium with a bulk solution of molality m_2^0. The comparison of expressions (6.56b) and (6.57b) indicates that the adsorbent is considered as thermodynamically inert, i.e., its parameters m_S, S, and h_S do not change during adsorption.

The law of mass conservation applied to the solute gives rise to the following relation:

$$\Gamma_2^s = \frac{\left(M_{1T} - M_{1C}^0\right)}{10^3 \cdot m_S \cdot S} m_2^0 \tag{6.58a}$$

This relation shows that the quantity of adsorption for the solute may be directly measured in the flow calorimetry experiment if the values of M_{1C}^0 and M_{1T} are known. When the solid bed in the cell contains a non-adsorbing solid, the difference

$M_{1T} - M_{1C}^0$ is equal to zero; the value of M_{1C}^0 can thus be measured in the appropriate "blank" run. Otherwise, this difference depends on how much longer the solute is retained by the adsorbent. In the model flow experiment, the concentration curves obtained with a non-adsorbing and adsorbing solid represent obviously square profiles (Fig. 6.18a).

If the feed solution flow rate d_{pump} is constant and the solvent migrates at the same velocity throughout the whole system, the mass of solvent entering the trap is directly proportional to the retention time τ_A:

$$M_{1T} = \frac{10^3 \cdot d_{pump}}{10^3 + m_2^0 \cdot M_{solute}} \cdot \tau_A \tag{6.58b}$$

where M_{solute} is the molar mass of the solute. The amount of solute adsorbed onto solid sample can be thus calculated from the *corrected retention time*, i.e., $\tau_A - \tau_{NA}$ (Fig. 6.18a). In a real calorimetry run, the shape of the m_2 *vs.* time plot depends on the underlying equilibrium isotherm of solute adsorption, as well as diffusion and mass-transfer kinetics. This results in a breakthrough curve, as shown in Fig. 6.18b. The quantity of solute adsorption is calculated from the difference between the areas over the breakthrough curves obtained with a non-adsorbing and adsorbing solid [79, 80].

When using a pen recorder, the retention measurements are made in terms of chart distances and the area difference is determined by graphical integration. In modern systems, this theoretically correct analysis is also easy to perform when the digitized signal is recorded on the computer hard disk. In practice, the calculation of the areas always includes signal noise and it is very sensitive to the integration limits (the mass transfer is often slow and the plateau concentration m_2^0 is reached slowly, thereby resulting in significant systematic errors). Therefore, it is easier and better to handle

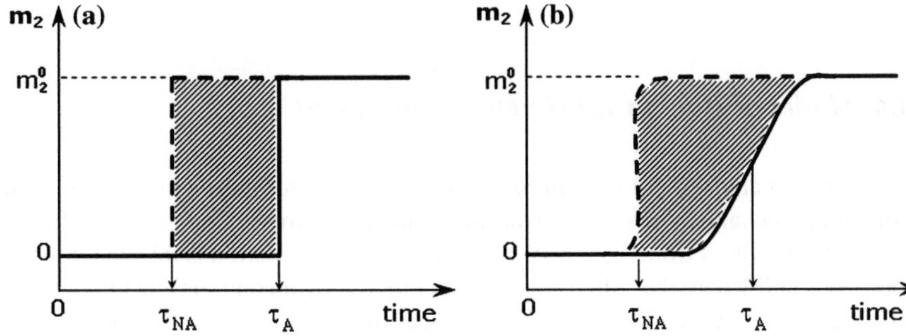

Fig. 6.18 Square profiles of the solute concentration obtained in the model flow experiments (**a**) and breakthrough curves of the solute registered during real flow calorimetry runs (**b**). The *dashed lines* refer to the profiles obtained with a non-adsorbing solid; τ_A and τ_{NA} are the retention times of the retained solute and the completely unretained solute, respectively. The *dashed areas* represent the amount of solute adsorbed (the calibration factor is needed)

the retention times (chart distances) defined by characteristic points method (e.g., the retention parameters derived from the inflection point or the half-height [84]).

The thermal balance in the flow calorimetry experiment can now be expressed in the following way:

$$\Delta H_{\exp} = H^{fin} - H^{init} = m_S \cdot S \cdot \left[\left(H_{SL}^s - H_{S0}^s \right) - \Gamma_2^s \cdot \Phi_H \left(m_2^0 \right) \right] + \\ - m_S \cdot S \cdot \left(H_{SL1}^s - H_{S0}^s \right) \quad (6.59a)$$

Taking into account the discussion in Sect. 6.4.4, the first term on the right hand side represents the enthalpy of immersion $\Delta_{imm} H_{12} \left(m_2^0 \right)$ of the solid in excess solution of molality m_2^0 and one obtains:

$$\Delta H_{\exp} = \Delta_{imm} H_{12} \left(m_2^0 \right) - \Delta_{imm} H_1 = \Delta_{dpl} H \quad (6.59b)$$

where $\Delta_{dpl} H$ denotes the *integral enthalpy of displacement* [64], which is the main enthalpy effect measured in the liquid-flow calorimetry experiment.

The *molar integral enthalpy* of displacement is calculated as follows:

$$\Delta_{dpl} h = \frac{\Delta H_{\exp}}{m_S \cdot S \cdot \Gamma_2^s} \quad (6.60)$$

This quantity provides information about the excess of component-adsorbent interactions averaged over all surface domains from which the solvent has been displaced by the adsorbing solute species. In consequence, it is not easy to monitor subtle changes in the adsorption mechanism based on usually small variations of the $\Delta_{dpl} h$ values with increasing quantity of adsorption. Compared to the differential molar enthalpy of displacement, the enthalpy $\Delta_{dpl} h$ is less sensitive to the energetic heterogeneity of the solid surface.

6.4.6 Model of Batch Calorimetry Experiment

Liquid titration calorimetry is a *microcalorimetric batch technique* most often used to study the mechanism of solute adsorption onto solids from binary dilute solution [83, 85, 86]. It differs from the flow variant in that a stock solution is injected into the calorimetric cell where the solid sample is kept in homogeneous suspension in a liquid (solvent or solution). The possibility of measuring the pseudo-differential molar enthalpy effects is the most important advantage of this technique. The differential enthalpy is very sensitive to various partial processes of adsorption occurring at the Solid-Liquid interface, thus allowing changes in the interfacial properties with surface coverage to be continuously monitored. From this standpoint, calorimetric data are very useful for theoretical consideration and modelling. Additionally, titration calorimetry may be an important analytical tool for determining the enthalpy effects accompanying the dilution of solutions with given compositions.

A typical operational procedure involves injection of a stock solution by small steps into the calorimetric cell containing either dilute solution (*dilution experiment*) or suspension of powder solid sample in a dilute solution (*adsorption experiment*). Serious drawbacks to the use of this calorimetry technique in studying the adsorption phenomena are due to the difficulty of direct evaluation of the related amounts adsorbed and to insufficient control of the environment of the liquid phase (e.g., the pH and ionic strength cannot be maintained constant through the whole run) [73, 87]. The progress in the adsorption quantity during successive injections of the adsorbate to the cell is quantified from the adsorption isotherm measured separately under *exactly the same* experimental conditions. For the adsorption of ionic solutes at electrified interfaces, it is impossible to well reproduce the charging behaviour of the adsorbent at a constant pH since this parameter cannot be re-adjusted during injections.

The model of dilution experiment is depicted in Fig. 6.19. Prior to injection sequence, there is M_1^0 grams of pure solvent in the calorimetric cell. Then, small aliquots of the stock solution of molality m_2^0 are injected by the syringe pump operating at a constant flow rate d_{pump}. When the time of injection t_{inj} is maintained constant, the amount of the solute n_2^{inj} introduced into the calorimetric cell is always the same. As a result, the equilibrium solution in the cell becomes more and more concentrated, i.e., $m_2^i \to m_2^0$. The dilution data may be further processed in two different ways, by calculating either *differential molar* or *cumulative molar* enthalpy changes.

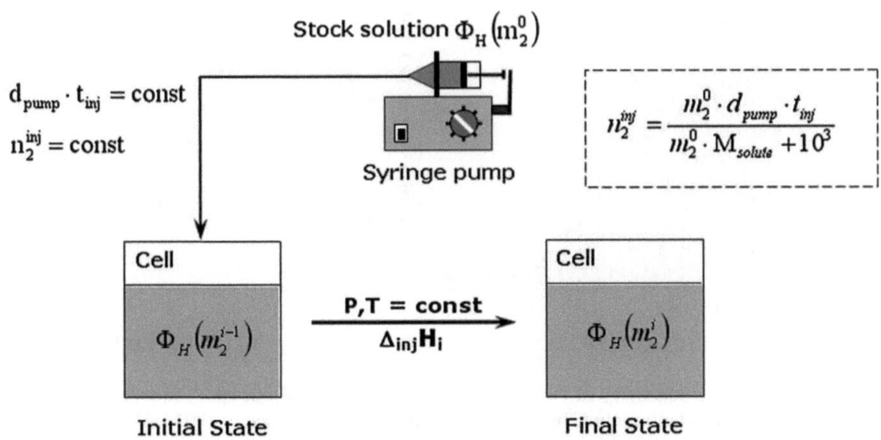

Fig. 6.19 Schematic representation of an injection step in the batch dilution experiment. The flow rate d_{pump} of the pump and the time t_{inj} of injection are kept constant during the experiment. The equation shows how to calculate the number of moles of the solute injected; m_2^0 and M_{solute} are the molality of the stock solution in the syringe and the molar mass of the solute, respectively; Φ_H (m_2) is the value of the apparent molal enthalpy of the solute in a binary solution corresponding to the molality m_2

The total enthalpy change during the ith injection *per mole of the solute* is expressed as follows [73, 74, 88]:

$$\frac{\Delta_{inj} H_i}{n_2^{inj}} = \Phi_H\left(m_2^i\right) + (i-1)\cdot n_2^{inj}\cdot\frac{\left[\Phi_H\left(m_2^i\right) - \Phi_H\left(m_2^{i-1}\right)\right]}{n_2^{inj}} + \\ - \Phi_H\left(m_2^0\right) \qquad (6.61a)$$

where n_2^{inj} depends on d_{pump}, t_{inj}, and m_2^0, according to the formula given in Fig. 6.19.

The equilibrium molality of the solution m_2^i in the cell after this injection is calculated by means of the following expression:

$$m_2^i = \frac{10^3 \cdot i \cdot n_2^{inj}}{M_1^0 + \frac{10^3 \cdot i \cdot n_2^{inj}}{m_2^0}} \qquad (6.61b)$$

The determination of all $\frac{\Delta_{inj} H_i}{n_2^{inj}}$ terms in function of m_2^{i-1} and m_2^i yields the results in the form of a histogram, as shown in Fig. 6.20. Referring to the general definition of the partial molal enthalpy $\overline{h}_2$ and apparent molal enthalpy Φ_H for any particular

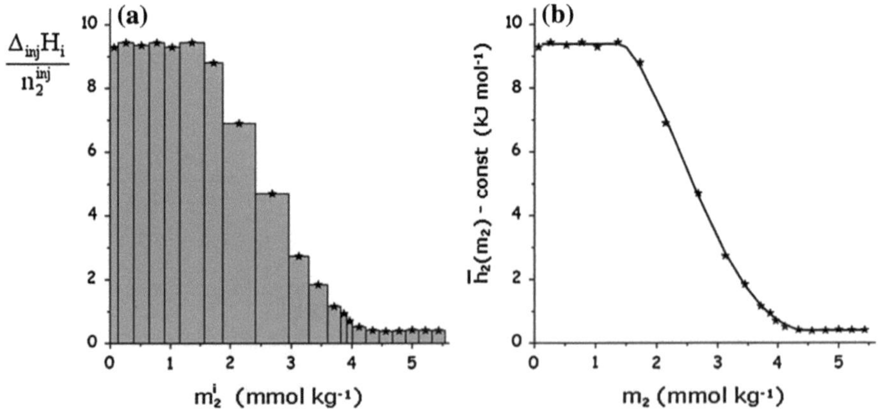

Fig. 6.20 Titration calorimetry of successive dilutions of a 0.028 mol kg^{-1} aqueous solution of benzyldimethyldodecylammonium bromide (BDDAB) in the presence of 0.01 mol kg^{-1} NaBr at 298 K [89]: (**a**) histogram of the molar enthalpy change during the *i*th injection as a function of the equilibrium solution molality, (**b**) plot of the partial molal enthalpy of the solute against its bulk molality. The solution is regarded as a binary system: the solute ions BDDAB$^+$ and Br$^-$ form together the mean solute, whereas NaBr is included in the mean solvent. The smooth curve of $\overline{h}_2$ against the equilibrium solution molality m_2 is achieved by choosing the value of m_2 to lie in the middle of interval $\left[m_2^{i-1}, m_2^i\right]$

solute in a solution of two components (Eqs. 6.36 and 6.39), it is to be noted that, for the sufficiently small quantities n_2^{inj} of the solute injected during each injection, Eq. (6.61a) can be transformed to give:

$$\frac{\Delta_{inj} H_i}{n_2^{inj}} \approx \lim_{n_2^{inj} \to 0} \frac{H\left(m_2^i\right) - H\left(m_2^{i-1}\right)}{n_2^{inj}} - const = \overline{h}_2\left(m_2\right) - const,$$
$$m_2^{i-1} < m_2 < m_2^i \tag{6.62}$$

where H denotes the enthalpy of a binary solution in the calorimetric cell.

Thus, values of $\overline{h}_2$ may be obtained experimentally by measuring both enthalpy changes $\Delta_{inj} H_i$ and very small quantities n_2^{inj} of the solute injected during successive dilution of the stock solution. It is clear from Eq. (6.62) that the above relation does not hold for the first injection: the first points in the calorimetric curves of dilution should not be taken into consideration.

The thermal effects of successive injections can be also summed up to obtain the molar cumulative enthalpy of dilution $\Delta_{dil} h_{cum}$. After k injections, one obtains:

$$\Delta_{dil} h_{cum} = \sum_{i=1}^{k} \frac{\Delta_{inj} H_i}{n_2^{inj}} = k \cdot \left[\Phi_H\left(m_2^k\right) - \Phi_H\left(m_2^0\right)\right],$$
$$\text{with } m_2 = m_2^k \tag{6.63}$$

The experimentally measured values of $\Delta_{dil} h_{cum}$ are further plotted against the equilibrium solute molality m_2 and only such a representation may have clear physical meaning from a thermodynamic standpoint. Nevertheless, it is sometimes more useful to present the enthalpy of dilution curve in terms of $\Delta_{dil} h_{cum}$ as a function of the injection number k, especially when the dependence of $\Delta_{dil} h_{cum}$ *vs* m_2 is linear or contains several linear portions (cf., Fig. 6.21).

In the adsorption calorimetry experiment, a small amount n_2^{inj} of the stock solution injected during a given injection is diluted in the supernatant liquid inside the cell and some of the resulting species subsequently adsorb onto solid particles. They displace a certain amount of solvent molecules and can exchange with some pre-adsorbed molecules or ions, because of the limited extent of the adsorption space. The effects of desolvation and re-solvation of various compounds taking part in the displacement process contribute to the competitive character of adsorption at the solid-solution interface. The flow chart of the batch displacement experiment is shown in Fig. 6.22. Since the enthalpy effects accompanying dilution of the stock solution inside the cell should be known, both the dilution and adsorption experiments are carried out under the same conditions (cf., Fig. 6.19).

When the adsorbent is initially immersed in M_1^0 grams of pure solvent and the final state corresponds to the formation of a Solid-Liquid interface in equilibrium with a bulk solution of molality m_2^k, the mass balance inside the cell leads to the following expression for the amount of solute adsorbed at the Solid-Liquid interface:

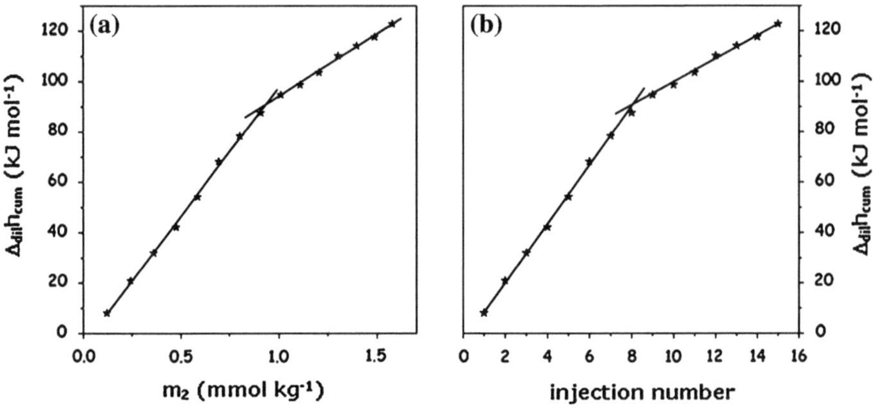

Fig. 6.21 Cumulative enthalpy of dilution of a 0.009 mol kg^{-1} aqueous solution of hexadecyltrimethylammonium bromide (HTAB) at 303 K as a function of (**a**) the solute molality and (**b**) the injection number [88]. The solution is regarded as a binary system: the solute ions HTA^+ and Br^- form together the mean solute. Plot **a**: the intersection of the two linear portions provides estimate of the critical micelle concentration for HTAB; Plot **b**: the enthalpy of HTAB micellisation is determined directly from the difference between the slopes of the two linear regression segments

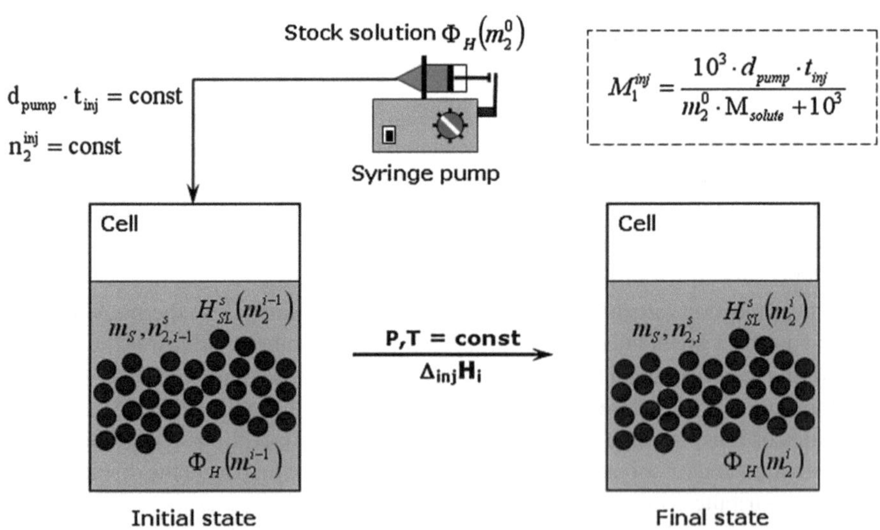

Fig. 6.22 Schematic representation of an injection step in the batch adsorption experiment. The values of d_{pump}, t_{inj}, n_2^{inj}, and m_2^0 are the same as those in Fig. 6.19. The equation shows how to calculate the mass of solvent injected with the stock solution. $H_{SL}^s(m_2)$ refers to the Solid-Liquid interface in equilibrium with a bulk solution of molality m_2

$$n^s_{2,k} = n^s_2\left(0 \to m^k_2\right) = \sum_{i=1}^{k} n^{inj}_2 - \frac{m^k_2}{10^3}\left(M^0_1 + \sum_{i=1}^{k} M^{inj}_1\right) \tag{6.64}$$

where M^{inj}_1 is the mass of solvent introduced into the calorimetric cell during one injection of the stock solution (see the appropriate equation in Fig. 6.22).

Equation 6.64 represents the so-called *calorimetric line*, since the amount adsorbed $n^s_{2,k}$ varies linearly with m^k_2 between two characteristic points: (P1) the whole amount of the solute injected is retained by the adsorbent surface and $m^k_2 = 0$; (P2) the whole amount of the solute injected remains in the bulk solution and $n^s_{2,k} = 0$. In practice, it is impossible to measure directly the values of n^s_2 for all injections (it would not be reasonable to interrupt the calorimetry run after each injection and remove the calorimetric cell in order to analyse its content). Nevertheless, the partition of the solute between the adsorbed and bulk phases is strictly determined by the adsorption equilibrium at a given temperature and does not depend on the path by which the adsorption system passes from its initial state to the equilibrium. Hence, the values of m^k_2 and $n^s_{2,k}$ are evaluated from the intersection between the calorimetric line (6.64) and the experimental isotherm $\Gamma^s_2 = \Gamma^s_2\left(m^\beta_2\right)$, obtained under *exactly the same* experimental conditions (Fig. 6.23).

The total change in enthalpy during the *i*th injection in the adsorption experiment may be written as follows:

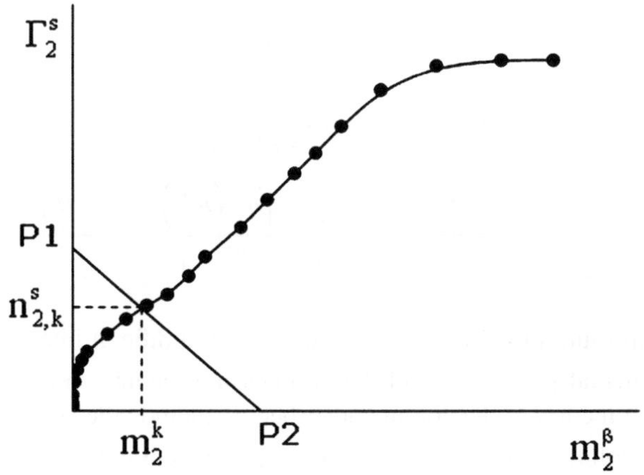

Fig. 6.23 Graphical determination of the amount adsorbed $n^s_{2,k}$ after a given injection step in the titration calorimetry run and the related molality of the equilibrium bulk solution m^k_2 from the intersection between the calorimetric line 6.64 (*passing through two characteristic points P1 and P2*) and the experimental adsorption isotherm measured under the same conditions

$$\Delta_{inj} H_i = m_S \cdot S \cdot \left[H^s_{SL}\left(m^i_2\right) - H^s_{SL}\left(m^{i-1}_2\right)\right] - n^s_{2,i} \cdot \Phi_H\left(m^i_2\right) + \\ + n^s_{2,i-1} \cdot \Phi_H\left(m^{i-1}_2\right) + \\ + n^{inj}_2 \cdot \left\{ \Phi_H\left(m^i_2\right) + (i-1) \cdot n^{inj}_2 \cdot \frac{\left[\Phi_H\left(m^i_2\right) - \Phi_H\left(m^{i-1}_2\right)\right]}{n^{inj}_2} - \Phi_H\left(m^0_2\right)\right\} \tag{6.65a}$$

where S is the specific surface area of the adsorbent; $H^s_{SL}(m_2)$ is the interfacial enthalpy for a Solid-Liquid interface in equilibrium with a bulk solution of molality m_2. The above equation has been developed by neglecting any changes in the enthalpy of the bulk of the adsorbent $m_S \cdot h_S$. The comparison with Eqs. (6.54a, 6.54b) and (6.59a, 6.59b) in Sects. 6.4.4 and 6.4.5 indicates that the first three terms on the right hand side of Eq.(6.65a) represent the difference in the enthalpy of displacement $\Delta_{dpl} H\left(m^i_2\right) - \Delta_{dpl} H\left(m^{i-1}_2\right)$ when passing from m^{i-1}_2 to m^i_2. The forth term has the same form as the overall enthalpy change recorded in the dilution experiment (cf., Eq. 6.61a). For small n^{inj}_2 values accompanied by small increments in the amount adsorbed n^s_2, the batch adsorption experiment allows measuring the *differential enthalpy of displacement per mole of the solute* adsorbed at the Solid-Liquid interface, since:

$$\Delta_{dpl} h_{diff} = \lim_{\Delta n^s_2 \to 0} \frac{\Delta_{dpl} H\left(m^i_2\right) - \Delta_{dpl} H\left(m^{i-1}_2\right)}{n^s_{2,i} - n^s_{2,i-1}} \\ \approx \frac{\Delta_{dpl} H\left(m^i_2\right) - \Delta_{dpl} H\left(m^{i-1}_2\right)}{n^s_{2,i} - n^s_{2,i-1}} = \\ = \frac{\Delta_{inj} H_i - n^{inj}_2 \cdot \left[\overline{h}_2\left(m^\beta_2\right) - const\right]}{n^s_{2,i} - n^s_{2,i-1}} \tag{6.65b}$$

where the equilibrium molality m^β_2 is located in the middle of interval $\left[m^{i-1}_2, m^i_2\right]$.

The experimental procedure and data processing leading to a smooth curve of $\Delta_{dpl} h_{diff}$ as a function of the amount adsorbed n^s_2 include the following stages: the quantity of the solute injected n^{inj}_2 and the successive enthalpy changes $\Delta_{inj} H_i$ are measured in the adsorption calorimetry experiment; the limit molality values m^{i-1}_2 and m^i_2 for each injection class and the related increments in the amount adsorbed $n^s_{2,i-1}$ and $n^s_{2,i}$ are evaluated with the aid of the adsorption isotherm; based on the "differential" enthalpy curve obtained in the dilution experiment, the correction terms for dilution, $n^{inj}_2 \cdot \left[\overline{h}_2\left(m^\beta_2\right) - const\right]$, are determined for the values of m^β_2 taken as the middle of appropriate intervals $\left[m^{i-1}_2, m^i_2\right]$; with small increments $\Delta n^s_2 =$

$n^s_{2,i} - n^s_{2,i-1}$, the differential displacement terms $\Delta_{dpl}h_{diff}$ are finally calculated and ascribed to the n^s_2 values in relation with the equilibrium molalities m^β_2. It should be noted again that the enthalpy value corresponding to the first injection cannot be considered as the differential molar enthalpy of displacement and has to be removed from the experimental curve.

6.5 Calorimetry Applied to Study Competitive Adsorption from Dilute Solution

The integral and differential enthalpies of displacement can be measured directly using calorimeters of the isothermal type. Such instruments either are produced and marketed by some manufacturers of high-performance calorimetric systems for different applications (e.g., SETARAM Instrumentation, LKB-ThermoMetric, TA Instruments, Microscal Ltd), or are home-mode prototypes based on original work carried out in university laboratories and therefore documented in considerable detail in scientific publications. This section focuses on the properties of two calorimetry systems used by the author to study the competitive adsorption from dilute solution.

6.5.1 Flow Calorimetry System

A liquid-flow calorimetry system (commercialised by Microscal Ltd) is represented schematically in Fig. 6.24 [79, 80]. Here the calorimetric cell is simply a cylindrical cavity inside the calorimetric bloc and its volume is limited by two removable tubes. The outlet tube is also used as a holder for a powdered solid sample. Prior to each calorimetry run, the inlet and outlet tubes are taken away from the calorimeter and the calorimetric cell is cleaned.

The measuring termistors detect the temperature of the cell content *via* a PTFE cell wall membrane which protects them from chemical attack. A separate temperature sensor is installed in the calorimetric block to permit the digital display of block temperature. The thermistor sensors are part of a Wheatstone bridge, which has the advantage that only temperature differences between the calorimetric cell and the block will put the bridge out of balance. When the solution containing a preferentially adsorbing solute or the pure solvent reaches the adsorbent bed there is an exothermic or endothermic displacement with resultant evolution or absorption of heat. This causes a small resistance change to the measuring thermistors and the effect is registered in the form of a thermal peak, the area of which is proportional to the heat measured.

It is necessary to have an evenly packed adsorbent bed in the calorimetric cell and a smooth, steady flow of liquid to pass through the bed so as to obtain reproducible results. Since the volume of the cell is limited (about 0.17 mL), a suitable volume

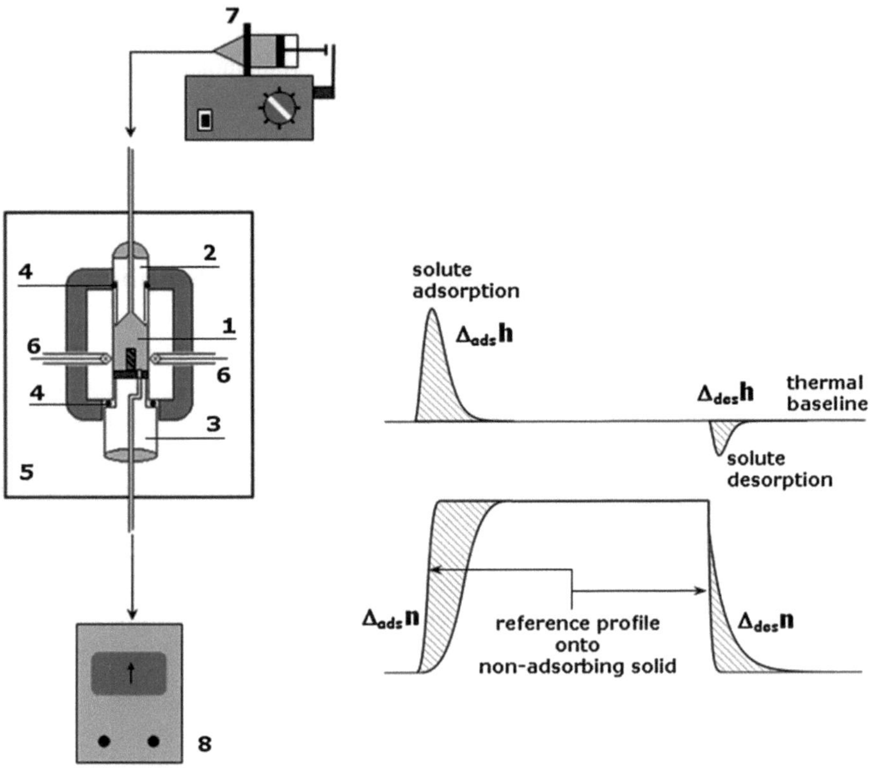

Fig. 6.24 Schematic representation of a liquid-flow microcalorimetry system operating in continuous-flow mode, together with traces showing the thermal and mass exchange profiles for adsorption of solute from its solution in the solvent, followed by the desorption of the solute by flow of pure solvent: *1* adsorbent bed, *2* inlet tube, *3* outlet tube, *4* toric seals, *5* aluminium block, *6* measuring thermistors, *7* syringe pump, *8* downstream detector

of powder or granular solid is weighed out in a special measuring tube provided by the manufacturer. Then the outlet tube is fitted in the calorimeter and the powder is poured into the sample holding space by means of an extended stainless steel funnel. Contrary to appearances, this operation is crucial for further measurements. Sometimes, the sample should be gently crushed to decrease the dead volume of the adsorbent bed. The use of monodispersed particles can be beneficial. When the particles are agglomerates of much finer material, they should not disintegrate in the liquid flow through the adsorbent. For solids having significant differences in their dry and wet packing densities (e.g., solids swelling in carrier liquids), the sample volume placed in the cell must take account of the volume change occurring on sample evacuation and immersion with the liquid. It is still possible to check whether the quantity of solid sample introduced and its packing in the cell are correct by fitting the inlet tube carefully. Any adsorbed species can be subsequently removed from the sample surface by evacuation, making use of a vacuum pump connected to the outlet tube. The temperature may be raised on evacuation by heating the calorimetric block with the aid of a stabilised D.C. power supply. The temperature limit depends on the thermal resistance of the materials of construction used. The efficiency of the evacuation step can be enhanced by flushing the adsorbent bed with the carrier liquid.

Further operating procedure for adsorption studies includes establishing a steady flow of the solvent used as a carrier liquid, awaiting thermal equilibrium, and setting the sensitivity controls. Different flow rates should be usually tested to optimise the operation for peak height, shape, sensitivity and duration. It should be remembered that the pump flow rate is limited by the syringe volume (normally 10 or 20 mL), compressibility of the solvent utilised, size of the solid particles, and their packing density in the sample holding space: the most commonly applicable flow rates are included in the range 0.05–0.1 mL per minute. Moreover, all occluded air should be carefully removed from the syringe and the whole line purged. As the equilibration proceeds, the thermal signal in function of time approaches a straight line asymptotically and this state is the criterion by which to judge the equilibration and to set the sensitivity of the calorimeter.

After attaining thermal equilibrium in the system, the flow of carrier liquid solvent is replaced with the identical flow rate of the solution of adsorbate at defined concentration by means of a changeover valve. The solution is fed by a syringe pump other than the one directing the solvent *via* the valve to the calorimeter. The progress of solute adsorption is monitored by the evolution or absorption of heat measured by thermistors sensing temperature changes in the calorimetric cell and simultaneously adsorbate transfer from the bulk phase to the Solid-Liquid interface is monitored by measuring composition changes in the effluent leaving the adsorbent and passing through the downstream detector. The principle of detection (UV, referactometry, thermal conductivity) and thus the choice of the detector depend on the chemical nature of both the solute and the solvent, the flow rates to be used, as well as the technical specification of the detector (e.g., operating parameters, sensitivity of detection, linearity of the signal with concentration, baseline stability, ease of operation). The adsorbent bed and the solution are allowed to remain in contact until heat evolution or absorption ceases (the signal returns to the thermal baseline) and no further change in the effluent composition is detected (the recorder trace shows a straight line), as illustrated in Fig. 6.24.

Since the calorimetric system is particularly well adapted to the study of the thermodynamic reversibility of the adsorption phenomenon, it is always worth carrying out the first desorption stage under exactly the same experimental conditions, just by returning to the flow of the carrier liquid through the adsorbent bed. The reader should be reminded that the heat effects of adsorption and desorption are opposite in sign. When the thermal and detector signals return to those of pure solvent, the adsorption-desorption cycle can be repeated to test the attainment of adsorption reversibility (e.g., the heat effects of adsorption and desorption are equal and repeatable). For some porous materials, the desorption process may last longer than the corresponding adsorption due to the slower mass transfer kinetics and the peak tailings effects are observed, thereby reducing the measurement accuracy and repeatability.

Calibration of the instrument is required at some stage for the particular operating conditions employed. Calibration of the areas under the thermal peaks is carried out by dissipating a known amount of energy in the adsorbed bed with the aid of a calibration probe incorporated into the outlet tube and encapsulated in PTFE. The related "exothermic" peak may be integrated making use of appropriate software facilities to process the digitized signal recorded on the computer hard disk. The data processing is the same as that described in Sect. 6.3.2 for immersional and wetting calorimetry. The downstream detector provides a plot of the effluent composition, the profile of this curve being influenced by the amount of solute molecules retained on the adsorbent surface during adsorption or released from the surface during desorption. To determine the related amount of solute adsorbed or desorbed, this profile is to be compared with composition changes obtained in a "blank" experiment with the use of a "non-adsorbing" solid through which pure solvent and the solution are passed at the same flow rate and temperature as those for the adsorbing sample. Glass or PTFE balls of low specific surface areas are usually used as non-adsorbing adsorbents, but the main difficulty here is to reproduce the same flow conditions of liquids through the adsorbent bed in the blank run (e.g., packing density of the solid, hydrophobic-hydrophilic character of its surface, pressure drop over the adsorbent bed). The two types of composition profile can be matched and presented in the form of net mass-transfer for adsorption and desorption. The areas of the segments, resulting from the subtraction of the peaks obtained on adsorbing and non-adsorbing solids, provide a direct measure of the quantity of the solute which either fails to emerge in the effluent solution due to its adsorption by the adsorbent bed in the calorimetric cell or is released to the carrier stream by the flow of pure solvent. The detector calibration factor is determined from the injection of a given volume of the solution into the stream of the solvent percolating through the detector (calibrated injection loop facility).

6.5.2 Measurements of Integral Enthalpy of Displacement

Despite the limited physical meaning of the integral enthalpy of displacement, measuring $\Delta_{dpl}H$ or $\Delta_{dpl}h$ may be very useful in several cases.

The first case certainly corresponds to the study of the thermodynamic reversibility of adsorption onto solids from binary solutions. Liquid-flow calorimetry measurements usually provide clear, unambiguous arguments for irreversible character of the phenomenon in numerous systems. An example of such systems is illustrated in Fig. 6.25. With non-porous Graphon possessing a very small number of surface polar sites, the adsorption of C_{60} fullerene from toluene is completely reversible. In the case of porous active carbons, the phenomenon is only partially reversible, the degree of reversibility being evaluated from the difference between the values of $\Delta_{dpl}h$ measured for the adsorption and desorption stage.

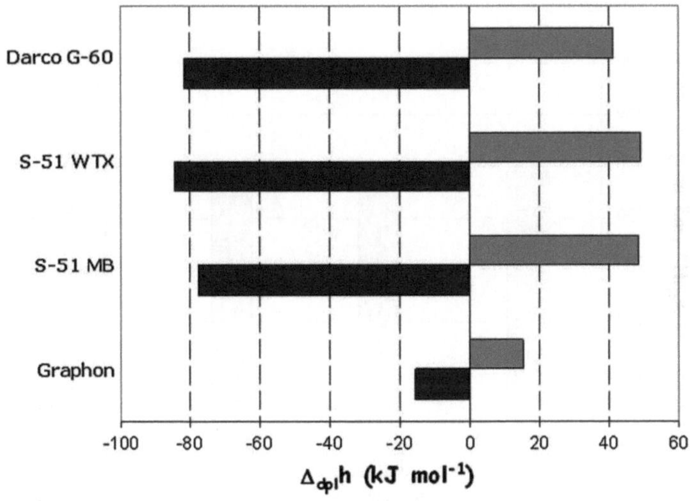

Fig. 6.25 Integral molar enthalpies (in kJ mol^{-1}) of C_{60} fullerene adsorption (*black bars*) and desorption (*grey bars*) obtained in one adsorption-desorption cycle from a 0.5 g L^{-1} toluene solution onto graphitised carbon *black* (Graphon) and three active carbons (S-51 MB, S-51 WTX, Darco G-60) at 293 K [75]

The irreversible enthalpy component may be calculated from the following formula:

$$\Delta_{dpl}h_{IR} = \frac{\Delta_{ads}n \cdot \Delta_{ads}h + \Delta_{des}n \cdot \Delta_{des}h}{\Delta_{ads}n - \Delta_{des}n} \tag{6.66}$$

where $\Delta_{ads}n$ and $\Delta_{des}n$ are positive changes in the number of moles of solute measured during the adsorption and desorption run, respectively; $\Delta_{ads}h$ and $\Delta_{des}h$ are the corresponding molar enthalpies of displacement observed in both stages.

Low sensitivity of the $\Delta_{dpl}H$ or $\Delta_{dpl}h$ values to the surface heterogeneity effects makes the integral enthalpy of displacement useful for probing specific sites on the surface of solid materials. The principle of this method lies in measuring the enthalpy of displacement per unit area of the adsorbent $\Delta_{dpl}H^*$ during adsorption of specific probe species (solute) capable of displacing non-specific solvent molecules from the targeted surface sites. For example, the polar contribution to the interfacial enthalpy H^s_{SL} may be thus approximated by determining the integral enthalpy of displacement of an apolar solvent by a polar solute. Prior to calorimetry measurements, the solution composition should be carefully optimised to ensure monomolecular adsorption of the solute on the solid surface [90, 91].

Figure 6.26 shows the effect of heteroatom incorporation into the framework of ordered mesoporous silica, as inferred from the adsorption of *1*-butanol from *n*-heptane.

When diluted in an apolar solvent, the molecules of *1*-butanol (BuOH) may be regarded as monomer species, potentially acting as both hydrogen-bond donors and

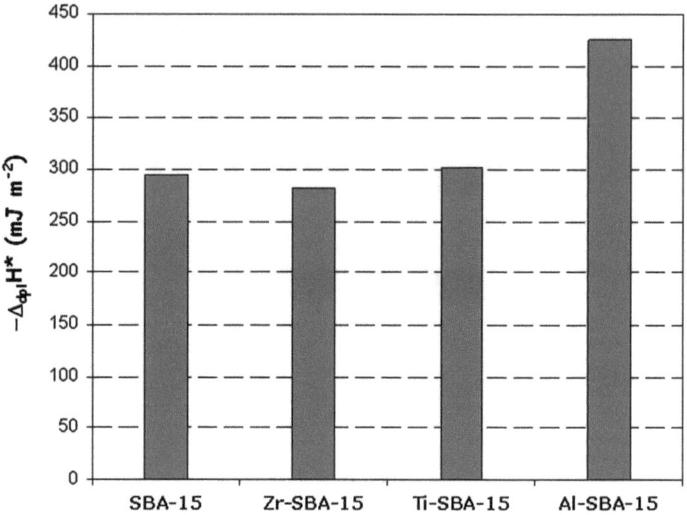

Fig. 6.26 Integral enthalpy of displacement per unit surface area $\Delta_{dpl}H^*$ (taken with the opposite sign) measured in the liquid-flow calorimetry experiment for adsorption of *1*-butanol from a 2 g L^{-1} solution in *n*-heptane at 298 K onto ordered mesoporous silica of the SBA-15 type and three mesoporous silica-based materials doped with various heteroatoms [92]

acceptors. Since the doping procedure aims at isomorphic substitutions of silicon by such heteroatoms as Al, Ti or Zr in the tetrahedral structures, the hydrophilic surface of the four mineral oxides should be dominated by surface hydroxyl groups. Butanol molecules can form hydrogen bonds with these silanols (Si-OH), aluminols (Al-OH), titanols (Ti-OH), or zirconols (Zr-OH). The monolayer adsorption of BuOH is accompanied by simultaneous desorption of apolar heptane molecules. In consequence, the integral enthalpy of displacement per unit area of the adsorbent surface $\Delta_{dpl}H^*$ is a function of the surface density of hydroxyl groups. In the case of materials doped with Ti(IV) and Zr(IV), there is no reason for a marked change in the surface density of hydroxyl groups.

This hypothesis is well illustrated by quite similar (within the experimental error) $\Delta_{dpl}H^*$ values obtained for SBA-15, Zr-SBA-15, and Ti-SBA-15, irrespective of differences in the surface area and porous structure among the samples. The partial replacement of Si(IV) by Al(III) results in additional surface hydroxyls related to the 'bridging' Si(OH)Al hydroxyl structures, thereby enhancing the value of $\Delta_{dpl}H^*$ (Fig. 6.26).

In the liquid-flow calorimetry experiment, the purified adsorbent bed remains in contact with a stock solution of constant composition. It is clear that the environment of the liquid phase does not change during the measurement. This is an important advantage of the flow calorimetry, especially in the case of solid-solution systems containing electrified interfaces. The study of ions adsorption from aqueous solutions

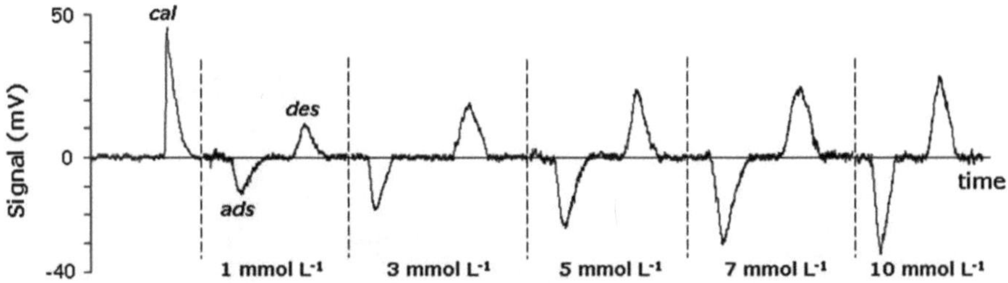

Fig. 6.27 Adsorption of Cd^{2+} cations from $Cd(NO_3)_2$ aqueous solutions of varying concentration at pH 7 onto Spherosil (S_{BET} = 25 m^2g^{-1}) at 298 K [94]: a record of successive saturation and desorption runs showing heat absorption and evolution: calibration (cal), adsorption (ads), and desorption (des). For each concentration of the stock solution m_2^0, the areas under the adsorption and desorption peaks are equal

onto mineral oxides bearing a pH-dependent surface charge always requires constant pH and ionic strength [71, 93].

Figure 6.27 presents the thermogram resulting from adsorption of a heavy metal cation from aqueous solution on the negatively charged surface of Spherosil registered during liquid-flow calorimetry measurements.

For a given concentration of the stock solution m_2^0, continuous percolation of the solution through the calorimetric cell containing the solid sample leads to saturation of the adsorbent with the solute giving rise to a negative heat effect (i.e., adsorption is endothermic) in the form of a peak in which the beginning and the end depend on the solution concentration, flow rate of solution throughout the adsorbent bed, and the kinetics of adsorption. Then the solute is removed from the adsorbent by exchanging the flow of the solution for that of pure solvent. The solute desorption is exothermic and complete in each cycle since the adsorption and desorption enthalpy effects have the same absolute value. Therefore, the successive saturation-desorption cycles can be performed without changing the solid sample in the cell. Although the thermogram in Fig. 6.27 shows that the resulting values of $\Delta_{dpl}H$ increases with increasing concentration, the molar enthalpy $\Delta_{dpl}h$ is proven to be a monotonously decreasing function of m_2^0 [94].

The comparison among thermal displacement effects accompanying the individual adsorption of an alkaline earth metal from aqueous solution is given in Fig. 6.28. The pH and ionic strength of the aqueous phase were identical in the four systems studied.

According to the appropriate speciation diagrams, each metal forms divalent cationic species in aqueous solution under the experimental conditions applied. In spite of the same electric charge of the four cations, the positive enthalpy values are very different, indicating that electrostatic attraction is not the only driving force of adsorption. Since there are Na^+ ions in the heterogeneous system, the total displacement effect should also include ion exchange between sodium and a given divalent cation. Modelling attempts to reproduce the positive displacement effects lead to the conclusion that metal cations may form multidentate complexes with oxygen atoms of ionised silanol groups and changes in the hydration layers of the adsorbing and

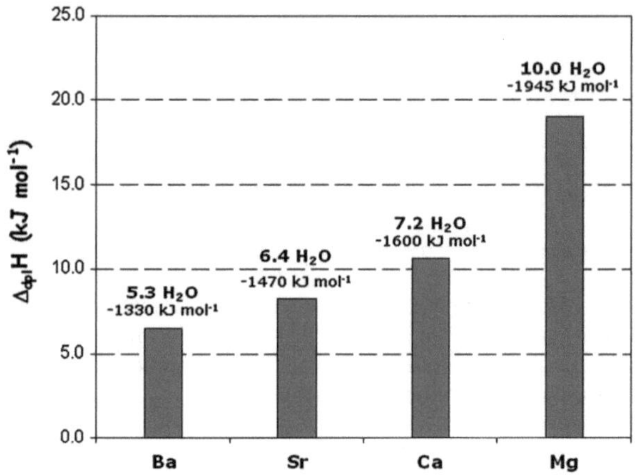

Fig. 6.28 Integral molar enthalpy of displacement related to the adsorption of a metal cation from 10^{-2} M nitrate solution onto Spherosil in the presence of 10^{-1} M $NaNO_3$ at pH 7 and 298 K for various divalent cations [95]. The number of water molecules in the hydration layer and the total enthalpy of hydration are given for each metal cation in aqueous solution

desorbing cations are the main reason for the endothermic character of the overall process [96].

6.5.3 Titration Calorimetry System

Liquid titration calorimeters contain a stirring device, which ensures the homogeneity of the liquid solution or solid suspension in the calorimetric cell, and an injection system permitting the controlled introduction of the reagents from outside the calorimetric cell. Contrary to the flow system, nothing flows out of the calorimeter to the surroundings since the reagents from a stock solution fed to the injection device are collected within the calorimetric cell. The heat detection is usually based on the principles of the isothermal batch or flow microcalorimeters. An example of home-made microcalorimeter [86] designed for study of the enthalpies of mixing of liquid and adsorption from dilute solution onto divided solids is represented schematically in Fig. 6.29.

The variations of temperature in the calorimetric cell induced by dilution or adsorption phenomena, as well as by the electrical calibration are recorded by thermistors arranged as a Wheatstone bridge. The two measuring thermistors, calibration coil and the inlet end of the injection tube are immersed in the solution or suspension. A precision syringe pump injects a stock solution into the cell at a constant rate ranging between 0.01 and 0.2 g min^{-1} without introducing any significant thermal perturbation. The injected solution flows through a heat exchanger tube where it is heated up to attain the temperature of the calorimetric block. The measurement of the

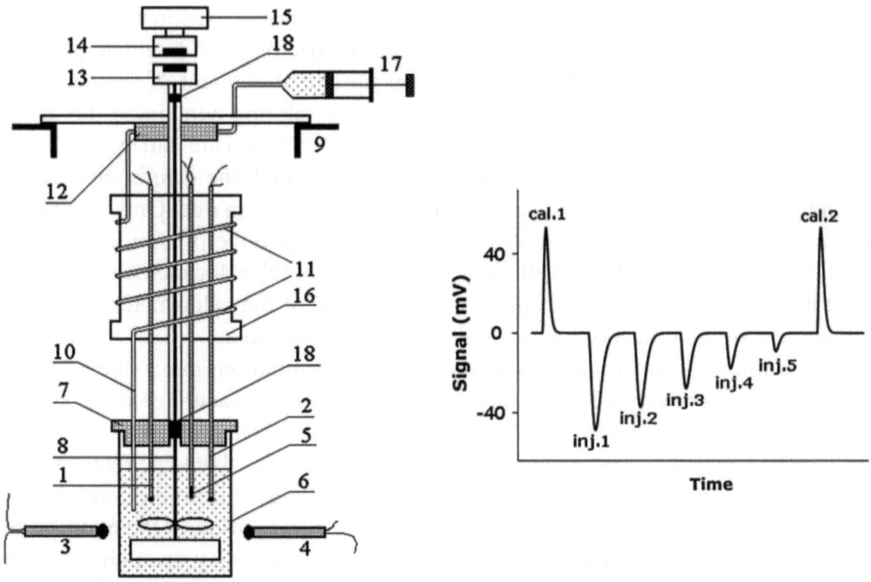

Fig. 6.29 Schematic representation of the calorimetric cell, stirring device, liquid injection and heat effect recording systems of the Montcal titration microcalorimeter [86], together with a thermal record of successive injections of a stock solution by small steps of 0.1 g mn^{-1} into the calorimetric cell: *1,2* measuring thermistors, *3,4* reference thermistors, *5* calibration coil, *6* stainless steel calorimetric cel (12–30 mL), *7* inert cover for calorimetric cell, *8* stirrer, *9* aluminium calorimetric block, *10* injection tube, *11* heat exchanger tube, *12* pre-heater, *13* magnet attached for stirrer, *14* magnet attached to electric motor, *15* electric motor, *16* aluminium cylinder supporting the exchanger tube, *17* syringe pump, *18* bearing; cal.1, cal.2—calibration peaks, inj.1, …, inj.5—injection peaks

enthalpy changes may be carried out also at higher temperatures (this microcalorimeter has been proven to give satisfactory results at temperatures ranging from 20 and 50°C [86]). For this purpose, an additional power unit provides a stabilised D.C. power supply for heating the calorimetric block to the desired temperature. To reduce the heat loss or temperature variations during injection, it is sometimes necessary to thermostate both the stock solution in the syringe and the injection tube outside the calorimetric block. A horizontal agitator with a variable speed of rotation is driven by a stepper motor fitted through a magnetic transmission.

The dissipation of a known amount of electrical energy inside the calorimetric cell by means of a calibration coil (i.e., the Joule effect) is used to relate the area of the thermal peaks recorded to the enthalpy effects which this represents. The difficulty with this type of calibration in the titration calorimetry systems is related to the fact that the mass of solution in the calorimetric cell is constantly increased by successive injections, thereby changing the calorific capacity of the cell. Therefore, thermal calibration should be regularly repeated after each series of injections in the same calorimetric run.

The calorimetric cell, together with the measuring termistors, stirring device and injection system, can be removed from the calorimetric block to facilitate cleaning and refilling the cell. After the instrument set up and attainment of thermal equilibrium, all steps of the run (injections of the stock solution containing the adsorbate, calibration processing, and recording of the heat effects) are carried out by the appropriate computer system. It is a best practice to weigh the calorimetric cell at the end of each run in order to calculate the real (mean) pump rate. One of the original features of the construction of this calorimeter is also the possibility of easily changing the thickness of the insulating barrier between the calorimetric cell and the metal block. This operation permits the modification of the calorimeter sensitivity and the time of the return of the thermal signal to the baseline.

The titration calorimetry technique presents some limitations due to the necessity of a strict correlation between the dilution and adsorption measurements. A special care must be taken to avoid experimental artefacts and erroneous interpretations in the study of solute adsorption onto solid supports which dissolve to a great extent in solutions or when the quantity of foreign substances released from the solid surface to the bulk solution is significant. In such cases, the composition of the equilibrium supernatant does not correspond to the pure solvent and it may even change constantly with increasing adsorption of the preferentially adsorbed solute [87, 93, 97]. Firstly, this evolution of the supernatant liquid should be monitored thoroughly during adsorption. Then, the evaluation of the correction term for dilution may pose serious problems since the stock solution has to be prepared by dissolving solute in the actual supernatant liquid and not in the pure solvent.

An example of the adsorption system investigated by means of titration calorimetry technique is given in Fig. 6.30 [74]. The experimental adsorption isotherm (Fig. 6.30b) has been determined separately based on the solution depletion method. The solute (dodecyldimethylammonio) butanoate (C12N3C) is a zwitterionic surfactant possessing a dipolar head-group and a linear aliphatic tail. In aqueous solution, the solute molecules self-assemble into aggregates called *micelles*. The concentration of the surrounding aqueous phase at which the surfactant monomers begin to form micelles is known as the *critical micelle concentration* (CMC).

When a micellar stock solution (i.e., its concentration is 10 times the CMC) is injected into a more dilute solution in the calorimetric cell, the constant value of partial molal enthalpy $\overline{h}_2$ in the premicellar region is due to destruction of micelles and dilution of unmicellized species (Fig. 6.30a); the constant $\overline{h}_2$ value in the postmicellar region is ascribed to dilution of micelles.

The molar change in partial molal enthalpy of the surfactant when monomers associate into a micelle at the cmc represents the *standard enthalpy of micellisation* $\Delta_{mic}h^o$ per mole of surfactant monomers [98]. In accordance with the variations of $\overline{h}_2$ as a function of m_2^β in Fig. 6.30a, the enthalpy of micellisation for C12N3C at 298 K is positive, indicating that the micellisation process is endothermic.

The experimental curves describing the adsorption of C12N3C onto Spherosil XOB015 from aqueous solution at 298 K (i.e., Figs. 6.30b and 6.30c) suggest that the phenomenon generally occurs in two stages. At very small quantities of adsorp-

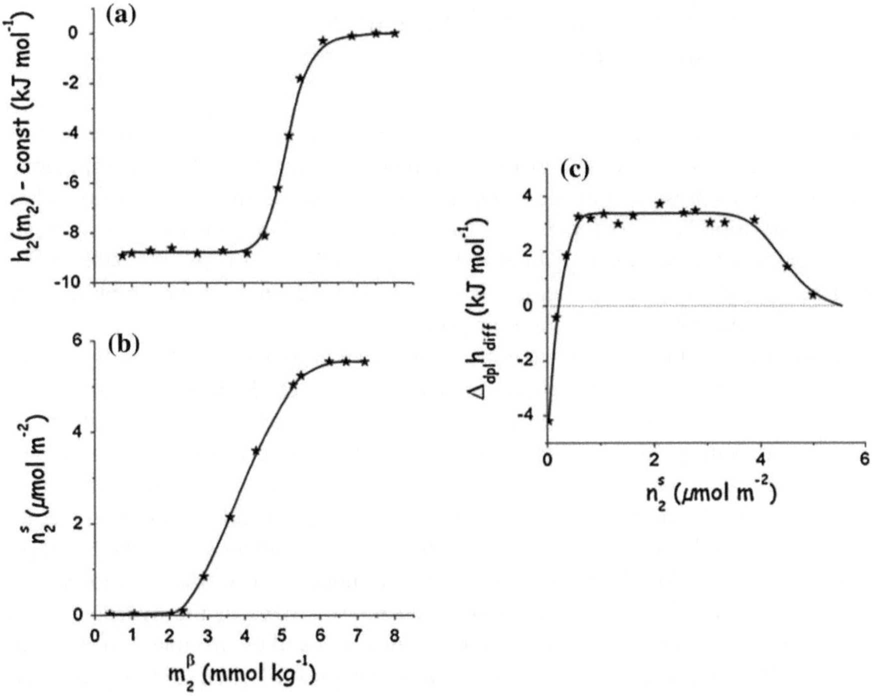

Fig. 6.30 Dilution of aqueous solution of (dodecyldimethylammonio) butanoate (C12N3C) and its adsorption onto Spherosil XOB015 ($S_{BET} = 25 m^2 g^{-1}$) at 298 K: (**a**) enthalpy of dilution, (**b**) adsorption isotherm, (**c**) differential molar enthalpy of displacement. In both types of titration calorimetry experiment, a 0.3 mol kg^{-1} C12N3C solution in pure H_2O was used

tion n_2^s, the values of $\Delta_{dpl}h_{diff}$ are negative so that the phenomenon is exothermic. The negative enthalpies of displacement are usually attributed to individual adsorption of surfactant molecules on an empty surface, where there may be only a few adsorbed molecules, and therefore lateral adsorbate-adsorbate interactions can be neglected [74]. The subsequent adsorption stage is dominated by adsorbate-adsorbate interactions. At moderate and great adsorption amounts, the driving force of adsorption derives from the hydrophobic effect, i.e., lateral chain-chain attractions and the tendency of hydrophobic tails to escape from an aqueous environment. This mode of adsorption is characterised by a constant, positive enthalpy of displacement, showing much similarity to micelle formation in the bulk solution [74]. This surface aggregation is likely controlled by a pseudo nucleation step, i.e., individual adsorption: the first adsorbed monomers act as nucleation centres for future surface-bound surfactant aggregates formed through chain-chain association.

6.5.4 Scanning of Surfactant Aggregation by Titration Calorimetry

Surface-active molecules or ions with an amphiphilic structure are known to have low solubility in water and self-assemble into large aggregates [61]. According to the enthalpy curves presented in Fig. 6.30, surfactant aggregation may occur not only in aqueous solution but also at the Solid-Liquid interface. Since the topic of surfactant aggregation is a well-developed field of research [62, 71, 73, 93, 99, 100], only a brief review of the broad principles is proposed in the present paragraph to better illustrate the contribution of titration calorimetry.

In aqueous solution, the formation of surfactant aggregates is driven by the tendency of the surfactant units to densely pack their tails. The hydrophobic tails remain inside the *liquid-like micellar core* due to unfavourable interactions with water molecules, whereas the polar head-groups, due to favourable interactions with the solvent, form a *hydrophilic outer layer* protecting the hydrophobic core. In the case of ionic surfactants, the Coulombic repulsion among the ionised head-groups is moderated by the specific adsorption of some counter-ions close to them within the Stern part of a curved ionic double layer surrounding the micelle. The micelle morphology and size depend primarily on the nature and relative sizes of the hydrophilic and hydrophobic moieties, as well as on the composition and environment of the aqueous phase [61, 62]. Titration calorimetry may be very useful when studying surfactant micellisation under different experimental conditions. The fundamental thermodynamic parameters, namely the CMC and the standard molar enthalpy of micellisation $\Delta_{mic}h^{\circ}$, can be easily inferred from calorimetric measurements of successive dilutions of a micellar stock solution injected by small steps into aqueous solution having a given composition [73, 74]. Several examples of cationic and zwitterionic surfactants are given in Table 6.5.

It is important to note that the formation of micelles in pure water may be an exothermic, endothermic or even athermic phenomenon, depending on the detailed molecular structure of the surfactant. The CMC value is related to the standard Gibbs energy of micellization, $\Delta_{mic}G^{\circ}$, which always takes negative values. This negative energy results rather from a large increase in entropy, which is ascribed either to structural changes in the solvent, associated with loss of hydration of the hydrophobic tail when the surfactant enters the micelle [61] or to increased freedom of the hydrophobic chain in the interior of the micelle compared to the bulk aqueous medium [101]. Besides the volume and length of the hydrophobic tail, the area per head-group at a curved interface between the micelle core and the aqueous solution σ_{mic} is a critical packing factor having an impact on the ultimate micelle structure [102]. For conventional ionic and zwitterionic surfactants with a single hydrocarbon chain, globular micelles are formed above but near the CMC. A significant increase in the overall surfactant content in aqueous solution may induce a change from spherical micelles to cylindrical (prolate) or disc-like (oblate) aggregates. This change is parralled by a decrease in σ_{mic} which allows a larger number of monomers to be inserted into each aggregate. The maximum cohesion is attained in large lamellar

Table 6.5 Critical micelle concentrations, CMC, and standard enthalpies of micellisation, $\Delta_{mic}h°$, per mole of surfactant monomers for selected quaternary ammonium surfactants in pure water at 298 K [73, 74, 88, 97, 103, 104]

Surfactant acronym and formula	CMC mmol kg^{-1}	$\Delta_{mic}h°$ kJ mol^{-1}
Zwitterionic surfactants		
C12N1C: $C_{12}H_{25}(CH_3)_2N^+(CH_2)CO_2^-$	1.9	4.6
C12N3C: $C_{12}H_{25}(CH_3)_2N^+(CH_2)_3CO_2^-$	4.6	8.8
C12N3S: $C_{12}H_{25}(CH_3)_2N^+(CH_2)_3SO_3^-$	3.0	3.6
Classical cationic surfactants		
BDDAB: $(C_6H_5)(CH_2)N^+(CH_3)_2(C_{12}H_{25})Br^-$	5.6	−5.3
TTAB: $C_{14}H_{29}N^+(CH_3)_3Br^-$	4.0	−4.7
DTAB: $C_{12}H_{25}N^+(CH_3)_3Br^-$	14.8	−1.6
Gemini cationic surfactants with a hydrophobic spacer		
C12S2C12:		
$C_{12}H_{25}(CH_3)_2N^+(CH_2)_2N^+(CH_3)_2C_{12}H_{25}Br^-$	0.84	−22
C12S6C12:		
$C_{12}H_{25}(CH_3)_2N^+(CH_2)_6N^+(CH_3)_2C_{12}H_{25}Br^-$	1.03	−8.5
C12S12C12:		
$C_{12}H_{25}(CH_3)_2N^+(CH_2)_{12}N^+(CH_3)_2C_{12}H_{25}Br^-$	0.37	−12.2
Gemini cationic surfactants with a hydrophilic spacer		
C12EO3C12:		
$C_{12}H_{25}(CH_3)_2N^+(C_2H_4O)_3N^+(CH_3)_2C_{12}H_{25}Br^-$	1.02	−6.9
C12EO7C12:		
$C_{12}H_{25}(CH_3)_2N^+(C_2H_4O)_7N^+(CH_3)_2C_{12}H_{25}Br^-$	1.58	0.0
C12EO12C12:		
$C_{12}H_{25}(CH_3)_2N^+(C_2H_4O)_{12}N^+(CH_3)_2C_{12}H_{25}Br^-$	1.93	6.3

sheets (flat bilayers) two molecules thick, though surfactant tails never attain such a close-packed arrangement in dilute solutions.

Dimeric or Gemini surfactants (composed of two surfactant units connected by a hydrophobic or hydrophilic chain—*spacer*) usually show a much stronger tendency for micellar growth and self-assemble into larger aggregates with a lower degree of curvature (e.g., linear thread-like and tree-like micelles, or spheroids) [105, 106].

The addition of solid particles (porous or non porous) into aqueous solution induces a decrease in the chemical (or electrochemical) potential of the surfactant solute and, in consequence, the adsorbing surfactant units form, at the solid surface, some periodic adsorbate self-assemblies closely related to the micellar structures encountered in the bulk solution at higher monomer concentrations. Such surface-bound aggregates are spoken of as *interfacial aggregates* or *solloids* [107]. In the general case, the shape and size of solloids are considered to be a compromise between the free curvature, as defined by the energetic, geometrical and packing factors arising from the molecular structure of the surfactant in a given environment, and some influences and constraints imposed, on the one hand, by direct solute-surface interactions and, on the other hand, by the porosity of the adsorbent. For example, the images of extended aggregate structures showing a closer registry with the underlying surface have been obtained by atomic force microscopy (AFM) on atomically smooth crystalline surfaces [99]. Such solloids have cylindrical and hemi-cylindrical morphologies, depending on the hydrophilic-hydrophobic character of the solid surface. In the case of powders or porous solids, where such microscopy techniques as AFM or ellipsometry have very limited applicability, the titration calorimetry measurements of the differential molar enthalpy of displacement as a function of the surfactant adsorption may provide important information about the self-assembled surfactant structures when compared with the thermal effects of surfactant micellisation in aqueous solution under the same experimental conditions. Nevertheless, calorimetry alone cannot be used to scan for the detailed solloid morphology and appropriate modelling of the adsorption system is necessary.

As far as the adsorption of ionic surfactants on the oppositely charged (and macroscopically flat) surfaces of mineral oxides is concerned, the following three types of solloid are frequently used in the empirical explanation or modelling of the experimental data:

(1) *monolayered hemimicelles*, composed of surfactant units oriented 'head-on' towards the surface, with the surfactant tails forming a hydrophobic film in contact with the equilibrium aqueous solution [108],
(2) *bilayered admicelles*, containing two adsorbed layers of the surfactant monomers directed 'head-on' and 'head-out' with respect to the surface [109],
(3) *small surface micelles*, i.e., spherical isolated aggregates anchored to certain surface sites with aggregation numbers markedly smaller than in bulk micelles [110].
Each of these solloid morphologies is claimed to have a micelle-like character, although the contact area between the water molecules and the hydrophobic surfactant moieties is not always reduced to the minimum.

Typical plots of the differential molar enthalpy of displacement against the amount of the surfactant adsorbed reveal significant variations in the enthalpy value when the adsorption progresses. An example of the enthalpy curve is given in Fig. 6.31. Such trends in $\Delta_{dpl}h_{diff}$ with increasing n_2^s suggest almost continuous evolution of the solloid morphology and size: the aggregates self-assembled from the adsorbing surfactant monomers at equilibrium concentrations in the bulk phase lower than the CMC may grow in the direction parallel and perpendicular to the solid surface. The most successful theoretical attempts to mimic complex shapes of both the experimental adsorption isotherms and enthalpy of displacement curves have been based on the assumption that the adsorbed phase at a given surface coverage can be seen as a mixture of mutually interacting surface-bound monomers, monolayered hemimicelles, and bilayered admicelles varying in size and number [65, 111]. The proportion between the various types of adsorbate species is shown to undergo significant changes with increasing surfactant adsorption, first monomers and monolayered aggregates and then bilayered admicelles dominating on the surface.

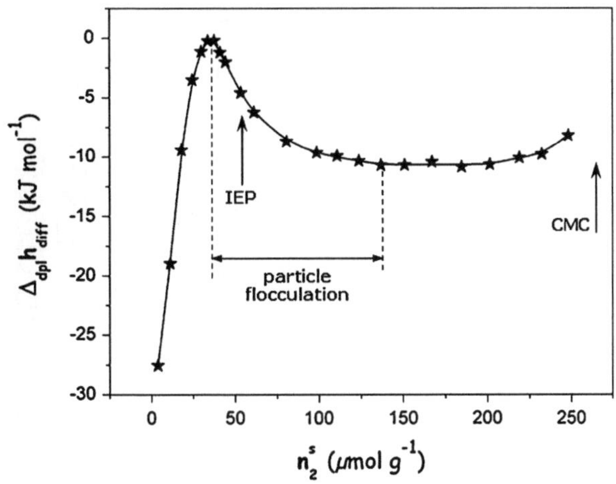

Fig. 6.31 Variations of the differential molar enthalpy of displacement as a function of the adsorption of benzyldimethylammonium bromide (BDDAB) onto silica powder S91-16 (Rhône-Poulenc, France) from aqueous solutions at 298 K at the initial pH 8 [89]. The arrows indicate the critical micelle concentration (CMC) and the isoelectric point (IEP) at which the effective charge of the silica particles together with the specifically adsorbed surfactant cations becomes equal to zero. The region of particle flocculation (where the silica particles covered with the adsorbed species are predominantly hydrophobic) is also shown

When surfactants are adsorbed onto fine-pore solids, the growth of solloids in the direction perpendicular to the pore walls should be limited by the pore volume [93]. For mesoporous ordered mineral oxides, it may be even that the surfactant monomers adsorb only head-on with respect to the hydrophilic surface and the hydrophobic tails of the surfactant units adsorbed on the opposite walls interpenetrate themselves in such a way as to produce "internal" aggregates, which fill the pore space. To check whether the head-out adsorption of surfactant monomers is to be excluded in such systems, one can refer to the micellar solubilisation of water-insoluble, hydrophobic materials as well as polar substances, which dissolve in water only to a limited extent [62, 69, 112, 113]. For low contents in aqueous solution, some small molecules may behave as molecular probes, occupying specific sites in surfactant aggregates without greatly disturbing their morphology and size. The exact locus of a given probe molecule in a micelle reflects the type of forces operating between the aggregate and the solubilised material. This justifies the use of titration calorimetry to study micellar solubilisation. One of the possible methodologies is to lump water and the additive together into the mean solvent and investigate the micellisation of the surfactant in this new medium. The molar enthalpies of micellisation for several cationic surfactants in the absence and the presence of phenol at various additive contents are compared schematically in Fig. 6.32.

On the addition of phenol to the aqueous phase, the enthalpy of micellisation per mole of the surfactant becomes more negative (and the CMC value is decreased), thereby indicating more favourable phenol-surfactant interactions after the transfer

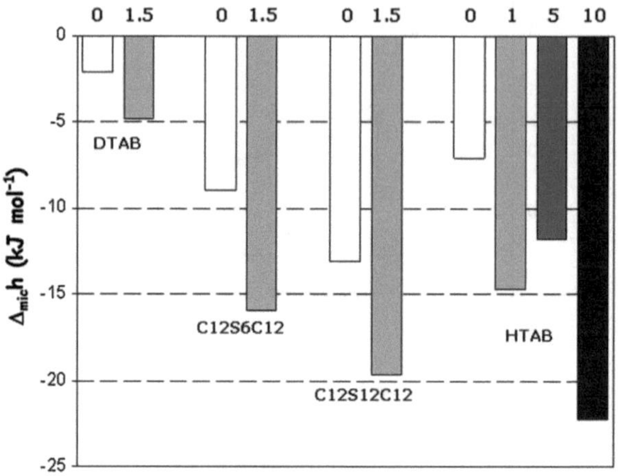

Fig. 6.32 Molar enthalpies of micellisation, $\Delta_{mic}h$, for selected quaternary ammonium surfactants in pure water and in the presence of phenol molecules in the aqueous phase at 298 K [88, 114, 115]. The overall phenol (PhOH) content (in mmol kg^{-1}) is reported on the X-axis. The surfactant acronyms are explained in Table 6.5. For hexadecyltrimethylammonium bromide (HTAB), the calorimetry measurements were carried out at 303 K safely above the Krafft point

of the additive to the micellar phase. Therefore, phenol molecules are preferentially located in the outer portions of cationic micelles close to the surfactant head-groups, without involving much rearrangement of the micelle structure. When the phenol concentration increases (e.g., PhOH-HTAB systems), the existence of an endothermic contribution to $\Delta_{mic}h$ may be deduced from the evolution of the enthalpy value. To better understand this positive enthalpy component, one may refer to the detailed analysis of ^{1}H NMR spectra recorded with the various PhOH-HTAB solutions: additional phenol units penetrate deeper into the micelle core producing unfavorable interactions with cationic micelles of HTAB [88]. Consequently, if phenol is to be used as a molecular probe for detecting the presence of the head-out adsorbed surfactants at the Solid-Liquid interface, the overall additive content in the system should remain low.

Based on the assumption that surfactant aggregation on the solid surface is a prerequisite for the uptake of phenol and the aromatic molecules can be located only close to the "free" head-groups of the surfactant units within the interfacial aggregates, it may be helpful to compare the curves presenting the differential enthalpy of displacement as a function of the surface coverage by the surfactant adsorbate in the absence and in the presence of the additive. Figure 6.33 illustrates such a comparison for a cationic surfactant adsorbed onto ordered mesoporous aluminosilicate of the MCM-41 type.

The difference between the two curves is clearly pronounced at higher surface coverage ratios where surface aggregation is considered to be the predominant sorption mode. With a small amount of phenol added to the aquous phase, the displacement phenomenon is more exothermic in this region: the related portion of the enthalpy

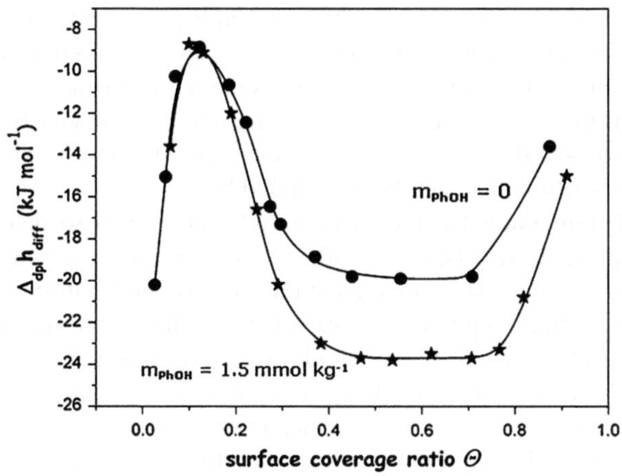

Fig. 6.33 Effect of phenol addition on the differential molar enthalpy of displacement upon adsorption of cationic Gemini C12C12C12 onto ordered mesoporous aluminosilicate of the MCM-41 type (S_{BET} = 860 m^2g^{-1}, mean pore diameter = 5 nm, Si:Al = 32) from aqueous solution at 298 K and the initial pH 8 [114]. The $\Delta_{dpl}h_{diff}$ enthalpy is plotted against the adsorption coverage of the solid surface by the surfactant cations

curve can be viewed as shifted towards more negative values by a constant value in comparison with the curve corresponding to the system without phenol. Similar enthalpy behaviour of this surfactant has been observed during its micellisation in aqueous solution (cf., Fig. 6.32). This analysis provides strong indication for phenol incorporation in the interfacial aggregates having their head-groups oriented outwards. Therefore, the image of all surfactant units interacting directly with the negatively charged surface and their hydrophobic tails filling the whole pore space available is rather to be excluded.

The coexistence of hydrophilic and hydrophobic nano-domains separated in space, with a local order and fluidity typical of liquids, confer to supramolecular surfactant structures remarkable properties, which are advantageous in applications involving molecular confinement within nanoscopic regions and reactivity in microheterogeneous media. Micelle-mediated reactions constitute the basis of the so-called micellar catalysis [62, 116], admicellar catalysis [117] or admicellar polymerisation [118] in which reaction mechanisms may be controlled at a molecular level to save energy and raw materials, as well as to avoid lengthy post-reaction purification and analytical steps.

6.6 Concluding Remarks

The intention of the present chapter was to present mostly the prospective advantages, but also some limitations, of the use of isothermal calorimetry at the Solid-Liquid interface as a powerful tool in the study of interactions between solid surfaces and the surrounding liquid phase. The operating principles and examples of applications were described for three calorimetry techniques frequently used to date in Surface Science and Technology: (i) wetting and immersional calorimetry, (ii) liquid flow calorimetry, (iii) titration batch calorimetry. The interested reader is encouraged to search for other outstanding examples of commercial or home-made instruments and their specific uses which have not been included here.

Wetting and immersional calorimetry may be employed to determine surface properties of catalysts, adsorbents and other solid materials in contact with liquids. Based on the calorimtery measurements of the various contributions to the total surface enthalpy of a solid, it is possible to evaluate the hydrophobic-hydrophilic character of its surface. The Harkins-Jura method for the evaluation of the specific surface area of a solid based on enthalpy changes in the so-called immersion-adsorption-wetting cycle gives the surface area of contact between this solid and a pure liquid or a solution, thereby shedding light on the availability of the solid surface under real experimental conditions.

Liquid flow or batch titration calorimetry techniques offer an opportunity for studying the macroscopic outcome of the various interactions involved in interfacial phenomena occurring at the Solid-Liquid Interface. The enthalpy changes appear very sensitive to the partial mechanisms through which a given phenomenon can occur. In particular, the effects of heterogeneity of a solid surface (i.e., surface sites with different adsorption energies, "confinement effects" due to adsorbent porosity) show up more clearly in heat quantities than in adsorption isotherms, thereby allowing easier interpretation of the phenomenon studied. The assessment of the thermodynamic reversibility of competitive adsorption from multicomponent solutions is one of the different possibilities of isothermal calorimetry. Self-assembled surfactant structures, defined by the regular assembly of small molecular entities into larger supra-molecular structures either in aquous solution or at the Solid-Liquid interface, may be thermodynamically described based on the results of titration calorimetry measurements. Here the next step would be to use high sensitivity isothermal calorimeters to determine the thermal effects of micelle-mediated reactions in micellar or admicellar catalysis.

In spite of many advantages, Solid-Liquid calorimetry alone is not capable of solving satisfactorily many detailed problems concerning the resulting interfacial mechanisms. It certainly cannot provide much information on entropy changes. In consequence, calorimetric measurements have to be always supplemented by other experimental studies reported on the system so as to obtain a more complete description of the phenomenon

References

1. R. Defay, I. Prigogine, A. Bellemans, D.H. Everett, *Surface Tension and Adsorption* (Longmans, London, 1966)
2. P.C. Hiemenz, *Principles of Colloid and Surface Chemistry*, 2nd edn. (Marcel Dekker, New York, 1986)

3. R.J. Hunter, *Foundations of Colloid Science,*, vol. 1 and 2, (Oxford University Press, Oxford, 1989)
4. A.W. Adamson, *Physical Chemistry of Surfaces*, 5th edn. (Wiley-Interscience, New York, 1990)
5. J. Lyklema, *Fundamentals of Interface and Colloid Science*, vol. 1–3, (Academic Press, London, 1991–2000)
6. D.H. Everett, Reporting data on adsorption from solution at the solid/solution interface (Recommendations 1986). Pure Appl. Chem. **58**(7), 967–984 (1986). doi:10.1351/pac198658070967
7. W. Rudzinski, D.H. Everett, *Adsorption of Gases on Heterogeneous Surfaces* (Academic Press, London, 1992)
8. E.A. Guggenheim, *Thermodynamics*, 5th edn. (North Holland Publishing Co., Amsterdam, 1967)
9. J.F. Padday, in *Surface Tension. II. The Measurement of Surface Tension*, vol 1, ed. by E. Matijevic, F. Eirich. Surface and Colloid Science, vol 1 (Wiley-Interscience, New York, 1969), pp. 101–149
10. N.R. Pallas, Y. Harrison, An automated drop shape apparatus and the surface tension of pure water. Colloids Surf. **43**(2), 169–194 (1990). doi:10.1016/0166-6622(90)80287-E
11. R. Cini, G. Loglio, A. Ficalbi, Temperature dependence of the surface tension of water by the equilibrium ring method. J. Colloid Interface Sci. **41**(2), 287–297 (1972). doi:10.1016/0021-9797(72)90113-0
12. G. Loglio, A. Ficalbi, R. Cini, A new evaluation of the surface tension temperature coefficients for water. J. Colloid Interface Sci. **64**(1), 198–198 (1978). doi:10.1016/0021-9797(78)90352-1
13. R.C. Weast (ed.), *Handbook of Chemistry and Physics*, 45th edn. (CRC, Cleveland, 1964)
14. A.R.C. Westwood, T.T. Hitch, Surface energy of 100 potassium chloride. J. Appl. Phys. **34**(10), 3085–3089 (1963)
15. S. Boffi, M. Ricci, On the cleavage energy of magnesium oxide. Mater. Chem. **1**(4), 289–296 (1976). doi:10.1016/0390-6035(76)90030-4
16. J.J. Gilman, Direct measurements of the surface energies of crystals. J. Appl. Phys. **31**(12), 2208–2218 (1960)
17. E. Orowan, Die Zugfestigkeit von Glimmer und das Problem der technischen Festigkeit. Z. für Phys. A Hadrons Nuclei **82**(3), 235–266 (1933). doi:10.1007/bf01341490
18. R.J. Good, Contact angle, Wetting, and Adhesion: a critical review, in *Contact Angle*, ed. by K.L. Mittal, Wettability and Adhesion (VSP, Utrecht, 1993), pp 3–36
19. C.J. Van Oss, M.K. Chaudhury, R.J. Good, Interfacial Lifshitz-van der Waals and polar interactions in macroscopic systems. Chem. Rev. **88**(6), 927–941 (1988). doi:10.1021/cr00088a006
20. A.C. Zettlemoyer, Hydrophobic surfaces. J. Colloid Interface Sci. **28**(3–4), 343–369 (1968). doi:10.1016/0021-9797(68)90066-0
21. J.N. Israelachvili, *Intermolecular and Surface Forces*, 2nd edn. (Academic Press, London, 1991)
22. J.C. Berg, The Role of Acid-Base Interactions in Wetting and Related Phenomena, in *Wettability*, ed. by J.C. Berg (Marcel Dekker, New York, 1993), pp. 75–148
23. C.J. Van Oss, M.K. Chaudhury, R.J. Good, Monopolar surfaces. Adv. Colloid Interface Sci. **28**, 35–64 (1987). doi:10.1016/0001-8686(87)80008-8
24. D.B. Hough, L.R. White, The calculation of Hamaker constants from Liftshitz theory with applications to wetting phenomena. Adv. Colloid Interface Sci. **14**(1), 3–41 (1980). doi:10.1016/0001-8686(80)80006-6
25. J. Visser, On Hamaker constants: a comparison between Hamaker constants and Lifshitz-van der Waals constants. Adv. Colloid Interface Sci. **3**(4), 331–363 (1972). doi:10.1016/0001-8686(72)85001-2
26. C.J. Van Oss, R.J. Good, M.K. Chaudhury, The role of van der Waals forces and hydrogen bonds in "hydrophobic interactions" between biopolymers and low energy surfaces. J. Colloid Interface Sci. **111**(2), 378–390 (1986). doi:10.1016/0021-9797(86)90041-X

27. R.S. Drago, Quantitative evolution and prediction of donor-acceptor interactions. Struct. Bond. (Berlin) **15**, 73–139 (1973)
28. F.M. Fowkes, Quantitative characterization of the acid-base properties of solvents, polymers, and inorganic surfaces. in *Acid-Base Interactions—Relevance to Adhesion Science and Technology* ed. by K.L. Mittal, H.R.J. Anderson (VSP, Utrecht, 1991), pp. 93–115
29. M.K. Chaudhury, Interfacial interaction between low-energy surfaces. Mater. Sci. Eng. R Rep. **16**(3), 97–159 (1996). doi:10.1016/0927-796X(95)00185-9
30. J.M. Douillard, T. Zoungrana, S. Partyka, Surface Gibbs free energy of minerals: some values. J. Pet. Sci. Eng. **14**(1–2), 51–57 (1995). doi:10.1016/0920-4105(95)00018-6
31. L.A. Girifalco, R.J. Good, A theory for the estimation of surface and interfacial energies. I. Derivation and application to interfacial tension. J. Phys. Chem. **61**(7), 904–909 (1957). doi:10.1021/j150553a013
32. C.J. Van Oss, Acid-base interfacial interactions in aqueous media. Colloids Surf. A Physicochemical Eng. Aspects **78**, 1–49 (1993). doi:10.1016/0927-7757(93)80308-2
33. P.G. De Gennes, Wetting: statics and dynamics. Rev. Mod. Phys. **57**(3), 827 (1985)
34. A.W. Neumann, R.J. Good, Technique of measuring contact angle, in *Surface and Colloid Science*, vol. 11, ed. by R.J. Good, R.R. Stromberg (Plenum Press, New York, 1979), pp. 31–91
35. C.J. Van Oss, R.F. Giese, Z. Li, K. Murphy, J. Norris, M.K. Chaudhury, R.J. Good, Determination of contact angles and pore sizes of porous media by column and thin layer wicking. J. Adhes. Sci. Technol. **6**, 413–428 (1992). doi:10.1163/156856192X00755
36. H.G. Bruil, J.J. van Aartsen, The determination of contact angles of aqueous surfactant solutions on powders. Colloid Polym. Sci. **252**(1), 32–38 (1974). doi:10.1007/bf01381692
37. J.M. Douillard, V. Médout-Marère, A new interpretation of contact angle variations in view of a recent analysis of immersion calorimetry. J. Colloid Interface Sci. **223**(2), 255–260 (2000). doi:10.1006/jcis.1999.6679
38. V. Médout-Marère, S. Partyka, G. Chauveteau, J.M. Douillard, R. Dutartre, Surface heterogeneity of passively oxidized silicon carbide particles: vapor adsorption isotherms. J. Colloid Interface Sci. **262**(2), 309–320 (2003). doi:10.1016/S0021-9797(03)00198-X
39. C.J. Van Oss, The Apolar and Polar Properties of Liquid Water and Other Condensed-Phase Materials. in *Interface Science and Technology*, vol 16 (Elsevier, Amsterdam, 2008), pp. 13–30. doi:10.1016/S1573-4285(08)00202-0
40. V. Médout-Marère, A. El Ghzaoui, C. Charnay, J.M. Douillard, G. Chauveteau, S. Partyka, Surface Heterogeneity of Passively Oxidized Silicon Carbide particles: Hydrophobic-Hydrophilic partition. J. Colloid Interface Sci. **223**(2), 205–214 (2000). doi:10.1006/jcis.1999.6625
41. J.M. Douillard, Concerning the thermodynamic consistency of the "Surface Tension Components" equations. J. Colloid Interface Sci. **188**(2), 511–515 (1997). doi:10.1006/jcis.1997.4768
42. J.M. Douillard, J. Zajac, H. Malandrini, F. Clauss, Contact angle and film pressure: study of a talc surface. J. Colloid Interface Sci. **255**(2), 341–351 (2002). doi:10.1006/jcis.2002.8611
43. M.J. Meziani, J. Zajac, J.-M. Douillard, D.J. Jones, S. Partyka, J. Rozière, Evaluation of surface enthalpy of porous aluminosilicates of the MCM-41 type using immersional calorimetry: effect of the pore size and framework Si:Al ratio. J. Colloid Interface Sci. **233**(2), 219–226 (2001). doi:10.1006/jcis.2002.8611
44. V. Médout-Marère, H. Belarbi, P. Thomas, F. Morato, J.C. Giuntini, J.M. Douillard, Thermodynamic analysis of the immersion of a swelling clay. J. Colloid Interface Sci. **202**(1), 139–148 (1998). doi:10.1006/jcis.1998.5400
45. M.A. Wilson, A. Pohorille, L.R. Pratt, Molecular dynamics of the water liquid-vapor interface. J. Phys. Chem. **91**(19), 4873–4878 (1987). doi:10.1021/j100303a002
46. W. Drost-Hansen, Structure of water near solid interfaces. Ind. Eng. Chem. **61**(11), 10–47 (1969). doi:10.1021/ie50719a005
47. W.D. Harkins, *The Physical Chemistry of Surface Films* (Reinhold, New York, 1952)
48. Proceedings of BP Symposium on the Significance of the Heats of Adsorption at the Solid-Liquid Interface. in A.J. Groszek, Sunbury-on-Thames, BP Research Centre (1971)

49. S. Partyka, J.M. Douillard, Nature of interactions between organic pure liquids and model rocks: a calorimetric investigation. J. Pet. Sci. Eng. **13**(2), 95–102 (1995). doi:10.1016/0920-4105(94)00065-C
50. J.M. Douillard, What can really be deduced from enthalpy of immersional wetting experiments? J. Colloid Interface Sci. **182**(1), 308–311 (1996). doi:10.1006/jcis.1996.0468
51. T.W. Healy, D.W. Fuerstenau, The oxide-water interface-Interrelation of the zero point of charge and the heat of immersion. J. Colloid Sci. **20**(4), 376–386 (1965). doi:10.1016/0095-8522(65)90083-8
52. D.A. Griffiths, D.W. Fuerstenau, The effect of pH and temperature on the heat of immersion of alumina. J. Colloid Interface Sci. **80**(1), 271–283 (1981). doi:10.1016/0021-9797(81)90181-8
53. M. El Wafir, Approche thermodynamique des interactions entre les liquides et les solides modeles issus des roches reservoirs de petrole. Ph.D. Thesis, (University of Montpellier 2, Montpellier 1991)
54. S. Partyka, F. Rouquerol, J. Rouquerol, Calorimetric determination of surface areas: Possibilities of a modified Harkins and Jura procedure. J. Colloid Interface Sci. **68**(1), 21–31 (1979). doi:10.1016/0021-9797(79)90255-8
55. J. Fripiat, J. Cases, M. Francois, M. Letellier, Thermodynamic and microdynamic behavior of water in clay suspensions and gels. J. Colloid Interface Sci. **89**(2), 378–400 (1982). doi:10.1016/0021-9797(82)90191-6
56. X.-C. Zeng, Y. Chen, X.-N. Chen, J.-Q. Xie, F.-B. Jiang, Thermo-kinetic research method for faster reactions: modifier method of distorted thermoanalytical curve. Thermochim. Acta **332**(1), 97–102 (1999). doi:10.1016/S0040-6031(99)00092-1
57. W. Hemminger, G. Höhne, *Calorimetry—Fundamentals and Practice* (Verlag Chemie, Weinheim/Basel, 1984)
58. W. Zielenkiewicz, E. Margas, *Theory of Calorimetry* (Kluwer Academic Publishers, Dordrecht, 2002)
59. E. Calvet, H. Prat, *Microcalorimetrie, Applications Physico-Chimiques et Biologiques* (Masson, Paris, 1956)
60. H. Malandrini, Une etude thermodynamique de l'energie superficielle des solides divises : Determination de la tension superficielle de poudres talco-chloriteuses. Ph.D. Thesis, (University of Montpellier 2, Montpellier, 1995)
61. C. Tanford, *The Hydrophobic Effect. Formation of Micelles and Biological Membranes*, 2nd edn. (Wiley, New York, 1980)
62. M.J. Rosen, *Surfactants and Interfacial Phenomena*, 2nd edn. (Wiley, New York, 1989)
63. J. Lyklema, Adsorption at solid-liquid interfaces with special reference to emulsion systems. Colloids Surf. A Physicochemical Eng. Aspects **91**, 25–38 (1994). doi:10.1016/0927-7757(94)02718-8
64. R. Denoyel, F. Rouquerol, J. Rouquerol, Thermodynamics of adsorption from solution: Experimental and formal assessment of the enthalpies of displacement. Journal of Colloid and Interface Science **136**(2), 375–384 (1990). doi:10.1016/0021-9797(90)90384-Z
65. W. Rudzinski, J. Narkiewicz-Michalek, R. Charmas MD, Piasecki W, Zajac J, Thermodynamics of adsorption at heterogeneous solid—liquid interfaces, in *Interfacial Dynamics*, ed. by N. Kallay, Surfactant Science Series (Marcel Dekker, New York , 1999), pp. 83–162
66. F.D. Rossini, *Chemical Thermodynamics*, 3rd edn. (Wiley, New York, 1961)
67. I. Prigogine, A. Bellemans, V. Mathot, *The Molecular Theory of Solutions* (North-Holland Publishing Company, Amsterdam, 1957)
68. H.S. Harned, B.B. Owen, *The Physical Chemistry of Electrolytic Solutions* (Reinhold, New York, 1958)
69. R.M. Garrels, C.L. Christ, *Solutions, Minerals and Equilibria* (Freeman, Cooper & Co., San Francisco, 1965)
70. M.J. Sparnaay, *The Electrical Double Layer*, vol 4, 1st edn. Properties of Interfaces, (Pergamon Press, Glasgow, 1972)

71. L.K. Koopal, Adsorption of ions and surfactants, in *Coagulation and Flocculation: Theory and Applications*, ed. by B. Dobias, vol. 47, Surfactant Science Series, (Marcel Dekker, New York, 1993), pp. 101–208
72. H.-H. Kohler, Surface charge and surface potential, in *Coagulation and Flocculation: Theory and Applications*, ed. by B. Dobias, vol. 47, Surfactant Science Series, (Marcel Dekker, New York, 1993), pp. 37–56
73. J. Zajac, Adsorption microcalorimetry used to study interfacial aggregation of quaternary ammonium surfactants (zwitterionic and cationic) on powdered silica supports in dilute aqueous solutions. Colloids Surf. A Physicochemical Eng. Aspects **167**(1–2), 3–19 (2000). doi:10.1016/S0927-7757(99)00479-3
74. J. Zajac, C. Chorro, M. Lindheimer, S. Partyka, Thermodynamics of Micellization and Adsorption of Zwitterionic Surfactants in Aqueous Media. Langmuir **13**(6), 1486–1495 (1997). doi:10.1021/la960926d
75. J. Zajac, A.J. Groszek, Adsorption of C60 fullerene from its toluene solutions on active carbons: Application of flow microcalorimetry. Carbon **35**(8), 1053–1060 (1997). doi:10.1016/S0008-6223(97)00058-4
76. W. Rudzinski, J. Zajac, I. Dekany, F. Szanto, Heats of immersion in monolayer adsorption from binary liquid mixtures on heterogeneous solid surfaces: equations for excess isotherms and heats of immersion corresponding to condensation approximation and Rudzinski-Jagiello approach. J. Colloid Interface Sci. **112**(2), 473–483 (1986). doi:10.1016/0021-9797(86)90115-3
77. D.H. Everett, Enthalpy and entropy effects in adsorption from solution. J. Phys. Chem. **85**(22), 3263–3265 (1981). doi:10.1021/j150622a012
78. G.W. Woodbury Jr, L.A. Noll, Heat of adsorption of liquid mixtures on solid surfaces: comparison of theory and experiment. Colloids Surf. **8**(1), 1–15 (1983). doi:10.1016/0166-6622(83)80068-7
79. A.J. Groszek, Flow adsorption microcalorimetry. Thermochim. Acta **312**(1–2), 133–143 (1998). doi:10.1016/S0040-6031(97)00447-4
80. A.J. Groszek, M.J. Templer, Innovative flow-adsorption microcalorimetry. ChemTech **29**(11), 19–26 (1999)
81. R. Denoyel, F. Rouquerol, J. Rouquerol, Interest and requirements of liquid-flow microcalorimetry in the study of adsorption from solution in the scope of tertiary oil recovery, in *Adsorption from Solution*, ed. by C. Rochester (Academic Press, London, 1982), pp. 1–10
82. G.W. Woodbury Jr, L.A. Noll, Heats of adsorption from flow calorimetry: relationships between heats measured by different methods. Colloids Surf. **28**, 233–245 (1987). doi:10.1016/0166-6622(87)80187-7
83. R. Denoyel, F. Rouquerol, J. Rouquerol, Adsorption of anionic surfactants on alumina: complementarity of the information provided by batch and liquid flow microcalorimetry. Colloids Surf. **37**, 295–307 (1989). doi:10.1016/0166-6622(89)80126-X
84. J.M. Miller, *Chromatography: Concepts and Contrasts*, 2nd edn. (Wiley, New York, 2005)
85. Z. Kiraly, R.H.K. Borner, G.H. Findenegg, Adsorption and Aggregation of C8E4 and C8G1 Nonionic Surfactants on Hydrophilic Silica Studied by Calorimetry. Langmuir **13**(13), 3308–3315 (1997). doi:10.1021/la9620768
86. S. Partyka, E. Keh, M. Lindheimer, A. Groszek, A new microcalorimeter for the study of solutions, adsorption and suspensions. Colloids Surf. **37**, 309–318 (1989). doi:10.1016/0166-6622(89)80127-1
87. M. Chorro, C. Chorro, O. Dolladille, S. Partyka, R. Zana, Adsorption mechanism of conventional and dimeric cationic surfactants on silica surface: effect of the state of the surface. J. Colloid Interface Sci. **210**(1), 134–143 (1999). doi:10.1006/jcis.1998.5936
88. R. Chaghi, L.-C. de Ménorval, C. Charnay, G. Derrien, J. Zajac, Interactions of phenol with cationic micelles of hexadecyltrimethylammonium bromide studied by titration calorimetry, conductimetry, and 1H NMR in the range of low additive and surfactant concentrations. J. Colloid Interface Sci. **326**(1), 227–234 (2008). doi:10.1016/j.jcis.2008.07.035

89. J.L. Trompette, Contribution de la calorimetrie a l'etude de l'interaction tensioactif cationique - solide divise. Ph.D. Thesis (University of Montpellier 2, Montpellier, 1995)
90. A.J. Groszek, Graphitic and polar surface sites in carbonaceous solids. Carbon **25**(6), 717–722 (1987). doi:10.1016/0008-6223(87)90140-0
91. A.J. Groszek, S. Partyka, Measurements of hydrophobic and hydrophilic surface sites by flow microcalorimetry. Langmuir **9**(10), 2721–2725 (1993)
92. K. Szczodrowski, B. Prélot, S. Lantenois, J.-M. Douillard, J. Zajac, Effect of heteroatom doping on surface acidity and hydrophilicity of Al, Ti, Zr-doped mesoporous SBA-15. Microporous Mesoporous Mater. **124**(1–3), 84–93 (2009). doi:10.1016/j.micromeso.2009.04.035
93. J. Zajac, Mechanism of ionic and zwitterionic surfactant adsorption from dilute solutions onto charged non-porous and porous mineral oxides inferred from thermodynamic studies, in *Recent Research Developments in Surface and Colloids*, ed. by S.G. Pandalai (Research Signpost, Kerala, 2004), pp. 265–300
94. S. Lantenois, B. Prélot, J.-M. Douillard, K. Szczodrowski, M.-C. Charbonnel, Flow microcalorimetry: experimental development and application to adsorption of heavy metal cations on silica. Appl. Surface Sci. **253**(13), 5807–5813 (2007). doi:10.1016/j.apsusc.2006.12.064
95. B. Prelot, S. Lantenois, M.-C. Charbonnel, F. Marchandeau, J.M. Douillard, J. Zajac, What are the main contributions to the total enthalpy of displacement accompanying the adsorption of some multivalent metals at the silica–electrolyte interface? J. Colloid Interface Sci. **396**, 205–209 (2013) . doi:10.1016/j.jcis.2012.12.049
96. B. Prelot, S. Lantenois, C. Chorro, M.-C. Charbonnel, J. Zajac, J.M. Douillard, Effect of nanoscale pore space confinement on cadmium adsorption from aqueous solution onto ordered mesoporous silica: a combined adsorption and flow calorimetry study. J. Phys. Chem. C **115**(40), 19686–19695 (2011). doi:10.1021/jp2015885
97. J.L. Trompette, J. Zajac, E. Keh, S. Partyka, Scanning of the cationic surfactant adsorption on a hydrophilic silica surface at low surface coverages. Langmuir **10**(3), 812–818 (1994)
98. R. De Lisi, C. Ostiguy, G. Perron, J.E. Desnoyers, Complete thermodynamic properties of nonyl- and decyltrimethylammonium bromides in water. J. Colloid Interface Sci. **71**(1), 147–166 (1979). doi:10.1016/0021-9797(79)90229-7
99. R. Atkin, V.S.J. Craig, E.J. Wanless, S. Biggs, Mechanism of cationic surfactant adsorption at the solid-aqueous interface. Adv. Colloid Interface Sci. **103**(3), 219–304 (2003). doi:10.1016/S0001-8686(03)00002-2
100. J.H. Clint, *Surfactant Aggregation* (Blackie, Glasgow/London, 1992)
101. R.H. Aranow, L. Witten, The environmental influence on the behavior of long chain molecules. J. Phys. Chem. **64**(11), 1643–1648 (1960). doi:10.1021/j100840a010
102. H. Hoffmann, Fascinating phenomena in surfactant chemistry. Adv. Mater. **6**(2), 116–129 (1994). doi:10.1002/adma.19940060204
103. A. Bendjeriou, G. Derrien, P. Hartmann, C. Charnay, S. Partyka, Microcalorimetric studies of cationic gemini surfactant with a hydrophilic spacer group. Thermochim. Acta **434**(1–2), 165–170 (2005). doi:10.1016/j.tca.2005.01.034
104. L. Grosmaire, M. Chorro, C. Chorro, S. Partyka, R. Zana, Alkanediyl-alpha, omega-bis(dimethylalkylammonium bromide) surfactants - 9. Effect of the spacer carbon number and temperature on the enthalpy of micellization. J. Colloid Interface Sci. **246**(1), 175–181 (2002). doi:10.1006/jcis.2001.8001
105. M. Pisarcik, M.J. Rosen, M. Polakovicova, F. Devinsky, I. Lacko, Area per surfactant molecule values of gemini surfactants at the liquid-hydrophobic solid interface. J. Colloid Interface Sci. **289**(2), 560–565 (2005). doi:10.1016/j.jcis.2005.03.092
106. R. Zana, Dimeric (Gemini) surfactants: effect of the spacer group on the association behavior in aqueous solution. J. Colloid Interface Sci. **248**(2), 203–220 (2002). doi:10.1006/jcis.2001.8104
107. P. Somasundaran, J.T. Kunjappu, In-situ investigation of adsorbed surfactants and polymers on solids in solution. Colloids Surf. **37**, 245–268 (1989). doi:10.1016/0166-6622(89)80123-4

108. P. Somasundaran, D.W. Fuerstenau, Mechanisms of Alkyl Sulfonate Adsorption at the Alumina-Water Interface. J. Phys. Chem. **70**(1), 90–96 (1966). doi:10.1021/j100873a014
109. M.A. Yeskie, J.H. Harwell, On the structure of aggregates of adsorbed surfactants: the surface charge density at the hemimicelle/admicelle transition. J. Phys. Chem. **92**(8), 2346–2352 (1988). doi:10.1021/j100319a048
110. H. Rupprecht, T. Gu, Structure of adsorption layers of ionic surfactants at the solid/liquid interface. Colloid Polym. Sci. **269**(5), 506–522 (1991). doi:10.1007/bf00655889
111. B. Li, E. Ruckenstein, Adsorption of Ionic Surfactants on charged solid surfaces from Aqueous solutions. Langmuir **12**(21), 5052–5063 (1996). doi:10.1021/la951559t
112. S.D. Christian, J.F. Scamehorn (eds.), *Solubilisation in Surfactant Aggregates, Surfactant Science Series*, vol. 55 (Marcel Dekker, New York, 1995)
113. R. Zana, Aqueous surfactant-alcohol systems: a review. Adv. Colloid Interface Sci. **57**, 1–64 (1995). doi:10.1016/0001-8686(95)00235-I
114. M.J. Meziani, H. Benalla, J. Zajac, S. Partyka, D.J. Jones, Adsorption of a cationic gemini surfactant from aqueous solution onto aluminosilicate powders of the MCM-41 type: effect of pore size and co-adsorption of phenol. J. Colloid Interface Sci. **262**(2), 362–371 (2003). doi:10.1016/S0021-9797(03)00204-2
115. H. Benalla, J. Zajac, S. Partyka, J. Rozière, Calorimetric study of phenol adsolubilisation by cationic surfactants adsorbed on a flat silica surface or confined within small mesopores of powdered MCM-41 aluminosilicates. Colloids Surf. A Physicochem. Eng. Aspects **203**(1–3), 259–271 (2002). doi:10.1016/S0927-7757(01)01109-8
116. M.N. Khan, *Micellar Catalysis. Surfactant Science Series*, vol. 133 (Taylor and Francis Group, Boca Raton, 2006)
117. C.-C. Yu, L. Lobban Lance, Admicellar catalysis. in *Surfactant Adsorption and Surface Solubilization*, vol. 615, ACS Symposium Series. American Chemical Society, (1996), pp 67–76. doi:doi:10.1021/bk-1995-0615.ch005.
118. A.D.W. Carswell, E.A. O'Rea, B.P. Grady, Adsorbed surfactants as templates for the synthesis of morphologically controlled polyaniline and polypyrrole nanostructures on flat surfaces: from spheres to wires to flat films. J. Am. Chem. Soc. **125**(48), 14793–14800 (2003). doi:10.1021/ja0365983

Part II
Applications and Case Studies

Chapter 7
Study of Selective Adsorption of Gases by Calorimetry

Jean-Pierre Bellat

Abstract This chapter is devoted to the study of coadsorption of gases in nanoporous solids by using the differential calorimetry. In the first part, the thermodynamic principles of adsorption of gases are recalled. Some of them have already presented in chapter one. However a special attention has been paid here to the determination of the adsorption enthalpies and entropies and we focused on the selective adsorption of binary mixtures. Then the specific experimental technique based on the combination of differential calorimetry with manometry and gas phase chromatography or mass spectrometry is shown in details. In the last part, the thermodynamic concepts on coadsorption are illustrated with experimental results taken from studies on gas separation by selective adsorption in microporous solids.

7.1 Introduction

Adsorption is a phenomenon, which systematically occurs when a solid is in contact with a liquid or a gas. The molecules of the adsorptive fluid are accumulated at the pore surface of the solid to form another phase, called the adsorbate, which has different physical and chemical properties than the fluid bulk. When several components are present in the adsorptive fluid, some components can be adsorbed in greater amount than the other ones. It results that the composition of the adsorbed phase is different that the composition of the adsorptive fluid. In this case, the adsorption process is called selective adsorption or competitive adsorption.

The adsorption process is all the more important as the specific surface area of the solid is large. So, this phenomenon will be predominant when the solid is highly divided as in nanometric powdered material or highly porous as in microporous or mesoporous materials.

J.-P. Bellat (✉)
Laboratoire Interdisciplinaire Carnot de Bourgogne, UMR 6303 CNRS-Université de Bourgogne,9 avenue A. Savary, BP 47870, 21078 Dijon, France
e-mail: jean-pierre.bellat@u-bourgogne.fr

A. Auroux (ed.), *Calorimetry and Thermal Methods in Catalysis*,
Springer Series in Materials Science 154, DOI: 10.1007/978-3-642-11954-5_7,

Adsorption is the first step in the mechanism of reactivity of solids: corrosion, oxidation of metals, catalytic reactions begin always by an adsorption process. Therefore, the understanding of the adsorption step is of a relevant interest from a fundamental point of view in the study of reactivity of solids. Moreover, adsorption finds many applications in various fields of investigations as in characterization of porous solids, gas separation or purification, heterogeneous catalysis, energy storage, toxic or green-house gas sequestration, gas detection, storage of active molecules... This concerns several industrial sectors as for example the petroleum chemistry, the environment, the renewable energy, the electronics, the nuclear energy, the pharmacology, the agriculture and food chemistry... This is certainly the reason why we observe today a real renewed interest for adsorption.

The aim of this chapter is to present, in a simple and didactic way, the bases of the selective gas adsorption theory and how we can experimentally study the coadsorption of gases by using the calorimetry technique. It focuses on the energetic aspect of adsorption and on the selective physisorption of gases in microporous solids in order to use the selective adsorption for gas separation or gas purification. Thus, the thermodynamics concepts presented in this chapter are illustrated at the end by two experimental case studies on the selective adsorption in faujasite zeolites. The former is devoted to the separation of xylenes isomers, the latter to the desulphurization of natural gas.

7.2 Definition of Selective Adsorption

7.2.1 Adsorption of Single Component

a. Definition

The adsorption of a single gaseous component on a solid is the accumulation process of the gas at the surface of this solid as illustrated in Fig. 7.1a [1, 2]. However this picture is restrictive. Indeed, if we can imagine that the molecules adsorbed on a homogeneous plane surface form a film with a given thickness, in the case of microporous solids with pores having a size close to that one of adsorptive molecules, the notion of adsorbed layer has no more physical meaning (Fig. 7.1b). Adsorption is different than absorption. In adsorption, the molecules are localized on the surface where they can form a film while in absorption the molecules can diffuse inside the solid structure. Adsorption occurs on a surface, absorption in a volume. The accumulation of the gas on the surface is due to the fact that the atoms have not the same surrounding at the surface than within the solids. The surface atoms can interact with the molecules coming from the outside by forming physical or chemical bonds in order to minimize their energy. When the molecules interact with the surface by specific or non-specific physical interactions (van der Waals interactions and hydrogen bonds), the adsorption process is called physisorption. When the molecules are bound to the surface by chemical bonds with electronic transfer and modification of chemical

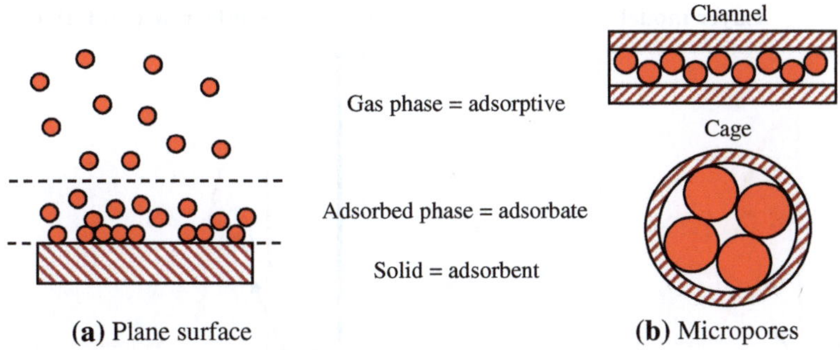

Fig. 7.1 Schematic representation of gas adsorption on a plane surface and inside micropores. **a** Plane surface. **b** Micropores

Table 7.1 Main differences between physisorption and chemisorption

	Physisorption	Chemisorption
Adsorbate-adsorbent interaction	Electrostatic	Electronic
Interaction energy	$E \sim E_{liquefaction}$	$E >> E_{liquefaction}$
Reversibility of the adsorption process	Reversible	Non reversible
Kinetics	Fast	Slow

natures of gas and solid surface, the adsorption process is called chemisorption or reactive adsorption. The main differences between physisorption and chemisorption are reported in Table 7.1. The term sorption is often used to describe in a general manner adsorption, absorption, physisorption and chemisorption without distinction.

b. Adsorbed amount and surface excess amount

As the adsorption process is an accumulation of gas at the solid surface, the gas concentration decreases progressively with increased distance z from the surface to the bulk phase (Fig. 7.2).

The *adsorbed amount*, n^a, is defined as the amount of matter present in the layer of thickness t adsorbed at the surface [3]. This amount is given by the relation:

$$n^a = A \int_0^t c\,dz \tag{7.1}$$

where A is the specific surface area of the solid. This value corresponds to the white area (a) below the curve $c = f(z)$ on Fig. 7.2a.

The *surface excess amount,* n^σ, is the amount of matter localized at the surface, in addition to the amount of gas present in the layer of thickness t. This amount is then defined by the relation:

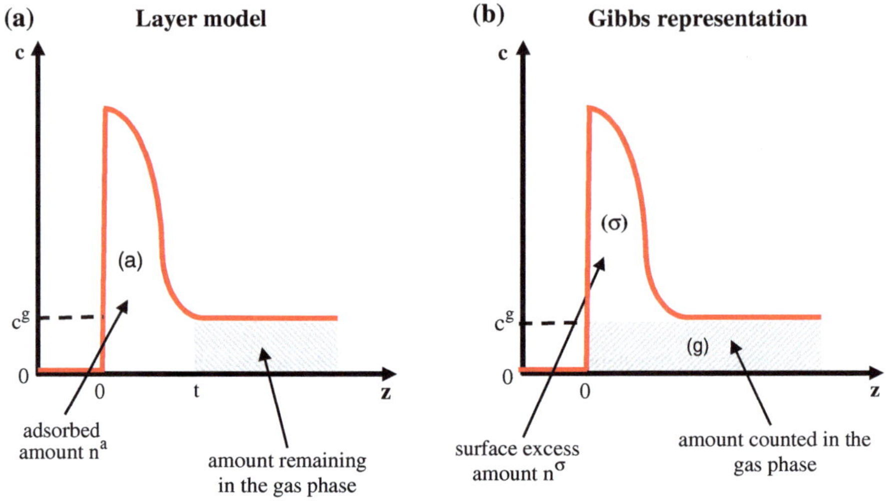

Fig. 7.2 Concentration of the gas c as a function of the distance z from the solid surface. Graphic representation of adsorbed amount and Gibbs surface excess amount. **a** Layer model **b** Gibbs representation

$$n^{\sigma} = A \int_{0}^{t} (c - c^{g}) dz \tag{7.2}$$

This value is represented by the area (σ) below the curve $c = f(z)$ on Fig. 7.2b.

Adsorbed amount and surface excess amount are two different quantities, which are often used without clear distinction. We will see later in which conditions these two values can be considered as equal.

c. Variance and representation of adsorption equilibria

The adsorption equilibrium of a single component i on a solid can be written:

$$i_{(g)} \leftrightarrow i_{(ads)}$$

The adsorption equilibrium constant is [4, 5]:

$$K(T) = \frac{\gamma \theta}{p/p^{\circ}} \tag{7.3}$$

with:

- γ the activity coefficient of the adsorbed phase,
- p the vapour pressure at equilibrium,
- p° the standard pressure (101 325 Pa),

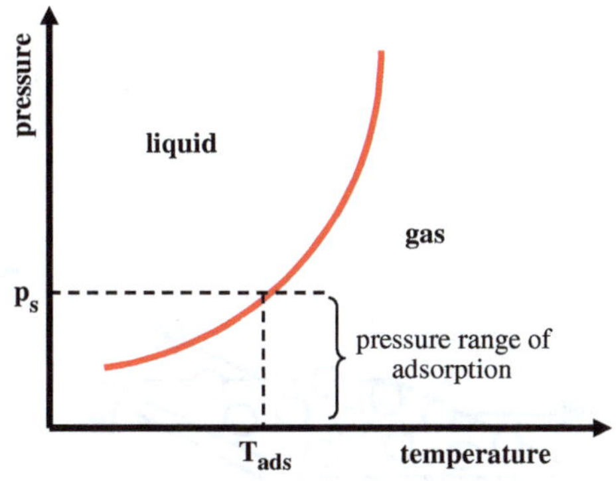

Fig. 7.3 Pressure range of adsorption at constant temperature in the liquid-gas phase diagram of the adsorptive. The pressure cannot exceed the saturated vapour pressure p_s of the fluid at the adsorption temperature T_{ads}

- $\theta = \frac{n^a}{n^a_s}$ the filling coefficient or surface coverage coefficient, ratio of the amount adsorbed under the pressure p on the amount adsorbed under the saturated vapour pressure p_s of the adsorptive at the adsorption temperature.

Thus K(T) is dimensionless. The adsorption process of a gas on a solid can occur only in a restricted pressure range limited by the saturated vapour pressure of the component i at the same temperature as the adsorption temperature. If the vapour pressure is above p_s then adsorption is replaced by liquefaction (Fig. 7.3). The saturation of the adsorbent is established when p is equal to p_s. We often use the notion of relative pressure defined by the ratio p/p_s. Its value lies between 0 and 1. When the relative pressure is equal to one, the filling coefficient is also equal to one.

The variance of the adsorption equilibrium can be regarding as the number of intensive variables that must be set to know completely the thermodynamic state of the system. It can be calculated by using the well-known phase rule of Gibbs [6]:

$$v = N - r + P - \varphi \tag{7.4}$$

where N is the number of components in the system, r the number of independent equilibria between phases, P the number of intensive variables and φ the number of phases.

In the case of adsorption of a single component, N, r and φ are easily determined:

N = 3 (adsorbent, gas and adsorbate),
r = 1 (adsorption equilibrium),
φ = 3 (adsorbent, gas and adsorbate).

The determination of the number of intensive variables P is less evident. Indeed, if the pressure and the temperature must obviously be taken into account, a new intensive variable must be introduced, the spreading pressure π. As a gas enclosed in a bulb of volume V creates a pressure due to the collision of gas molecules against the wall, the molecules adsorbed at the surface. A exert a bi-dimensional pressure (Fig. 7.4). As for the gas the elementary mechanical work is given by −pdV, the

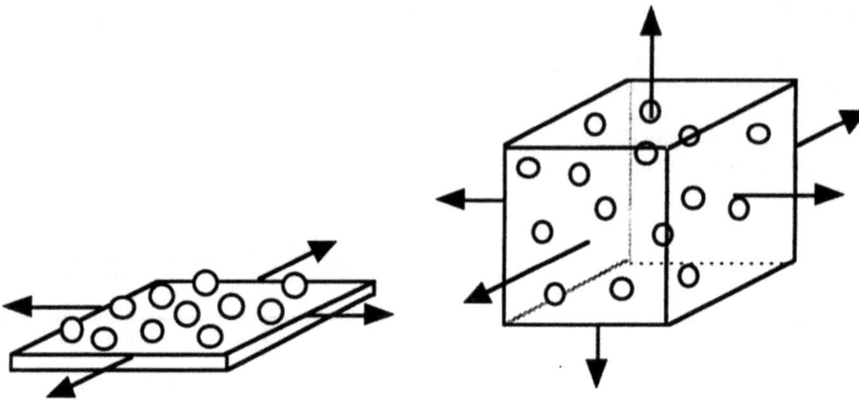

Adsorbed phase
π : spreading pressure
(bi-dimensional space)
Mechanical work of surface strength:
$\delta W = -\pi dA$

Gas phase phase
p: pressure
(tri-dimensional space)
Mechanical work of volume strength:
$\delta W = -pdV$

Fig. 7.4 Illustration of the spreading pressure exerted by molecules adsorbed on a surface compared to the pressure exerted by a gas in a volume

elementary mechanical work of surface strength is given by the term $-\pi dA$. The physical meaning of spreading pressure is analogous to that of surface tension of a monomolecular film at the gas–liquid interface.

The spreading pressure defines the lowering of the surface tension at the gas–solid interface upon adsorption $\pi = \sigma^\circ - \sigma$, where σ° and σ are the surface tensions of the clean and monolayer covered surfaces, respectively. We will show below that the spreading pressure is related to the adsorbed amount. Thus, the number of intensive variables is $P = 3$ (T, p and π) and the variance of the adsorption equilibrium is equal to 2. The system is "bivariant" and has two degrees of freedom. This means that the spreading pressure or the adsorbed amount depends on the temperature and the pressure. Thus, as illustrated in Fig. 7.5, there are three manners to represent an adsorption equilibrium by plotting:

- an adsorption isotherm $n^a = f(p)_T$
- an adsorption isobar $n^a = f(T)_p$
- an adsorption isoster $p = f(T)_{n^a}$ or rather $Ln(p) = f\left(\frac{1}{T}\right)_{n^a}$.

The adsorption isotherm is the representation, which is the more often used.

7.2.2 *Adsorption of Gas Mixtures*

a. Definition

Selective adsorption so-called competitive adsorption occurs when a gas containing more than one component is in interaction with a solid in such a way that the

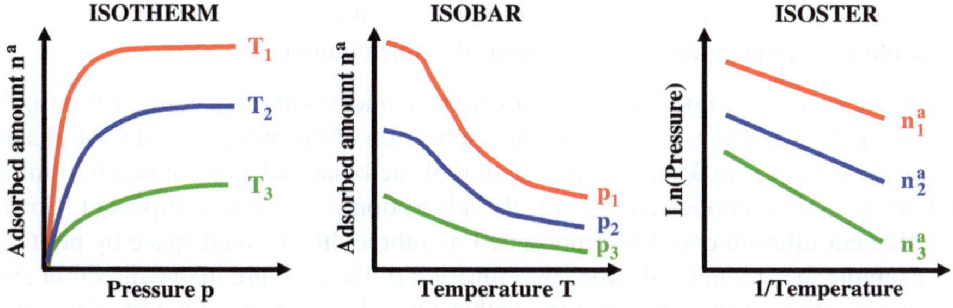

Fig. 7.5 The three representations of adsorption equilibrium of gas on solid at constant temperature (isotherm), constant pressure (isobar) and constant adsorbed amount (isoster)

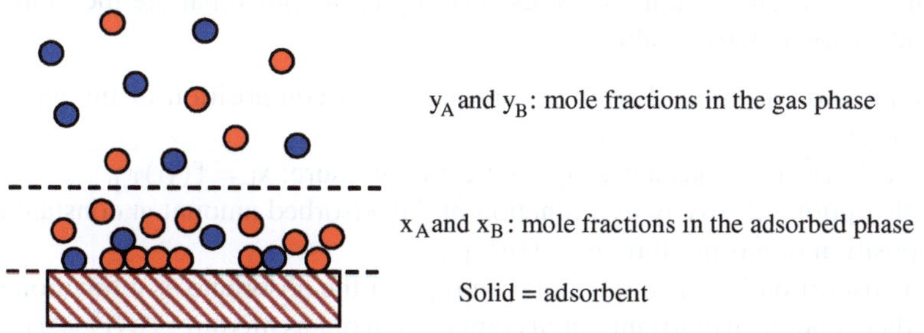

Fig. 7.6 Schematic representation of the adsorption of a binary mixture A (*red*)+B (*blue*) on a solid. The adsorbate is rich in component A, the gas phase in component B. The adsorption process is in favour of component A

composition of the gas is different than the composition of the adsorbate. This implies that one component is more adsorbed by the solid than the other ones. For example, the Fig. 7.6 is an illustration of a selective adsorption process of two components A and B where the component A is favourably adsorbed. The adsorption process is called to be "selective for A".

b. Variance and representation of coadsorption equilibria

The adsorption equilibrium of binary mixture A + B on a solid can be written:

$$A_{(g)} + B_{(ads)} \leftrightarrow A_{(ads)} + B_{(g)}$$

The equilibrium constant is defined by:

$$K(T) = \frac{\gamma_A\,\theta_A\,p_B}{\gamma_B\,\theta_B\,p_A} = \frac{K_A}{K_B} \tag{7.5}$$

with:

- γ_i the activity coefficient of component i in the adsorbed phase,
- θ_i the partial filling coefficient of component i

- p_i the partial pressure of component i in the gas phase.
- K_i the adsorption equilibrium constant of single component i.

There are seven thermodynamics parameters characterizing this equilibrium (p, T, π, y_A, y_B, x_A, and x_B). In that case, the application of the phase rule of Gibbs leads to a variance equal to 3. The representation of the binary adsorption equilibrium is obviously more complicated than for the adsorption of a single component. Coadsorption equilibrium could be represented in a three dimensional space by plotting for example the amounts adsorbed as a function of the pressure in one direction and the composition of the adsorbate in another one. However, this kind of plot is not common because it requires a lot of data measured at different composition and pressure. We prefer to represent the coadsorption equilibria in a two-dimension space by maintaining constant certain variables. The representations that are the more often used and easily understandable are:

- the total and partial isotherms at constant initial composition of the gas phase: $n^a = f(p)_{T,y_i^o}$ and $n_i^a = f(p)_{T,y_i^o}$
- the x-y diagram at constant temperature and pressure: $x_i = f(y_i)_{T,p}$
- the adsorption selectivity as a function of the adsorbed amount at constant initial composition of gas mixture: $\alpha = f(n^a)_{T,y_i^o}$
- the coadsorption heat and the molar entropy of the adsorbate as a function of the adsorbed amount at constant initial composition of gas mixture: $\Delta_r H^a = f(n^a)_{T,y_i^o}$ and $S^a_{m,i} = f(n^a)_{T,y_i^o}$, respectively.

These different representations are illustrated in Fig. 7.7.

c. Adsorption selectivity

The adsorption selectivity is the more appropriate parameter to describe the preferential adsorption of one component with respect to another one. For the adsorption equilibrium of a binary mixture,

$$A_{(g)} + B_{(ads)} \leftrightarrow A_{(ads)} + B_{(g)}$$

the adsorption selectivity is defined by the ratio:

$$\underset{A/B}{\alpha} = \frac{x_A/y_A}{x_B/y_B} \tag{7.6}$$

with x_i and y_i the mole fractions of component i in the adsorbed phase and the gas phase, respectively. This parameter is analogous to a separation factor of two components between two phases. It may be pointed out that the selectivity is the equilibrium constant of the binary adsorption equilibrium in the case of an ideal adsorbed solution. We have only to replace in Eq. 7.5 the partial filling coefficients and pressures by their relation with the mole fractions of each component in the adsorbed and gas phases ($\theta_i = x_i\,\theta$ and $p_i = y_i p$) and to take the activity coefficients equal to unity.

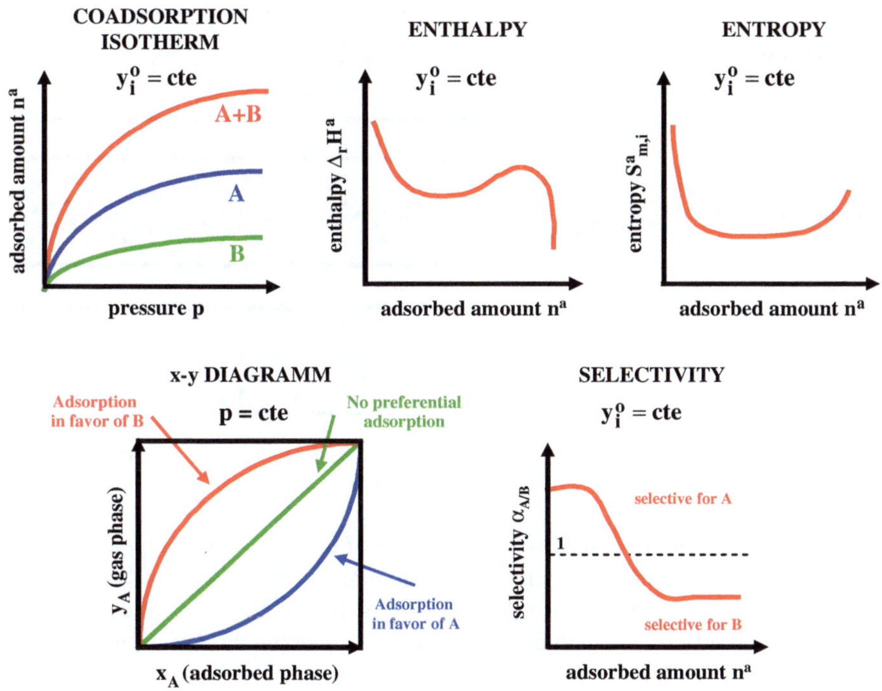

Fig. 7.7 Different representations of coadsorption equilibrium which are often used for a binary mixture

When the solid adsorbs two components A and B in the same proportions, the composition of the mixture in the gas and adsorbed phases are the same and the selectivity is equal to unity. The adsorption process is not selective. If the solid adsorbs preferentially the component A, the mole fraction of this component is higher in the adsorbed phase than in the gas phase and the selectivity is higher than unity. On the opposite, if the adsorption process is in favour of component B, the selectivity is lower than unity. Seeing that the selectivity is the equilibrium constant in the case of an ideal adsorbed solution, it must be constant whatever the composition. The fact that the selectivity changes with the composition indicates that the adsorption process is not ideal.

d. Different types of selective adsorption

The selective adsorption of several components on a solid is a very complex phenomenon. The reason why, an adsorption process is in favour of one component rather than another one is far to be elucidated. In practical applications, we consider there are three types of separation by selective adsorption (Fig. 7.8):

- the steric exclusion, which occurs when the size of one component in the mixture is higher than the pore opening of the adsorbent. The small molecules can be adsorbed in the microporosity while the biggest remain in the gas phase. This is a real molecular sieving with a selectivity that tends to infinity.

Steric exclusion: (molecular sieving) $\alpha_{A/B} \rightarrow \infty$

Kinetic separation: $\alpha_{A/B} = \frac{K_A}{K_B}\sqrt{\frac{D_A}{D_B}}$

Energetic separation: $\alpha_{A/B} = \frac{K_A}{K_B} = \frac{x_A y_B}{y_A x_B}$

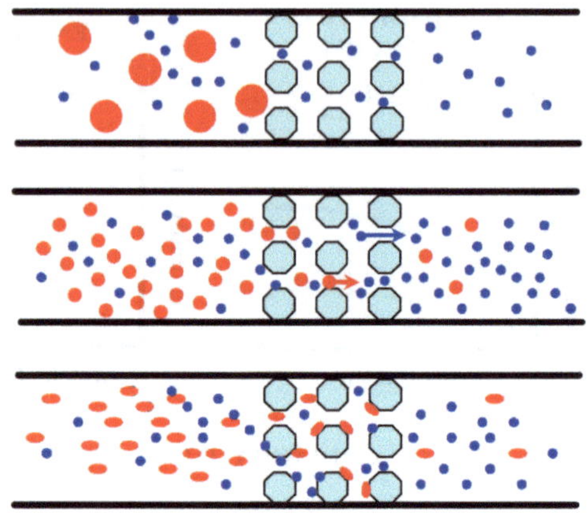

Fig. 7.8 The three types of separation by selective adsorption [K_i is the adsorption equilibrium constant and D_i is the diffusion coefficient of component i]

- the kinetic separation based on the difference of diffusivity of components in the microporosity. All the molecules of the mixture can pass through the adsorbent but some of them diffuse faster than the others. In that case the expression of the selectivity is different than in Eq. 7.6. It takes into account the diffusion coefficient of each component in the microprosity.
- the energetic separation which results of difference in adsorption affinity of the adsorbent towards the components of the mixture. All the molecules can penetrate the microporosity but are adsorbed with different energies. At equilibrium, the component adsorbed in favour is that one having the strong adsorption energy with the solid.

7.3 Adsorption Enthalpies and Entropies

Adsorption is a spontaneous and exothermic process. The adsorption Gibbs energy and the adsorption enthalpy are then negative. As the molecules are adsorbed in micropores and bond to the solid by physical or chemical interactions, they lose degrees of freedom. The adsorbate forms a phase, which is more ordered compared to the gas phase. Therefore the adsorption entropy is negative too. Thus the sign of the Gibbs energy depends on the enthalpic term which is always negative and the entropic term which is positive: $\Delta_r G^a = \Delta_r H^a - T\Delta_r S^a$. In most cases, the adsorption process is spontaneous owing to the enthalpic term, which is generally more important than the entropic one. However, the weight of the entropic term must not be neglected because it can compensate the enthalpic term, especially when adsorption occurs in very confined spaces like nanopores. One of particularities of the adsorption process is that the adsorption enthalpy and entropy depend on the adsorbed amount and, sometimes, the composition of gas mixture. The knowledge of these two thermodynamics values, as a function of the adsorbed amount, is then of a relevant interest for the understanding of adsorption mechanisms.

7.3.1 Adsorption Enthalpy

Adsorption enthalpy versus adsorbed amount gives interesting information about molecular interactions between the solid surface and the adsorbed molecules (adsorbate-adsorbent interactions) and between the adsorbed molecules themselves (adsorbate-adsorbate interactions). The adsorbate-adsorbent interactions depend on the energetic properties of the surface. If all the adsorption sites have the same energy, the adsorbate-adsorbent interactions are almost constant whatever the surface coverage. Each molecule interacts in the same way with each site. On the opposite, if the adsorption sites are very different in energy, the adsorbate-adsorbent interactions decrease with the surface coverage. First, the adsorption occurs on the more energetic sites and then goes on the less energetic ones. As for the adsorbate-adsorbate interactions, they are always increasing with the surface coverage. If they can be neglected at the beginning of the adsorption process, they are maximal at complete surface coverage and can be very important especially when adsorption occurs in nanopores in which the molecules are very confined. The adsorption enthalpy is the sum of these two kinds of interactions and the shape of the plot adsorption enthalpy versus adsorbed amount will be different according to the adsorption process. As illustrated on Fig. 7.9, adsorption on a heterogeneous surface is characterized by a sharp decrease of the adsorption enthalpy at very low filling while this value is almost constant when the surface is homogeneous. An increase of the adsorption enthalpy at high filling indicates the presence of adsorbate-adsorbate interactions. In all cases, at saturation, when the adsorption process is complete, the adsorption enthalpy tends to the liquefaction enthalpy. It is not possible to distinguish adsorbate-adsorbent and adsorbate-adsorbate interactions from the adsorption enthalpy. However, if we know that the surface is homogeneous, the adsorbate-adsorbate interactions can be estimated by subtracting to the adsorption enthalpy its value extrapolated at zero filling. Moreover, in the case of coadsorption, the adsorption enthalpy curve can show additional changes resulting from endothermic or exothermic effects consecutive to displacement of a species by another one or specific interactions between components in the adsorbed phase.

7.3.2 Adsorption Entropy and Molar Adsorbate Entropy

Adsorption entropy and molar adsorbate entropy are related by the relation:

$$S_m^a = \Delta_r S^a + S_m^{g^\circ} \tag{7.7}$$

where $S_m^{g^\circ}$ is the standard molar entropy of the gaseous adsorptive.

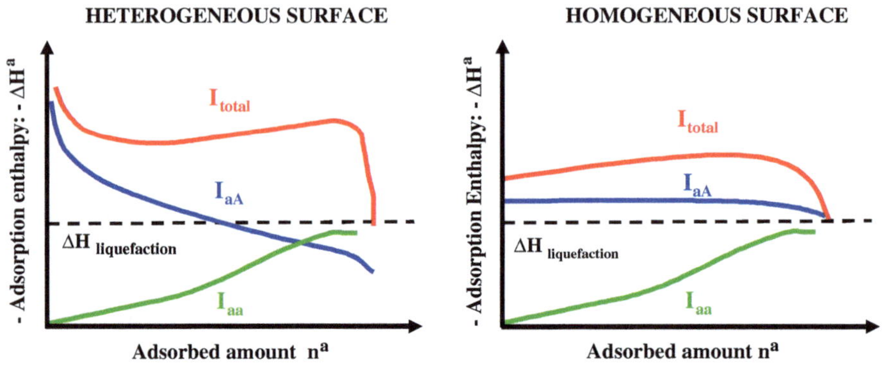

Fig. 7.9 Change of the adsorption enthalpy as a function of the adsorbed amount for an adsorption process energetically homogeneous or heterogeneous. I_{aa} is the adsorbate-adsorbate interaction, I_{Aa} is the adsorbate-adsorbent interaction and $I_{total} = I_{aa} + I_{aA}$ corresponds to the adsorption enthalpy

The knowledge of molar adsorbate entropy compared to molar entropies of solid, liquid or gaseous adsorptive is particularly interesting when represented as a function of the adsorbed amount. These values give interesting information about the physical state of the adsorbate and allow us to know if the molecules adsorbed on the solid surface flutter or are rather frozen around their adsorption sites. It is currently admitted that the adsorbed phase is in a physical state similar to a liquid particularly at complete filling. Thus, to calculate the adsorbed volume we used the density of the liquid fluid. However the determination of the molar adsorbate entropy shows that this hypothesis is not always true. Indeed, if in the case of a mobile adsorption, the molar adsorbate entropy is often close to the molar entropy of the liquid adsorptive, for a localized adsorption, it can be near to that one of the solid adsorptive, the molecules having lost numerous modes of vibration. This result is particularly true when the molecules are adsorbed in narrow nanopores. Figure 7.10 shows two examples of molar entropy curve corresponding to these two cases. However, the entropy curves have often more complicated shapes that make their interpretation very difficult.

7.4 Calculation of Adsorption Enthalpy and Entropy from Single Adsorption Isotherms

The adsorption enthalpie and entropie of single component can be extracted from adsorption data obtained under isothermal conditions by using two different methods. The first one is based on the van't Hoff equation and the second one on the Clausius-Clapeyron equation. These methods only need a set of adsorption isotherms measured at various temperatures.

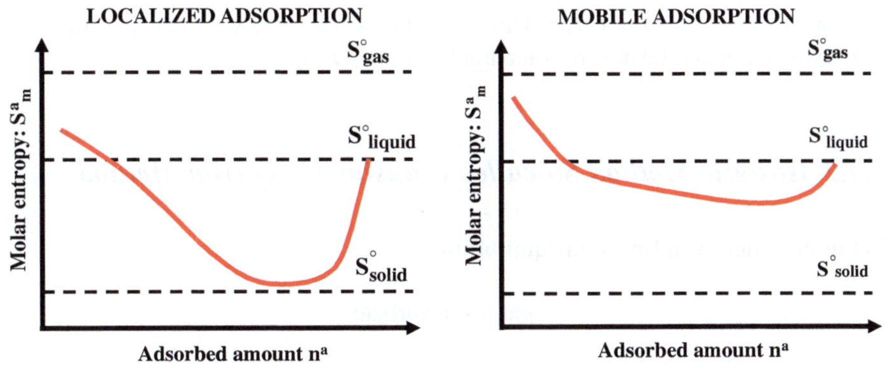

Fig. 7.10 Change of the molar adsorbate entropy as a function of the adsorbed amount for localized and mobile adsorption process. S° is the standard molar entropy of the adsorptive in the gas, liquid or solid state

7.4.1 Van't Hoff Method

This method consists to find the thermodynamic model (Henry, Langmuir, Fowler, Freundlich…) which gives the best fit with adsorption isotherms measured at various temperatures [7]. The parameter of the model K_{model} is temperature dependent. It can be related to the dimensionless equilibrium constant of adsorption K(T) by the relation:

$$K_{model} = \frac{K_{(T)}}{p^\circ} \tag{7.8}$$

where p° is the standard pressure ($p^\circ = 101325 \sim 10^5$ Pa).

At equilibrium we have:

$$\Delta_r G^a = \Delta_r H^a - T\Delta_r S^a = -RT \mathrm{Ln} K_{(T)} \tag{7.9}$$

and

$$\mathrm{Ln}K_{(T)} = -\frac{\Delta_r H^a}{R}\frac{1}{T} + \frac{\Delta_r S^a}{R} \tag{7.10}$$

After differentiation of this equation we retrieve the well-known van't Hoff equation:

$$\Delta_r H^a = RT^2 \left(\frac{d\mathrm{Ln}K_{(T)}}{dT}\right) \tag{7.11}$$

Thus, by plotting the curve $LnK_{(T)} = f(1/T)$ it is easy to determine the adsorption enthalpy and entropy from the slope and the intercept of this straight-line (Fig. 7.11a). In this method we have to keep in mind that we consider that the adsorption enthalpy and entropy are constant i.e. independent on the temperature and the adsorbed amount. In fact this is not true. However, this is a simple way that is often used to roughly estimate the adsorption enthalpy and entropy.

7.4.2 Isosteric Method So-called Clausius-Clapeyron Method

Let us consider the adsorption equilibrium:

$$\text{gas} \leftrightarrow \text{adsorbate}$$

If equilibrium exists between gas and adsorbed phases we have, for a given adsorbed amount, equality in chemical potentials:

$$\mu^{g}(T,p) = \mu^{a}(T,p,n^{a}) \tag{7.12}$$

The chemical potential of the gas assumed as ideal is defined by:

$$\mu^{g}(T,\ p) = \mu^{g^{\circ}}(T) + RT\,Ln\frac{p}{p^{\circ}} \tag{7.13}$$

where $\mu^{g^{\circ}}$ is the standard chemical potential of the gas at the temperature T.

The chemical potentials can be expressed as a function of molar enthalpies and entropies as follows:

$$\mu^{g^{\circ}}(T) = H_{m}^{g^{\circ}}(T) - TS_{m}^{g^{\circ}}(T) \tag{7.14}$$

$$\mu^{a}(T,\ p,\ n^{a}) = H_{m}^{a}(T,\ p,\ n^{a}) - TS_{m}^{a}(T,\ p,\ n^{a}) \tag{7.15}$$

By combining Eqs. 7.13, 7.14 and 7.15 we get:

$$Ln\left(\frac{p}{p^{\circ}}\right)_{n^{a}} = \frac{H_{m}^{a}(T,p,n^{a}) - H_{m}^{g^{\circ}}(T)}{R}\frac{1}{T} - \frac{S_{m}^{a}(T,p,n^{a}) - S_{m}^{g^{\circ}}(T)}{R} \tag{7.16}$$

If we define the isosteric enthalpy and entropy by:

$$\Delta H_{iso}(n^{a}) = H_{m}^{a}(T,p,n^{a}) - H_{m}^{g^{\circ}}(T) \tag{7.17}$$

$$\Delta S_{iso}(n^{a}) = S_{m}^{a}(T,p,n^{a}) - S_{m}^{g^{\circ}}(T) \tag{7.18}$$

we obtain:

$$Ln\left(\frac{p}{p^{\circ}}\right)_{n^{a}} = \frac{\Delta H_{iso}(n^{a})}{R}\frac{1}{T} - \frac{\Delta S_{iso}(n^{a})}{R} \tag{7.19}$$

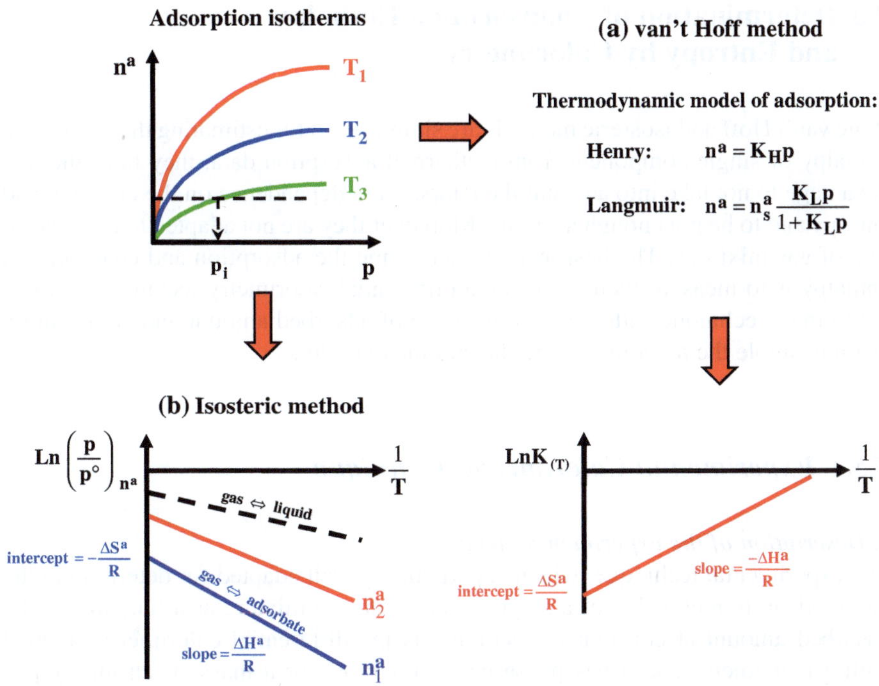

Fig. 7.11 Graphic determination of adsorption enthalpy and entropy from isothermal adsorption data measured at various temperatures. **a** Van't Hoff method applied to the adsorption equilibrium constant deduced from the parameter of a thermodynamic model like Henry or Langmuir fitting well the adsorption isotherms. **b** Clausius-Clapeyron or isosteric method: the isosters lie *below* a limit line, which corresponds to the liquefaction equilibrium of the adsorptive

Thus the adsorption enthalpy and entropy can be determined by plotting $\mathrm{Ln}\left(\frac{p}{p^\circ}\right)_{n^a} = f\left(\frac{1}{T}\right)$ at constant adsorbed amount. The slope and the intercept of this plot called "adsorption isoster" give $\Delta H_{iso}(n^a)$ and $\Delta S_{iso}(n^a)$ as illustrated in Fig. 7.11b. We need to apply this method a series of adsorption isotherms in a large range of pressure and temperature in order to explore the complete domain of adsorbed amounts. This second method has the advantage to take into account the dependence of the adsorbed amount on the adsorption enthalpy and entropy and allows a more accurate determination of these values than the van't Hoff method.

However we consider that the enthalpy and entropy are not function of the temperature. It may be noticed that by differentiating Eq. 7.19 we obtain:

$$\frac{d\mathrm{Ln}p}{dT} = -\frac{\Delta H_{iso}(n^a)}{RT^2} \tag{7.20}$$

This relation is analogous to the Clausius-Clapeyron equation characterizing the equilibrium between a liquid and a gas.

7.5 Determination of Coadsorption Enthalpy and Entropy by Calorimetry

If the van't Hoff and isosteric methods are simple ways for estimating the adsorption enthalpy of single component from isothermal adsorption data, they have the disadvantage to not take into account the temperature dependence on the enthalpy and entropy and to be not enough accurate. Moreover they are not adapted to the adsorption of gas mixtures. The best mean to determine the adsorption and coadsorption enthalpy is to measure them by using a differential calorimetry technique coupled with others techniques allowing the measure of adsorbed amount and composition as for example the manometry and the chromatography.

7.5.1 Experimental Calorimetric Technique

a. Description of the experimental device
The experimental technique, which is particularly well adapted for determining the molar adsorption enthalpy of a single component or a mixture as a function on the adsorbed amount at constant temperature, is the differential calorimetry coupled with a manometry and a gas phase chromatography or a mass spectrometry [8]. A schematic representation of the experimental device is shown on Fig. 7.12. The calorimeter is a Setaram C80 differential calorimeter type Calvet. Composed of a sample cell containing the adsorbent and a reference cell surrounded by a vast number of thermocouples, it allows measuring the heat flow being given off during the adsorption of a small dose of gas put in contact with the adsorbent. The amount of gas introduced in the calorimeter is controlled with the manometric device by measuring the pressure of gas mixture in glass bulbs with calibrated volumes V_S and V_F. The composition of the gas is analyzed with the gas chromatograph or the mass spectrometer. A bypass system connected to the vacuum line (turbomolecular pump and dry primary pump) and composed of a six-way gas injection valve (GIV) with sample loops allows the collection of a small dose of gas under low pressure for analysis. The pressure is measured with several MKS Baratron pressure sensors in the range 10^{-4}–10^{3} hPa. This apparatus can be used to study the adsorption of single component as well as mixtures. It simultaneously allows the determination of the total and partial adsorption isotherms at constant initial gas composition, the coadsorption enthalpy and the adsorption selectivity as a function of the total adsorbed amount.

b. Operating procedure
After activation by heating under dynamic vacuum, the sample is brought to the adsorption temperature. The gas mixture is prepared in the flask of volume V_s by introducing each component of the mixture, one after the other one, while controlling the partial pressures. The flask is then slightly heated in order to homogenize the mixture by convection. The composition of the mixture is controlled by GPC or MS. Coadsorption isotherms are obtained by successive introduction of small doses of

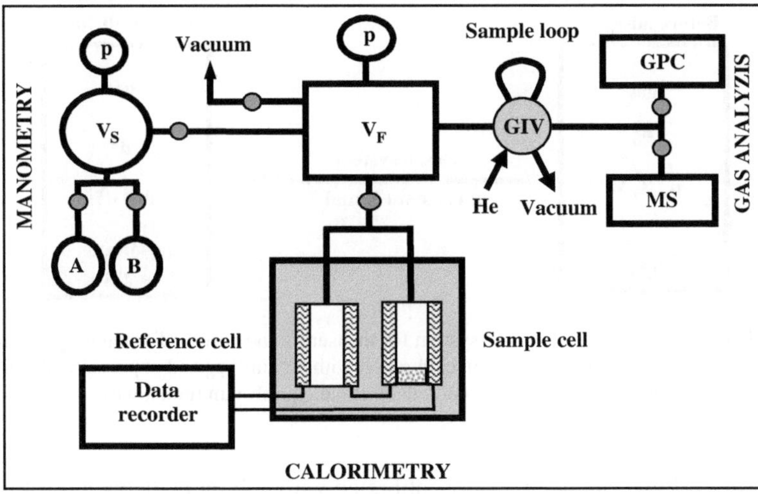

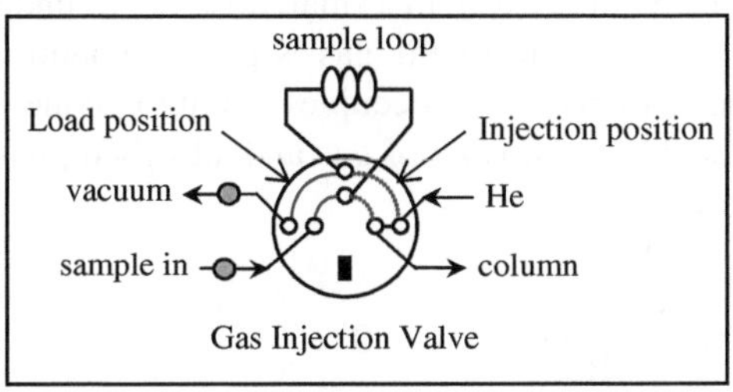

Fig. 7.12 Schematic representation of the calorimeter coupled with a manometric device and a gas analysis technique used to study the coadsorption of gas. Details of the connection of the six-way valve used for GPC or MS sampling

gas mixture stored in bulb V_S first into the flask V_F, previously evacuated and then in the adsorption cells. The pressure is recorded and the gas composition is analyzed before and after each step.

7.5.2 Measurement of Adsorbed Amounts

In the manometry technique coupled with CPG or MS, the determination of the amount of each component i adsorbed (n_i^a) consists to perform a mass balance on the gas phase from the total pressure and the mole fraction of component i in the gas phase, measured before and after adsorption [9]. As the system is closed, we consider that the amount of matter leaving the gas phase is equal to that one adsorbing on the solid.

First, let us consider the adsorption of a single component in a closed system (Fig. 7.13). The volume of the system containing the adsorbent previously evacuated

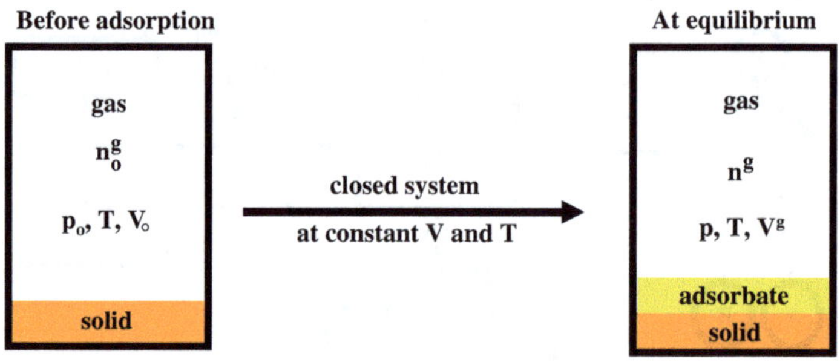

Fig. 7.13 Equivalent thermodynamic system for measuring the adsorbed amount by manometry. V_o: dead volume of reactor (known from calibration with helium); p_o and p: pressure of gas before and after adsorption (measured); V^g = volume of gas at equilibrium (unknown)

is V_o. This volume can be precisely calibrated with helium by using a pycnometric technique. We introduced in the system a small dose of gas under the initial pressure p_o. At equilibrium, the pressure of the gas is p. The volume of the gas phase is $V^g < V_o$ because the adsorbed phase occupies a finite volume in the system. Let us assume the gas as ideal, the amounts of gas and adsorbed phase before adsorption and at equilibrium are:

	$A_{(g)}$	$\rightarrow$	$A_{(ads)}$
Initial condition	$n_o^g = \dfrac{V_o}{RT} p_o$		0
Equilibrium	$n^g = \dfrac{V^g}{RT} p$		n^a

The adsorbed amount is given by:

$$n^a = \frac{V_o}{RT} p_o - \frac{V^g}{RT} p \tag{7.21}$$

As for an ideal gas the concentration is defined by $c = \dfrac{p}{RT}$, Eq. 7.21 becomes:

$$n^a = c_o^g V_o - c V^g \tag{7.22}$$

c°, c and V_o are experimentally measured but V^g is unknown. As the volume is constant,

$$V_o = V^a + V^g \tag{7.23}$$

we can write:

$$(c_o^g - c^g) V_o = n^a - c^g V^a \tag{7.24}$$

By introducing Eq. 7.1 in the right hand side and as the volume of the adsorbed phase is

$$V^a = A\int_o^t dz \tag{7.25}$$

Eq. 7.24 becomes:

$$(c_o^g - c^g)V_o = n^\sigma \tag{7.26}$$

In the manometric technique, the measure of the concentration (pressure) of the gas before and after adsorption does not lead to the adsorbed amount but to the surface excess. However, if the volume of the adsorbed phase is small compared to that one of the system ($V^a << V^g$) and if the concentration of the gas is not too much high, Eq. 7.24 can be reduced to:

$$n^a = V_o(c_o^g - c^g) \tag{7.27}$$

Thus, the surface excess can be assimilated to the adsorbed amount if the volume of the reactor is enough large and the pressure enough low.

In the case of the adsorption of a gas mixture, we have just to introduce in Eq. 7.27 the composition of the gas before adsorption and after adsorption. Let us define by y_{io} and y_i the mole fraction of the component i in the gas phase before adsorption and at equilibrium, respectively, the adsorbed amount of each component i is given by:

$$n_i^a = V_o(y_{io}c_o^g - y_i c^g) \tag{7.28}$$

with obviously:

$$\sum_i n_i^a = n_{total}^a \tag{7.29}$$

Practically, the system composed of a volume V_o containing the adsorbent under the initial pressure p_o, as illustrated in Fig. 7.13, is not conceivable because the gas is adsorbed as soon as it is in contact with the solid. It is the reason why the manometric device shown in Fig. 7.12 is composed of two volumes: the volume V_c of the calorimetric cells, which contains the adsorbent and the volume V_F, which is used to prepare the dose of gas. By applying the Eq. 7.28 to this system, we find that the amount of each component of gas mixture adsorbed on the solid at the adsorption step j is given by the relation:

$$n_{i,j}^a = \frac{V_F}{RT_F}\left(y_{io,j}p_{o,j} - y_{i,j}p_j\right) - \frac{V_c}{RT_c}\left(y_{i,j}p_j - y_{i,j-1}p_{j-1}\right) \tag{7.30}$$

T_F and T_C being the temperature of volume V_F and V_C, respectively.

The total amount of component i adsorbed since the first step and expressed per unity of mass of solid is:

$$N_{i,j}^a = \frac{1}{m}\sum_{j=1}^{j} n_{i,j}^a \tag{7.31}$$

and the total amount of mixture adsorbed after j adsorption steps is:

$$N^a_{total,j} = \sum_{i=1}^{i} N^a_{i,j} \quad (7.32)$$

where m is the mass of the adsorbent.

Then, the mole fraction of component i in the adsorbed phase can be calculated by mean of the relation:

$$x_{i,j} = \frac{N^a_{i,j}}{N^a_{total,j}} \quad (7.33)$$

Finally, the adsorption selectivity at the adsorption step j can be calculated with Eq. 7.6.

7.5.3 Measurement of Differential Adsorption Enthalpy

The calorimetry technique coupled with manometry shown in Fig. 7.12 allows the measure of the heat emanated by the adsorption of a given amount of gas on the solid. In thermodynamics it is essential to know at what function of state corresponds this adsorption heat. This depends on the calorimetric technique and experimental conditions used.

When an amount of gas is introduced in the calorimetric cells, two energetic exchanges must be taken into account: thermal energy due to the adsorption and mechanical energy due to the expansion of the gas in the cells. Thermodynamically, the operation, which consists to introduce a given amount of gas into a vessel containing or not the adsorbent, is equivalent to the compression of gas enclosed in a cylinder by moving the piston [10]. This equivalent thermodynamic system is represented in Fig. 7.14.

Let us consider the adsorption of a single component in the sample cell. The amount of matter present in the system and their molar internal energies are defined as below:

	gas	$\leftrightarrow$	adsorbate
before adsorption	n^g_o, u^g, p_o		0
after adsorption	n^g, u^g, p		n^a, u^a

with u^g and u^a the molar internal energy of gas and adsorbate, respectively. The change in internal energy is:

$$\Delta U_s = n^a u^a + n^g u^g - n^g_o u^g \quad (7.34)$$

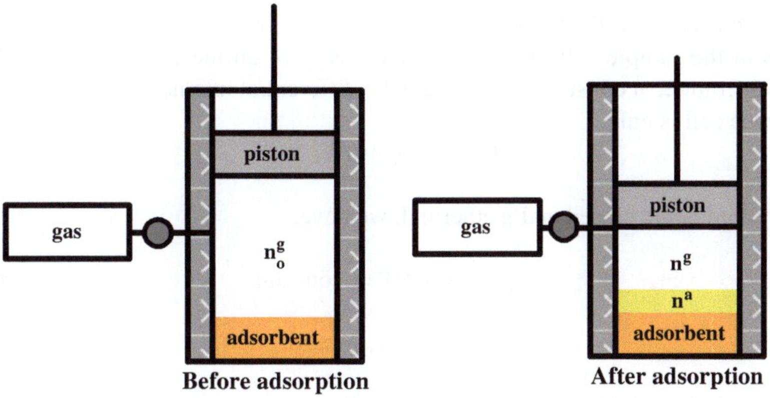

Fig. 7.14 Thermodynamic system composed of a cylinder closed by a frictionless piston equivalent to the sample cell of the manometric system. For the reference cell, the equivalent system is the same but without adsorbent

As the system is closed:

$$n^g_o = n^a + n^g \tag{7.35}$$

we have:

$$\Delta U_s = n^a(u^a - u^g) \tag{7.36}$$

Then, by derivation of Eq. 7.36 we obtain the differential energy:

$$dU_s = (u^a - u^g)dn^a + n^a du^a = \delta Q_s + \delta W_s \tag{7.37}$$

The differential mechanical work exchanged with the outside world, assuming the transformation as reversible, is expressed by:

$$\delta W_s = -pdV_s \tag{7.38}$$

Let us consider the gas as ideal ($pV_s = n^g RT$) and as the system is close ($dn^g = -dn^a$), this equation becomes:

$$\delta W_s = V_s dp - dn^g RT = V_s dp + dn^a RT \tag{7.39}$$

Thus the differential heat in the sample is:

$$\delta Q_s = dU_s - \delta W_s \tag{7.40}$$

$$\delta Q_s = (u^a - u^g)dn^a + n^a du^a - RTdn^a - V_s dp \tag{7.41}$$

Let us consider now the reference cell in which we introduced the same amount of gas as in the sample cell. As the internal energy of an ideal gas depends only on the temperature, at constant temperature the differential internal energy of gas in the reference cell is null:

$$dU_r = \delta Q_r + \delta W_r = 0 \tag{7.42}$$

The system being closed and isothermal, we have:

$$pV_r = n^g RT = \text{constant} \tag{7.43}$$

$$pdV_r + V_r dp = 0 \tag{7.44}$$

Thus the differential heat in the reference cell is:

$$\delta Q_r = -V_r dp \tag{7.45}$$

The adsorption heat measured by differential calorimetry is equal to the difference between the heat emanated in the sample cell and that one emanated in the reference cell.

$$\delta Q^a = \delta Q_s - \delta Q_r \tag{7.46}$$

By introducing Eqs. 7.41 and 7.45 in this relation we obtain:

$$\delta Q^a = (u^a - u^g)dn^a + n^a du^a - RTdn^a + (V_r - V_s)dp \tag{7.47}$$

Let us define now the molar internal energy of the adsorbate at the adsorbed amount n^a by:

$$u^a(n^a) = u^a + n^a \frac{du^a}{dn^a} \tag{7.48}$$

We obtain:

$$\delta Q^a = u^a(n^a)dn^a - u^g dn^a - RTdn^a + (V_r - V_s)dp \tag{7.49}$$

If the volume of sample and reference cells are identical and larger than that one of the adsorbent we can assume:

$$(V_r - V_s)dp = 0 \tag{7.50}$$

Therefore the molar heat of adsorption measured by differential calorimetry is equal to:

$$Q_m^a = \frac{\delta Q^a}{dn^a} = u^a(n^a) - u^g - RT = \Delta_r U_m^a - RT \tag{7.51}$$

Q_m^a is called the differential calorimetric heat of adsorption and $\Delta_r U_m^a$ the differential molar internal energy of adsorption. They depend on the adsorbed amount n^a.

The differential molar internal energy is related to the differential molar enthalpy by:

$$\Delta_r H_m^a = \Delta_r U_m^a + p\Delta_r V \tag{7.52}$$

For the adsorption equilibrium gas ↔ adsorbate, the change in molar volume is equal to:

$$\Delta_r V = V_m^a - V_m^g \tag{7.53}$$

Let us consider the molar volume of the adsorbed phase as negligible compared to that one of the ideal gas we have:

$$\Delta_r V \approx -V_m^g = -\frac{RT}{p} \tag{7.54}$$

Finally, by combining Eqs. 7.51, 7.52 and 7.54 we obtain:

$$Q_m^a = \Delta_r H_m^a \tag{7.55}$$

Thus, the molar adsorption heat measured by differential calorimetry at constant temperature corresponds to the differential molar adsorption enthalpy. This value is equivalent to the isosteric enthalpy of adsorption [11].

Practically, when an adsorption step j is performed at constant temperature between the initial state j-1 defined by the couple of parameters $p_{j\text{-}1}$ and N_{j-1}^a and the final state j defined by the couple p_j and N_j^a as illustrated on Fig. 7.15, the calorimetric molar adsorption heat at the mean total adsorbed amount $\frac{N_{j-1}^a + N_j^a}{2}$ is calculated by the relation:

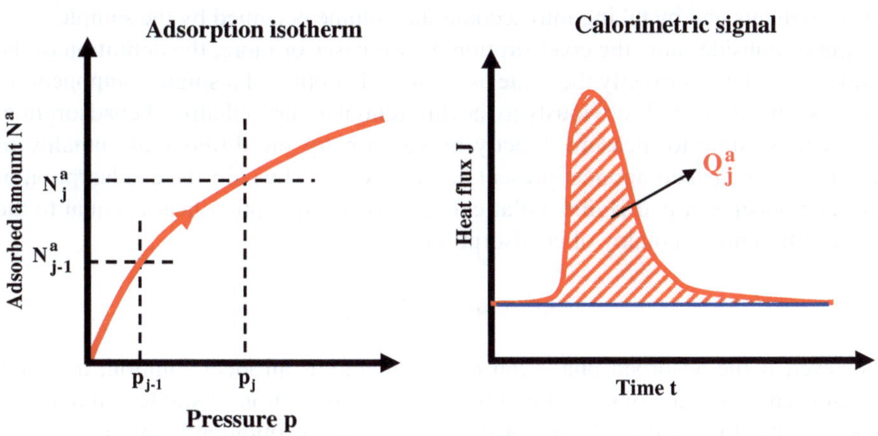

Fig. 7.15 Schematic representation of an isothermal adsorption step from initial equilibrium state j-1 to final equilibrium state j and corresponding thermal effect measured by differential calorimetry

$$Q^a_{m,j} = \frac{Q^a_j}{N^a_j - N^a_{j-1}} = \frac{N^a_j u^a(N^a_j) - N^a_{j-1} u^a(N^a_{j-1})}{N^a_j - N^a_{j-1}} - u^g - RT \quad (7.56)$$

where Q^a_j is the heat emanated by the adsorption of the $N^a_j - N^a_{j-1}$ moles of gas per unity of mass of solid. Q^a_j is measured by integration of the calorimetric signal giving the heat flux J as a function of time (Fig. 7.15):

$$Q^a_j = \int_0^t J_j dt \quad (7.57)$$

Experimentally, the difference $N^a_j - N^a_{j-1}$ must be as small as possible in order that the calorimetric adsorption heat can be considered as a differential value. This implies to have accurate pressure sensors particularly for exploring the low adsorbed amount range where the pressure variations between the initial and final states can be very small.

It may be noted that some authors [3] use a different procedure to determine the molar differential enthalpy of adsorption by mean of a differential calorimeter coupled with a manometric device. They prefer to work with only one calorimetric vessel connected to the manometric device, the reference cell being maintained under vacuum. In these conditions we obtain from Eq. 7.48 that the molar heat of adsorption is equal to:

$$Q^a_m = \frac{\delta Q^a}{dn^a} = \Delta_r H^a_m - V_s \frac{dp}{dn^a} \quad (7.58)$$

This procedure needs to know precisely the volume V_s of the sample cell, which is located within the calorimetric detector and to reduce the dead volume, which is outside the detector. V_s can be estimated by liquid weighting or from geometrical measurements and by taking into account the volume occupied by the sample.

Let us consider now the coadsorption of two gases or more, the definition of the calorimetric heat is exactly the same as for the adsorption of a single component. In this case, it corresponds obviously to the differential molar enthalpy of coadsorption. It is not possible to measure directly by calorimetry the differential enthalpy of adsorption of each component present in the mixture. Thus, for the coadsorption of two components A and B, the molar calorimetric coadsorption heat is equal to the molar differential enthalpy of coadsorption:

$$Q^a_{m,A+B} = \Delta_r H^a_{m,A+B} \quad (7.59)$$

However, if the adsorbed phase can be considered as an ideal solution, in which the molecular interactions are the same as for the adsorption of single components, it is possible to calculate the molar differential coadsorption enthalpy by mean of the relation:

$$\Delta_r H^a_{m,\,A+B} = x_A \Delta_r H^{a^*}_{m,\,A} + x_B \Delta_r H^{a^*}_{m,\,B} \quad (7.60)$$

where $\Delta_r H^{a^*}_{m,i}$ is the molar differential enthalpy of adsorption of single component i at the same adsorbed amount as in the mixture.

7.5.4 Adsorption Gibbs Energy

a. Gibbs adsorption isotherm

For a three-dimensional fluid phase like, for example, a gas enclosed in a volume, the free enthalpy so-called Gibbs energy depends on the temperature, the pressure and the amount of gas: $G^g = f(T, p, n^g)$. The differential Gibbs energy is:

$$dG^g = \left(\frac{\partial G^g}{\partial T}\right)_{p,n^g} dT + \left(\frac{\partial G^g}{\partial p}\right)_{T,n^g} dp + \left(\frac{\partial G^g}{\partial n^g}\right)_{T,p} dn^g \tag{7.61}$$

and we obtain by expressing the partial derivatives according to the Maxwell relations [12]:

$$dG^g = -S^g dT + V^g dp + \mu^g dn^g \tag{7.62}$$

For the adsorbed phase, which is a two-dimensional phase, we have to take into account the specific surface area A of the solid on which the molecules are adsorbed. Thus, the Gibbs energy of the adsorbed phase is a function: $G^a = f(T, p, A, n^a)$. The differential Gibbs energy is given by:

$$dG^a = \left(\frac{\partial G^a}{\partial T}\right)_{p,A,n^a} dT + \left(\frac{\partial G^a}{\partial p}\right)_{T,A,n^a} dp + \left(\frac{\partial G^a}{\partial A}\right)_{T,p,n^a} dA + \left(\frac{\partial G^a}{\partial n^a}\right)_{T,p,A} dn^a \tag{7.63}$$

and the Maxwell relations give:

$$dG^a = -S^a dT + V^a dp - \pi\, dA + \mu^a dn^a \tag{7.64}$$

We retrieve here the spreading pressure p, which corresponds to the change of Gibbs energy due to a change of the interface area between solid and adsorbate. At constant temperature and pressure Eq. 7.64 is reduced to:

$$dG^a = -\pi\, dA + \mu^a dn^a \tag{7.65}$$

The application of the Euler theorem gives:

$$G^a = -\pi\, A + n^a \mu^a \tag{7.66}$$

By derivation of this equation we have:

$$dG^a = -\pi\, dA - A d\pi + n^a d\mu^a + \mu^a dn^a \tag{7.67}$$

The subtraction of Eqs. 7.65 and 7.67 gives:

$$\mathrm{A d}\,\pi = \mathrm{n^a d}\mu^{\mathrm{a}} \tag{7.68}$$

The equilibrium condition between the adsorbed phase and the gas phase is:

$$\mu^{\mathrm{g}}(\mathrm{T, p}) = \mu^{\mathrm{a}}(\mathrm{T, p, n^a}) \tag{7.69}$$

$$\mathrm{d}\mu^{\mathrm{g}} = \mathrm{d}\mu^{\mathrm{a}} \tag{7.70}$$

with the chemical potential of the gas assumed as ideal:

$$\mu^{\mathrm{g}}(\mathrm{T, p}) = \mu^{\mathrm{g}^{\circ}}(\mathrm{T}) + \mathrm{RTLn}\frac{\mathrm{p}}{\mathrm{p}^{\circ}} \tag{7.71}$$

$$\mathrm{d}\mu^{\mathrm{g}} = \mathrm{RTdLn}\frac{\mathrm{p}}{\mathrm{p}^{\circ}} \tag{7.72}$$

Thus we obtain with Eqs. 7.68, 7.70 and 7.72:

$$\mathrm{d}\,\pi = \frac{\mathrm{n^a}}{\mathrm{A}}\mathrm{RTLn}\frac{\mathrm{p}}{\mathrm{p}^{\circ}} \tag{7.73}$$

This is the Gibbs adsorption isotherm. The spreading pressure cannot be directly measured but can be calculated owing to this equation. The integration of Eq. 7.73 from 0 to p gives:

$$\pi = \frac{\mathrm{RT}}{\mathrm{A}}\int_{0}^{\mathrm{p}} \mathrm{n^a dLn}\frac{\mathrm{p}}{\mathrm{p}^{\circ}} \tag{7.74}$$

Thus the spreading pressure can be calculated from the integration of the adsorption isotherm in the form $\mathrm{n^a} = \mathrm{f}\left(\mathrm{Ln}\frac{\mathrm{p}}{\mathrm{p}^{\circ}}\right)_{\mathrm{T}}$ (Fig. 7.16).

It may be noted that a problem of integration appears because $\mathrm{Ln}\frac{\mathrm{p}}{\mathrm{p}^{\circ}}$ tends to infinity when p tends to zero. To overcome this problem we can use the Henry law, which states that, in the low pressure region, the adsorbed amount is proportional to the pressure:

$$\mathrm{n^a} = \mathrm{K_H p} \tag{7.75}$$

$\mathrm{K_H}$, the Henry constant is given by the slope of the adsorption isotherm in the very low pressure region (Fig. 7.16).

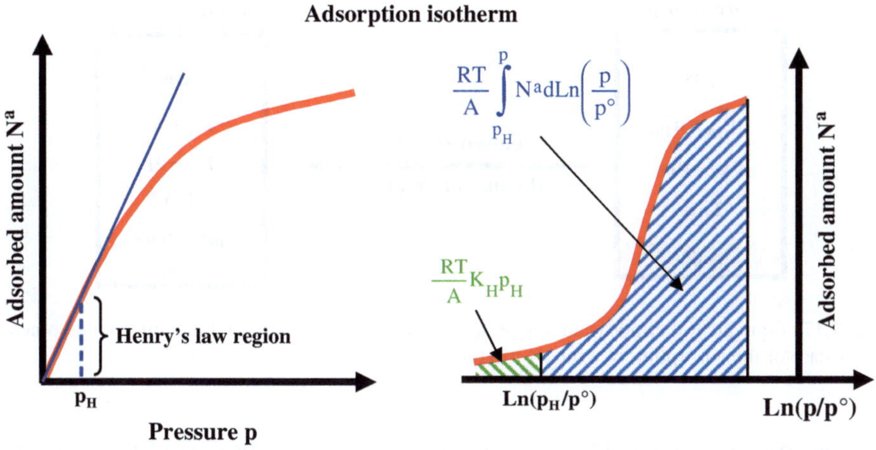

Fig. 7.16 Determination of the spreading pressure by integration of the adsorption isotherm

Thus the spreading pressure is calculated by mean of the relation:

$$\pi = \frac{RT}{A}\left(K_H p_H + \int_{p_H}^{p} n^a d\mathrm{Ln}\frac{p}{p^\circ}\right) \tag{7.76}$$

where p_H is the pressure up to which Henry's law is valid.

In the case of multi-components adsorption the total spreading pressure is:

$$\pi = \sum_i \pi_i = \frac{RT}{A}\sum_i\left(\int_0^p n_i^a d\mathrm{Ln}\frac{y_i p}{p^\circ}\right) \tag{7.77}$$

It can be calculated from partial adsorption isotherms in the same way as for a single component.

In the case of microporous solids like zeolites for example, where the curvature of the surface can be very important, the notion of specific surface area is questionable. It is the reason why some authors recommend replacing the term πA by Φ that they call the surface potential.

b. Differential adsorption Gibbs energy

Let us consider the coadsorption of two components A and B in a closed system at constant temperature. As for the adsorption of a single component, we can define an equivalent thermodynamic system composed of a closed volume containing the gas mixture in contact with the sample (Fig. 7.17). The pressure of gas mixture is $p_o = p_{Ao} + p_{Bo}$ in the initial condition and $p = p_A + p_B$ at the equilibrium. The molar Gibbs energy of each component is G_{mi}.

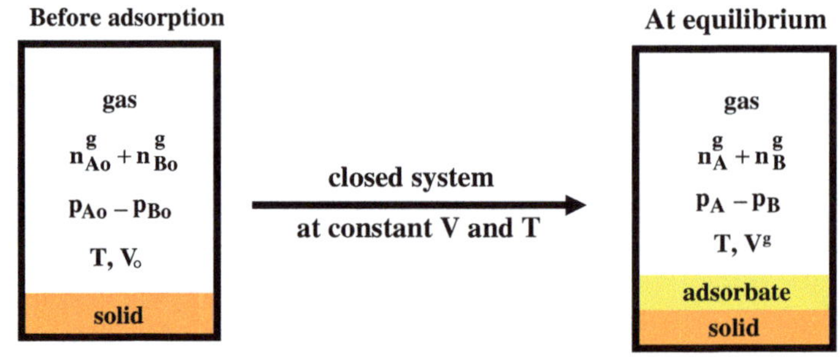

Fig. 7.17 Equivalent thermodynamic system for measuring the adsorbed amount by manometry in the case of the adsorption of a binary mixture A + B

The coadsorption equilibrium corresponding to the adsorption of a dose of gas mixture A+B at a given composition is:

	$A_{(g)}$	$+\ B_{(g)}$	$\leftrightarrow\ A_{(ads)}$	$+\ B_{(ads)}$
before adsorption	$n^g_{Ao}, G^g_{mAo}, p_{Ao}$	$n^g_{Bo}, G^g_{mBo}, p_{Bo}$	0	0
after adsorption	n^g_A, G^g_{mA}, p_A	n^g_B, G^g_{mB}, p_B	n^a_A, G^a_{mA}	n^a_B, G^a_{mB}

The total Gibbs energy of the system in the initial conditions is:

$$G_o = \sum_i n^g_{io} G^g_{mio} = n^g_{Ao} G^g_{mAo} + n^g_{Bo} G^g_{mBo} \tag{7.78}$$

For each component i in the gas phase we can write:

$$G^g_{mio} = \mu^g_{io}(T,p) = \mu^{g^\circ}_i(T) + RTLn\frac{p_{io}}{p^\circ} \tag{7.79}$$

Thus we have before adsorption:

$$G_o = \sum_i n^g_{io}(\mu^{g^\circ}_i(T) + RTLn\frac{p_{io}}{p^\circ}) \tag{7.80}$$

$$G_o = n^g_{Ao}\left(\mu^{g^\circ}_A(T) + RTLn\frac{p_{Ao}}{p^\circ}\right) + n^g_{Bo}\left(\mu^{g^\circ}_B(T) + RTLn\frac{p_{Bo}}{p^\circ}\right) \tag{7.81}$$

The total Gibbs energy of the system after adsorption is:

$$G = \sum_i n^a_i G^a_{mi} + \sum_i n^g_i G^g_{mi} \tag{7.82}$$

with:

$$n_i^g = n_{io}^g - n_i^a \tag{7.83}$$

and

$$G_{mi}^g = \mu_i^g(T,p) = \mu_i^{g^\circ}(T) + RTLn\frac{p_i}{p^\circ} = \mu_i^{g^\circ}(T) + RTLn\frac{y_i p}{p^\circ} \tag{7.84}$$

For each component i adsorbed, we can write from Eq. 7.66:

$$G_{mi}^a = \frac{-\pi_i^a A}{n_i^a} + \mu_i^a(T, \pi) \tag{7.85}$$

with $\mu_i^a(T, \pi)$ the chemical potential of the adsorbed phase defined by:

$$\mu_i^a(T, \pi) = \mu_i^{a^*}(T, \pi) + RTLn\gamma_i x_i \tag{7.86}$$

where $\mu_i^{a^*}(T, \pi)$ is the chemical potential of the single component i adsorbed at the same spreading pressure as the mixture.

Thus the expression of the total Gibbs energy after adsorption becomes:

$$G = -\sum_i \pi_i A + \sum_i n_i^a \mu_i^a(T, \pi) + \sum_i (n_{io}^g - n_i^a)\left(\mu_i^{g^\circ}(T) + RTLn\frac{p_i}{p^\circ}\right) \tag{7.87}$$

As at equilibrium we have equality of the chemical potentials of each component present in the gas phase and in the adsorbed phase:

$$\mu_i^a(T, \pi) = \mu_i^g(T,p) \tag{7.88}$$

the total Gibbs energy after adsorption is:

$$G = -\sum_i \pi_i A + \sum_i n_i^{g^\circ}\left(\mu_i^{g^\circ}(T) + RTLn\frac{p_i}{p^\circ}\right) \tag{7.89}$$

The change of coadsorption Gibbs energy between the final and initial states is then obtained by subtracting Eq. 7.89 with Eq. 7.80:

$$\Delta G = G - G_o = -\sum_i \pi_i A + \sum_i n_{io} RTLn\frac{p_i}{p_{io}} \tag{7.90}$$

For the adsorption of the mixture A+B we obtain:

$$\Delta G = -\pi_A A - \pi_B A + n_{Ao} RTLn\frac{p_A}{p_{Ao}} + n_{Bo} RTLn\frac{p_B}{p_{Bo}} \tag{7.91}$$

ΔG is an integral value. This is the change of Gibbs energy of the system for the coadsorption of the $\sum_i (n^g_{io} - n^g_i)$moles of gas mixture in a given proportion. ΔG can be easily calculated by measuring the pressure of gas mixture in the system before and after adsorption and by determining the partial spreading pressures from integration of the partial adsorption isotherm of each component i:

$$\pi\, A = \sum_i \pi_i\, A = \sum_i RT \int_0^{pi} RTdLn\frac{p_i}{p^\circ} \tag{7.92}$$

The differential molar coadsorption Gibbs energy can be calculated at given adsorbed amount by derivation of the integral coadsorption Gibbs energy:

$$\Delta_r G^a_{m,A+B} = \left(\frac{\partial \Delta G}{\partial n^a}\right)_{(T,\,A)} \tag{7.93}$$

Thus we can obtain the dependence of the adsorbed amount on the differential coadsorption Gibbs energy in a similar way as for the differential coadsorption enthalpy measured by calorimetry. Practically, $\Delta_r G^a_{m,A+B}$ can be calculated by numerical derivation of the plot $\Delta G = f(n^a)$. In the case of the adsorption of a single component, the equations for the calculation of the Gibbs adsorption energy are exactly the same, we have just to take i equal to one.

7.5.5 Differential Adsorption Entropy and Molar Entropy of Adsorbate

The differential molar coadsorption entropy can be easily calculated knowing the differential molar coadsorption enthalpy measured by calorimetry and the differential molar coadsorption Gibbs energy calculated as explained above by mean of the relation:

$$\Delta_r S^a_{m,A+B} = \frac{\Delta_r H^a_{m,A+B} - \Delta_r G^a_{m,A+B}}{T} \tag{7.94}$$

The molar entropy of the adsorbate at the total adsorbed amount n^a_{A+B} and for a given composition of the gas mixture is given by:

$$S^a_{m,A+B} = \Delta_r S^a_{m,A+B} + S^{g^\circ}_{m,A+B} \tag{7.95}$$

where $S^{g^\circ}_{m,A+B}$ is the molar entropy of the gas mixture. Its value can be calculated assuming the mixture as ideal with the relation:

$$S^{g^\circ}_{m,A+B} = y_{Ao} S^{g^\circ}_{m,A} + y_{Bo} S^{g^\circ}_{m,B} \tag{7.96}$$

y_{io} being the initial mole fraction of component i in the gas phase and $S^{g^\circ}_{m,i}$ the molar entropy of gaseous single component i.

7.5.6 Partial Adsorption Enthalpy and Entropy

In the case of multi-components adsorption, the partial molar differential adsorption enthalpies and entropies of each component i present in the gas mixture cannot be directly measured by experiments. However, it is possible to estimate them by mean of the "tangent method" based on the well-known Gibbs-Duhem relation.

Let us consider the adsorption of a binary gas mixture A+B and Z a function of state, which denotes either the enthalpy or the entropy. The differential molar coadsorption value $\Delta_r Z^a_{m,A+B}$ is equal to:

$$\Delta_r Z^a_{m,A+B} = x_A \Delta_r Z^a_{m,A} + x_B \Delta_r Z^a_{m,B} \tag{7.97}$$

where $\Delta_r Z^a_{m,i}$ is the partial molar differential adsorption value of component i and x_i is the mole fraction of component i in the adsorbed phase for a given adsorbed amount of mixture. As $x_A + x_B=1$, we obtain:

$$\Delta_r Z^a_{m,A+B} = \Delta_r Z^a_{m,B} + (\Delta_r Z^a_{m,A} - \Delta_r Z^a_{m,B}) x_A \tag{7.98}$$

The application of the Gibbs-Duhem relation $\sum_i x_i dZ^a_{m,i} = 0$ leads to:

$$x_A d\Delta_r Z^a_{m,A} + x_B d\Delta_r Z^a_{m,B} = 0 \tag{7.99}$$

The derivation of Eq. 7.96 with respect to x_A gives:

$$\frac{\partial \Delta_r Z^a_{m,A+B}}{\partial x_A} = \Delta_r Z^a_{m,A} + x_A d\Delta_r Z^a_{m,A} - \Delta_r Z^a_{m,B} + (1 - x_A) d\Delta_r Z_{m,B} \tag{7.100}$$

By combining with Eq. 7.99 and by reporting in Eq. 7.98 we have:

$$\Delta_r Z_m, B^a = \Delta_r Z^a_{m,A+B} - \frac{\partial \Delta_r Z^a_{m,A+B}}{\partial x_A} x_A \tag{7.101}$$

In the same way, by derivation of Eq. 7.97 with respect to x_B we obtain:

$$\Delta_r Z^a_{m,A} = \Delta_r Z^a_{m,A+B} - \frac{\partial \Delta_r Z^a_{m,A+B}}{\partial x_B} x_B \tag{7.102}$$

As $x_A + x_B = 1$ and $dx_A = -dx_B$, this equation can be written as:

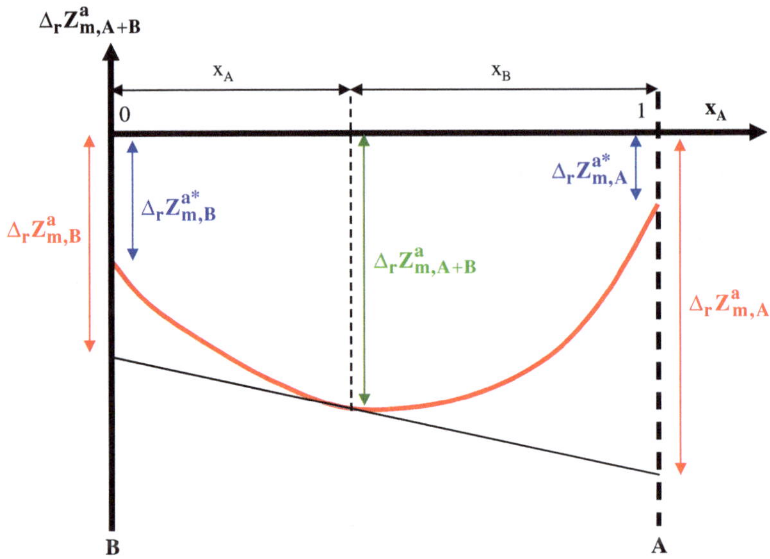

Fig. 7.18 Graphical method for the determination of the partial molar differential adsorption enthalpies or entropies in the case of a binary mixture A + B

$$\Delta_r Z^a_{m,A} = \Delta_r Z^a_{m,A+B} + \frac{\partial \Delta_r Z^a_{m,A+B}}{\partial x_A}(1 - x_A) \tag{7.103}$$

Thus, the partial molar differential values $\Delta_r Z^a_{m,A}$ and $\Delta_r Z^a_{m,B}$ can be graphically determined from the curve $\Delta_r Z^a_{m,A+B} = f(x_a)$ plotted at a constant total adsorbed amount. As illustrated on Fig. 7.18, they are given by the intercepts of the straight-line tangent to the curve at $x_A = 1$ and $x_A = 0$. We just need for that to have a lot of accurate measurements of $\Delta_r Z^a_{m,A+B}$ in a large range of composition and total adsorbed amount of mixtures.

It may be noticed that in the case of an ideal adsorbed solution the partial molar differential values $\Delta_r Z^a_{m,i}$ is constant and equal to the partial molar value of single component $\Delta_r Z^{a^*}_{m,i}$. If Z represents the enthalpy, we retrieve from the Eq. 7.97 the Eq. 7.60, which allows the prediction of the coadsorption enthalpy from the adsorption enthalpies of single components.

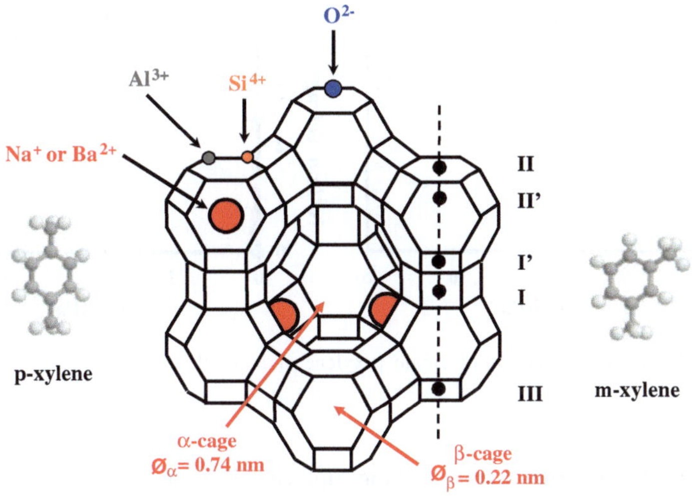

Fig. 7.19 Structure of the Y type zeolite. For clarity reason the aluminium, silicon and oxygen atoms are not represented. Only the compensation cations (in *red*) located on sites II inside the supercages accessible to xylenes isomers are shown. *Black points* represent the cationic sites

7.6 Case Studies

7.6.1 Separation of Xylenes Isomers by Selective Adsorption on FAU Type Zeolite

This first example of application concerns the separation of xylenes isomers (o-, m- and p-dimethylbenzene) by selective adsorption on FAU type zeolites exchanged with various compensation cations. As these isomers have very close boiling points, they cannot be separated by distillation. So, they are separated by competitive adsorption using the liquid chromatography technique and the simulated moving bed technology [13, 14]. The aim of this work was to perform a complete thermodynamic study of the adsorption of xylenes in the gas phase in order to obtain a detailed understanding of the selective adsorption process. We focused here only on the separation of p-xylene (pX) and m-xylene (mX), which is particularly difficult. This basic study was essential for the development of a more efficient separation process [15] and, in particular, for the optimization of adsorbent formulation and operating conditions of the process.

The FAU type zeolite is a microporous aluminosilicate [16]. Its chemical formula can be written as: $M^{n+}_{x/n}(AlO_2)_x(SiO_2)_y$, mH_2O where M is a compensation cation (alkaline or alkaline-earth metals) at the oxidation state +n. The simplified structure of FAU type zeolite is shown on Fig. 7.19. It is composed of SiO_4 and AlO_4 tetrahedrons arranged in such a way that they form sodalite cages (so-called β-cages) and supercages (so-called α-cages) having a free aperture of 0.22 and 0.74 nm

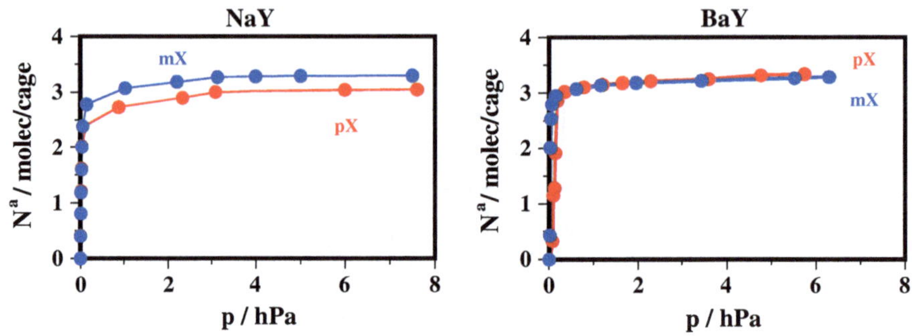

Fig. 7.20 Adsorption isotherms of single p-xylene and m-xylene on NaY and BaY zeolites at 423 K

respectively. The internal diameter of the supercage is around 1.3 nm. As the kinetic diameters of p-xylene and m-xylene are about 0.67 and 0.71 nm respectively, these isomers can only enter the supercages. FAU zeolite is usually called X zeolite if the Silicon/Aluminium ratio is lower than 1.5 and Y zeolite if this ratio lies between 1.5 and 3. According to the Si/Al ratio, the compensation cation can be located on different sites inside supercages (sites II and III), sodalite cages (sites I' and II') or hexagonal prisms linking two sodalite cages (sites I) [17, 18]. In this study, the Si/Al ratio was equal to 2.4 and the compensation cations were sodium or barium. They were essentially located at the center of the hexagonal window of each sodalite cage but some of them are also present inside sodalite cages and hexagonal prims.

First, the adsorption of single xylenes has been studied on NaY and BaY zeolites at 423 K by calorimetry coupled with manometry [8, 19]. The adsorption isotherms are given in Fig. 7.20. No significant difference is observed between p-xylene and m-xylene. Adsorption isotherms are perfectly superimposed on BaY while the amount of m-xylene is slightly higher for m-xylene on NaY. This indicates that whatever the compensation cation, the adsorption affinities for p-xylene and m-xylene are very close and the supercage can accommodate around 3 aromatic molecules [20]. The adsorption enthalpies as a function of supercage filling are shown on Fig. 7.21. The values measured by calorimetry are higher than those estimated by using the van't Hoff method as shown for the NaY zeolite in Table 7.2 [21]. Relevant information can be extracted from these calorimetric curves. Firstly, on both zeolites, the adsorption enthalpy at zero filling is higher with m-xylene than with p-xylene. However, the difference is rather low (5 kJ.mol^{-1}). This means that the aromatic molecule-zeolite interactions are almost the same for the two isomers and consequently this suggests that the adsorption sites for p-xylene and m-xylene are probably very similar. This result has been confirmed by a detailed structural study on the location of xylenes adsorbed in the supercages by neutron diffraction [22, 23] and molecular simulation [24, 25]. Figure 7.22 shows the location of xylene molecules inside the supercage. Both isomers have a very similar position. Their aromatic rings are centred on compensation cation in site II and their methyl groups are in interaction with

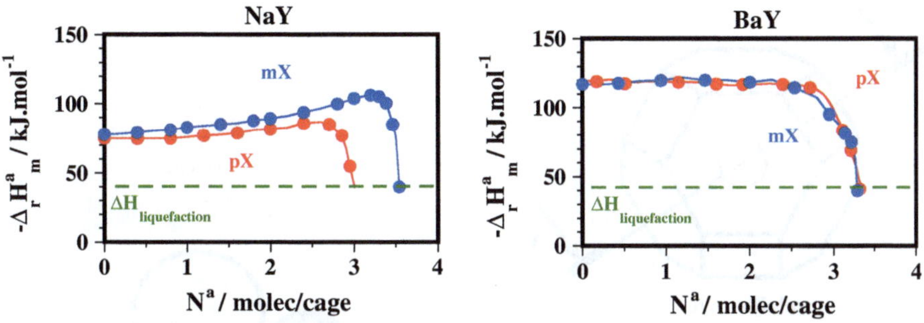

Fig. 7.21 Differential molar adsorption enthalpies of single p-xylene and m-xylene as a function of filling in NaY and BaY zeolites at 423 K

Table 7.2 Comparison of adsorption enthalpies of p-xylene and m-xylene on NaY measured by calorimetry and estimated from adsorption isotherms using the Van't Hoff method

$\Delta_r H^a$/kJ.mol^{-1}		p-xylene	m-xylene
Experimental	Calorimetry	75	80
	Henry	50	58
Van't Hoff	Langmuir	64	67

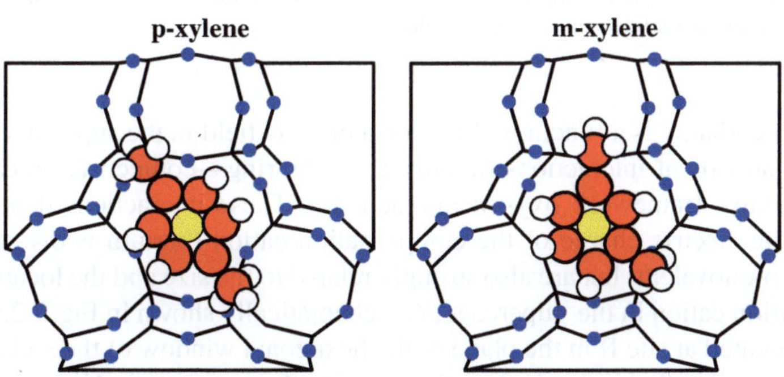

Fig. 7.22 Schematic representation of the location of p-xylene and m-xylene molecules adsorbed in the supercage of FAU zeolite [*blue*:oxygen atoms; *yellow*: compensation cation; *red*: carbon atoms; *white*: hydrogen atoms]

oxygen atoms of the double 6-ring unities joining two sodalite cages. The fact that the adsorption enthalpies are higher with m-xylene could be explained by a better interaction of methyl groups with the oxygen atoms of hexagonal prisms, which are also in the meta position to the hexagonal window.

Secondly, the adsorption enthalpy at zero filling is largely higher on BaY than on NaY (around 75 against 120 kJ.mol^{-1}). This indicates that the adsorbate–adsorbent interactions are stronger with barium than with sodium. This important difference in adsorption enthalpies can be explained from structural considerations. The

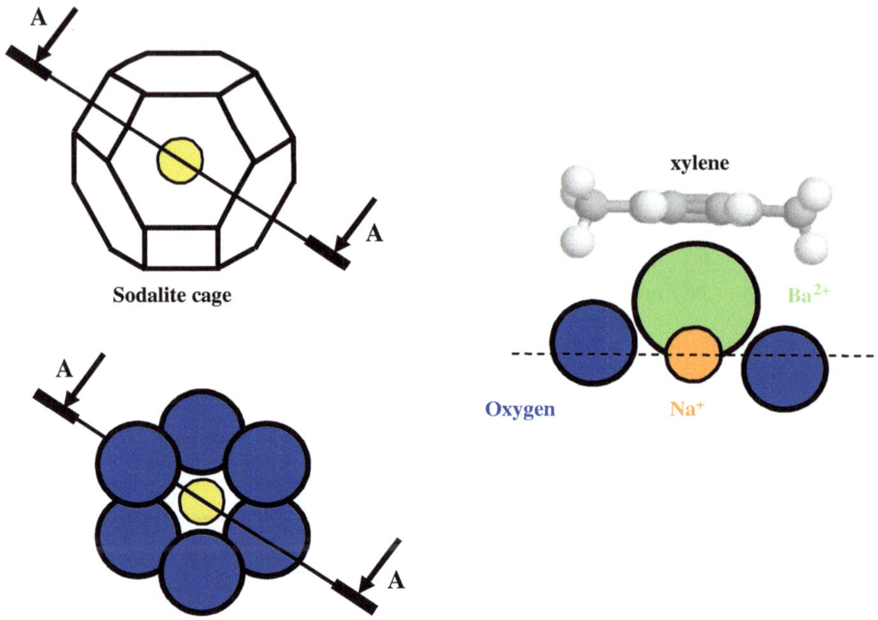

Fig. 7.23 Schematic representation of the location of sodium and barium cation in site II in the supercage and in interaction with a xylene molecule

adsorption enthalpy is a function of the local electric field in the supercages and the adsorbate-adsorbent interactions (mainly aromatic ring—compensation cation and methyl group—framework oxygen interactions). These interactions depend obviously of the electric charge of the compensation cation (barium is divalent while sodium is monovalent) but are also strongly related to the size and the location of the compensation cation in the supercage. As schematically shown in Fig. 7.23, sodium cation is located at site II in the plane of the hexagonal window of the sodalite cage. The xylene molecule is adsorbed on this site, involving aromatic-Na^+ and methyl–oxygen interactions. As for the barium cation, it lies out of this plane, inside the supercage, because of its bigger size ($r_{Na^+} = 0.095\text{n}m$ and $r_{Ba^{2+}} = 0.143\,\text{n}m$). It is more accessible to xylene molecule than the sodium cation, which is rather hidden by the oxygen atoms. With barium, the aromatic—ring cation distance is shorter than with sodium, leading thus to stronger aromatic-ring cation interaction and weaker methyl-oxygen interaction because the methyl groups, in this case, are farther from the framework oxygen atoms. But this weakening is however in a less extent and it results that the adsorption enthalpy at zero filling is higher with barium than with sodium. Thirdly, as the filling increases, above 2 molec/cage, the adsorption enthalpy increases slightly for NaY while it decreases for BaY. This means that the xylene molecules do not interact in the same way with both zeolites. For NaY, the increase of the adsorption enthalpy is due to the increase of adsorbate-adsorbate interactions, which is not compensated by the decrease of the adsorbate-adsorbent interactions. As in a supercage, each xylene molecules interact with one sodium cation, the adsorbate-adsorbent interactions are almost constant during the adsorption. But when the last molecule is adsorbed, because of steric hindrance, the molecular interactions become important. For BaY, if the adsorbate-adsorbate interactions increase also with the filling, the adsorbate-adsorbent interactions are not constant. As there are only two cations in the supercage, contrary to the two first molecules, the last xylene molecule cannot interact with one barium cation and is adsorbed on a non cationic site with a lower energy. Finally, once the supercages are completely filled, the adsorption enthalpy decreases sharply down to the liquefaction enthalpy of xylenes. This last part of the enthalpy curve corresponds to the adsorption of some xylene molecules on the external surface of zeolite particles.

Then, the coadsorption of p-xylene and m-xylene has been studied on NaY and BaY zeolites. From the study on adsorption of single components indicating no important difference in adsorption capacities and enthalpies between p-xylene and m-xylene, we could expect that the coadsorption of these two isomers on these materials will not be selective. Nevertheless, results show that the FAU zeolites can selectively adsorb the xylene isomers. However these systems are very particular because the selectivity appears only at high filling of supercages [26, 27]. Adsorption isotherms of equimolar mixture of p-xylene and m-xylene on NaY and BaY zeolites at 423 K are displayed on Fig. 7.24. For both zeolites, in the low pressure range the amounts of p-xylene and m-xylene adsorbed are quite the same, but at the plateau of the partial isotherms, a small difference in adsorption capacities appears. NaY adsorbs more m-xylene while BaY adsorbs more p-xylene. The adsorption selectivities of p-xylene with respect to m-xylene versus filling are given in Fig. 7.25. Below 2 molec/cage and whatever the zeolite, the selectivity is equal to one. Both isomers are adsorbed in the same proportions. The adsorption process is not selective. Above this filling, during the adsorption of the third and last molecule in the supercages, the selectivity decreases slightly for NaY (m-xylene is preferentially adsorbed) while it increases sharply for BaY (p-xylene is preferentially adsorbed).

Figure 7.26 shows the coadsorption enthalpies as a function of the filling. For NaY, the calorimetric curve measured with the equimolar mixture is very similar to

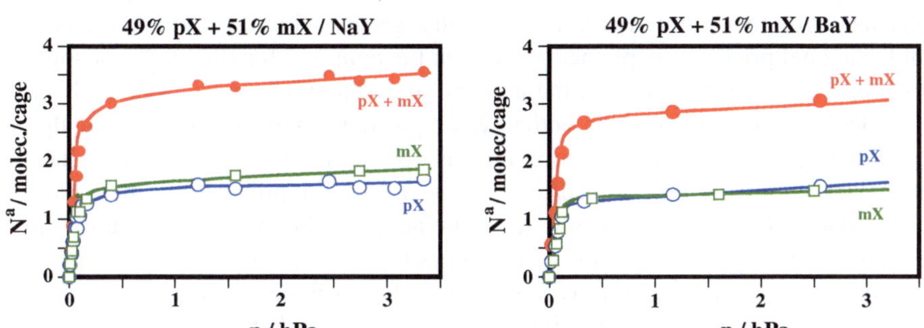

Fig. 7.24 Coadsorption isotherms of p-xylene and m-xylene on NaY and BaY zeolites at 423 K

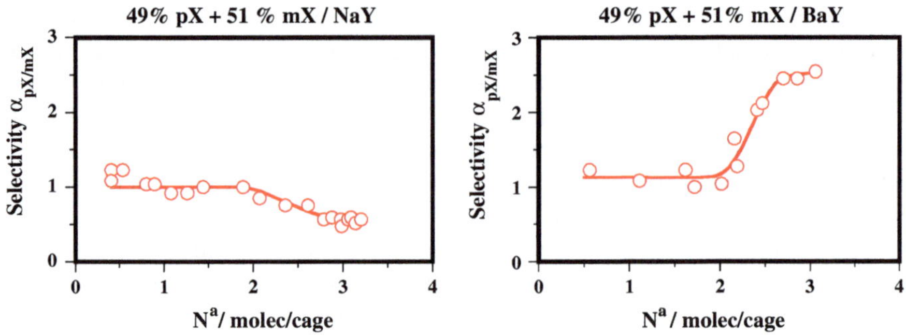

Fig. 7.25 Dependence of the total filling on the adsorption selectivity of p-xylene with respect to m-xylene on NaY and BaY zeolites at 423 K

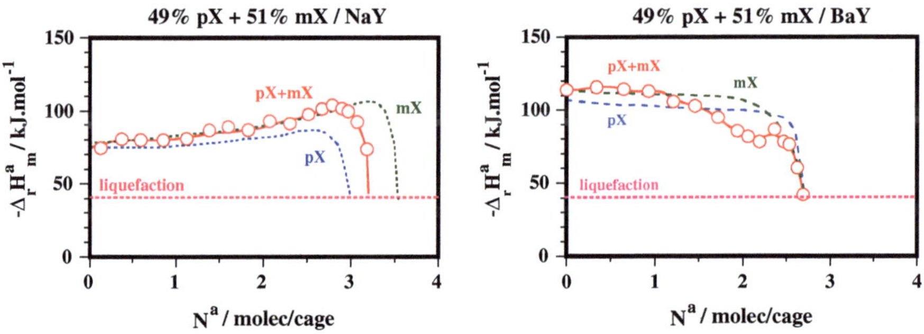

Fig. 7.26 Coadsorption enthalpies of p-xylene and m-xylene on NaY and BaY zeolites at 423 K as a function of the total filling [*Doted lines* adsorption enthalpies of single components]

those obtained with single xylenes. The coadsorption enthalpy is even very close to the adsorption enthalpy of single m-xylene, the isomer, which is adsorbed in favour at high filling. No particular change in the adsorption enthalpy during the filling is observed. The fact that the NaY adsorbs preferentially the m-xylene is certainly due to the adsorbate-adsorbent interactions which are stronger with m-xylene owing to a better orientation of methyl groups with respect to oxygen atoms of hexagonal prisms as explained above. On the contrary, for BaY the coadsorption enthalpy is very different than the adsorption enthalpy of single xylenes. If up to 1 molec/cage the coadsorption enthalpy is equivalent to the adsorption enthalpy of single m-xylene, it decreases during the adsorption of the second molecule. Then it increases slightly above 2 molec/cage, as the adsorption process becomes selective for p-xylene, before to go down to the liquefaction enthalpy when the saturation of supercages is approaching. This result shows that during the coadsorption process, the xylene-zeolite interactions as well as the xylene-xylene interactions are changed compared to the adsorption of single components. The fact that the BaY adsorbs selectively the p-xylene at high filling, while not the NaY, is attributed to steric hindrances caused by the barium cation. Although they are two times less numerous than sodium, the free space available to xylene isomers in the supercages is reduced because of their bigger size and their location. Thus, at high filling of supercages, the last xylene molecule, which is adsorbed is extremely confined and cannot directly interact with a barium cation. Moreover, this last molecule cannot be accommodated in the supercage without molecular rearrangement of the others. These phenomena are accompanied by exo- or endothermic effects, which are at the origin of the changes observed in coadsorption calorimetric curve. In these conditions, this is the isomer having the smallest size and the most symmetric geometry, i.e. the p-xylene, which is preferentially adsorbed. The compensation cation appears as the key parameter. The selectivity depends strongly of its size, number, electric charge and location. In such systems, the selective adsorption process does not result of differences in adsorbate-adsorbent interactions between the isomers. These ones are too much low to induce a competitive adsorption. The selectivity is essentially governed by the entropic effects, consecutive to the steric hindrance caused by the compensation cations and leading to important molecular rearrangement of xylenes confined in the zeolitic cavities.

7.6.2 Desulphurization of Natural Gas by Selective Adsorption on FAU Type Zeolite

This second example of application is taken from a complete thermodynamic study devoted to the adsorption of light mercaptans, paraffins and aromatics on FAU type zeolite [28–30]. The aim of this work was to find a good adsorbent able to perform a deep desulphurization of natural gas by selective adsorption. Results presented below concern the adsorption and coadsorption of ethylmercaptan (ESH), n-heptane (HEP) and toluene (TOL) on the NaX zeolite a 298 K. NaX zeolite has exactly the same structure as NaY excepted that it contains more compensation cations located in sites II and also in sites III around the dodecagonal window (Fig. 7.27).

Adsorption isotherms of single components are shown in Fig. 7.28. In order to compare the three adsorbates, they are represented by using the loading rate θ and the relative pressure p/p_s. In the low relative pressure range, isotherms show that the adsorption affinity increases according to the sequence: ESH > TOL > HEP, which is the same as for the dipolar moment (Table 7.3). Moreover, the adsorption affinity of the NaX zeolite for the sulphur compound is strong enough to consider this adsorbent as a good candidate for a deep desulphurisation of natural gas. Regarding the adsorption capacities, they are very similar. The volume of each molecule adsorbed is very close to the crystallographic volume of supercages (Table 7.3). Figure. 7.29 shows the adsorption enthalpy of ethylmercaptan measured by differential calorimetry and calculated with the isosteric method from a set of isotherms measured at different temperatures (Fig. 7.30). A very good agreement is obtained between the two methods. The filling range investigated with the isosteric method is however limited to moderate fillings, the determination of equilibrium pressure at fixed adsorbed amount on

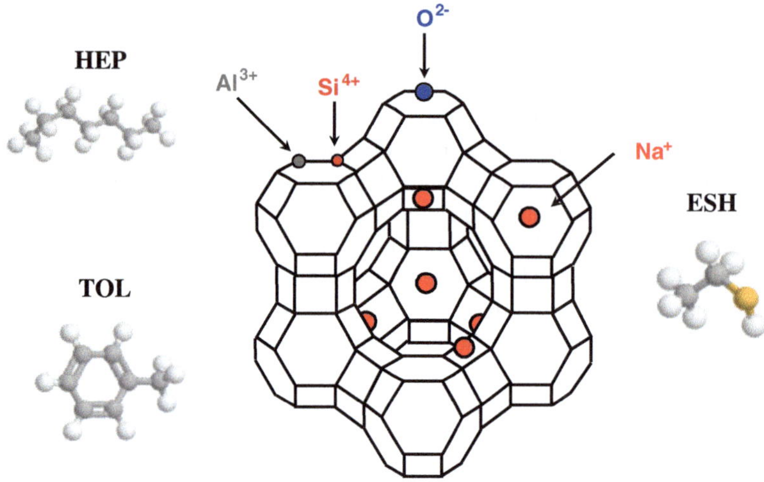

Fig. 7.27 Structure of the NaX zeolite. The compensation cations inside the supercage are located on sites II and III

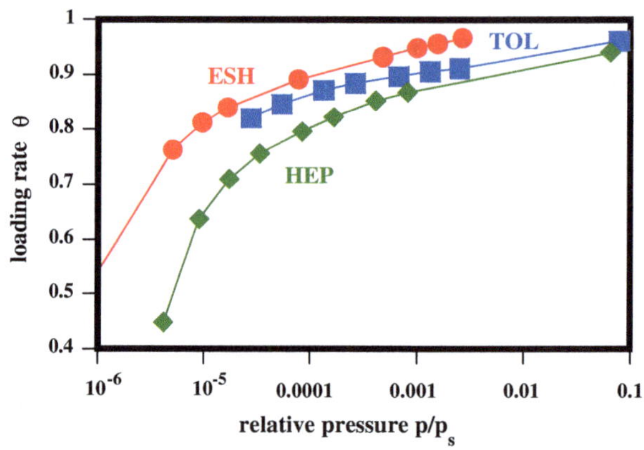

Fig. 7.28 Adsorption isotherms of ethylmercaptan, n-heptane and toluene on NaX zeolite at 298 K

Table 7.3 Dipolar moment and volume adsorbed in NaX at 298 K of ethylmercaptan, n-heptane and toluene

	ESH	TOL	HEP
μ / D	1.58	0.36	0
$V^a/cm^3.g^{-1}$	0.294	0.265	0.270

Crystallographic volume of supercages: $V_\alpha = 0.294\,cm^3.g^{-1}$

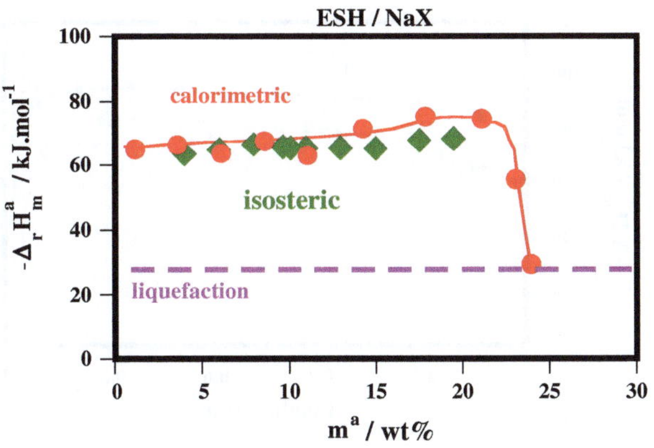

Fig. 7.29 Comparison of adsorption enthalpies of ethylmercaptan on NaX zeolite at 298 K measured by calorimetry *(red points)* and calculated with the isosteric method *(green points)*

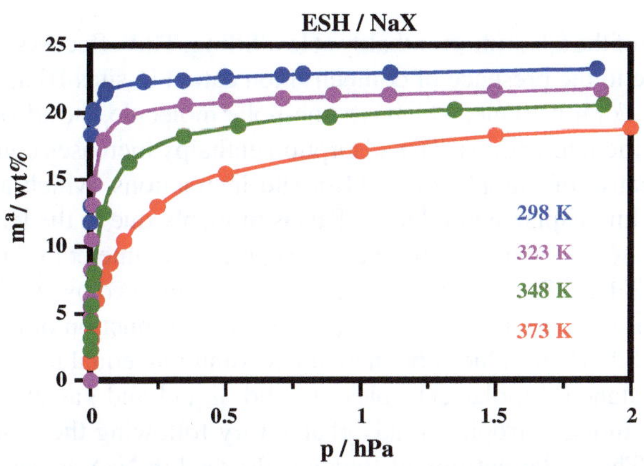

Fig. 7.30 Adsorption isotherms of ethylmercaptan on NaX zeolite measured at different temperatures

the adsorption isotherms being difficult in low and high pressure ranges. Adsorption enthalpies of three single adsorptives are compared in Fig. 7.31. At zero filling the adsorption enthalpies of ESH and TOL are equivalent and slightly higher than with HEP. This is attributed to the fact that ESH and TOL have a dipolar moment contrary to ESH. Their adsorption involves specific interactions. As the filling increases, the adsorption enthalpies of toluene and ethylmercaptan are rather constant. The adsorption process of these two molecules is relatively homogeneous from the energetic point of view. The molecules are in interaction with the compensation cation in the supercages without important interactions between them. The adsorption enthalpy is higher with toluene. This could be explained by a stronger interaction with the compensation cation owing to the aromatic ring. It may be pointed out with this molecule a

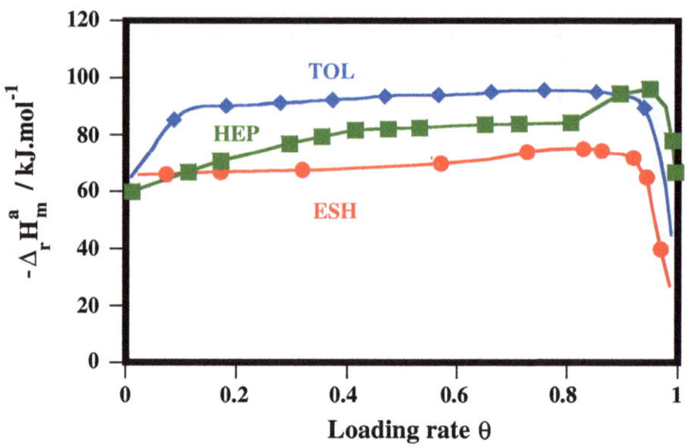

Fig. 7.31 Adsorption enthalpies of ESH, HEP and TOL on NaX zeolite determined by differential calorimetry at 298 K as a function of the loading rate

sharp increase of the adsorption enthalpy at low filling. This effect is still unexplained. It could be due to the presence of compensation cation in sites III at the entrance of the supercage. For n-heptane, which is a non-polar molecule, the adsorption involves only non-specific interactions. The adsorption enthalpy increases continuously with the filling because of the adsorbate-adsorbate interactions, which appear stronger than with ethylmercaptan and toluene. This is probably due to the length of this linear paraffin (1.16 nm), which is close to the supercage diameter (1.3 nm). The molar entropies of ESH, HEP and TOL adsorbed in NaX, calculated by combining isosteric method and calorimetric measurements, are given as a function of the filling coefficient in Fig. 7.32. Their values obtained at a loading rate equal to 0.5 are compared with those of standard molar entropies of solid, liquid and gas phases at 298 K in Table 7.4. The molar entropies of adsorbates vary following the sequence: TOL < ESH < HEP. The molar entropy of toluene adsorbed in NaX is very low. Its value is lower than that one of the solid state. This means that the toluene is extremely confined in the supercage. The molecule is "frozen" on its adsorption site. The molar entropies of ethylmercaptan and n-heptane are higher and lie between the standard molar entropies of the liquid and solid states. These molecules have lost less degrees of freedom and are more flexible and mobile on their adsorption sites than toluene.

The coadsorption of ESH with HEP or TOL has then been studied in order to check if the presence of long paraffins or aromatics in the natural gas can perturb the capture of sulphur compound. Although the study on the adsorption of single components shows that the strongest adsorption affinity of NaX is for ESH, in the case of coadsorption, the behaviour of NaX can be different. The adsorption selectivity of ESH with respect to HEP or TOL as a function of the total filling, for an initial gas mixture containing 25 % of sulphur compound, is given in Fig. 7.33. Similar selectivity curves are obtained for other compositions of gas mixtures. When ESH is in the presence of HEP in the gas phase, the adsorption selectivity is largely higher than one. Whatever the filling, NaX adsorbs preferentially the mercaptan.

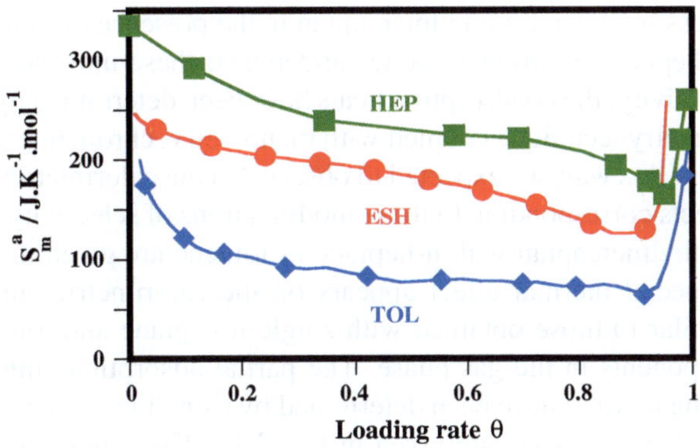

Fig. 7.32 Molar entropies of ESH, HEP and TOL adsorbed in NaX zeolite at 298 K as a function of the loading rate

Table 7.4 Comparison of molar entropies of ESH, HEP and TOL adsorbed in NaX at 298 K at a loading rate equal to 0.5 with the standard molar entropies of solid, liquid and gas states at 298 K

$S/J.K^{-1}.mol^{-1}$	ESH	HEP	TOL
S^a_m at $\theta = 0.5$	128	182	71
S^o_{solid}	70	127	81
S^o_{liquid}	207	305	192
S^o_{gas}	296	428	320

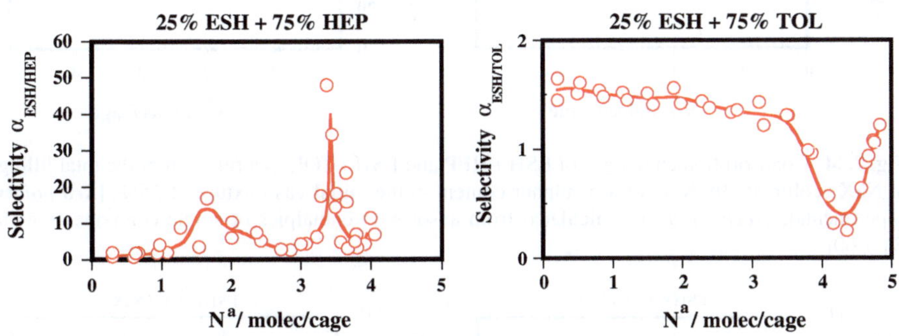

Fig. 7.33 Adsorption selectivity of ESH with respect to HEP or TOL as a function of the total filling of NaX zeolite at 298 K and for a sulphur content in the initial gas mixture of 25 %

However, the selectivity curve exhibits some peaks of selectivity at certain fillings which are less or more intense according to the initial gas composition. When ESH is in the presence of TOL, the selectivity is in favour of the mercaptan in a large range of filling but decreases sharply as the complete filling of supercages is approaching. Then, NaX adsorbs preferentially the toluene and we observe, in the partial adsorbed amounts, a displacement of ESH by TOL. It may be noticed that

NaX is much less selective for ethylmercaptan in the presence of toluene than in the presence of n-heptane. In order to better understand these unusual changes in the adsorption selectivity, the coadsorption heats have been determined by using the differential calorimetry technique coupled with manometry, chromatography and mass spectrometry. By this way, we expected to observe on the calorimetric curve particular thermal effects corresponding to these modifications of selectivity. Coadsorption enthalpies of ethylmercaptan with n-heptane or toluene are given in Fig. 7.34. Surprisingly, no special thermal effect appears on the calorimetric curves. They are rather very similar to those obtained with single n-heptane and toluene which are the major components in the gas phase. The partial adsorption enthalpies of each component in the mixture have been determined by using the "tangent method" with Eqs. 7.101 and 7.102. Results are given in Fig. 7.35. They show that whatever the composition, the partial adsorption enthalpies are very close to those of single component excepted for the toluene for which, the partial molar enthalpy increases slightly when the mixture becomes very reach in mercaptan. As the partial enthalpies are constant and equal to the adsorption enthalpies of single components in a large range

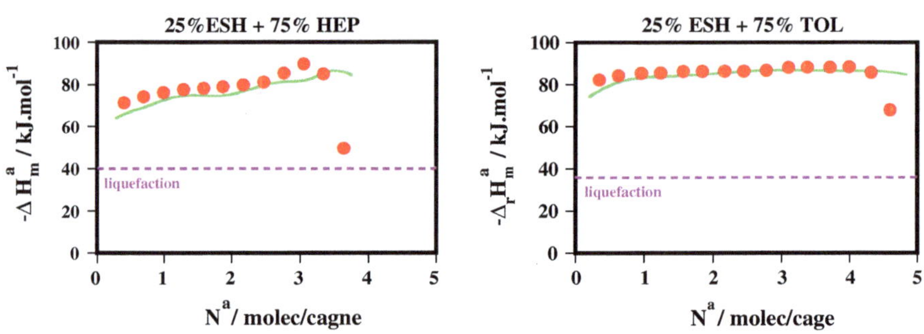

Fig. 7.34 Coadsorption enthalpies of ESH + HEP and ESH + TOL as a function of the total filling of NaX zeolite at 298 K and for a sulphur content in the initial gas mixture of 25%. [*Red points* experimental; *green solid line* calculated from adsorption enthalpies of single components with Eq. 7.60]

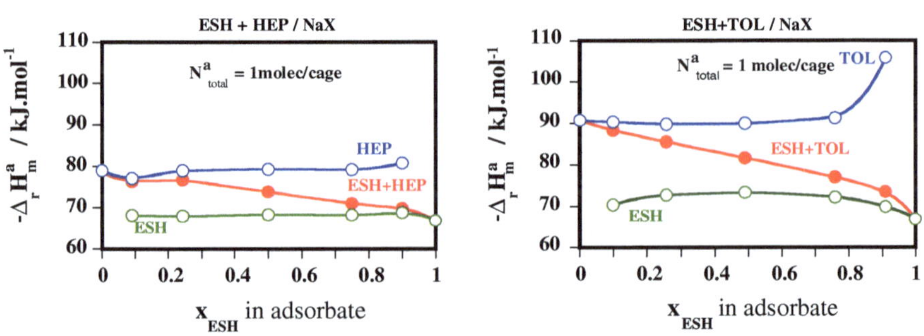

Fig. 7.35 Partial adsorption enthalpies of ESH, HEP and TOL coadsorbed in NaX zeolite at 298 K and for a total amount of mixture adsorbed equal to 1 molec/cage

of composition, the coadsorption enthalpies of ESH + HEP and ESH + TOL can be calculated by the adsorption enthalpies of single components using the Eq. 7.60. As shown on Fig. 7.34, a very good agreement is obtained between the calculation and the calorimetric measurements. Therefore the variations of the adsorption selectivity observed as filling increases are not related to particular thermal effects. At least these effects are not important enough to be detected by our calorimetry technique. These results suggest that the coadsorption affinity is more dependent on the entropic effects than the enthalpic ones. As in the case of the adsorption of xylenes, in the supercages of NaX, the molecules of ESH, HEP and TOL are probably confronted to steric hindrance leading to molecular rearrangements. In these systems the selectivity is also essentially governed by entropic effects.

7.7 Conclusions

The development of new gas separation or purification processes by selective adsorption requires the elaboration of more and more efficient adsorbents. New solids having large adsorption capacity, strong adsorption affinity, good hydrothermal and mechanical stability and high selectivity must be discovered. This imposes to have a detailed knowledge of the adsorption process at macroscopic and molecular levels, especially when the adsorption involves several components. Since the last two decades, the numerical molecular simulation methods (molecular dynamics and Monte Carlo simulations) have shown fantastic advances in the field of gas adsorption in nanoporous solids. It is possible now to predict with acceptable precision the adsorption capacities, energies and selectivities in shorter time than by experiments. Moreover these methods give interesting information at the microscopic level as the location of the adsorbed species, the energy of molecular interactions, the diffusion coefficients that are very difficult or impossible to measure by experiments. Simulations are becoming more numerous than experiments. Nevertheless, adsorption experiments are essential for checking the validity of numerical simulations codes and to find the real physico-chemical phenomena involved in the adsorption process. It is the reason why the experimental approach is essential in any thermodynamic study of gas adsorption. In this field, the differential calorimetry coupled with manometry and gas chromatography or mass spectrometry is a powerful technique. It allows performing a complete and detailed thermodynamic study of gas adsorption on solids. The knowledge of the adsorption enthalpies and entropies are now very useful to design the porous structure and the surface chemical properties of efficient selective adsorbents. This is a complementary technique, which can give rich information even at the molecular level. The use of this technique remains however delicate and is time consuming. It needs a lot of calibrations and adsorption experiments must be repeated several times. The calorimetry requires the greatest care and patience but this technique is really indispensable to have a detailed knowledge of adsorption mechanisms that take place at the gas-solid interface.

References

1. S.J. Gregg, K.S.W. Sing, *Adsorption, Surface Area and Porosity* (Academic Press, London, 1982)
2. K.S.W. Sing, D.H. Everett, R.A.W. Haul, L. Moscou, R.A. Pierotti, J. Rouquerol, T. Siemienewska, Pure Appl. Chem. **57**(4), 603 (1985)
3. F. Rouquerol, J. Rouquerol, K. Sing, *Adsorption by Powders and Porous Solids* (Academic Press, London, 1999)
4. R.M. Barrer, *Zeolites and clay minerals as sorbents and molecular sieves* (Academic Press, New York, 1978)
5. F. Thibault-Starzyk, M.H. Simonot-Grange, Les matériaux micro et mésoporeux—Caractérisation (EDP Sciences, Les Ulis, 2004)
6. P.W. Atkins, *Physical Chemistry* (Oxford University Press, Oxford, 1990)
7. D.D. Do, *Adsorption Analysis: Equilibria and Kinetics* (Imperial College Press, London, 1998)
8. M.H. Simonot-Grange, O. Bertrand, E. Pilverdier, J.P. Bellat, C. Paulin, J. Thermal Anal. **48**, 741 (1997)
9. V. Cottier, J.P. Bellat, M.H. Simonot-Grange, A. Méthivier, J. Phys. Chem. B **101**(24), 4798 (1997)
10. S. Sircar, J. Chem. Soc. Faraday Trans. 1 **81**, 1527 (1985)
11. S. Ross, J.P. Olivier, *On Physical Adsorption* (Interscience Publishers John Wiley & Sons, New York USA, 1964)
12. A. Boutin, G. Dosseh, A. Fuchs, *Eléments de Thermodynamique* (Masson, Paris, 1997)
13. D.B. Broughton, R.W. Neuzil, J.M. Pharis, C.S. Breatly, Chem. Eng. Prog. **66**(9), 70 (1970)
14. R.T. Yang, *Gas separation by adsorption processes* (Imperial College Press, London, 1987)
15. G. Hotier, B. Balannec, Revue de l'Institut Français du Pétrole **46**, 803 (1991)
16. D.W. Breck, *Zeolite Molecular Sieves* (Wiley, New York, 1974)
17. C. Beauvais, A. Boutin, A.H. Fuchs, Chem. Phys. Chem. **5**, 1791 (2004)
18. C. Beauvais, A. Boutin, A.H. Fuchs, C.R. Chimie **8**, 485 (2005)
19. V. Cottier, E. Pilverdier, M.H. Simonot-Grange, J.P. Bellat, J. Thermal Anal. **58**, 121 (1999)
20. J.P. Bellat, M.H. Simonot-Grange, S. Jullian, Zeolite **15**, 124 (1995)
21. J.P. Bellat, M.H. Simonot-Grange, Zeolite **15**, 219 (1995)
22. M. Czjzek, H. Fuess, T. Vogt, J. Phys. Chem. **95**, 5255 (1991)
23. C. Mellot, M.H. Simonot-Grange, E. Pilverdier, J.P. Bellat, D. Espinat, Langmuir **11**(5), 1727 (1995)
24. V. Lachet, A. Boutin, B. Tavitian, A. Fuchs, J. Phys. Chem. **102**, 9224 (1998)
25. V. Lachet, S. Buttefey, A. Boutin, A. Fuchs, Phys. Chem. Chem. Phys. **3**, 80 (2001)
26. V. Cottier, J.P. Bellat, M.H. Simonot-Grange, J. Phys. Chem. B **101**(24), 4798 (1997)
27. J.C. Moïse, J.P. Bellat, J. Phys. Chem. B **109**, 17239 (2005)
28. G. Weber, J.P. Bellat, F. Benoit, C. Paulin, S. Limborg-Noetinger, M. Thomas, Adsorption **11**, 183 (2005)
29. G. Weber, F. Benoit, J.P. Bellat, C. Paulin, P. Mougin, M. Thomas, Microporous Mesoporous Mater. **109**, 184 (2008)
30. F. Benoit, G. Weber, J.P. Bellat, C. Paulin, P. Mougin, M. Thomas, Adsorption **14**, 501 (2008)

Chapter 8
Characterization of Acid–Base Sites in Oxides

Antonella Gervasini

Abstract The acidity and basicity of the oxides and metal oxide surfaces is one among the most peculiar properties of this group of inorganic products that have been exploited in catalysis from decades. The different types of oxides (single oxides, doped and modified oxides, supported oxides, mixed and complex oxides) are here described with emphasis on their acid-base functionalities in relation with the catalytic activity. The prediction of the acidity insurgence from the oxide composition is discussed. The possibility to have a deep knowledge of the amount of acid sites and of the acid strength and strength distribution at the oxide surface is here presented. Adsorption by suitable probes with basic/acidic characteristics can be realized in a microcalorimetry of adsorption giving information on the heterogeneity of the surface. Alternatively, probe desorption can be realized by a temperature increasing program to obtain the acid-site strength distribution. In any case, quantitative information on the acid oxide surfaces can be obtained with the relevant thermodynamic of kinetic parameters which determine the amount and strength of the acid sites.

8.1 Introduction

Oxides are relevant products of the inorganic chemistry that are extensively used in many applied fields such as adsorption technology, pigment technology, ceramics, and heterogeneous catalysis. In any applied field, the oxide functionality is governed by surface interaction with the environment. In particular, the interactions of the oxide surfaces with gases and liquids, active in the fields of adsorption and catalysis, are mainly governed by the acid-base properties of the oxide. These properties are tightly connected to the catalytic functionality of several divided, wide surface area oxide surfaces. Acid oxide catalysts have great importance in petroleum chemistry

A. Gervasini (✉)
Dipartimento di Chimica, Università degli Studi di Milano, Milan, Italy
e-mail: antonella.gervasini@unimi.it

A. Auroux (ed.), *Calorimetry and Thermal Methods in Catalysis*,
Springer Series in Materials Science 154, DOI: 10.1007/978-3-642-11954-5_8,

(e.g., amorphous silica-alumina cracking catalyst) and in any reaction involving hydrocarbons and they have been extensively investigated, while basic oxide catalysts have been studied to a lesser extent. The nature of the acid and base sites on oxide surfaces is an intriguing and moot subject. Surface exposed unsaturated cations and hydroxyl groups give rise to Lewis and Brønsted acid sites, respectively, while basic sites could be associated with surface oxygen atoms able to interact attractively with a proton. Severe pretreatment conditions of the oxides are needed for generating active basic sites by removal of carbon dioxide, water, and in some cases, oxygen from the surface.

Among the various types of heterogeneous acid and basic catalysts that found real application in various reactions and catalytic processes (Table 8.1, [1–4]), various types of oxide structures are found. The surface properties, nature and structure of acid and base sites of oxides have been investigated for many decades by various approaches, comprising spectroscopic, thermal, and theoretical studies, and newly developed measurement methods using modern techniques are nowadays used for this purpose. Other studies are aimed at developing novel acid and base oxide formulations with improved functionality. In the following, the acidity-basicity of oxide materials of catalytic importance by adsorption calorimetry by conventional and novel methodologies of study is presented. Beside the conventional gas-solid adsorption calorimetry studies, liquid-solid adsorption calorimetry has been developed to disclose sound relations between acid-base properties and activity of catalysts working in liquid-solid heterogeneous processes.

8.2 The Surface Acido–Basicity of Metal Oxides

To introduce the concept of acidity and basicity of surface sites of any solid, it is necessary to start with the very general definition of acids and bases. Two different definitions are used today; they are the most relevant and accepted ones among all the others. Brønsted and Lowry in 1923 defined that an acid (HA) is any hydrogen-containing species able to release a proton and a base (B) is any species capable of combining with a proton. In this view, the acid-base interaction consists in an equilibrium exchange of a proton and the final product of the exchange is the conjugated base (A^-) and acid (BH^+), respectively. Alternatively, Lewis proposed in the same year (1923) a different view to define the acidity-basicity concept. An acid (A) is any species possessing an incomplete electronic grouping that can accept an electron pair to form a dative or coordination bond (A^{δ^-}). A base (B) is any species possessing a non-bonding electron pair that can be donated to form a dative or coordination bond ($B^{\delta+}$). This second definition is more general than the first one and can be applied to any species, even if hydrogen is not concerned in the exchange. The Brønsted and Lewis acid and basic strength can be quantitatively determined. According to the Brønsted theory, pK_a measures the position of the equilibrium in water solution or in gas-phase so allowing the quantitative determination of the relative strength

Table 8.1 Types of heterogeneous acid and basic catalysts [1–4]

Acid catalyst		Basic catalyst	
(1)	Alumina chloridrated alumina sulfated alumina	(1)	Single component metal oxides alkaline earth oxide alkali metal oxide rare earth oxide
(2)	Silica silicalite-1 solid phosphoric acid (kiselghur)	(2)	Zeolites alkali ion-exchanged zeolite alkali ion-added zeolite
(3)	Silica-alumina …aluminated silica	(3)	Supported alkali metal ions alkali metal ions on alumina alkali metal ions on silica alkali metal ions on alkaline earth oxide alkali metals and alkali metal hydroxides on alumina
(4)	H-zeolites	(4)	Clay minerals Hydrotalcite chrysotile sepiolite
(5)	SAPOs (silicoaluminophosphates)	(5)	Non-oxide KF supported on alumina Lanthanide imide and nitride zeolites
(6)	Pure/mixed/supported metal oxides (titania, zirconia, tungsta, molibdena, etc.)		
(7)	Zirconia sulfated zirconia tungstated zirconia		
(8)	Resins Sulfonic acid resins		
(9)	Clays acid-treated montmorillonite		
(10)	Niobic acid and niobium phosphate		
(11)	Heteropolyacids		

of the acids and bases. The pK_a of the various acids/bases is usually reported with respect to water, while in the gas-phase, where solvation effects are absent, the acidity strength scale of given acids can be greatly different. If HA bond is fully ionic, the equilibrium of the acid-base couple is modified and now the *Hammett acidity function* (H_o) is introduced. H_o holds in concentrated solutions where the pH concept is no longer applicable. The *Hammett* function is determined by the extent to which an indicator (weak or very weak organic base) is protonated, by measuring the $[BH^+]/[B]$ ratio by UV-vis spectroscopy. The acid strength concept of the Lewis acids is more difficult to be quantified. A simple scale of acid strength for Lewis acids is not feasible because in this case the acidity depends upon the nature of the base involved in the acid-base equilibrium. Only few quantitative measurements of acid strength are available compared to Brønsted acids.

Other definitions of the acidity concept have been proposed in the literature; the concept of "*Hard and Soft Acids and Bases (HSAB)*" initially proposed by R. Pearson in 1963 [5] and extended by G. Klopman merits to be mentioned. Pearson classified Lewis acids and Lewis bases as *hard*, *borderline*, or *soft and* G. Klopman attempted to quantify Pearson's HSAB principle using frontier molecular orbital (FMO) theory [6]. *Hard* acids are defined as small-sized, highly positively charged, and not easily polarizable electron acceptors. *Hard* bases are species holding their electrons tightly as a consequence of large electronegativities, low polarizabilities, and difficulty of oxidation of their donor atoms. *Hard* acids prefer to associate with *hard* bases, while *soft* acids prefer to associate with *soft* bases. The formed complexes are thermodynamically more stable than mixed complexes and also form faster. According with this theory, protons and hydroxides and oxide ions are all hard species, as are most O- and N-terminated anions. C- and S- terminated anions, sulphides, phosphines, and aromatic hydrocarbons are soft bases.

It may be taken into consideration that the proton affinity, *PA*, of a anion or of a neutral atom or molecule is a measure of its gas-phase basicity. It is the energy released in the following reactions: $A^- + H^+ \rightarrow HA$ or $B + H^+ \rightarrow BH^+$. For any species, the higher the proton affinity, the stronger the base and the weaker the conjugate acid in the gas phase. Recently, the relationships between the proton affinity and atomic charge and with the transfer of charge have been reexamined [7].

When getting on to the acidity-basicity of solids, we face with a different problem. The quantitative concepts that hold in dilute and concentrated solutions (pK_a and H_o) cannot be extended to solid surfaces. In general terms, a solid acid may be viewed as a surface on which a base is chemically adsorbed and, if the base acts as an indicator, its color changes upon adsorption. Following both the Brønsted and Lewis definitions, a solid acid shows a tendency to donate protons or to accept electron pairs, respectively, while a solid base should tend to accept protons or to donate electron pairs. For a complete evaluation of the solid acidity/basicity, the knowledge of the nature, amount and strength of the acid/base sites is of crucial importance, besides the knowledge of the distribution and density of the sites in nature and strength. This last point is of great importance when the solid acts as catalyst since the catalyst activity mainly depends on the amount of active sites but the catalyst selectivity may be different as a function of the site nature and their distribution and density on the surface. The amount of acid/base on a solid is usually expressed as the number of moles of acid/base sites per unit mass or per unit surface area of the solid, and it is obtained by measuring the amount of a base which is adsorbed on it. The extent of the adsorption reaction can be studied with different analytical techniques, among these UV-vis spectroscopy, volumetric and gravimetric apparatus. Others techniques of study have to be selected to investigate on the nature, strength, and distribution of the acid/base sites. The need for integrated approach of study clearly emerges to successful face the solid acidity/basicity question.

Confining our interest on metal oxides, knowing that they are solid compounds of oxygen and metal or semimetal elements, their surfaces are constituted by positive and negative arrays of coordinated ions which are bound through ionic or covalent bonds, depending on the nature of the cations and structure of the oxide. The positive and negative centers can behave as Lewis acids and Lewis bases that con be described as a bidimensional organization of AB acid-base pairs $(AB)_n$. Above that, hydroxyl groups variably cover any real oxide surfaces, as these groups are potential proton donors or acceptors, it is evident that Brønsted acidity and basicity should be considered to appear on such surfaces. Of course Lewis and Brønsted species are often simultaneously present on the same surface where they form densely populated patches. Consequently due to the complexity of any real metal oxide surface, the quantitative study of their acid-base properties is not a trivial matter and it deserves considerable work.

8.3 Acid, Basic, and Amphoteric Oxides

Before providing an overview of the acid-base properties of metal oxide surfaces, a brief systematic description of their bulk and structural properties and surface chemistry is needed. According to basic inorganic chemistry concepts, the oxides of non-metals as well as the oxides of the metals in very high oxidation states are defined as acidic oxides and *anhydrides*, respectively, the oxides of metals as basic oxides, and the oxides having both acidic and basic characters are denoted as amphoteric.

Basic oxides possess low-valency metals (oxidation number lower than or equal to +4) and they are typically ionic compounds obtained in crystalline forms. In such oxides the coordination of the cations (from 4 to 8) is generally higher than the valency (from 1 to 4) and the same occurs for the coordination of the oxide ions (from 3 to 6). The basic nature of the ionic oxides is associated with the strong polarization of the metal-oxygen bond and with the formation of basic products (metal hydroxides) formed from the reaction with water. Examples of relevance in catalysis are: Cu_2O, CuO, Ag_2O, MgO, BaO, MnO, FeO, La_2O_3, etc. They found application in various base-catalyzed reactions (alcohol dehydrogenation, olefines, aromatic carboxylic acids hydrogenation, amination of primary and secondary amines, Meerwein-Ponndorf-Verley reduction, side-chain alkylation of aromatics, aldol addition and condensation, Tischenko reaction to produce esters by aldehydes dimerization, Michael addition, Knoevenagel condensation, etc.) [2].

Acidic oxides are oxides of non-metal elements with high electronegativity characterized by a nearly covalent bond between the non-metal and oxygen. In normal conditions, some of these oxide structures are gaseous or liquids while others are solids (e.g., Se_4O_{12}, P_4O_{10}). Decreasing the electronegativity of the element, covalent network structures are formed (e.g., SiO_2, B_2O_3). In this case, the covalent nature of the non-metal oxides is evidenced by the coordination number of the non-metal element which corresponds to its valency or even lower and coordination of oxygen is two or one. They can form crystalline phases but they usually exist in very stable amorphous states that are utilized for wide applications (e.g., silica) and convert into stable crystalline phases after treatment at high temperatures. The acid nature of these oxides is associated with the acidity of the products of their reaction with water: oxo-acids. Different types of acid oxides, denoted as anhydrides, are constituted of metal elements in very high oxidation states (e.g., Mn_2O_7, CrO_3, MoO_3, WO_3, Sb_2O_5, Nb_2O_5, etc.). The very high electronegativity of the metal imparts them nearly the same properties of non metal oxides. Concerning the structure of these oxides, ii depends on the position of the metal in the period (from group 4 to group 7); network structure, as TiO_2, layered structure as V_2O_5, linear polymeric structure, asCrO_3, and molecular structure, as Mn_2O_7. These metal oxides react with water giving rise to acid and polyoxoacid products.

Some oxides exhibit both acidic and basic properties, that is, they have amphoteric properties. In a strongly acidic environment, they act as bases while in a strongly basic environment they act as acids. In general the electropositive character of the oxide central atom determines whether the oxide will be acidic or basic. The more electropositive the metal is the more basic the oxide will be and the more electronegative the central atom is, the more acidic oxide will be. Electropositive character increases from right to left across the period and increases down the column for each period column. A resultant borderline between basic and acidic oxides occurs along the diagonal of the period including oxides of Zn, Al, Ga, Sb, etc. Alumina represents the most well-known example of amphoteric oxide able to react with acids and d bases:

$$Al_2O_{3(s)} + 6\ HCl_{(aq)} = 2\ Al^{3+}_{(aq)} + 6Cl^{-}_{(aq)} + 3\ H_2O_{(aq)} \quad (8.1)$$

$$Al_2O_{3(s)} + 2\ NaOH_{(aq)} + 3\ H_2O_{(aq)} = 2\ Na^{+}_{(aq)} + 2\ Al(OH)^{-}_{4(aq)} \quad (8.2)$$

Taking into account a given series of metal oxides of different acid-base properties, including Al_2O_3, BeO, CaO, Cr_2O_3, Ga_2O_3, La_2O_3, MgO, MoO_3, Nb_2O_5, Nd_2O_3, Pr_6O_{11}, SiO_2, Ta_2O_5, ThO_2, TiO_2, V_2O_5, WO_3, ZnO, ZrO_2 [8, 9], those able to adsorb at their surface both carbon dioxide (an acid molecular probe) and ammonia (a basic molecular probe) with evolution of an appreciable amount of heat, typically in the range of chemisorbed reactions (from 50 to 200 $kJ\cdot mol^{-1}$) are classed as amphoteric oxides, while if only ammonia or carbon dioxide are retained on a given surface, the complete acidic or basic character of the oxide, respectively, can be evinced. Figure 8.1 shows a quantitative description of the acid base characteristics of the above cited series of oxides by using a three-dimensional picture in which each experimental point of each curve refers to three axes: the amount of adsorbed ammonia and the amount of adsorbed carbon dioxide (expressed per unit surface area of oxide sample), and the relevant evolved heat of adsorption (expressed by the differential heat, $kJ\cdot mol^{-1}$). The pure acidic oxides are settled completely in the right hand side plane of the diagrams, the basic ones which reacted noticeably with carbon dioxide only stayed completely in the left hand side of the diagrams, and the amphoteric oxides can be found in the middle of the diagram planes with their experimental points in the forehead space. For instance, ZnO showed high acidic tendency

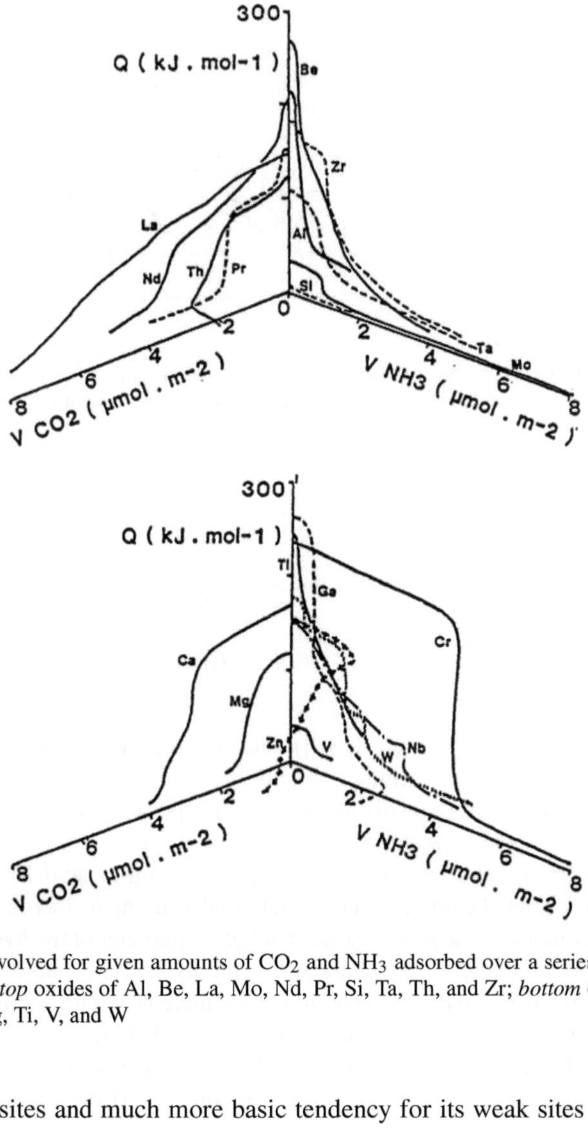

Fig. 8.1 Heat evolved for given amounts of CO_2 and NH_3 adsorbed over a series of metal oxides (from Ref. [9]); *top* oxides of Al, Be, La, Mo, Nd, Pr, Si, Ta, Th, and Zr; *bottom* oxides of Cr, Ca, Ga, Mg, Nb, Mg, Ti, V, and W

for its strong sites and much more basic tendency for its weak sites (starting from $100\,kJ{\cdot}mol^{-1}$).

8.4 Heterogeneous Character of Oxides

The heterogeneous character of oxides is due to (i) irregularity in the surface crystallographic structure (different types of crystal planes, growth steps, crystal edges and corners, Fig. 8.2); (ii) presence of various functional groups (M–O$^-$, M–O–M, M=O,

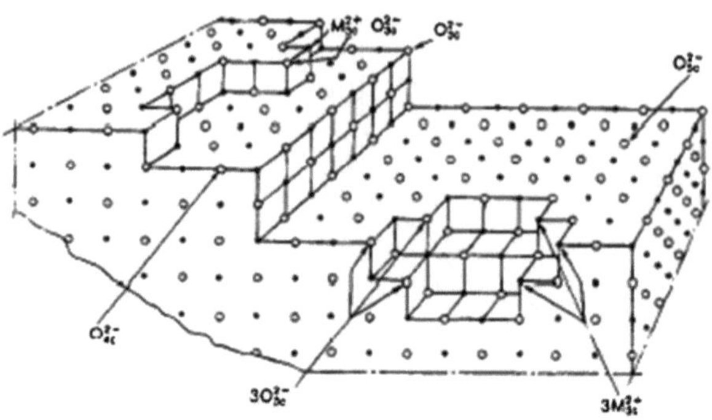

Fig. 8.2 Schematic representation of a real oxide particle with irregularity in crystal planes, growth steps, edges, and corners which causes the presence of surface species with different coordinative unsaturation (from Ref. [12])

Scheme 8.1 Hydroxyl groups of metal oxide surfaces [11]: Type 1, mono bridging hydroxyl group; Type II, doubly bridging hydroxyl group; Type III, triply bridging hydroxyl group

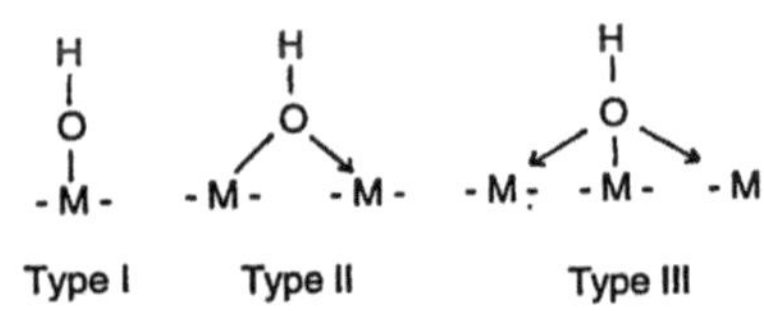

$(M)_x$–OH, M(–OH)$_2$, M–[ ], etc. [8, 10]); and (iii) presence of impurities strongly bounded to the surface. Concerning the acid and basic functionalites of any oxide surface, different species can be distinguished: (i) exposed oxide species (mono-oxo or di-oxo surface species), potentially acting as basic oxides; (ii) exposed hydroxyl groups arising from water adsorption, potentially acting as acid Brønsted sites or basic sites; (iii) exposed cationic centers, potentially acting as Lewis acid sites; and (iv) cationic or anionic vacancies. The acid or basic character of the hydroxyl species is dependent on the nucleophilic character of the hydroxyl group that is governed by the M–OH bond strength; the proton acidity increases with the increase of the coordination number of the OH group (Scheme 8.1 [10, 11]).

Other than the intrinsic heterogeneous character of any surface, it is possible to induce surface heterogeneity by several actions as (i) the mode of formation of the solid (e.g., by precipitation from liquid phase, flocculation of colloidal suspension, sintering of powder, partial/total re-crystallization, sputtering/condensation from vapor phase, surface chemical reaction, etc.); (ii) the thermal treatment of the solid; in fact, heating causes roughness of initially smooth surface by atom displacement and migration to sites of higher energy on top of the surface; and (iii) the introduction of additional phase (active phase or promoter) that imparts functionality to the surface.

The complex picture of oxide surfaces can be studied from both qualitative and quantitative point of view by spectroscopic techniques, conventionally infrared (FT-IR) but also Raman and NMR spectroscopies [11, 13–15], spectrophotometric techniques, in particular XPS [16, 17], by directly observing the surface and by the help of suitable probe molecules adsorption [2, 4, 18–21] which give the possibility to study the molecular adduct formed between the acid/base site and the probe [2, 10, 11]. Reasons for the choice of one or another probe are essentially linked with their size (i.e., kinetic diameter) and basicity [10, 22].

The most used molecular probes are among different chemical families: cyclic amines (piperidine), alkyl amine (n-butylamine), heterocyclic amines (pyridine, 2-phenylethylamine), phosphines (trimethyl-phosphine), ammonia, carbon monoxide. The probe basicity can be expressed as pK_a of the conjugated acid and as proton affinity (PA). The former parameter, pK_a, is relevant with respect to acid-base interactions occurring in water solution, also implying solvation with water of both the base and its conjugated acid. The latter parameter, PA, is more appropriate for gas-phase interactions. The two basic scales are not coincident, for example ammonia is associated with a higher pK_a and a lower PA than pyridine: This suggests that ammonia is more basic than pyridine in water but it is less basic than pyridine in gas phase. The inversion of the relative acidity in water with respect to the gas phase observed for the ammonia and pyridine probes is due to the greater solvation of ammonium ion in water compared with the pyridinium ion. At last, the choice of the probe is also dependent on the analytical technique and the method used to study the adsorbed molecular acid-base complex formed. For example, a high molecular mass probe is preferred when the acid-sites of a surface are determined by gravimetric analysis (by adsorption or thermal desorption), in this case 2-phenylethylamine is a better choice than pyridine [23, 24].

The quantitative evaluation of the amount of acid/base sites and relative acid/base strength of a surface can be more easily determined by gravimetric, volumetric, microcalorimetric, and thermal techniques (Temperature Programmed Desorption, TPD, among others) [8, 24–35]. Moreover, calorimetric and thermal techniques are able to provide quantitative information on the strength of interaction between the acid/base site and the probe, that is, the energy of the bond formed, through the direct determination of the heat of adsorption or desorption, depending on the experimental method adopted for the measurements. Such information let us to classify the acid/base sites of any solid following a scale of acid/base strength with important relationship with the solid reactivity [8, 36, 37]. When a heterogeneous surface is concerned, the acid/base site strength distribution can be determined by mathematical analysis of the experimental data obtained from adsorption and desorption thermal techniques [30–32, 34, 38]. In this context, the studies of adsorption calorimetry can give an estimation of the energy distribution of acid or base sites of heterogeneous surfaces by the direct inspection of the collected calorimetric profiles which show a more or less pronounced decreasing trend of the differential heat values as a function of the degree of coverage of the used probe (q_{diff} versus adsorbed moles). The decreasing trend reflects heterogeneity of the surface sites but also induced heterogeneity, due to lateral interactions between the adsorbed molecules [38, 39], leads to the observed decreasing trends, so making complicated the quantitative evaluation of the energy distribution of any surface. Following a common and simplified viewing, the different types of site are assumed to be covered by the probe successively starting from the strongest ones. This assumption can hold only if large differences among the adsorption constants are involved [40, 41].

Carniti and Gervasini [42] presented the possibility to study the heterogeneity of oxide surfaces by performing adsorption calorimetry experiments, by using a suitable base probe, at different adsorption temperatures. If on a same sample, adsorption calorimetry measurements are performed at different temperatures, different distributions of sites could be observed [42–44]. If no modification of the surface occurs in the temperature range considered, the different observed distributions can be due to either kinetic or thermodynamic effects. Regarding the kinetic effects, at low adsorption temperature, the strongest sites may not be activated or, even if sufficiently activated, may be covered by the probe at an adsorption rate too low to achieve equilibrium during the experiment. Moreover, surface diffusion could not be accomplished if enough time does not elapse in the equilibration. When the temperature and the time employed in the measurements ensure that the adsorption equilibrium has been obtained on all sites, the differences among the observed distributions have to be ascribed to the influence of temperature on the thermodynamic parameters of adsorption, particularly on the adsorption constants. Accurate adsorption calorimetric measurements carried out at different temperatures can give the *true* site distribution; besides, the thermodynamic parameters of adsorption can be obtained as functions of temperature. The *true* distribution can be different from the observed one obtained by using the above-mentioned oversimplification from a single adsorption calorimetric curve. Concerning the temperature range employed, temperatures should be sufficiently high that the adsorption process is activated on all types of site, but low enough that the adsorption equilibrium constants are favorable.

Once obtained the data (sets of adsorption and calorimetric isotherms collected at different temperatures, n and Q, adsorbed moles and evolved heat, respectively), a computation fitting procedure has to be applied [42]. In the model, the adsorption isotherms are interpreted as summations of single Langmuir isotherms, each of them relevant to the sites of a definite type (with $n_{max,i}$ representing the maximum number of moles adsorbed on sites of ith type and having a defined adsorption enthalpy with the probe, $\Delta_a H_i$). The temperature dependence of the Langmuir adsorption constant of each site-type (b_i) was obtained through the integration of the vant'Hoff differential equation, considering the adsorption enthalpy either independent or dependent on temperature by introducing the heat capacity of adsorption ($\Delta_a C_{p,i}$). On the basis of the experimental values of n and Q obtained at the investigated temperatures as a function of equilibrium pressure and surface coverage, respectively, the characteristic parameters of adsorption ($n_{max,i}$, $\Delta_a H_i$, number of acid sites and adsorption enthalpy for each type of ith site, and any other parameters related to them) can be evaluated by a computational program including an optimization subroutine. When the vant'Hoff differential equation is considered,

$$b_i = b_{0,i} \cdot \exp(-\Delta_a H_i/RT) \quad \text{or} \quad b_i = b_{0,i} \cdot \exp(-\Delta_a H_i/RT + (\Delta_a C_{p,i}/R) \cdot \ln T) \tag{8.3}$$

the $b_{0,i}$ pre-exponential parameter has to be optimized too, in order to obtain the energy distribution of the surface, that is $n_{max,i}$ versus $\Delta_a H_i$. The validity of the $b_{0,I}$ parameter is difficult to evaluate because it does not have a simple physical meaning and the range of its numerical values is far from being clearly defined. A new parameter [45] related to $b_{0,i}$ and defined as *half-coverage temperature at unit pressure*, $T^0_{1/2,i}$, can be optimized instead of $b_{0,i}$. The new parameter corresponds to the temperature at which the adsorption constant, b_i, becomes unity. In the case of a Langmuir isotherm, this condition corresponds to half-coverage; therefore the sites of *i*th type are half-covered at $T = T^0_{1/2}$. Then $b_{0,I}$ can be easily calculated from the optimized $T^0_{1/2,i}$, values.

$$b_i = b_{0,i} \cdot \exp(-\Delta_a H_i / R \cdot T^0_{1/2,i}) = 1 \quad \text{or} \quad b_i = b_{0,i} \cdot \exp(-\Delta_a H_i / R \cdot T^0_{1/2,i} + (\Delta_a C_{p,i}/R) \cdot \ln \cdot T^0_{1/2,i}) = 1 \tag{8.4}$$

$$b_{0,i} \cdot = \exp(\Delta_a H_i / R \cdot T^0_{1/2,i}) \quad \text{or} \quad b_{0,i} = \exp(\Delta_a H_i / R \cdot T^0_{1/2,i} - (\Delta_a C_{p,i}/R) \cdot \ln \cdot T^0_{1/2,i}) \tag{8.5}$$

Following the approach above described, the energy distribution of several acidic simple metal oxides has been determined by using ammonia as probe [42]. The obtained total number of acid sites ($n_{max} = \Sigma_i n_{max,i}$) allowed to disclose the following acidity scale: WO_3 (5.55 μmol/m^2) > ZrO_2 (5.50 μmol/m^2) > Nb_2O_5 (4.76 μmol/m^2) < BeO(4.56 μmol/m^2) > TiO_2 (3.20 μmol/m^2) > Al_2O_3 (1.98 μmol/m^2) (Fig. 8.3). No more than three or four different types of site were found for each oxide. All the oxide surfaces showed a significant quantity of sites

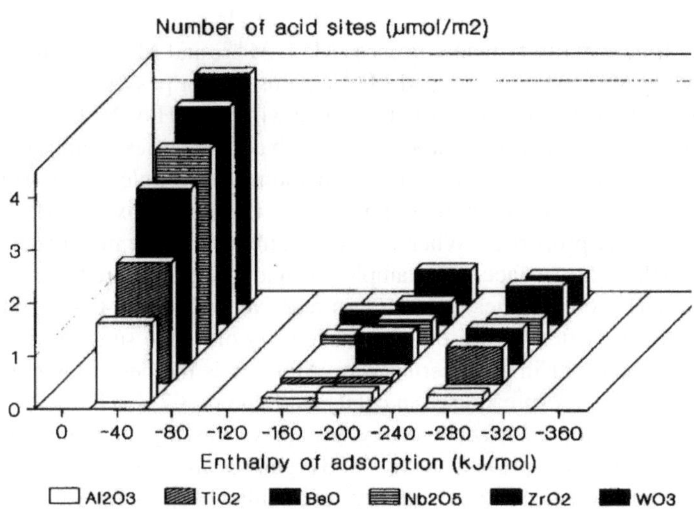

Fig. 8.3 Acid site energy distribution for several oxides of catalytic relevance (from Ref. [42])

(70–80 %) associated with $\Delta_a H_i$ of −40 kJ/mol, corresponding to hydrogen-bonded ammonia (the weakest mode of interaction). The other 20–30 % of sites was distributed in a $\Delta_a H_i$ range of −160 to −280 kJ/mol, corresponding to more energetic interactions of ammonia with the sites.

8.5 Single Oxides, Doped and Modified Oxides, Supported Oxides, Mixed Oxides, and Complex Oxides

In catalysis, oxides with well defined acidic and basic properties are used in different forms that have found application in numerous catalytic applications in the gas-solid and liquid-solid heterogeneous catalysis [3, 46, 47]. Among the most used oxide materials in catalysis, we find: (i) bulk oxides (one component metal oxides); (ii) doped and modified oxides; (iii) supported metal oxides (dispersed active oxide component onto a support oxide component); (iv) bulk and supported binary metal oxides to quaternary metal oxides (mixed oxide compositions); (v) complex oxides (e.g., spinels, perovskites, hexa-aluminates, bulk and supported hydrotalcites, pillared clays, bulk and supported heteropolyacids, layered silicas, etc.).

One component metal oxides can crystallize with different morphologies (isotropic, anisoptropic, or remain amorphous) and local coordination. All one component metal oxide phases crystallize at high temperatures, providing detectable crystalline XRD spectra, and many phases can remain amorphous at moderate calcinations temperatures (SiO_2, Al_2O_3, Nb_2O_3, MoO_3, etc.). Their surfaces terminate with various functionalities; for example, amorphous silica terminates with isolated Si–OH, hydroxylpairs of $(Si–OH)_2$, and $Si(–OH)_2$, bridging Si–O–Si bonds [48]; niobia terminates with Nb–OH, Nb–O–Nb, and Nb=O bonds [49, 50]. In general, surfaces terminating with M=O functionality typically possess M^{+7}, M^{+6}, and M^{+5} cations, while surfaces showing M^{+4}, M^{+3}, M^{+2}, and M^{+1} cations do not possess enough electrons to form terminal bonds and typically terminate with M–OH or M–O–M bonds. The literature reports several investigations on acid-base properties of amorphous metal oxides by adsorption calorimetry in research and review articles [8, 9, 41].

The presence of the different functionalities on the various oxide surfaces leads to different reactivity properties. When molecules, metal salts, or metal complexes are adsorbed on the SiO_2 surface, for example, the anchoring sites are the more reactive isolated Si–OH groups rather than the less reactive bridged Si–O–Si groups. Acid and base sites can often be present simultaneously on oxide surfaces and they can work independently or in a concerted way. Alumina is the best known example of amphoteric oxide widespread used in catalysis (Scheme 8.2) [51].

In this case, multiple surface hydroxyl groups varying by the number of Al sites and Al coordination are present as well as oxygen vacancies of defects, such determining the characteristic surface chemistry of alumina [48]. As for SiO_2, the isolated Al–OH sites are the most reactive towards molecules, metal salts, and metal complexes that have to be adsorbed on the alumina surface, these sites are basic, for

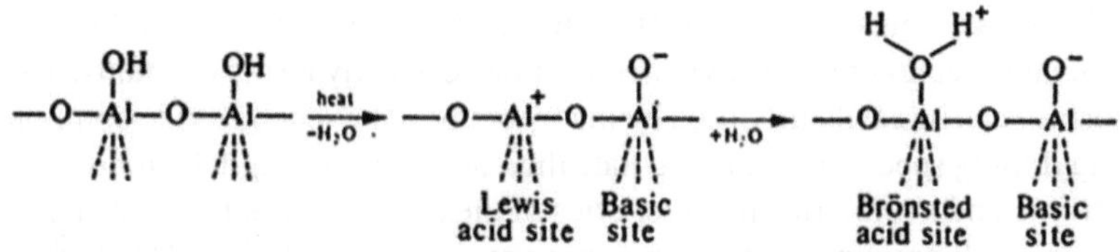

Scheme 8.2 Picture of acid and base sites creation on the alumina surface (from Ref. [51])

the most part, while the bridging Al–OH–Al sites are the most acidic ones, and the Al–[ ] defects possess Lewis acidic character.

There are some other metal oxides crystallize with an anisotropic morphology, such as platelets [52]. Both MoO_3 and V_2O_5 crystallize with platelet morphology. In these structures, the terminal M=O and bridging M–O–M groups are present on the basal planes and the terminal M–OH predominate on the edge plains. Concerning surface reactivity of such surfaces, it was demonstrated that the MoO_3 basal planes do not chemisorb methanol while the flat portions of surface containing Mo–OH sites on the edges react with methanol to form Mo–OCH_3 and water [53]. The same behavior was observed for V_2O_5 and ZnO when exposed to methanol [54–56]. On such materials, the number of catalytically active sites, i.e., acid sites, are only a small fraction of the total exposed metal oxide sites and we assist to structure sensitivity of active sites to different crystalline planes.

The recent discovery of mesoporous silica or in general mesoporous molecular sieves (MMS) has attracted a great deal of attention [57, 58]. The adjustable porosity of silica-based MMS allows large reactant molecules to penetrate into the internal void space to be processed at the active-acid sites and then diffuse out freely as products. Because of the low acidity of the silanol groups of such materials non-silica mesoporous transition-metal oxide materials have been recently prepared for catalytic purposes.

Many early transition metal oxides possess high acidity, especially in sulfated and phosphated forms; titanium, niobium, and tantalum oxides demonstrated very high activity for large series of acid-catalytic reactions such as benzylation, alkilation, and isomerisation [57]. For example, sulfated mesoporous Nb oxide has excellent activity (*ca.* 200 times greater than that of sulfated bulk Nb oxide) in the benzylation of anisole or toluene with benzyl alcohol in liquid phase. The mesoporous structure is maintained even after acid treatment (both with sulfuric or phosphoric acids), Brønsted and Lewis acid sites coexist in a roughly 50:50 mixture on the surface of mesoporous Nb oxide while a strong dominance of Brønsted acid sites is present on the sulfated and phosphated materials [59]. The very high surface area of these mesoporous materials results in a great increase in the exposed acid sites per mass, so justifying the observed excellent catalytic activity.

Doped and modified oxides are a wide family of samples which are synthesized with the aim of modifying some surface property of a given oxide. In particular, when an acid or base solid has to be used as catalyst, it can happen that the average acid/base strength of its surface active sites is not useful for the studied reaction.

The acidic or basic sites can be either too strong causing some irreversible adsorption of the reactant species or too weakly acidic or basic to activate the reactants. Then, the oxide surface can be easily modified by incorporation of a second oxide component or by adding doping species that can regulate their acid-base strength by modification of the electronic and geometric characteristics of the acid or base sites. The effect of the addition of small amounts of various ionic species (Ca^{2+}, Li^{+}, Nd^{3+}, Ni^{2+}, Zr^{4+}, and SO_4^{2-},) on the acid-base properties of γ-alumina, silica, and magnesia surfaces was studied by ammonia and sulfur dioxide probe molecules, respectively, in an adsorption calorimeter [60].

The added metal ion concentration was in the range from 0.1 to 0.3 μmol·m^2 corresponding to a surface coverage of about 0.5–1 % of moles of metal ion per mole of support. It was found that the modification of γ-Al_2O_3 surface properties by the ion dopants only moderately changed its amphoteric properties, the surfaces of the doped alumina samples remained amphoteric with acid/base pairs independent of the additives. More substantial changes were observed on magnesia concerning its basic properties, formation of even more basic sites in the domain of medium weak sites, but not in the strong sites domain, were revealed (Fig. 8.4 upper). On the weakly acid silica sample, the number of acid-base centers was strongly affected by the introduction of the doping ions, as shown in Fig. 8.4 down.

Concerning the relationships between the surface acid-base properties and more general intrinsic properties of the ions, such as electronegativity, ionicity character of the cation-oxygen bond, etc., it was found that the acidity of the samples correlated with the charge/radius ratio of the ionic species and with the electronegativity of the doping ions. The basicity correlated well with the partial oxygen charge associated with the cationic dopant [60]. Cardona-Martínez and Dumesic [61] arrived at the same conclusion studying the acid properties of a series of doped silicas with small amounts (0.2–0.3 μmol·m^2) of Mg^{2+}, Sc^{3+}, Fe^{2+}, Fe^{3+}, Al^{3+}, Zn^{2+}, and Ga^{3+} ions by adsorption microcalorimetry using pyridine as probe. The new acidity created by the ion introduction could be correlated with the Sanderson electronegativity of the corresponding oxide formed on the surface (Fig. 8.5). In particular, the Ga-, Al-, and Sc-silica samples were found to have both Brønsted and Lewis acidity while all the other samples showed only Lewis acidity.

Besides the doped oxides, modified oxides are gained a prominent role in the catalytic *scenario*. Many different oxide materials were synthesized chemically modifying in particular, but not exclusively, silica structures with alumina, titanium, niobium, tantalum, and zirconium giving rise to many successful materials used in catalysis as active phases or support phase [23, 24, 34, 62–67].

The possibility of modifying the acid properties of the silica surface by chemical modification, for example, by covalent anchoring of acid groups or by introducing elements of other valence [68–71] which create a defect of charge causing the formation of a Brønsted site to balance it, is well known from long time and it has been well exploited in catalysis to synthesize solid acids with modulated acidity strength. The integral and differential heats evolved from ammonia adsorption on a series of oxides comprising silica, alumina, and three modified silicas with amounts of Al, Ti, and Zr, are shown in Fig. 8.6. The curves for alumina (A), silica-alumina (SA),

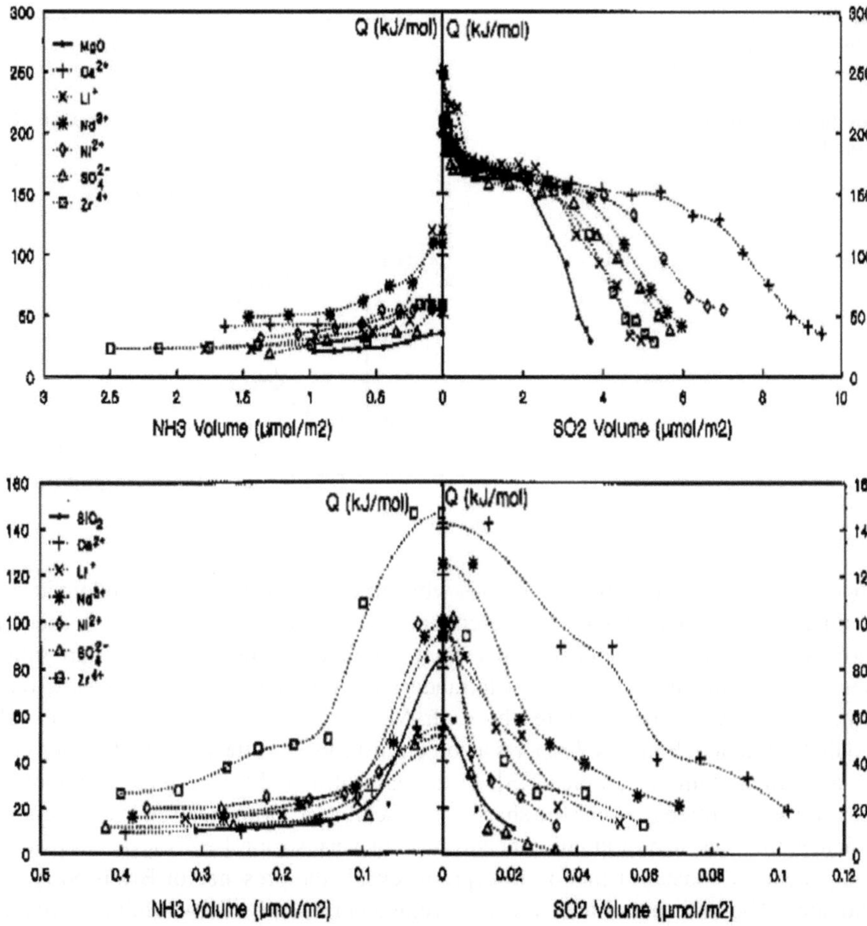

Fig. 8.4 Heat of adsorption for NH_3 and SO_2 probes on doped magnesia samples (*upper*) and on doped silica samples (*bottom*) (from Ref. [60])

and silica-zirconia (SZ) run together while that for silica-titania (ST) lays well below those of the other samples and the curves for all the samples are very higher than that of silica. This suggests that by chemical modification of the silica structure it is possible to enhance the surface acidity of the resultant samples and in particular that alumina, silica-alumina, and silica-zirconia samples have surfaces with the highest acidity in terms of number and strength of sites with high heterogeneity of the surfaces as indicated by the continuous decreasing heat values as a function of coverage [34].

Among these kinds of oxides, the most important example is silica-alumina that gained success as substitute for acidic zeolites in many catalytic processes of petrol chemistry and refining and in general of hydrocarbon chemistry [72]. Synthetic

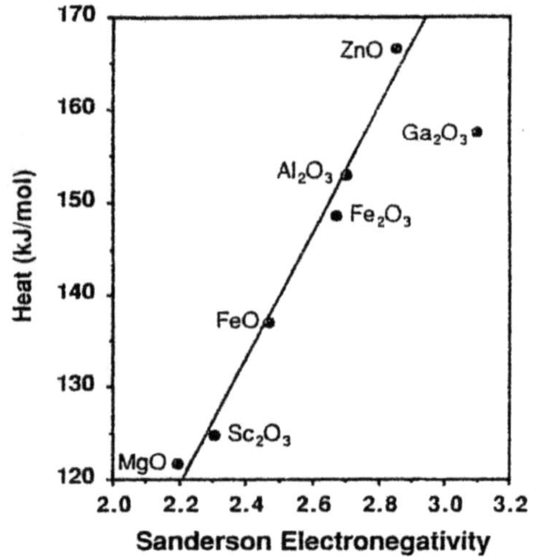

Fig. 8.5 Integral heat of pyridine adsorption as a function of the Sanderson electronegativity of the cations introduced on silica surface (from Ref. [61])

silica-aluminas are amorphous materials with structure consisting of a random array of silica and alumina tetrahedral interconnected over three dimensions. In order to maximize the number of acid sites in a silica-alumina it is important to prepare samples with the maximum amount of tetrahedrally coordinated Al; in this way, one acidic proton would be generated by each Al atom, while formation of Al–O–Al should be avoided. Figure 8.7 reports a comparative view of the acidity of some zeolite and silica-alumina samples [33]. In the case of a zeolite, a typical differential heat plot shows three regions; the sharp decrease of adsorption heats at low coverage indicates the presence of small concentration of very strong Lewis type-acid sites. The plateau of constant heats of adsorption reveals the presence of Brønsted type-acid sites. The differential heat versus coverage curves for silica-aluminas reveal a more heterogeneity of the acid sites with a more balanced presence of Brønsted and Lewis acid sites than zeolite materials that have a predominant Brønsted acidity.

Not only the number and strength of the acid sites but also the accessibility to the reactant molecules is a target in the case of the solid acids when reaction forecasts large reactants to transform. A new family of mesoporous aluminosilicates with a regular array of uniform pores of 20–100 Å diameter has been discovered [73]. These MCM-41 materials can be synthesized over a large range of framework Si/Al ratios developing acidity properties [74]. The presence of these very large uniform pores combined with acidic properties opened new possibilities for cracking heavier feeds and other hydrocarbon transformation.

Supported metal oxides obtained by spreading an active metal oxide component over an oxide support, in prevalence of ceramic type (Al_2OI_3, SiO_2, ZrO_2, TiO_2, etc.), represent a very large family of oxides widely used in heterogeneous catalysis. The driving force for this surface wetting of metal oxides is the lower surface free energy of the final supported metal oxide system. The hydroxylated oxide support surface possesses a much higher surface free energy than the terminal M=O bonds

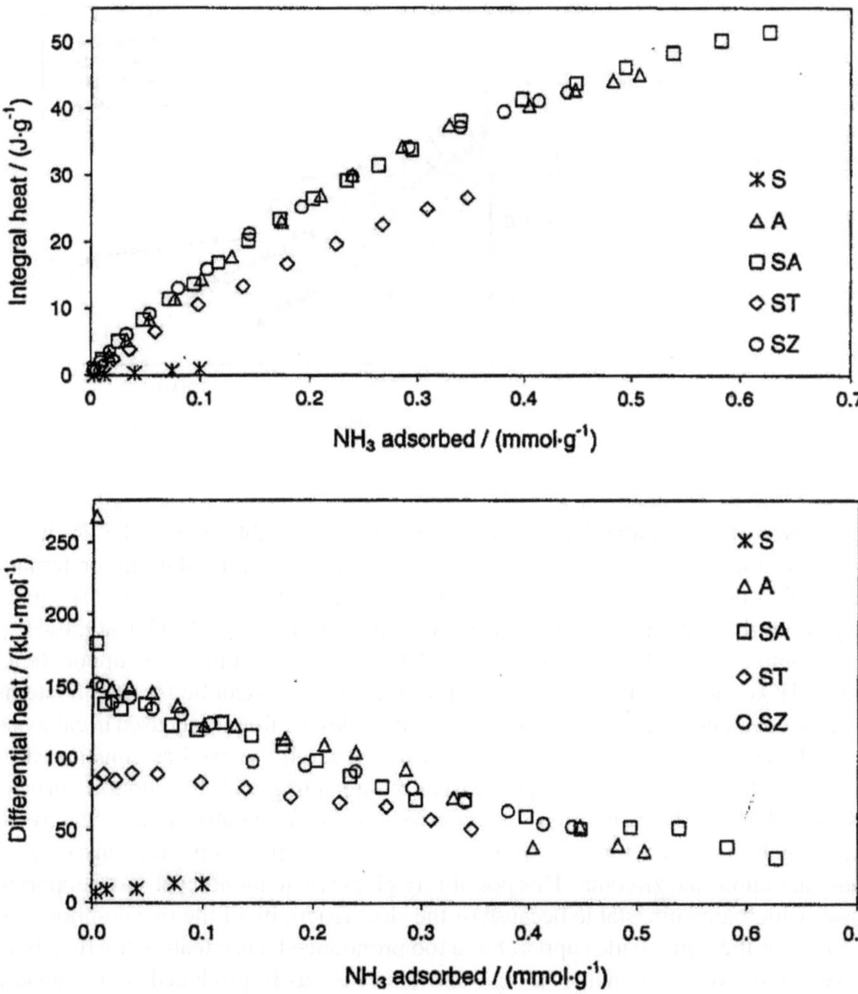

Fig. 8.6 Integral (*upper*) and differential (*bottom*) heats of ammonia adsorption versus coverage for a series of modified silicas with alumina, SA, titania, ST, and zirconia, SZ; for comparison the curves for silica, S, and alumina, A, are reported (from Ref. [34])

formed in the surface metal oxide monolayer. The electronic and molecular structures of the surface metal oxide species dispersed on oxide supports have received enormous attention over the past decades because (i) the industrial significance as catalysts for numerous applications and (ii) the ability to serve as model mixed metal oxides systems due to the essentially completed dispersed state.

Depending on the mutual characteristics and nature of the supported oxide and oxide components, supported systems with different properties and stability can be formed [10]. As a general trend, it is possible to support ionic oxides on ionic

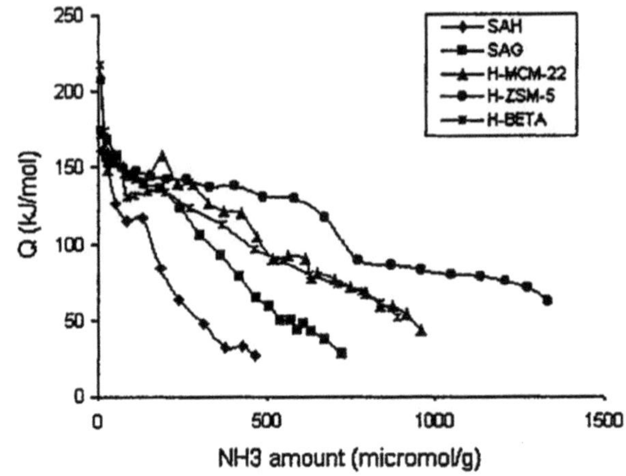

Fig. 8.7 Differential heat of ammonia adsorption versus coverage for a series of zeolites (H-ZSM-5, H-BETA, and H-MCM-22) and two silica-aluminas (SAH and SAG) (from Ref. [33])

oxides even if the resultant compositions are quite unstable due to the ability of the two oxides to react with each other giving rise to solid solutions or ternary phases. This occurs, for example, for CuO on alumina which forms $CuAl_2O_4$ spinel. Only when the supported oxide is formed by big cations (e.g., K_2O, CaO, La_2O_3, etc.) unable to enter the close packing of the oxide ions forming the support (e.g., Al_2O_3, TiO_2, etc.), fairly stable oxide-on-oxide structures can be formed. Systems based on covalent oxides deposited on covalent oxides are frequently used in catalysis (e.g., MoO_3/SiO_2, WO_3/SiO_2, etc.). They are generally viewed as constituted of small particles of the supported phase weakly interacting with the support surface. Also covalent oxides supported on ionic oxides are frequently used in catalysis. Oxides such as vanadia, tungsta, niobia, rhenia can be usefully supported on oxides as alumina, titania, and zirconia. This possibility gives rise to monolayer-like supported phases which are quite stable because of the slow reactivity of the two components. If, however, the ionic oxide support has a too pronounced basic feature the reactivity between the two components is high and a salt is easily produced (e.g., vanadia on magnesia gives rise to Mg vanadate). Ionic oxides supported on covalent oxides are sometimes used in catalysis. The covalent support is almost amorphous silica, on which many ionic oxides can be supported. The stability of such compositions appears to be quite limited, due to salt formation.

The acidic metal oxides (e.g., CrO_x, MoO_x, WO_x, VO_x, NbO_x, TaO_x, etc.) usually anchor to the oxide substrate by preferentially titrating the basic surface hydroxyls of the support surfaces [47]. The active basic metal oxides (e.g., NiO_x, CoO_x, InO_x, CuO, etc.) usually anchor to the oxide substrate by preferentially titrating the surface Lewis acid sites of the oxide supports [47, 75]. For many supported metal oxide systems, the active component can be present as a 100 % dispersion (typically when Al_2O_3, TiO_2, ZrO_2 are used as oxide supports) when its loading on the support is not high and the support coverage is below the monolayer. Less than 100 % dispersion is usually obtained for metal oxides at high loading and when the support surface

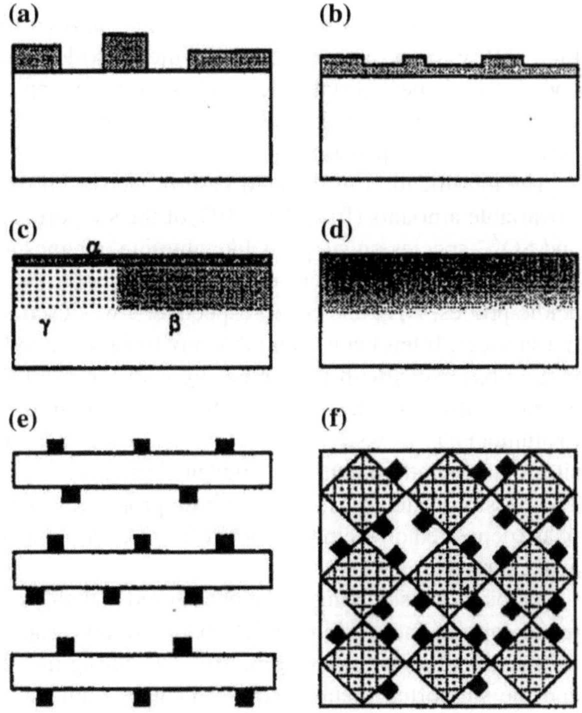

Fig. 8.8 Examples of different supported oxide systems concerning the two oxide component interaction (from Ref. [77]): **a** weak interaction; **b** medium interaction; **c** strong interaction, with formation of α-surface compound, β-bulk compound, and γ-solid solution; **d** composition with gradient-concentration; **e** layered support with adsorbed particles, **f** diamond layered structure

presents lower reactivity of the surface hydroxyls. In addition, it is known that some active basic metal oxides do not interact strongly with the different oxide functionalities of the oxide supports and, consequently, do not disperse to form well-dispersed phases (e.g., MnO_x, CeO_x, [47, 76]).

Another important distinction has to be made based on the strength of the interaction between the support and the supported phase, designated as active phase, in view of its catalytic role [77]. By varying the support and the active oxide nature, different situations can be met which cover a continuous range of interaction strength between the two phases (Fig. 8.8). When weak interaction strength occurs (Fig. 8.8a), the supported phase gives rise to isolated crystallites deposited on, but not necessarily covering, the support phase. For medium interaction strength (Fig. 8.8b), the active phase tends to spread over the support surface if its loading is low, while, for increasing concentrations, multiple layers, and even distinct crystallites, of active oxide are built. For strong interaction strength (Fig. 8.8c), a spreading of the supported phase id favored, but a strong interaction can also lead to a new surface compound (Fig. 8.8c, α) or to bulk compound (Fig. 8.8c, β) or to solid solution (Fig. 8.8c, γ). Figure 8.8d–f illustrate other situations with formation of a gradient composition compound, the layered support with adsorbed particles, and a diamond-layered model, respectively.

The electronic and molecular structures of surface metal oxide species present on oxide supports have received enormous attention over the past three decades because of their industrial significance as catalysts for numerous applications and their ability to serve as model mixed metal oxide catalytic systems due to the quite completely dispersed state. Here below, some examples of supported oxide catalysts are illustrated with emphasis on the possibility to judiciously develop surfaces with defined acid-base properties.

The modification of the acid-base properties of oxides following the deposition of variable amounts (from 1 to 50% of the support surface coverage) of Li^+, Ni^{2+}, and SO_4^{2-} species on supports like alumina, magnesia, and silica have been studied [78]. Once again the adsorption microcalorimetric experiments using a basic and an acidic probes, NH_3 and SO_2, respectively, was used to study the acid properties of the surfaces. It has been found that any linear changes in the amount and in strength of acid-base sites with the increasing addition of lithium, nickel, or sulfate ions to alumina, silica, or magnesia were found. The addition of small amounts of additives to alumina has very slight impact on the heat of adsorption and density of acid/base sites until the attainment of a sudden change when half of its surface was covered. The possibility to adjust the acid-base properties of bulk oxides by a second oxide phase deposited on them was more effective on silica owing to its very weak intrinsic acid character.

Niobia is a fascinating oxide phase used in catalysis because it is a typical strong-metal-support-interaction (SMSI) oxide; for this property it is used as a support phase for metal and metal oxides [50, 79]. The strong interaction between niobia surface and the supported metal species enhances both the redox of the reducible metal oxides species and the acid properties of the supported phases. Such modifications could be favorable for given catalytic processes; for example, some group III oxides (B_2O_3, Al_2O_3, Ga_2O_3, and In_2O_3) are frequently studied as semiconductor materials used as sensor devices for the detection of NO_x, O_2, H_2O, CO [80–82], and more recently as de-NO_x catalysts [83, 84], or as catalysts for the dehydrogenation or aromatization of light alkanes [85] in the form of supported phases. When niobia is concerned as support phase for Al_2O_3 (Al/Nb), Ga_2O_3 (Ga/Nb), and In_2O_3 (In/Nb) oxides, amphoteric surfaces are obtained giving interesting catalytic properties in the reaction of dehydrogenation of propene [86]. The Ga/Nb and In/Nb samples presented differential heat curves of ammonia adsorption clearly different from that of niobia. Up to a certain adsorbed amount (ca. 25 $\mu mol \cdot g^{-1}$) the samples have adsorption heat values higher than those of niobia, attributed to newly created centers of guest oxide; then the adsorption heats become lower than those of the bare support (Fig. 8.9, left). This means that gallium and indium oxides are preferentially deposited on the acid sites of niobia. Concerning the basicity of the surfaces, as determined by SO_2 adsorption, the differential heat curves obtained (Fig. 8.9, right) do not reflect the basicity order of the group III bulk oxides. Despite the well assessed basic properties of In_2O_3, the poor In-dispersion on niobia (In/Nb sample) was responsible for the very lower adsorbed amount of SO_2 than on Ga/Nb and, in particular on Al/Nb (possessing the highest surface area). At this subject, the influence of the support nature in promoting the dispersion of the supported phase is of fundamental importance. Still considering the dispersed In_2O_3 active phase, it was shown [84] that among a series of oxide supports (Al_2O_3, Nb_2O_5, SiO_2, and TiO_2), Al_2O_3 and, to a lower extent,

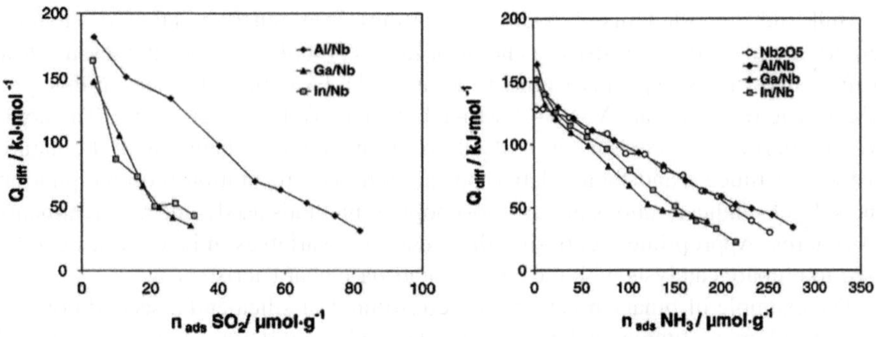

Fig. 8.9 Sulfur dioxide adsorption (*left*) and ammonia adsorption (*right*) on some supported group III oxides (Al_2O_3, Al/Nb, Ga_2O_3, Ga/Nb, and In_2O_3, In/Nb) on niobia. Bulk niobia does not show any basicity [84]

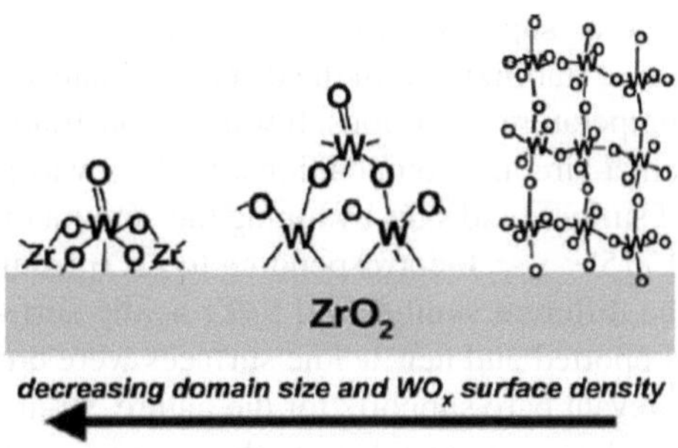

Fig. 8.10 Representation of the evolution of supported WO_x domains from isolated mono tungstates to two-dimensional polytungstates and three-dimensional WO_3 clusters. The increasing W–O–W connectivity with increasing domain size is related to the extent of oligomerization (from Ref. [87])

TiO_2 were found to be the best ones for obtaining active de-NO_x catalysts, since the good In-dispersion and high active surface area.

A peculiar characteristic of the highly dispersed oxides in nanometer size is that their local structure and electronic properties vary with domain size [87]. For example, in the catalysis applications, turnover rates vary as oxide domains evolve from isolated monomers to two-dimensional oligomers, and into three-dimensional clusters with bulk-like properties. Many examples on dispersed metal oxide domains in relation to acid catalyzed reactions can be recalled. For example, WO_x domains dispersed on ZrO_2 constitute an excellent acid catalyst for xylene isomerization at low temperatures [87]. The maximum rates of reaction, expressed *per* W-atom, were observed at WO_3 surface densities of *ca.* $10\,\mathrm{atom_W{\cdot}nm^{-2}}$, corresponding to the two-dimensional polytungstate monolayer formation. Raman evidences suggested that W–O–W connectivity (vibrational mode at $807\,\mathrm{cm^{-1}}$) which predominate over the terminal W=O groups (vibrational mode at $1019\,\mathrm{cm^{-1}}$) in two-dimensional and three-dimensional extended WO_x oligomers constitute the active acid sites, they could stabilize the cationic transition state involved in the xylene isomerization. Figure 8.10 reports the evolution of WO_x species on zirconia surface from the smallest to most wide aggregates.

Bulk and supported mixed oxide compositions, from binary metal oxides to quaternary metal oxides, consist in general of large crystalline phases possessing low surface area values (typically from 1 to $10\,m^2\cdot g^{-1}$). Examples of oxides of this type of catalytic relevance are V–Nb–O, Mo–Nb–O, Co–Ti–O, Ni–Ti–O, etc. The acid-base properties of mixed metal oxides have been found to change with the nature of the constituents and their relative concentrations, preparation (co-precipitation and sol-gel synthesis among are the most popular methods used), and pre-treatments procedures. Appropriately choosing the mentioned variables, mixed oxides can be used to prepare catalysts with the desired-acid-base characteristics.

The example of binary mixed oxides constituted of silica and a second component like alumina, titania, and zirconia has been above reported. The so constituted surfaces have different acidity properties in terms of number of sites and site strength distribution, depending on the nature of the second component (Al, Ti, or Zr), the Si/Al, Si/Ti, and Si/Zr ratios, and the final sample structure [23, 34]. In catalysis such oxides can be used as support phases as well as acid catalysts. When they are used as support phases, the acid sites serve as anchoring points for the supported phase stabilizing it in the dispersed state, even when the samples operate under high temperature conditions. It was demonstrated by XPS spectrophotometry [23] that for silica-zirconia compositions, Si/Zr, it was possible to regularly increase the amount of surface acid sites by tuning the amount of Zr during preparation ($Si_xZr_{1-x}O_2$ with $0.715 < x < 1$, corresponding to ZrO_2 from 5 to 45 wt% in the silica) (Fig. 8.11). On the different synthesized Si/Zr acidic surfaces, the iron oxide dispersed phase was supported and new acidic surfaces were created, $Fe/Si_xZr_{1-x}O_2$, differing from the relevant bare supports for the nature, strength, and amount of the acid sites.

The average acid strength of the Fe-samples was higher than that of $Si_xZr_{1-x}O_2$ supports and tuned towards a prominent Lewis acidity. Concerning the amount of acid sites of the Fe-catalysts, they were higher or lesser compared with that of the relevant supports depending on the acidity of the bare support. When a portion of highly acidic surface of support (Si/Zr samples at high Zr content) was covered by iron oxide having lower surface area than its support, a decrease of the number of surface acid sites was observed (Fig. 8.12, see the SZ-15, SZ-30, and SZ-45 supports and relevant Fe-catalysts) while when low acidic surfaces, as pure silica or low containing zirconium samples were covered by iron oxide an increase of the amount of acid sites was observed (Fig. 8.12, see the S and SZ-5 supports and relevant Fe-catalysts). It is then possible to tune not only the amount of the acid sites but also the Brønsted and Lewis nature and the acid strength of a surface by a judicious presence of different components at the surface.

Silica-alumina has been used to support the CuO coupled with Ga_2O_3, and SnO_2 dispersed phases to enhance the catalytic properties of CuO-based catalysts in reactions of environmental importance (hydrocarbon combustion, NO and N_2O decomposition and reduction [88]). The acidic properties of such oxide systems were studied from a qualitative (nature of the acid sites) and quantitative (number, acid strength, and strength distribution of acid sites) points of view through the adsorption and desorption of two basic probes (ammonia and pyridine) by coupled volumetric-calorimetric technique and XPS and FT-IR spectroscopy.

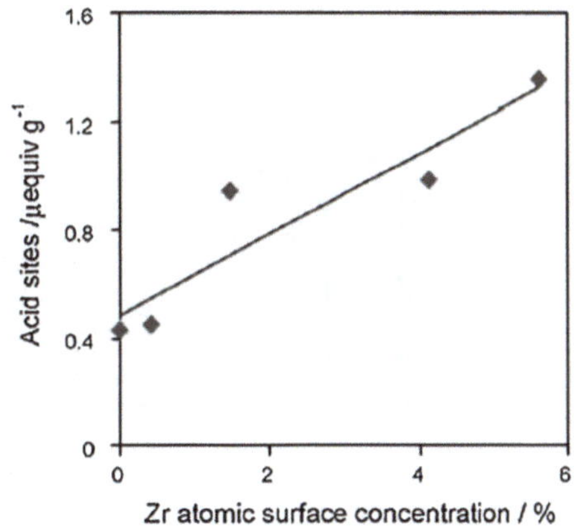

Fig. 8.11 Relation between the amount of acid sites of $Si_xZr_{1-x}O_2$ oxides determined by 2-phenylethylamine probe and the amount of surface Zr concentration determined by XPS (from Ref. [23])

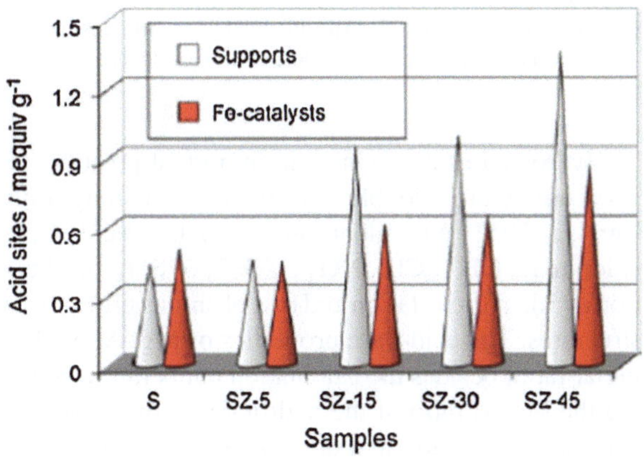

Fig. 8.12 Amount of acid sites of the $Si_xZr_{1-x}O_2$ supports in comparison with the Fe oxide covered surfaces, determined by 2-phenylethylamine probe (S, pure silica, and SZ-X, silica-zirconia oxides with 5 < x < 45 zirconia mass concentration) (from Ref. [23])

The Brønsted acidity of the silica-alumina support was converted into predominantly Lewis acidity upon metal oxide deposition, in particular that of the CuO supported phase. The highest amount of Lewis acid centers were formed on the CuO sample, the partial coverage of the Cu phase by Ga-phase caused a marked decrease of the amount of Lewis acid centers of the Cu/Ga surfaces. The Lewis to Brønsted acid site ratio of the Cu/Sn samples was higher compared to that of Cu/Ga samples, accounting for a higher surface Sn-dispersion. Concerning the acid strength of the surfaces, moderate acidity was associated with the Cu sites ($100\,kJ\cdot mol^{-1} < q_{diff} < 150\,kJ\cdot mol^{-1}$) whereas the most acidic fraction of the sites ($q_{diff} > 150\,kJ\cdot mol^{-1}$) increased with the presence of Ga and Sn (Fig. 8.13). The results have been discussed in the light of the intrinsic acidity of the Cu, Ga, and Sn metal ions derived from electronic parameters (in particular electronegativity).

Complex oxides comprise a large variety of structures which gained importance in the catalysis field due to their acidity or basicity properties (e.g., spinels, perovskites, hexa-aluminates, bulk and supported hydrotalcites, pillared clays, bulk and supported heteropolyacids, etc.).

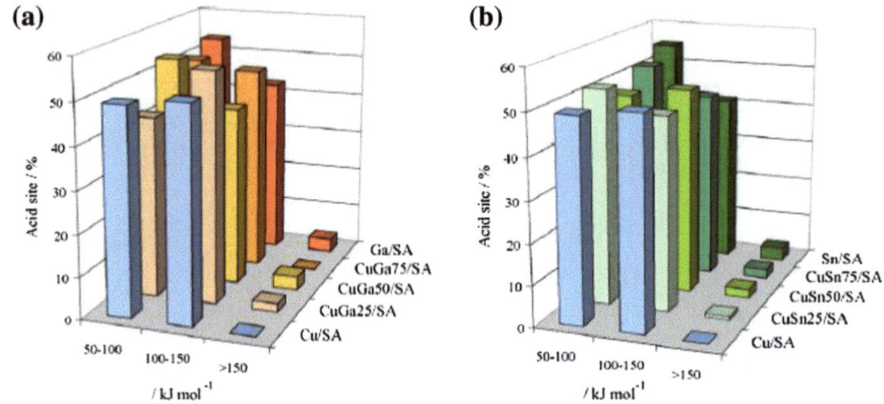

Fig. 8.13 Acid site strength percent distributions of the Cu/Ga (**a**) and Cu/Sn (**b**) series samples supported on silica-alumina (SA) as obtained from ammonia adsorption by calorimetric measurements (the Ga/Cu and Sn/Cu atomic ratios, 25, 50, or 75, are indicated) (from Ref. [88])

Hydrotalcites are gained an important position, in particular, in basic catalysis. They are layered double hydroxides containing a divalent ion (e.g., Mg^{2+}, Ca^{2+}, Zn^{2+}, or Ni^{2+}) a trivalent ion (e.g., Al^{3+}, Fe^{3+}, Cr^{3+}) and a charge compensating anion (e.g., OH^-, Cl^-, NO_3^-, CO_3^{2-}, or SO_4^{2-}). Thermal treatments cause dehydration and dehydroxylation and loss of anions giving rise to mixed oxide with MgO-type structures. The acid-base properties of Mg–Al oxides are governed by the Mg to Al molar ratio, besides the calcination temperature and preparation procedure. Increasing the Mg:Al ratio of the hydrotalcites, increase of the total number of basic sites is observed. The thermal decomposition of hydrotalcites gives rise to mixed oxides with basic properties and they can act as precursors for the preparation of basic oxides with catalytic activity [89]. Several reviews have been reported recently concerning the application of Mg–Al or calcined hydrotalcites as basic materials in particular in the field of fine organic chemistry [90–92].

Among the mixed oxides, those with perovskite-like structure are among the most widely studied. They have general formula ABO_3 where A is usually a large-radius metallic cation (e.g., lanthanide, alkali-earth metal ion) and B is a lower size transition metal ions. The principal advantages of this oxide family are the high versatility of their composition that can give compounds with widely different properties and their exceptional stability of the structures. The crystal field force of the perovskite structure can force some ions to assume unusual oxidation states. A typical example is Cu in Ba-substituted lanthanum cuprate, in which Cu is present in both 2+ and 3+ oxidation states [93]. Another interesting property of perovskites is their nonstoichimetry. As the whole structure has to be neutral, so cation charge have to be distributed to give an overall +6 sum: $A^{1+}B^{5+}O_3$, or $A^{2+}B^{4+}O^3$, or $A^{3+}B^{3+}O_3$. The presence of defects, like anionic or cationic vacancies, is very common in real structures. Anionic vacancies are the most frequent and they can extend up to a whole oxygen layer, leading to a different family of compounds, the brownmillerites [94], like $Ca_2Fe_2O_5$ and $La_2Ni_2O_5$ in which one sixth of the oxygen atoms are vacant. In other cases, an excess oxygen can be found; this is the case of $LaMnO_{3+\delta}$.which shows very good catalytic activity for total oxidation reactions. This defect or excess of charge is connected with acid or basic properties of these structures which found wide application in the hydrocarbon catalytic combustion processes rather than in the acid-base catalytic reactions, due to their excellent thermal and chemical stability.

Another interesting family of oxide compounds are the heteropolycompounds possessing Keggin-type structure. They consist of heteropolyanions and counter-cations such as NH^{4+}, Cs^+, H^+. When the counter-cations are protons, the compounds are called heteropolyacids, they can have very strong Brønsted acid sites, like $H_3PW_{12}O_{40}$. They can have different acid strength depending on the nature of the compounds but always presents a plateau of sites of the same strength. For the catalytic application, they are often supported on high-surface area oxides or activated carbons to increase the surface contact with the reactants of the fluid phase. They are used in several acid-catalyzed reactions, like the hydration of alkanes, synthesis of methacrolein, isobuturric acid, etc. Various studies are reported in the literature [95–97] concerning the acid strength of heteropolyacids, influence of the support on acidity, influence of the acid site distribution on the substitution of protons with cationic species, change in acidity upon heat treatment, etc.

8.6 Acidity Prediction from Composition

From practical and theoretical points of view concerning binary metal oxides, it is interesting to find oxide combinations having well defined and tunable acid or basic properties. On a catalytic oxide surface, the acid or basic sites can be either too strong causing some irreversible adsorption of the substrate species or the sites can be too weak to activate the substrate species. Therefore, the possibility to regulate the acid-base strength, besides the acid site amount of the oxide surfaces, appears a necessary tool for catalytic purposes.

The acid/base strength of an oxide surface may be enhanced or decreased by the addition of a secondary component which modifies the electronic and/or geometric characteristics of the parent acid/base sites or creates new acid/base centers. The oxides belonging to the silica-alumina system, wherein the Al introduction generates new Brønsted or Lewis acid sites, are representative examples of acid binary oxides. Several authors have found generation of acidity also for other mixed oxide systems (silica-zirconia, silica-niobia, silica-titania, [23, 51, 98]). Then, several models appeared in the literature (Thomas [99], Tanaka [100], Tanabe [101], Seiyama [102], Dumesic [103], and Gervasini [60]) which attempted to generalize the experimental observations picturing acidity generation mechanisms.

The Tanabe model for acidity prediction in mixed oxide compositions is based on the interaction of the oxides at molecular level, the interaction generates an excess of negative or positive charge in the mixed composition localized around the guest element. According to this model, the substitution/introduction of a metal ion into the structure of a host oxide, whether the charge is an excess or not, and whether it is positive or negative, may generate acid sites. They are determined by the coordination numbers, C, and valences, V, of the positive and negative elements in the model oxide structure pictured according to two postulates: (i) the coordination numbers of the positive elements of the metal oxides are maintained even when mixed; (ii) the coordination number of oxygen of the major component oxide is retained for all the oxygen atoms in the mixed composition. It is then possible to explain the mechanisms of the acidity generation of mixed oxides and to predict whether the generated acid sites will be of the Brønsted or the Lewis type.

$$\Delta = (V_A/NC_A - V_O/NC_O) \cdot NC_A \tag{8.6}$$

where Δ is the excess/ of charge, V/NC is the ionic valence to coordination number ratio, and the A and O subscripts concern the positive element of the added ion and the negative element (oxygen) of the major component oxide, respectively. In any case, Lewis acidity is assumed to appear upon the presence of an excess of positive charge and Brønsted acidity for an excess of negative charge.

Some model oxide structures pictured according to Tanabe's model are here below reported.

Silica-titania, with silica the major component oxide:

$\Delta \;= (+4/6 - 2/2) \cdot 2 = -2$

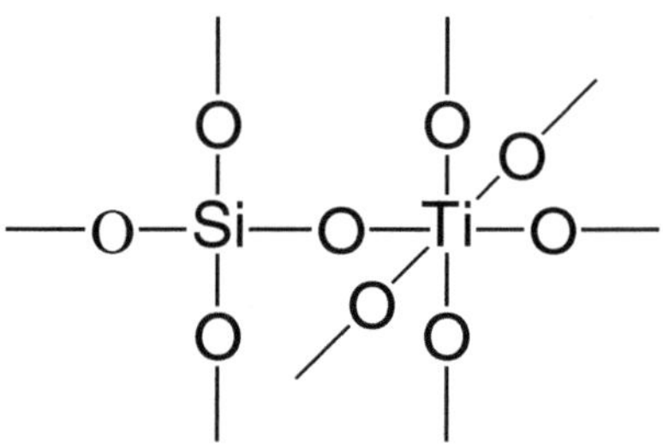

Titania-silica, with titania the major component oxide:

$\Delta \;= (+4/4 - 2/3) \cdot 4 = +4/3$

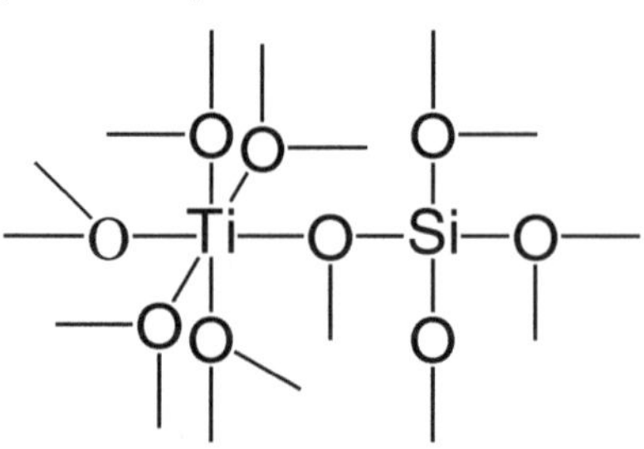

Silica-niobia with Nb tetrahedral coordination (NbO_4):

$\Delta \;= (+5/4 - 2/2) \cdot 4 = +1$

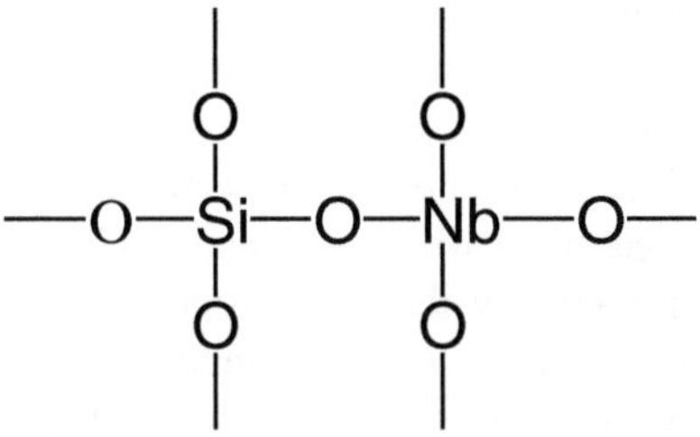

Silica-niobia with Nb octahedral coordination (NbO_6):
$\Delta \;= (+5/6 - 2/2) \cdot 6 = -1$

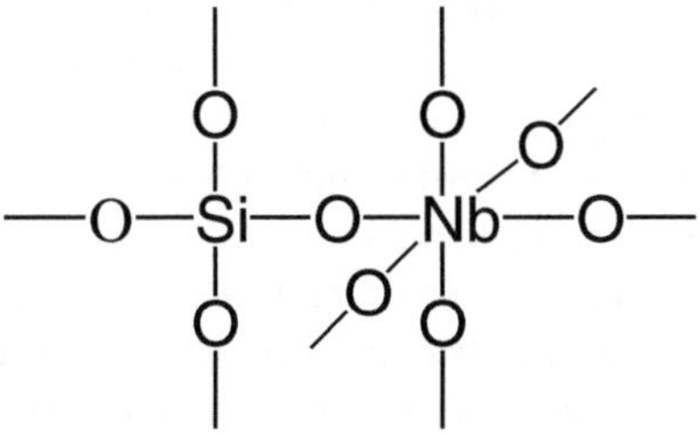

Silica-zirconia wirh 8-fold coordination of Zr (fluorite-like structure):
$\Delta \;= (+4/8 - 2/2) \cdot 8 = -4$

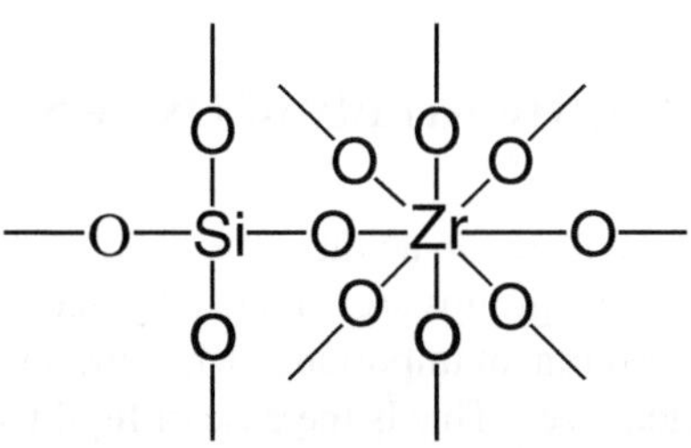

Seiyama has presented a different model in which the oxygen bridging between the two different metal ions develops a positive or negative charge as a consequence of the different coordination of the two cations. In this case too, it is possible to calculate the effective charge of oxygen as the sum of the boundary charges of the two oxides:

$$\Delta = (V_A/NC_A + V_S/NC_S) \cdot -2 \tag{8.7}$$

where the S subscript concerns the positive element of the major component oxide (support oxide).

There is not very often accordance among the different models for the acidity prediction of a given oxide structure. Moreover, it is also hard to justify the predicted acid properties with the catalytic activity of the oxide composition; this is the case of titania-silica system [67, 104]. $TiO_2{-}SiO_2$ mixed oxide is a very important industrial material and catalyst in both the amorphous and crystalline phases which found several industrial applications (e.g., isomerization of olefins, epoxidation of olefins with hydroperoxides, selective oxidation of a number of organic compounds, etc.). The TiO_2-SiO_2 activity in the 1-butene isomerization, phenol ammination, and ethane hydration was attributed to the formation of Brønsted acidity [105]. While for the oxidation reactions there is a general consensus that the active sites are tetrahedrally coordinated Ti^{4+} isolated in SiO_2 matrix, less clear is the source of the activity in the other reactions which demand acid sites. Because the catalytic activity could not be explained by the acid properties of the two pure oxides: in fact TiO_2 has only Lewis acid sites and the silanol groups of SiO_2 are too weakly acidic for any reaction requiring an acid catalysis. The performances of TiO_2-SiO_2 as acid catalyst were at first explained by Tanabe on the basis of his theory of acidity generation in mixed compositions. From his first hypothesis, many authors studied the TiO_2-SiO_2 system and support the Tanabe hypothesis. Notari et al. [67] did not agree with the literature evidences on the acidity of TiO_2-SiO_2 and confuted them by selected experiments. In particular, he proved that highly divided TiO_2 operated the same acid reactions that TiO_2-SiO_2 did with a radical or anion mechanism.

In some other different examples, acidity prediction of an oxide system correlates well with its activity. This is the case of TiO_2-SnO_2 which generated strong new acid sites. The number of acid sites become maximum at the composition TiO_2 50 %. At this composition, the oxide shows the maximum catalytic activity in the butane isomerization reaction [106].

8.7 Intrinsic and Effective Acidity of Oxide Surfaces

The distinction between the *intrinsic* acidity of a solid and the *effective* acidity displayed when the surface acidic groups are screened by interaction with solvent molecules becomes a topic of prominent importance when the solid has to work in contact with liquids for its practical uses. This is the case of liquid-solid heterogeneous catalysts in which the activity of the catalyst can be modified by the presence of solvent which may establish physical or chemical interactions with the acid sites of the surface. For reactions carried out in liquid phase, the knowledge of the *effective* acidity (in terms of number and strength of the sites) of the catalyst in given liquids allows determining sound relationships between the catalytic activity and surface acidity.

Adsorption calorimetry in liquid is a complex matter of study because many variables are simultaneously present: solids with different acidity in terms of distributions of acid sites (nature and acid strength); liquids of different polarity, proticity, and solvating ability; and probes with different basicity (pKa scale). Therefore, results permit comparing for a given solid and probe couple, the influence of the solvent; for a given solid and liquid couple, the influence of the probe; and for a given liquid and probe couple, the properties of different solids.

One of the main goals of the calorimetric experiments of acid-base titrations in liquid is to evaluate the *effective* acid strengths of the surfaces when suspended in liquids of different nature. As for the gas-solid acid-base titrations, solutions of base probe in liquids of various natures are used for titrating the acid sites of the solid which is placed in a reaction calorimeter equipped with a stirring system. Successive pulse injections of dosed amounts of the probe solution are sent onto the sample maintained at constant temperature up to surface saturation with the probe. In this case, the evolved heat from the adsorption reaction derives from two different contributions: the exothermic enthalpy of adsorption, $\Delta_{ad}H_{probe}$, and the endothermic enthalpy of displacement of the liquid, $\Delta_{dpl}H_{liquid}$, the enthalpy effects describing dilution and mixing phenomena can be neglected depending on the differential design and preheating of the probe solution.

Various papers have recently been published concerning the liquid-phase titration of acid solids, such as acidic polymeric resins [107–109]. Example is here presented on the study of the *intrinsic* and *effective* acidities of two catalysts based on niobium: niobium oxide (NBO) and niobium phosphate (NBP) [110] which found application in reaction of acid transformation of monosaccharides (fructose, glucose, in particular) to useful chemicals, like 5-hydroxymethyl-2-furaldehyde (HMF) [111, 112].

When a strongly basic probe, like 2-phenylethylamine (PEA), in the decane solvent is used for titrating the NBO and NBP surfaces (Fig. 8.14), the information obtained on the two acid surfaces support the conclusions obtained from the more classical determinations of acidity using gas-solid calorimetric adsorption and thermal desorption approaches [110]. The calorimetric curve of acid-base titration in decane for NBP always lies above that of NBO, and the difference between the two curves increases as the titration progresses further. This seems confirming that the main difference between the NBO and NBP surfaces has to be ascribed to the medium and weak strength acid sites.

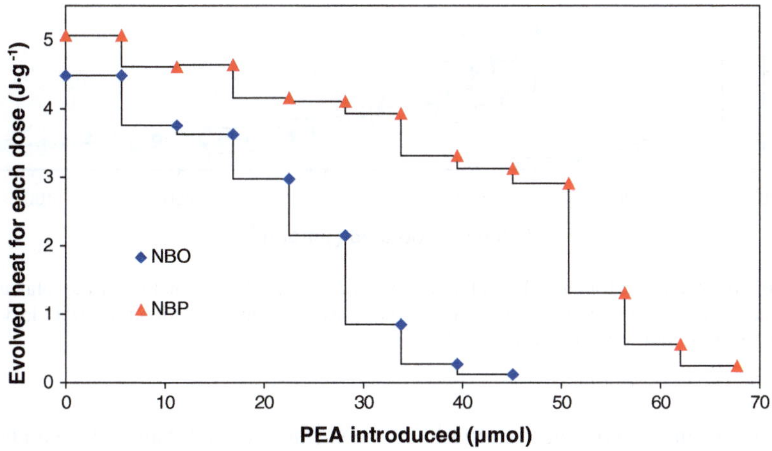

Fig. 8.14 Evolved heat for each dose of PEA solution injected as a function of the total amount of PEA injected in the liquid-solid adsorption experiments performed at 40 °C in decane (from Ref. [110])

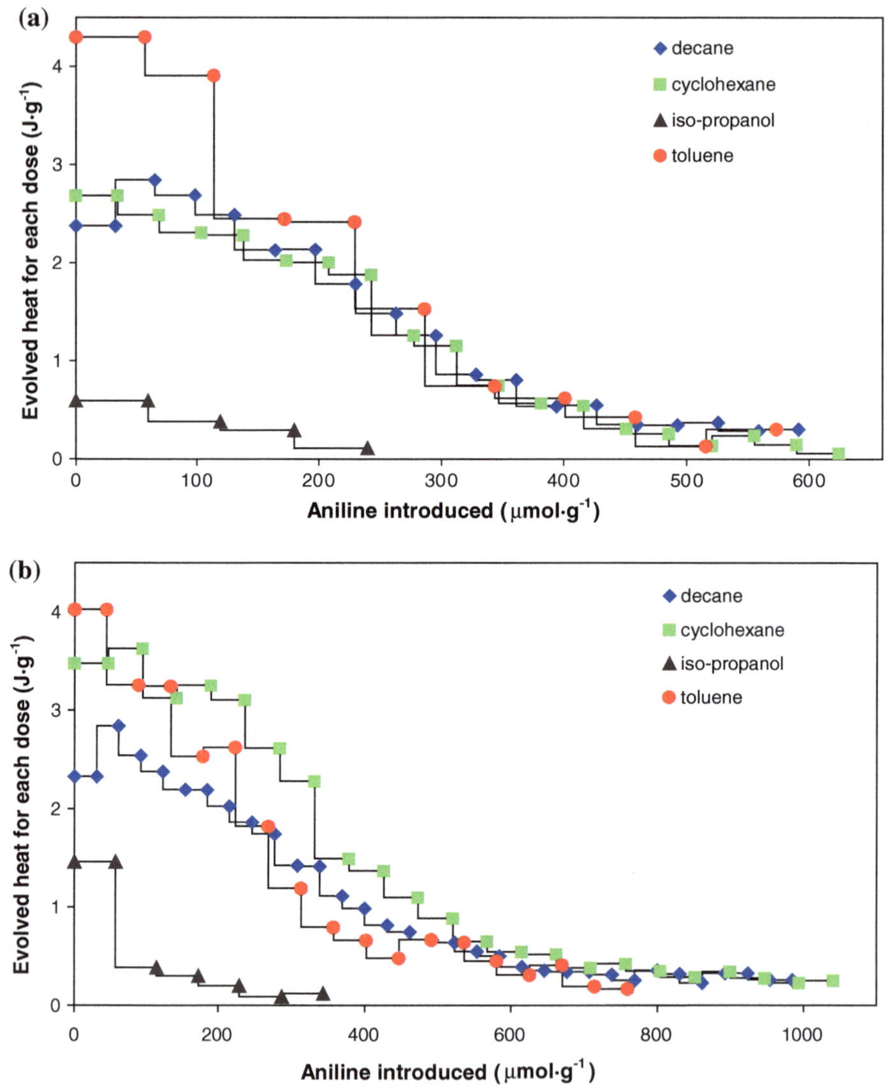

Fig. 8.15 Evolved heat for each dose of aniline solution injected as a function of the total amount of aniline injected during the liquid-solid adsorption experiments performed at 40 °C in various liquids: NBO (**a**) and NBP (**b**) (from Ref. [110])

The different *effective* acid strengths of the surfaces of NBO and NBP can be evidenced by the comparative experimental results summarized in Fig. 8.15a,b, respectively, obtained by using aniline as base probe. In these figures, the areas of the heat flow peaks obtained during aniline titration in cyclohexane, decane, toluene, and isopropanol are plotted as functions of the amount of aniline introduced. The higher acidity of NBP compared to that of NBO, in terms of both the number and strength of the sites, is clearly evident. The total heat evolved following the introduction of a definite amount of aniline, and the amount of aniline necessary to complete the adsorption reaction, are in all cases higher for NBP than for NBO, independently of the nature of the liquid. Since the evolved heat is a sum of two contributions with opposite signs ($\Delta_{ad}H_{probe}$, adsorption enthalpy, exothermic, and $\Delta_{dpl}H_{liquid}$, enthalpy of displacement of the liquid, endothermic), the higher the endothermic contribution, the lower the experimental heat flow measured. For a given solid and probe, $\Delta_{ad}H_{probe}$ depends only on the solid acidity, while $\Delta_{dpl}H_{liquid}$ can be different, depending on solid-solvent interactions. It may then be argued that the surface acid sites of NBO (Fig. 8.15a) present a stronger interaction with cyclohexane and decane than with toluene; the $\Delta_{dpl}H_{liquid}$ contribution is then higher in cyclohexane and decane, which results in lower measured heats than in toluene.

A completely different picture emerges when aniline titration is carried out in isopropanol, a protic solvent with high polarity. On both NBO and NBP, very low heats evolved have been measured; the initial value of evolved heat obtained for NBP is much higher than that for NBO (Fig. 8.15a,b). This confirms the ability of NBP to maintain a highly acidic character in solvents of high polarity, like alcohols or water.

The acid performances of the NBO and NBP catalysts were compared in the dehydration of fructose to 5-hydroxymethyl-2-furaldehyde (HMF). The reaction runs with high conversion and good selectivity to HMF in organic solvents but in agreement with a green chemistry process development, friendly solvents have to be used, like water or alcohol mixtures. In water, NBP shows higher fructose conversion and HMF selectivity than NBO, so justifying its higher effective acidity in polar and protic liquids [110].

The calorimetric titration of the acid sites in liquid phase makes it possible to discriminate the acid site strength more accurately than the more conventional gas-solid phase titration in order to understand the *effective* acidity of the solids measured in various liquids of very different polarities and proticities. The obtained calorimetric results are of not easy interpretation since solute-solvent and solid-solvent interactions have to be taken into account, besides the solute-site interaction.

References

1. G. Busca, Chem. Rev. **107**, 5366 (2007)
2. H. Hattori, Chem. Rev. **95**, 537 (1995)
3. K. Tanabe, W.F. Hölderich, Appl. Catal. A **181**, 399 (1999)
4. G. Busca, Chem. Rev. **110**, 2217 (2010)
5. R.G. Pearson, J. Am. Chem. Soc. **85**, 3533 (1963)
6. G. Klopman, J. Am. Chem. Soc. **90**, 223 (1968)
7. J.L. Reed, J. Phys. Chem. **98**, 10477 (1994)
8. A. Auroux, A. Gervasini, J. Phys. Chem. **94**, 6371 (1990)
9. A. Gervasini, A. Auroux, J. Therm. Anal. **37**, 1737 (1991)
10. G. Busca, Phys. Chem. Chem. Phys. **1**, 723 (1999)
11. J.C. Lavalley, Catal. Today **27**, 377 (1996)

12. S. Coluccia, A.J. Tench, *Proceedings of the 7th International Congress on Catalysis*, Tokyo, Japan, Kodansha, 1980
13. I.E. Wachs, Catal. Today **27**, 437 (1996)
14. G. Busca, Catal. Today **41**, 191 (1998)
15. F. Haw, I.S. Chuang, B.L. Hawkins, G.E. Maciel, J. Am. Chem. Soc. **105**, 7206 (1983)
16. M. Johansson, K. Klier, Top. Catal. **4**, 99 (1997)
17. R. Borade, A. Sayari, A. Adnot, S. Kaliaguine, J. Phys. Chem. **94**, 5989 (1990)
18. D. Barthomeuf, G. Coudurier, J.C. Vedrine, Mater. Chem. Phys. **18**, 553 (1988)
19. R.J. Gorte, D. White, Top. Catal. **4**, 57 (1997)
20. A. Boréave, A. Auroux, C. Guimon, Microporous Mater. **11**, 275 (1997)
21. J. Kotrla, L. Kubelková, C.-C. Lee, R.J. Gorte, J. Phys. Chem. B **102**, 1437 (1998)
22. G. Busca, in *Metal Oxide Catalysis*, ed. by S.D. Jackson, J.S.J. Hargreaves. The Use of Infrared Spectroscopy Methods in the Field of Heterogeneous Catalysis by Metal Oxide, vol 1 (Wiley-VCH, Weinheim, 2009), p. 95
23. A. Gervasini, C. Messi, D. Flahaut, C. Guimon, Appl. Catal. A **367**, 113 (2009)
24. P. Carniti, A. Gervasini, S. Bennici, J. Phys. Chem. B **109**, 1528 (2005)
25. V. Solinas, I. Ferino, Catal. Today **41**, 179 (1998)
26. A. Auroux, Mol. Sieves **6**, 45 (2008)
27. A. Auroux, Top. Catal. **19**, 205 (2002)
28. A. Auroux, in *Catalyst characterization: physical øøøtechniques for solid materials*, ed. by B. Imelik, J.C. Vedriine (Plenum Press, New York, 1994), p. 611
29. M. Fadoni, L. Lucarelli, in *Adsorption and its Applications in Industry and Environmental Protection*, ed. by A. Dabrowski. Applications in Industry, vol 1 (Stud. Surf. Sci. Catal. **120A**) (1999)
30. H.G. Karge, V. Dondur, J. Phys. Chem. **94**, 765 (1990)
31. H.G. Karge, V. Dondur, J. Weitkamp, J. Phys. Chem. **95**, 283 (1991)
32. F. Arena, R. Dario, A. Parmaliana, Appl. Catal. A **170**, 127 (1998)
33. B. Dragoi, A. Gervasini, E. Dumitriu, A. Auroux, Thermochim. Acta **420**, 127 (2004)
34. A. Gervasini, P. Carniti, A. Auroux, Thermochim. Acta **434**, 42 (2005)
35. S. Bennici, A. Auroux, in *Metal Oxide Catalysis*, ed. by S.D. Jackson, J.S.J. Hargreaves. Thermal Analysis and Calorimetric Methods, vol 1 (Wiley-VCH, Weinheim, 2009), p. 391
36. A. Auroux, Top. Catal. **4**, 71 (1997)
37. A. Auroux, A. Gervasini, L. Nemeth, G. Gati, I.S. Pap, G. Mink, Surf. Interface Anal. **19**, 529 (1992)
38. W. Rudzinski, T. Borowiecki, T. Panczyk, A. Dominko, Langmuir **16**, 8037 (2000)
39. M.V. Gargiulo, J.L. Sales, M. Ciacera, G. Zgrablich, Surf. Sci. **501**, 282 (2002)
40. W. Rudzinski, D.H. Everett, *Adsorption of Gases on Heterogeneous Surfaces* (Academic Press, London, 1992)
41. N. Cardona-Martinez, J.A. Dumesic, Adv. Catal. **38**, 149 (1992)
42. P. Carniti, A. Gervasini, A. Auroux, J. Catal. **150**, 274 (1994)
43. K. Tsutsumi, Y. Mitani, H. Takahashi, Bull. Chem. Soc. Jpn. **56**, 1912 (1983)
44. N. Cardona-Martinez, J.A. Dumesic, J. Catal. **125**, 427 (1990)
45. P. Carniti, A. Gervasini, React. Kinet. Catal. Lett. **52**, 285 (1994)
46. S.D. Jackson, J.S.J. Hargreaves (eds.), Wiley-VCH, Weinheim, vol I and II, 2009 (ISBN 978-3-527-31815-5).
47. I.E. Wachs, Catal. Today **100**, 79 (2005)
48. H.-P. Boehnm, H. Knozinger, in *Nature and Estimation of Functional Groups on Solid Surfaces*, ed. by J.R. Anderson, M. Boudart. Catalysis, Science and Technology, vol 4 (Springer, Berlin, 1984)
49. I. Nowak, M. Ziolek, Chem. Rev. **99**, 3603 (1999)
50. M. Ziolek, Catal. Today **78**, 47 (2003)
51. K. Tanabe, in *Solid Acid and Base Catalysts*, ed. by J.R. Anderson, M. Boudart. Catalysis, Science and Technology, vol 5 (Springer, Berlin, 1987), p. 232

52. J. Haber, in *Crystallography of Catalyst Types*, ed. by M. Boudart, J.R. Anderson. Catalysis, vol 2 (Springer, Berlin, 1984)
53. W.E. Farneth, F. Ohuchi, R.H. Staley, U. Chowdhry, A.W. Sleight, J. Phys. Chem. **89**, 2493 (1985)
54. M. Badlani, I.E. Wachs, Catal. Lett. **75**, 137 (2001)
55. I.E. Wachs, Y. Chen, J.-M. Jehng, L.E. Briand, T. Tanaka, Catal. Today **78**, 13 (2003)
56. L.E. Briand, J.-M. Jehng, L. Cornaglia, A.M. Hirt, I.E. Wachs, Catal. Today **78**, 257 (2003)
57. Y. Rao, D.M. Antonelli, J. Mater. Chem. **19**, 1937 (2009)
58. Y. Rao, J. Kang, M. Trudeau, D.M. Antonelli, J. Catal. **266**, 1 (2009)
59. Y. Rao, M.L. Trudeau, D.M. Antonelli, J. Am. Chem. Soc. **128**, 13996 (2006)
60. A. Gervasini, G. Bellussi, J. Fenyvesi, A. Auroux, J. Phys. Chem. **99**, 5117 (1995)
61. N. Cardona-Martínez, J.A. Dumesic, J. Catal. **127**, 706 (1991)
62. E. Cano-Serrano, G. Blanco-Brieva, J.M. Campos-Martin, J.L.G. Fierro, Langmuir **19**, 7621 (2003)
63. A. Gervasini, C. Messi, A. Ponti, S. Cenedese, N. Ravasio, J. Phys. Chem. C **112**, 4635 (2008)
64. C. Flego, L. Carluccio, C. Rizzo, C. Perego, Catal. Commun. **2**, 43 (2001)
65. T. Klimova, M.-L. Rojas, P. Castello, R. Cuevas, J. Ramírez, Microporous Mesoporous Mater. **20**, 293 (1998)
66. G. Guiu, P. Grange, J. Catal. **156**, 132 (1995)
67. B. Notari, R.J. Willey, M. Panizza, G. Busca, Catal. Today **116**, 99 (2006)
68. A. Desmartin-Chomel, J.L. Flores, A. Bourane, J.M. Clacens, F. Figueras, G. Delahay, A. Giroir-Fendler, C. Lehaut-Burnouf, J. Phys. Chem. **110**, 858 (2006)
69. S.Y. Kim, J.G. Goodwin, S. Hammache, A. Auroux, D. Galloway, J. Catal. **201**, 1 (2001)
70. M. Occelli, D.A. Schiraldi, A. Auroux, K.A. Keogh, B.H. Davis, Appl. Catal. A **209**, 165 (2001)
71. D. Deutsch, V. Quaschning, E. Kemnitz, A. Auroux, H. Ehwald, H. Lieske, Top. Catal. **13**, 281 (2000)
72. A. Corma, Chem. Rev. **95**, 559 (1995)
73. J.S. Beck, C.W. Chu, I.D. Johnson, C.T. Kresge, M.E. Leonowicz, W.J. Roth, J.C. Vartuli, JWO 91/11390 (1991)
74. A. Corma, V. Fornés, M.T. Navarro, J. Pérez-Pariente, J. Catal. **148**, 569 (1994)
75. M.A. Vuurman, D.J. Stufkens, A. Oskam, G. Deo, I.E. Wachs, J. Chem. Soc. Faraday Trans. **92**, 3259 (1996)
76. F. Kapteijn, A.D. Vanlangeveld, J.A. Moulijn, A. Andreini, M.A. Vuurman, A.M. Turek, J.-M. Jehng, I.E. Wachs, J. Catal. **150**, 94 (1994)
77. A. Cimino, D. Gazzoli, M. Valigi, J. Electron Spectrosc, Relat. Phenom. **104**, 1 (1999)
78. A. Gervasini, J. Fenyvesi, A. Auroux, Langmuir **12**, 5356 (1996)
79. T. Uchijima, Catal. Today **28**, 105 (1996)
80. C. Rizzo, A. Carati, M. Tagliabue, C. Perego, Stud. Surf. Sci. Catal. **128**, 137 (2000)
81. R. Pohle, M. Fleischer, H. Meixner, Sens. Actuators B **68**, 151 (2000)
82. M. Kudo, T. Kosaka, Y. Takahashi, H. Kokusen, N. Sotani, S. Hasegawa, Sens. Actuators B **69**, 10 (2000)
83. J.A. Perdigón-Melón, A. Gervasini, A. Auroux, J. Catal. **234**, 421 (2005)
84. A. Gervasini, J.A. Perdigón-Melón, C. Guimon, A. Auroux, J. Phys. Chem. B **110**, 240 (2006)
85. K. Nakagawa, C. Kajita, Y. Ide, M. Okamura, S. Kato, H. Kasuya, N. Ikenaga, T. Kobayashi, T. Suzuki, Catal. Lett. **64**, 215 (2000)
86. A.L. Petre, J.A. Perdigón-Melón, A. Gervasini, A. Auroux, Catal. Today **78**, 377 (2003)
87. J. Macht, E. Iglesia, Phys. Chem. Chem. Phys. **10**, 5331 (2008)
88. S. Bennici, A. Auroux, C. Guimon, A. Gervasini, Chem. Mater. **18**, 3641 (2006)
89. D. Tichit, M.H. Lhouty, A. Guida, B.H. Chiche, F. Figueras, A. Auroux, D. Bartolani, E. Garrone, J. Catal. **151**, 50 (1995)
90. A. Vaccari, Appl. Clay Sci. **14**, 161 (1999)
91. D. Tichit, B. Coq, CATTECH **7**, 206 (2003)
92. A. Corma, S. Iborra, Adv. Catal. **49**, 239 (2006)

93. M.A. Peña, J.L.G. Fierro, Chem. Rev. **101**, 1981 (2001)
94. M.J. Sayagues, M. Vallet-Regi, A. Caneiro, J.M. Gonzales Calbet, J. Solid State Chem. **110**, 295 (1994)
95. F.X. Liu-Cai, B. Sahut, E. Faydi, A. Auroux, G. Hervé Appl, Catal. A **185**, 75 (1999)
96. L. Damjanovic, V. Rakic, U.B. Mioc, A. Auroux, Thermochim. Acta **434**, 81 (2005)
97. T. Nakato, M. Kimura, S. Nakata, T. Okuhara, Langmuir **14**, 319 (1998)
98. P. Carniti, A. Gervasini, M. Marzo, J. Phys. Chem. C **112**, 14064 (2008)
99. C.L. Thomas, Ind. Eng. Chem. **41**, 2564 (1949)
100. K.I. Tanaka, A. Ozaki, J. Catal. **8**, 1 (1967)
101. K. Tanabe, T. Sumiyoshi, K. Shibata, T. Kiyoura, J. Kitagawa, Bull. Chem. Soc. Japan **47**, 1064 (1974)
102. T. Seiyama, *Metal Oxides and Their Catalytic Actions* (Kodansha, Tokyo, 1978)
103. G. Connell, J.A. Dumesic, J. Catal. **102**, 216 (1986)
104. B. Notari, Adv. Catal. **41**, 253 (1996)
105. M. Itoh, H. Hattori, K. Tanabe, J. Catal. **35**, 225 (1974)
106. M. Itoh, H. Hattori, K. Tanabe, J. Catal. **43**, 192 (1976)
107. M. Hart, G. Fuller, D.R. Brown, J.A. Dale, S. Plant, J. Mol. Catal. A: Chem. **182–183**, 439 (2002)
108. S. Koujout, B.M. Kiernan, D.R. Brown, H.G.M. Edwards, J.A. Dale, S. Plant, Catal. Lett. **85**, 33 (2003)
109. S. Koujout, D.R. Brown, Catal. Lett. **98**, 195 (2004)
110. P. Carniti, A. Gervasini, S. Biella, A. Auroux, Chem. Mater. **17**, 6128 (2005)
111. P. Carniti, A. Gervasini, S. Biella, A. Auroux, Catal. Today **118**, 373 (2006)
112. P. Carniti, A. Gervasini, M. Marzo, Catal. Today **152**, 42 (2010)

Chapter 9
Characterization of Acid–Base Sites in Zeolites

Dušan Stošić and Aline Auroux

Abstract The review of the use of adsorption microcalorimetry for the characterization of acid-base properties of various types of zeolites is provided. Factors influencing these properties are introduced and explained. Furthermore, the relationship between the data obtained by this technique and catalytic activity of investigated materials is discussed.

9.1 Introduction

Zeolites are natural or synthetic microporous materials with an ordered structure: zeolite frameworks consist of 4-fold connected TO_4 tetrahedra (T = Si, Al) that form three-dimensional networks. In these networks, each oxygen atom is shared between two neighbouring tetrahedra; in other words, the linkage in between tetrahedra is accomplished through T-O-T bridges [1–3]. It should be mentioned that aluminate tetrahedra cannot be neighbours in the frameworks of zeolites. In other words Al-O-Al linkages are forbidden; this requirement being known as the Loewenstein rule [1, 2]. In zeolitic network, the binding capability in between tetrahedra reaches its maximum; thus, ideal zeolite crystals should have terminal silanol groups only on their external surface [4].

The linkage of tetrahedra within zeolites leads to open network structures: the tetrahedra (primary building units) form rings of various sizes which are linked to form complex units (secondary building units).These secondary building units may be connected in many different ways to give a large number of different zeolite structure types. Thus formed network of interconnected tetrahedra results in the zeolite

D. Stošić (✉) · A. Auroux
Institut de Recherches sur la Catalyse et l'Environnement de Lyon,
UMR 5256 CNRS/Université Lyon 1, 2 avenue Albert Einstein, 69626 Villeurbanne
Cedex, France
e-mail: dusan.stosic@ircelyon.univ-lyon1.fr; aline.auroux@ircelyon.univ-loyn1.fr

A. Auroux (ed.), *Calorimetry and Thermal Methods in Catalysis*,
Springer Series in Materials Science 154, DOI: 10.1007/978-3-642-11954-5_9,

framework. From the other point of view, it can be told that zeolite crystals contain pore or channel systems of molecular dimensions with fixed geometry and size [3].

According to the pore size, zeolites are classified into small- (pore size up to 5 Å), medium-(pore size 5–6 Å), and large-pore types (pore size 7–8 Å). Typical representatives of the different types are zeolite A, ZSM-5, and the faujasites (X and Y), respectively [5]. According to the International Zeolite Association Website, about 190 different framework types of zeolites, zeolitic silicates and phosphates with precisely estimated structures are known to date (2010) [6].

In zeolite network, any tetrahedral T atom (Si^{4+} or Al^{3+}) is surrounded by four O^{2-} ions. Network constructed of SiO_2 units should be therefore, neutral, because each O^{2-} ion is shared by two tetrahedra. However, the isomorphous substitution of Si^{4+} by Al^{3+} causes a negative excess charge of the framework. This framework anionic charge is compensated by charge-balancing cations located in the channels [7–9]. In contrast to the semipolar character of bonds within the zeolite framework, the interaction between cations and the framework is ionic. Oxygen atoms with radii of ca. 1.36 Å surround the smaller central atoms of the tetrahedra nearly completely [3]. Consequently, the interior surface of zeolites is nearly entirely composed of oxygen atoms. Nonframework charge-balancing cations and molecules within the pore system coordinate framework oxygen atoms. Water, if present, can coordinate these cations and interact with other sorbed molecules [1]. Finally, the overall composition of zeolitic unit cell is given by the formula:

$$1/\mathrm{m}\,\mathrm{Me}^{\mathrm{m}+}[(\mathrm{SiO_2})_{\mathrm{n}}\mathrm{AlO_2^-}](\mathrm{H_2O})\mathrm{z}$$

where n denotes the Si/Al ratio and z is the number of water molecules adsorbed per framework Al atom.

Previously mentioned isomorphous substitution of Si^{4+} ions with Al^{3+} that provokes a negative excess charge of the framework is at the origin of their acidity [1, 2], one of the most important characteristics of these materials which explains their numerous catalytic applications. Unlike framework T atoms, the charge-balancing cations can be exchanged by other cations from aqueous solutions [10, 11]. If cations are exchanged by protons, the zeolite acquires considerable Brönsted acidic properties and for this reason they can be viewed as solid acids. Importantly, both Lewis and Brönsted acid sites can be found in the structure of zeolites [12]. The way to generate Brönsted acid sites (proton donor sites) is based either on the thermal decomposition of ammonium form of as-synthesized zeolites or on the direct ion exchange of Na^+ (usually) by H^+ (with mineral or organic acids). Brönsted acid sites can be converted into Lewis acid sites by dehydroxylation at elevated temperatures [5]. These processes are presented in Fig. 9.1. However, it is important to notice that charge-balancing cations can also play a role of Lewis acidity.

Altogether, the structural diversity of zeolites discussed above is responsible for a wide range of interesting zeolite properties such as ion-exchange capacity, specific adsorption behaviour, and catalytic activity due to acidity, shape selectivity caused by size and polarity of molecules, high thermal stability and resistance against solvents, and wide flexibility for adjustments by post synthesis modification.

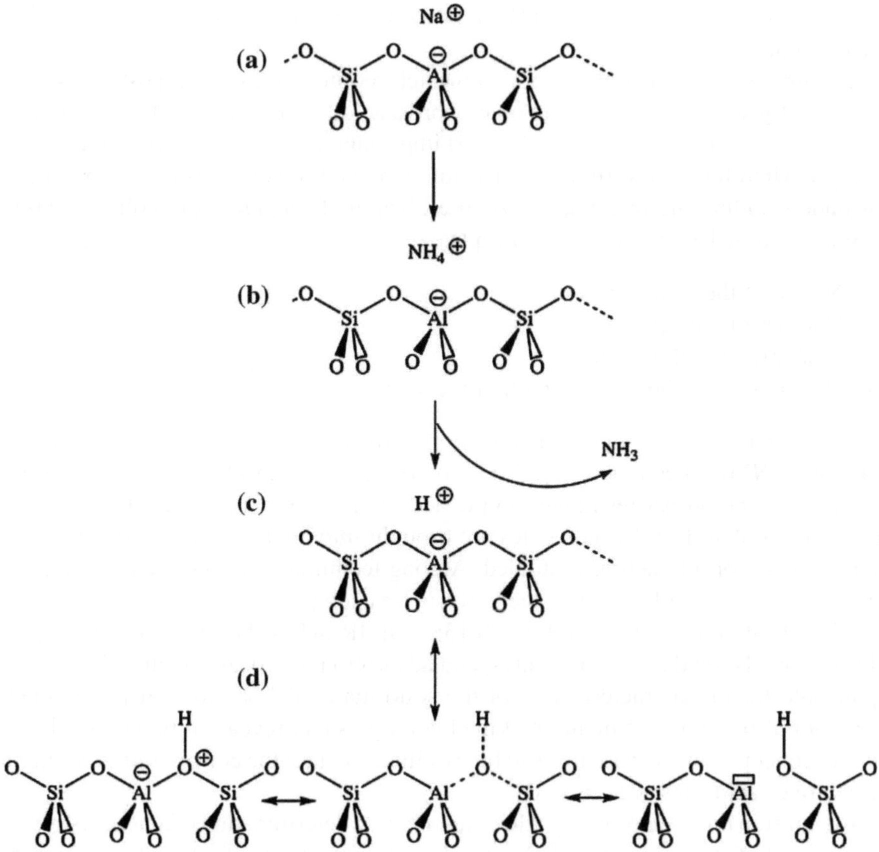

Fig. 9.1 Shematic presentation of generation of Brönsted and Lewis acidity in zeolites. **a** the presentation of charge-balancing cation; **b** cation-exchange Na^+ to NH_4^+; **c** from NH_4^+ cation, isolated proton (proton-donor or Brönsted acid site) arises by calcination; **d** Lewis acid site (Al atom with the empty electron orbital, electron acceptor) is formed by dehydroxylation

Because their thermal and mechanical stabilities are not enough to be used for the industrial purposes, and the zeolite synthesis requires too much cost and time, so many kinds of zeolite species are not available industrially. In fact, only about ten kinds of zeolites have been applied in industrial catalytic processes. In practice, Y-zeolite, ZSM-5, mordenite, MCM-22, and β zeolites are most typical zeolite catalysts [13]. Accordingly to the needs of specific catalytic reaction, they can be modified appropriately. For example, steaming, dealumination by HCl, cation-exchange or metal loading, are processes that can be applied to tune zeolites' features for their applications as industrial catalysts. In addition, since the remarkable developments of the mesoporous materials such as SBA-15, MCM-41 and FSM-16, many kinds of mesoporous materials are also synthesized and studied. Evidently, many opportunities to develop a new catalyst based on zeolites and zeolite like materials are opened.

To summarize, industrial processes in which zeolites are used are mostly reactions catalysed by acid sites [14–26]. In other words, acidity with a controlled distribution of acid sites strengths is probably the most important property observed in the zeolite catalyst. Therefore, the estimation of features related to acidity is evidently crucial for understanding the reactivity of zeolites. The understanding of zeolites' acidity comprises knowledge of related concepts:

1. Nature of the acid sites.
2. Number of the acid sites.
3. Acid strength of the sites.
4. Distribution of the acid strength of the sites.

Experimentally, the physicochemical properties of the solid *per se*, obtained by solid-state NMR spectroscopy [27–31], or IR [32, 33] spectroscopy, give insight into the surface properties related to the acidity. However, more detailed information is accessible if probe molecules are brought into contact with the surface sites and the mode of interaction is studied. Among techniques funded on adsorption of specific probe molecules, several methods such as: temperature-programmed desorption (TPD) of amines [34–37]; UV-Vis [38–40], IR [41–43] or XPS spectroscopies [44, 45] of adsorbed probe molecules; and adsorption microcalorimetry [46–52] are applicable for the characterization of the solid materials' acidities. It is important to point out that none of mentioned methodologies can reveal all previously listed acidity concepts. In any case it has to be specified whether the concentration, strength or nature of acid sites is estimated.

Adsorption microcalorimetry allows an accurate determination of amount, strength and strength distribution of surface sites, based on the heats of adsorption of appropriate probe molecules and differential heats as a function of surface coverage. Following text discuss the applicability of adsorption microcalorimetry for the determination of zeolites' acidity, and the estimation of different factors that can influence this important property.

9.2 Factors Influencing the Acid Properties of Zeolites

The strategy for estimation of zeolites' acidity comprises (includes) the adsorption and subsequent desorption of chosen probe molecules; both events can be studied by different physical methods. From the obtained results, the facts about acidic sites (such as: the strength, strength distribution...) can be derived. As it has been already discussed (Chap. 3), the characteristics of probe molecule (its nature and size), as well as the temperature dependence of the heats of adsorption have to be considered during the estimation of zeolite' acidity. However, additional parameters that are related to zeolite of interest might influence its acidity, and have to be taken into consideration. Here, the influence of:

(1) zeolite topology;
(2) pre-treatment parameters;
(3) proton (cation) exchange level;
(4) Si/Al ratio and dealumination;
(5) the nature of framework T atom;

on acidity of zeolites and their determinations by the method of adsorption microcalorimetry will be discussed.

9.2.1 Influence of the Zeolite Topology

Topology of zeolite structures plays an important role in the number of phenomena related to this important group of solid catalysts, such as selective adsorption and diffusion. It is well known that in the cases of adsorptions performed on the molecular sieves that possess same chemical composition but different pore systems, the topology of microporous solid (the structure of the pore system and the diameters of pores openings) is responsible for the phenomenon of selective adsorption [53–60]. Although other factors have additional impact on selective adsorption (the intensity of overall electrostatic field; outer surface acidity which can diminish adsorption and even cause pore blocking by coke formation), topology of zeolite structure is the most important factor responsible for the fact that in acid-catalazyd reactions, zeolites often show shape selectivity [61–67].

The topology of the zeolite host is also crucial for the diffusion of the guest through the interior; it controls the maximum uptake achievable. Zeolites with tridirectional structures are more accessible and allow diffusion readily, whereas diffusion in monodirectional zeolites is seriously affected [68].

Besides its significant role on the catalytic behavior of zeolites through the so called ‘shape selectivity effects’, the zeolite geometry plays an important role in determining the acidity of zeolites. From the overall knowledge concerning zeolitic structures, it is known that the distances and bond angles in the Al-OH-Si group can affect the acid strength of the hydroxyl group; besides, the zeolite structure may also cause preferential location of acid sites [69, 70]. From the molecular orbital calculations for protonated Si-O-Al bridge it was suggested that the deprotonation energy decreases with the increase of the T-O-T bond angle, and consequently, the corresponding acidity increases [71–75]. However, in the attempts to estimate the influence of zeolite structure on acidity, the difficulties in experimental work arise from the fact that it is not easy to obtain samples with the same chemical composition but with different homogenous distribution of Al atoms through the zeolitic particles. Nevertheless, there are results of investigation of zeolites’ topology influence on the acidity reported in the literature. Here, the examples of such investigations done by adsorption microcalorimetry are summarized.

Influence of structure on the number and strength of available Brönsted acid sites was investigated on a series of dealuminated zeolites (HY, mordenite, ZSM-5 and

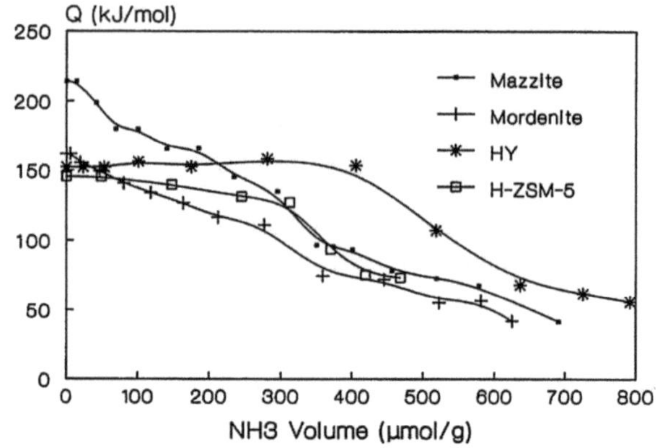

Fig. 9.2 Differential heat of ammonia adsorption versus the adsorbed amount for various zeolites [46]

mazzite) with similar framework composition (Si/ Al ratios ≈ 15). It is known that in the case of dealuminated zeolites, microporosity does not influence importantly their acidity; while the bond angles play an important role. Differential heat curves plotted as a function of coverage, represented in Fig. 9.2, shows that the number of sites is higher on HY, but the strength of the sites is greater for the mazzite sample [46]. It seems therefore that these systems contain domains of more extensive dealumination than the other samples, including less accessible zones, thus leading to strength distributions of acid sites dependent on the samples.

Contrary to this behavior, various structures with high Si/Al ratio (H-ZSM-5, H-ZSM-12, HY and H-mordenite) were found to exhibit almost the same adsorption energies measured at 480 K, suggesting that the isolated Brönsted acid sites in these materials are identical [76]. Authors observed that the strength of Brönsted acid sites associated with framework Al atoms does not vary in a given zeolite, and they suggest that Brönsted acid sites in unsteamed zeolites are independent of the sample (what means, independent of Si/ Al ratio) and equal in concentration to the framework aluminum content [77].

A study performed by adsorption microcalorimetry of NH_3 and SO_2 on a series of milled Na zeolites [78, 79] with different degrees of crystal collapse, revealed that losses in zeolite crystallinity cause a decrease in the number and strength of both Lewis acid sites and Lewis basic sites. It was observed that Lewis acid strength is strongly cristallinity dependent while Lewis basic strength is less so. Combined XPS and calorimetric results have shown that the extraframework Lewis acid sites (Na cations) are remarkably influenced by the long range stabilization effect of the zeolite lattice which gives a higher strength of Lewis acidity.

A similar study performed on crystals of HA, HX and HY zeolites [80] revealed that milling caused the destruction of the long-range symmetry of the crystal while the primary building units, namely Si(Al)-O_4 tetrahedra, remained intact. Microcalorimetry showed that the collapse in zeolite structure also caused decrease in the number and strength of acid sites, and that the Brönsted acid strength is strongly crystallinity-dependent. The destruction of long range crystal symmetry induces an

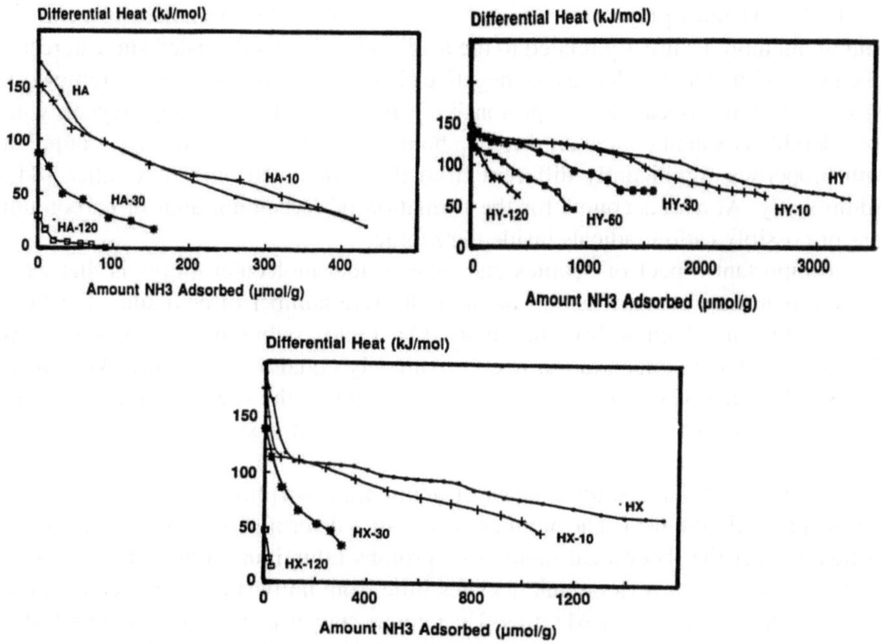

Fig. 9.3 Differential heat NH_3 of adsorption on initial and ball-milled zeolite samples (each milled sample is designated by milling time noted behind the sample name) [80]

increase of the population of weak Brönsted acid sites and a decrease of the population of strong Brönsted acid sites (Fig. 9.3).

It can be concluded that the heat curves should be similar only when the solids contain exlusively isolated sites where the next-nearest neighbors (NNN) are identical. It appears that the NNN determines Brönsted acidity, in fact, it can be said that zeolite geometry has an influence on acidity due to long-range effects as well as on the local structure of the acid sites.

9.2.2 *Influence of the Si/Al Ratio*

The Si/Al ratio has an important influence on the properties of zeolites. The Al content determines the number of cations in the framework and the properties such as the thermal and chemical stability or the polarity of internal species. Typically, zeolites with high Al content are thermally and chemically less stable, so that dehydratation at high temperature may cause partial dealumination that happens together with water desorption, what results in a certain decrease of cristallinity[81]. The hydrophilicity / hydrophobicity of zeolites are related to the polarity of the pores. Zeolites without framework Al are the most hydrophobic [82, 83].

The Si/ Al ratio plays also a significant role on acidic properties of zeolites. Aluminum atom is directly related to the acidic site, since a Brönsted site can result when the cation which balances the negative charge associated with each framework Al is a "proton". Because the "protons" are bonded to the bridging oxygen with considerable covalent character, the site should be viewed as a hydroxyl group, but with properties significantly different from that associated with pure silica [84]. Additionally, Al also accounts for the formation of carbenium and/ or carbonium ions or possibly cation radicals inside the zeolite.

An important aspect of zeolites and other acidic molecular sieves is that each of these materials contains a well defined, discrete number of acid sites. For high silica zeolites in which all bridging hydroxyls are accessible to adsorbates, such as H-ZSM-5, the site concentration is approximately equal to framework Al content [85, 86]. In some other materials, such as HY, many of the sites are inaccessible to adsorbate molecule, so the measured site concentration is significantly lower than the Al content [87].

Tsutsumi et al. have studied heats of ammonia adsorption on mordenites with different Si / Al ratio [88]. The number of acid sites determined from the heat curves plotted against the theoretical number of protons (aluminum atoms) per unit cell of mordenites, shows a slope that deviates little from unity (Fig. 9.4); the deviation being smaller for higher Si / Al ratios. The authors suggested that the framework aluminium atoms participate in the formation of acid sites and that some of the generated Brönsted acid sites are converted to Lewis acid sites. Displacement of aluminium atoms from the zeolite framework may also be responsible for the observed deviation.

To investigate how Si/Al ratio influences the acidity for a given zeolite structure, several H-ZSM-5 samples have been prepared, with Si/ Al values varying between 50 and 14 [89]. Auroux et. al. found that the Al atoms progressively incorporated in larger proportion not only create new acid sites, but also modify the strength of pre-existing acid sites as well. The initial heat of ammonia adsorption goes through a maximum. When the Al content increases, the initial heat of adsorption, and thus the

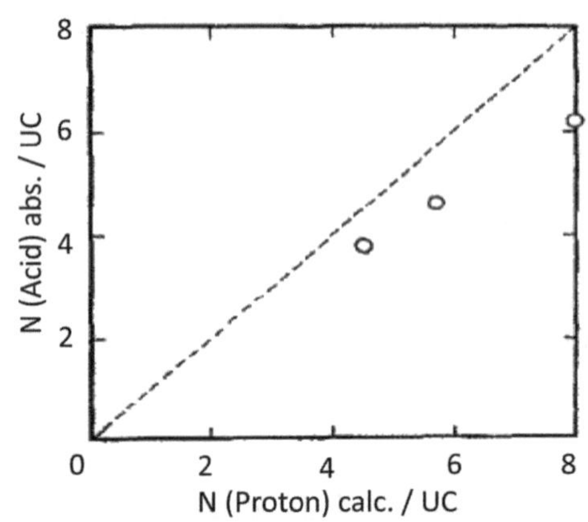

Fig. 9.4 Relationship between the number of acid sites calculated from calorimetric heat curves and the calculated number of protons on mordenite samples with different Si/Al ratio [88]

strength of the strongest acid sites increases, then remains constant (for Si/ Al between 35 and 18), and finally strongly decreases for low Si/ Al ratios. However, the integral heats of irreversible adsorption, representative of the overall acidity, increase sharply when the Al content increases from 0 to 2 atoms per unit cell, and then moderately for higher Al loadings. These results show that when relating the acidity of material to its catalytic properties, it is important to discriminate the strength of individual acid sites and the total acidity. It has been discovered that appropriate Si/Al ratio has to be discovered for a given reaction; so the importance of a careful selection of adequate zeolite sample is emphasized. For example, to prevent an excessive polymerization, the presence of very strong acid sites must be avoided, while the total acidity must be sufficient; for that purpose, ZSM-5 zeolites with a Si/Al ratio between 10 and 40 have been found to be best suited for this purpose.

The chemical composition of synthetic zeolites can be controlled either during synthesis (by varying the composition of mother gels) or by postsynthetic modifications. Dealumination process can promote modifications of porous structure, which may improve some important properties of zeolites, like: thermal and hydrothermal stability, acidity, catalytic activity, resistance to aging and low coking rate, and material transfer. Different dealumination processes have been proposed: steaming and acid treatments, as well as reactions with $SiCl_4$ or SiF_6^{2-}. From many theoretical and experimental investigations on the acidity of faujasites with different framework aluminium content, it was concluded that the number of bridging hydroxyl groups increases with the number of Al atoms in the lattice [90]. In contrast, the dependence of the acid strength on aluminium content is more complicated. From theoretical considerations, the acid strength is expected to increase with decreasing number of aluminium atoms, whereas from the studies of Al topology in the framework, a curve often presenting a maximum was derived from the number of strong sites [90].

Figure 9.5 shows that on aluminium-deficient HY zeolites, dealumination generally causes a decrease in the acid sites concentration followed by an increase in strength [91].

Mitani at al., in their work related to the investigation on the acidity of faujasites with different framework aluminium content, reported that the extent of the decrease in acid sites concentration varies with the kind of basic probe used in microcalorimetry experiments (see Fig. 9.6) [92]. The experimental results were explained by differences in molecular diameters of ammonia and pyridine, and the fact that pore size of zeolite becomes widely distributed as the dealumination proceeds; so OH groups in supercages have become available for adsorption of such large molecule as pyridine is.

In another study performed on steamed deluminated faujasites presenting both framework and non framework Al, authors found no evidence for the presence of a small concentration of very strong acid sites [93]. For low coverage, differential heats of adsorption did not show any dependence on the Si/Al ratio of the samples or on the sample preparation. Since the samples with different degrees of dealumination showed considerable different catalytic activities, it was suggested that factors other than acid strength are responsible for this behaviour. It was concluded that when

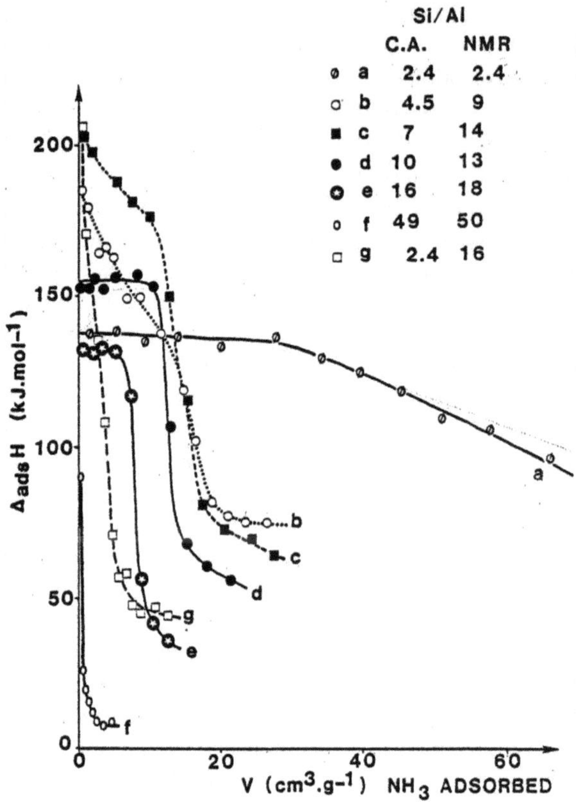

Fig. 9.5 Differential heats of ammonia adsorption over HY zeolites with various Al content as a function of coverage [91]

heterogeneity in the strength of sites is revealed by microcalorimetry (i.e. a fall of differential heat with coverage), this is the result of molecular interactions of molecules adsorbed at neighbouring sites, rather than a true indication of the differences between the sites. Strong repulsive interactions between adsorbate molecules arising from ionic repulsion of two positively charged adsorbates, from steric interactions between neighbouring adsorbates, or from changes in a protonic site due to adsorption on adjacent sites, could lead to a large decrease in measured adsorption heats with coverage. The slope of this decrease depends on the adsorbate.

Generally, it can be seen that, when differential heats of adsorption of ammonia or pyridine are plotted as a function of surface coverage for the samples with different aluminium content, at least three domains of acidity are revealed, whose relative importance depends on the aluminum content, and whether aluminium is exclusively located in framework positions (Fig. 9.6). First domain, with a sharp initial decrease in q_{diff}, is generally associated with non-framework Al and a very strong Lewis sites (heats of pyridine adsorption above ~150 kJ/mol). A region of intermediate

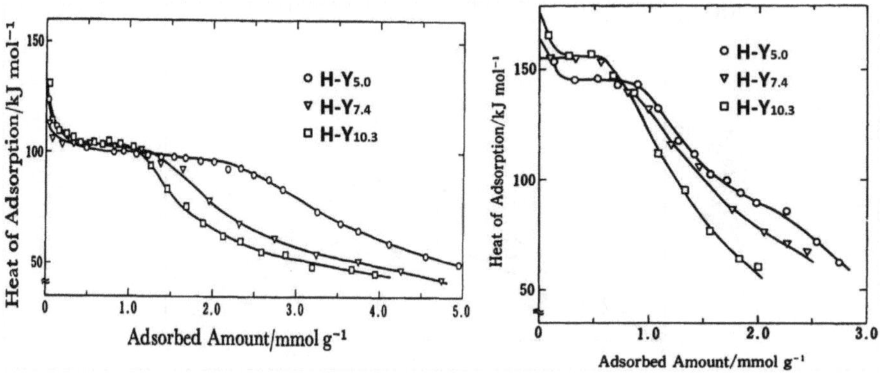

Fig. 9.6 Differential heats of adsorption of ammonia (*left*) and pyridine (*right*) at 473 K [92]

heats, represented by a plateau, follows. This plateau corresponds to strong Brönsted acid sites associated with framework Al around 140–150 kJ/mol depending on the zeolite. At the end there is again a decrease of heat, corresponding to weak Lewis acid sites associated with alkali cations and weak Brönsted sites associated with non-framework Al or neighboring species.

The heats of adsorption of ammonia and the catalytic activity in isooctane cracking are reported by Mishin et al. [94] for high-silica Y zeolites with Si/Al ratios varying from 1.25 to 100. The aluminium-deficient Y zeolites have been found to possess stronger acid sites than the parent zeolites. It has been revealed using adsorption microcalorimetry measurements that even a minor increase in the Si/Al framework ratio results in an increase of the bond strength between NH_4^+ ions and the lattice. In agreement with the changes in acidity (number of sites with $q_{diff} \approx 130$ kJ/mol), progressive dealumination of Y zeolites results firstly in an increase followed by a decrease in catalytic activity, with a maximum for 25–30 Al atoms per unit cell. The increase in catalytic activity is associated with the appearance of strong acid sites ($q_{NH_3} > 120$ kJ/mol). The effect of acidity on activity of dealuminated zeolites is explained in terms of isolated acid-site generation. It is postulated that nonframework Al atoms do not contribute to catalytic activity.

The effect of steaming on the number and strength of the acid sites is obvious from a comparison of differential heat curves for samples with different degrees of dealumination. The microcalorimetric curves show also that the strength of the sites corresponding to the intermediate plateau region first increases and then progressively decreases with steaming severity. The dependence of the acid strength distribution of dealuminated mordenites and dealuminated faujasites on the Si/Al ratio is depicted in Fig. 9.7. The number of strong acid sites presents a distinct maximum. The abscissa of the maximum corresponds to an Al content of 4.6 per unit cell (u.c.) (and Si/ Al ratio equal to 9.5) for dealuminated mordenites and an Al content of about 29 / u.c. (and Si / Al ratio around 5.5) for dealuminated Y zeolites. These values are in good agreement with the values predicted theoretically [95], where the limit values, m_{lim},

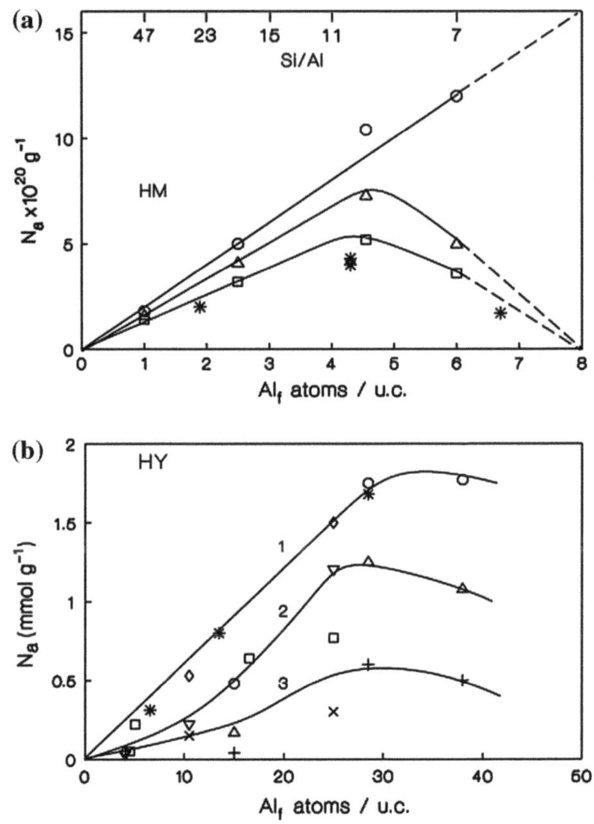

Fig. 9.7 **a** Acid strength distribution dependence on Al(IV) content per unit cell of dealuminated H-mordenites. *N*a number of acid sites in molecules g^{-1} with Q >80 kJ mol^{-1} (o), with Q >100 kJ mol^{-1} (Δ), with Q > 120 kJ mol^{-1} (□, *) from [96]. Differential molar heats of ammonia adsorption were measured at 423 K. **b** Acid strength distribution dependence on the framework aluminum content per unit cell in dealuminated Y zeolites. *Curve 1* total number of acid sites with Q >80 kJ mol^{-1} (*,o ,◇). *Curve 2*: number of acid sites with Q > 100 kJ mol^{-1} (Δ,▼). *Curve 3* number of acid sites with Q >120 kJ mol^{-1} (+,×,□) from [90]. Differential molar heats of ammonia adsorption were measured at 423 K

for mordenite and H-Y zeolites were calculated. It was found that they correspond to m_{lim} = 0.096 (or Si/ Al = 9.4) for mordenite and m_{lim} = 0.150 (Si/ Al = 5.8) for H-Y zeolites. According to Barthomeuf [95], below this maximum no Al atom has another Al atom as a next-nearest neighbour, and therefore all acid sites show a high acid strength. Above these maxima, there are Al atoms in the second coordination sphere, with the consequence that some Brönsted sites show weak acidity (although the total number of acid sites is still increasing), and their concentration decreases with increasing of framework Al content. Measurements from Stach et al. [90, 96] and Macedo et al. [97] were found to confirm this model and to be in very good agreement, though the dealumination and the measurements were not performed under the same conditions.

9.2.3 Influence of the Pre-treatment Parameters

It is known that increasing of the pre-treatment temperature modifies the surface acidity of the solids. High-temperature calcination is a well-known method of reducing total acidity of zeolites via dehydroxylation and dealumination. Although for the estimation of the real nature of an active site other techniques (spectroscopic investigation of adsorption/desorption of probe molecules) are needed, it can be told that, in principle, by using adsorption calorimetry technique it should be possible to determine the number of "strong" Brönsted sites, assuming that activation at 673K before NH_3 adsorption gives rise to a maximum in H^+, i.e. that no dehydroxylation occurs at that temperature. This is a crude approximation but can be considered as valid for comparison of different samples [48]. After calcination at increasing temperatures, dehydroxylation of the zeolite is observed: above 675K, the number of Brönsted acid sites decreases, while that of strong Lewis acid sites increases. However, a limited dealumination occurs and the constraining character of the intracrystalline voids increases. Microcalorimetric studies of ammonia adsorption confirm the very strong acidic character of the acid sites and show their dependence in strength and heterogeneity upon calcination temperature [89, 98, 99].

In a ZSM-5 zeolite (Si/Al = 19), a thermal treatment at high temperatures leads to condensation of two OH groups (representative of two protons) to form a Lewis site. Changes in pre-treatment temperature modify the differential heat curve and thus allow identification of the nature (Lewis or Brönsted) of acid sites. Figure 9.8 presents the acidity spectra (the values of dn_a/dq ratios) obtained for the adsorption

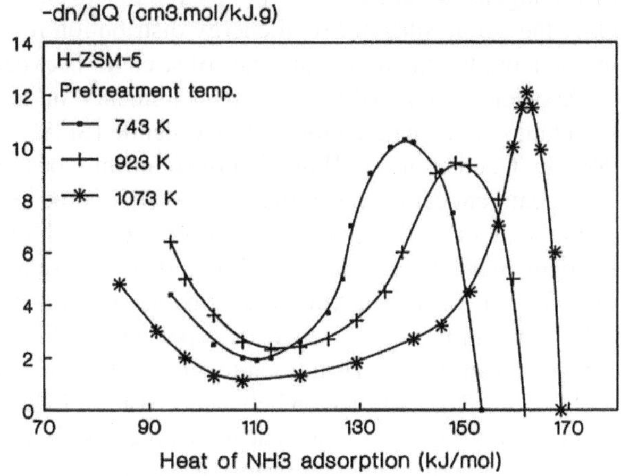

Fig. 9.8 Acidity spectra ($-dn_a/dq$ versus q_{diff}) of a ZSM-5 sample for various pretreatment temperatures [46]

of ammonia at 416 K on these ZSM-5 zeolites after the samples were pre-treated at various temperatures. The initial heat of adsorption increases as a function of outgassing temperature. From the results presented by Fig. 9.8, it is evident that the main band of the spectrum is progressively shifted towards higher adsorption heats, while its area (and thus the total number of sites) simultaneously decreases, when the sample is dehydrated at increasing temperatures (743, 923 or 1073 K). It follows that a high activation temperature (1073 K) decreases the total number of acid sites, while the acid strength of some of them is increased. The hypothesis of formation of Lewis sites stronger than Brönsted ones is thus confirmed [46].

However, the formation of Lewis acid sites cannot be revealed only from this shifting of the band in the acidity spectrum, since as already mentioned, calorimetry alone is not able to provide a simple procedure for identification of the nature of sites, and needs to be associated with another technique like infrared spectroscopy.

It is important to notice that the dehydroxylation at high temperatures produces an irreversible transformation; a sample outgased at 1073 K, rehydrated at room temperature and outgased again at 673 K, does not exhibit the initial acidity spectrum [46].

A similar microcalorimetric study has been performed by Karge et al. [100, 101] on a well-crystallized H-ZSM-5 activated at 673 K which contains very few Lewis sites, i.e. with aluminium located almost exclusively in framework positions. The microcalorimetric curve of ammonia adsorption at 423 K showed a nearly horizontal plateau around 150 kJ/mol of homogeneous acidic strength due to the interaction of ammonia with Brönsted acid sites. After high temperature dehydroxylation in vacuum at 1073 K, true Lewis sites were created, inducing a large increase in the number of sites evolving differential heat over 150 kJ/mol. These were considered to be heats which have a contribution both from the Lewis sites induced by the high activation temperature and from the remaining Brönsted sites. The overall number of titrated strong and medium acidic sites had simultaneously decreased from 2.7 to 1.6 NH_3 molecules per u.c. while the width and the homogeneity of the plateau were considerably reduced. After dehydroxylation at high temperature, no Brönsted sites of homogeneous energy distribution remained on dealuminated ZSM-5 samples. Only the acidic sites of broad energy distribution remained unaffected by the high temperature treatment, i.e. the sites of strength between 140 and 80 kJ/mol.

Another example of pre-treatment influence on acidity is the study concerning high temperature calcination of H-mordenite (Si/Al = 13) in air at 1008 K, that was done by Chen et al. [102] using adsorption microcalorimetry of pyridine at 473 K. This treatment caused a significant reduction in the total number and in the strength of the acid sites. The plateau characteristic of a large number of sites of uniform strength (near 200 kJ mol^{-1}), observed on the sample pre-treated at 673 K, also disappeared. High-temperature calcination is known to induce dealumination and dehydroxylation, both of which are expected to reduce the number of acid sites [102]. Dehydroxylation of an H-mordenite sample at 923 K caused the appearance of centres with heats of NH_3 adsorption between 170 and 175 kJ mol^{-1} which were not present in the same sample pre-treated at 703 K. Increasing the dehydroxylation temperature to 1023 K provoked the increasing of the concentration of centers characterized by a heat of ammonia adsorption of 175–170 kJ mol^{-1} to 0.2 mmol g^{-1}, and sharply

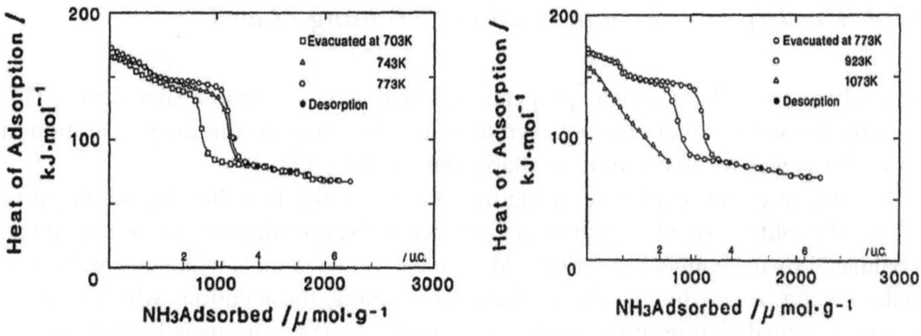

Fig. 9.9 Calorimetrically determined molar heats of adsorption of ammonia at 473 K on HM-20 evacuated at various temperatures. Filled symbols represent heats of re-adsorption on samples which were evacuated at 473 K after the first run of the adsorption measurement [88]

decreased the concentration of centres generating heats of 160–130 kJ mol^{-1} [103]. It was shown that the latter values are typical of the dissociation of ammonia over Lewis acid centers.

The acidic properties of a mordenite zeolite with a Si/Al ratio of 10 were analyzed by Tsutsumi et al. [88] through calorimetric measurements of the heats of ammonia adsorption done at 473 K, on samples previously evacuated at temperatures varying from 703 to 1073 K, as can be seen from Fig. 9.9. Drastic energy changes were observed between low and high coverage, with the exception of the sample evacuated at 1073 K. Both the number of the more energetic sites and their energy increased with an increase in the evacuation temperature, and reached a maximum at 773 K (around 170–175 kJ mol^{-1}); then a decline was observed. The heat curve of the sample evacuated at 1073 K did not exhibit a plateau, and its initial value was much less than those of the other samples (the value of 160 kJ mol^{-1} was obtained, instead of 175 kJ mol^{-1}). This may be attributed to the breakdown of part of the mordenite structure and/or to the local formation of amorphous species [88].

A sample of mordenite (98 % degree of ammonium-ion exchange) deammoniated at various temperatures from 693 to 923 K was studied at 573 K by NH_3 adsorption microcalorimetry by Bankos et al. [104]. On increasing the pre-treatment temperature, the number of acid sites passed through a maximum at 753 K as a result of simultaneous decationation and dehydroxylation. The heat of adsorption of NH_3 on Brönsted acid sites formed by decationation was 110–160 kJ mol^{-1}. During dehydroxylation, two types of Lewis sites were formed, characterized by heats of NH_3 adsorption of 170–185 kJ mol^{-1} and 95–100 kJ mol^{-1} respectively, on which dissociative chemisorption of ammonia was evidenced by IR [103].

9.2.4 The Effect of Proton (Cation) Exchange Level

Ion exchange is a characteristic property manifested by most molecular sieves. This property is used routinely for post synthesis modification during the preparation of molecular sieves for major industrial applications [10, 11].

The origin of ion exchange properties resides in the fact that the net negative charge of zeolite network, equal to the number of the constituent aluminium atoms, is balanced by exchangeable cations, M^{n+}, located in channels which normally also contain water. According to the synthetic procedures, these cations which neutralize the electrical charge of the aluminosilicate framework, are usually Na^+, K^+ or template cations. This charge-compensating entities can further be exchanged for other cations. Conventionally, ion exchange is carried out by suspending the zeolite powder in aqueous solutions of salts which contain the desired in-going cation, or in mineral acids to introduce protons into acid resistant zeolites [10, 11]. The same can be done by solid state ion exchange, where dry powders of zeolites and salts or oxides containing the exchangeable cations are reacting [105].

Evidently, ion exchange capacities can be considered as functions of the quantity and distribution of aluminium atoms within the structure of zeolites. In addition, it is worth to notice that zeolites often possess high ion exchange selectivity for certain cations, and this can be used for their concentration and isolation. Molecular sieving and acid properties of zeolites are modified in this way, what gives possibility to tailor their catalytic activity. The framework anions and exchangeable cations form electrostatic fields which strongly interact with electronic structure of adsorbate molecules [106]. The reactivity of adsorbed species depends strongly on the number and kind of cations of the zeolite under investigation.

The exchange degree plays an important role in the heterogeneity of the acid sites of zeolites. From the study performed on NaY parent sample and NH_4NaY zeolites exchanged to 29, 56, 80, 90, and 94 % respectively, it was observed that acidity appears slowly at the early stages of replacement, and that up to about 60 % exchange, the number of acid sites present in the zeolites is four times lower than the amount of sodium replaced [107, 108]. At higher extents of decationation, the density of acid sites increases rapidly and approaches a theoretical value [108]. A question is how the framework charge balance can be achieved at low exchange degree, if the number of cations that are removed is greater than that of the acid sites formed. Two alternative interpretations were offered: either the hydroxyls generated at low exchange levels are unstable and can be eliminated during pretreatment, or an incomplete decomposition of NH_4^+ ions occurs upon thermal activation, the latter hypothesis being supported by TPD spectra. To rationalize the observed acidity-exchange level profile a model was developed, extending the shielding power of a single Na^+ ion to several neighboring AlO_4^- tetrahedra. The catalytic activity of these decationated faujasites in the cracking of isooctane and the disproportionation of ethylbenzene was also found to vary in the same manner as the number of acid sites, increasing rapidly beyond 70 % exchange degree. However, the observed discrepancy between the two increases (the catalytic activity increased much faster) prevented

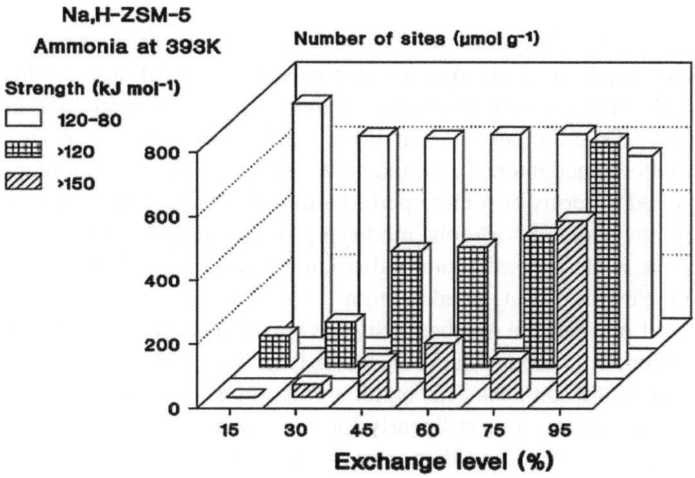

Fig. 9.10 Acid site strength distribution of Na,H-ZSM-5 zeolites as a function of the exchange level. Ammonia adsorption at 393 K, pretreatment at 673 K (111)

direct interpretation of the catalytic activity in terms of acid sites population. It is reasonable to assume that both the number and the strength of the acid sites are affected by the exchange level.

Similar study was performed on decationated mordenites [107], which have narrower pores and a higher Si/Al ratio than Y zeolites. From microcalorimetric measurements it was found that the number of acid sites increases nearly linearly with an increase of decationation level. Growth of the exchange level increased the acid strength, and removal of the last NH_4^+ ions resulted in the formation of strong sites ($q_{diff} > 130\,kJ\,mol^{-1}$). This increased concentration of strong acid sites explained that mordenites are more active in the cracking of n-octane than HY zeolites. However with bulkier reactants such as isooctane and ethylbenzene, the difference in catalytic activity decreases due to shape selectivity effects.

In a study by Muscas et al., a variety of NaHZSM-5 zeolites with varying extent of Na exchange was subjected to adsorption of various probe molecules in order to determine the selectivity of adsorption of these molecules on their acid centres of variable nature, such as Na^+ in Na forms and acidic OHs in the corresponding H forms [109]. Using ammonia as a basic probe it has been shown that the number of acid sites also increased quasi-linearly with the exchange level, while the acid strength increased monotonously and moderately up to about 40 % of exchange level, remained almost constant at intermediate exchange level, and then dramatically increased when 80 % of the original Na^+ ions were removed. The removal of residual Na^+ ions not only generated much stronger sites but resulted also in a general increase of the acid site population already present. Figure 9.10 clearly shows that at low exchange levels, most of the acid sites were rather weak sites. While this population of weak sites remained almost constant as the exchange level increased, the population of stronger sites increased progressively up to the point where, for extensively exchanged samples, the strongest sites became predominant. The population of sites, the heat of adsorption which was above 150 kJ mol^{-1}, illustrated the effect of removal of the very last sodium ions on the acid strength, not only on that of newly created sites but also on that of pre-existing ones. This phenomenon, fairly common to those encountered with faujasites [96, 108, 109] and mordenites [107], was interpreted in terms of remote perturbation of the structure of the acid centers by the very few residual Na^+ ions modifying significantly the T-O-T bond angles. In addition, possible general cation redistribution induced by NH_3 adsorption may also affect the evolved heats of adsorption.

Figure 9.11 presents the number of sites evolving heats above 110 kJ mol^{-1} upon NH_3 adsorption, as a function of the exchange level; for ZSM-5 [109], faujasite [107] and mordenite [108] samples. This number of strong sites increases in all cases, but in different proportions; almost linearly for mordenite and ZSM-5, and first slowly (for low exchange levels) and then sharply (at high exchange level) for faujasite. Concomitantly, the catalytic efficiency increases as does the acid strength. But it is not clear yet how the acid strength increases with increased exchange level.

The nature of the exchanged cation is one of the key points that determine acidity in zeolites. The acid-base properties of alkali-exchanged X and Y zeolites have been studied using microcalorimetry of ammonia, pyrrole or SO_2 probes. Generally, the alkali-exchanged X zeolites are more basic than the corresponding Y zeolites. It was shown that the basic strength increases quasi-linearly with the negative charge of oxygen atom, calculated from the Sanderson electronegativity equivalence method. Pyrrole chemisorbed on the basic site creates a bonding between the framework oxygen (Lewis basic site) and the H atom of the NH group of the pyrrole molecule, so the stronger base (Cs) will produce a larger heat of adsorption. Briefly, a thermodynamic scale of the Lewis basic strength distribution of alkali-exchanged X and Y zeolites was obtained, supporting the conclusion that the Lewis basicity in alkali-exchanged zeolites is a local property, strongly influenced by the adjacent alkali cation [110].

The acidic properties of a series of X faujasites exchanged with Li, Na, K, Rb and Cs have been studied by adsorption microcalorimetry, using ammonia as acidic probe. The heats of NH_3 adsorption were found to decrease in the sequence from Li to Cs (Fig. 9.12).

Li and Na zeolites presented much higher heats of NH_3 adsorption and greater coverage at the same pressure in comparison with the other zeolites. The acid-base properties of alkali-metal ion exchanged X and Y zeolites have also been investigated by ammonia and sulphur dioxide adsorption microcalorimetry, in parallel with the study of a catalytic reaction, viz. 4-methylpentan-2-ol conversion [111].

9.2.5 The Influence of the Framework T Atom

Isomorphous substitution is an important way to modify zeolite properties for practical applications and has achieved considerable interest in the field of zeolite chemistry. The thermodynamics of this process have been considered by Barrer [112]. The

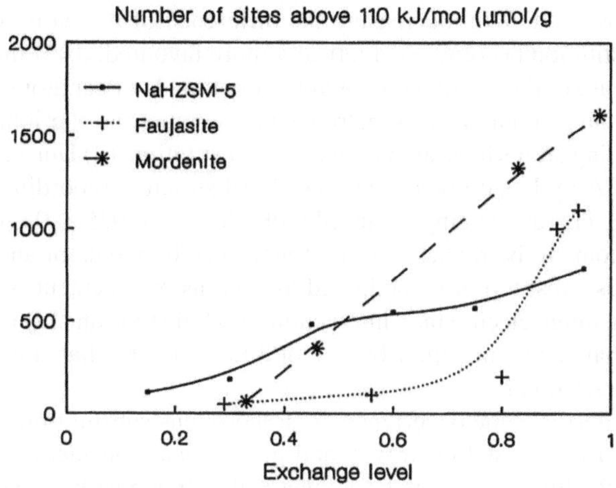

Fig. 9.11 Number of acid sites evolving heats above 110 kJ mol^{-1} upon NH_3 adsorption as a function of exchange level for HNa-ZSM-5 (■), HNaY (+) and HNaM (*) zeolites

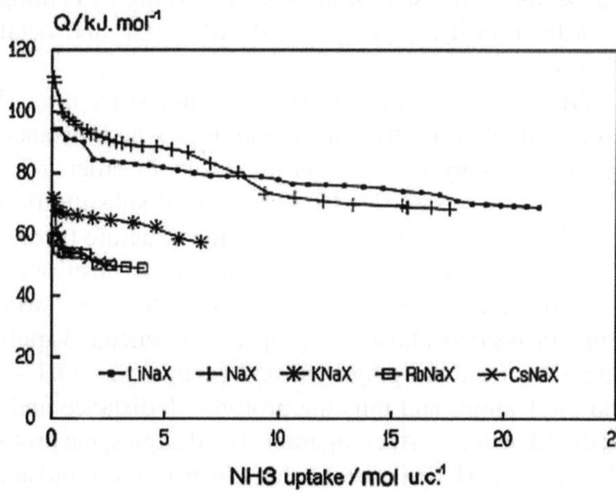

Fig. 9.12 Differential heats of NH_3 adsorption on alkaline X zeolites versus coverage [90]

phenomenon of isomorphous substitution is well-known in the field of mineralogy [1, 113]. By isomorphous substitution, framework atoms of crystalline compounds are replaced by atoms of other elements without changing the type of the crystal structure. Most elements can undergo substitution, at least to a very low degree. Even chromium, which prefers octahedral coordination, may substitute silicon in tetrahedral framework positions, to a certain extent [114, 115]. Extremely low degrees of substitution (0.2 atom %) can give rise to a remarkable change of properties.

The main factors governing isomorphous substitution are summarized in the following way [116]. (a) The tendency of substitution depends on the ratio of radii of the atoms involved. With decreasing the difference of the radii Δr of atoms A and B, the substitution becomes energetically more favoured. (b) If the replacement leads to a decrease of the coordination number of atom A, larger atom A will replace more easily smaller atom B. (c) Electronegativity, ratio and the ionization potentials of exchanging atoms have an influence on substitution. (d) During isomorphous replacement of A by B with preservation of local structure (coordination number), a minimum free energy is being achieved with $\Delta r/r = 0.025$–0.03, r denoting the radius of the atoms to be replaced in the framework by atoms of another element. (e) Isomorphous substitution is facilitated as long as replacement of A by B does not change long range electrostatic interactions. (f) Substitution may occur when the charges of exchanged atoms differ by 1, 2 or 3 units. (g) Exchanging atoms should not react with each other.

The main criteria for the occurrence of isomorphous substitution, which are primarily derived from crystal chemistry and geometrical considerations, have been formulated by Pauling [117]. The basic idea is that framework of crystals, i.e., silicates including zeolites, consist of packages of negatively charged oxygen anions (O^{2-}). Therefore, tetrahedral and octahedral vacancies are formed. The size of these vacancies depends on the size of anions. According to Pauling, cations prefer tetrahedral coordination if $r_{Me}/r_{O2-} = 0.214 - 0.4$, and octahedral sites if $r_{Me}/r_{O2-} = 0.4 - 0.6$.

It should be said that all rules cited above are not strict laws. The ability of the framework to change its fine structure (bond angles and distances) by rotation of tetrahedra, by tilting, or by inversion is important in order to relax the strain resulting from the substituting atom [118]. The aim of substitution is to integrate the new element in the structure while preserving the structure type. Often a certain percentage of the modifier remains in the extra-framework positions.

The motivation to replace aluminium in aluminosilicate zeolite structures by other elements arose from the need to adjust their properties to intended applications. Since the nature and strength of the bridging hydroxyl groups (Si – OH – T, T = Al, Fe, Ga, B, ...) depend on T atom, and thus the proton – T distance and resulting acid strength of the modified material. At comparable bond angles, the proton – T distance decreases in order Fe, Ga, Al. This means that electrostatic repulsion between the proton and T increases in the same way, and that the acidity is expected to increase in the same order [48]. These changing in the acidity are easily accessible by adsorption calorimetry technique.

The use of organic templates has rendered possible the substitution of many elements, including other trivalent (Cr^{3+}), bivalent (Be^{2+}), and tetravalent ions (Ge^{4+}, Ti^{4+}). Most of these metallosilicate compositions have been synthesized with ZSM-5 crystal structure. Adsorption microcalorimetry enabled the studying of acidic properties of thus obtained materials, as in the case of parent zeolites.

A co-incorporation of aluminium and boron in the zeolite lattice has revealed weak acidity for boron-associated sites [119] in boron-substituted ZSM-5 and ZSM-11 zeolites. Ammonia adsorption microcalorimetry gave initial heats of adsorption of about 65 kJ mol^{-1} for H-B-ZSM-11 and showed that B-substituted pentasils have only very weak acidity [120]. Calcination at 1073 K increased the heat of NH_3 adsorption to about 170 kJ mol^{-1} by creating the strong Lewis acid sites. The lack of strong Brönsted acid sites in H-BZSM-11 was additionally confirmed by poor catalytic activity in methanol conversion and in toluene alkylation with methanol.

Gallium has been successfully introduced into numerous zeolite frameworks (β, MFI, offretite, faujasite, ...). The Ga^{3+} ions in zeolites can occupy both tetrahedral framework sites (T) and non-framework cationic positions. The isomorphous substitution of gallium into aluminosilicate zeolites results in modified acidity and subsequently modified catalytic activity such as enhanced selectivity towards aromatic hydrocarbons.

Microcalorimetric experiments with ammonia and pyridine as probe molecules have been used to investigate the effects of framework Ga on the acidic properties of several zeolites [121–127]. Experiments of NH_3 adsorption microcalorimetry, together with FTIR results from pyridine thermodesorption, have shown that the isomorphous substitution of Al by Ga in various zeolite frameworks (offretite, faujasite, β) leads to reduced acid site strength, density, and distribution [122–125]. To a lower extent, a similar behaviour has also been observed in the case of a MFI framework [126, 127]. A drastic reduction in the acid site density of H,Ga-offretites has been reported, while the initial acid site strength remained high [122, 124].

For Ga-β zeolite it was found that, when the Si/Ga ratio increased from 10 to 40, the number of strong sites decreased drastically for Si/Ga between 10 and 25 and then reached a plateau above Si/Ga = 25 [121]. The strength and density of acid sites in H-(Ga,La)-Y have also been found to be lower than those in H-Y crystals of the type used in FCC preparation (LZY-82) [123].

Parrillo et al. [128] have used microcalorimetric measurements of ammonia and pyridine adsorption to compare the acid sites in H-[Fe]ZSM-5, H-[Ga]ZSM-5, and H-[Al]ZSM-5. On each of the molecular sieves, the differential heats of adsorption for both ammonia and pyridine were constant up to coverage of one molecule per Brönsted site. The differential heats at a coverage below 1 : 1 were identical on each of the materials, with values for ammonia of 145 ± 5, 150 ± 5, and 145 ± 5 kJ mol^{-1} on H-[Fe]ZSM-5, H-[Ga]ZSM-5, and H-[Al]ZSM-5, respectively; and for pyridine of 195 ± 5, 200 ± 5, and 200 ± 5 kJ mol^{-1} on H-[Fe]ZSM-5, H-[Ga]ZSM-5, and H-[Al]ZSM-5, respectively. The authors [128] concluded that the microcalorimetric heats of adsorption for ammonia and pyridine at Brönsted acid sites formed by framework Fe(III) and Ga(III) were very similar to heats of adsorption at Al(OH)Si sites, and that the three samples were effectively equivalent proton donors. By contrast, they found very different reactivity measurements for *n*-hexane cracking and propene oligomerization on the same materials. The authors claimed that heats of adsorption for strong bases do not reflect differences in inherent acid strength and may not be related to catalytic activity in a simple manner.

Iron silicates of MFI structure with various Si/Fe ratios have also been studied by NH_3 adsorption microcalorimetry [127, 129], and compared with the Al and Ga analogues. The intermediate strength sites (predominantly Brönsted sites) were found to correspond to a plateau around 145 kJ mol^{-1} for H-[Al]-ZSM-5 (Si/Al = 19), 140 kJ mol^{-1} for H-[Ga]-ZSM-5 (Si/Ga = 22), and 135 (Si/Fe = 41), 125 (Si/Fe = 26), or 120 (Si/Fe = 12) kJ mol^{-1} for H-[Fe]-ZSM-5, respectively.

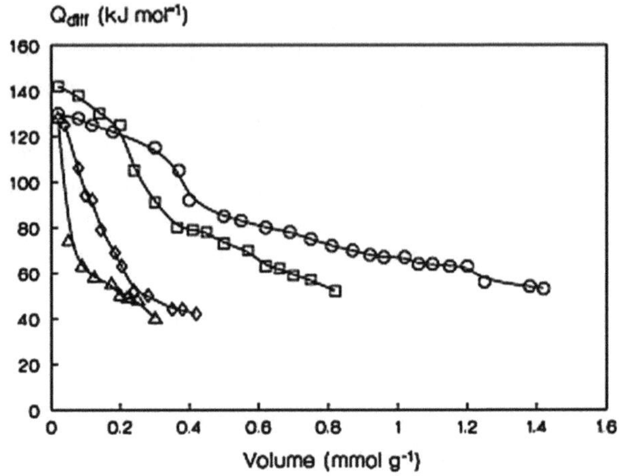

Fig. 9.13 Differential molar heats of ammonia chemisorption at 423 K on MFI zeolites as a function of the adsorbed amount: □ Al-Sil, ○ Fe-Sil, ◇ In-Sil, Δ silicalite. Pretreatment temperature 673 K ([132])

Besides, the adsorption of acetonitrile as a probe molecule has been studied using microcalorimetry at 400 K, in the case of H-[Fe] ZSM-5 and H-[Al] ZSM-5 [130]. The heats of formation of the complexes were found to differ slightly, ca. 95 kJ mol^{-1} on H-[Fe] ZSM-5 compared to ca. 110 kJ mol^{-1} on H-[Al] ZSM-5; suggesting that the hydrogen bonds at FeOHSi sites may be slightly weaker.

Ga- and Fe-substituted MFI zeolites have been investigated using adsorption microcalorimetry of different alkanes at 353 K by Auroux et al. [131]. The acid strength of the zeolite protons decreased following the sequence H-[Al]MFI > H-[Ga]MFI >H-[Fe]MFI. The heats of adsorption decreased with the basicity of the alkane in the order *n*-butane > isobutane > propane.

The active sites of isomorphously substituted MFI structures activated at 673 K have been characterized by Jänchen et al. [132] using microcalorimetric measurements carried out at 423 K with ammonia as a probe. In accordance with decreasing heats of NH_3 adsorption, the Brönsted acid site strength of the modified MFI was reported to decrease in the sequence Al > Fe > In > silicalite. In addition to these strong sites, weaker Lewis centres due to the non-framework material were found (Fig. 9.13).

Crystalline MFI-type indosilicates containing indium ions in framework positions were hydrothermally synthesized by Vorbeck et al. [133] and characterized by adsorption microcalorimetry of NH_3 at 423 K (after activation of the sample at 670 K). They exhibited rather weak acidity attributed to Brönsted sites. However, the number of acid sites with a sorption heat between 80 and 110 kJ mol^{-1} was significantly higher than for silicalite [133].

Titanium-substituted silicalite can be prepared with a homogeneous distribution of Ti ions in the crystal. The Ti^{4+} ions seem to be all surrounded by four NNN Si ions and, thus, the catalytic site is an isolated Ti^{4+} ion. For TS-1 (titanium silicate molecular sieve, zeolite of pentasil family), relatively low heats of adsorption due to coordinatively bonded ammonia were detected by Jänchen et al. [132]. Moreover, the amounts of adsorption with heats higher than found for silicalite correlated with the amount of Ti in the sample [132].

Similarly, the acidity of titanium-silicalites with different titanium contents was characterized by adsorption calorimetry at 353K of various probe molecules by Muscas et al. [134]. These molecular sieves had a molar composition $xTiO_2$ $(1 - x)SiO_2$, where x ranged from 0 to 0.02. Subjected to ammonia adsorption, these solids showed an acidic character compared to a pure silicalite-1 sample. A small amount of titanium induced a high increase in the strong acid sites (Q_{init} = 160–170 kJ mol^{-1} instead of 75 for silicalite-1). The integral heat and the total amount of acid sites increased with increasing titanium loading and then reached a plateau for $x \geq 0.014$. All the curves showed a sharp decrease in Q_{diff} at very low coverage. The next region corresponded to a plateau with heats evolving around 70 kJ mol^{-1}, instead of 40 kJ mol^{-1} found for silicalite-1. Other basic probes such as pyridine and substituted pyridines (DMP) were also used in an attempt to identify, by selective adsorption, the different sites of these catalysts [134]. The heats and amounts adsorbed were in the order: pyridine > 3.5 lutidine > 2.6 lutidine.

The adsorption properties of titanium silicalites-1 synthesized via two different routes (in the presence or the absence of sodium in the precursor gel) have been compared by Auroux et al. [135]. Adsorption calorimetric measurements of a basic probe (NH_3) and an acidic probe (SO_2) showed that these solids were very acidic compared to a silicalite-1 sample. The presence of Na in the different samples decreased the number and the strength of the acid sites. The modification strongly depended on the synthesis procedure [135].

The treatment of Ti-silicalite-1 (TS-1) and silicalite by aqueous solutions of ammonium acetate has been shown to suppress the most energetic sites on these two catalysts, as evidenced by the heats of adsorption which were much lower for the treated samples than for the untreated ones [136], while the number of NH_3 molecules absorbed per Ti atom was unaffected by the treatment in the case of TS-1.

In another study, the heterogeneity of framework Ti (IV) atoms in Ti-silicalite (TS-1) was studied by NH_3 adsorption calorimetry [137] and compared to a Ti-free silicalite taken as reference material. The evolution of the heat of adsorption with coverage was found to be typical of heterogeneous surfaces, not only due to the presence of sites active towards ammonia on the silicalite matrix, but also due to the heterogeneous distribution of Ti (IV) sites. This suggests that a considerable number of framework sites (among the 12 available in the MFI framework) is occupied in a nearly equally distributed manner.

9.3 Correlation Between Adsorption Heat and Catalytic Activity

Finding correlations in between catalysts surface properties and catalytic behaviour is the main reason to study adsorption of probe molecules on catalysts surfaces. If the probe molecules are chosen to resemble possible reaction intermediates of the catalytic cycle, measurements of heat of adsorption of such probe molecules can provide essential information about reaction mechanisms.

Finding correlations between the heats of adsorption and the catalytic behaviour (activity and/ or selectivity) of the solid catalysts is not an easy task. Surface sites may exist in different configurations what typically brings to the distribution of adsorption sites' strengths. Among the active sites present on the surface of catalyst certain number may possess adequate strength to activate the adsorbed molecule and to form a reactive intermediate [52]. In other words, some of the sites may be too weak to activate the reactants or they can be too strong, leading to strongly held species which block and deactivate these sites or cause excessive fragmentation of reactants or products.

Therefore, the determination of the strength and strength distribution of surface active sites is of fundamental importance, because in this way one can distinguish the undesired sites from the desired ones. This further gives possibility of creating novel catalysts, which properties can be properly tuned for the target reaction. In revealing these very important features, adsorption microcalorimetry plays one irreplaceable role.

In a study performed by Auroux et. al. [111] microcalorimetry experiments of ammonia and sulfur dioxide were performed in order to analyze the possible correlations between the acidity and basicity of the alkali-metal ion-exchanged X and Y zeolite structures and their catalytic properties. The catalytic results for the 4-methylpentan-2-ol conversion show that activity and selectivity are both affected to some extent by the acid-base character of the catalysts. The activity was found to increase in order Cs > Rb > K > Na > Li for both X and Y zeolites. The dehydrogenation reaction occurs only on $CsX + Cs_2O$, which presents very strong basicity. The product selectivity of the reaction depends on both Lewis acidity and basicity; Lewis basic or acidic sites of zeolites can be considered as acid-base pairs, in which both framework basic oxygens and neighboring cations are important. The selectivity ratio between the 1-alkene and (2-alkene+isomers) increases linearly with the ratio between basic and acid sites number, n_{SO2}/n_{NH3}, for both X and Y zeolites, as shown in Fig. 9.14.

In another study [138], authors presented evidence that heats of adsorption of ammonia on three different kinds of zeolite structure (Y, mordenite, ZSM-5) can be used to obtain the correlation plots that describe relationships between acidic and catalytic properties of these catalysts. Catalytic properties were tested in simple carbonium ion reactions. An approach to search "acidity-catalytic activity" correlations was based on the straightforward concept which implies that the heat of adsorption of the base is directly related to the energy needed for the protonation of hydrocarbon molecules leading to the carbocation formation. For example it was

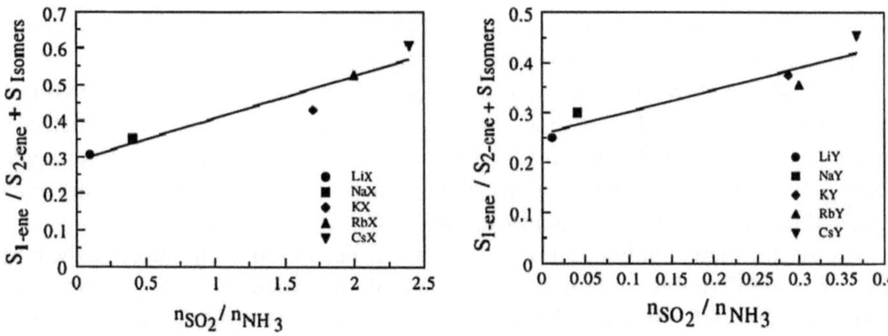

Fig. 9.14 1-ene/(2-ene+isomers) selectivity ratio versus the basic/acidic site number for X samples left and for Y samples right

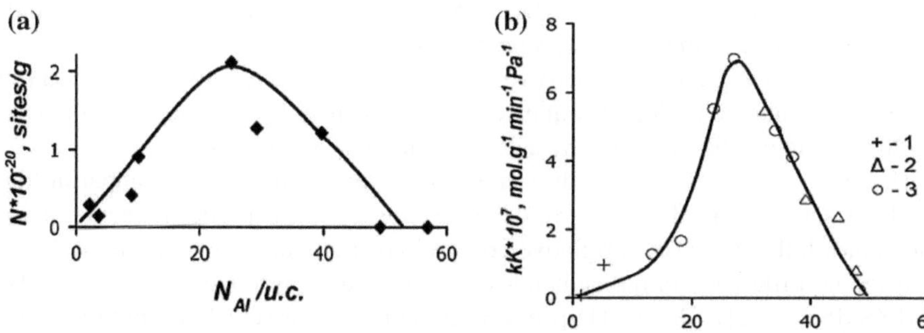

Fig. 9.15 Variation in the number of strong acid sites with $Q_{NH3} = 122–136$ kJ/mol (**a**) and activity in cracking of isooctane as a function of N_{Al} for *Y* zeolites (**b**) prepared by treating the stabilized *Y* zeolites with HCl (1), prepared by thermal dealumination at 600–700°C (2) and produced by treatment of sodium faujasites with $SiCl_4$ (3)

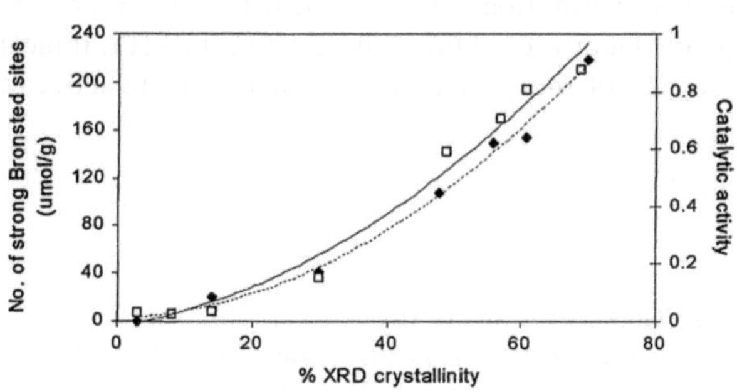

Fig. 9.16 Plots of number of strong Brønsted acid sites (◇) and catalytic activity for *n*-hexane cracking (□) as a function of percentage XRD crystallinity of ZSM-5-based samples[139]

found that the variation of strong acidity and the change of activity with progressing dealumination of Y zeolites show similar trend (Fig. 9.15).

In a study on ZSM-5 based materials [139], Nicolaides et. al. have shown that the number of Brönsted acid sites, determined by adsorption microcalorimetry of ammonia, increases with increasing XRD cristallinity. A strong correlation was observed between the catalytic activity of these materials, in n-hexane cracking reaction, and the number of strong acid sites, as presented in Fig. 9.16.

Microcalorimetry study of NH_3 adsorption has been used to characterize the acid sites of H-USY zeolite and another USY sample in which the strong Lewis sites were poisoned with ammonia. Poisoning of Lewis acid sites did not affect the rate of deactivation, the cracking activity, or the distribution of cracked products during 2-mehylpentene cracking. From these findings it was concluded that strong Lewis sites does not play any important role in cracking reactions [140].

The effect of the Si/Al ratio of HZSM-5 zeolite-based catalysts on surface acidity and on selectivity in the transformation of methanol into hydrocarbons has been studied using adsorption calorimetry of ammonia and tert-butylamine. The observed increase in light olefin selectivity and decrease in methanol conversion with increasing Si/Al ratio can be explained by a decrease in total acidity [141].

However, it is worth noting that experiments aimed at directly comparing the cracking activity and the enthalpy of adsorption of basic probe molecules have sometimes failed to make correlative conclusions. For example, Gorte et al. [128] detected no difference in the enthalpies of ammonia or pyridine adsorption on H-[Al] ZSM5, H-[Ga] ZSM5 and H-[Fe] ZSM5 samples whose cracking activities were quite different. Likewise, in a study on dealuminated Y zeolites, no significant differences in the highest enthalpies of adsorption could be detected by microcalorimetry for samples of different cracking activities [93]. Kuehne et al. [142] confirm these conclusions by the results of another study where a 57 kJ mol^{-1} difference in the enthalpy of ammonia adsorption was detected between HY and (H, NH4)- USY sample, although the latter was 33 times more active than HY. It means that there is only a certain category of sites with a given strength that are active in the reaction.

References

1. D.W. Breck, *Zeolite Molecular Sieves: Structure Chemistry and Use* (Wiley, New York, 1974)
2. P.A. Jacobs, *Carboniogenic Activity of Zeolites* (Elsevier Science Ltd., Amsterdam, 1977)
3. J.V. Smith et al., Topochemistry of zeolites and related materials. 1. Topology geometry. Chem. Rev. **88**, 149–182 (1988)
4. W.M. Meier, *Molecular Sieves* (Society of Chemical industry, London, 1968)
5. R. Fricke, H. Kosslick, G. Lische, M. Richter et al., Incorporation of gallium into zeolites: syntheses, properties and catalytic application. Chem. Rev. **100**, 2303–2405 (2000)
6. Database of zeolite structures (2000) The International Zeolite Association, http://www.iza-structure.org/databases/, Accessed 24 Sept. 2011
7. A. Dyer, *An Introduction to Zeolite Molecular Sieves* (Wiley, New York, 1988)
8. W.M. Meier, Zeolites and zeolite like materials. Stud. Surf. Sci. Catal. **28**, 13–22 (1986)

9. J.B. Uytterhoeven, L.C. Christner, W.K. Hall et al., Studies of the hydrogen held by solids. VIII. The decationated zeolites. J. Phys. Chem. **69**, 2117–2126 (1965)
10. R.P. Townsend, R. Harjula, in *Ion Exchange in Molecular Sieves by Conventional Techniques*, vol. 3, ed. by H. Karge, J. Weitkamp. Molecular Sieves–Science and Technology, (Springer, Berlin, 2002), pp. 2–42
11. R.P. Townsend, Ion exchange in zeolites. Stud. Surf. Sci. Catal. **58**, 359–390 (1991)
12. W.E. Farneth, R.J. Gorte et al., Methods for characterizing zeolites acidity. Chem. Rev. **95**, 615–635 (1995)
13. M. Niwa, N. Katada, K. Okumura, *Characterization and Design of Zeolite Catalysts* (Springer, Berlin, 2010)
14. H. Ichihashi, M. Ishida, A. Shiga, M. Kitamura, T. Suzuki, K. Suenobu, K. Sugita et al., The catalysis of vapor-phase Beckmann rearrangement for the production of ε-caprolactam. Catal. Surv. Asia. **7**, 261–270 (2003)
15. S. Shimizu, N. Abe, A. Iguchi, H. Sato et al., Synthesis of pyridine bases: general methods and recent advances in gas phase synthesis over ZSM-5 zeolite. Catal. Surv. Jpn. **2**, 71–76 (1998)
16. H. Ishida et al., Liquid-phase hydration process of cyclohexene with zeolites. Catal. Surv. Jpn. **1**, 241–246 (1997)
17. H. Tsuneki, M. Kirishiki, T. Oku et al., Development of 2,2'-iminodiethanol selective production process using shape-selective pentasil-type zeolite catalyst. Bull. Chem. Soc. Jpn. **80**, 1075–1090 (2007)
18. M. Iwamoto, H. Yahiro, K. Tanda, N. Mizuno, Y. Mine, S. Kagawa et al., Removal of nitrogen monoxide through a novel catalytic process. 1. Decomposition on excessively copper-ion-exchanged ZSM-5 zeolites. J. Phys. Chem. **95**, 3727–3730 (1991)
19. X.B. Feng, W.K. Hall et al., FeZSM-5: a durable SCR catalyst for NOx removal from combustion streams. J. Catal. **166**, 368–376 (1997)
20. Y.J. Li, J.N. Armor et al., Catalytic reduction of nitrogen oxides with methane in the presence of excess oxygen. Appl. Catal. B Environ. **1**, L31–L40 (1992)
21. D.J. Wang, J.H. Lunsford, M.P. Rosynek et al., Characterization of a Mo/ZSM-5 catalyst for the conversion of methane to benzene. J. Catal. **169**, 347–357 (1997)
22. Y. Ashina, T. Fujita, M. Fukatsu, K. Niwa, J. Yagi et al., Manufacture of dimethylamine using zeolite catalyst. Stud. Surf. Sci. Catal. **28**, 779–786 (1986)
23. A. Corma, V. Martinez-Soria, E. Schnoeveld et al., Alkylation of benzene with short-chain olefins over MCM-22 zeolite: catalytic behaviour and kinetic mechanism. J. Catal. **192**, 163–173 (2000)
24. U. Freese, F. Heinrich, F. Roessner et al., Acylation of aromatic compounds on H-Beta zeolites. Catal. Today. **49**, 237–244 (1999)
25. T.R. Hughes, W.C. Buss, P.W. Tamm, R.L. Jacobson et al., Aromatization of hydrocarbons over platinum alkaline earth zeolites. Stud. Surf. Sci. Catal. **28**, 725–732 (1986)
26. R.E. Jentoft, M. Tsapatsis, M.E. Davis, B.C. Gates et al., Platinum clusters supported in zeolite LTL: influence of catalyst morphology on performance in n-hexane reforming. J. Catal. **179**, 565–580 (1998)
27. H. Pfiefer, D. Freude, J. Kärger, *Catalysis and Adsorption by Zeolites* (Elsevier, Amsterdam, 1991)
28. M. Hunger, D. Freude, D. Fenzke, H. Pfeifer et al., ^{1}H solid-state NMR studies of the geometry of Brönsted acid sites in zeolites H-ZSM-5. Chem. Phys. Lett. **191**, 391–395 (1992)
29. T.J. Gluszak, D.T. Chen, S.B. Sharma, J.A. Dumesic, T.W. Root et al., Observation of Brönsted acid sites of D-Y zeolites with deuterium NMR. Chem. Phys. Lett. **190**, 36–41 (1992)
30. P. Batamack, C. Doremieux-Morin, J. Fraissard et al., Broad-line ^{1}H NMR: a new application for studying the Brönsted acid strength of solids. J. Phys. Chem. **97**, 9779–9783 (1993)
31. A.L. Blumenfeld, J.J. Fripiat et al., Characterization of Brönsted and Lewis acidity in zeolites by solid-state NMR and the recent progress in the REDOR technique. Magn. Reson. Chem. **37**, S118–S125 (1999)

32. A.G. Pelmenshchikov, E.A. Paukshtis, V.G. Stepanov, V.I. Pavlov, E.N. Yurchenko, K.G. Ione, S. Beran et al., Scattering vector dependence of mutual diffusion coefficients for rodlike mlcelles in aqueous sodlum halide solutions. J. Phys. Chem. **93**, 6720–6725 (1989)
33. K.P. Schröder, J. Sauer, M. Leslie, C.R.A. Catlow, J.M. Thomas et al., Bridging hydroxyl groups in zeolitic catalysts: a computer simulation of their structure, vibrational properties and acidity in protonated faujasites (H-Y zeolites). Chem. Phys. Lett. **188**, 320–325 (1992)
34. K. Suzuki, T. Noda, N. Katada, M. Niwa et al., IRMS-TPD of ammonia: Direct and individual measurement of Brönsted acidity in zeolites and its relationship with the catalytic cracking activity. J. Catal. **250**, 151–160 (2007)
35. N. Katada, H. Igi, J.H. Kim, M. Niwa et al., Determination of the acidic properties of zeolite by theoretical analysis of temperature-programmed desorption of ammonia based on adsorption equilibrium. J. Phys. Chem. B **101**, 5969–5977 (1997)
36. J.R. Anderson, K. Foger, T. Mole, R.A. Rajadhyaksha, J.V. Sanders et al., Reactions on ZSM-5-type zeolite catalysts. J. Catal. **58**, 114–130 (1979)
37. V.S. Nayak, V.R. Choudhary et al., Acid strength distribution and catalytic properties of H-ZSM-5: effect of deammoniation conditions of NH_4-ZSM-5. J. Catal. **81**, 26–45 (1983)
38. D. Atkinson, G. Curthoys et al., The acidity of solid surfaces and its determination by amine titration and adsorption of coloured indicators. Chem. Soc. Rev. **8**, 475–479 (1979)
39. B.S. Umansky, W.K. Hall et al., A spectrophotometric study of the acidity of some solid acids. J. Catal. **124**, 97–108 (1990)
40. D. Farcasiu, A. Ghenciu et al., Acidity functions from ^{13}C-NMR. J. Am. Chem. Soc. **115**, 10901–10908 (1993)
41. E.P. Parry, An infrared study of pyridine adsorbed on acidic solids. Characterization of surface acidity. J. Catal. **2**, 371–379 (1963)
42. C.A. Emeis, Determination of integrated molar extinction coefficients for infrared absorption bands of pyridine adsorbed on solid acid catalysts. J. Catal. **141**, 347–354 (1993)
43. O. Cairon, T. Chevreau et al., Quantitative FTIR studies of hexagonal and cubic faujasites by pyridine and CO adsorption. J. Chem. Soc. Faraday Trans. **94**, 323–330 (1998)
44. R. Borade, A. Sayari, A. Adnot, S. Kaliaguine et al., Characterization of acidity in ZSM-5 zeolites: an X-ray photoelectron and IR spectroscopy study. J. Phys. Chem. **94**, 5989–5994 (1990)
45. C. Guimon, A. Boreave, G. Pfister-Guillouzo et al., Study of the bulk and surface acidity of protonated Y zeolites by TPD and XPS. Surf. Interface Anal. **22**, 407–411 (1994)
46. A. Auroux, Acidity characterization by microcalorimetry and relationship with reactivity. Top. Catal. **4**, 71–89 (1997)
47. A. Auroux, Microcalorimetry methods to study the acidity and reactivity of zeolites, pillared clays and mesoporous materials. Top. Catal. **19**, 205–213 (2002)
48. A. Auroux, in *Acidity and Basicity: Determination by Adsorption Microcalorimetry*, vol. 6. Molecular Sieves–Science and Technology: Acidity and Basicity, (Springer, Berlin, 2000), pp. 46–154
49. N. Cardona-Martinez, J.A. Dumesic et al., Applications of adsorption microcalorimetry to the study of heterogeneous catalysis. Adv. Catal. **38**, 149–244 (1992)
50. V. Solinas, I. Ferino et al., Microcalorimetric characterization of acid-basic catalysts. Catal. Today **41**, 179–189 (1998)
51. A. Auroux, Thermal methods: calorimetry, differential thermal analysis and thermogravimetry, in *Thermal Methods in Catalysts Characterization*, ed. by B. Imelik, J.C. Vedrine (Plenum Press, New York, 1994), pp. 611–650
52. Lj. Damjanović, A. Auroux, Heterogeneous catalysis on solids, in *Handbook of Thermal Analysis and Calotimetry*, vol. 5, ed. by M.E. Brown, P.K. Gallagher (Elsevier, Amsterdam, 2008), pp. 387–438
53. A. Corma, Inorganic solid acids and their use in acid-catalyzed hydrocarbon reactions. Chem. Rev. **95**, 559–614 (1995)
54. P.B. Weisz, Zeolites-new horizons in catalysis. Chemtech **3**, 498–505 (1973)

55. P.B. Weisz, V.J. Frilette et al., Intercrystalline and molecular-shape-selective catalysis by zeolite salts. J. Phys. Chem. **64**, 382 (1960)
56. W.O. Haag, R.M. Lago, P.B. Weisz et al., Transport and reactivity of hydrocarbon molecules in a shape-selective zeolite. Faraday Discuss. Chem. Soc. **72**, 317–331 (1982)
57. V.J. Frilette, W.O. Haag, R.M. Lago et al., Catalysis by crystalline aluminosilicates: characterization of intermediate pore-size zeolites by the "Constraint Index". J. Catal. **67**, 218–222 (1981)
58. S.M. Csicsery, In: J.A. Rabo (eds), Zeolite chemistry and catalysis. ACS Symposium series **171**, 680–701 (1976)
59. S.M. Csicsery, Shape-selective catalysis in zeolites. Zeolites **4**, 202–213 (1984)
60. S.M. Csicsery, The reactions of 1-methyl-2-ethylbenzene I. Exploring the structure of intracrystalline void space and the catalytic properties of molecular sieves and other catalysts. J. Catal. **108**, 433–443 (1987)
61. E.G. Derouane, Shape selectivity in catalysis by zeolites: the nest effect. J. Catal. **100**, 541–544 (1986)
62. E. G. Derouane, New aspects of molecular shape-selectivity: catalysis by zeolite ZSM - 5. Stud. Surf. Sci. Catal. **4**, 5–18 (1980)
63. E.G. Derouane, Molecular shape-selective catalysis in Zeolites - Selected topics. Stud. Surf. Sci. Catal. **19**, 1–17 (1984)
64. J. Dwyer, Uses of natural zeolites. Chem. Ind. **7**, 241–245 (1984)
65. P.A. Jacobs, J.A. Martens et al., Exploration of the void size and structure of zeolites and molecular sieves using chemical reactions. Stud. Surf. Sci. Catal. **28**, 23–32 (1986)
66. U. Hammon, G.T. Kokotailo, L. Riekert, J.Q. Zhou et al., Deactivation of dealuminated zeolite Y in the catalytic cracking of n-hexane. Zeolites **8**, 338–339 (1988)
67. E.G. Derouane, Z. Gabelica et al., A novel effect of shape selectivity: molecular traffic control in zeolite ZSM-5. J. Catal. **65**, 486–489 (1980)
68. J. Kärger, D.M. Ruthven, *Diffusion in Zeolites and Other Microporous Solids* (Wiley, New York, 1992)
69. J. Datka, M. Boczar, P. Rymarowicz et al., Heterogeneity of OH groups in NaH-ZSM-5 zeolite studied by infrared spectroscopy. J. Catal. **114**, 368–376 (1988)
70. A. Chatterjee, T. Iwasaki, T. Ebina, H. Tsuruya, T. Kanougi, Y. Oumi, M. Kubo, A. Miyamoto et al., Effects of structural characteristics of zeolites on the properties of their bridging and terminal hydroxyl groups. Appl. Surf. Sci. **130–132**, 555–560 (1998)
71. J. Sauer, Molecular models in ab initio studies of solids and surfaces: from ionic crystals and semiconductors to catalysts. Chem. Rev. **89**, 199–255 (1989)
72. J. Sauer, C.M. Kölmel, J.R. Hill, R. Ahlrichs et al., Brönsted sites in zeolitic catalysts. An ab initio study of local geometries and of the barrier for proton jumps between neighbouring sites. Chem. Phys. Lett. **164**, 193–198 (1989)
73. J.B. Nicholas, R.E. Winans, R.J. Harrison, L.E. Iton, L.A. Curtiss, A.J. Hopfinger et al., Ab initio molecular orbital study of the effects of basis set size on the calculated structure and acidity of hydroxyl groups in framework molecular sieves. J. Phys. Chem. **96**, 10247–10257 (1992)
74. J. Limatrakul, S. Hannongbua et al., Catalytic properties of a free hydroxyl on a silica, a zeolite, and modified zeolites: quantum-chemical calculations. J. Mol. Struc. (Theochem) **280**, 139–145 (1993)
75. R. Carson, E.M. Cooke, J. Dwyer, A. Hinchliffe, P.J. O'Malley, Molecular orbital calculations of structural and acidic characteristics of aluminophosphates (AlPO), silicoaluminophosphates (SAPO) and metaluminophosphate (MeAPO) based molecular sieves. Stud. Surf. Sci. Catal. **46**, 39–48 (1989)
76. D.J. Parrillo, R.J. Gorte et al., Characterization of acidity in H-ZSM-5, H-ZSM-12, H-mordenite, and H-Y using microcalorimetry. J. Phys. Chem. **97**, 8786–8792 (1993)
77. D.J. Parrillo, C. Lee, R.J. Gorte et al., Heats of adsorption for ammonia and pyridine in H-ZSM-5: evidence for identical Brönsted-acid sites. Appl. Catal. A **110**, 67–74 (1994)

78. M. Huang, S. Kaliaguine, A. Auroux et al., Crystallinity dependence of Lewis acidity and Lewis basicity in Na zeolites. J. Phys. Chem. **99**, 9952–9959 (1995)
79. M. Huang, S. Kaliaguine, A. Auroux et al., Lewis basic and Lewis acidic sites in zeolites. Stud. Surf. Sci. Catal. **97**, 311–318 (1995)
80. M. Huang, A. Auroux, S. Kaliaguine et al., Crystallinity dependence of acid site distribution in HA, HX and HY zeolites. Micropor. Mater. **5**, 17–27 (1995)
81. C.V. McDaniel, P.K. Maher et al., In: J.A. Rabo (eds). Zeolite chemistry and catalysis. ACS Symposium series **171**, 171–180 (1976)
82. M.A. Camblor, A. Corma, S. Valencia et al., Spontaneous nucleation and growth of pure silica zeolite-P free of connectivity defects, Chem. Commun. 2365–2366 (1996)
83. M.A. Camblor, A. Corma, A. Mifsud, J. Perez-Pariente, S. Valencia et al., Synthesis of nanocrystalline zeolite beta in the absence of alkali metal cations. Stud. Surf. Sc. Catal. **105**, 341–348 (1997)
84. P.A. Jacobs, R. von Ballmoos et al., Framework hydroxyl groups of H-ZSM-5 zeolites. J. Phys. Chem. **86**, 3050–3052 (1982)
85. T.J. Gricus Kofke, R.J. Gorte, W.E. Farneth et al., Stoichiometric adsorption complexes in H-ZSM-5. J. Catal. **114**, 34–45 (1988)
86. T.J. Gricus Kofke, R.J. Gorte, G.T. Kokotailo, W.E. Farneth et al., Stoichiometric adsorption complexes in H-ZSM-5, H-ZSM-12, and H-mordenite zeolites. J. Catal. **115**, 265–272 (1989)
87. A.I. Biaglow, D.J. Parrillo, R.J. Gorte et al., Characterization of H, Na-Y using amine desorption. J. Catal. **144**, 193–201 (1993)
88. K. Tsutsumi, K. Nishimiya et al., Differential molar heats of adsorption of ammonia on silicious mordenites at high temperature. Thermochim. Acta. **143**, 299–309 (1989)
89. A. Auroux, P.C. Gravelle, J.C. Védrine, M. Rekas, in *Proc 5th Int Zeolite Conf, Naples, Italy, June 2–6, 1980*, ed. by L.V.C. Rees (LV, Heyden, London, 1980), pp. 433–438
90. H. Stach, J. Jänchen, U. Lohse et al., Relationship between acid strength and framework aluminium content in dealuminated faujasites. Catal. Lett. **13**, 389–393 (1992)
91. A. Auroux, Y. Ben Taarit et al., Calorimetric investigation of the effect of dealumination on the acidity of zeolites. Thermochim. Acta. **122**, 63–70 (1987)
92. Y. Mitani, K. Tsutsumi, H. Takahashi et al., Direct measurement of the interaction energy between solids and gases. XI. Calorimetric measurements of acidities of aluminum deficient H-Y zeolites. Bull. Chem. Soc. Jpn. **56**, 1921–1923 (1983)
93. A.I. Biaglow, D.J. Parrillo, G.T. Kokotailo, R.J. Gorte et al., A study of dealuminated faujasites. J. Catal. **148**, 213–223 (1994)
94. I.V. Mishin, A.L. Klyachko, T.R. Brueva, O.P. Tkachenko, H.K. Beyer et al., Composition, acidity and catalytic activity of high silica fujasites. Kinet. Catal. **34**, 502–508 (1993)
95. D. Barthomeuf, Aluminium topological density and correlations with acidic and catalytic properties of zeolites. Stud. Surf. Sci. Catal. **38**, 177–186 (1987)
96. H. Stach, J. Jänchen, H.G. Jerschkewitz, U. Lohse, B. Parlitz, M. Hunger et al., Mordenite acidity: dependence on the silicon/aluminum ratio and the framework aluminum topology. 2 Acidity investigations. J. Phys. Chem. **96**, 8480–8485 (1992)
97. A. Macedo, A. Auroux, F. Raatz, E. Jacquinot, R. Boulet et al., Strong acid sites of dealuminated Y zeolites prepared by conventional treatments and isomorphous substitution. ACS Symp. Ser. **368**, 98–104 (1988)
98. J.C. Védrine, A. Auroux, G. Coudurier, Catalytic materials, relationship between structure and reactivity. ACS Symp. Ser. **248**(13), 254 (1984)
99. J.C. Védrine, A. Auroux, V. Bolis, P. Dejaifve, C. Naccache, P. Wierzchowski, E.G. Derouane, J.C.H. van Hoff et al., Infrared, microcalorimetric, and electron spin resonance investigations of the acidic properties of the H-ZSM-5 zeolite. J. Catal. **59**, 248–262 (1979)
100. H.G. Karge, L.C. Jozefowicz et al., A comparative study of the acidity of various zeolites using the differential heats of ammonia adsorption as measured by high-vacuum microcalorimetry. Stud. Surf. Sci. Catal. **84**, 685–692 (1994)
101. L.C. Jozefowicz, H.G. Karge, E.N. Coker et al., Microcalorimetric investigation of H-ZSM-5 zeolites using an ultrahigh-vacuum system for gas adsorption. J. Phys. Chem. **98**, 8053–8060 (1994)

102. D.T. Chen, S.B. Sharma, I. Filimonov, J.A. Dumesic et al., Microcalorimetric studies of zeolite acidity. Catal. Lett. **12**, 201–211 (1992)
103. G.I. Kapustin, L.M. Kustov, G.O. Glonti, T.R. Brueva, V.Y. Borovkov, A.L. Klyachko, A.M. Rubinshtein, V.B. Kazanskii et al., A microcalorimetric study of NH_3 adsorption on H, Na-mor catalysts. Kinet. Katal. **25**, 959–964 (1984)
104. I. Bankós, J. Valyon, G.I. Kapustin, D. Kallo, A.L. Klyachko, T.R. Brueva et al., Acidic and catalytic properties of hydrogen sodium mordenite. Zeolites **8**, 189–195 (1988)
105. H.G. Karge, H.K. Beyer, Solid-state ion exchange in microporous and mesoporous materials, in *Molecular Sieves–Science and Technology*, vol. 3, ed. by H.G. Karge, J. Weitkamp (Springer, New York, 2002)
106. H. Garcia, H.D. Roth et al., Generation and reactions of organic radical cations in zeolites. Chem. Rev. **102**, 3947–4008 (2002)
107. I.V. Mishin, A.L. Klyachko, G.I. Kapustin, H.G. Karge et al., The effect of exchange degrees on the heterogeneity of acid sites in decationated mordenites, Kinet. Katal . **34**, 828–834 (1993)
108. I.V. Mishin, A.L. Klyachko, T.R. Brueva, V.D. Nissenbaum, H.G. Karge et al., Effect of the exchange degree on the heterogeneity of acid sites of decationed zeolites. Kinet. Catal. **34**, 835–840 (1993)
109. M. Muscas, J.F. Dutel, V. Solinas, A. Auroux, Y. Ben Taarit et al., A dynamic assessment of MFI acidity using microcalorimetric techniques. J. Mol. Catal. A **106**, 169–175 (1996)
110. M. Huang, S. Kaliaguine, M. Muscas, A. Auroux et al., Microcalorimetric characterization of the basicity in alkali-exchanged X zeolites. J. Catal. **157**, 266–269 (1995)
111. A. Auroux, P. Artizzu, I. Ferino, R. Monaci, E. Rombi, V. Solinas et al., Conversion of 4-methylpentan-2-ol over alkali-metal ion-exchanged X and Y zeolites: a microcalorimetric and catalytic investigation. Micropor. Mater. **11**, 117–126 (1997)
112. R.M. Barrer, *Hydrothermal Chemistry of Zeolites* (Academic Press, London, 1982)
113. R. Szostak, *Molecular Sieves: Principle of Synthesis and Identification* (Van Nostrand Reinhold, New York, 1989)
114. T. Chapus, A. Tuel, Y. Ben Taarit, C. Naccache et al., Synthesis and characterization of a chromium silicalite-1. Zeolites **14**, 349–355 (1994)
115. K.G. Ione, L.A. Vostrikova, V.M. Mastikin et al., Synthesis of crystalline metal silicates having zeolite structure and study of their catalytic properties. J. Mol. Catal. **31**, 355–370 (1985)
116. K.G. Ione, L.A. Vostrikova et al., The isomorphism and catalytic properties of silicates with the zeolite structure. Uspechi. Chimii. **56**, 231–252 (1987)
117. L. Pauling, *The Nature of the Chemical Bond*, 3rd ed. (Cornell University Press, Ithaka, New York, 1967)
118. F. Liebau, *Structural Chemistry of Silicates* (Springer, Berlin, 1985)
119. M. Sayed, A. Auroux, J.C. Védrine et al., The effect of boron on ZSM-5 zeolite shape selectivity and activity: II. Coincorporation of aluminium and boron in the zeolite lattice. J. Catal. **116**, 1–10 (1989)
120. G. Coudurier, A. Auroux, J.C. Védrine, R.D. Farlee, L. Abrams, R.D. Shannon et al., Properties of boron-substituted ZSM-5 and ZSM-11 zeolites. J. Catal. **108**, 1–14 (1987)
121. M.L. Occelli, H. Eckert, A. Wölker, A. Auroux et al., Crystalline galliosilicate molecular sieves with the beta structure. Micropor. Mesopor. Mat. **30**, 219–232 (1999)
122. M.L. Occelli, H. Eckert, C. Hudalla, A. Auroux, P. Ritz, P.S. Iyer et al., Acidic properties of galliosilicate molecular sieves with the offretite structure. Stud. Surf. Sci. Catal. **105**, 1981–1987 (1997)
123. M.L. Occelli, G. Schwering, C. Fild, H. Eckert, A. Auroux, P.S. Iyer et al., Galliosilicate molecular sieves with the faujasite structure. Micropor. Mesopor. Mat. **34**, 15–22 (2000)
124. H. Eckert, C. Hudalla, A. Wölker, A. Auroux, M.L. Occelli et al., Synthesis and structural characterization of mixed aluminum-gallium-offretites. Solid State NMR **9**, 143–153 (1997)
125. M.L. Occelli, A.E. Schweizer, C. Fild, G. Schwering, H. Eckert, A. Auroux et al., Gallioaluminosilicate molecular sieves with the faujasite structure. J. Catal. **192**, 119–127 (2000)

126. B. Ducourty, M.L. Occelli, A. Auroux et al., Processes of ammonia adsorption in gallium zeolites as studied by microcalorimetry. Thermochim. Acta **312**, 27–32 (1998)
127. E. Dumitriu, V. Hulea, I. Fechete, C. Catrinescu, A. Auroux, J.F. Lacaze, C. Guimon et al., Prins condensation of isobutylene and formaldehyde over Fe-silicates of MFI structure. Appl. Catal. A Gen. **181**, 15–28 (1999)
128. D.J. Parrillo, C. Lee, R.J. Gorte, D. White, W.E. Farneth et al., Comparison of the acidic properties of H-[Al]ZSM-5, H-[Fe]ZSM-5, and H-[Ga]ZSM-5 using microcalorimetry, hexane cracking, and propene oligomerization. J. Phys. Chem. **99**, 8745–8749 (1995)
129. E. Dumitriu, V. Hulea, I. Fechete, A. Auroux, J.F. Lacaze, C. Guimon et al., The aldol condensation of lower aldehydes over MFI zeolites with different acidic properties. Micropor. Mesopor. Mat. **43**, 341–359 (2001)
130. J. Kotrla, L. Kubelkova, C.C. Lee, R.J. Gorte, Calorimetric and FTIR studies of acetonitrile on H-[Fe]ZSM-5 and H-[Al]ZSM-5. J. Phys. Chem. B **102**, 1437–1443 (1998)
131. A. Auroux, A. Tuel, J. Bandiera, J.M. Guil et al., Calorimetric and catalytic investigation of alkanes reactivity over a variety of MFI structures. Appl. Catal. **93**, 181–190 (1993)
132. J. Jänchen, G. Vorbeck, H. Stach, B. Parlitz, J.H.C. van Hooff et al., Adsorption calorimetric and spectroscopic studies on isomorphous substituted (Al, Fe, In, Ti) MFI zeolites. Stud. Surf. Sci. Catal. **94**, 108–115 (1995)
133. G. Vorbeck, J. Jänchen, B. Parlitz, M. Schneider, R. Fricke et al., Synthesis and characterization of crystalline indosilicates with the MFI structure. J. Chem. Soc. Chem. Com. 123–124 (1994)
134. M. Muscas, V. Solinas, S. Gontier, A. Tuel, A. Auroux et al., Microcalorimetry studies of the acidic properties of titanium-silicalites-1. Stud. Surf. Sci. Catal. **94**, 101–107 (1995)
135. A. Auroux, A. Gervasini, E. Jorda, A. Tuel et al., Acidic properties of titanium-silicalites-1. Stud. Surf. Sci. Catal. **84**, 653–659 (1994)
136. V. Bolis, S. Bordiga, C. Lamberti, A. Zecchina, A. Carati, F. Rivetti, G. Spano, G. Petrini et al., A calorimetric, IR, XANES and EXAFS study of the adsorption of NH_3 on Ti-silicalite as a function of the sample pre-treatment. Micropor. Mater. **30**, 67–76 (1999)
137. V. Bolis, S. Bordiga, C. Lamberti, A. Zecchina, A. Carati, F. Rivetti, G. Spano, G. Petrini et al., Heterogeneity of framework Ti(IV) in Ti-silicalite as revealed by the adsorption of NH_3. Combined calorimetric and spectroscopic study. Langmuir **15**, 5753–5764 (1999)
138. I.V. Mishin, T.R. Brueva, G.I. Kapustin et al., Heats of adsorption of ammonia and correlation of activity and acidity in heterogeneous catalysis. Adsorption **11**, 415–424 (2005)
139. C.P. Nicolaides, H.H. Kung, N.P. Makgoba, N.P. Sincadu, M.S. Scurrell et al., Characterization by ammonia adsorption microcalorimetry of substantially amorphous or partially crystalline ZSM-5 materials and correlation with catalytic activity. Appl. Catal. A Gen. **223**, 29–33 (2002)
140. S.M. Babitz, M.A. Kuehne, H.H. Kung, J.T. Miller et al., Role of Lewis acidity in the deactivation of USY zeolites during 2-methylpentane cracking. Ind. Eng. Chem. Res. **36**, 3027–3031 (1997)
141. A.G. Gayubo, P.L. Benito, A.T. Aguayo, M. Olazar, J. Bilbao et al., Effect of Si/Al ratio and of acidity of H-ZSM5 zeolites on the primary products of methanol to gasoline conversion. J. Chem. Tech. Biotechnol. **65**, 183–191 (1996)
142. M.A. Kuehne, H.H. Kung, J.T. Miller et al., Effect of steam dealumination on H-Y acidity and 2-methylpentane cracking activity. J. Catal. **171**, 293–304 (1997)

Chapter 10
Adsorption/Desorption of Simple Pollutants

Vesna Rakić

Abstract This text reviews the possibilities to apply microcalorimetry and thermo-analytical methods of analysis in the fields related to the environment pollution. At the beginning, short overview of chemical species that can be found as common pollutants in the atmosphere, waters and soils is given. Further, it is shown how the mentioned techniques can be applied for direct investigation of some event that includes specific pollutants. The possibilities to use calorimetry and thermo-analytical methods for the characterization of substances used either as adsorbents or catalysts in the processes of pollutants' abatement are presented. Besides, it is shown how all mentioned methods can provide data useful in the removal of certain pollutant. The importance of microcalorimetry and thermo-analytical methods in environment protection is underlined.

10.1 Introduction

Nowadays, environment is in the middle of our concerns. The pollution of the environment is very complex problem: numerous chemicals that have been recognized as poisonous or dangerous in any other sense come in the environment; mainly as a result of anthropogenic actions. Atmosphere, waters and soils may contain a number of different organic and inorganic residues present in a wide range of concentrations. Therefore, their de-pollution has become one of the most important tasks worldwide, imposed also by legislative procedures for environment protection.

The drastic increase of fossil fuel consumption linked to the industrialization of the world is at the origin of the emission of the major part of **atmospheric pollution**. The energy stored in fossil fuels is freed mostly in air-combustion processes by flame. The operative temperatures thermodynamically favour the formation of

V. Rakić (✉)
Faculty of Agriculture, University of Belgrade, Nemanjina 6, 11080 Zemun, Serbia
e-mail: vesna.rakic@ffh.bg.ac.rs

A. Auroux (ed.), *Calorimetry and Thermal Methods in Catalysis*,
Springer Series in Materials Science 154, DOI: 10.1007/978-3-642-11954-5_10,

nitrogen oxides: N_2O, and in particular NO and NO_2, jointly known as NO_x. Besides, flame combustion of fossil fuels is a common source of CO_2 and SO_x. The oxides of nitrogen (NO_x) and sulphur (SO_x) are responsible for occurrence of the acid-rain phenomenon. The increasing N_2O and CO_2 emissions have been identified as main contributors to global warming what is commonly known as the "greenhouse effect" [1–7]. In addition, there are the other pollutants which make the problem of atmospheric pollution even more complicated. For example, the emissions of volatile organic compounds (VOCs) into the atmosphere have the potential to cause adverse effects on human health and the environment [8]. Coal-burning power plants, iron and steel plants, paper mills, internal-combustion engines are the major sources of these pollutants. Depending on the chemical nature and sources of contaminants, the concentrations of VOCs in the resulting gas stream may vary from tens of ppb to a few percent by volume. Therefore, not only the elimination of pollutants present in trace amounts is important; even their detection in the environment can be a real challenge.

During the past decades, widespread chemical contamination of **lakes, streams and groundwater**, and consequently, **soils**; have been documented worldwide [9]. A large variety of compounds such as polyaromatic and heterocyclic compounds, pesticides, pharmaceuticals (for example, antibiotics) have been detected. More recently, other pharmaceuticals (nonsteroidal and steroidal anti-inflammatory drugs, antihypertensive drugs, etc.) and personal care products (PPCPs) have been discovered in various surface and ground waters, some of which have been linked to ecological impacts even at trace concentrations [10, 11].

Besides, there are reports that certain synthetic and natural compounds could mimic natural hormones in the endocrine systems of animals. These substances are now collectively known as endocrine-disrupting compounds (EDCs), and have been linked to a variety of adverse effects in both humans and wildlife. EDCs are released into aquatic environments as a result of industrial, agricultural and sewage runoff. Phenolic compounds, such as alkylphenols and bisphenol-A, many pesticides and steroidal hormones (17α-estradiol, synthetic agent in contraceptives with estrogenic activity of about tenfold that of female hormone 17β-estradiol, which is found in wastewaters coming from pharmaceutical industries) are among the other compounds recognized as hormone disruptors [12].

Particular problem is a frequent occurrence of phenol and phenolic compounds (phenols) in rivers, what occasionally disturbs the production of drinkable water [13]. Phenol often originates from coal power plants or as a result of pesticides degradation. In addition, large quantities of wastewater are generated during tobacco processing, and they can be toxic due to the presence of alkaloids such as nicotine, the strongest addictive agent from tobacco [14]. Nicotine and the majority of EDCs and PPCPs are polar molecules - they possess acidic or basic functional groups. Consequently, they are usually soluble in water. These properties, coupled with occurrence at trace levels, create unique challenge for both analytical detection and removal processes. All previously mentioned indicate that the production of drinkable water could be a very complex problem.

Table 10.1 The systematization of possible environment pollutants

Eco system	Pollutant(s)	Comment
Atmosphere	NOx	Cause acidic rains. N_2O is responsible for greenhouse effect (global warming).
	SOx	Cause acidic rains.
	Cox	CO_2 is responsible for greenhouse effect.
	VOCs.	
Waters	Polyaromatic compounds (PAHs, among them - phenolic compounds ...); Heterocyclic compounds (alcaloides, nicotine ...); Pesticides; Heavy metals; Pharmaceuticals and personal care products (PPCPs); Endocrine disrupting compounds (EDCs).	
Soils	May contain all chemical species that arrive in surface and ground waters.	

Table 10.1 summarizes previously stated facts and gives a systematization of atmospheric, waters' and soils' pollutants.

Environment pollution control is presently one of major thrust areas of both scientific research and technological development. Over the last few decades, concerns about the potential ecological and health impacts of exposure to substances recognized as atmospheric, water or soil pollutants have led to the establishment of new, multi-stakeholder research and testing initiatives, expert groups, committees, databases, etc. throughout the world. The combined efforts of scientists, governments and communities as a whole gave noticeable results in this domain. However, emission of pollutants coming from industrial and anthropogenic sources is a persistent problem. It has to be pointed out additionally that new pollutants in the environment can be expected, due to the continuous development of new technologies. Evidently, the pollution of environment presents a complex problem which has to be permanently controlled.

10.1.1 Possible Abatement Procedures

Several main strategies have been established for the abatement of environment pollution. In general, there are four classes of possibilities to decrease the presence of pollutants in the environment:

- reduction of pollutant emissions,
- reuse of pollutants (if possible),
- their recycling,
- treatment of waste(s) in order to obtain benign or less dangerous substances.

For the sake of environment remediation, the procedures that decrease pollutant emissions or those that enable the elimination of pollutants already discharged into the environment are still of the highest interest. For that purposes, physical (filtration, adsorption), chemical (oxidation, catalytic conversions, ozonization in the presence of a catalyst or without one) or biological methods are available. The choice of particular method depends on the nature and the concentration of pollutant, and on its physical state. All mentioned processes are permanently objects of further development.

In the domain of atmospheric de-pollution, still the most important is to remove several main classes of pollutants (CO_2, NO_x, SO_x and VOCs) from the gas-phase effluents. The scientific efforts have been, and still are, directed towards several classes of procedures that can be generalized in a following way: the usage of various solid materials as adsorbents, chemical treatment of effluent gases either in heterogeneous catalytic processes or with ozone, the treatment of pollutant gases with non-thermal plasmas, and plasma-assisted adsorption and/or catalytic processes.

For all processes that involve the adsorption step, such as physical processes of separation or catalytic transformations, the usage of solid materials with optimised activity as adsorbents and catalysts is necessary. Various solids, such as porous materials (zeolites—molecular sieves with hierarchical porosities and natural clays), activated carbons, mesoporous silica-based materials, pillared clays and metal oxides, have shown the ability to act as adsorbents or as catalysts for the conversions of previously mentioned atmospheric pollutants. Solid materials are also used for the removal of pollutants that can be found in wastewaters. The possibilities to remove polyaromatic hydrocarbons (PAHs) and heavy metal particles using the adsorptive characteristics of activated carbon and porous materials from wastewaters have been proven [15–17]. The same classes of solids are used for the elimination of organic pollutants form wastewaters by heterogeneous catalytic oxidation processes; one of the most important tasks is to eliminate phenolic compounds [13].

Evidently, the synthesis, characterizations and the investigations of adsorption or catalytic capabilities of particular materials are necessary for enabling the complete processes of pollutant eliminations. Besides, there is a necessity to find new or to improve the existing methods for the de-pollution of all mentioned eco-systems. It can be inferred that the permanent investigation of de-pollution procedures as a whole is imposed to the researchers. Numerous scientific methods are available for investigations of materials and for development of procedures applied in the previously mentioned domains: from the methods that enable the investigation of materials' structure (based primarily on the diffraction of X-rays) or texture (such as the low-temperature adsorption of different gases, or microscopies) up to the application of different spectroscopies that can give information on both de-pollution procedures and on the characteristics of applied adsorbents or catalysts.

This text will discuss the possibilities to use calorimetry, temperature-programmed techniques, particularly, temperature-programmed desorption (TPD), and thermo-analytical methods (such as thermogravimetry or differential scanning calorimetry) in the fields of environment protection and remediation.

10.2 The Application of Calorimetry in Environment Protection

Calorimetry is the science of measuring the amount of heat which is involved in some chemical or physical processes. Heat may be exchanged between some system and the environment (generated in the case of exothermic or consumed in the case of endothermic process), or dissipated by the sample. Since calorimetry's advent in the late eighteenth century, a large number of calorimetric techniques have been developed. Initially they were based on simple measurements of temperature, but more recently, it became possible to collect data and maintain samples under conditions that enable a wide range of applications [18].

Any process that results in heat being exchanged with the environment is a candidate for a calorimetric study; therefore, calorimetry has a very broad range of applicability. The classification of calorimeters may be based on three different criteria: the measuring principle (heat conduction, heat accumulation or heat exchange calorimeters have been constructed), the mode of operation (isothermal, isoperibol or adiabatic calorimeters exist) and the principle of construction (single or differential measuring system is possible). Additionally, calorimeters may be classified as those that enable measurements at constant pressure, or those working at constant volume [18]. Besides, another classification of these apparatuses can be done according to the amounts of samples used and the heat effects monitored. Microcalorimetry is the calorimetry of small samples, specifically microgram samples, designed to work in the microwatt range. Recently, a new generation of miniaturized calorimeters has been developed with highly reduced dimensions compared to the state-of-the-art microcalorimeters. Accordingly to the amounts of sample that can be used, this improved version of calorimetry is denoted as nanocalorimetry [19].

The wide range of different constructions allow measurements in many different fields: examples can be ranged from drug design in the pharmaceutical industry, to quality control of processes in the chemical industry and the study of metabolic rates in biological (people included) systems. Therefore, this methodology can be also very useful in the domain of environment protection - pollutants detection and abatement. There are two main ways in which calorimetry can be useful in this domain.

10.2.1 Calorimeters Can be Applied for Direct Investigation of Some Event that Includes Specific Pollutant(s)

In those cases, the heat of pollutants' degradation or their chemical transformations can be measured; the obtained results can be useful either in the process of monitoring of dangerous substances, or for discovering the abatement possibilities. As it has been already stated, the elimination of pollutants can be performed using physical processes of adsorption, or catalytic procedures can be applied for their chemical transformations. These processes may be monitored and characterized through the measurement of the heats that they exchange with the environment.

It may be inferred generally that calorimeters are used to register the heat evolved/consumed as a result of a contact between the adsorbent (reactant) and adsorbate (the other reactant) which can be admitted from either gas or liquid phase. The enthalpy of adsorption provide an insight into the mechanism of adsorption and hence, into the mechanism of chemical transformation if it happens after the adsorption step. These values can be determined from the isosteric measurements, by the application of Clausius-Clapeyron equation—isosteric heats of adsorption are determined from the variation of adsorption isotherms with temperature. Another way to determine the enthalpy of adsorption is to perform direct calorimetric measurements in order to obtain the desired values as $-\Delta H = f(n_a)_T$ (n_a—the adsorbed amount, T—temperature). In the case of gaseous adsorbate, its adsorbed amount can be determined in several different ways: from the variations of gas pressure in a known volume (volumetric measurements), from the variations of solid sample's weight in a static or continuous-flow lines (gravimetric measurements) or from the variations of a mass spectrometer signal [20]. The insight into the available literature data gives evidence that calorimetry linked to the volumetric lines is the most commonly used method.

Being the methodology that reveals the heat exchanged as a result of any physical or chemical event, calorimetry can help to obtain the information about the effects that versatile pollutants provoke in the environment, in a number of different ways. It can be stated that there is a growing application of calorimetric techniques that enables the monitoring of pollutant's presence as well as the investigation of effects that they produce in the environment. Several examples of the application of calorimetry in obtaining the informations concerning soils', waters' and atmosphere' pollution will be presented here.

10.2.1.1 Calorimetry Can be Applied to Assess and Compare the Toxic Effect of Different Pollutants on the Chosen Ecosystem

As and example, the case of soil microbial community' pollution will be presented. Soil microorganisms form an essential component of most terrestrial ecosystems and play a critical role in the environment due to their role in cycling mineral compounds and in the decomposition of organic material. Environmental stress caused by the presence of versatile chemicals generally decreases the diversity and activity of soil microbial populations. Their biological functionality is mostly affected by the presence of heavy metals or agrochemicals (such as pesticides) [21–25]. Heavy metal contamination of soil is widespread due to metal processing industries, combustion of wood, coal and mineral oil. Different pesticides are used on a vast and expanding scale in modern agriculture, with the aim of eliminating undesirable weeds, insects, and diseases. Although some of these compounds are directly applied to soil, the great majority can reach plants and animals.

Soil microorganisms release heat when decomposing the substrate. It is well known that the heat released closely correlates with the CO_2 evolution. Therefore, CO_2 is considered as one of the most important indicator of soil microbial activity [26]. Because it is independent on the organisms and reaction pathways [27], it is enough only to know the initial and final energetic states of investigated soil, which means that application of calorimetry can provide specific information on soil microbial activity. Currently, microcalorimetry is sensitive enough to detect very low values of heat and it is applied in studies of soil microbial activity [28–30]. In calorimetric measurements, the microbial activity which is affected by the pollutants' presence produces typical power vs. time curves. The variation in thermal effects that occurred after programmed applications of particular pollutant to the soil containing microorganisms can be monitored as important information. Apart from these values, microcalorimetry permits to reach several parameters simultaneously, such as microbial growth rate constant, the heat yield of the microbial growth and the metabolic enthalpy.

In order to study the effects of contamination by agrochemicals, selected pesticide can be applied to soil containing a degradable source of carbon. Then, the variation in thermal effects that occurred after programmed applications of this pesticide to an appropriate, nutrient-containing system can be monitored as important and valuable information. For example, the pesticide denoted as picloram is toxic for the microorganisms residing in the soil and causes the inhibition of their metabolism. Consequently, the progressive addition of pesticide picloram affects the metabolism of the microflora in the soil, what can be recognized through the changes in the profiles of power–time curves, as presented in Fig. 10.1. The profile of each recorded curve reflects the thermal effect generated by the activity of the microorganisms present in the system. The effects caused by distinct doses of picloram on the total thermal effect (Q) evolved by microorganisms are evident.

Similar results are obtained in the cases of soil contamination by other agrochemicals. Figure 10.2 presents the variation of enthalpy with time, obtained for samples of Red Latosol soil that contained certain ammount of glucose as degradable source of carbon and variable amounts of herbicide diquat. These results are presented as the integral heats vs. time. The curves show that the increase of the applied agrochemicals' doses produces the decreases in activity generated by microbes.

The toxic effects of heavy metals on soil microorganisms arrive mainly from the interaction of metals with proteins (enzymes) what provoke the inhibition of metabolic processes. These effects of heavy metals on natural soil microbial communities are complex and difficult to characterize [21]. However, similarly to the

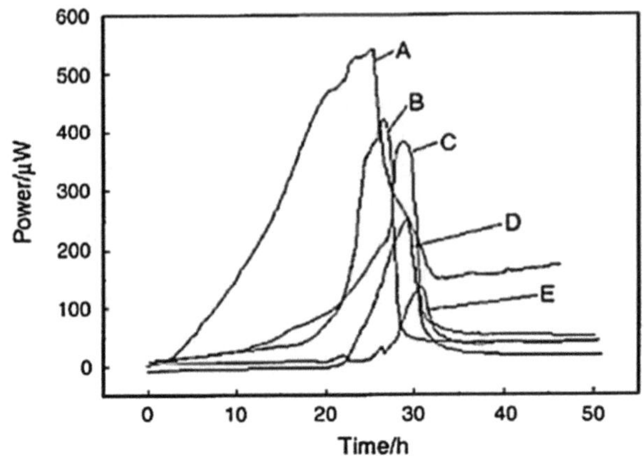

Fig. 10.1 Power–time curves of the microbial metabolism of 6.0 mg of glucose plus 6.0 mg of ammonium sulfate at 34.8 % humidity in a sample of 1.50 g red Latosol soil, with different doses of picloram: (*A*) 0.00, (*B*) 1.00, (*C*) 2.50, (*D*) 5.00, (*E*) 10.00 mg (g soil) [24]

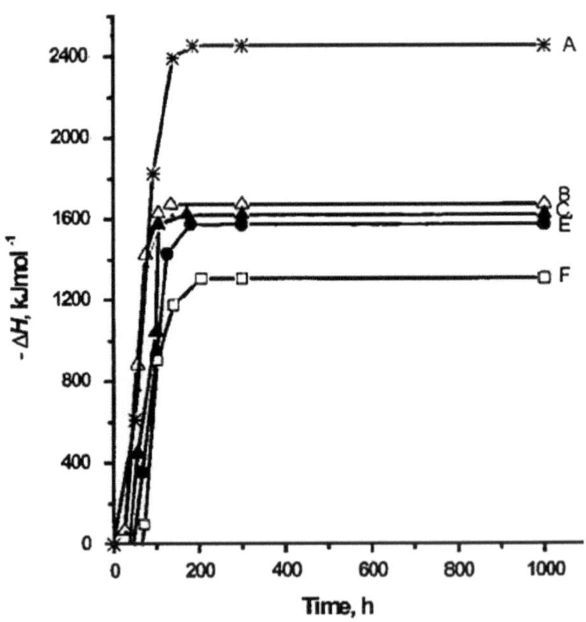

Fig. 10.2 Variation of enthalpy with time for samples with 1.50 g of Red Latosol soil, 6.0 mg of glucose, 6.0 mg of ammonium sulfate with 35 % of moisture content, control (*A*), and variable amounts of diquat: 1.30 (*B*); 2.70 (*C*), 5.30 (*D*), 6.70 (E), and 8.00 (F) mg [25]

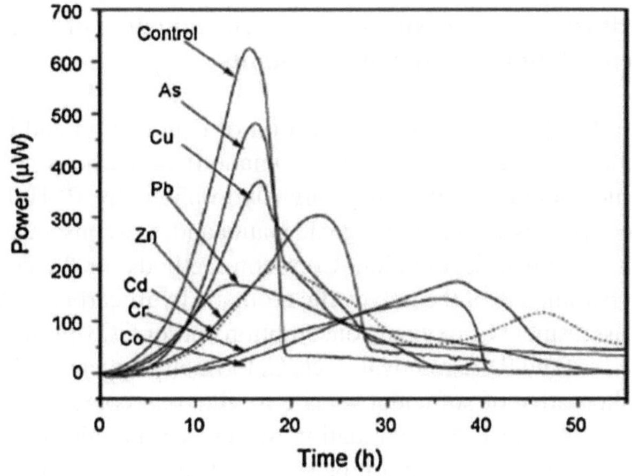

Fig. 10.3 Power–time curves for soil amended with glucose and ammonium sulphate and different selected heavy metals (1.0 mg g^{-1} As, Cu, Pb, Zn, Cd, Co and 0.01 mg g^{-1} Cr) [21]

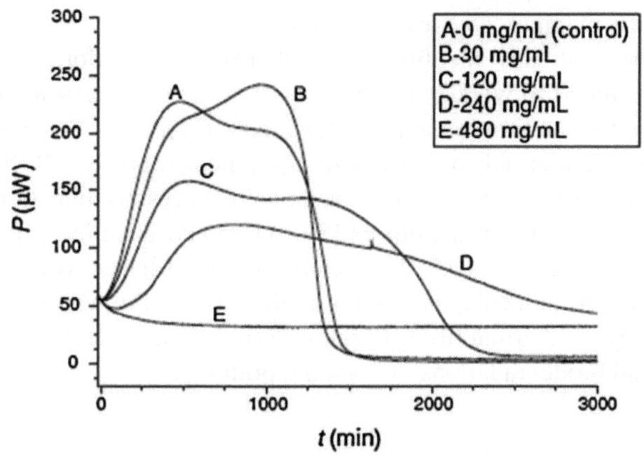

Fig. 10.4 The thermogenic curves of the growth of pure culture of Bacillus subtilis for the different concentrations of Cd^{2+} at 28–30°C [31]

pollution by different agrochemicals, the inhibition of metabolic processes that arrive from the presence of heavy metals can be revealed using microcalorimetric measurements. Figure 10.3 presents power vs. time curves, obtained for one "clean" and several polluted soils. Different thermogenic curves obtained for different metals clearly indicate the influence of heavy metal character on microbial activity. However, the influence of metal content is also obvious, as revealed by microcalorimetric investigation of toxic effects of cadmium on pure microbes (see Fig. 10.4) [31]. From the results presented in Figs. 10.1–10.4, and being aware of all other results reported in the literature, it can be inferred that microcalorimetry is a powerful tool to provide qualitative and quantitative data on microbial activities in soils.

10.2.1.2 Calorimetry Can Be Applied in the Investigation of Degradation and Biogedradation of Different Pollutants

Calorimetry can be applied in the investigation of degradation and biogedradation of different pollutants such as polycyclic aromatic hydrocarbons (PAH's) or other persistent organic pollutants. Due to their high hydrophobicity, PAHs tend to interact with non-aqueous phases and natural organic matter and, as a consequence, are poorly bioavailable for microbial degradation. Common methods for the determination of biodegradation parameters of the polycyclic aromatic hydrocarbon (PAH) either rely on continuous assessment of oxygen consumption and/or CO_2-production or on the chemical analysis of the PAH-removal rates. Calorimetry appears to be a convenient and powerful alternative of sufficient sensitivity that overcomes some of the problems, since the rates of PAH biodegradation by appropriate microorganisms can be directly followed as absolute values of the reaction heat production in the bioreactor. For example, in the case of biodegradation of anthracene, it has been shown that the achievable thermal sensitivity between few nanowatts and 1 μW corresponds to a degradation rate of few $\mu g L^{-1} h^{-1}$ and is sufficient to follow the biodegradation of ultra traces in the magnitude of few nanograms [29].

As it has been already mentioned, "isothermal microcalorimeters" are those calorimeters designed to work in the microwatt range under essentially isothermal conditions. Isothermal titration microcalorimetry (ITμC) is designed to connect extremely sensitive thermal measurement equipment (approx. 20–100 nanowatts) with an automatic syringe able to add reactants in successive injections to the solution with a precision of few nanoliters [32, 33]. Each injection produces specific heat effect, as shown in Fig. 10.5. The determination of heats evolved as a result of interaction between molecules is a main application of this variation of calorimetry. Consequently, isothermal titration calorimetry is a suitable method for studying degradations and biodegradations of versatile pollutants.

10.2.1.3 Calorimetry Can Be Applied in the Development of Different Bio-Sensors

The presence of versatile pollutants in different mediums (soils, water, and atmosphere) can be noticed by the application of microcalorimetry. For example, lichen tissue can be used to detect the presence of benzene, because it can metabolise benzene derivatives in high quantities. Therefore, lichen tissue can be used to detect traces of benzene in aqueous solutions. It has been shown that by measuring the enthalpy changes the interaction that occurs between lichen and 2-Chlorophenol can be recognized and quantified [34].

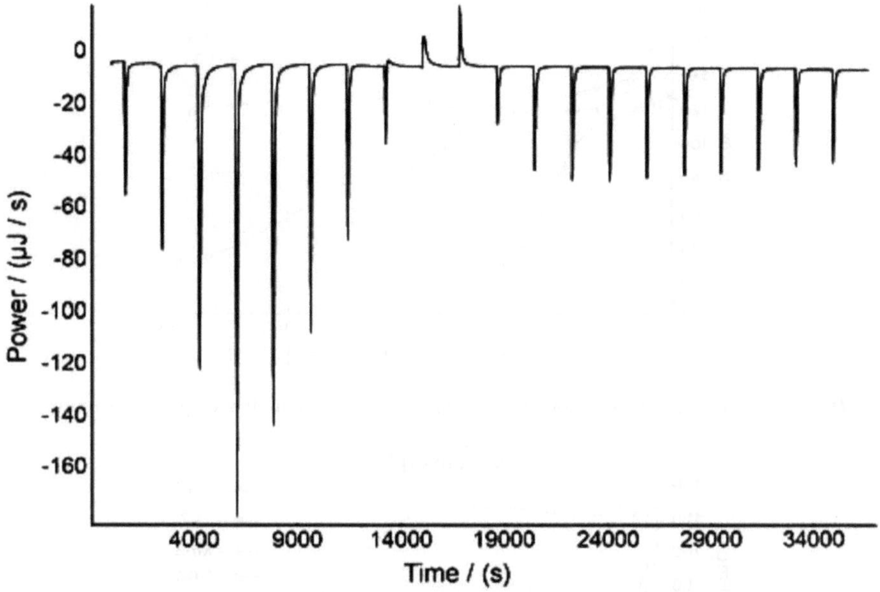

Fig. 10.5 Typical data obtained from an isothermal titration microcalorimeter [33]

10.2.1.4 Calorimetry Can Be Applied to Obtain the Data Concerning the Adsorption of Pollutants, From Gas or Liquid Phase

As mentioned previously, calorimetry linked to the volumetric technique is the most commonly used. The adsorbate (admitted either from gas or liquid phase) can be added in desired increments (doses) and the heat can be measured after the equilibration is reached, for each dose. In such a way, differential heats, defined as Qdiff = ∂Qint/∂n_a (the ratio of the amount of heat evolved for each dose to the number of moles adsorbed, or molar adsorption heat for each dose of adsorbate), become available. The results are most commonly presented as differential heat variations in relation to the adsorbed amount; in the form of histograms, or by a continuous curve connecting the centers of the steps (increments or doses). In this variation of calorimetric technique, the calorimeter is connected to a gas handling and a volumetric system equipped with manometers for precision pressure measurement. In the case of liquid adsorbates, the calorimeter is connected with a stirring system and a system for the introduction of liquids (a programmable syringe pump). Successive small doses of gas (or liquid) are introduced until appropriate final equilibrium pressure (or saturation of adsorbent with liquid adsorbate) is achieved [20, 35, 36].

The obtained differential heats of adsorption allow the estimation of adsorption affinity toward the gas-phase pollutant; the amounts adsorbed are also available. This possibility is illustrated by example presented in Fig. 10.6: the values of differential heats and the amounts adsorbed reveal that alumina-supported tin oxide expresses

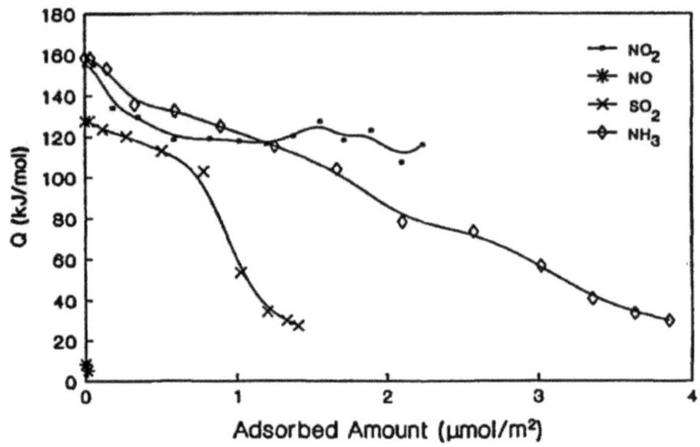

Fig. 10.6 Differential heats vs. adsorbate coverage obtained for alumina-supported tin oxide [37]

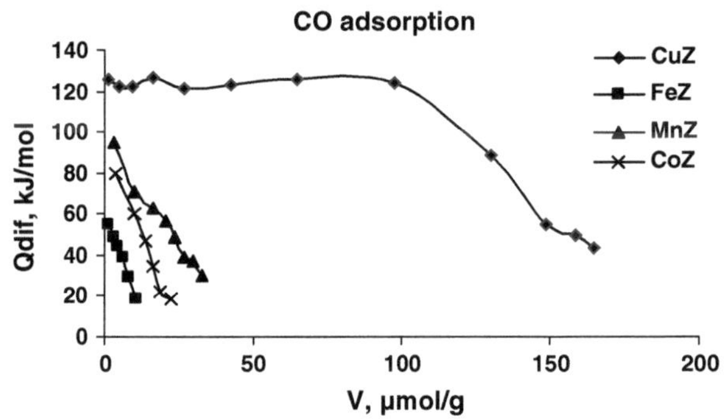

Fig. 10.7 Differential heats vs. adsorbate coverage obtained for CO adsorption on different cation exchanged ZSM-5 zeolites [38]

significant affinity toward ammonia adsorption, moderate affinity toward NO_2 and SO_2 adsorption, and does not express the affinity toward NO adsorption [37].

The values of differential heats of adsorption and the profile that they form plotted versus the adsorbed amounts, allow to chose an efficient adsorbent (among several investigated) for the adsorption of particular gas. Figure 10.7 illustrates the abilities of different cation-exchanged ZSM-5 zeolites toward CO adsorption. Evidently, the values of differential heats clearly indicated that Cu(II)-containing ZSM-5 zeolite expresses the highest possibility to adsorb carbon monoxide, among investigated ones.

The values of differential heats of adsorption enable the knowledge about the pollutant which will be preferably adsorbed on a given solid material. The example presented in Fig. 10.8 show that ZSM-5 zeolite which contain both Cu(II) and Fe(III) as charge-balancing cations is better adsorbent for CO then for N_2O (both the amount adsorbed up to equilibrium and the values of differential heats are higher for CO then for N_2O adsorption).

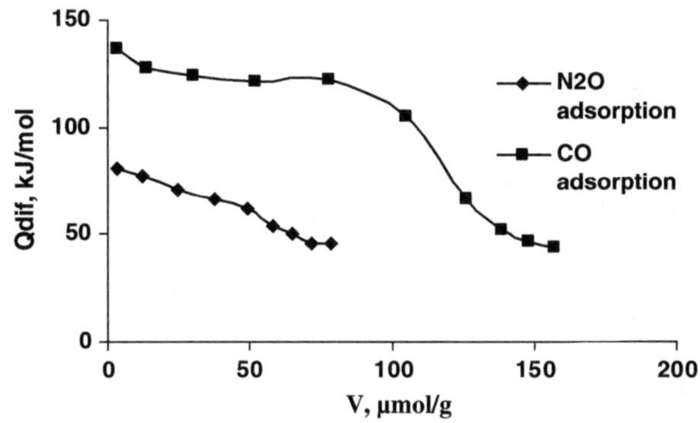

Fig. 10.8 Differential heats vs. adsorbate coverage obtained for CO and N_2O adsorption on Cu, FeZSM-5 zeolite [39]

In the case of water pollution, the estimation of adsorption affinity of potential solid adsorbent toward the specific pollutant can be done using the so-called liquid microcalorimetry. The instruments used for this purpose are differential heat flow microcalorimeters modified to allow continuous stirring of liquid samples. The adsorbate is added to both sample and reference cells simultaneously using a programmable twin syringe pump, linked to the calorimeter. The heat evolved as a result of adsorption can be obtained by integration of the area under the calorimeter signal, for each particular injection (dose). The output of typical microcalorimetric experiment of this type is shown in Fig. 10.9.

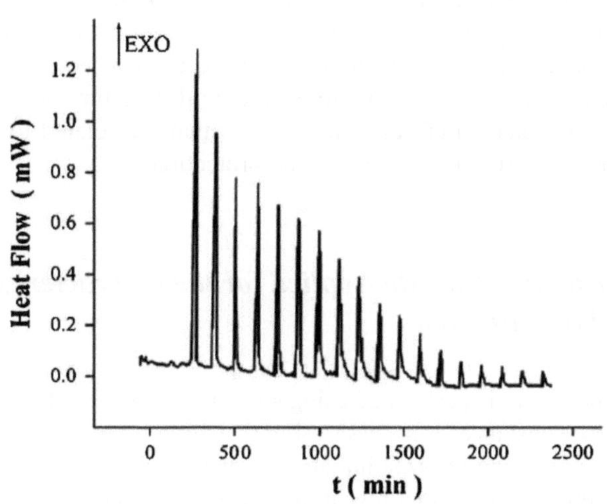

Fig. 10.9 The output of typical liquid-phase microcalorimetric experiment. Each pulse injection of dosed amount of particular solution gives as a result the exothermic signal that formed a specific peak related to the heat of adsorption [14]

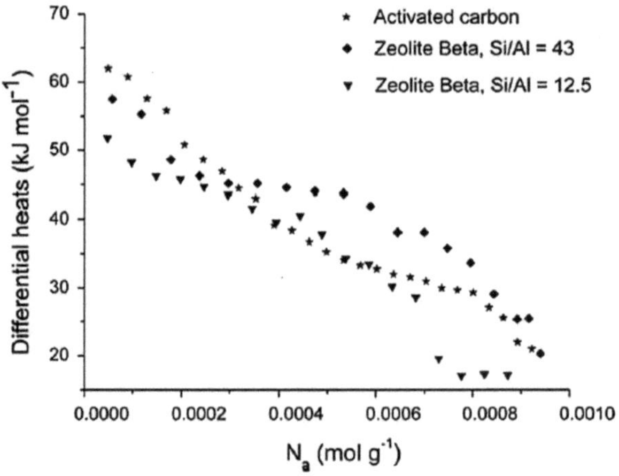

Fig. 10.10 Differential heats vs. adsorbate coverage obtained for nicotine adsorption from the aqueous solution on activated carbon and β-zeolites [14]

Figure 10.10 shows the profiles of differential heats revealed as a function of the amounts of aqueous solution of nicotine adsorbed on different solids. Based on the values and profile of differential heats, it can be concluded that adsorption capabilities of β-zeolites are comparable with that of activated carbon, solid known to be effective adsorbent for remediation of wastewaters. There is evidence in the literature on the studies of other water pollutants adsorption, such as phenol, aldehydes and the ketones, done by the application of microcalorimetry [40].

Evidently, calorimetry can bring noticeable and valuable data concerning soil, water and atmosphere pollution. Having in mind numerous modifications of this technique that have been done in the past decades; it can be expected that new versions of apparatuses and related equipments will give further improvement in the possibilities to detect small amount of heats that are related to the processes significant in both environment pollution and protection.

10.2.2 Calorimeters Can Be Applied for the Characterization of Solid Materials

If pollutants' abatement is performed using solid adsorbent or catalysts, the characterization of these substances is necessary. For that purpose, numerous methods have to be used: those enabling information about structure; the others that provide data concerning texture and those that reveal the existence of active sites for adsorption (conversion) of particular pollutants. Evidently, multi-technique approach has to be applied in order to perform full characterization of materials applied for pollutants abatement. Among the other techniques, calorimetry may be applied for the characterization of materials that can be used for capturing of pollutants (as adsorbents) or as catalysts (for conversion of pollutants) [20, 35, 36].

It is important to point out that acidic/basic character may be decisive for the reactivity of a solid material in the reactions important for pollutants abatement. The data concerning acidity/basicity can be obtained using several methods, such as IR, XPS and Raman spectroscopies; temperature-programmed desorption, and calorimetry. Among them, calorimetry gives multi-indicative data: the adsorption of successive injections (doses) of so-called "probe" molecules happens onto the sample's surface while it is kept at a constant temperature and a heat-flow detector notice the amount of heat transferred per time.

An experiment of adsorption from the gas-phase, performed in microcalorimeter coupled with volumetric line can give: a profile of Q_{dif} versus the amount adsorbed, integral heats of adsorption, adsorption isotherms (adsorbed amounts vs. equilibrium pressure) and irreversibly absorbed amount of a chemisorbed gas; the same stands for the adsorption from the liquid-phase, where the adsorbate (titrant) is added to both sample and reference cells simultaneously. The profile of differential heats versus the uptake of "probe" gives the data concerning the amount, strength and distribution of the active sites. Besides, the values of initial heats of adsorption characterize the strongest sites active in adsorption process. For the sake of acidic/basic characterization of solids' surface, the most commonly used gas-phase probes are ammonia, pyridine or some amines for the interaction with acidic sites. SO_2 and CO_2 are the probes used to notice and characterize the basic sites. In microporous solids, the accessibility of active sites is not the same for the molecules of different sizes. Therefore, many different probes can be applied to study acidity or basicity of same solid materials; this approach brings additional information. For example, acidity of zeolites can be characterized by adsorption of ammonia, but also by adsorption of pyridine (from the gas phase) and aniline (from the liquid phase) [20–22]. Liquid microcalorimetry can be also used for the determination of acidic character of solid adsorbent; the common liquid-phase probe is aniline dissolved in n-decane [40].

The obtained information about the acidic/basic character can be correlated with the adsorption possibilities. The possibility of some solid material to adsorb (or modify) the atmospheric or water pollutant depends very often on these features. Therefore, there are literature sources that report the correlations between acidity/basicity of different materials and their abilities in pollutants' abatement. For example, balanced acidic/basic properties are indispensable for an efficient conversion of NOx, that is, high activity and selectivity, over a broad temperature range. In case of alumina-supported indium oxide catalysts, microcalorimetry of ammonia and SO_2adsorption helped to find effective formulations [41].

Finally, it can be summarized that, in the domain of environment protection microcalorimetry gives possibilities to:

1. correlate the acidic/basic character with the adsorption capacities and thus, to find the appropriate formulations of adsorbents/catalysts;
2. determine the adsorption capacities toward different gas-phase or liquid phase pollutants, and thus:

3. choose an efficient adsorbent, among several investigated, for the adsorption of:
   - particular gas-phase pollutant,
   - particular water pollutant;
4. recognize the pollutant which will be preferably adsorbed on a given solid material.

10.3 The Application of Temperature-Programmed Techniques in Environment Protection

Temperature-programmed desorption (TPD) can be described as the measurement of the rate of desorption of preadsorbed molecules as a function of temperature. It involves heating of sample while contained in a sample holder—as the temperature rises, certain absorbed species will have enough energy to desorb and will be detected simultaneously by means of specific detector (for example, mass spectrometer). In TPD experiments, temperature increases linearly and the concentration of desorbed gas is recorded as a function of temperature. Therefore, TPD profile is traced as desorption rate versus time (or temperature). If detector signal is properly calibrated, concentration of desorbed gas can be plotted as a function of temperature.

TPD provides the estimation (or precise determination in the case of quantified detector response) of the amount of adsorbed species; the enthalpy of adsorption (i.e. the strength of binding to the surface) and the kinetics of desorption. The temperatures of peak maxima in the TPD profile are influenced by the strength of interaction adsorbate molecule—active site, while the area under this profile is affected by the population of active sites. Therefore, this method is an excellent tool for determination of surface coverage and for studying the adsorption states of adsorbate molecules, their binding energies and surface concentration, and desorption kinetics [20]. Two other temperature-programmed techniques: temperature-programmed oxidation (TPO) and temperature-programmed reduction (TPR) provide the red-ox features of solid materials: the amount of reducible/oxidable species, their types and kinetic parameters concerning reduction-oxidation [42].

One typical TPD experiment is designed in a way to enable the pre-treatment of the sample, *in situ*; the admission of specific adsorbate (probe) up to some specific surface coverage or up to the saturation; and subsequently, desorption which is performed in a temperature-controlled regime. Many different chemical species can be used as probes: if a chosen probe can titrate acid or basic sites at the surface, TPD can be used for the characterization of acidity/basicity of some adsorbent. In fact, temperature-programmed desorption and adsorption calorimetry are most commonly used for the study of acid/base properties of solid materials [20, 35, 36]. The same probe molecules that are used for adsorption calorimetry experiments are applicable in the case of TPD; while the investigation of acidic/basic character of solids is perhaps the most common application of this technique. TPD method has been used to study the acidity/basicity of wide variety of materials [20].

In the domain of environment protection, temperature-programmed techniques provide: (i) the characterization of important features of materials that should be employed for pollutants' abatement (as adsorbents or catalysts), (ii) the detection of environment contamination, (iii) the estimation of the possibility to remove the pollutant.

Often, there is a correlation between acidic/basic or red-ox properties of some solid material and its ability to adsorb or catalytically convert certain pollutant. For example, acidity of different zeolites and mesoporous materials is important for their ability to adsorb aldehydes and ketones (from the gas phase) or phenol and nicotine (from the aqueous phase) [40, 43]. Red-ox properties are often correlated with catalysts' efficiency; for example, red-ox features of ceria-based mixed oxides are of importance for their ability to catalyse direct conversion of methane to synthesis gas, and they can be affected by incorporation of ZrO_2 [44]. The oxidability of mixed oxides is an important feature, which determines the possibility of their use as catalysts for combustion reactions; and it can be also determined using TPR-TPO techniques [45].

Apart from investigation of adsorption-desorption of probe molecules used to estimate acidic/basic character, TPD can be applied to study all relevant parameters concerning thc adsorption of any other adsorbate. Among other possibilities, TPD can be applied to get the information about adsorption/desorption of many different pollutants of soils, waters and atmosphere. In that way, the contamination of environment by different pollutants can be detected and quantified using TPD procedures; besides, the possibility to remove it can be estimated. Figure 10.11a presents the desorption profiles of sulphur dioxide from activated carbon [46]; while TPD profiles presented in Fig. 10.11b prove the pollution of investigated soil by agrochemical phenantrene [47].

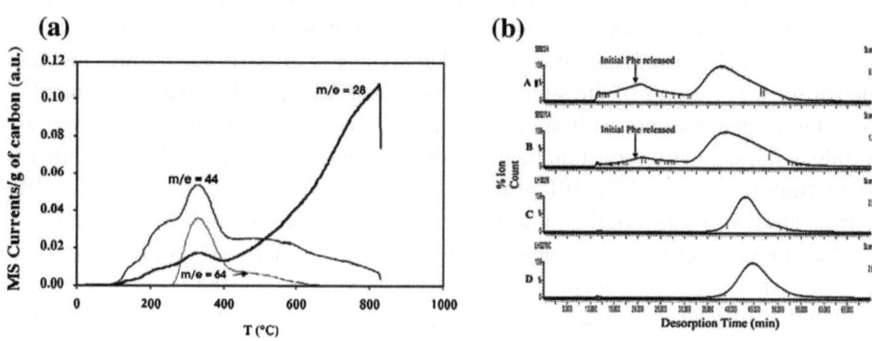

Fig. 10.11 **(a)** The profiles of temperature-programmed desorption of SO_2 (m/e = 64) as atmospheric pollutant from activated carbon [46]; and **(b)** the profiles of phenantrene (soil pollutant) released during temperature programmed heating [47]

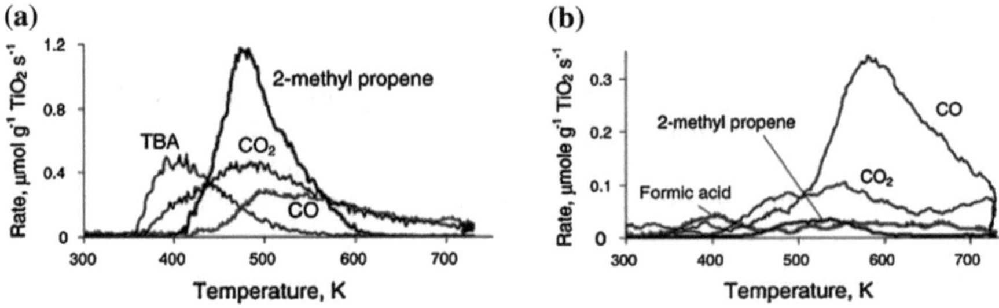

Fig. 10.12 Temperature programmed techniques allow to monitor the adsorption and chemical transformation of pollutants: **(a)** TPD spectra of adsorbed tert-butyl alcohol on TiO_2 in He stream; **(b)** TPD spectra after 300 s of photocatalytic oxidation of tert-butyl alcohol on TiO_2 (*right*) [48]

It is important to point out that the products of catalytic conversion can be monitored in temperature-programmed regime, if appropriate detector is chosen. Hence, temperature-programmed decomposition of some pollutant can be studied using the same methodology. Figure 10.12 presents TPD profiles of tert-butyl alcohol desorption and catalytic conversion on TiO_2 [48].

Evidently, the possibility to make a choice of appropriate detection technique makes possible to adapt temperature programmed techniques for many different applications. From all presented examples, it becomes evident that these methods can be of big help in the field of investigations related to environment protection.

10.4 The Application of Thermo-analytical Methods in Environment Protection

The possibility to use any other technique which belongs to the group of thermo-analytical methods for detection of pollutants or for the development of abatement procedures can be also discussed. Thermal analysis is a group of techniques in which one (or more) property of the sample is studied while it is subjected to a controlled temperature programme. Although it can take many different forms, most often, the sample is subjected to a constant heating (or cooling) rate (β). Here, the possibility to use thermogravimetry (TG), differential scanning calorimetry (DSC) and differential thermal analysis (DTA) in the field of environment protection will be exposed.

Thermogravimetry (TG) is defined by ICTAC (International Confederation for Thermal Analysis and Calorimetry) as a technique in which the mass change of a substance is measured as a function of temperature whilst the substance is subjected to a controlled temperature programme. In TG experiment, the sample is placed in a crucible which is positioned in a furnace; balance detects the mass loss, while the results are presented as a plot of mass against temperature (T) or time (t). The alternative presentation is the derivative of the original experimental curve: dm/dt (or dm/dT) plotted against temperature T or time t.

Mass loss is only noticed if a volatile component is lost in the investigated process. Therefore, the processes which happen without mass loss can not be investigated by TG. For these kinds of processes, other thermal techniques such as DTA or DSC are available. The term "differential" emphasises an important feature of these techniques—two identical measuring sensors are used: one for the sample and the other for the reference; the signal from the instrument depends on the difference between the responses of the two sensors.

The practical distinction between DTA and DSC is in the nature of the signal obtained from the equipment. In the case of DTA it is proportional to the temperature difference established between the sample and an inert reference when both are subjected to the same temperature program: $\Delta T = T_S - T_R$ (subscripts S and R indicate the sample and reference respectively). DSC signal can be regarded as proportional to the difference in thermal power between the sample and reference: $d\Delta q/dt$. The results from DTA and DSC experiments are displayed as a thermal analysis curve in which the instrument signal is plotted against temperature (usually the sample temperature) or time. Combining different heating or cooling rates with isothermal periods is also possible and often employed.

Often, TG is coupled with DTA or DSC technique. In case of all these techniques, the investigated processes are usually performed in a controlled atmosphere. In cases where gaseous products are evolved, their identification can be performed if a system is coupled with a detector (e.g. mass spectrometer - MS, infrared spectrophotometer with Fourier transformation—FTIR or gas chromatograph—GC). So, the combined techniques such as: TG-FTIR, TG/DSC-FTIR, DSC-FTIR, DSC-MS, DSC-GC or TG/DSC-MS exist, and are employed in versatile investigations.

Similarly to previously exposed possibilities of microcalorimetry and temperature-programmed techniques, TG, DSC, DTA and related coupled techniques can be also helpful in the domain of environment protection. They enable to: (i) obtain data useful in the preparation of materials applied in the abatement of pollutants; (ii) investigate the processes of pollutant removal or degradation; (iii) identify the nature of the decomposition products; (iv) monitor the processes that indicate the presence of pollutants and to investigate their effects. Here, several typical applications will be shown.

In the processes of preparation of solid materials that would be used as adsorbents or catalysts, TG is usually applied to determine the temperature of calcinations of crude samples. Similarly TG coupled with DSC or DTA is often applied to reveal the stability of solid catalysts to heat. In addition, TG can be used to get the adsorption capacities of some adsorbents toward specific gas or liquid phase pollutants.

By observing the difference in heat flow between the sample and reference, differential scanning calorimeter is able to measure the amount of heat adsorbed or released during some event. The enthalpy changes found as a result of certain process are always significant information which elucidates its mechanism. In the field of environment protection, the enthalpy changes that appear as a result of pollutant adsorption, degradation or interaction with some living system, can be monitored. Therefore, DSC can play an important role in elucidation of abatement mechanisms for particular pollutants.

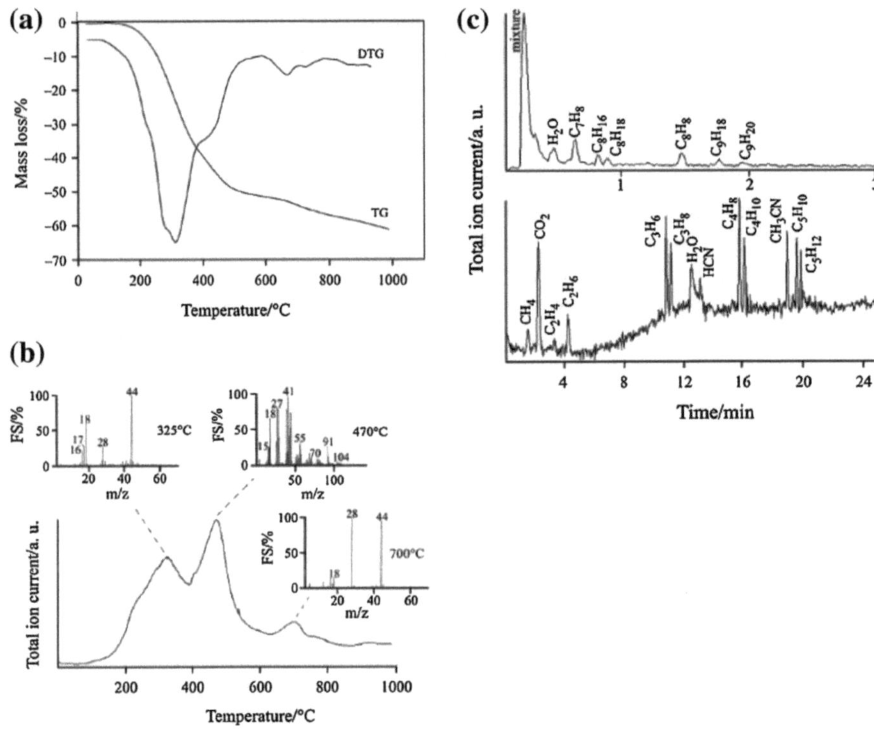

Fig. 10.13 Thermal analysis of the dried sewage sludge. **(a)** Thermogravimetric analysis of dried seawage sludge; **(b)** total ion current (TIC) curve of the evolved gas phase in the TG-MS analysis of the dried sewage sludge; **(c)** gas chromatographic elutions of the gas phase evolved during the pyrolysis of the sewage sludge [49]

Besides, TG or TG/DSC coupled with appropriate detection systems such are MS, GC or FTIR are often used to investigate the processes of pollutant removal or degradation. As an example, Fig. 10.13 presents the results obtained using coupled TG-MS and TG-GC-MS systems for studying the products of an urban plant sewage sludge pyrolisis carried out under helium atmosphere. Gas chromatographic analyses of the gas released allow the identification of various chemical species.

TG and DTG curves recorded in the TG-MS measurement (Fig. 10.13a) show the profile of mass loss along temperature increase. The profile of total ion current (TIC) curve of the evolved gas phase in the TG-MS analysis of dried sewage sludge (Fig. 10.13b) indicate the temperature regions where the majority of products appear; while isothermally performed mass spectrometric analysis gives precise data about mass fragmentation that appear as a result of pyrolisis (Fig. 10.13 b, in the insets mass spectra recorded corresponding to selected pyrolysis temperatures are presented). Finally, a detailed identification of gaseous species, evolved during the thermal treatment up to 1000 ° C, has been enabled by gas chromatographic analysis (Fig. 10.13c) [49].

DSC or DSC coupled with appropriate detector can be useful in investigation of effects that some particular pollutant can provoke on humans, animals or plants. Especially interesting is the case of persistent pollutants and effects that they can cause. For example, perfluorooctanesulfonic acid (which belongs to the group of perfluorinated surfactants - PFOS) is a persistent pollutant of the environment, which extreme inertness towards chemical and biological degradation, as well as its hydrophobic character, contributes to their worldwide distribution in the environment and their accumulation in the food chain. Recent bio-monitoring studies have shown a widespread prevalence of PFOS in humans, thus giving raise to human health concerns. One example of DSC application in this field is the investigation of interaction of perfluorooctane-1-sulfonic acid (PFOS) with biological systems. It has been shown that DSC can be useful in studying the changes in the structure of lipid bi-layers caused by pollutant presence [50]. Also, DSC can be employed to notice the enthalpy changes that arise as a result of altered protein denaturation in polluted seawater. For that purpose, the denaturation of proteins in mussels Mytilus galloprovincialis collected from polluted and non-polluted sites can be compared. In that way, DSC helps finding a reliable biomarker of seawater pollution [51].

10.5 Conclusion

It can be inferred that all discussed techniques: calorimetry, temperature-programmed techniques, as well as TG, DSC, DTA and coupled techniques, have important place in the domain of environment science: in characterisation of materials used for the abatement of pollutants; in investigation of processes of pollutant removal or degradation; and generally in monitoring the processes that include numerous environment pollutants. Having in mind the versatility of possible atmospheric, water or soil pollutants, we can estimate that these techniques will be further developed and adjusted for the applications in domain of environment protection.

References

1. P.G. Robertson, Abatement of nitrous oxide, methane, and the other non-CO_2 greenhouse gases: the need for a systems approach, in *Global Carbon Cycle, Scientific Committee on Problem of the Environment - SCOPE 62*, ed. by C.B. Field, M.R. Raupach (Island Press, Washington, DC, 2004)
2. R.G. Prinn, Non-CO_2 greenhouse gases, in *Global Carbon Cycle, Scientific Committee on Problem of the Environment - SCOPE 62*, ed. by C.B. Field, M.R. Raupach (Island Press, Washington, DC, 2004)
3. M. Dalla Valle, E. Codato, A. Marcomini, Climate change influence on POPs distribution and fate: a case study. Chemosphere **67**, 1287–1295 (2007). doi:10.1016/j.chemosphere.2006.12.028

4. P.F. Verdes, Global warming is driven by anthropogenic emissions: a time series analysis approach. Phys. Rev. Lett. **99**, 048501-1–048501-4 (2007). doi:10.1103/PhysRevLett.99.048501
5. J.C. Kramlich, W.P. Linak, Nitrous oxide behavior in the atmosphere, and in combustion and industrial systems. Prog. Energy. Combust **20**, 149–202 (1994). doi:10.1016/0360-1285(94)90009-4
6. S. Seitzinger, C. Kroeze, R. Styles, Global distribution of N_2O emissions from aquatic systems: natural emissions and anthropogenic effects. Chemosphere Glob. Change Sci. **2**, 267–279 (2000). doi:10.1016/S1465-9972(00)00015-5
7. Y. Kamata, A. Matsunami, K. Kitagawa, N. Arai, FFT analysis of atmospheric trace concentration of N_2O continuously monitored by gas chromatography and cross-correlation to climate parameters. Microchem. J. **71**, 83–93 (2002). doi:10.1016/S0026-265X(01)00140-0
8. D.A. Sarigiannis, S.P. Karakitsios, A. Gotti, I.L. Liakos, A. Katsoyiannis, Exposure to major volatile organic compounds and carbonyls in European indoor environments and associated health risk. Environ. Int. **37**, 743–765 (2011). doi:10.1016/j.envint.2011.01.005
9. D.J. Lapworth, N. Baran, M.E. Stuart, R.S. Ward, Emerging organic contaminants in groundwater: a review of sources, fate and occurrence. Environ. Pollut. **163**, 287–303 (2012). doi:10.1016/j.envpol.2011.12.034
10. S.T. Glassmeyer, D.W. Kolpin, E.T. Furlong, M.J. Focazio, Environmental presence and persistence of pharmaceuticals: an overview, in *Fate of Pharmaceuticals in the Environment and in Water Treatment Systems*, ed. by D.S. Aga (CRC Press, Taylor & Francis Group, Boca Raton, 2008)
11. S.K. Khetan, T.J. Collins, Human pharmaceuticals in the aquatic environment: a challenge to green chemistry. Chem. Rev. **107**, 2319–2364 (2007). doi:10.1021/cr020441w
12. J.D. Meeker, Exposure to environmental endocrine disrupting compounds and men's health. Maturitas **66**, 236–241 (2010). doi:10.1016/j.maturitas.2010.03.001
13. G. Busca, S. Berardinelli, C. Resini, L. Arrighi, Technologies for the removal of phenol from fluid streams: a short review of recent developments. J. Hazard. Mater. **160**, 265–288 (2008). doi:10.1016/j.jhazmat.2008.03.045
14. V. Rakic, L.J. Damjanovic, The adsorption of nicotine from aqueous solutions on different zeolite structures. Water. Res. **44**, 2047–2057 (2010). doi:10.1016/j.watres.2009.12.019
15. C.Y. Yin, M.K. Aroua, W.M.A.W. Daud, Review of modifications of activated carbon for enhancing contaminant uptakes from aqueous solutions. Sep. Purif. Technol. **52**, 403–415 (2007). doi:10.1016/j.seppur.2006.06.009
16. K.G. Bhattacharyya, S.S. Gupta, Adsorption of a few heavy metals on natural and modified kaolinite and montmorillonite: a review. Adv. Colloid Interfac **140**, 114–131 (2008). doi:10.1016/j.cis.2007.12.008
17. N. Rajic, D.J. Stojakovic, Removal of aqueous manganese using the natural zeolitic tuff from the Vranjska Banja deposit in Serbia. J. Hazard. Mater. **172**, 1450–1457 (2009). doi:10.1016/j.jhazmat.2009.08.011
18. R.J. Willson, Calorimetry, in *Principles of thermal analysis and calorimetry*, ed. by P.J. Haines (The Royal Society of Chemistry, Cambridge, 2002)
19. G.A. Urban, T. Weiss, Hydrogels for biosensors, in *Hydrogel Sensors and Actuators, Springer Series on Chemical Sensors and Biosensors Volume 6*, ed. by G. Gerlach, K.F. Arndt (Springer-Verlag, Berlin, 2010)
20. Lj Damjanovic, A. Auroux, Determination of acid/base properties by temperature programmed desorption (TPD) and adsorption calorimetry, in *Zeolite Chemistry and Catalysis: An Integrated Approach and Tutorial*, ed. by A.W. Chester, E.G. Derouane (Springer, Berlin, 2009)
21. F. Wang, J. Yao, Y. Si, H. Chen, M. Russel, K. Chen, Y. Qian, G. Zaray, E. Bramanti, Short-time effect of heavy metals upon microbial community activity. J. Hazard. Mater. **173**, 510–516 (2010). doi:10.1016/j.jhazmat.2009.08.114
22. A.G.S. Prado, C. Airoldi, The effect of the herbicide diuron on soil microbial activity. Pest. Manag. Sci. **57**, 640–644 (2001). doi:10.1002/ps.321

23. K.E. Giller, E. Witter, S.P. McGrath, Toxicity of heavy metals to microorganisms and microbial processes in agricultural soils: a review. Soil Biol. Biochem. **30**, 1389–1414 (1998). doi:10.1016/S0038-0717(97)00270-8
24. A.G.S. Prado, C. Airoldi, Toxic effect caused on microflora of soil by pesticide picloram application. J. Environ. Monit. **3**, 394–397 (2001). doi:10.1039/b103872a
25. S.A.M. Critter, C. Airoldi, Application of calorimetry to microbial biodegradation studies of agrochemicals in oxisols. J. Environ. Qual. **30**, 954–959 (2001)
26. A. Tancho, R. Merckx, F. Schoovaerts, K. Vlassak, Relation between substrate induced respiration and heat loss from soil samples amended with various contaminants. Thermochim. Acta **251**, 21–28 (1995). doi:10.1016/0040-6031(94)02044-O
27. A.G.C. Prado, C. Airoldi, Effect of the pesticide 2, 4-D on microbial activity of the soil monitored by microcalorimetry. Thermochim. Acta **349**, 17–22 (2000). doi:10.1016/S0040-6031(99)00492-X
28. N. Barros, S. Feijóo, S. Fernández, Microcalorimetric determination of the cell specific heat rate in soils: relationship with the soil microbial population and biophysic significance. Thermochim. Acta **406**, 161–170 (2003). doi:10.1016/S0040-6031(03)00255-7
29. N. Barros, S. Feijóo, S. Fernández, J.A. Simoni, C. Airoldi, Application of the metabolic enthalpy change in studies of soil microbial activity. Thermochim. Acta **356**, 1–7 (2000). doi:10.1016/S0040-6031(00)00495-0
30. N. Barros, M. Gallego, S. Feijóo, Sensitivity of calorimetric indicators of soil microbial activity. Thermochim. Acta **458**, 18–22 (2007). doi:10.1016/j.tca.2006.12.020
31. H. Chen, J. Yao, Y. Zhou, H. Chen, F. Wang, N. Gai, R. Zhuang, B. Ceccanti, Th Maskow, G. Zaray, The toxic effect of cadmium on pure microbes using a microcalorimetric method and a biosensor technique. J. Environ. Sci. Health Part A **43**, 1639–1649 (2008)
32. F. Buchholz, L.Y. Wick, H. Harms, Th Maskow, The kinetics of polycyclic aromatic hydrocarbon (PAH) biodegradation assessed by isothermal titration calorimetry (ITC). Thermochim. Acta. **458**, 47–53 (2007). doi:10.1016/j.tca.2007.01.028
33. A.C. Dos Santos Pires, N. de Fátima Ferreira Soares , L.H.M. da Silva, N.J. de Andrade, Silveira MFA, A.F. de Carvalho, Polydiacetylene as a biosensor: fundamentals and applications in the food industry. Food Bioprocess Tech. **3**, 172–181 (2010). doi:10.1007/s11947-008-0171-x
34. M.L. Antonelli, L. Campanella, P. Ercole, Lichen-based biosensor for the determination of benzene and 2-chlorophenol: microcalorimetric and amperometric investigations. Anal. Bioanal. Chem. **381**, 1041–1048 (2005). doi:10.1007/s00216-004-3014-2
35. V. Solinas, I. Ferino, Microcalorimetric characterisation of acid-basic catalysts. Catal. Today **41**, 179–189 (1998). doi:10.1016/S0920-5861(98)00048-0
36. A. Auroux, Microcalorimetry methods to study the acidity and reactivity of zeolites, pillared clays and mesoporous materials. Top. Catal. **19**, 205–213 (2002). doi:10.1023/A:1015367708955
37. B. Gergely, A. Auroux, Calorimetric study of the adsorption of air pollutants on alumina-supported tin and gallium oxides. Res. Chem. Intermed. **25**, 13–24 (1999)
38. V. Rakic, V. Rac, A. Auroux, unpublished results
39. V. Rakic, V. Rac, V. Dondur, A. Auroux, Competitive adsorption of N_2O and CO on CuZSM-5, FeZSM-5, CoZSM-5 and bimetallic forms of ZSM-5 zeolite. Catal. Today **110**, 272–280 (2005). doi:10.1016/j.cattod.2005.09.027
40. B. Dragoi, V. Rakic, E. Dumitriu, A. Auroux, Adsorption of organic pollutants over microporous solids investigated by microcalorimetry techniques. J. Therm. Anal. Calorim. **99**, 733–740 (2010). doi:10.1007/s10973-009-0353-4
41. J.A. Perdigon-Melon, A. Gervasini, A. Auroux, Study of the influence of the In_2O_3 loading on γ -alumina for the development of de-NOx catalysts. J. Catal. **234**, 421–430 (2005). doi:10.1016/j.jcat.2005.07.001
42. S. Bennici, A. Auroux, Thermal analysis and calorimetric methods, in *Metal Oxide Catalysis*, ed. by S.D. Jackson, J.S.J. Hargreaves (Wiley, New York, 2009)
43. B. Dragoi, E. Dumitriu, C. Guimon, A. Auroux, Acidic and adsorptive properties of SBA-15 modified by aluminum incorporation. Microporous Mesoporous Mater. **121**, 7–17 (2009). doi:10.1016/j.micromeso.2008.12.023

44. K. Otsuka, Y. Wang, M. Nakamura, Direct conversion of methane to synthesis gas through gas-solid reaction using CeO_2-ZrO_2solid solution at moderate temperature. Appl. Catal. A **183**, 317–324 (1999). doi:10.1016/S0926-860X(99)00070-8
45. T. Yuzhakova, V. Rakić, C. Guimon, A. Auroux, Preparation and characterization of Me_2O_3-CeO_2 (Me = B, Al, Ga, In) mixed-oxide catalysts. Chem. Mater. **19**, 2970–2981 (2007). doi:10.1021/cm062912r
46. C. Martin, A. Perrard, J.P. Joly, F. Gaillard, V. Delecroix, Dynamic adsorption on activated carbons of SO_2 traces in air I. Adsorption capacities. Carbon **40**, 2235–2246 (2002)
47. A. Abu, S. Smith, Mechanistic characterization of adsorption and slow desorption of Phenanthrene aged in soils. Environ. Sci. Technol. **40**, 5409–5414 (2006). doi:10.1021/es060489h
48. S. Preis, J.L. Falconer, R. del Prado Asensio, N.C. Santiago, A. Kachina, J. Kallas, Photocatalytic oxidation of gas-phase methyl tert-butyl ether and tert-butyl alcohol. Appl. Catal. B-Environ. **64**, 79–87 (2006). doi:10.1016/j.apcatb.2005.11.002
49. M. Ischia, C. Perazzolli, R. Dal Maschio, R. Campostrini, Pyrolysis study of sewage sludge by TG-MS and TG-GC-MS coupled analyses. J. Therm. Anal. Calorim. **87**, 567–574 (2007). doi:10.1007/s10973-006-7690-3
50. H.J. Lehmler, W. Xie , G.D. Bothun, P.M. Bummer, B.L. Knutson (2006) Mixing of perfluorooctanesulfonic acid (PFOS) potassium salt with dipalmitoyl phosphatidylcholine (DPPC). Colloid. Surf. B **51**: 25–29. doi:10.1016/j.colsurfb.2006.05.013
51. S. Gorinstein, S. Moncheva, F. Toledo, P. Arancibia-Avila, S. Trakhtenberg, A. Gorinstein, I. Goshev, J. Namiesnik, Relationship between seawater pollution and qualitative changes in the extracted proteins from mussels mytilus galloprovincialis. Sci. Total. Environ. **364**, 251–259 (2006). doi:10.1016/j.scitotenv.2005.06.013

Chapter 11
Hydrogen and Calorimetry: Case Studies

Simona Bennici and Aline Auroux

Abstract In this chapter the use of calorimetric data is shown to be of primary interest for the determination of the reaction/diffusion and reaction/absorption mechanisms involved in hydrogen storage or hydrogen production reactions. Examples of calorimetric measurements (alone or coupled to volumetric devices working at atmospheric or high pressure) and thermal analysis experiments applied to different hydrogen storage systems are reported. In particular, applications of calorimetric tools to the study of irreversible and reversible hydrogen storage systems are detailed, such as borohydrides hydrolysis and hydrogen desorption from magnesium hydrides.

11.1 Introduction

Hydrogen as energy vector is one of the most promising approaches to the challenges posed by global warming, and hydrogen-based fuel cell technologies are expected to become one of the most prevalent energy sources in the near future [1, 2].

Now that technologies to use hydrogen as a clean fuel are readily available, like the Proton Exchange Membrane Fuel Cell (PEMFC), and can be developed at an industrial scale, research mainly focuses on the barrier of development which is hydrogen storage for delayed use. In fact, if nowadays the H_2 production methods are well known and controlled, the storage and transportation of the fuel remain major obstacles to its use [3].

During the last decade a lot of research effort has been put into the development of suitable and safe technologies for hydrogen storage, such as materials for high-pressure cylinders, liquefaction processes, hydrogen adsorption materials, and metal hydrides [4, 5], adaptable to a wide range of applications from stationary and automotive to portable devices. Although H_2 adsorption capacities have recently been

S. Bennici (✉) · A. Auroux
Institut de recherches sur la catalyse et l'environnement de Lyon, UMR5256,
CNRS-Université Lyon1, 2 avenue Einstein, 69626 Villeurbanne, France
e-mail: simona.bennici@ircelyon.univ-lyon1.fr

A. Auroux (ed.), *Calorimetry and Thermal Methods in Catalysis*,
Springer Series in Materials Science 154, DOI: 10.1007/978-3-642-11954-5_11,

brought up to values near 6–8 wt%, this storage method requires a high pressure and low temperature.

Portable applications in particular require the implementation of H_2 technologies in the short term, due to a rapidly growing market and ever increasing energy demand. In this case, conception of a H_2 storage/generation system operating under ambient conditions is a primary requirement. The ultra-pure hydrogen produced by hydrolysis of metal borohydrides, possessing the right humidity level, can be directly used in PEM fuel cells, which represents a very interesting alternative to lithium-ion battery. At parity of mass, PEM based systems are 5–6 times more performing than lithium batteries [6]. Moreover, chemical hydrides have an excellent potential for high energy density storage at room temperature and ambient pressure [4, 5, 7].

To enable the commercialisation of fuel cells and hydrogen technologies, governmental institutions worldwide are working to identify the most promising hydrogen storage systems in order to promote the future R&D activities [7–9]. U.S Department of Energy in particular has excluded the use of $NaBH_4$ for on-board vehicle hydrogen storage due to its drawbacks in terms of exothermic reaction and formation of very stable dehydrogenated products. The high energy consumption required for the regeneration of these stable products back into hydrogen-containing fuel contributes to an inefficient overall energy balance [10]. Despite this, $NaBH_4$ remains a promising candidate for mobile, portable and stationary isolated site applications in which the targets are less demanding than for automotive applications and the final product price can be relatively high [11].

An appropriate catalyst is necessary to carry out the hydrolysis reaction at a high rate. Noble metal based catalysts were initially developed and studied for this purpose [5, 12–15], but the high cost of these materials associated with the low supply available shifted the focus of research towards cheaper catalytic materials. In fact, non-noble metals that form boride compounds such as Ni-B or Co-B alloys are efficient and low-cost catalysts for this reaction [6, 8, 11, 12, 16–19].

Besides borohydrides, solid state complex hydrides [20] such as magnesium, sodium, and lithium aluminium hydrides [21, 22], lithium amides [23], and their combination with MgH_2 [24, 25] have also proven to be potential candidates, in this case for reversible hydrogen storage. The goal of the recent research is to overcome the kinetic limitation and/or thermodynamic stability of Mg-based hydrogen storage materials adding additives and catalysts to these mixtures [26, 27].

A third class of materials potentially useful in hydrogen storage is nanostructured materials (i.e. carbon nanofibers, carbon nanotubes, zeolites, clathrate hydrates, metal-organic frameworks (MOFs)) that by virtue of their large surface-to-volume ratios can adsorb considerable amounts of hydrogen in the molecular state via weak molecular-surfacic interactions (physisorption) [4, 28–31]. For these materials the main challenge is to find ways by which they can be engineered to store reversibly high quantity of hydrogen close to room temperature.

For all the cited hydrogen storage systems, a precise determination of the heat of reaction is needed for an industrially applicable system design and evaluation of feasibility. The measurement of the heat evolved during a catalytic reaction or a hydrogen desorption process is important from both the practical and fundamental points of view.

Firstly it is an essential tool for the assessment of thermal risks related to the performance of the reacting system at industrial scale (i.e. the capability of a system to enter into a runaway reaction). This type of safety data is particularly important for reactions like hydrogen generation by hydrolysis of borohydrides, where the rapid increase in temperature may result in a sharp pressure increase. Secondly the thermodynamic data are of primary interest for the determination of the reaction/diffusion and reaction/absorption mechanisms.

Therefore, calorimetry and thermal analysis techniques alone or preferentially combined to volumetric systems play a fundamental role in the study of hydrogen storage materials and related phenomena.

11.2 First Case Study: Irreversible H_2 Storage (Borohydrides)

The potential applications of sodium borohydride have been widely studied and most of the research efforts and industrial devices developed up to now are based on the catalyzed hydrolysis of sodium borohydride solutions (stabilized with NaOH) [32], Eq. 11.1.

$$\begin{aligned} &NaBH_4(aq) + 4H_2O(l) \rightarrow NaB(OH)_4(aq) + 4H_2(g) \\ &\Delta_r H^\circ = -250.5\,kJ.mol_{NaBH4}{}^{-1} \end{aligned} \tag{11.1}$$

Sodium borohydride hydrolysis in aqueous solution can be represented in terms of the overall stoichiometric equation (Eq. 11.1) where $NaBH_4$ reacts with 4 molecules of water to produce 4 molecules of H_2 [33]. Although the reaction of $NaBH_4$ hydrolysis has been studied since the discovery of sodium borohydride by Stock in 1933 [34], the theoretical, calculated energy of the reaction is often cited in an incompatible manner to the application, and real experimental data are very scarce [35–38]. The thermodynamic features of the catalyzed hydrolysis in fact are not yet well understood, as the evolved energy depends on the physical state and the hydration degrees of borohydride and metaborate and on-side reactions.

Calorimetric techniques, and liquid phase calorimetry in particular, are promising methods to study catalytic reactions [39]. Notably, the use of a differential reaction calorimeter (DRC) makes it possible to determine the most important thermodynamic data such as the heat of reaction and heat capacity of the system [40–42].

11.2.1 Hydrolysis of $NaBH_4$ *Stabilized Solutions*

The study of Garron et al. [43] is one of the first investigations by liquid-phase calorimetry of the mechanism of hydrogen generation by hydrolysis of sodium borohydride catalysed by Co_2B nanoparticles generated in situ. The differential reaction calorimeter (Fig. 11.1) was coupled with a hydrogen volumetric measurement, allowing simultaneous thermodynamic and kinetic studies of the reaction. In a typical experiment the "sample" vessel was filled with the reactants and the "reference" cell with the solvent (NaOH solution). The oil circulating in the jackets maintains the surroundings of the reactor at a constant temperature. The used calorimeter principle is based on the continuous measurement of the temperature difference between the two vessels during the experiment. In order to determine the heat capacity (Cp) of the system and correlate the temperature difference (ΔT) to the heat flow, a Joule-effect calibration is performed before and after each reaction step (see Fig. 11.2), and the corresponding energy is obtained by the $Q = Cp \times \Delta T$ equation. The measured heat corresponds to the global energy evolved during the $NaBH_4$ hydrolysis reaction after each addition of fresh reactants ($NaBH_4$ solution stabilized by NaOH) (Fig. 11.2). The endothermic effects observed at the beginning of each peak correspond to the addition of the reactive solution at a temperature slightly lower than that of the thermostated vessel. In their paper [43] the authors show that the total enthalpy of the catalytic process is strongly influenced by the evolution of the catalyst during the hydrolysis reaction and by some water evaporation which is related to the $NaBH_4$ concentration. For the first time the calorimetric

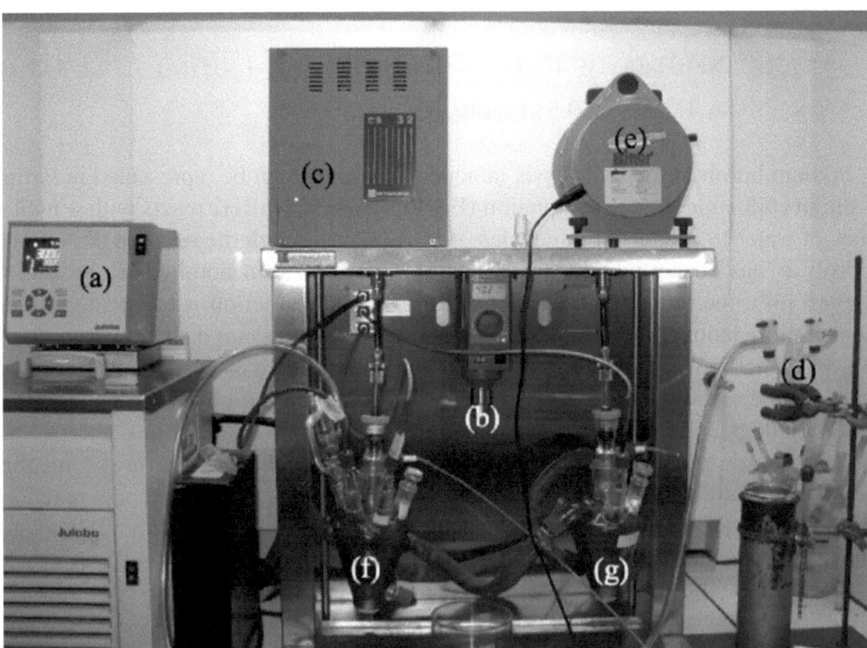

Fig. 11.1 Picture of the DRC system coupled with a volumetric gas-meter. **a** Julabo Thermostat. **b** IKA parallel stirrers. **c** DRC CS 32 control unit. **d** liquid N_2 trap. **e** Gas volume-meter Ritter TG01. **f** measurement cell. **g** reference cell

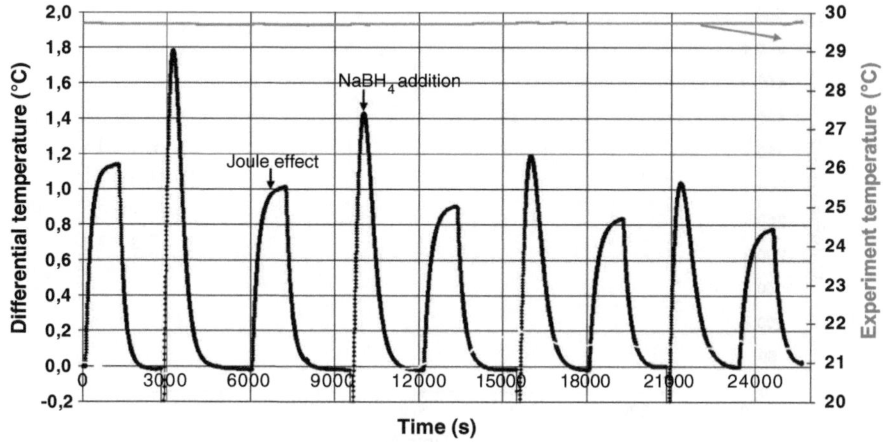

Fig. 11.2 Typical profile of a DRC analysis with 4 injections (10 mL) of NaOH-stabilized solution of $NaBH_4$ (19 wt.%) on 1 mmol Co-nanoparticles suspended in 20 mL water. Before and after each injection peak a Joule effect is performed for determining the relative heat of reaction

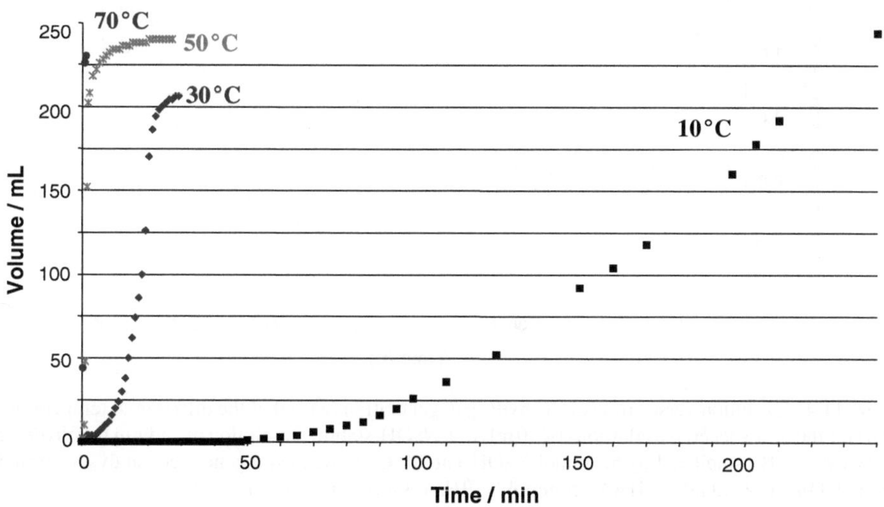

Fig. 11.3 Evolution versus time of the hydrogen generation at different temperatures for the $NaBH_4$ hydrolysis reaction (Injection of 0.2 mL of water on a solid mixture of 200 mg $NaBH_4$ + 20 mg of $Co(H_2O)_6Cl_2$

technique was used to study the catalytic reaction in aqueous phase as complement of "ex situ" characterization of the solid catalyst and the reacting solution, thus providing "operando" information about the reaction thermodynamics and kinetics and consequently the mechanism.

The setup shown in Fig. 11.1 was thermostated so that experiments reported in Fig. 11.3 could be performed at constant temperature between −5 °C and 70 °C, with

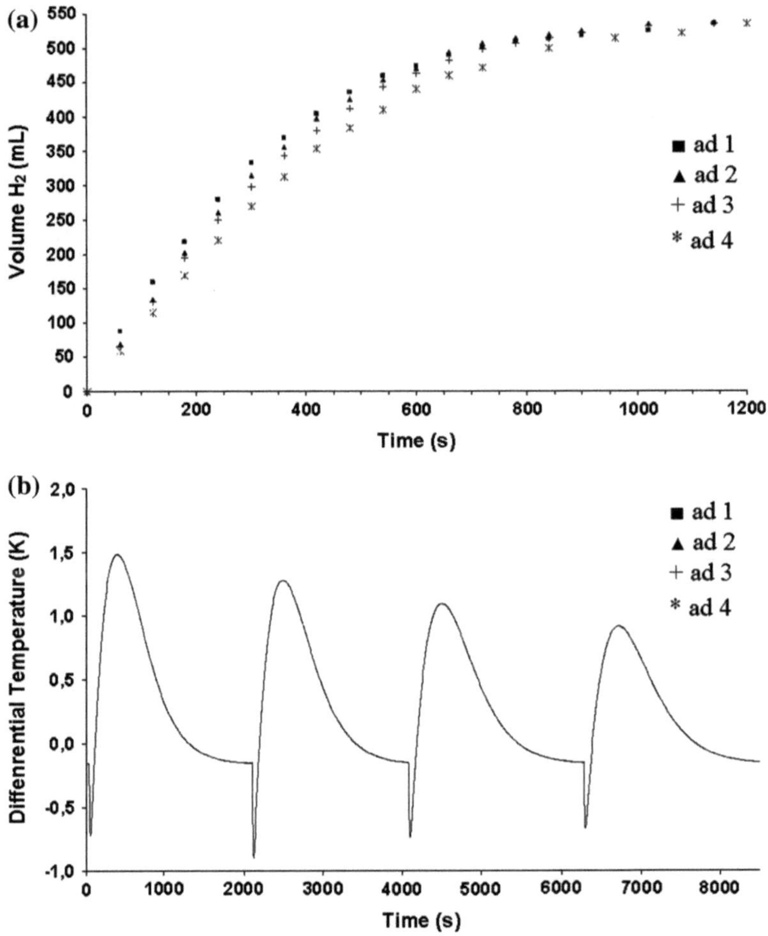

Fig. 11.4 Evolution versus time of the hydrogen generation (**a**) and of the differential temperature (**b**) for the four successive additions of 10 mL of a NaOH-stabilized solution of sodium borohydride (2 wt% $NaBH_4$ stabilized by 5.7 mmol NaOH) at 30 °C on the in situ generated catalyst (1 mmol $Co(H_2O)_6Cl_2$ + 20 mL of H_2O + 5 mmol $NaBH_4$ dissolved in 5 mL of H_2O)

the purpose of determining the activation energy by studying the dependence of the rate of hydrogen production on the temperature. The activation energy determined in this manner was $-42.7\,kJ.mol^{-1}$ [43].

Figure 11.4 shows the evolution of the hydrogen volume and differential temperature *versus* time upon four successive additions of $NaBH_4$ solution (Fig. 11.4a, b). The curves of the amounts of evolved hydrogen (Fig. 11.4a) for each addition show similar profiles, with a kinetic response that slightly decreases with each successive addition as the concentration of metaborate and the total volume of the reactant solution increase. In fact, the presence of residual sodium metaborate during the reaction lowers the ability of borohydride to reach the catalyst surface [43, 44]. The four differential temperature peaks obtained after successive additions of solution (Fig. 11.4b) present very similar shapes, but the increasing of the width at half-height (thermokinetic parameter), confirmed that thermal transfer within the solution becomes increasingly difficult as its viscosity increases (Fig. 11.4b). Nonetheless, the total area of the peak, and hence the total energy evolved during the reaction, remain constant.

The activity towards $NaBH_4$ hydrolysis of metals which can be readily oxidized and form stable boride species was also studied by liquid reaction calorimetry [45]. Co, Ni, and Fe are stable in alkaline aqueous media and they are not affected by the concentration of residual metaborate in the solution. In Ref. [45] the use of liquid-phase calorimetric methods provided deep insight on the catalytic behaviour of the samples by studying the energy production and its evolution as a function of time. The volume of hydrogen generated at a given time is reported in Fig. 11.5 for Co, Ni and Fe salts acting as catalysts in the $NaBH_4$ hydrolysis reaction.

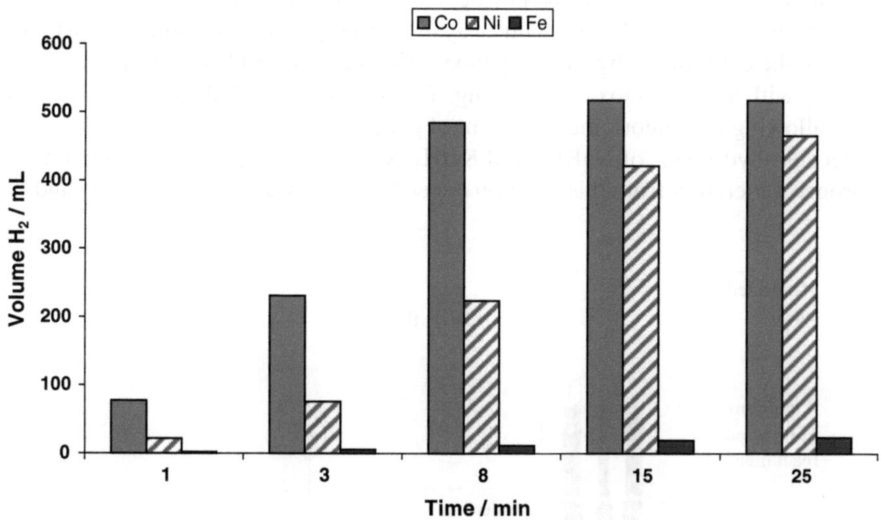

Fig. 11.5 Hydrogen volumes generated at 1, 3, 8, 15, and 25 min for Co, Ni and Fe salts acting as catalysts. Catalyst generated in situ by addition of 5 mL of aqueous solution containing 200 mg of $NaBH_4$ on the chloride salt (1 mmol equivalent of metal in 20 mL water), pH = 14 by addition of 5 mL NaOH 6 M. Test: additions of 10 mL of $NaBH_4$ (2 wt.% solution) stabilized by 10 mmol of NaOH (4 wt.%) [45]

11.2.2 *Hydrolysis of* $NaBH_4$ *and* KBH_4 *Powders*

The hydrogen storage capacity of sodium borohydride depends on the quantity of water involved in the whole storage system. For example, for a standard commercial solution containing 20 wt% of $NaBH_4$ the storage capacity corresponds to 4.2 wt% H_2. Higher hydrogen storage capacities can be achieved by the reaction of stoichiometric quantity of water with solid sodium borohydride. If the water content in the Na,KBH_4(s)/catalyst(s)/H_2O(l) system is high enough to dissolve the metaborate product at room temperature (25 °C), the advantages of this configuration are the possibility to store the reagents separately and to remove the products of the reaction from the system in liquid form, avoiding the formation of a borate solid crust that will decrease the diffusion of water in the solid reactants [46].

To mimic and study this storage configuration the instrument used was a Calvet type differential heat flow microcalorimeter which allows continuous stirring of liquid-solid mixtures (Fig. 11.6). After stabilization of the calorimeter baseline, the water (or the hydrolyzing solution) was added to the sample cell which contains the potassium or sodium borohydride powders (eventually activated with a catalyst) and to the empty reference cell in the same way using a programmable twin syringe pump linked to the calorimeter by capillary tubes. The reaction cell had a home-made air-tight cap with attached inox outlet tubing. The inox tube was linked to a gas volume meter allowing continuous measurement of generated hydrogen.

Because hydrolysis of $NaBH_4$ and KBH_4 is a very slow process and the rate of hydrogen generation from this reaction depends on the addition of a catalyst and on

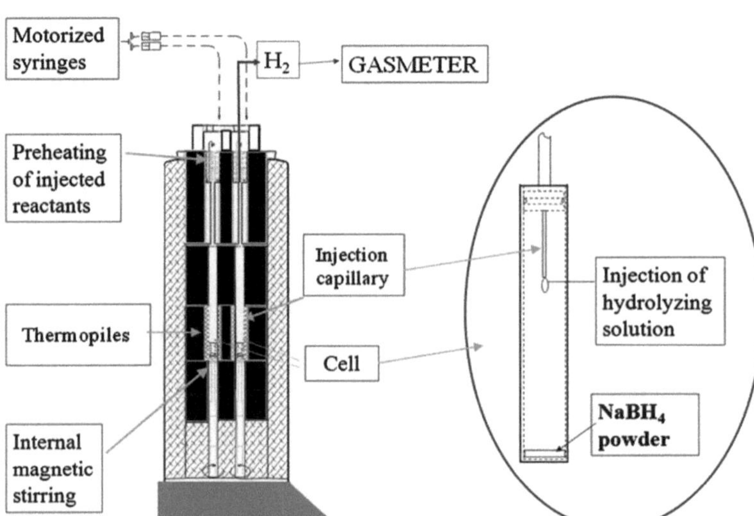

Fig. 11.6 Scheme of the Titrys calorimeter coupled with a volumetric gas-meter and a programmable twin syringe pump

its preparation method, the study of non-catalyzed hydrolysis can be very difficult as well as the experimental determination of the reaction heat. In fact, long time scale experiments are challenging in terms of accuracy of the total evolved heat (integral method) determination.

Self-hydrolysis occurs spontaneously but slowly when $NaBH_4$ and KBH_4 are in contact with water, and liquid calorimetry coupled with a gas-meter (Fig. 11.6) permits to precisely determine the heats of the hydrolysis reaction for $NaBH_4$ and KBH_4 that are respectively of -236 and $-220\,kJmol_{Na,KBH4}{}^{-1}$ [46]. These values were obtained by integration of the calorimetric peaks shown in Fig. 11.7 that display the heat flow evolved during the hydrolysis reactions of $NaBH_4$ and KBH_4 without catalyst. The endothermic signal visible on the heat flow signal versus time curves for both investigated borohydrides is attributed to the water injection. The hydrolysis reaction is an exothermic process and, after an initial increase, the heat flow signal reaches an almost constant value until the end of the hydrolysis reaction, when it exhibits an abrupt decrease. The value of the measured heat includes the enthalpy of dissolution of the borohydride, the enthalpy of the hydrolysis reaction, the enthalpy of dissolution of the by-product ($NaBO_2$ or KBO_2) and the evaporation of some water during the exothermic reaction.

Therefore, to increase the decomposition rate of $NaBH_4$ by hydrolysis a catalyst is needed. As already pointed out, cobalt-based catalysts have shown promising activity, and acids are also suitable catalytic accelerators [47, 48].

In Fig. 11.8 the heat flow peaks for the $NaBH_4$ hydrolysis reaction in presence of malic acid (a), Co-nanoparticules (b) and pure water (c) are reported, and the heats of reaction obtained by integration of these peaks were respectively of -298, -244, and $-236\,kJ \cdot mol_{NaBH4}{}^{-1}$. By analyzing the shape of the calorimetric peaks a perception of the hydrogen release kinetics can be obtained; the most rapid

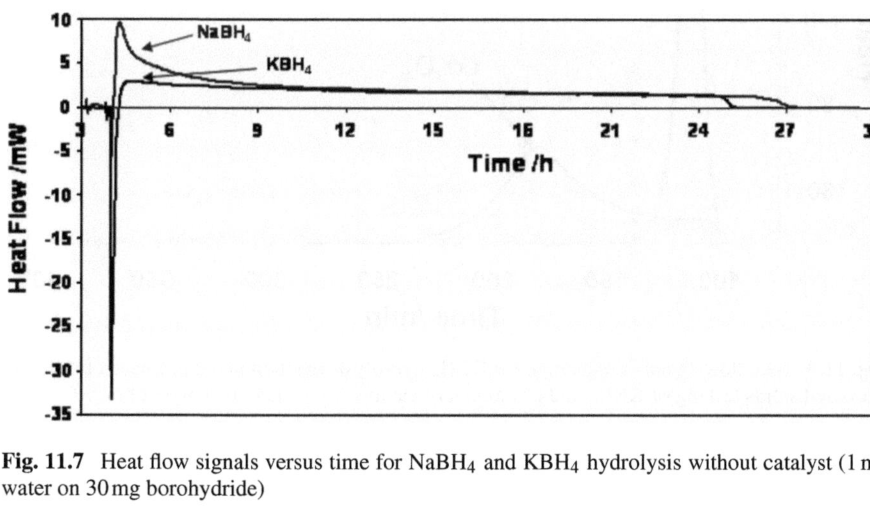

Fig. 11.7 Heat flow signals versus time for $NaBH_4$ and KBH_4 hydrolysis without catalyst (1 mL water on 30 mg borohydride)

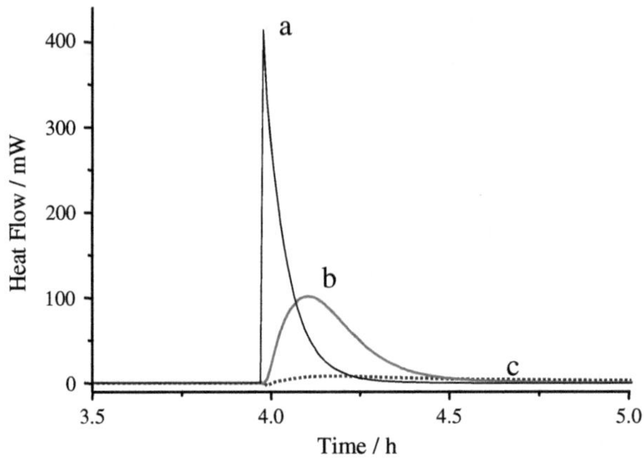

Fig. 11.8 Heat flow generated during $NaBH_4$ hydrolysis by addition of **a** 1 mL of 1 M maleic acid, **b** 1 mmol Co-nanoparticules dispersed in 1 mL of water, and **c** pure water

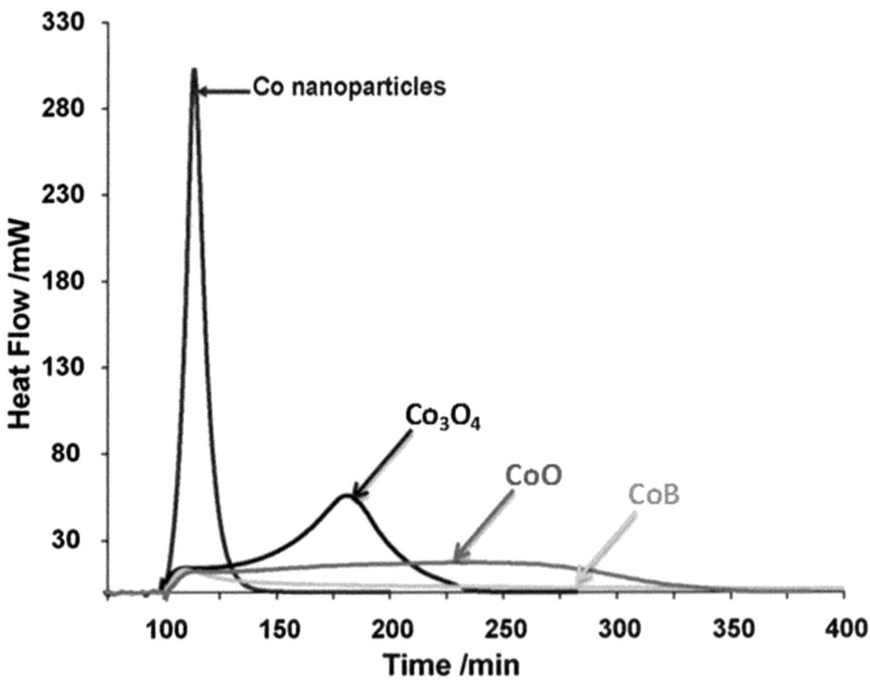

Fig. 11.9 Heat flow signals versus time for KBH_4 hydrolysis reaction in the presence of different Co-based solids (30 mg of KBH_4 and an amount of catalyst equivalent to 10 wt% of pure Co)

hydrogen production corresponds to the use of malic acid and it can be noticed that the faster reaction corresponds to the higher heat generated.

The experimental calorimetric curves obtained for the hydrolysis reaction of $NaBH_4$ in the presence of different catalysts (Co nanoparticles, CoO and Co_3O_4) are shown in Fig. 11.9, as well as the curve obtained for the $NaBH_4$ hydrolysis reaction without catalyst reported as reference. The measured heats of the $NaBH_4$ hydrolysis reaction were −243, −235, and $-\ 236\,kJ \cdot mol^{-1}{}_{NaBH4}$ in the presence of Co nanoparticles, CoO, and Co_3O_4, respectively [46]. The reaction times for the complete hydrogen generation were: 33 min for Co nanoparticles, 120 min for Co_3O_4 and 240 min for CoO. In all cases, 100 % of the stoichiometric amount of hydrogen was generated during the stated reaction time [46]. Much higher hydrogen production rates (6000 mL $min^{-1}g_{Co}$, corresponding to a reaction time of 7 min) were measured in the same way on Co based catalysts with the cobalt phase dispersed on acidic heteropolyanions supports [49].

By means of liquid calorimetry coupled with a volumetric gas-meter Damjanovic et al. [50] have shown that addition of solid NaOH to the reacting mixture ($NaBH_4$ + catalyst) increases the rate of $NaBH_4$ hydrolysis reaction in the presence of Co-based catalysts. Comparing the results available in the literature it is clear that the effect of NaOH concentration on the hydrolysis generation rate greatly depends on the type of catalysts [51]. Consequently, it is important to find an optimum range for NaOH concentration to improve the hydrogen generation rate in the presence of Co based catalysts. Figure 11.10 shows the experimental calorimetric curves obtained for hydrolysis of 30 mg of $NaBH_4$ in the presence of 4.1 mg of Co_3O_4 (corresponding to 10 wt% of pure Co with respect to sodium borohydride), and different amounts of solid NaOH. The measured heats after injection of 0.5 ml of water were

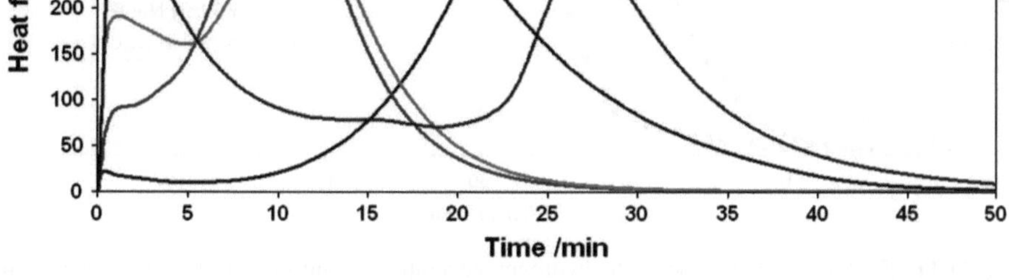

Fig. 11.10 Heat flow signals versus time for injection of 0.5 mL of water on 30 mg of $NaBH_4$ hydrolysis in the presence of 4.1 mg of Co_3O_4, and different amounts of solid NaOH (as reported on the figure)

−225, −420, −390 and − 250 $kJmol^{-1}$ for 5.5, 30, 66 and 125 mg of solid NaOH, respectively [52]. As it can be seen on Fig. 11.5, a clear shift from 20 to about 10 min of the peak maximum occurs when increasing the amount of solid NaOH added from 5.5 to 30 mg. Adding a balanced amount of solid NaOH (around 15–20 mg) to the "$NaBH_4$ + catalyst" mixture improves the kinetics of the reaction and creates a real benefit in terms of reaction kinetics. In parallel to the calorimetry measurements, the volume of generated H_2 versus time during the $NaBH_4$ hydrolysis reaction in the presence of Co_3O_4, and different amounts of solid NaOH was determined. Full conversion was achieved after 35, 12, 15 and 40 min for 5.5, 30, 66 and 125 mg of solid NaOH, respectively. The maximum rates reached over Co_3O_4 were 8, 16, 16 and 10 ml min^{-1} for 5.5, 30, 66 and 125 mg of solid NaOH, respectively, as presented in Fig. 11.11 [50].

Another experimental solution is to inject very small amounts of water (that will act as limiting reactant) and to avoid any separation step involving reactants and reacted products. In this simpler configuration the issues to solve are the improvement of the water diffusion in the solid (once the first injection performed and consequently further $NaBO_2 \cdot xH_2O$ formed) and the optimisation of water consumption (by reducing the $NaBO_2$ hydration ratio).

To give an answer to these issues, interesting studies about the thermal properties of $NaBH_4$ solid mixtures were obtained with other thermal techniques, as reported in Ref. [16], where an infrared camera has been used to evaluate the temperature

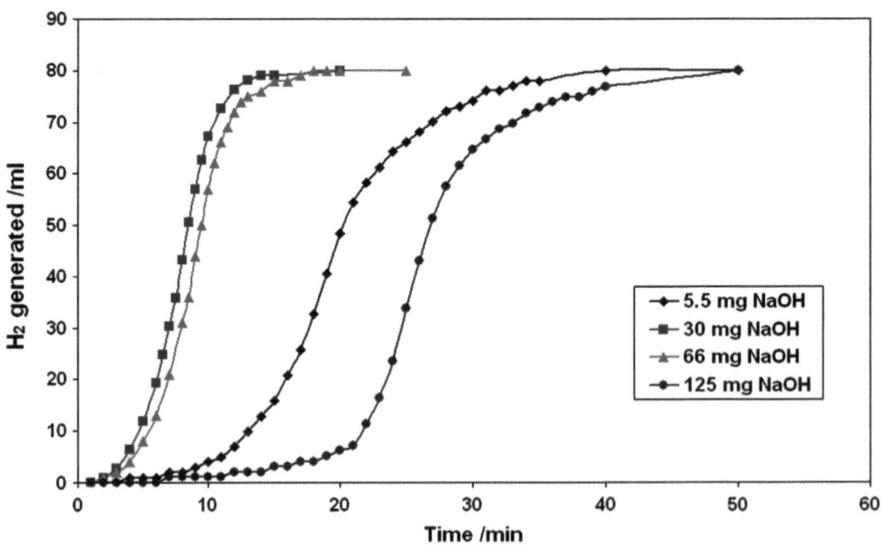

Fig. 11.11 Evolution versus time of the hydrogen generation for injection of 0.5 mL of water on 30 mg of $NaBH_4$ in the presence of 4.1 mg of Co_3O_4, and different amounts of solid NaOH (as reported on the figure)

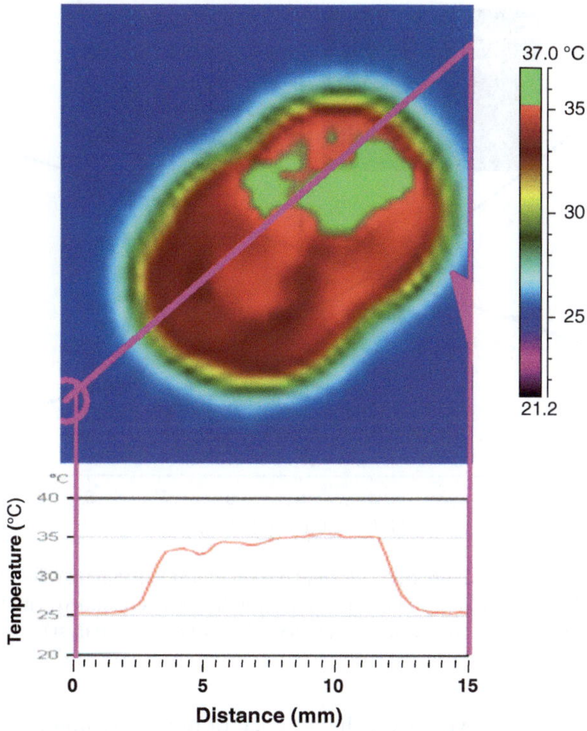

Fig. 11.12 Thermal imaging and temperature profile during the hydrolysis of $NaBH_4$ + 10 wt% nCo at 25 °C in presence of a less-than-stoichiometric amount of water

profile in a $NaBH_4$ + Co-nanoparticles solid mixture (with or without addition of solid NaOH) when a drop of water is added (Fig. 11.12).

Knowing that the major limitation to reach the theoretical 10.9 wt% H_2 storage density by hydrolysis of $NaBH_4$ powder with water is the hydration level of the metaborate product, to attain a higher H_2 storage density, a higher temperature of the system is required for dehydrating the metaborate products, and an easy way to reach this objective is to utilize the thermal effects of hydrolysis.

These thermal effects associated to the hydrolysis reaction have been studied on a fully dehydrated $NaBH_4$ powder by means of an IR imaging camera and a differential titration calorimeter. Various amounts of solid sodium hydroxide were added to the system ($NaBH_4$ + metallic nanoCobalt catalyst) allowing an increase of the maximum reaction temperature (up to 140 °C). The reaction maximum temperature and the hydrogen yield were considerably modified by varying the amount of NaOH and the amount of catalyst (Fig. 11.13). At a temperature of more than 140 °C, it is reasonable to expect the formation of low hydration borate phases. In fact, at temperatures above 105 °C water is expected to participate preferentially in the hydrolysis reaction rather than in the hydration of the

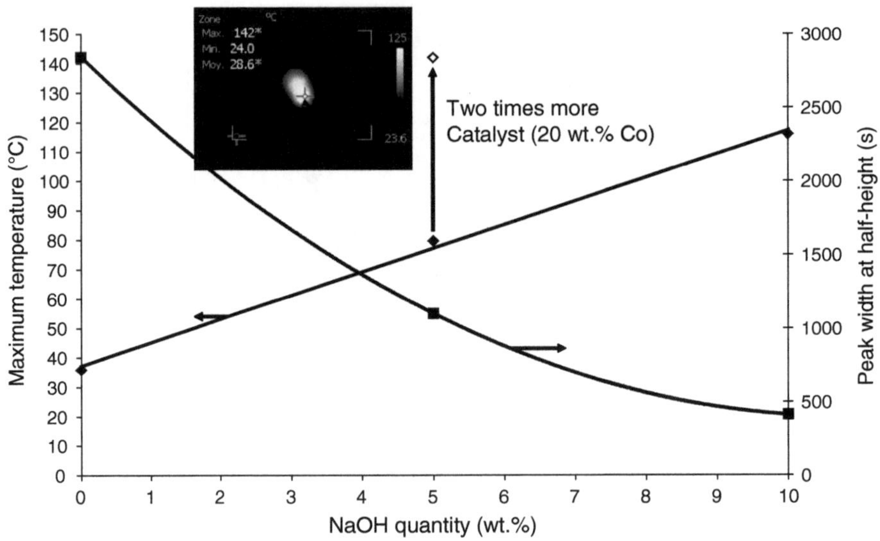

Fig. 11.13 Maximum temperature (*left side*) and peak width at half-height (*right side*) as functions of the sodium hydroxide loading, and comparison with the results obtained with twice the amount of catalyst (the two curves are obtained in presence of 10 wt% nCo catalyst)

formed metaborate [52]. Moreover the absence of highly hydrated borate favors the diffusion of water by avoiding the formation of the hydrophobic $NaBO_2 \cdot 2H_2O$ layer.

11.3 Second Case Study: Reversible H_2 Storage (Mg-Based Materials)

As mentioned in the introductory section of this chapter, for automotive applications the requirements for on-board hydrogen storage are very severe, as mainly stated by the United States Department of Energy [7] and the Japanese Government. The same kind of restrictions are also associated to the storage of intermittent energies where the amount of energy involved can be extremely high and consequently a stable storage system is required. To reach the demanded targets for the application in the automotive field research efforts have been made to develop interstitial, binary or even more complex hydrides capable to store and release hydrogen at temperature and pressure compatible with the different applications. For example, Mg (non toxic and inexpensive) shows a hydrogen storage capacity of 7.7 wt%, but the major impediment is its high H_2 desorption temperature (>300 °C). In order to decrease the desorption temperature of MgH_2 many research groups have tried to add other hydrides to MgH_2 in order to form complex hydrides with lower hydrogen desorption

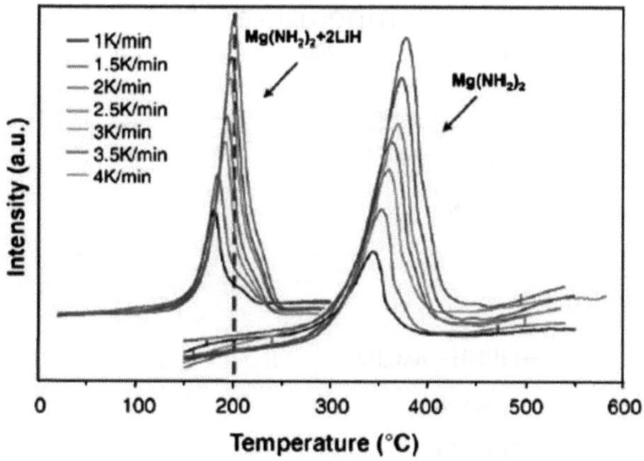

Fig. 11.14 DSC profiles for ($Mg(NH_2)_2 + 2LiH$) and $Mg(NH_2)_2$ at various scan rates [30]

temperature. Temperature programmed desorption techniques (TPD, DSC, TG) are the right tools to follow the hydrogen desorption process, determining the desorption temperature, evaluating the amount of desorbed H_2 and in DSC measuring the heat associated to the hydrogen release.

As example, the Li-Mg-N-H system was studied by differential scanning calorimetry by Gross's group [53]. The major issue for metal-N-H storage systems is the formation of NH_3, that takes place in parallel with H_2 release, and that acts as a

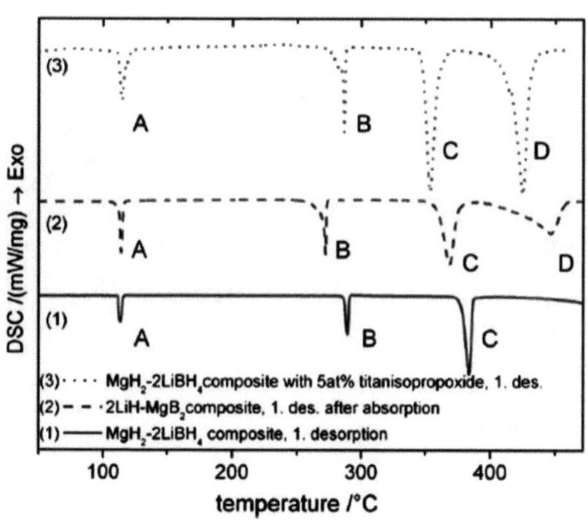

Fig. 11.15 HP-DSC of the desorption reaction of MgH_2–$2LiBH_4$ composites after milling (*1*), after initial hydrogenation (*2*) and after milling with additive (*3*). Heating rate $5°C \cdot min^{-1}$; 20 $mL_{H2} \cdot min^{-1}$; 3 bar H_2 [57]

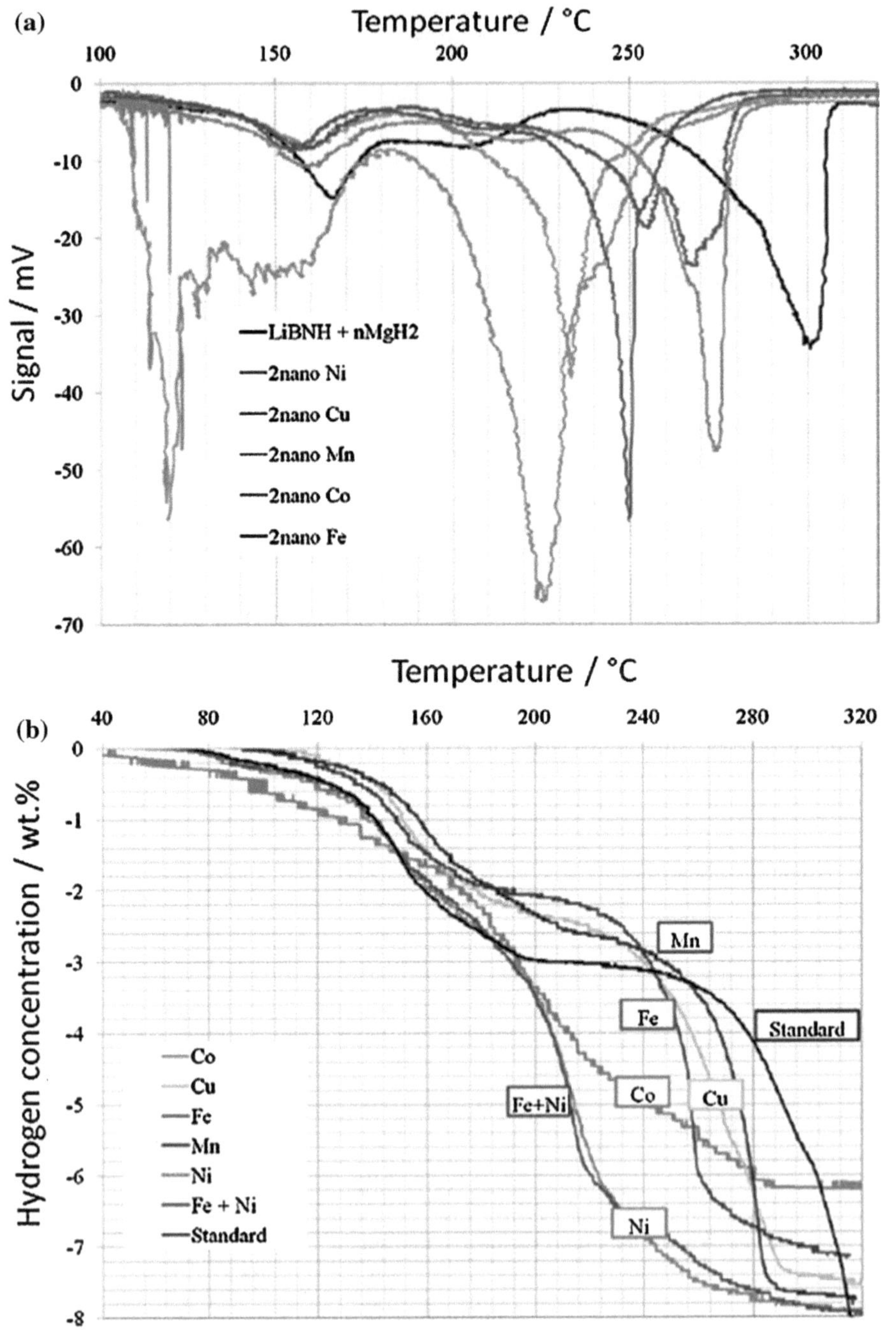

Fig. 11.16 **a** TPD comparison of LiBNH + $nMgH_2$ without additive and with 2 mol% Ni, Cu, Mn, Co and Fe at a constant ramping rate of 1°C/min and **b** Ramping kinetics measurements of LiBNH + $nMgH_2$ without and with 2 mol% nano Mn, Fe, Co, Cu, Ni and Fe + Ni [24]

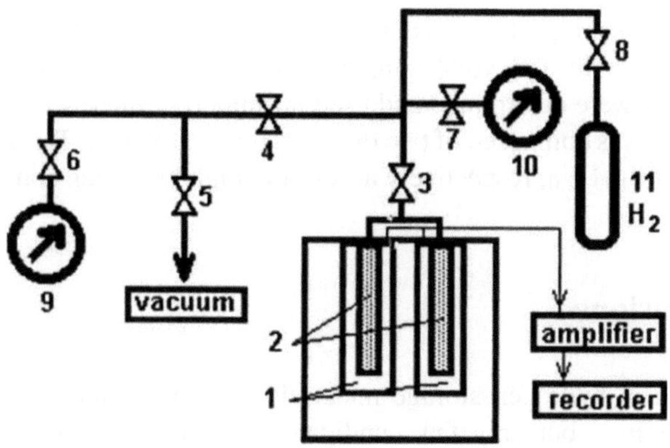

Fig. 11.17 Scheme of the calorimetric device: (*1*) calorimetric chambers; (*2*) calorimetric cells (reaction and blank); (*3–8*) needle valves; (*9*) vacuum gauge; (*10*) pressure gauge; (*11*) hydrogen source

poison for the PEM fuel cell catalysts [54]. As reported in Fig. 11.14, the lower temperature peak obtained by DSC method and centered at 200 °C is related to hydrogen release by the $Mg(NH_2)_2 + 2LiH \rightarrow Li_2Mg(NH_2)_2 + H_2$ reaction, while the higher temperature peak corresponds to the self-decomposition of Mg amide (irreversible reaction) related to NH_3 production. The identification of these phenomena and of their relative temperatures helps to identify the maximum working temperature necessary to avoid the $Mg(NH_2)_2$ decomposition and improve the cyclic stability of the material.

Addition of $LiBH_4$ to Mg allows to decrease the hydrogen desorption temperature by approximately 30 °C by formation of MgB_2 intermediary species [55, 56]. New techniques as high-pressure differential scanning calorimetry (HP-DSC) have been used to verify the mechanism of decomposition of MgH_2–$LiBH_4$ composites as well as the influence of the addition of different additives such as titanium isopropoxide and VCl_3, under 3 bar hydrogen pressure [57]. As shown in Fig. 11.15, the absorption-desorption cycling and the addition of additives deeply influence the kinetics of hydrogen release. As consequence, the peaks C and D shift at lower temperature for the MgH_2–$LiBH_4$ materials activated by the absorption reaction (curve 2) or by addition of titanium isopropoxide (curve 3).

Novel complex hydrides ($LiBH_4/LiNH_2/MgH_2$) are also under study and the addition of various nanosized additives as nickel, cobalt, iron, copper, manganese was followed by thermoprogrammed desorption technique (TPD) [24].

The TPD tool was used in Ref. [24] to identify the optimal hydrogen release temperature and to get an indication of the H_2 desorption rate by measuring the peak width. In Fig. 11.16, the addition of the different additives has a big effect on reducing the temperature of the second H_2 releasing peak.

Unusual are set-ups combining Tian-Calvet calorimetry and high pressure volumetric lines. One of the few examples is reported in the work of Anikina and Verbetsky in Ref. [58]: the schematic diagram of their set-up is reported on Fig. 11.17. This technique was used to study TiZrMnV compounds used as H_2storage system. Desorption molar enthalpies were determined and used in support to the researcher assumption that the material was composed of two different hydride phases, $Ti_{0.9}Zr_{0.1}Mn_{1.1}H_{1.0}$ and $Ti_{0.9}Zr_{0.1}Mn_{1.1}H_{2.0}$, respectively at low and high hydrogen content.

11.4 Conclusions

Different kinds of hydrogen storage materials have been studied and reported in the recent literature, but only few candidates are able to answer to the requirements for portable and on-board applications. The main problems still to be solved are the continuous control of the reaction for safety reasons, the increasing of the kinetics of hydrogen release, and the reversibility of the reaction to perform adsorption/desorption cycling.

For an extensive use of these materials it is important to deeply understand the thermal phenomena associated to hydrogen release reactions and adsorption/desorption cycles.

The thermodynamic features of these systems (enthalpies, entropies) as well as the thermal properties of the involved compounds (thermal conductivity, heat capacity) are crucial data to predict the thermal behavior of large quantity of material, as those used in real applications.

Calorimetry and thermal analysis techniques (alone or coupled to other instruments) have been developed and applied to the study of hydrogen storage systems. As example, information about the heat evolved during the hydrogen release as a function of time has given the possibility to contemporarily evaluate the kinetics of reaction and to point out the presence of diffusion problems, connected to borohydride hydrolysis. In other showcases the identification of specific thermal phenomena helped in understanding the reaction mechanism and the material structure.

It is certain that calorimetry and thermal analysis are indispensable tools in the study of hydrogen storage and further developments and applications of these techniques have to be expected in the near future.

References

1. G. Marban, T. Valdes-Solis, Int. J. Hydrogen Energy **32**, 1625 (2007)
2. D.A.J. Rand, R.M. Dell, *Hydrogen Energy* (The Royal Society of Chemistry, Cambridge, 2008)
3. F. Jackow, J. Loughead, European Technology Platform for Hydrogen and Fuel Cells (HFP), Implementation plan-status 2006 http://www.fch-ju.eu/sites/default/files/documents/hfp-ip06-final-apr2007.pdf/
4. L. Schlappbach, A. Züttel, Nature **414**, 353 (2001)
5. B. Sakintuna, F. Lamari-Darkrimb, M. Hirscher, Int. J. Hydrogen Energy **32**, 1121 (2007)
6. S.F.J. Flipsen, J. Power Sour. **162**, 927 (2006)

7. U.S. Department of Energy Hydrogen Program, http://www.hydrogen.energy.gov/
8. U.B. Demirci, O. Akdim, P. Miele, Int. J. Hydrogen Energy **34**, 2638 (2009)
9. S. Bakker, Int. J. Hydrogen Energy **35**, 6784 (2010)
10. C. Cakanyildirim, M. Guru, Int. J. Hydrogen Energy **33**, 4634 (2008)
11. S.U. Jeong, R.K. Kim, E.A. Chob, H.-J. Kim, S.-W. Nam, I.-H. Oh, S.-A. Hong, S.H. Kim, J. Power Sour. **144**, 129 (2005)
12. B.H. Liu, Z.P. Li, J. Power Sour. **187**, 527 (2009)
13. S.C. Amendola, S.L. Sharp-Goldman, M.S. Janjua, N.C. Spencer, M.T. Kelly, P.J. Petillo, Int. J. Hydrogen Energy **25**, 969 (2000)
14. Y. Kojima, K.I. Suzuki, K. Fukumoto, M. Sasaki, T. Yamamoto, Y. Hawai, Int. J. Hydrogen Energy **27**, 1029 (2002)
15. J.S. Zhang, W.N. Delgass, T.S. Fisher, J.P. Gore, J. Power Sour. **164**, 772 (2007)
16. S. Bennici, A. Garron, A. Auroux, Int. J. Hydrogen Energy **35**, 8621 (2010)
17. B.H. Liu, Z.P. Li, S. Suda, J. Alloys Compd. **415**, 288 (2006)
18. P. Krishnan, K.L. Hsueh, S.D. Yim, Appl. Catal. B **77**, 206 (2007)
19. R. Chen, X. Wang, L. Xu, L. Chen, S. Li, C. Chen, Mater. Chem. Phys. **124**, 83 (2010)
20. A. Züttel, P. Sudan, P. Mauron, P. Wenger, Mater. Sci. Eng. B **108**, 9 (2004)
21. N. Eigen, F. Gosch, M. Dornheim, T. Klassen, R. Bormann, J. Alloys Compds. **465**, 310 (2008)
22. M. Ismail, Y. Zhao, X.B. Yu, S.X. Dou, Int. J. Hydrogen Energy **35**, 2361 (2010)
23. K.R. Graham, T. Kemmitt, M.E. Bowden, Energy Environ. Sci. **2**, 706 (2009)
24. S.S. Srinivasan, M.U. Niemann, J.R. Hattrick-Simpers, K. McGrath, P.C. Sharma, D.Y. Goswami, E.K. Stefanakos, Int. J. Hydrogen Energy **35**, 9646 (2010)
25. P. Chen, Z. Xion, J. Luo, J. Lin, K.L. Tan, Nature **420**, 302 (2002)
26. W.N. Yang, C.X. Shang, Z.X. Guo, Int. J. Hydrogen Energy **35**, 4534 (2010)
27. C.X. Shang, Z.X. Guo, Int. J. Hydrogen Energy **32**, 2920 (2007)
28. D. Dybtsev, C. Serre, B. Schmitz, B. Panella, M. Hirscher, M. Latroche, P.L. Llewellyn, S. Cordier, Y. Molard, M. Haouas, F. Taulelle, G. Férey, Langmuir **26**, 11283 (2010)
29. M. Latroche, S. Suble, C. Serre, C. Mellot-Draznieks, P.L. Llewellyn, J.-H. Lee, J.-S. Chang, S.H. Jhung, G. Ferey, Angew. Chemie Int. Ed. **45**, 8227 (2006)
30. H.W. Langmi, D. Book, A. Walton, S.R. Johnson, M.M. Al-Mamouri, J.D. Speight, P.P. Edwards, I.R. Harris, P.A. Anderson, J. Alloys Compds. **637**, 404 (2005)
31. G.E. Ioannatos, X.E. Verykios, Int. J. Hydrogen Energy **35**, 622 (2010)
32. V.G. Minkina, S.I. Shabunya, V.I. Kalinin, V.V. Martynenko, A.L. Smirnova, Int J. Hydrogen Energy **33**, 5629 (2008)
33. D.R. Lide, *CRC Handbook of Chemistry and Physics*, 82nd edn. (CRC Press, Boca Raton, 2001)
34. A. Stock, *Hydrides of Boron and Silicon* (Cornell University Press, Ithaca, 1933)
35. J.C. Ingersoll, N. Mani, J.C. Thenmozhiyal, A. Muthaiah, J. Power Sour. **173**, 450 (2007)
36. J. Zhang, T.S. Fisher, J.P. Gore, D. Hazra, P.V. Ramachandran, Int. J. Hydrogen Energy **31**, 2292 (2006)
37. N.O. Gonzales, M.E. Levin, L.W. Zimmerman, J. Hazard. Mater. **142**, 369 (2007)
38. J. Li, B. Li, S. Gao, Phys. Chem. Miner. **27**, 342 (2000)
39. S. Bennici, A. Auroux, Thermal Analysis and Calorimetric Methods, in *Metal Oxide Catalysis*, vol. 1, ed. by S.D. Jackson, S.J. Hargreaves (Wiley-VCH Verlag GmbH & Co, Weinheim, 2009), p. 391
40. H. Nogent, X. Le Tacon, L. Vincent, N. Sbirrazzuoli, J. Loss Prev. Process Ind. **18**, 43 (2005)
41. R. Andre, L. Bou-Diab, P. Lerena, F. Stoessel, M. Giordano, C. Mathonat, Organ. Process Res. Dev. **6**, 915 (2002)
42. J. Andrieux, D. Swierczynski, L. Laversenne, A. Garron, S. Bennici, C. Goutaudier, P. Miele, A. Auroux, B. Bonnetot, Int. J. Hydrogen Energy **34**, 938 (2009)
43. A. Garron, D. Swierczynski, S. Bennici, A. Auroux, Int. J. Hydrogen Energy **34**, 1185 (2009)
44. S.J. Kim, J. Lee, K.Y. Kong, C.R. Jung, I.G. Min, S.-Y. Lee, H.-J. Kima, S.W. Nam, T.-H. Lim, J. Power Sour. **170**, 412 (2007)
45. A. Garron, S. Bennici, A. Auroux, Appl. Catal. A **378**, 90 (2010)

46. L. Damjanovic, S. Bennici, A. Auroux, J. Power Sour. **195**, 3284 (2010)
47. O. Akdim, U.B. Demirci, P. Miele, Int. J. Hydrogen Energy **34**, 4780 (2009)
48. H.I. Schlesinger, H.C. Brown, A.E. Finholt, J.R. Gilbreath, H.R. Hoekstra, E.K. Hyde, J. Am. Chem. Soc. **75**, 215 (1953)
49. S. Bennici, H. Yu, E. Obeid, A. Auroux, Int. J. Hydrogen Energy **36**, 7431 (2011)
50. L. Damjanovic, S. Bennici, A. Auroux, Int. J. Hydrogen Energy **36**, 1991 (2011)
51. H. Tian, Q. Guo, D. Xu, J. Power Sour. **195**, 2136 (2010)
52. E.Y. Marrero-Alfonso, J.R. Gray, T.A. Davis, M.A. Matthews, Int. J. Hydrogen Energy **32**, 4723 (2007)
53. W. Luo, J. Wang, K. Steward, M. Clift, K. Gross, J. Alloys Compds. **336**, 446 (2007)
54. G. Postole, A. Auroux, Int. J. Hydrogen Energy **36**, 6817 (2011)
55. S. Orimo, Y. Nakamori, G. Kitahara, K. Miwa, A. Ninomiya, H. Li, N. Ohba, S. Towata, A. Zuttel, J. Alloys Compd. **427**, 404 (2005)
56. J.J. Vajo, S.L. Skeith, F. Mertens, J. Phys. Chem. B Lett. **109**, 3719 (2005)
57. U. Bosenberg, S. Doppiu, L. Mosegaard, G. Barkhordarian, N. Eigen, A. Borgschulte, T.R. Jensen, Y. Cerenius, O. Gutfleisch, T. Klassen, M. Dornheim, R. Bormann, Acta Materialia **55**, 3951 (2007)
58. E.Yu. Anikina, V.N. Verbetsky, J. Alloys Compd. **45**, 330 (2002)

Chapter 12
Adsorption Microcalorimetry as a Tool to Study the CO–Pt Interaction for PEMFC Applications: A Case Study

Georgeta Postole and Aline Auroux

Abstract To date, microcalorimetry of CO adsorption onto supported metal catalysts was mainly used to study the effects induced by the nature and the particle size of supported metallic clusters, the conditions of pretreatment and the support materials on the surface properties of the supported metallic particles. The present chapter focuses on the employ of adsorption microcalorimetry for studying the interaction of carbon monoxide with platinum-based catalyst aimed to be used in proton exchange membrane fuel cells (PEMFCs) applications.

12.1 Evolution and Types of Fuel Cell

Fuel cells are one of the oldest electrical energy conversion technologies known to man for more than 160 years. A fuel cell is a galvanic cell, in which the free energy of a chemical reaction is converted into electrical energy. Their development lacked a drive at the beginning, as primary energy sources were abundant, unrestricted, and inexpensive. After the second world war, this technology became the subject of intense research; one of the major factors that have influenced their development have been the increasing concern about the environmental consequences of fossil fuel use in production of electricity, and for the propulsion of vehicles. The historical development of fuel cells has been described by Carrete et al. [1], Litster et al. [2] and Boudghene Stambouli and Traversa [3]. In spite of the attractive system efficiencies and environmental benefits associated with fuel-cell technology, it is still a difficult task to transform the early scientific experiments into commercially viable industrial products. These problems have been often associated with the lack of appropriate

G. Postole (✉) · A. Auroux
Université Lyon 1, CNRS, UMR 5256, IRCELYON, Institut de recherches sur la catalyse et l'environnement de Lyon, F-69626 Villeurbanne Cedex, France
e-mail: georgeta.postole@ircelyon.univ-lyon1.fr

A. Auroux (ed.), *Calorimetry and Thermal Methods in Catalysis*,
Springer Series in Materials Science 154, DOI: 10.1007/978-3-642-11954-5_12,

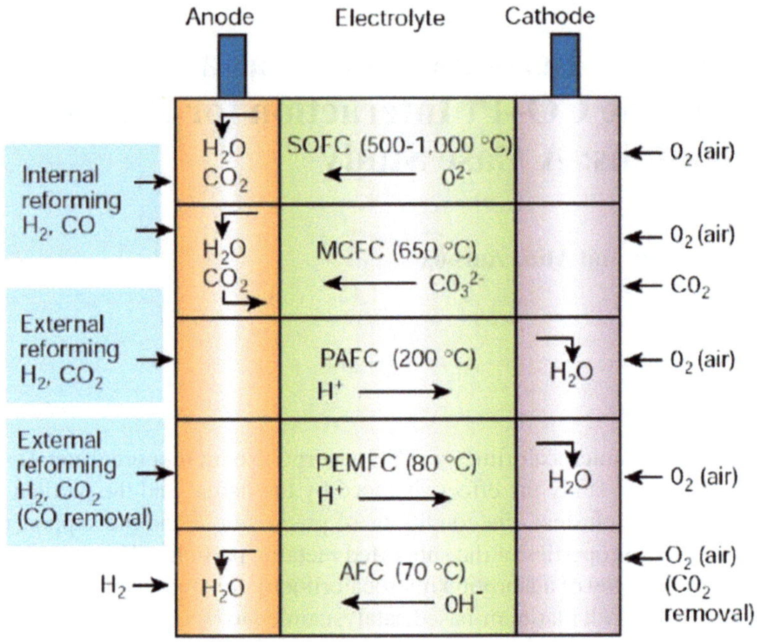

Fig. 12.1 Summary of fuel-cell types [5]

materials or manufacturing routes that would enable the cost of electricity per kWh to compete with the existing technology [4].

However, fuel cells may help to reduce our dependence on fossil fuels and diminish poisonous emissions into the atmosphere and that is the reason for their continuous development. For example, fuel cells using pure hydrogen as fuel produce only water, thus eliminating locally all noxious byproducts otherwise caused by electricity production.

Different types of fuel cells under active development are usually classified by the electrolyte employed as the ionic conductor in the cell, or by their operating temperature. Figure 12.1 summarizes the types of fuel cells in order of increasing operating temperature [5], while an overview of their characteristics is given in Table 12.1 [1, 3].

As it can be seen, low-temperature and high-temperature fuel cells can be distinguished. Low-temperature fuel cells are the Alkaline Fuel Cell (AFC), the Polymer Electrolyte Fuel Cell (PEMFC), and the Phosphoric Acid Fuel Cell (PAFC). The high-temperature fuel cells operate in the temperatures region from 500 to 1000 °C; two different types have been developed: the Molten Carbonate Fuel Cell (MCFC) and the Solid Oxide Fuel Cell (SOFC). They have the ability of using methane as fuel and thus present high inherent generation efficiency (45–60 % for common fuels such as natural gas, 90 % with heat recovery [3]).

Table 12.1 The characteristics of different fuel cells that have been realized and are currently in use and development

Fuel cells	AFC (Alkaline)	PEMFC (Polymer electrolyte membrane)	DMFC (Direct methanol)	PAFC (Phosphoric acid)	MCFC (Molten carbonate)	SOFC (Solid oxide)
Operating temperature (°*C*)	<100	60–120	60–120	160–200	600–800	800–1000 low temperature (500–600) possible
Anode reaction	$H_2 + 2OH^- \rightarrow 2H_2O + 2e^-$	$H_2 \rightarrow 2H^+ + 2e^-$	$CH_3OH + H_2O \rightarrow CO_2 + 6H^+ + 6e^-$	$H_2 \rightarrow 2H^+ + 2e^-$	$H_2 + CO_3^{2-} \rightarrow H_2O + CO_2 + 2e^-$	$H_2 + O^{2-} \rightarrow H_2O + 2e^-$
Cathode reaction	$1/2O_2 + H_2O + 2e^- \rightarrow 2OH^-$	$1/2O_2 + 2H^+ + 2e^- \rightarrow H_2O$	$3/2O_2 + 6H^+ + 6e^- \rightarrow 3H_2O$	$1/2O_2 + 2H^+ + 2e^- \rightarrow H_2O$	$1/2O_2 + CO_2 + 2e^- \rightarrow CO_3{}^{2-}$	$1/2O_2 + 2e^- \rightarrow O^{2-}$
Electrolyte	Potassium hydroxide (KOH)	Polymer, proton exchange membrane	Polymer	Phosphoric acid	Molten salt such as nitrate, sulphate, carbonates…	Ceramic as stabilized zirconia and doped perovskite
Charge carrier in the electrolyte	OH^-	H^+	H^+	H^+	$CO_3{}^{2-}$	O^{2-}

(continued)

Table 12.1 (continued)

Fuel cells	AFC (Alkaline)	PEMFC (Polymer electrolyte membrane)	DMFC (Direct methanol)	PAFC (Phosphoric acid)	MCFC (Molten carbonate)	SOFC (Solid oxide)
Fuel	Hydrogen or hydrazine	Hydrogen	Liquid methanol	Hydrogen	Hydrogen, carbon monoxide, natural gas, propane	Hydrogen, hydrocarbons, natural gas, renewable fuels
Applications	Transportation, space, military, energy storage system			Combined heat and power for decentralized stationary power systems	Combined heat and power for decentralized stationary power systems and for transportation (train, boats, …)	
Realized power	Small plants 5–150 kW modular	Small plants 5–250 kW modular	Small plants 5 kW	Small-medium plants 50 kW–11 MW modular	Small power plants 100 kW–2 MW modular	Small power plants 100–200 kW
Efficiency (%)	50–55	40–50	40–55	40–50	50–60	45–60

Not mentioned in this classification is the Direct Methanol Fuel Cell (DMFC) working at 60–120 °C and using methanol as a fuel. There are also other types of fuel cells (e.g. air-depolarised cells, sodium amalgam cells, alkali metal-halogen cells, etc.) which are less employed, but that can possibly find a specific application in the future.

The basic structure of all fuel cells is similar: the cell consists of two electrodes which are separated by the electrolyte and which are connected in an external circuit. The electrodes are exposed to gas or liquid flows to supply the electrodes with fuel or oxidant (e.g. hydrogen or oxygen). As it can be seen in Table 12.1, the anode reaction in fuel cells is either the direct oxidation of hydrogen (low temperature fuel cells) or the oxidation of methanol (DMFC). An indirect oxidation via a reforming step can also occur in the case of high temperature operation fuel cells. The cathode reaction is oxygen reduction, in most cases from air.

Both the low-temperature and the high-temperature fuel cells have their advantages and disadvantages depending on the application. For example, among the fuel cells used for transportation, the high-temperature fuel cells (such as solid oxide or molten carbonate cells) present advantages for the use in ships or locomotives in which frequent on/off cycling is not required. They allow more flexibility in fuel selection and may be used without a reformer. The high-grade waste heat is more easily used in a thermally integrated system. Conversely, the low-temperature fuel cells, such as polymer electrolyte or alkaline, may be a better choice for passenger cars to which rapid start-up and wide power range are important [6].

By far the greatest research interest throughout the world has focussed on proton exchange membrane (PEM) and solid oxide (SO) cell stacks. PEMFCs are well advanced type of fuel cells that has found widespread area of use, especially in transportation applications, distributed generation (DG) units and portable electronic equipments [7]. Some of the key advantages of PEMFC systems over the other competitive types of fuel cells for its potential market competitiveness arise from [8]:

- PEMFCs can operate at relatively low temperatures.
- PEMFCs are tolerant to CO_2; so they can use the atmospheric air.
- PEMFCs have high voltage, current and power density.
- PEMFCs can work at low pressure (1 or 2 bars), which enhances security.
- PEMFCs have a good tolerance to the difference of pressure of the reactants.
- PEMFCs are compact and robust and have a simple mechanical design.
- PEMFCs use stable building materials.

12.2 Proton Exchange Membrane Fuel Cells

PEM fuel cells use a proton exchange membrane as an electrolyte and operates at low-temperatures, between 60 and 120 °C. From various types of fuel cells, PEMFCs are suitable choice for both stationary and portable applications due to their fast start up, high power density, suitability for discontinuous operation as well as low operating temperature [8–10]. Their components and related functions are well described in a recent review [9]. The first application of a PEM fuel cell was in the 1960s as an auxiliary power source in the Gemini space flights. It was also used to provide the astronauts with clean drinking water. The membrane used was a polystyrene sulfonate (PSS) polymer, which has been proven do not be enough stable. Advances in the PEMFCs technology were stagnating until the late 1980s when the fundamental design underwent significant reconfiguration. A major breakthrough in this field came with the use of Nafion® or Dow® membranes, possessing a higher acidity and conductivity and being more stable than the polystyrene sulfonate membranes [1, 2].

Figure 12.2 presents the principle of a single proton exchange membrane fuel cell fed with hydrogen which is oxidized at the anode, and oxygen that is reduced at the cathode [1]. The protons released during the oxidation of hydrogen are conducted through the proton exchange membrane (the electrolyte) to the cathode. Since the membrane is not electrically conductive, the electrons released from the hydrogen travel along the electrical detour provided and an electrical current is generated. The reaction product is water, which is formed at the cathode [2].

In order to be efficient enough, the electrochemical reactions that take place in fuel cell must be catalyzed. To date, platinum has proven to be the best catalyst for both the hydrogen oxidation (anode) and the oxygen reduction (cathode) reactions. For PEM fuel cell applications, platinum is usually implemented in the form of Pt/C catalysts because of its significantly higher surface area compared with that of platinum black catalysts resulting in its cost reduction [5, 11].

The oxidation of hydrogen occurs readily on Pt-based catalysts involving the adsorption of the gas onto the catalyst surface followed by a dissociation of the molecule and electrochemical reaction to two hydrogen ions as follows:

$$2Pt_{(s)} + H_2 \rightarrow Pt{-}H_{ads} + Pt{-}H_{ads} \tag{12.1}$$

$$Pt{-}H_{ads} \rightarrow H^+ + e^- + Pt_{(s)} \tag{12.2}$$

where $Pt_{(s)}$ is a free surface site and $Pt{-}H_{ads}$ is an adsorbed H-atom on the Pt active site.

The overall reaction of hydrogen oxidation is:

$$H_2 \rightarrow 2H^+ + 2e^- \tag{12.3}$$

The rate of the hydrogen oxidation process at the Pt-based catalyst at 80 °C is very high, as long as the catalyst surface is not contaminated by adsorbed impurities.

The highest performing PEMFC systems employ pure hydrogen as the fuel, but for many applications, especially mobile, pure hydrogen is not yet a viable option due to the technical difficulty of on-board storage and refuelling. Currently, the most viable technology for on-site H_2 generation is reforming technology (e.g. steam reforming, partial oxidation or autothermal reforming) of hydrocarbons such as methanol,

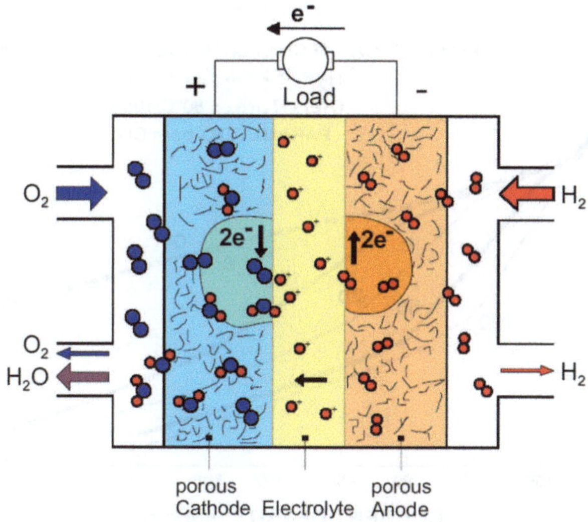

Fig. 12.2 Schematic presentation of a single typical proton exchange membrane fuel cell [1]. On the anode, hydrogen gas is catalytically disassociated according to the reaction $H_2 \rightarrow 2e^- + 2H^+$. The hydrogen ions pass through the polymer electrolyte to the cathode where oxygen, in most cases from air, is reduced ($O_2 + 4e^- \rightarrow O^{2-}$). The overall reaction is: $H_2 + 1/2O_2 \rightarrow H_2O$

gasoline or natural gas and alcohols that are readily available through existing distribution channels [12]. Accordingly, the use of hydrocarbon or alcohol fuels requires an external fuel processor (reformer) to be incorporated into the system. This item not only increases the complexity and cost of the fuel cell system, but also decreases the overall efficiency. The use of reformate fuel presents special challenges for a PEM fuel cell in terms of system efficiency, reliability, and durability. Many of these issues can be attributed to the detrimental effects of impurities such as CO, CO_2, NH_3 and H_2S obtained as the by-products of the reforming process [13]. The hydrogen impurities due to manufacturing process are brought along with the fuel feed stream into the anode of a PEMFC stack, causing performance degradation, and sometimes permanent damage of the membrane electrode assemblies [14–16]. The effect of different contaminants including: fuel impurities (CO,CO_2, H_2S, and NH_3), air pollutants (NO_x, SO_x, CO, and CO_2), and cations resulting from the corrosion of fuel cell stack system components (such are Fe^{3+} and Cu^{2+}) on the efficiency of PEM fuel cells was recently reviewed by Cheng et al. [17]. It was found that even trace amounts of impurities present in either fuel or air streams or fuel cell system components could severely poison the anode, membrane, and cathode, particularly at low-temperature operation, what results in dramatic performance drop. Thus, elucidation of the degradation mechanism of anode or cathode reactions by impurity materials in polymer electrolyte fuel cells (PEMFCs) is a crucial topic, to attain its longevity.

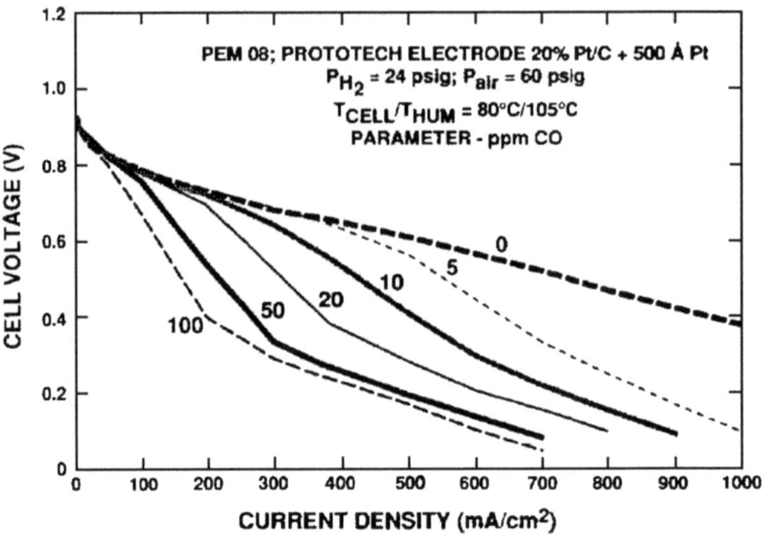

Fig. 12.3 The performance losses caused by trace amounts (from 5 to 100 ppm) of carbon monoxide in the fuel stream [6]

Poisoning of the anode catalyst is caused primarily by carbon monoxide, either brought into the cell with the fuel feed stream or generated in situ by the reduction of CO_2 [9, 17–19]. The hydrogen produced by steam reforming, contains more than 1 % of CO [6, 13, 20]. Since a PEMFC cannot tolerate such high CO levels, the carbon monoxide content is reduced through a series of high- and low-temperature water-gas-shift (WGS) and preferential oxidation (PROX) reactions to bring the CO level to less than 10 ppm [21]. However, due to the typical low temperature of operation (ca. 80 °C) and the choice of Pt as the electrocatalyst, the CO-poisoning effect could significantly affect the long-term performance of the PEMFC stack, even at this low CO level as it can be seen in Fig. 12.3 [6, 22, 23]. CO poisoning on Pt electrocatalysts becomes more severe with increases in CO concentration and exposure time. For example, Fig. 12.3 illustrates typical fuel cell stack polarization curves obtained at 80 °C in both the absence and presence of various concentrations of CO [6]. The figure indicates that the CO impurities from fuel streams, even at a level of a few ppm, can cause a substantial degradation in cell performance, especially at high current densities.

This problem is therefore particularly severe with fuel feed streams derived from the steam reforming of hydrocarbons, methanol or other liquid fuels, and is of lesser concern when the PEMFC operates on neat hydrogen. Nevertheless, the level of CO which brings about significant poisoning at the hydrogen anode in a PEMFC is so small (several parts per million) that even in the case of relatively pure hydrogen feeds, e.g., bottled hydrogen of nominal 99.99 % purity, some long-term platinum anode performance loss has been observed [9]. Overcoming the CO poisoning problem is thus of paramount interest and needs to be addressed in order to make reformate gas a viable fuel for PEM fuel cells. Extensive research has focused on gaining a better understanding of poisoning effect that CO has on PEMFC performance. Baschuk and Li [24] reviewed the CO poisoning of platinum electrocatalyts used in PEM fuel cells in terms of characteristics, mechanism, mitigation, and theoretical models. Many electrochemical, spectroscopic and theoretical studies [23, 25–31] have been employed to investigate the CO adsorption process on platinum and Pt-based electrocatalysts, including in situ attenuated total reflectance-Fourier transform infrared (ATR-FTIR) study [32, 33], CO isotope exchange experiments [34], cyclic and stripping voltammetry [35], etc. Although these studies provide valuable information for the design of new catalysts, extrapolations of the conclusions with respect to the CO tolerance to the real systems are not straightforward, mainly because of reaction conditions that are different from the actual PEM fuel cell operating environment. However, it has been shown without doubts that the CO poisoning effect was strongly related to the concentration of CO, the exposure time to CO, the cell operation temperature, and anode catalyst types.

12.3 CO Adsorption Microcalorimetry on Pt-Based Materials: Literature Survey

Gas-solid interactions are fundamental for the understanding of adsorption, which is the first step in a variety of processes in surface science. The characterization of many gas-solid surface processes and the study of the gas interaction with heterogeneous surfaces (surfaces composed of different kinds of sites) are therefore important aspects. One of the challenges in the field of gas-solid interactions is to envisage methods for the determination of the energetic topography of heterogeneous substrates from adsorption experiments. In this context, the adsorption microcalorimetry technique coupled to volumetry is a powerful tool able to supply information about the strength of gas-surface interactions [36–39]. Moreover, the adsorption microcalorimetry also provides a direct measurement of heat of adsorption and their evolutions with the coverage and can contribute in the study of all phenomena which can be involved in one catalyzed process, e.g. activation/deactivation of the catalyst, coke production, pore blocking, sintering, and adsorption of poisoning in the feed gases [40, 41].

Chemisorption of carbon monoxide on transition metal surfaces (either single crystals or supported clusters) is a tool of general use to study the active sites present over this type of solid surfaces [42]. CO adsorption on noble metals has been the subject of a large number of papers. For instance, the adsorption behaviour of CO on surfaces of Pt single crystals, polycrystalline Pt films, and supported Pt catalysts has been discussed in terms of adsorbed species or adsorption structures formed during interaction of CO with the metal surface.

The adsorption of CO on a series of platinum based catalysts was also carried out by microcalorimetry technique, supplying information about the number, the strength distribution and the heat associated to the carbon monoxide adsorption on available Pt surface sites. Table 12.2 lists some literature reports on the average heats of CO adsorption over different platinum supported catalysts. In this table, the experimental data on powdered catalysts were collected from the vicinity of room temperature up to 130 °C. Prior to CO adsorption, the samples were reduced at 200 or 500 °C under hydrogen flow.

In these literature reports [43–58], the data obtained by adsorption microcalorimetry are considered together with those obtained from complementary techniques (i.e. infrared spectroscopy, temperature programmed desorption, X-ray photoelectron spectroscopy) in order to elucidate the influence of loading and dispersion of Pt, type of support material and the reduction temperature, on energetics and mechanism of CO adsorption on supported Pt catalysts for a better understanding of their catalytic performances.

For example, by examining the initial and differential heats of adsorption measured on Pt/Al_2O_3 powders calcined at different temperatures, Uner and Uner [50] concluded that CO adsorption processes is not structure-sensitive. CO heats of adsorption values obtained by the authors are plotted against carbon monoxide coverage in Fig. 12.4. The heat of adsorption data for all catalysts fell on the same curve,

Table 12.2 Literature data of CO adsorption over supported Pt surfaces

Catalyst	Pt (wt%)	Temperature of adsorption (°C)	Average heat of adsorption (kJ/mol)	Reference
Pt/C (Norit RX-3)	0.22	27	115	43
Pt/C (Vulcan)	5.00	27	110	44
Pt/graphite	2.00	57	115	45
Pt/C	1.00	25	80	46
PtSn/TiO_2	2.00	25	105	47
Pt/TiO_2	2.00	27	90	48
Pt/A_2O_3	1.88	25	120	49
Pt/Al_2O_3	2.00	30	135	50
Pt/Al_2O_3 nano-fibre	2.91	30	120	51
Pt/η-Al_2O_3	2.10	47	100	48
Pt/Al_2O_3	3.00	57	130	42,52
Pt/Al_2O_3	5.00	50	130	53
Pt/SiO_2	10.00	27	135	54
Pt/SiO_2	2.10	47	113	48
Pt/SiO_2	1.20	130	130	55,56
Pt/SiO_2–Al_2O_3	1.50	27	115	48
Pt/NaX zeolite	1.00	27	130	57
Pt/K-L zeolite	1.00	130	120	58
Pt/K-ZSM5 zeolite	1.00	130	110	58

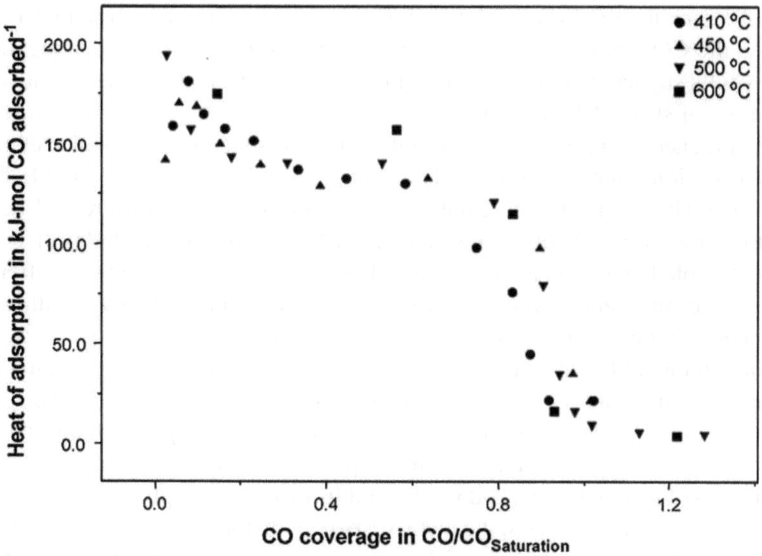

Fig. 12.4 Differential heat of carbon monoxide adsorption at 30 °C over 2 % Pt/γ-Al_2O_3 reduced under hydrogen at 270 °C. The temperatures indicate the calcination temperature of the catalysts [50]

but the adsorbed gas amounts decreased with increasing the calcinations temperature in agreement with the decrease in metal dispersions.

Serrano-Ruiz et al. [49] used the adsorption microcalorimetry of CO at room temperature, X-ray photoelectron spectroscopy (XPS), and ^{119}Sn Mössbauer spectroscopy to study the effect of adding Sn to Pt/CeO_2–Al_2O_3 and Pt/Al_2O_3 catalysts. Microcalorimetric analysis indicated that adding cerium caused the appearance of a more heterogeneous distribution of active sites, whereas adding tin led to a higher homogeneity of these sites. The influence of reduction conditions on the Pt–CO adsorption strength have also been investigated in detail for Pt/Al_2O_3. The authors [49] observed that the catalyst reduction at higher temperature caused a decrease of the initial CO adsorption heat (from 140 to 120 kJ/mol) and of CO saturation coverage from 55 to 45 $\mu mol_{CO}/g_{cat}$. Thus a higher homogeneity of the surface metal atoms for CO adsorption was reported when catalyst was reduced at 500 °C. The higher initial heat of adsorption obtained in the sample reduced at low temperature (200 °C) was attributed to the interaction of CO with highly unsaturated metal atoms at corners and edges. The decrease of the total amount of CO adsorbed with increasing the pretreatment temperature was explained to be due to sintering of the Pt particles after high-temperature reduction.

The CO adsorption microcalorimetry was also used to explain the promoting effect of Pt in bimetallic Ni–Pt catalysts supported on alumina nano-fibre (Alnf) tested for the liquid phase reforming of sorbitol to produce hydrogen [51]. The differential heat of adsorption for Ni–Pt/Alnf reduced to around 111 kJ/mol, which was 12 and

6 kJ/mol lower than Pt/Alnf and Ni/Alnf respectively. This is substantial because reducing the CO binding strength can avoid the poisoning of the active metal sites. These results suggest that in the case of bimetallic catalysts there was a reduction in the number of strong CO-adsorption sites.

The influence of support on CO–Pt interaction was studied by Vannice et al. [48] using adsorption microcalorimetry. The supports used were SiO_2, η-Al_2O_3, SiO_2–Al_2O_3, and TiO_2. The studied catalysts possessed a range of differential heats of CO adsorption at 27 °C between 88 and 135 kJ/mol (21 and 32 kcal/mol) with the Pt/TiO_2 sample having the lowest values. This variation in the values of differential heats of adsorption appears to be a strong function of crystallite size, with weaker Pt–CO bonding occurring as Pt dispersion increases.

Although one of the most commonly used catalysts today is Pt/Al_2O_3, the use of carbonaceous materials as catalyst supports is continuously increasing. Their porous texture can be easily tailored, yielding high surface-area supports where the active phase can be well dispersed, and with the required pore-size distribution to facilitate the diffusion of reactants and products to, and from, the active phase [59]. In the case of noble-metal-based catalysts, the metal dispersion in the final catalyst depends on a number of factors: porous texture of the support, nature of the metal precursor, types and amount of surface complexes on the support, etc. [60]. Guerrero-Ruiz et al. [45] applied CO adsorption microcalorimetry to study the factors affecting the Pt dispersions over high surface area graphite. Using different carbon supports (e.g. with, and without, oxygen surface groups) and different platinum precursors (H_2PtCl_6 and $Pt(NH_3)_4(OH)_2$) for the catalysts preparation, the authors showed that adsorption microcalorimetry is one useful tool to provide some evidence concerning the specific interaction that takes place between graphite carbon and metal particles. Thus, microcalorimetry of CO adsorption evidences that the presence of oxygen surface groups diminishes the metal-support interaction having as results lower differential heats of CO adsorption.

Heat of adsorption of a gas on a solid is, in general, composed of several contributions, including energy of the formed surface bond, energy associated with perturbation (or even dissociation) of the adsorbate, energy of interactions between the adparticles and energy associated with the surface relaxation or rearrangement [61]. Thus, in the basic research of adsorption, it is frequently desirable to separate the individual effects. In order to facilitate such a separation, and to minimize the complicating effects of the polycrystallinity and contamination of the surface, the adsorption microcalorimetry is often carried out for the surfaces of well defined crystal structure. Such surfaces can be obtained in the forms of metal filaments, vacuum-evaporated thin films, and single crystals of metals. The comparison of adsorption heats obtained by microcalorimetry over supported metallic clusters with those determined over metallic single crystals can help to understand the actual surface structure of the metal aggregates, and then their catalytic properties. Moreover, the adsorptive properties of CO on platinum single crystals can provide a better understanding of CO–Pt interaction, what can bring important information for designing of new catalysts and for the purpose of understanding the activity of Pt metal nanoparticles employed in fuel cells in the size range of few nanometers. Information concerning the CO adsorption on single platinum crystals has been mainly gained from kinetic studies and from spectroscopic techniques, whereas the direct determination of adsorption heats has been less applied. Karmazyn and coworkers [62] studied, for example, the chemisorption of CO on the stepped Pt{2 1 1} surface using first-principles density functional theory (DFT) and microcalorimetry experiments. The heat of adsorption of CO on Pt{2 1 1} were measured at room temperature as a function of coverage and an initial heat of adsorption of 185 kJ/mol was found. Yeo et al. [63, 64] reported the values of 180 and 215 kJ/mol for the initial CO heats of adsorption on unsupported Pt{1 1 1} and Pt{1 0 0}, respectively. It can be observed that the heat of adsorption data obtained for Pt single crystal surfaces with two different orientations is different, beyond experimental errors.

The insight into the data presented in Table 12.2 gives evidence that it is difficult to elucidate the structure dependency of CO chemisorptions from the heats of adsorption found over supported catalysts. Even though, based on literature reports it can be concluded that the adsorption microcalorimetry, although not widely used, is a powerful technique for surface characterization of supported metal clusters, because it enables obtaining data concerning the strength of interaction and population of active sites. Concerning CO–Pt interaction, it is evident that CO is chemisorbed on all Pt-based catalysts (average heats of adsorption is higher than 100 kJ/mol, see Table 12.2). The differential heat profiles are usually characterised by a plateau of nearly constant heat of adsorption at low CO coverage, followed by an abrupt decrease as the surface saturation limit is reached.

12.4 The CO Poisoning Effects on Pt/C Studied by Adsorption Microcalorimetry: A Case Study

However, although the determination of the heat of adsorption of carbon monoxide on platinum have been reported in the literature (Table 12.2), the possible use of adsorption microcalorimetry as an available technique for giving valuable information in PEM fuel cells studies has not been reported. In what follows, recent results from our work in the field of microcalorimetry of CO adsorption on Pt/C catalysts are presented [65–67]. The catalysts used in these studies were different commercial carbon-supported platinum, with high Pt loading, aimed to be used in PEMFCs applications. Particular emphasis on the sample history (preparation method, the support material, the metal loading) and the pre-treatment of the catalysts on the CO poisoning effect on supported platinum is paid.

The results of CO adsorption microcalorimetry described in this overview were collected with a differential and isothermal microcalorimeter (Tian–Calvet Microcalorimeter) linked to a static volumetric system. The equipment permits the introduction of successive small doses of CO onto the catalyst. Both the calorimetric and the volumetric data were stored and analyzed by microcomputer processing. The obtained data are presented as differential heats versus the amount of CO adsorbed while the amount of gas adsorbed at constant temperature are plotted as a function of the equilibrium pressure, thus giving adsorption isotherms. By using microcalorimetry, the structure sensitivity of hydrogen and carbon monoxide adsorption was investigated by using the Pt/C commercial catalysts with different Pt loading on carbon support. The studied catalysts were thus Pt/C powders provided by E-Tek (lot#E 1280702, 16.8 wt% Pt), Tanaka (lot 103-1341R, 24.5 wt% Pt), and Johnson Mathhey (lot 128372001, 16.6 wt% Pt) companies. As in a PEM fuel cell CO comes at the anode surface as an impurity in the hydrogen flow, H_2 adsorption microcalorimetry was also applied for a better understanding of CO effect on Pt based catalysts.

Due to higher CO poisoning effect at lower temperatures, it is important to know its effect at room temperature at which the start-up of the fuel cell system takes place. It is why the adsorption studies were carried out at two different temperatures: near room temperature (30 °C) and at 80 °C (the PEMFCs operation temperature). Prior to the adsorption experiments, the catalysts were reduced under static hydrogen (27 kPa) at 25, 100 or 200 °C.

The obtained microcalorimetric results showed that both H_2 and CO can be chemisorbed on all Pt/C catalysts. Indifferent of their provenience (different carbon support and different method of preparation), Pt/C exhibited significantly higher differential heats of carbon monoxide adsorption in comparison with hydrogen adsorption as can be seen in Fig. 12.5.

It is thus evident that in case of co-adsorption of these two gases (similar situation to that in fuel cell), CO will be primarily adsorbed. In addition, the adsorbed amount of CO is much higher than the amount of adsorbed H_2, which means that the Pt based catalysts present a larger number of sites active for carbon monoxide adsorption, in comparison with those active for hydrogen adsorption. By comparison with results

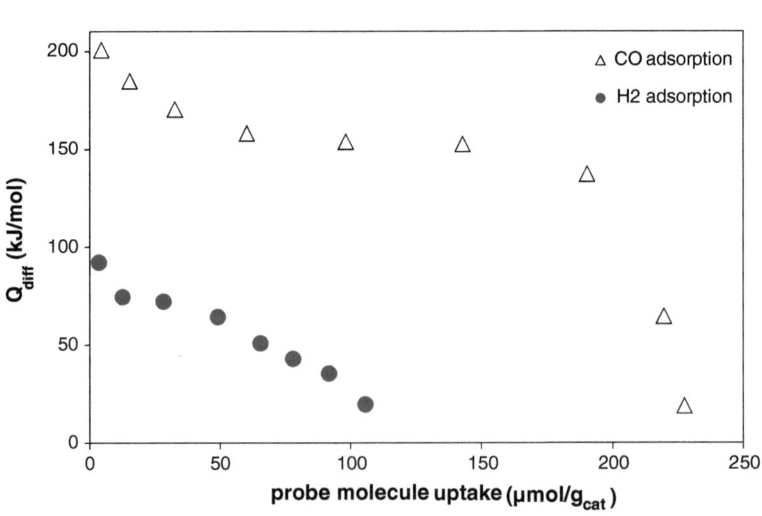

Fig. 12.5 Differential heats of H_2 and CO adsorption at 30 °C as a function of surface coverage for Pt/C sample (E-Tek, lot#E 1280702) pre-treated at 25 °C in static hydrogen (27 kPa)

obtained for different Pt/C catalysts [65, 67], no structure dependency was observed for hydrogen initial heats of adsorption. Similar values were obtained for hydrogen adsorption for all Pt/C catalysts, in good agreement with previously published values for supported and unsupported platinum catalysts [56, 58, 68–70]. Higher values of CO adsorption heats (average values ranged from 135 to 155 kJ/mol for the three tested catalysts) where observed when compared with those reported in Table 12.2.

The difference observed in the adsorption heats values could be attributed to different experimental conditions used, different loading of Pt and possibly to some remaining contaminants species (H_2, O_2, H_2O...) on the platinum, due to a poor cleaning of its surface when the sample is pre-treated at room temperature (25 °C). As reported in the literature, platinum shows a preference for a linear mode of carbon monoxide adsorption [71]. Anyway, for low surface coverage, two IR bands are observed from spectroscopic data of CO adsorbed on Pt supported powders: at around 1870 and 2060 cm^{-1} [72–74]. These bands have been attributed either to CO adsorbed on various sites of the metal [71] or to linear (2060 cm^{-1}) and bridged (1870 cm^{-1}) forms of adsorbed CO [18, 75]. The microcalorimetric results, with rather similar adsorption heat values in a large domain, can suggest that the main adsorbed species, over all Pt/C catalysts, are linear, as may also be inferred from spectroscopic data [71, 73, 74]. However, the presence of surface oxygen functional groups (SOFGs) on the support could modify the strength of the Pt–CO bonds. Thus, the differences observed in between Pt/C catalysts for CO adsorption were attributed to the contribution of the support to the surface properties, knowing that electron-donating supports produce an enrichment of the electron density of metal atoms interacting with them [76], inducing changes in the chemisorption properties of the metal particles. Nørskov et al. [77] have also demonstrated that the heats of adsorption of a species is directly related to the local structure of the catalysts, the step sites are more active unless poisoned, and they bind the adsorbates more strongly. The heats of adsorption are closely related to the adsorbate-substrate bond strength. Furthermore, the differential heat of adsorption can be dependent on the surface coverage of the adsorbate due to the lateral adsorbate-adsorbate interactions or due to the surface heterogeneity [50].

In contrast to other techniques for studying adsorption, heat-flow calorimetry yields both kinetic and thermodynamic information. The kinetics of heat release during adsorption can be monitored by changes in the thermokinetic parameter, τ. The calorimetric signal decreases exponentially with the adsorption time after the maximum of each adsorption peak. This can be expressed in the form $D(t) = D_m \exp(-t/\tau)$, where $D(t)$ and D_m are the deviation of the calorimetric signal at time t and the maximum deviation, respectively. The thermokinetic parameter τ in this expression can thus be computed as the inverse of the slope of the logarithm of the evolved heat curve during the return to equilibrium, and depends mainly on the accessibility of the adsorption sites in the samples. Thus, it was found [65] that the pore architecture of Pt/C catalysts (e.g. pore volume and pore size distribution) influenced the kinetics of heat release during CO adsorption.

The accessibility of CO molecules to the adsorption sites increased with the mesoporosity of the catalyst. When the Pt/C catalyst is predominantly mesoporous, the diffusion is not limited by movement of CO through micropores. More mesoporosity facilitates the creation of larger paths for diffusion, leading to an increased accessibility of CO molecules to the adsorption sites in the samples.

Different behaviour was also observed for Pt/C powders when their CO adsorption properties were studied [66, 67] after the catalysts were reduced at various temperatures. When the pre-treatment temperature was increased from 25 to 100 and further to 200 °C, the adsorptive properties of high loading Pt/C catalysts were influenced as shown for example in Fig. 12.6 [66]. Increasing the pre-treatment temperature provoked a decrease in the amount of chemisorbed carbon monoxide as well as in the values of differential heats of CO adsorption for the sample provided by E-Tek. These results were related with the changes in Pt particle sizes and dispersion as well as to a better cleaning of the catalyst surface when a higher pre-treatment temperature was used. This demonstrates once more that adsorption microcalorimetry is a powerful tool allowing the detection of different changes in the catalyst surface.

It has to be pointed out that not all the tested Pt/C samples presented the same behaviour for CO adsorption with increasing temperature of pre-treatment. Observation of the calorimetric profiles (Fig. 12.7) reveals a plateau at around 135 kJ/mol for Pt/C Tanaka, which did not change with increasing of reduction temperature.

The constant value of the plateau extends over a wide range of surface coverage, thus indicating a high homogeneity of the surface metal atoms of this sample for the CO adsorption. The drop in the differential heats at higher coverage is indicative of saturation of the accessible platinum metallic surface sites. The changing of the reduction conditions influenced only the total capacity (the length of plateau) of CO adsorption, which decreased with increasing pre-treatment temperature, due to the changes in Pt particle sizes and dispersion. Moreover, no differences in the adsorptive

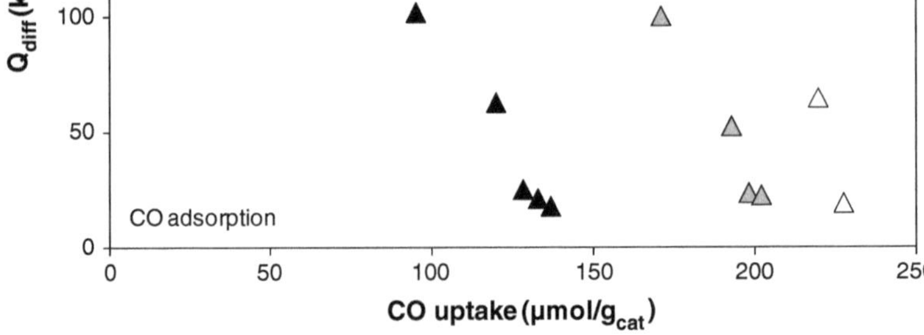

Fig. 12.6 Differential heats of CO adsorption at 30 °C as a function of surface coverage for Pt/C catalyst (E-Tek, lot#E 1280702) pre-treated at 25, 100 and 200 °C in static hydrogen (27 kPa) [66]

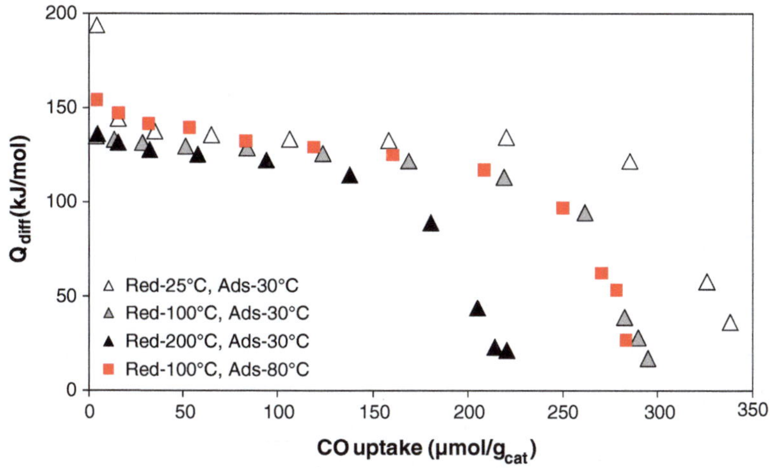

Fig. 12.7 Differential heats of CO adsorption at 30 and 80 °C as a function of surface coverage for Pt/C catalyst (Tanaka, lot 103-1341R) pre-treated at 25, 100 and 200 °C in static hydrogen (27 kPa)

properties of Pt/C catalyst for CO were observed when the adsorption temperature was increased from 30 to 80 °C and the same pre-treatment procedure was employed, which was the case for all studied powders.

When the thermokinetic parameter was measured, a more detailed characterization of the available sites for CO adsorption on the Pt/C surface was provided. For example, it was possible to differentiate reversible and irreversible adsorption processes. Because chemisorption may be a slow, irreversible process, involving activation of the adsorbate, a longer time and, therefore, a broader thermogram would distinguish such a process from a faster, reversible physisorption process. This feature was thus exploited to monitor the change in adsorption with coverage. As presented in Fig. 12.8, the adsorption process was initially slow and became slower, reaching a minimum of rate, before a significant acceleration of the process which was observed on approaching the physisorbed state at high coverage. The minimum rate appears as a maximum in a plot of the thermokinetic parameter as a function of the surface coverage, being indication of a change from irreversible to reversible adsorption.

As it can be seen in Fig. 12.8, the kinetic of CO adsorption on Pt based catalysts did not change when the adsorption and the pre-treatment were performed at higher temperature. As expected, the thermokinetic parameter firstly increases to reach a maximum and then slowly decreases showing that CO is almost completely irreversibly adsorbed on the surface of Pt-based catalyst. The amount held by the strong chemisorption sites at a certain adsorption temperature gives valuable information about the catalysts behaviour towards poisoning. Indeed, as deduced from volumetric data, 93 and 85 % of the total amount of CO was irreversibly adsorbed on Pt/C Tanaka catalyst for the powder pre-treated at 25 °C and CO adsorbed at 30 °C, and for the same sample pre-treated at 100 °C and CO adsorbed at 80 °C, respectively. Similar

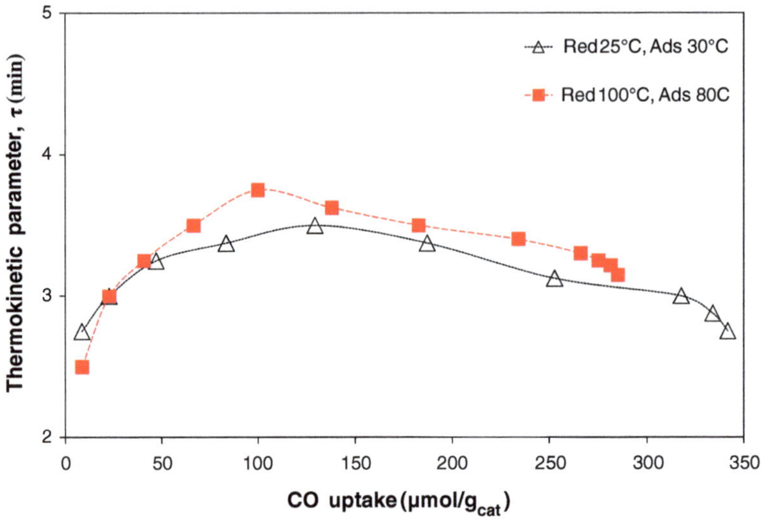

Fig. 12.8 Variation of the thermokinetic parameter, τ, versus CO uptake for Pt/C sample (Tanaka, lot 103-1341R)

results were also obtained for the catalysts provided by E-Tek and Johnson Matthey companies, with more than 90 % of CO irreversibly adsorbed for both powders.

A different and unexpected behavior was observed for the Pt/C catalyst provided by Johnson Matthey showing that the results obtained by direct adsorption calorimetric measurements provide accurate values of the heats of CO adsorption, and that they are directly related with the catalyst history (carbon used as support, the method of preparation). The calorimetric results obtained for this sample are given in Fig. 12.9 and Table 12.3.

The differential heat profiles of CO adsorption presented in Fig. 12.9 are very similar at low CO coverage (up to around 100 $\mu mol_{CO}/g_{cat}$) in spite of different reduction temperatures used. After a CO coverage of 100 $\mu mol_{CO}/g_{cat}$, when the catalyst was pretreated at 25 and 100 °C, an increase in the adsorption heats could be observed with increasing CO uptake, before a drastic drop to the values corresponding to weakly adsorbed CO (40 kJ/mol). The total metal stoichiometries at saturation are not influenced by increasing the pre-treatment temperature.

As can be seen in Table 12.3, the irreversibly chemisorbed amount of CO (n_{irrev}) can also be measured from the calorimetric studies. Obviously, this volume corresponds to the total amount held by the strong sites at the adsorption temperature over the catalysts. In order to accurately determine the chemisorbed amount from the overall adsorption isotherm, the catalyst was outgassed after the first adsorption run at the same temperature to remove the physically adsorbed amount, after which a new adsorption procedure was carried out to obtain a second isotherm. The difference between the first and second isotherm gives the extent of irreversible adsorption (n_{irrev}) at a given temperature. However, in the first approximation, the magnitude of

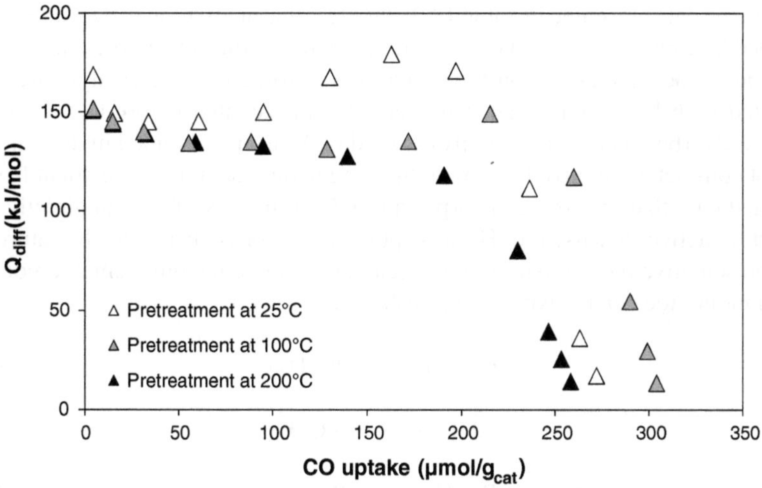

Fig. 12.9 Differential heats of CO adsorption at 30 °C as a function of surface coverage for Pt/C catalyst (Johnson Matthey, lot 128372001) pre-treated at 25, 100 and 200 °C in static hydrogen (27 kPa) [67]

Table 12.3 Data obtained from microcalorimetric measurements for Pt/C Johnson Matthey sample (lot 128372001)

Temperature of reduction (°C)	H_2 adsorption at 30 °C				CO adsorption at 30 °C			
	D (%)	Q_{init}[a] (kJ/mol)	H_2 uptake (μmol/g_{Pt})		D (%)	Q_{init}[a] (kJ/mol)	CO uptake (μmol/g_{Pt})	
			n_{total}[b]	n_{irrev}[c]			n_{total}[b]	n_{irrev}[c]
25	26.12	89	551	72	31.95	169	1413	1253
100	30.98	89	669	207	35.75	152	1744	1587
200	31.56	94	653	291	30.31	151	1490	1322

[a] Initial differential heats of H_2 and CO adsorption; [b] Amount of H_2 and CO adsorbed under an equilibrium pressure of 27 Pa; [c] Amount of irreversibly chemisorbed H_2 and CO under an equilibrium pressure of 27 Pa

the heat of adsorption can be considered as a simple criterion to distinguish between physical and chemical adsorption. As can be deduced from Table 12.3 which summarizes n_{irrev}, n_{total} as well as the dispersion obtained from the total amounts of H_2 and CO uptake at the monolayer, CO is almost completely irreversibly adsorbed on the surface of the Pt/C catalyst (i.e. 87, 91 and 89 % for the sample pre-treated at 25, 100 and 200 °C, respectively).

It was found that the available hydrogen adsorption sites have a broader site energy distribution, while the intermediate and low adsorption sites are almost not observed for carbon monoxide. The broader site energy distribution monitored by H_2 adsorption microcalorimetry can be attributed to the higher surface mobility of hydrogen.

From the data presented in Table 12.3, it is clear that at 30 °C an irreversible form of adsorbed hydrogen is detected on platinum, together with a reversible one for higher coverage. It means that carbon monoxide was adsorbed on the catalyst previously contacted with hydrogen. Even if the sample is evacuated before CO adsorption, CO was adsorbed on a surface partly covered with strongly bonded hydrogen which was not completely removed during outgassing under vacuum. It is evident from the presented data that the strong adsorption of CO at the Pt surface can directly block the surface active sites used for H_2 adsorption. As reported also in the literature [17], the representative mechanism of hydrogen adsorption containing small amounts of carbon monoxide can be expressed as follows:

$$H_2 + 2Pt \rightarrow 2Pt{-}H_{ads} \tag{12.4}$$

$$CO + Pt \rightarrow Pt{-}CO_{ads} \tag{12.5}$$

$$2CO + 2Pt{-}H_{ads} \rightarrow 2Pt{-}CO_{ads} + H_2 \tag{12.6}$$

The adsorption of CO occurs not only at bare Pt sites through reaction (12.5) but also at Pt hydride sites via reaction (12.6). This is not surprising, since the heats of adsorption of CO on Pt are much higher than that of irreversibly adsorbed hydrogen.

As observed in Fig. 12.9, up to coverage of about 100 $\mu mol_{CO}/g_{cat}$, the CO molecules are adsorbed on unoccupied active sites on the catalyst surface. At higher CO coverage, the noticed increase of differential heats can be attributed to the replacement of hydrogen irreversibly adsorbed by CO (reaction 12.6) which explains the different platinum dispersion found from H_2 and CO adsorption.

When the catalyst was pre-treated at a temperature of 200 °C, no increase of the heats was observed and similar values for the hydrogen and CO dispersion were found. This means that with increasing activation temperature, CO adsorption occurs on a reduced and clean surface. Therefore, for this catalyst, the temperature of pre-treatment does not seem to influence the adsorption capacity, but plays a role in the type of available sites on the surface. It has to be noted that the bell shaped profile was not observed for the other investigated catalyst which means that the adsorption properties are strongly dependent on the catalyst surface structure. Thus, the history of the investigated materials is of importance: platinum is supported on different carbons, in different quantities and possibly different preparation procedures have been used; which influenced its dispersion and orientation. All these features can be responsible as a whole for the different adsorptive properties of the studied materials. But, in spite of different differential heats profiles observed, there is no doubt that CO adsorbs quickly and essentially irreversibly on Pt even in the presence of pre-adsorbed hydrogen on the surface of catalyst. In the absence of CO, hydrogen adsorbs onto active platinum sites. Nevertheless, when only traces of CO are presented in the anode gas mixture, it will gradually accumulate on the platinum surface through a replacement reaction or a free site adsorption. Thus, the CO-poisoning arises from the adsorption of CO molecules on the platinum catalyst sites, which avoids the hydrogen to reach the platinum particles.

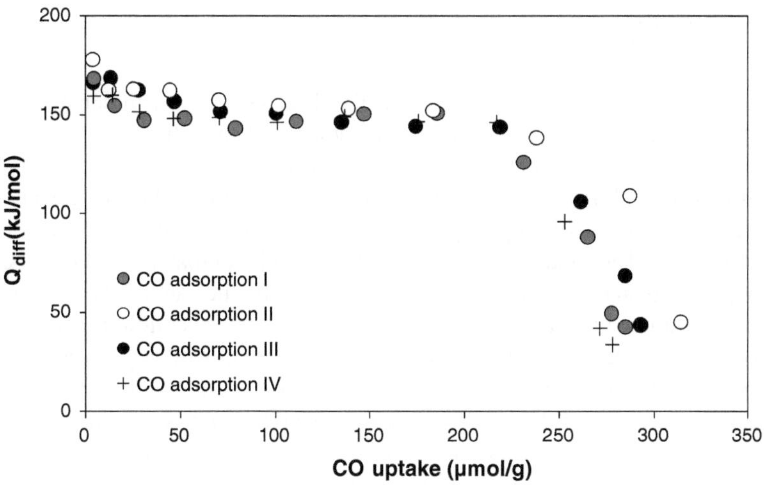

Fig. 12.10 Differential heats of CO adsorption at 80 °C as a function of surface coverage for Pt/C powder (Johnson Matthey, lot 128372001) pre-treated at 100 °C in static hydrogen (27 kPa) [67]

Finally, the response to several cycles poisoning-recovering has also been assessed by using the adsorption microcalorimetry. The poisoning degree of Pt/C and the catalysts regeneration was followed during four successive CO adsorption/desorption/readsorption cycles on the same bath of sample [65, 67]. For this, in between two successive CO adsorption runs, the catalysts were kept under air overnight and again reduced under static hydrogen before a new adsorption cycle. The results obtained for Pt/C Johnson Matthey, are given as example in Fig. 12.10. Here, the CO adsorption was performed at 80 °C after the catalyst was reduced at 100 °C. As it can be seen, this sample seems to be tolerant to CO poisoning in spite of the fact that CO binds strongly on Pt and cannot be desorbed at the operation temperature of PEM fuel cells (lower than 100 °C). The same result was obtained when a similar study was performed at room temperature [65]. This behaviour suggests that the platinum morphology, structure and adsorbed CO species are restored or remain unchanged, independently of the repeated exposure to air/H_2/CO. It was not the case for the other studied samples as shown in the same reference [65]. Differences appeared in the total amount of CO adsorbed after each cycle, which decreases considerably at the end of the experiments. The degree of catalyst poisoning by CO upon successive air/H_2/CO cycles varied between 2 and 30 % for different studied samples. The decrease in CO adsorption capacity, which means a decrease in the number of adsorption sites, can be attributed to an irreversible poisoning of the surface. No relation in between the loading of Pt and the tolerance to CO poisoning could be found. These results confirm that the surface chemistry of the catalyst affects the surface site energy distribution and consequently the adsorptive properties towards H_2 and CO. Thus it can be concluded that if the catalyst characteristics are well controlled, the reformate gas can be used as fuel for operation in PEMFC but an exhaustive control of CO must be anyway taken into account.

12.5 Summary

The CO poisoning effect on PEM fuel cells efficiency is still under active study. The detailed mechanism is still not well understood and there are contradictory observations that need to be clarified. The goal of our work was to verify if the adsorption microcalorimetry is an available technique for further understanding of the effects of impurities on hydrogen activation and hydrogen surface coverage of Pt/C catalysts used as anode in PEM fuel cells. It is evident that a correlation between poisoning effects and the adsorptive properties exists, the rate of the former being catalyst structure sensitive. The CO-poisoning can be a reversible process through air bleed at the anode and the injection of clean hydrogen. Anyway it was shown that the degree of recovery performances is directly related to the specific structure of the surface metallic atoms of the catalyst. It was proven that microcalorimetry technique is quite well developed and very useful in providing information on the strength and distribution of active sites for CO adsorption on the catalyst surface.

References

1. L. Carrette, K.A. Friedrich, U. Stimming, Fuel cells-fundamentals and applications. Fuel Cells **1**, 5–39 (2001)
2. S. Litster, G. McLean, PEM fuel cell electrodes. J. Power Sources **130**, 61–76 (2004)
3. A. Boudghene Stambouli, E. Traversa, Solid oxide fuel cells (SOFCs): a review of an environmentally clean and efficient source of energy. Renew. Sustain. Energy Rev. **6**, 433–455 (2002)
4. B.C.H. Steele, Material science and engineering: the enabling technology for the commercialisation of fuel cell systems. J. Mater. Sci. **36**, 1053–1068 (2001)
5. B.C.H. Steele, A. Heinzel, Materials for fuel-cell technologies. Nature **414**, 345–352 (2001)
6. R.A. Lemons, Fuel cells for transportation. J. Power Sources **29**, 251–264 (1990)
7. T. Yalcinoz, M.S. Alam, Improved dynamic performance of hybrid PEM fuel cells and ultracapacitors for portable applications. Int. J. Hydrogen Energy **33**, 1932–1940 (2008)
8. O. Erdinc, M. Uzunoglu, Recent trends in PEM fuel cell-powered hybrid systems: investigation of application areas, design architectures and energy management approaches. Renew. Sustain. Energy Rev. **14**, 2874–2884 (2010)
9. A. Bıyıkoğlu, Review of proton exchange membrane fuel cell models. Int. J. Hydrogen Energy **30**, 1181–1212 (2005)
10. M.L. Perry, R.M. Darling, S. Kandoi, T.W. Patterson, C. Reiser, in *Polymer Electrolyte Fuel Cell Durability*, ed. by F.N. Buchi, M. Inaba, T.J. Schmids (LLC, Springer Science and Business Media, New York , 2009), p. 405
11. H. Liu, F.D. Coms, J. Zhang, H.A. Gasteiger, A.B. LaConti, in *Polymer Electrolyte Fuel Cell Durability*, ed. by F.N. Buchi, M. Inaba, T.J. Schmids (LLC, Springer Science and Business Media, New York , 2009), p. 80
12. A. Midilli, M. Ay, I. Dincer, M.A. Rosen, On hydrogen and hydrogen energy strategies I: current status and needs. Renew. Sustain. Energy Rev. **9**, 255–271 (2005)
13. R. Jiang, H.R. Kunz, J.M. Fenton, Influence of temperature and relative humidity on performance and CO tolerance of PEM fuel cells with Nafion®-Teflon®-$Zr(HPO_4)_2$ higher temperature composite membranes. Electrochim. Acta **51**, 5596–5605 (2006)
14. S. Gottesfeld, J. Pafford, A new approach to the problem of carbon monoxide poisoning in fuel cells operating at low temperatures. J. Electrochem. Soc. **135**, 2651–2652 (1988)

15. J. Wu, X.Z. Yuan, J.J. Martin, H. Wang, J. Zhang, J. Shen, S. Wu, W. Merida, A review of PEM fuel cell durability: degradation mechanisms and mitigation strategies. J. Power Sources **184**, 104–119 (2008)
16. H.-F. Oetjen, V.M. Schmidt, U. Stimming, F. Trila, Performance data of a proton exchange membrane fuel cell using H_2/CO as fuel gas. J. Electrochem. Soc. **143**, 3838–3842 (1996)
17. X. Cheng, Z. Shi, N. Glass, L. Zhang, J. Zhang, D. Song, Z.-S. Liu, H. Wang, J. Shen, A review of PEM hydrogen fuel cell contamination: impacts, mechanisms, and mitigation. J. Power Sources **165**, 739–756 (2007)
18. J.Z. Zhang, Z. Liu, The effect of low concentrations of CO on H_2 adsorption and activation on Pt/C. Part 1: in the absence of humidity. J. Power Sources **195**, 3060–3068 (2010)
19. F.A. De Bruijn, D.C. Papageorgopoulos, E.F. Sitters, G.J.M. Janssen, The influence of carbon dioxide on PEM fuel cell anodes. J. Power Sources **110**, 117–124 (2002)
20. A. Pitois, A. Pilenga, G. Tsotridis, CO desorption kinetics at concentrations and temperatures relevant to PEM fuel cells operating with reformate gas and PtRu/C anodes. Appl. Catal. A **374**, 95–102 (2010)
21. B. Du, R. Pollard, J.F. Elter, M. Ramani, in *Polymer Electrolyte Fuel Cell Durability*, ed. by F.N. Buchi, M. Inaba, T.J. Schmids (LLC, Springer Science and Business Media, New York , 2009), pp. 341–366
22. L. Li, G. Wu, B.Q. Xu, Electro-catalytic oxidation of CO on Pt catalyst supported on carbon nanotubes pretreated with oxidative acids. Carbon **44**, 2973–2983 (2006)
23. H. Igarashi, T. Fujino, M. Watanabe, Hydrogen electro-oxidation on platinum catalysts in the presence of trace carbon monoxide. J. Electroanal. Chem. **391**, 119–123 (1995)
24. J.J. Baschuk, X.G. Li, Carbon monoxide poisoning of proton exchange membrane fuel cells. Int. J. Energy Res. **25**, 695–713 (2001)
25. K. Wang, H.A. Gasteiger, N.M. Markovic, On the reaction pathway for methanol and carbon monoxide electrooxidation on Pt-Sn alloy versus Pt-Ru alloy surfaces. Electrochim. Acta **41**, 2587–2593 (1996)
26. G. Garciá, J.A. Silva-Chong, O. Guillén-Villafuerte, J.L. Rodríguez, E.R. González, E. Pastor, CO tolerant catalysts for PEM fuel cells. Spectroelectrochemical studies. Catal. Today **116**, 415–421 (2006)
27. J. Sobkowski, A. Czerwinski, Voltammetric study of CO and CO_2 adsorption on smooth and platinized platinum electrodes. J. Phys. Chem. **89**, 365–369 (1985)
28. B. Beden, C. Lamy, N.R. de Tacconi, J. Arvia, The electrooxidation of CO: a test reaction in electrocatalysis. Electrochim. Acta **35**, 691–704 (1990)
29. M. Hachkar, T. Napporn, J.-M. LeÂger, B. Beden, C. Lamy, An electrochemical quartz crystal microbalance investigation of the adsorption and oxidation of CO on a platinum electrode. Electrochim. Acta **41**, 2721–2730 (1996)
30. S.J. Lee, S. Mukerjee, E.A. Ticianelli, J. McBreen, Electrocatalysis of CO tolerance in hydrogen oxidation reaction in PEM fuel cells. Electrochim. Acta **44**, 3283–3293 (1999)
31. H. Steininger, S. Lehwald, H. Ibach, On the adsorption of CO on Pt(111). Surf. Sci. **123**, 264–282 (1982)
32. H. Hanawa, K. Kunimatsu, H. Uchida, M. Watanabe, In situ ATR-FTIR study of bulk CO oxidation on a polycrystalline Pt electrode. Electrochim. Acta **54**, 6276–6285 (2009)
33. T. Sato, K. Kunimatsu, H. Uchida, M. Watanabe, Adsorption/oxidation of CO on highly dispersed Pt catalyst studied by combined electrochemical and ATR-FTIRAS methods. Part 1: ATR-FTIRAS spectra of CO adsorbed on highly dispersed Pt catalyst on carbon black and carbon un-supported Pt black. Electrochim. Acta **53**, 1265–1278 (2007)
34. A. Pitois, J.C. Davies, A. Pilenga, A. Pfrang, G. Tsotridis, Kinetic study of CO desorption from PtRu/C PEM fuel cell anodes: temperature dependence and associated microstructural transformations. J. Catal. **265**, 199–208 (2009)
35. A.-K. Meland, S. Kjelstrup, Three steps in the anode reaction of the polymer electrolyte membrane fuel cell. Effect of CO. J. Electroanal. Chem. **610**, 171–178 (2007)
36. A. Auroux, in *Catalyst Characterization: Physical Techniques for Solid Materials*, ed. by B. Imelik, J.C. Vedrine (Plenum Press, New York, 1994), p. 611

37. P.C. Gravelle, Application of adsorption calorimetry to the study of heterogeneous catalysis reactions. Thermochim. Acta **96**, 365–376 (1985)
38. B.E. Spiewak, J.A. Dumesic, Microcalorimetric measurements of differential heats of adsorption on reactive catalyst surfaces. Thermochim. Acta **290**, 43–53 (1996)
39. V.E. Ostrovskii, Molar heats of chemisorption of gases at metals: review of experimental results and technical problems. Thermochim. Acta **489**, 5–21 (2009)
40. P.L. Llewellyn, G. Maurin, Gas adsorption microcalorimetry and modelling to characterise zeolites and related materials. C. R. Chim. **8**, 283–302 (2005)
41. G. Postole, A. Auroux, in Advances, in *Fluid Catalytic Cracking: Testing, Characterization, and Environmental Regulations*, ed. by M.L. Occelli (CRC Press Taylor & Francis Group, Boca Raton, 2010), pp. 199–256
42. A. Guerrero-Ruiz, A. Maroto-Valiente, M. Cerro-Alarcon, B. Bachiller-Baeza, I. Rodriguez-Ramos, Surface properties of supported metallic clusters as determined by microcalorimetry of CO chemisorption. Top. Catal. **19**, 303–311 (2002)
43. J. Silvestre-Albero, J.C. Serrano-Ruiz, A. Sepulveda-Escribano, F. Rodriguez-Reinoso, Modification of the catalytic behaviour of platinum by zinc in crotonaldehyde hydrogenation and iso-butane dehydrogenation. Appl. Catal. A **292**, 244–251 (2005)
44. J.C. Serrano-Ruiz, A. Lopez-Cudero, J. Solla-Gullon, A. Sepulveda-Escribano, A. Aldaz, F. Rodriguez-Reinoso, Hydrogenation of α, β unsaturated aldehydes over polycrystalline, (111) and (100) preferentially oriented Pt nanoparticles supported on carbon. J. Catal. **253**, 159–166 (2008)
45. A. Guerrero-Ruiz, P. Badenes, I. Rodriguez-Ramos, Study of some factors affecting the Ru and Pt dispersions over high surface area graphite-supported catalysts. Appl. Catal. A **173**, 313–321 (1998)
46. J.C. Serrano-Ruiz, A. Sepulveda-Escribano, F. Rodriguez-Reinoso, Bimetallic PtSn/C catalysts promoted by ceria: application in the nonoxidative dehydrogenation of isobutene. J. Catal. **246**, 158–165 (2007)
47. J. Ruiz-Martinez, A. Sepulveda-Escribano, J.A. Anderson, F. Rodriguez-Reinoso, Influence of the preparation method on the catalytic behavior of $PtSn/TiO_2$ catalysts. Catal. Today **123**, 235–244 (2007)
48. M.A. Vannice, L.C. Hasselbring, B. Sen, Direct measurements of heats of adsorption on platinum catalysts. J. Catal. **97**, 66–74 (1986)
49. J.C. Serrano-Ruiz, G.W. Huber, M.A. Sanchez-Castillo, J.A. Dumesic, F. Rodriguez-Reinoso, A. Sepulveda-Escribano, Effect of Sn addition to Pt/CeO_2-Al_2O_3 and Pt/Al_2O_3 catalysts: an XPS, ^{119}Sn Mössbauer and microcalorimetry study. J. Catal. **241**, 378–388 (2006)
50. D. Uner, M. Uner, Adsorption calorimetry in supported catalyst characterization: adsorption structure sensitivity on Pt/γ-Al_2O_3. Thermochim. Acta **434**, 107–112 (2005)
51. A. Tanksale, J.N. Beltramini, J.A. Dumesic, G.Q. Lu, Effect of Pt and Pd promoter on Ni supported catalysts-a TPR/TPO/TPD and microcalorimetry study. J. Catal. **258**, 366–377 (2008)
52. A. Maroto-Valiente, I. Rodriguez-Ramos, A. Guerrero-Ruiz, Determination of the surface states of metallic clusters supported on alumina using microcalorimetry of CO adsorption. Thermochim. Acta **379**, 195–199 (2001)
53. H. Kivrak, A. Mastalir, Z. Kiraly, D. Uner, Determination of the dispersion of supported Pt particles by gas-phase and liquid-phase measurements. Catal. Commun. **10**, 1002–1005 (2009)
54. R. Alcala, J.W. Shabaker, G.W. Huber, M.A. Sanchez-Castillo, J.A. Dumesic, Experimental and DFT studies of the conversion of ethanol and acetic acid on PtSn-based catalysts. J. Phys. Chem. B **109**, 2074–2085 (2005)
55. R.D. Cortright, J.A. Dumesic, Microcalorimetric, spectroscopic, and kinetic studies of silica supported Pt and Pt/Sn catalysts for isobutene dehydrogenation. J. Catal. **148**, 771–778 (1994)
56. R.D. Cortright, J.A. Dumesic, Effect of potassium on silica-supported Pt and Pt/Sn catalysts for isobutene dehydrogenation. J. Catal. **157**, 576–583 (1995)
57. N.D. Gangal, N.M. Gupta, R.M. Iyer, Microcalorimetric study of the adsorption and reaction of CO, O_2, CO+O_2, and CO_2 on NaX zeolite, Pt/NaX, and platinum metal: effect of oxidizing and reducing pre-treatment. J. Catal. **140**, 443–452 (1993)

58. S.B. Sharma, J.T. Miller, J.A. Dimesic, Microcalorimetric study of silica- and zeolite-supported platinum catalysts. J. Catal. **148**, 198–204 (1994)
59. F. Rodríguez-Reinoso, I. Rodríguez-Ramos, A. Guerrero-Ruiz, J.D. López-González, Platinum catalysts supported on activated carbons. Part I: Preparation and characterization. J. Catal. **99**, 171–183 (1986)
60. A. Sepúlveda-Escribano, F. Coloma, F. Rodríguez-Reinoso, Platinum catalysts supported on carbon blacks with different surface chemical properties. Appl. Catal. A **173**, 247–257 (1998)
61. S. Ćerný, Adsorption calorimetry on filaments, vacuum-evaporated films and single crystals of metals. Thermochim. Acta **312**, 3–16 (1998)
62. A.D. Karmazyn, V. Fiorin, S.J. Jenkins, D.A. King, First-principles theory and microcalorimetry of CO adsorption on the 211 surfaces of Pt and Ni. Surf. Sci. **538**, 171–183 (2003)
63. Y.Y. Yeo, L. Vattunoe, D.A. King, Calorimetric heats for CO and oxygen adsorption and for the catalytic CO oxidation reaction on Pt111. J. Chem. Phys. **106**, 392–403 (1997)
64. Y.Y. Yeo, L. Vattunoe, D.A. King, Energetics and kinetics of CO and NO adsorption on Pt100: restructuring and lateral interactions. J. Chem. Phys. **104**, 3810–3821 (1996)
65. G. Postole, S. Bennici, A. Auroux, Calorimetric study of the reversibility of CO pollutant adsorption on high loaded Pt/carbon catalysts used in PEM fuel cells. Appl. Catal. B **92**, 307–317 (2009)
66. G. Postole, S. Bennici, A. Auroux, Poisoning by CO of Pt/C commercial catalysts used in PEMFCs. A microcalorimetric study, in Proceedings of International Symposium on Fundamentals and Developments of Fuel Cells (FDFC) 2011 Conference, ISBN: 978-2-7466-2970-7 (2011).
67. G. Postole, A. Auroux, The poisoning level of Pt/C catalysts used in PEM fuel cells by the hydrogen feed gas impurities: the bonding strength. Int. J. Hydrogen Energy **36**, 6817–6825 (2011)
68. S. Cerny, M. Smutek, F. Buzek, The calorimetric heats of adsorption of hydrogen on platinum films. J. Catal. **38**, 245–256 (1975)
69. J.B. Lantz, R.D. Gonzalez, Development of a new isothermal calorimeter; heats of hydrogen adsorption on supported platinum versus crystallite size. J. Catal. **41**, 293–302 (1976)
70. J.M. Herrmann, M. Gravelle-Rumeau-Maillot, P.C. Gravelle, A microcalorimetric study of metal-support interaction in the Pt/TiO_2 system. J. Catal. **104**, 136–146 (1987)
71. P. Hollins, The influence of surface defects on the infrared spectra of adsorbed species. Surf. Sci. Rep. **16**, 51–94 (1992)
72. M. Primet, J.M. Basset, M.V. Mathieu, M. Prettre, Infrared study of CO adsorbed on Pt/Al_2O_3. A method for determining metal-adsorbate interactions. J. Catal. **29**, 213–223 (1973)
73. S.D. Jackson, B.M. Glanville, J. Willis, G.D. McLellan, G. Webb, R.B. Moyes, S. Simpson, P.B. Wells, R. Whyman, Supported metal catalysts: preparation, characterization, and function. Part II: carbon monoxide and dioxygen adsorption on platinum catalysts. J. Catal. **139**, 207–220 (1993)
74. D. Liu, G.H. Que, Z.X. Wang, Z.F. Yan, In situ FT-IR study of CO and H_2 adsorption on a Pt/Al_2O_3 catalyst. Catal. Today **68**, 155–160 (2001)
75. M. Primet, J.M. Basset, M.V. Mathieu, M. Prettre, Infrared investigation of hydrogen adsorption on alumina-supported platinum. J. Catal. **28**, 368–375 (1973)
76. P. Gallezot, D. Richard, Selective hydrogenation of α, ß-unsaturated aldehydes. Catal. Rev. Sci. Eng. **40**, 81–126 (1998)
77. J.K. Nørskov, T. Bligaard, A. Logadottir, S. Bahn, L.B. Hansen, M. Bollinger, H. Bengaard, B. Hammer, Z. Slivancanin, M. Mavrikakis, Y. Xu, S. Dahl, C.J.H. Jakobsen, Universality in heterogeneous catalysis. J. Catal. **209**, 275–278 (2002)

Chapter 13
Biodiesel: Characterization by DSC and P-DSC

Rodica Chiriac, François Toche and Christian Brylinski

The use of vegetable oils as engine fuel may seem negligible today. Nevertheless, such oils may become, in the passing years, as important as oil and coal tar presently.

Rudolf Diesel, 1912

Abstract Thermal analytical methods such as differential scanning calorimetry (DSC) have been successfully applied to neat petrodiesel and engine oils in the last 25 years. This chapter shows how DSC and P-DSC (pressurized DSC) techniques can be used to compare, characterize, and predict some properties of alternative non-petroleum fuels, such as cold flow behavior and oxidative stability. These two properties are extremely important with respect to the operability, transport, and long-term storage of biodiesel fuel. It is shown that the quantity of unsaturated fatty acids in the fuel composition has an important impact on both properties. In addition, it is shown that the impact of fuel additives on the oxidative stability or the cold flow behavior of biodiesel can be studied by means of DSC and P-DSC techniques. Thermomicroscopy can also be used to study the cold flow behavior of biodiesel, giving information on the size and the morphology of crystals formed at low temperature.

13.1 Introduction

Nowadays, the world's primary energy consumption is based on fossil fuels. However, the rising cost of petroleum oil, together with global warming, has led researchers to develop alternative non-petroleum fuels such as biofuels. These are divided in two main classes: bioethanol and biodiesel. The former is obtained from

R. Chiriac (✉) · F. Toche · C. Brylinski
Laboratoire des Multimatériaux et Interfaces, Université Lyon 1, CNRS, UMR 5615,
43 boulevard du 11 Novembre 1918, 69622 Villeurbanne, France
e-mail: Rodica.chiriac@univ-lyon1.fr

A. Auroux (ed.), *Calorimetry and Thermal Methods in Catalysis*,
Springer Series in Materials Science 154, DOI: 10.1007/978-3-642-11954-5_13,

Table 13.1 Fatty acids profiles of the most common VOs and animal fats

Fatty acid	Structure	Common acronym	Methyl Ester
Palmitic acid	$CH_3-(CH_2)_{14}-COOH$	C16 : 0	Methyl palmitate
Stearic acid	$CH_3-(CH_2)_{16}-COOH$	C18 : 0	Methyl stearate
Oleic acid	$CH_3-(CH_2)_7-CH=CH-(CH_2)_7-COOH$	C18 : 1	Methyl oleate
Linoleic acid	$CH_3-(CH_2)_7-CH=CH-CH_2-CH=CH-(CH_2)_4-COOH$	C18 : 2	Methyl linoleate
Linolenic acid	$CH_3-(CH_2)_7-(CH=CH-CH_2)_3-COOH$	C18 : 3	Methyl linolenate

sugar and starch crops and is used only in gasoline motors in its pure form (ethanol) or as a gasoline additive. Bioethanol is mostly produced in the USA and Brazil. The second class is obtained from vegetable oils, animal fats or recycled greases and is used only in diesel engines mostly in Europe whether in its pure form, transformed by transesterification, or as a diesel additive. More recently, it has been found that micro-algae can also be a source of biofuel. Algae can produce bioethanol (from biomass via hydrolysis and fermentation) and biodiesel (from algal oil). Algae fuel, also called *oilgae* is a promising field because it does not require fertile land or food crops [13, 14, 34, 40].

This chapter will however treat only biodiesel from vegetable oil (VO), the use of which in internal combustion engines is not a recent innovation. Indeed, in 1900, at the World Fair in Paris, the new diesel engine created by Rudolph Diesel (1858–1913) was tested for the first time with peanut oil.

There are four main ways to make biodiesel: direct use and blending, microemulsions, thermal cracking (pyrolysis) and transesterification. Raw VO or recycled greases (also called waste cooking oils) are not registered by the US Environmental Protection Agency (EPA) as biodiesel for legal use in vehicles. Biodiesel is in fact defined as a mixture of mono-alkyl esters of fatty acids derived from VOs or animal fats. Fatty acids are represented by two numbers: the first denotes the total number of carbon atoms in the fatty acid chain and the second is the number of double bonds present in the chain. For example, C18:1 designates oleic acid, which has 18 carbon atoms and one double bond. Table 13.1 lists the fatty acid profiles of a number of common VOs and animal fats.

Pure VO has both a higher viscosity and auto-ignition temperature than those of diesel fuel. Thus, one of the most common methods used to reduce oil viscosity in the biodiesel industry is transesterification. Biodiesel is the product obtained when a VO or animal fat is chemically reacted with an alcohol to produce fatty acid alkyl esters. These reactions are often catalyzed by a base or acid [6, 7, 18, 37]. Bases can catalyze the reaction by removing a proton from the alcohol thus making it more reactive, while acids catalyze the reaction by donating a proton to the carbonyl group, also making it more reactive. Glycerol is produced as a co-product and typical "contaminants" are water, free glycerin, bonded glycerin, free fatty acids, catalysts, soaps and oxidation products [31].

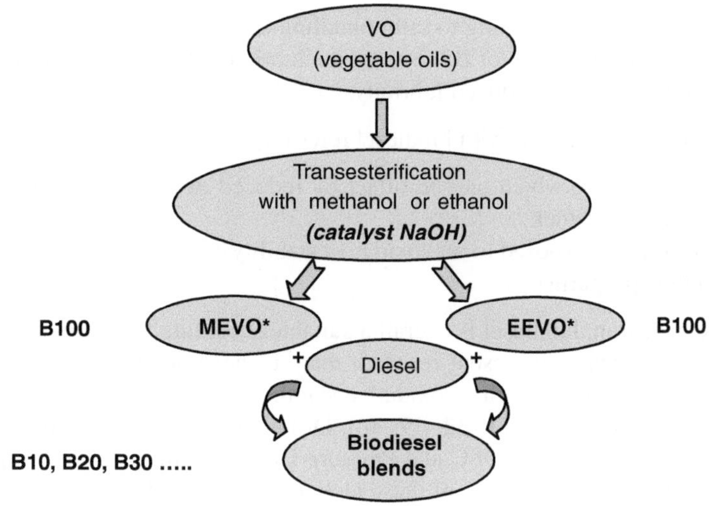

**MEVO = Methyl Ester of Vegetable Oil*
**EEVO = Ethyl Ester of Vegetable Oil*

Fig. 13.1 Simplified schematic of how to obtain biodiesel and biodiesel blends

Figure 13.1 summarizes the most common way to obtain biodiesel and biodiesel blends. Blends are denoted by acronyms such as B10, meaning 10 % of alkyl ester of VO and 90 % petrodiesel. The B designates biodiesel. B100 means 100 % alkyl ester of VO. Biodiesel blends are usually found at low percentages of alkyl ester of VO rather than 100 %.

The suitability of biodiesel as a fuel comes from similarities to petrodiesel or at least to some of its properties. The main properties are: the cetane number, flash point, heat of combustion, and lubricity, etc. For example, the cetane number (CN) is one of the most important properties of a diesel fuel. This indicator measures how quickly the fuel ignites under diesel engine conditions. The CN test (ASTM D 613) compares the test fuel with a blend of two reference fuels, n-cetane (hexadecane, with CN = 100, that ignites very easily under compression) and hepta-methyl-nonane (iso-cetane, CN = 15), on their tendency to auto-ignite. Fuels with a high CN value will have short ignition delays. Generally, diesel engines run well with fuel with a CN value from 40 to 55. Most biodiesel fuels have a CN higher than that of petroleum-based diesel fuels. Biodiesel from VO sources have been recorded as having a range of 46 to 52, and animal-fat based biodiesels (more saturated esters) a CN range from 56 to 65.

The **advantages** of biodiesel over petrodiesel are summarized as follows [31]:

- derived from a renewable domestic source → reducing dependence on and preserving petroleum;
- biodegradability;
- reduction of most exhaust emissions, except NO_x;

- higher flash point, leading to safer handling and storage;
- excellent lubricity: B1 or B2 restores the lubricity of low-sulfur petrodiesel fuels, which have greatly reduced lubricity.

Problems associated with biodiesel refer to:

- its higher price, which can be offset by reduced excise taxes or the use of less expensive feedstock;
- stability when exposed to air (oxidative stability);
- cold flow properties.

In conclusion, biodiesel is overall a suitable substitute for petrodiesel fuel.

Many publications present research that has been done on biodiesel from production to properties, such as oxidative stability, cold flow properties, and thermal decomposition, etc. [2, 11, 32, 39, 46]. Differential Scanning Calorimetry (DSC), Modulated Temperature-DSC and Pressure-DSC are among the techniques used to characterize biodiesel. Some of these techniques will be discussed in the following chapter.

13.2 Study of the Cold Flow Behavior of Biodiesel by DSC and Thermomicroscopy

13.2.1 Introduction

The cold flow properties of biodiesel and conventional petrodiesel are extremely important. Unlike gasoline, petrodiesel and biodiesel can both start to gel as the temperature gets colder. If this happens, it can clog filters or even become too thick to be pumped from the fuel tank to the engine. The high molecular weight paraffins (or n-alkanes with C18–C30) of a diesel fuel crystallize with decreasing temperature. Firstly, paraffin crystals appear and the liquid becomes cloudy, then with the decreasing temperature crystals grow, agglomerate and stop flowing. The same occurs for a liquid fatty material which becomes cloudy due to formation of crystals and solidification of the saturated fatty acid methyl esters (chains with no double bonds) present in biodiesel. This can be explained by the fact that these saturates have significantly higher melting points than the unsaturated fatty compounds (chains with double bonds), also present in biodiesel, and in a mixture they crystallize at a higher temperature than unsaturates. Thus the cold flow behavior of biodiesel greatly depends on the quantity and chemical structure of saturates.

The tendency of a (petrodiesel) fuel to gel or solidify at low temperature can be quantified by several experimental parameters such as Cloud Point (CP), Cold Filter Plugging Point (CFPP), and Pour Point (PP) [23, 31]. They are described in more detail below. The American and corresponding French and European standard methods are also given.

Cloud point is the temperature at which haziness is observed (small wax crystals start to form and the fuel appears cloudy) when the sample is cooled at a specified rate and examined at 1 °C intervals. CP is an index of the lowest temperature of the fuel's utility under certain applications. Operating at temperatures below the CP for a diesel fuel can result in fuel filter clogging due to the wax crystals. The repeatability of the CP test is <0.5 °C and the reproducibility is <2.6 °C (using the standards: *ASTM D-2500 (or AFNOR T60-105))*.

Cold filter plugging point is the lowest temperature at which a 20-mL sample passes through a 45-μm wire mesh filter under cooling at a specified rate and 0.0194 atm vacuum within 60 s (using the standard: *ASTM D6371 (EN 116))*.

Pour point represents the lowest temperature at which a fuel sample will still flow, when the temperature is decreased at a specified rate below the CP and the crystals are large enough (and agglomerate) to plug fuel filter systems. Examination of the sample is made at 3 °C intervals. The PP provides an index of the lowest temperature of the fuel's utility for certain applications. In addition, the PP has implications for the handling of fuels at cold temperature. The repeatability of the PP test is <3 °C and the reproducibility is <6 °C (*ASTM D-97* (*AFNOR T60-105*)). Biodiesel fuels derived from fats or oils which have significant amounts of saturated fatty compounds will display high CP and PP values.

The definition of a maximum value for the CP, PP, and CFPP properties depends on the fuel purpose, the region/country and season. In Europe, diesel fuels can be classified in 10 classes as a function of their cold flow characteristics: six classes for "temperate climates" with CFPP limits comprised between 5 and −20 °C; and four classes for "arctic regions" with the CFPP between −20 and −44 °C. Each country chooses one or more classes as a function of its climatic conditions. Thus, in France, there are 3 classes for summer, winter and "extremely cold" periods with CFPP limits of 0, −15 and −20 °C, respectively [23]. Only a few approaches are possible in order to achieve a given value of these properties for a fuel. One of them [40] combines the high CP or PP fuel with fuels of lower CP. The second approach to improve the cold flow behavior of a diesel fuel is either to decrease the *distillation final point* in order to reduce the heavy fractions containing n-paraffins in fuel, or to choose more naphthenic and aromatic fractions. However, these options are restrictive and costly [23]. The third option [17] is to decrease the total saturated ester concentration in biodiesel via *winterization*, but the yields are sometimes poor. Another approach uses ***cold flow improvers*** or ***additives*** such as ***CFPP and PP depressants.*** These additives are known to improve the paraffin crystals dispersion in order to avoid agglomerates forming and so the plugging of filter pores. Generally, copolymers of ethylene and vinyl acetate are used as CFPP and PP depressants at concentrations between 200 and 600 ppm; this diesel fuel treatment cost is relatively low at only a few cents per liter of fuel [23].

There are also additives having an impact only on the CP. These are called ***cloud point depressants*** and are obtained by a radical polymerization between an olefin and acrylic or vinyl compound. These macromolecules prevent the formation of paraffin crystals in solution. The use of all these additives necessitates new fundamental knowledge to be acquired relating to nucleation and crystal growth in biodiesel and biodiesel blends [17].

Different standard methods are used to determine or predict the low temperature behavior of a fuel in terms of CP, CFPP and PP. The repeatability and reproducibility tests are not always very good (see above: <6 °C) and other "non-standardized" methods can be used such as DSC and thermomicroscopy. These thermal analytical methods were successfully applied to neat petrodiesel and crude oils [9, 19, 26, 33] but also to engine oils [12, 16, 17, 27, 35].

As the crystallization and melting of paraffins, or saturated fatty compounds in the case of biodiesel, are accompanied by a low release of heat and adsorption of heat, respectively, calorimetry can therefore be used to accurately detect and study the process of crystallization and melting. Furthermore, DSC has the advantage of the rapid determination of cold flow parameters from heating and cooling scans. Among these parameters are glass transition temperatures and percentage of crystallizable fraction from the heating scan, crystallization onset temperature, PP and percentages of crystals at different temperatures (i.e. $-10\,^{\circ}$C, $-20\,^{\circ}$C) from the cooling scan. Claudy [9] and Heino [26] have shown that CP, PP and CFPP of diesel fuels can be obtained by DSC with more accuracy, better reproducibility and more rapidly than with standard tests. It has been found that the CP determined with standard tests corresponds to the onset temperature of the crystallization peak obtained by DSC, the latter being a few degrees lower than the CP.

Thermomicroscopy is useful for correlating morphological or structural changes with the thermal effects observed by DSC as has been shown in other studies of diesel and alternative fuels [12, 27, 33].

13.2.2 Experimental Procedures

13.2.2.1 Materials

Prior to DSC and thermomicroscopic measurements, gas chromatography analyses were carried out for both palm oil and rapeseed oil methyl esters.

Blends of palm oil methyl ester and rapeseed oil methyl ester with diesel fuel (DF) have been realized on a gravimetric basis. Blend ratios were 0, 10, 20, 30 and 100 %.

Palm and rapeseed biodiesels were mixed with cold flow improvers (additives) after adding them to DF. The additives were mixtures of ethylene vinyl acetate (EVA) copolymers and naphthenic distillates.

13.2.2.2 DSC Analyses

DSC measurements were carried out using a DSC 820 Mettler-Toledo apparatus. Calibrations for temperature and enthalpy were performed using the melting point and enthalpy of high purity metals and volatile organic compounds. The precision of these temperature and enthalpy measurements was within ±0.2 °C and 0.4 J/g, respectively. The instrument was flushed with nitrogen.

For each scan, approximately 20 mg of sample was hermetically sealed in a 40 μL aluminum pan and tested against an empty pan. Two scans (heating and cooling) have been run for each sample. For the heating run, the samples were cooled and held isothermally at −150 °C for 2 min before heating to 50 °C at 5 °C/min, while for the cooling run the samples were cooled from 25 to −30 °C at a rate of 0.5 °C/min.

13.2.2.3 Thermomicroscopic Analyses

The equipment used for the thermomicroscopic analyses was composed of: a Zeiss Axioplan microscope using transmitted light and equipped with polarized light and phase contrast devices; a LTS350 Heating/Freezing Stage unit which could operate in the temperature range +350 to −50 °C, controlled by a LINKAM TMS 94 control unit; and an AxioCam MRc5 Zeiss colour camera and an AxioVision 4.1 software. A Dewar flask filled with liquid nitrogen was used for cooling.

Samples were placed in an 8 μL glass crucible and analyzed in the temperature range +20 to −15 °C at a cooling rate of 0.5 °C/min, the same as for DSC analyses.

13.2.3 DSC and Thermomicroscopy for the Study of Biodiesel and Biodiesel blends

This section shows how DSC and thermomicroscopy can be used to study the cold flow behavior of two biodiesels (palm oil methyl ester (ME1) and rapeseed oil methyl ester (ME2)) and their blends with a conventional diesel fuel (DF). The impact of a cold flow improver on the quantity and size of crystals is also presented.

13.2.3.1 Pure Compounds

Figures 13.2 and 13.3 show the heating and cooling curves obtained by DSC for pure DF, ME1 and ME2.

The heating scan of DF (Fig. 13.2a) firstly shows an increase in heat capacity, ΔCp, of 0.4 J g^{-1} K^{-1}, corresponding to the glass transition of the hydrocarbon matrix at about −120 °C (midpoint value), followed by a small exothermal effect due to the crystallization of species that cannot crystallize on cooling. The same effect is observed in Fig. 13.2b for ME1 and ME2 at −38 °C and −34 °C, respectively, and on the DF scan, a broad endothermic effect is also seen, corresponding to the dissolution of wax (n-alkanes + iso-alkanes) in the liquid matrix. This effect on the heating curves for both biodiesels (Fig. 13.2b) corresponds to the dissolution of FAMEs (fatty acid methyl esters) in the composition of the ME1 and ME2 biodiesels.

Determination of the amount of paraffin, i.e. the crystallizable fraction (CF), on the heating of DF, requires the use of the equation $\Delta H_{diss} = f(T\,°C)$. The way this equation is established is explained elsewhere [5, 8]. This relation between the paraffin dissolution enthalpy and temperature should of course be reviewed when paraffins and FAME precipitate together. For the latter, a new equation needs to be assessed.

The cooling scan of the pure compounds (Fig. 13.3) shows a broad exothermal peak corresponding to the crystallization of paraffins (for DF) or FAME (for biodiesels). The first cold flow parameters obtained by this analysis are the onset crystallization temperature (CT) and the slope of crystallization. The CT is determined by the intersection of the tangents to the base line and the exothermal peak. The cooling rate should be as low as possible to get close to the equilibrium temperature: solid–liquid.

The analysis of each of the most abundant FAMEs by DSC at 5 °C/min has permitted us to identify the peaks obtained on the cooling and heating scans of rapeseed and palm biodiesels. Table 13.2 lists the FAME composition of rapeseed (ME2) and palm (ME1) biodiesels as well as the melting and crystallization onset and enthalpy. Methyl oleate and elaidate are the predominant FAMEs in ME1 and ME2 at 53.2 % and 57.1 %, respectively. For palm biodiesel, methyl palmitate was the next most abundant FAME (27.4 %), followed by methyl stearate (9.4 %) and methyl linoleate (6.4 %). Concerning the rapeseed biodiesel, methyl linoleate was the next most abundant FAME (24.5 %) after methyl oleate, followed by methyl linolenate (8.6 %) and methyl palmitate (4.6 %). Rapeseed biodiesel (ME2) has a higher amount of unsaturated fatty acids (UFA, ≈92 %) than palm biodiesel (UFA, ≈61 %). The total saturated fatty acid (SFA) content for palm ester was ≈38 %, and ≈7 % for rapeseed ester.

As shown in Table 13.2, the unsaturates present lower melting and crystallization points (temperatures and enthalpies) than saturates. Therefore, it can be assumed that the cold flow parameters will be better (i.e. low CPs and PPs) for the rapeseed oil methyl ester as it contains more unsaturated compounds compared to the palm oil methyl ester. This affirmation is also supported by Knothe [29] who stated that

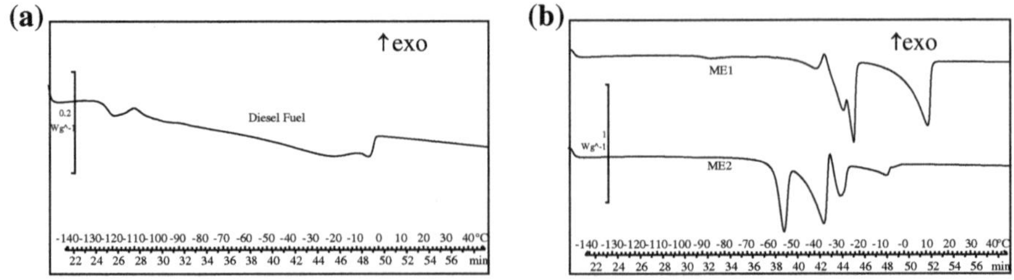

Fig. 13.2 Heating scans for pure compounds: **a** DF; **b** ME1 & ME2

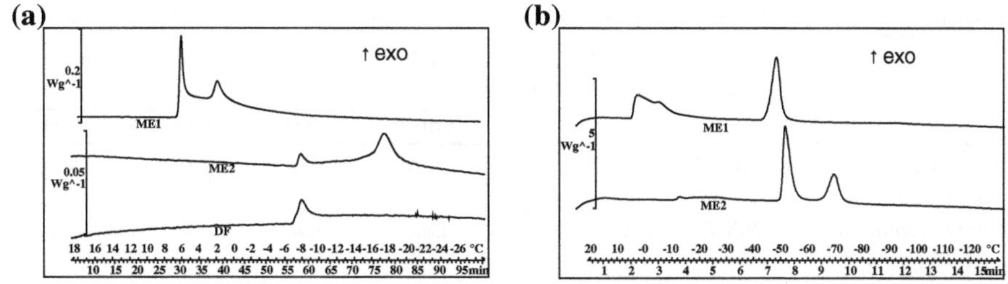

Fig. 13.3 Cooling scans for pure compounds. **a** DF, ME1, ME2 cooled at 0.5 °C/min until −30 °C. **b** ME1 & ME2 cooled at 5 °C/min until −130 °C

biodiesel fuels derived from fats or oils with significant amounts of saturated fatty compounds will display higher CPs and PPs.

The cooling scan of pure compounds (ME1 and ME2) at 5 °C/min (Fig. 13.3b) shows two broad exothermic peaks. As the melting and crystallization properties of FAME have been obtained and presented in Table 13.2, the identification of saturated and unsaturated compounds on these DSC cooling curves can therefore be made. The saturated ones start crystallizing at 6 and −7 °C for ME1 and ME2, respectively, while the unsaturates crystallize at very low temperatures (about −40/−50 °C).

Palm oil methyl ester presents an onset of crystallization above 0 °C (i.e. 6 °C) and a steep crystallization slope (Fig. 13.3a). Thus, its cold flow properties will be poorer than those of the rapeseed methyl ester. The crystallized fractions (CFs) determined at −15 °C for ME1, ME2 and DF are: 2.3, 1.5 and 25.9 %, respectively. These percentages were calculated using the enthalpy of crystallization of DF (210 J/g) obtained elsewhere [9]. The same linear relation obtained for diesel fuels has been used for biodiesel and biodiesel blends:

$$\Delta \text{Hcryst.} = 1.5167\text{T} + 211.7\,\text{Jg}^{-1}(\text{with T in}^\circ\,\text{C})$$

Even if the enthalpy of 210 J/g is not the corresponding value for the crystallization enthalpy of ME1 and ME2 it has been chosen in order to compare the variation of the crystallized fraction for all studied blends. The ME2 and DF have similar low percentages of crystals at −15 °C. However, ME1 shows a high percentage of crystals at −15 °C. The cold flow parameters of pure compounds are listed in Table 13.3. The CFPP and PP were determined by DSC at 0.45 and 1 % of crystals, respectively, using the same ΔHcryst as mentioned above. Claudy et al. [9] realized a study on 40 different diesel fuels and established correlations between CFPP and PP determined by ASTM standards and CFPP and PP obtained by DSC measurements. The best correlations were obtained at 0.45 and 1 % of crystals, respectively. In our study, the same values have been retained in order to compare the samples.

Palm oil methyl ester has a high % of crystals at −15 °C and high CT, PP and CFPP values compared with DF and ME2. This is due to the presence, in high quantity, of saturated fatty esters (no double bonds) which have high melting and crystallization points as shown in Table 13.2. Thus, from these results, we can conclude that ME1

Table 13.2 FAME compositions and DSC parameters for palm and rapeseed methyl esters

Compounds	MW (g/mol)	ME1[a] (wt)%	ME2[a] (wt)%	Melting		Crystallization	
				Onset °C	ΔH J/g	Onset °C	ΔH J/g
Methyl Palmitate (C16 : 0)	270.45	27.4	4.6	28.6	214.9	24.3	−214.1
Methyl Palmitoleate (C16 : 1n9c)	*268.43*	*0.0*	*0.2*	*n.a.*	*n.a.*	*n.a.*	*n.a.*
Methyl Heptadecanoate (C17 : 0)	284.48	0.1	0.0	*n.a.*	*n.a.*	*n.a.*	*n.a.*
Methyl Stearate (C18 : 0)	298.5	9.4	1.8	38.3	229.9	34.8	−219.1
Methyl Oleate (cis) & Elaidate (trans) (C18 : 1n9c)	*296.49*	*53.2*	*57.1*	*−20.4* *8.1*	*160.2* *189.8*	*−38.3* *2.8*	*−127.7* *−192.0*
Methyl Linoleate (C18 : 2n6c)	*294.47*	*6.4*	*24.5*	*−42.7*	*107.2*	*−77.5*	*−92.3*
Methyl Linolenate (C18 : 3n3c)	*292.46*	*0.5*	*8.6*	*−52.9*	*80.4*	*−78.8*	*−70.8*
Methyl Arachidate (C20 : 0)	326.56	0.5	0.5	47.1	242.8	42.0	−228.2
Methyl cis-11-Eicosenoate (C20 : 1)	*324.54*	*0.4*	*1.4*	*−15.6*	*138.7*	*−21.1*	*−143.4*
Methyl Behenate (C22 : 0)	354.61	0.2	0.4	*n.a.*	*n.a.*	*n.a.*	*n.a.*
Methyl Erucate (C22 : 1n9c)	*352.59*	*0.1*	*0.2*	*n.a.*	*n.a.*	*n.a.*	*n.a.*
Methyl Lignocerate (C24 : 0)	382.66	0.4	0.1	*n.a.*	*n.a.*	*n.a.*	*n.a.*
Total non identified compounds	–	**1.5**	**0.5**	–	–	–	–
Total saturated compounds	–	38	7.4	–	–	–	–
Total unsaturated compounds	–	*60.6*	*92*	–	–	–	–

[a] determined by GC; *n.a.* non-analyzed

cannot be used as a conventional diesel fuel in its pure form in spite of its high cetane number (54).

To visualize the crystallization evolution, the pure compounds were analyzed by thermomicroscopy. During cooling, three photos were taken for each compound at different temperatures (Fig. 13.4). ME2 and DF show continuous and homogenous crystallization of about 20 and 80 μm, respectively, but with no crystal agglomeration. In contrast, ME1 behaves differently with crystals agglomerating and freezing over only a 0.4 °C period (from 5.5 to 5.1 °C). These observations show that ME1 cannot be used as a diesel fuel as confirmed by the DSC results.

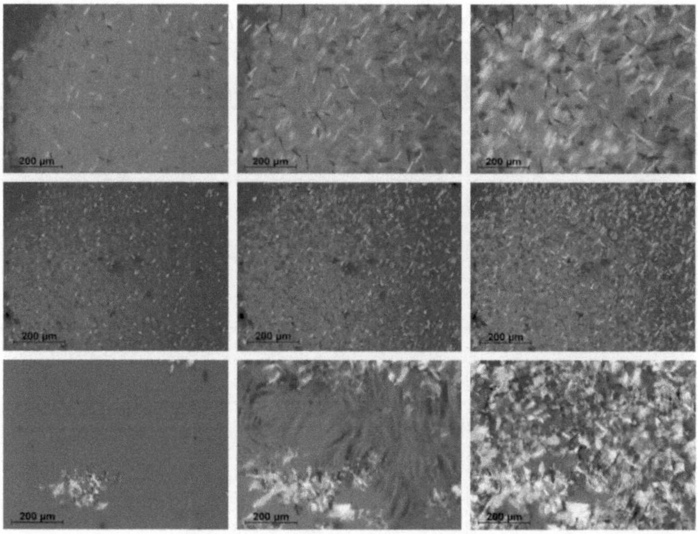

Fig. 13.4 Evolution of crystallization of pure compounds with decreasing temperature by thermomicroscopy

Table 13.3 Cold flow parameters of pure compounds (ΔHcryst. = 210 J/g)

	CT (°C)	CFPP (°C) at 0.45 % crystals	PP (°C) at 1 % crystals	% of crystals at −15 °C
DF	−6.9	−8.1	−8.5	2.3
ME2	−7.0	−10.4	−13.5	1.5
ME1	6.0	5.9	5.8	25.9

13.2.3.2 Diesel Fuel and Biodiesel Blends

The cold flow properties of diesel fuel and biodiesel blends are given in Table 13.4. As can be seen from these results, there is almost no influence when ME2 is blended with DF. This situation changes when ME1 is blended with DF, where the quantity of precipitated crystals and the cold flow parameters are higher than those of neat DF and increase when ME1 is increased from 10 to 30 %. As was expected after the analyses of the pure compounds, blends of ME1 with DF have lower cold flow parameters compared with blends of ME2 with DF.

The blends of ME1 and ME2 with DF have also been tested by thermomicroscopy. The results show that there is no impact on the morphology and the size of crystals when ME2 is added to DF (Fig. 13.5).

Concerning the blends of ME1 with DF, the crystals size increases and there is agglomeration as the quantity of ester increases from 10 to 30 % (see Fig. 13.6). In addition, the morphology of crystals is not uniform. The DF/ME1 70/30 shows a large increase in crystal size of about 120–150 µm, knowing that in DF they are

Table 13.4 Cold flow parameters of DF and biodiesel blends by DSC (ΔHcryst. = 210 J/g)

	CT (°C)	CFPP (°C) at 0.45 % crystals	PP (°C) at 1 % crystals	% of crystals at −15 °C
Diesel fuel	−6.9	−8.1	−9.5	2.3
DF/ME1 90/10	−5.3	−6.3	−8.7	2.7
DF/ME1 80/20	−5.9	−6.4	−7.9	3.7
DF/ME1 70/30	−5.0	−5.3	−5.9	5.8
DF/ME2 90/10	−6.3	−7.1	−9.6	2.0
DF/ME2 80/20	−5.6	−6.8	−10.0	1.9
DF/ME2 70/30	−6.3	−7.1	−10.4	1.7

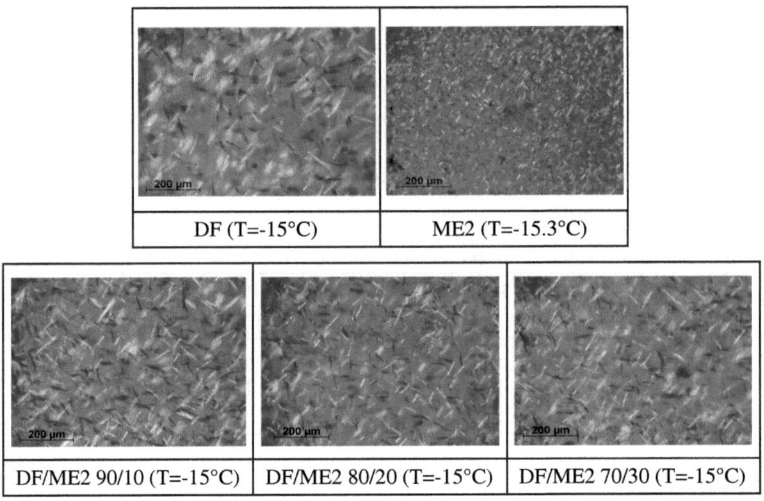

Fig. 13.5 Visualization of crystallization of pure compounds (DF and ME2) and corresponding biodiesel blends at −15 °C by thermomicroscopy

about 80 µm. Thus, this blend cannot be used in diesel engines unless a cold flow additive is used to reduce the crystal size.

13.2.3.3 DF and Biodiesel Blends: Impact of Cold Flow Improver

For any diesel fuel, the crystal size at low temperature is a critical parameter. Indeed, the standard tests for the determination of cold flow parameters are done on a 45 µm diameter filter. In reality, the injection pump filter pores can have different sizes depending on the vehicle make, varying from approximately 1–45 µm. Thus, the size of paraffin or FAME crystals needs to be drastically reduced.

The same analytical techniques as those presented before are used to study the impact of 200 ppm of a cold flow improver (additive "A") on the morphology of

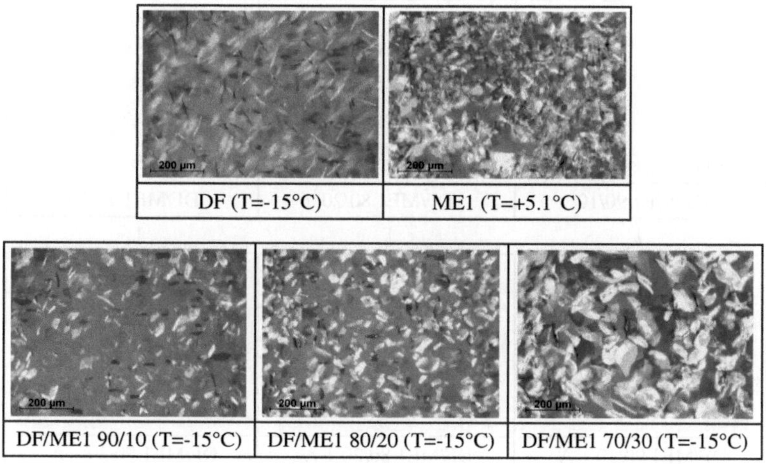

Fig. 13.6 Visualization of crystallization of pure compounds (DF and ME1) and corresponding biodiesel blends at −15 °C by thermomicroscopy

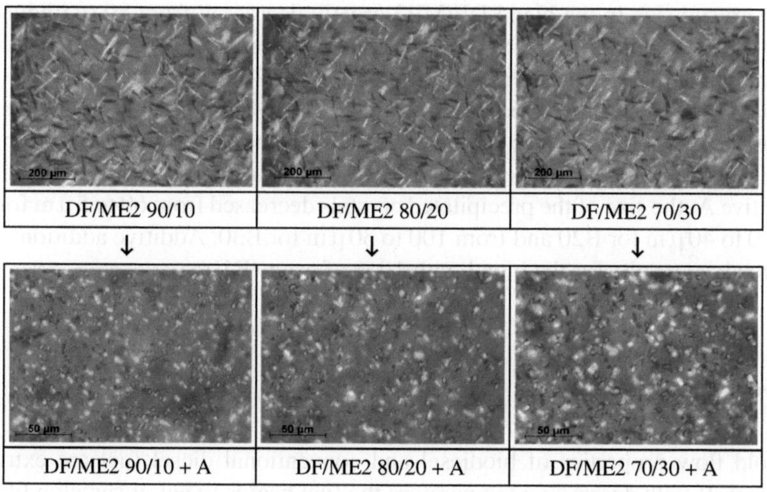

Fig. 13.7 Comparison of crystal morphology at −15 °C between biodiesel blends (DF/ME2) with and without cold flow improver (additive A)

precipitated crystals. The DSC results permitted the following conclusions to be drawn:

- no significant impact on the quantity of crystals was obtained in the presence of A;
- a slight decrease of the crystallization slope was observed for all studied blends, except for the blend DF/ME1 70/30, knowing that a small crystallization slope signifies the formation of crystals with small size. The latter result is confirmed by thermomicroscopy as shown in Figs. 13.7 and 13.8, where the photos were taken at −15 °C.

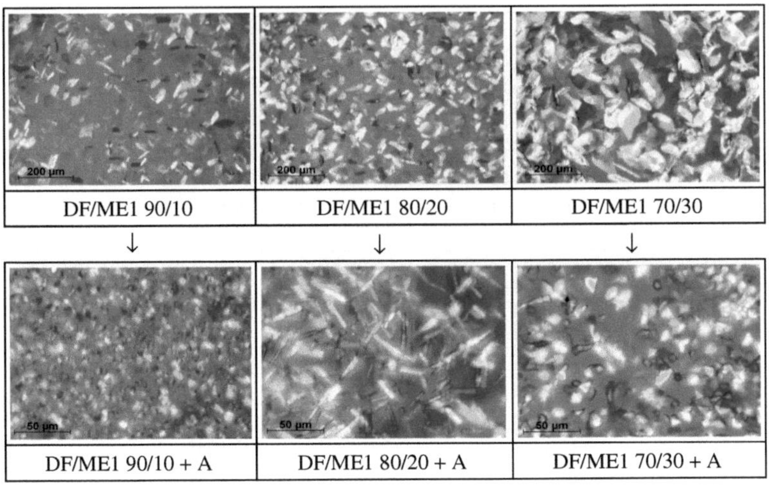

Fig. 13.8 Comparison of crystal morphologies at −15 °C between biodiesel blends (DF/ME1) with and without cold flow improver (additive A)

The crystal size in the blend DF/ME2 greatly decreased from 80 (without A) to 5 µm (with 200 ppm of A) for B10, but it slowly increased from 5 to 20 µm when the percentage of ME2 was increased from 10 to 30 %.

It has been shown that the addition of additive to the DF/ME1 blends led to a decrease in crystal size in the three studied blends, but less significantly for those with 20 and 30 % of ester. The crystals were also more heterogeneous. In the presence of additive A, the size of the precipitated crystals decreased from 60 to 5 µm for B10, from 60 to 40 µm for B20 and from 100 to 30 µm for B30. Additive addition clearly gives the best results for the blend with 10 % of ester (B10).

13.2.4 Conclusion

The cold flow properties of biodiesel and conventional diesel fuel are extremely important. If with decreasing temperature the fuel begins to gel, it can clog filters or even become too thick to be pumped from the fuel tank to the engine. We therefore characterized the cold flow behavior of biodiesel and biodiesel blends by two complementary techniques: DSC and thermomicroscopy. The former has permitted us to quantitatively determine the precipitated crystals and other cold flow parameters, whereas, the size and morphology of crystals were obtained by thermomicroscopy. We observed that the rapeseed methyl ester (ME2) with *a high percentage of unsaturated compounds, presents very good cold flow behavior*, because these compounds crystallize at lower temperatures than saturates. The latter are responsible for poor cold flow properties, thus their proportion should be controlled.

The palm oil methyl ester (ME1) and its blends with DF have shown poor cold flow behavior especially for B30 and this can be explained by the fact that this ester naturally contains more saturated fatty acids (38 %) than, for example, ME2 (7 %) . These saturated compounds have higher melting and crystallization points.

By adding 200 ppm of a cold flow improver, *the size of crystals was greatly reduced* for all the biodiesels and biodiesel blends studied. Therefore, the cold flow behavior was improved, but *no impact on the quantity of crystals* was obtained. This means that the additive used only had an impact on the nucleation and growth of particles.

13.3 Oxidative Stability of Biodiesel by P-DSC

13.3.1 Introduction

Although biodiesel is an excellent source of fuel, it has some disadvantages which include relatively poor cold flow properties and difficulty in maintaining fuel quality during long-term storage. The latter is a matter of oxidation stability and will be discussed in this part.

Degradation of the quality of this alternative fuel leads to an increase of deposits on injectors and pump parts and therefore an increase in pressure across filters [31]. Biodiesel is easily subject to oxidation under ambient conditions. This is mainly due to the presence of double bonds in the chains of fatty compounds. However, these unsaturated chains with low melting points lead to a good cold flow behavior (see Sect. 13.1), which is essential for a biodiesel.

The reaction rate of oxidation depends on the number and position of double bonds [24, 31, 44]. The greater the level of unsaturation in a fatty oil or ester, the more susceptible it will be to oxidation. The CH_2, allylic or bis-allylic to the double bonds, is the site of the first attack. After hydrogen is removed from such a carbon, a carbon-based free radical is formed. Then, oxygen quickly attacks to form a peroxy radical. This is sufficiently reactive to remove another hydrogen from a carbon chain to form another radical and a hydroperoxide. This is a chain reaction including a set of reactions such as initiation, propagation, and termination that can proceed rapidly after the initial induction period has occurred. Once the hydroperoxides have formed (*primary oxidation*), they decompose and react forming numerous *secondary oxidation* products such as aldehydes, alcohols, shorter chain carboxylic acids (which increase biodiesel acidity) and polymers (which increase biodiesel viscosity).

Other factors can have an impact on the oxidative stability of fatty oils and/or esters, such as, water, metals and light, etc. The presence of *water* can cause hydrolysis reactions and damage the fuel quality. Moreover, water can be dispersed in the biodiesel by mono-, di-glycerides and glycerol (co-products of the esterification). These compounds can play the role of emulsifier [3]. Furthermore, the presence of *metals*, even trace levels, can catalyze the oxidation reaction. Copper [30] has the strongest catalytic effect, but iron, nickel and tin also reduce the oxidative stability. Besides this, the presence of a commonly-used additive, fuel acid corrosion inhibitor, can greatly increase the formation of secondary oxidation products. When only 20 ppm of additive in a B50 blend was added, it has been found that polymeric gums were formed [45]. In addition, some polymers are soluble in biodiesel, due to its polar nature, and become insoluble when mixed with diesel fuel [4]. Some polymers are also produced in the biodiesel when the temperature reaches 250–300 °C (thermal polymerization). The mechanism is a Diels-Alder reaction [48]: a conjugated diene group from one fatty acid chain reacts with a single olefin group from another fatty acid chain to form a cyclohexene ring. Exposure to *light* is another factor accelerating oxidation of fatty oils and esters. This process, called photo-oxidation involves a different mechanism whereby oxygen directly attacks the olefinic carbons [44].

In order to prevent oxidation reactions, chemicals called antioxidants (AOs) are added to the fuel. There are two types depending on their action during the reaction process: some decompose hydroperoxides (e.g. zinc dithiophosphates, organophosphates, and organothio derivatives), and others break the reaction chain (phenolic and amine compounds). Most publications treat the latter type and they will also be discussed in this study. Such compounds contain highly labile hydrogen that is more easily removed by a peroxy-radical than the hydrogen of a fatty oil or ester. The resulting antioxidant free radical is either stable or reacts further to form a stable molecule that does not contribute to the chain oxidation process, as shown here:

$$\mathrm{ROO}\bullet + \mathrm{AH} \rightarrow \mathrm{ROOH} + \mathrm{A}\bullet$$

$$\mathrm{A}\bullet \rightarrow \text{stable products}$$

Some AOs can be naturally present in vegetable oils (VOs), such as tocopherols, or chemically synthesized. Tocopherol is a phenolic compound that is found in four isomeric forms (α, β, γ, and δ). They all occur naturally in VOs but not in animal-derived fats (except at trace levels). Several studies have shown that the naturally-occurring levels of tocopherols are optimized with respect to AO ability. In general, further addition of tocopherols has either no further benefit or may even decrease the oxidative stability [42]. The commonly synthesized AOs are:

- 3-t-butyl-4-hydroxyanisol (BHA)
- 2,6-di-t-butyl-4-methylphenol (BHT)
- 2-t-butylhydroquinon (TBHQ)
- 1,2,3-trihydroxybenzene (Pyrogallol or PY)

AO effectiveness is generally measured by stressing a fatty oil or ester both with and without AO and comparing the results. The methods used will be discussed in the following paragraphs.

A literature study has shown that AOs have different effects depending on the origin of the biodiesel [43]. The volatility of the AOs is also a critical parameter because part of the additive may be lost depending on the procedures used. The volatility of 3 AOs can be ranked as follows: BHT > BHA > TBHQ [44]. One study

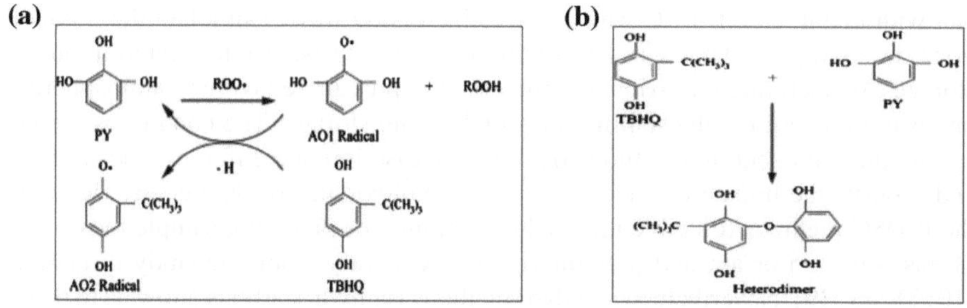

Fig. 13.9 Mechanisms of synergism between two antioxidants as proposed by Guzman [25]

reports that the addition of AO drastically improves the oxidative stability of B100 biodiesel up to a concentration of 1000 ppm [15]. Further additions do not reveal any significant benefit. Moreover, solids appeared in a B10 blend when 3000 ppm of BHT were added.

Cooperative effects (positive synergy) between two phenolic antioxidants in biodiesel have been brought to light. One of the mechanisms proposed by Guzman et al. [25], points the regeneration of PY by TBHQ (Fig. 13.9a). PY being more reactive, it easily donates the hydrogen from its hydroxyl group to fatty acid free radicals creating an antioxidant radical in the process. TBHQ then transfers hydrogen to the antioxidant radical to regenerate PY. As TBHQ was converted to a radical, it can then form stable products with other free radicals. The other mechanism (Fig. 13.9b) suggests the formation of a heterodimer. The as-made phenolic compound may have a better efficiency than the parent antioxidants.

In the same way, a study of lubricating oils [1] reported a synergy between amine and phenolic AOs. The same mechanisms could be applied, replacing PY by the amine AO. The proportions for optimum synergy should be a low amount of amine AO compared to that of the phenolic one, e.g. 1:4 ratio for binary blends. The same study also showed a positive influence of some detergent on the oxidative stability of the oil.

The most common method used to determine the oxidative stability of fatty compounds is Rancimat (EN14214) also called OSI (Oxidation Stability Index). This test method consists in bubbling air through a sample maintained at 110 °C. This air is then passed through deionized water while the conductivity is measured. The sudden increase in water conductivity is due to soluble secondary oxidation products (volatile organic acids). The time at which the conductivity drastically increases is the induction time.

The oxidative stability of diesel fuel is determined by the standard method ASTM D2274 or NF M 07-047, which consists in bubbling air through the sample maintained at 95 °C for 16 h. The solids formed are then dried and weighed. A maximum weight is set by the specification. These tests are however time-consuming. Thus, it has become necessary to develop new oxidation stability tests. In preliminary investigations, numerous pressurized DSC (P-DSC) methods have been developed to determine the oxidation stability of engine oil lubricants [47]. As lubricants are subject to harsh conditions, i.e. high temperature, pressure, atmosphere, and contact with metal, etc., P-DSC was found to be a good way to simulate these conditions. In addition, P-DSC can be used to assess the resistance to oxidation during storage. Accelerating the oxidation process by a pressurized oxygen atmosphere is one way to determine the stability of a fuel during storage. The conditions are not strictly the same, but this method could help to discriminate different types of fuels and observe the impact of additives, such as antioxidants or detergents. However, the P-DSC methods (crucible material—aluminum or platinum, sample mass, type of gas—oxygen or air, and pressure applied) can vary from one study to another [10, 21, 38, 39]. Nevertheless, standard methods for hydrocarbons allow us to determine two critical parameters by P-DSC: the oxidation onset temperature (OOT using the ASTM E 1858-08 standard) and the oxidation induction time (OIT with the ASTM E 2009-08 standard). These methods recommend using aluminum crucibles and sample masses comprises between 3 and 3.30 mg. Therefore, the use of P-DSC is a suitable technique to determine oxidation stability of biodiesel.

13.3.2 Experimental Procedure

Samples of conventional diesel fuel (DF), 3 rapeseed oil methyl esters and 3 antioxidants, BHT, TBHQ and EAP (mixture of N, N'-di-sec-butyl-p-phenylenediamine (40–60 %) 2, 6-di-t-butylphenol (<40 %) and tri-t-butylphenol (<10 %)) were used for this study.

P-DSC experiments were carried out using a High Pressure DSC827 from Mettler-Toledo. Approximately 3.5 ± 0.3 mg of sample was placed in a sealed 40 μL aluminum pan with a pinhole. The analysis cell was pressurized (20 bars) under static oxygen atmosphere.

To determine the OOT, a dynamic mode was used with a temperature range from 25 to 230 °C and heating rate of 5 °C min^{-1}. On the curve, the intersection of the tangents of the baseline and the exothermic peak corresponds to the OOT.

To determine the OIT, an isothermal mode was used where the isothermal temperature (140 °C) was reached at a heating rate of $\cong$ 120 °C min^{-1}. On the curve, the time between the beginning of the isothermal step and the onset of the oxidation peak corresponds to the OIT.

The P-DSC module was calibrated in terms of heat flow and temperature using naphthalene ($T_{fusion} = 80.1°$ C, $\Delta H_{fusion} = 147.64$ Jg^{-1}), indium ($T_{fusion} = 156.6°$ C, $\Delta H_{fusion} = 28.45$ Jg^{-1}), and tin ($T_{fusion} = 231.9°$ C, $\Delta H_{fusion} = 60.10$ Jg^{-1}) standards.

Repeatability was assessed by testing the same sample X, 3 times under the conditions described previously for both methods. Using Eq. (13.1), the average ($\overline{X}$) and standard deviation (σ) were calculated while the Eq. (13.2) shows the variation coefficient (VC). The results are given in Table 13.5 for both the OOT and OIT.

Table 13.5 Repeatability results for both P-DSC methods

	Dynamic method OOT (°C)	Isothermal method OIT (min)
Test 1	193.1	86.0
Test 2	192.4	85.9
Test 3	192.6	86.3
Average	192.7	86.1
Standard deviation	0.4	0.2
VC (%)	0.2	0.3

$$\sigma = \sqrt{\frac{1}{n}\sum_{i=1}^{n}(X_i - \overline{X})^2} \quad \text{with} \quad \overline{X} = \frac{1}{n}\sum_{i=1}^{n} X_i \quad \text{and} \quad n = 3 \tag{13.1}$$

$$VC = \frac{\sigma \times 100}{\overline{X}} \tag{13.2}$$

From Table 13.5, one can conclude that both methods are repeatable when they are carried out under identical conditions and by the same operator and with a freshly prepared sample (i.e. not more than a few days between tests).

13.3.3 Results and Discussions

P-DSC was used to compare different methyl esters of vegetable oil with a conventional diesel fuel, and to evaluate the efficiency of AO. In this study both methods, dynamic and isothermal, were used for every sample in order to determine OOT and OIT values.

First, P-DSC experiments were carried out on two rapeseed oil methyl esters (RME1 and RME2) and a conventional diesel fuel (DF). OOT and OIT values are given in Table 13.6. As expected, due to its chemical composition, DF is much more stable than both RME1 and RME2. Furthermore, RME1 is more stable to oxidation than RME2 (Fig. 13.10).

The two RMEs were blended with DF to make B7 blends. As shown in Table 13.6, the OOT of DF decreases in the presence of methyl esters (also illustrated in Fig. 13.11). The OIT of B7_1 is double that of B7_2. Therefore, B7_1 has a better stability to oxidation than B7_2, and both B7s are less stable than neat DF. Furthermore, the exothermal peak of DF shows a smaller oxidation slope compared with those of the two B7s. This is due to the kinetic of reaction which is slower for the oxidation of DF than in the case of B7s. This is expected because of their different chemical composition.

AOs were added to the ester and then blended with DF to obtain the B7 mixture with a concentration of 1000 ppm of AO. These mixtures were analyzed by P-DSC in terms of their OOT and OIT. Two AOs were tested: BHT and TBHQ. The presence

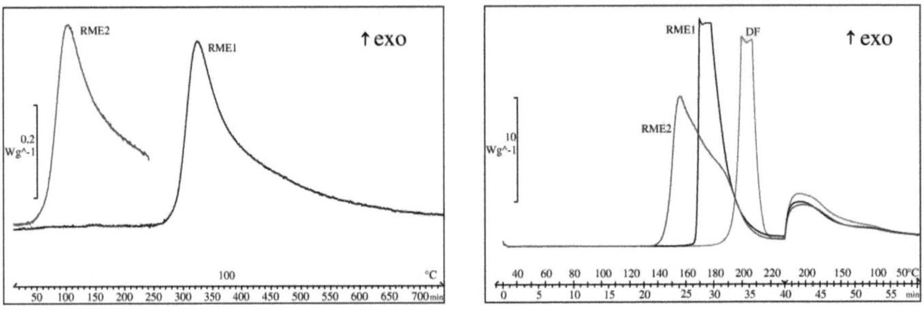

Fig. 13.10 P-DSC curves of DF and pure esters under dynamic and isothermal conditions

Table 13.6 OOT and OIT values of neat and blended compounds

	OOT °C	OIT min (at 140 °C)
DF	195.0	–
RME1	168.4	20.2
RME2	143.4	*n.a.*
DF+7%RME1	186.6	41.3
DF+7%RME2	181.9	20.6

Note n.a. = non analyzed because of its OOT value of 143.4 °C, quite similar with that of measurement.

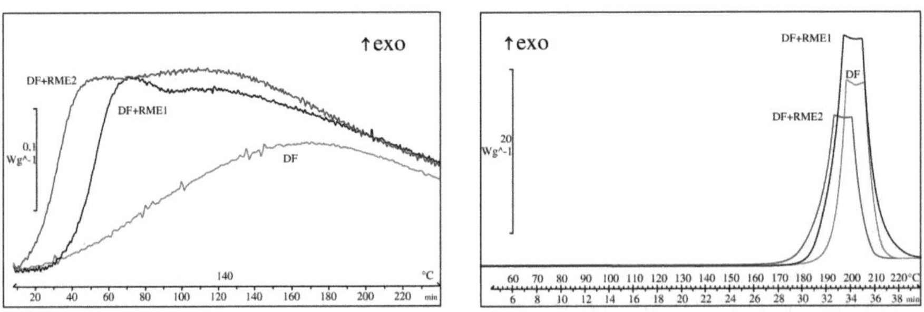

Fig. 13.11 P-DSC curves of diesel fuel and B7 blends under dynamic and isothermal conditions

of AO (with BHT better than TBHQ) greatly improves the oxidative stability of both biodiesel blends, as the OOT and OIT values have increased (Table 13.7).

An interesting point in this study is the relatively poor performance of TBHQ compared to BHT. It has been well-established that TBHQ is highly effective for fatty oils, esters and animal fats while BHT is more effective for diesel fuel and lubricants [15, 22, 36, 44]. Two reasons for this are that, first, the greatly different structures of fatty oils compared to non-polar hydrocarbons (fuels, mineral oil or lubricants) may interact with the highly hindered polar phenol group of BHT to

Table 13.7 OOT and OIT values of B7 with and without AOs

	OOT °C		**OIT min** (**at** 140 °C)	
	RME1	RME2	RME1	RME2
B7	186.6	181.9	41.3	20.6
B7+TBHQ 1000 ppm	191.5	186.1	57.3	31.2
B7+BHT 1000 ppm	196.5	192.6	79.9	74.4

Table 13.8 OOT and OIT values of a B7 with and without AOs

	OOT° **C**	**OIT min** (**at** 140 °C)
B7	186.5	22.0
B7+TBHQ 1000 ppm	193.2	35.8
B7+BHT 1000 ppm	196.5	103.5
B7+EAP 1000 ppm	211.1	319.9

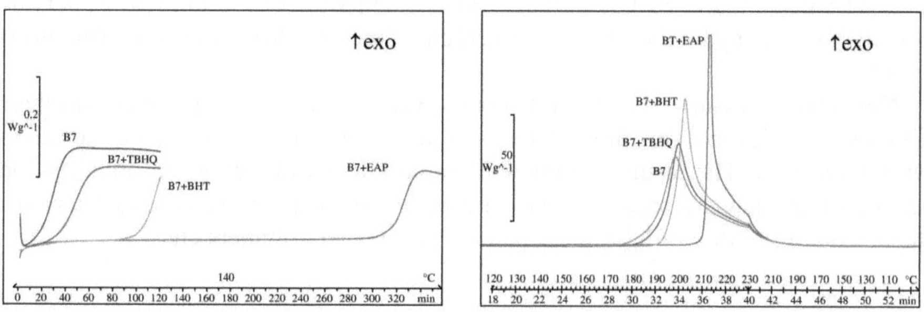

Fig. 13.12 P-DSC curves of a B7 with different AOs under dynamic and isothermal conditions

reduce its AO capacity. Second, BHT is relatively volatile, and under the conditions used for many of the procedures much of the additive may be lost during the early parts of the tests [22, 28, 36]. The blends used in this study being only 7 % ester in diesel fuel, it is understandable that BHT works better as AO than TBHQ.

Table 13.8 gives the results comparing 3 AOs (BHT, TBHQ and EAP) tested on a B7 (also illustrated in Fig. 13.12). B7 was different from the two other blends previously presented, but was still a rapeseed oil methyl ester. Addition of BHT and TBHQ led to improvements in OOT and OIT values. It is clear that EAP significantly improves oxidative stability as the OOT increased by more than 24 °C and the OIT by almost 300 min. This could be explained by its composition: one amine and two phenolic AOs, the whole mixture giving a positive synergy [1]. Thus, the three AOs tested can be classified as a function of their performance as follows:

$$\text{EAP} >> \text{BHT} > \text{TBHQ}$$

13.3.4 Conclusion

During long-term storage biodiesel can be easily subject to oxidation under ambient conditions due to the presence of double bonds in the chains of fatty compounds. The reaction of oxidation increases with the increasing of unsaturation level (polyunsaturated compounds are many times more reactive that monounsaturated ones). Some parameters such as acid value, peroxide value or kinematic viscosity increase when autoxidation occurs, thus the fuel's quality is affected. Degradation of the quality of this alternative fuel leads to an increase of deposits on injectors and pump parts and therefore an increase in pressure across filters.

The work herein has shown the suitability of P-DSC as an accelerated oxidation stability test method for biodiesel. Different biodiesel blends (B7) were compared in terms of their oxidative stability by P-DSC using dynamic (linear heating up to oxidation) and isothermal modes (maintaining constant temperature at a fixed value up to oxidation). The OOT and OIT parameters were determined for each studied blend and the positive influence of "chain-breaking" type additives (such as amine and/or phenolic compounds) has been shown. BHT showed better results (high OOT and OIT values) than TBHQ, but EAP, which is a mixture of 1 amine and 2 phenolic compounds, was by far the best AO. Synergetic effects could explain these very good results.

Nowadays, pressurized DSC has become a very adapted technique to discriminate different samples in terms of oxidative stability or to study the impact of additives on this property. The results are obtained rapidly and with better repeatability and accuracy than standard methods. Furthermore, P-DSC can also be used to determine the oxidative stability of lubricants, paintings, rubbers, polymers etc.

13.4 Overall Conclusion

In this chapter, it has been shown that DSC and P-DSC are both rapid and accurate methods to compare, characterize and predict some properties of alternative biodiesel fuels, such as cold flow behavior and oxidative stability which are closely related. These two properties are extremely important with respect to the operability, transport and long-term storage of biodiesels and diesel fuels. Therefore, much attention should be paid to them. Indeed, the quantity of unsaturated fatty acids (compounds with double bonds and very low melting points) has an important impact on both properties. A high proportion of these compounds *improves the cold flow behavior*, but also *increases the reactivity and thus reduces the stability to oxidation*. A good compromise must therefore be found, and therefore the selection of the appropriate cold flow improvers and antioxidants must be done with great care. Other techniques such as thermomicroscopy can be employed to study the cold flow behavior of biodiesel (i.e. the study of the evolution of paraffin and FAME crystallization with information about the size and the morphology of crystals), such analytical technique being complementary with DSC.

The chemical composition of a fuel is very complex and generally a lot of additives are added in order to improve some properties such as viscosity, oxidative stability, cold flow behavior, cetane number, color or smell etc. The impact on fuel properties of a complex mixture of miscellaneous additives is not always very well known. It has been already shown that it can bring a positive or negative synergy. Therefore, DSC and P-DSC can be very useful in evaluating with high accuracy and in a reasonable time the impact of a package of additives on the oxidative stability or the cold flow behavior of an alternative or conventional diesel fuel.

References

1. J.Z. Adamczewska, C. Love, Oxidative stability of lubricants measured by PDSC CEC L-85-T-99 test procedure. J. Therm. Anal. Calorim. **80**, 753–759 (2005)
2. Y. Ali et al., Fuel properties of tallow and soybean oil esters. J. Am. Oil Chem. Soc. **72** (1995)
3. P. Bondioli, L. Folegatti, Evaluating the oxidation stability of biodiesel. An experimental contribution. Riv. Ital. Sostanze Grasse **73**, 349–353 (1996)
4. P. Bondioli, A. Gasparoli et al., Evaluation of biodiesel storage stability using reference methods. Eur. J. Lipid Sci. Technol. **104**, 777–784 (2002)
5. F. Bosselet, *Application de l'analyse calorimétrique différentielle à l'étude des gazoles moteurs et de leur amélioration par dopage*. Thesis, Science/Chimie-Saint-Étienne, (1984), pp. 1–85
6. L. Bournay et al., New heterogeneous process for biodiesel production: A way to improve the quality and the value of the crude glycerine produced by biodiesel plants. Catal. Today **106**, 190–192 (2005)
7. D.G. Cantrell, L.J. Gillie et al., Structure-reactivity correlations in MgAl hydrotalcite catalysts for biodiesel synthesis. Appl. Catal. A General **287**, 183–190 (2005)
8. P. Claudy, J.M. Létoffé et al., Crude oils and their distillates: characterization by differential scanning calorimetry. Fuel **67**, 58–61 (1988)
9. P. Claudy, J.M. Létoffé et al., Diesel fuels: determination of onset crystallization temperature, pour point and filter plugging point by differential scanning calorimetry. Correlation with standard test methods. Fuel **65**, 861–864 (1986)
10. M.M. Conceiçao, M.B. Dantas et al., Evolution of the oxidative induction time of the ethilic castor biodiesel. J. Therm. Anal. Calorim. **97**, 643–646 (2009)
11. J. Cvengros et al., Production and treatment of rapeseed oil methyl esters as alternative fuels for diesel engines. Bioresour. Technol. **55** (1996)
12. M. De Torres, G. Jiménez-Osés et al., Fatty acid derivatives and their use as CFPP additives in biodiesel. Bioresour. Technol. **102**, 2590–2594 (2011)
13. A. Demirbas, Biodiesel from oilgae, biofixation of carbon dioxide by microalgae: A solution to pollution problems. Appl. Energy **88**, 3541–3547 (2011)
14. M.F. Demirbas, Biofuels from algae for sustainable development. Appl. Energy **88**, 3473–3480 (2011)
15. R.O. Dunn, Effect of antioxidants on the oxidative stability of methyl soyate (biodiesel). Fuel Process. Technol. **86**, 1071–1085 (2005)
16. R.O. Dunn, M.O. Bagby, Low-temperature filterability properties of alternative diesel fuels from vegetable oils, in *Proceedings of the third liquid fuel conference: Liquid fuel and industrial products from renewable resources*, American Society of Agricultural Engineers, St. Joseph, MI, (1996), ed. by J. S. Cundiff et al., pp. 95–103
17. R.O. Dunn, Thermal analysis of alternative diesel fuels from vegetable oils. J. Am. Oil Chem. Soc. **76**, 109–115 (1999)
18. A.K. Endalew et al., Heterogeneous catalysis for biodiesel production from Jatropha curcas oil (JCO). Energy **36**, 2693–2700 (2011)

19. H.R. Faust, The thermal analysis of waxes and petrolatums. Thermochim. Acta **26**, 383–398 (1978)
20. C.S. Foon, Y.C. Liang et al., Crystallisation and melting behavior of methyl esters of palm oil. Am. J. Appl. Sci. **3**, 1859–1863 (2006)
21. L.M.S. Freire, T.C. Bicudo et al., Thermal investigation of oil and biodiesel from Jatropha curcas L. J. Therm. Anal. Calorim. **96**, 1029–1033 (2009)
22. M.E. Gordon, E. Mursi, Comparison of oil stability based on the metrohm rancimat with storage at 20 C. J. Am. Oil Chem. Soc. **71**, 649–651 (1994)
23. J.C. Guibet, Chapitre 4.5. In: Carburants et moteurs (Tome 1). (Editions OPHRYS, Paris, 1997)
24. F.D. Gunstone, T.P. Hilditch, The union of gaseous oxygen with methyl oleate, linoleate and linolenate. J Chem Soc, 836–841 (1945)
25. R. Guzman, H. Tang et al., Synergistic effects of antioxidants on the oxidative stability of soybean oil- and poultry fat-based biodiesel. J. Am. Oil Chem. Soc. **86**, 459–467 (2009)
26. E.L. Heino, Determination of cloud point for petroleum middle distillates by differential scanning calorimetry. Thermochim. Acta **114**, 125–130 (1987)
27. J.C. Hipeaux, M. Born et al., Physico-chemical characterization of base stocks and thermal analysis by differential scanning calorimetry and thermomicroscopy at low temperature. Thermochim. Acta **348**, 147–159 (2000)
28. A.W. Kirleis, C.M. Stine, Retention of synthetic phenolic antioxidants in model freeze-dried food systems. J. Food Sci. **43**(5), 1457–1460 (1978)
29. G. Knothe, Dependence of biodiesel fuel properties on the structure of fatty acid alkyl esters. Fuel Process. Technol. **86**, 1059–1070 (2005)
30. G. Knothe, R.O. Dunn, Dependence of oil stability index of fatty compounds on their structure and concentration and presence of metals. J. Am. Oil Chem. Soc. **80**, 1021–1026 (2003)
31. G. Knothe, J. Van Gerpen, J. Krahl, *The Biodiesel Handbook* (AOCS Press, Urbana, 2005)
32. I. Lee et al., Reducing the crystallization temperature of biodiesel by winterizing methyl soyate. J. Am. Oil Chem. Soc. **73**(5), 631–636 (1996)
33. J.M. Létoffé, P. Claudy et al., Antagonism between cloud point and cold filter plugging point depressants in a diesel fuel. Fuel **74**, 1830–1833 (1995)
34. R. Maceiras et al., Macroalgae: Raw material for biodiesel production. Appl. Energy **88**, 3318–3323 (2011)
35. F. Noel, Thermal analysis of lubricating oils. Thermochim. Acta **4**, 377–392 (1972)
36. M. Peled, T. Gutfinger, A. Letan, Effect of water and BHT on stability of cottonseed during frying. J. Sci. Food Agric. **26**, 1655–1666 (1975)
37. G.R. Peterson et al., Rapeseed oil transesterification by heterogeneous catalysis. J. Am. Oil Chem. Soc. **61**(10), 1593–1597 (1984)
38. F.M.G. Rodrigues, A.G. Souza, I.M.G. Santos et al., Antioxidative properties of hydrogenated cardanol for cotton biodiesel by PDSC and UV/Vis. J. Therm. Anal. Calorim. **97**(2), 605–609 (2009)
39. N.A. Santos, J.R.J. Santos et al., Thermo-oxidative stability and cold flow properties of Babassu biodiesel by PDSC and TMDSC technique. J. Therm. Anal. Calorim. **97**(2), 611–614 (2009)
40. P.M. Schenk, S.R. Thomas-Hall et al., Second generation biofuels: High-efficiency microalgae for biodiesel production. Bioenergy Res. **1**, 20–43 (2008)
41. Y. Serpemen, F.W. Wenzel et al., Blending technology key to making new gasolines. Oil Gas J. **89**(11) (1991)
42. E.R. Sherwin, Oxidation and antioxidants in fat and oil processing. J. Am. Oil Chem. Soc. **55**, 809–814 (1978)
43. H. Tang, A. Wang et al., The effect of natural and synthetic antioxidants on the oxidative stability of biodiesel. J. Am. Oil Chem. Soc. **85**, 373–382 (2008)
44. J.A. Waynick, Characterisation of biodiesel oxidation and oxidation products. The Coordinated Research Council, Southwest Research Institute Project N° 08–10721, San Antonio, Texas (2005)
45. J.A. Waynick, Evaluation of the stability, lubricity, and cold flow properties of biodiesel fuel, in *Proceedings of the 6th International Conference on Stability and Handling of Liquid Fuels*, Vancouver, B.C., Canada, pp. 805–829 (1997)

46. G. Wenzel et al., Boiling properties and thermal decomposition of vegetable oil methyl esters with regard to their fuel suitability. J. Agric. Food Chem. **45** (1997)
47. A. Zeman, H.P. Binder, Determination of spontaneous ignition temperatures (SITs) of aviation lubricants using pressure differential calorimetry (PDSC). Thermochim. Acta **98**, 159–165 (1986)
48. O. Zovi, L. Lecamp et al., Stand reaction of linseed oil. Eur. J. Lipid Sci. Technol. **113**, 616–626 (2011)

Chapter 14
CO_2 Capture in Industrial Effluents. Calorimetric Studies

Jean-Yves Coxam and Karine Ballerat-Busserolles

Abstract In order to reduce environmental impact of CO_2 emissions, one possible option is the decarbonation of the effluents coming from fixed sources. A description of the different techniques proposed for a separation of CO_2 from gaseous effluents is explained with a focus on post-combustion processes. The design of specific separation units will require studies of gas dissolution in various selective absorbent solutions. The thermodynamic approach for CO_2 dissolution in aqueous solutions of amine is depicted, showing the physicochemical background and the main properties required in this domain. An overview of the main experimental developments for determining the enthalpy of solution of carbon dioxide in absorbent solutions is presented together with some representative results.

14.1 Introduction

Carbon dioxide is considered as a main greenhouse gas. Its contribution to global warming can be pointed up by correlations between the evolution of CO_2 in atmosphere and the average earth temperature [1]. Since the eighteenth century the concentration of CO_2 has approximately risen from 280–380 ppm when simultaneously temperature has increased by 0.6 K. Anthropogenic sources of carbon dioxide represents currently about 25 Gt per year. If we consider constant industrial development and population growth in the world, the emissions of carbon dioxide will

J.-Y. Coxam (✉) · K. Ballerat-Busserolles
Institut de Chimie de Clermont-Ferrand,
Clermont Université, Université Blaise Pascal,
BP 10448, 63000 Ferrand, France
e-mail: j-yves.coxam@univ-bpclermont.fr

K. Ballerat-Busserolles
e-mail: karine.ballerat-busserolles@univ-bpclermont.fr

A. Auroux (ed.), *Calorimetry and Thermal Methods in Catalysis*,
Springer Series in Materials Science 154, DOI: 10.1007/978-3-642-11954-5_14,

not decrease without an international mobilization. In order to avoid irreversible environmental impact, CO_2 emissions have to be drastically reduced [2].

The anthropogenic carbon dioxide emissions can roughly be classified in 3 main categories: industrial, transportation and residential. The key solutions to reduce these emissions will be found in an ecologic production and rational utilization of energy. One of them will be to reduce CO_2 emissions from industrial site called"fixed sources". It mainly concerns industries using fossil energy such as cement factory, metallurgy, glass industry or power plant. For this purpose pre-combustion and post-combustion processes have been proposed. The first ones consist in using specific comburant and technology for the combustion of coal, fuel or natural gas in order to release small quantity of CO_2 in the fumes. The second ones refer to techniques of separation of carbon dioxide from industrial effluent. In this last case carbon dioxide will be valorized or stored in specific secured sites.

Capture processes were developed in the past to remove acid gases such as carbon dioxide and hydrogen sulfide (H_2S) from natural gas. They are mainly based on chemical and physical dissolution of the acid gas in aqueous solutions of amines. The technique is considered as mature enough to be adapted in next future to the treatment of post combustion effluents. The new processes should take into account difference from natural gas treatment such as temperature, pressure or composition of the effluent. However one major barrier for the integration into industrial sites in the few coming years is the economical cost of the so called *ton of* CO_2 *avoided*. Specific researches are then carried out on both the technology and the choice of the absorbent solutions.

Future absorbent solutions have to combine high carbon dioxide loading charges (moles of dissolved carbon dioxide per mole of amine) with low energies of regeneration. Characterization of new absorbent solution can be performed by calorimetric studies of gas dissolution. The experimental data collected are essential to develop thermodynamic models representative of the {CO_2-absorbent solution} systems that will be used to design the future capture units. The dissolution properties required are the mainly the gas solubility and the enthalpy of solution. However some other properties also have to be studied, such as heat capacity, vapor pressure, chemical and thermal degradations. Then specific calorimetric techniques were set up to provide the essential experimental data.

14.2 Presentation of Techniques for CO_2 Separation from Gaseous Effluents

Adsorption on micro-porous solids is a common industrial technique used for removing impurities (e.g. CO_2 from gaseous streams such as hydrogen-rich gases resulting from gasification, steam reforming or shift of fossil hydrocarbons. The molecules to be adsorbed interact with the internal surface of microporous solids. Polar and/or condensable species like CO_2 will be strongly retained whereas "light components" such as N_2, O_2, CO will be less or not retained on the adsorbent. The regeneration, or gas desorption, is performed by running thermal cycles (Temperature Swing Adsorption, TSA) and/or pressure cycles (Pressure-Swing-Adsorption, PSA or Vacuum-Swing-Adsorption, VSA). The energy required is thus analogous to that of absorption. The most common adsorbents are activated carbons and zeolite molecular sieves, mostly used in combination [3, 4].

For separation processes, solid adsorbents are packed as fixed beds and submitted to successive adsorption and desorption steps. The operation is therefore transient instead of steady-state. To achieve a continuous production, multiple packed-beds are needed in parallel with shifted time-cycles. The technique necessitates a drastic pretreatment of gas effluent, especially for particle removal. A new class of nanoporous crystalline solids, called metal-organic framework (MOF) materials, is intensively studied as potential adsorbent materials for CO_2 capture. The interest of such material is their high internal surface area, large pore volume, and the possibility to adapt the pores with a variety of functionalities.

The membrane processes proposed for CO_2 separation consist in selective permeation of molecules through a thin layer of polymeric or mineral material. The two intrinsic criteria determining their performance are the selectivity and the permeability. Industrial membrane processes have been particularly developed for recovery of relatively concentrated CO_2 (>20 %) at high pressures from natural gas. [5, 6]. Selectivity and permeability of membranes depend on their chemical nature and their structure (microporosity, active layer thickness). Modern membranes are layered composites, with a thin dense selective layer supported by thicker, more permeable and less-selective or inert supports. There are basically two families of membrane materials: inorganic ceramic type membranes, and organic polymeric membranes. The former can stand high temperatures and pressures, and are corrosion resistant. They can be made selective for H_2/CO_2 separations, with hydrogen permeating preferentially. However the CO_2/N_2 separation selectivity remains relatively low in case of CO_2 capture from post combustion effluent.

The cryogenic process is used on a large scale for air separation but cryogenic distillation does not seem to be currently considered as an alternative for CO_2 capture, probably because of anticipated high costs when diluted streams are used [7]. However in the case oxy-fuel combustion technique where cryogenic air separation unit is implemented, it may be conveniently integrated with c ondensation units for separating water from CO_2.

Absorption in alkaline solution is a common principle used in acid gases capture processes operating for decades in natural gas treatment. The reference absorbent is aqueous solution of monoethanolamine (MEA). The mechanism of capture is a combination of chemical and physical dissolution. The chemical dissolution is based on an acido-basic reaction. The reaction must be reversible in order to regenerate the absorbent solution and recover carbon dioxide for storage. The physical dissolution, usually observed fot high partial pressure of CO_2, can be improved by addition of specific physical solvent.

Some other more emerging techniques are proposed. It includes gas hydrate crystallization [8], enzymatic capture [9] or dissolution in ionic liquids [10]. However the separation process by dissolution in amine solutions (amine washing process) is presently the most advanced technology proposed to remove carbon dioxide from industrial effluents. However the dilemma of the economic costs still constitutes a barrier for integration such processes into industrial sites.

14.3 Industrial Processes Proposed for CO_2 Capture in Post Combustion Effluents

The CO_2 capture processes will concern the industries that emit large volumes of carbon dioxide i.e. those using fossil energy. For near future integration, the most mature technologies are those based on gas separation by selective dissolution in appropriate absorbent solutions. The industrial process is schematically represented in Fig. 14.1. It consists in cycles of gas absorption and desorption that implicate a reversible reaction of gas dissolution [11]. In the first step the gaseous effluent flows counter-currently to the absorbent solution in an absorber unit. In the second step the "rich" absorbent solution containing dissolved CO_2 is pumped to the top of a stripper (or regeneration vessel). The regeneration of the chemical solvent is carried out by heating the solution using water vapor. The "lean" solvent is pumped back to the absorber.

Industrial pilots based on this principle are now under study. The design of the future installations must be tailored to the composition, flow rate, temperature and pressure of the fumes. These effluent parameters largely depend on the type of industry considered. The incoming gas flow rates will vary from 15,000 to 3,00,000 Nm^3/h

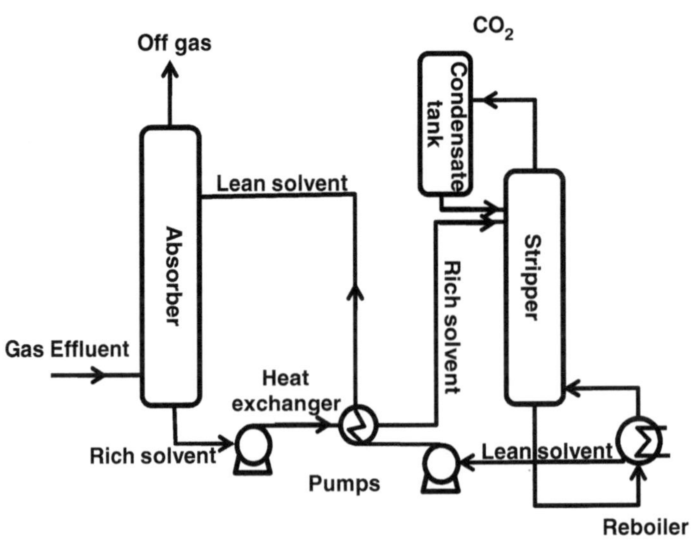

Fig. 14.1 Schematic representation of capture process

as in the same way, CO_2 compositions in the effluents will vary from 3 to 30 moles percent.

The regeneration of the solvent solution is planned to be performed at temperatures between 100 and 140 °C and at pressures not very much higher than atmospheric pressure. The costs of this step represent a major barrier to the integration of capture process into industrial sites. Then active researches are conducted to elaborate better absorbent solutions that will require less energy for regeneration.

The characterization of the new absorbent solutions will be achieved by a determination of its physical and chemical properties. We will focus here on some properties that can be investigated by calorimetric techniques.

For example, the energy of regeneration, related to the opposite of the energy of gas dissolution, will be studied through the determination of the enthalpy of solution of CO_2 in the absorbent solution. This enthalpy represents the energy needed to dissolve one mole of gas and will contribute in a great extend to the calculi of economical costs of carbon dioxide capture. Heat capacities will also be key properties for heat exchange calculations as well as the gas solubility, required to estimate solvent flux for given gas stream. In addition, the solvent characterization should be completed by studies on vapor pressure, transport properties, chemical and thermal degradations, kinetic of dissolution, corrosion...

14.4 Thermodynamic Approach of CO_2 Dissolution in Aqueous Solutions of Amine

14.4.1 Mechanism of CO_2 Dissolution

The dissolution of CO_2 in aqueous solution of amine is combination of chemical reactions and physical dissolution. The chemical reactions involved depend on the type of the amine.

In case of primary or secondary amine, the presence of hydrogen(s) branched on nitrogen atom allows formation of carbamate. The carbon dioxide reacts first with one molecule of amine to form a zwitterion (Eq. 14.1). Then the zwitterion reacts with a second amine molecule to form a carbamate (Eq. 14.2). The stoechiometry for chemical dissolution in primary or secondary amines is then limited to 2 molecules of amine per molecule of carbon dioxide (Eq. 14.3).

$$CO_2 + RR'NH = RR'N + HCOO^- \tag{14.1}$$

$$RR'N^+HCOO^- + RR'NH = RR'NCOO^- + RR'N^+H_2 \tag{14.2}$$

$$CO_2 + 2RR'NH = RR'NCOO^- + RR'N^+H_2 \tag{14.3}$$

In case of sterically hindered amine the carbamate is instable and primary or secondary amines will react as the tertiary amines. The instability of carbamate also increases with temperature. With tertiary amines, carbon dioxide cannot form carbamate because of no available free proton. The reaction mechanism then results in hydration of carbon dioxide by alkaline catalysis to form hydrogenocarbonate (Eq. 14.4).

$$RR'R''N + H_2O + CO_2 = RR'R''NH^+ + HCO_3^- \tag{14.4}$$

The increase of partial pressure of carbon dioxide forces the physical dissolution. This mechanism is purely mechanical and results in the apparition of molecular CO_2 in the solution. This physical dissolution is particularly considerate in the treatment of natural gas in which partial pressure of carbon dioxide can reach hundreds of bars. In the case of CO_2 capture in industrial effluent, partial pressure of CO_2 remains below atmospheric pressure and carbon dioxide is mainly chemically absorbed with formation of carbamate and hydrogen carbonate.

The solution concentration and the choice of the amine will be adapted to the conditions of temperature, pressure and composition of the gas stream to be treated. It has been shown that amines can also be combined to improve kinetic of absorption and quantity of dissolved gas [12]. In some cases, physical solvents can also be added to the solution in order to increase physical dissolution.

14.4.2 Selection of Amines for CO_2 Capture Processes

The design of specific amines for CO_2 capture in industrial effluent is a difficult task as no model really exists to predict the dissolution properties from structural properties. Then current studies consist in a screening of commercial or synthesized molecules, implicating numerous experimental investigations. The absorbent solution must be characterized by properties such as density, viscosity, heat capacity, thermal and chemical stability, or corrosion aspects. The gas dissolution can be described by the determination of dissolution properties such as the limits of gas solubility, enthalpies of solution or kinetic of gas dissolution. The experimental data make it possible the development of theoretical models representative of the {CO_2-water-amine} systems. These models can then be used for engineering calculations to estimate the pre-cited properties of dissolution of the absorbent solution as function of temperature, pressure or amine concentration.

First studies were carried out on alcanolamine such as monoethanolamine (*MEA*), diethanolamine (*DEA*) or methyl diethanolamine (*MDEA*). Investigations are actually extended to "new" amines chosen within families such as aliphatic, cyclic, polyfunctional ... Recently demixing amines solutions were proposed for new capture process. Aqueous solutions containing such amines present a liquid-liquid phase separation with a temperature change depending on the composition of the solution. The corresponding process DMX™ has been proposed and patented by IFPEN [13]. A demixing unit was added at the output of the absorber unit in which the temperature of the gas charged solution is increased until reaching demixing of the ternary solution CO_2-amine-water in two phases. Only the aqueous phase that contains dissolved carbon dioxide is sent to the stripper while the amine phase is directly returned to the absorber. In these conditions the energy required for the regeneration of the absorbent solution is significantly reduced. The characterization of these demixing amine solutions is comparable to the classic one excepted an additional determination of phase diagrams of binary {water-amine} and ternary {CO_2-water-amine} systems.

14.4.3 Calorimetric Experimental Data Required

The properties of dissolution as gas solubility and enthalpy of solution can be derived from vapor liquid equilibrium models representative of {CO_2-H_2O-amine} systems. The developments of such models are based on a system of equations related to phase equilibria and chemical reactions: electro-neutrality and mass balance. The non ideality of the system can be taken into account in liquid phase by the expressions of activity coefficients and by fugacity coefficients in vapor phase. Non ideality is represented in activity and fugacity coefficient models through empirical interaction parameters that have to be fitted to experimental data. Development of efficient models will then depend on the quality and diversity of the experimental data.

Regarding literature data reported in Tables 14.1 and 14.2 collected for classical amines (MEA-DEA-AMP), it appears that solubility data are quite available while enthalpy data remains very scarce. Solubility data in Table 14.1 are usually obtained using *pVT* measurements and provide gas solubility expressed as loading charge α (moles of gas per mole of amine) as function of temperature and pressure (CO_2 partial pressure or total pressure). The enthalpy data in Table 14.2 are obtained from measurements of heat of mixing gas with absorbent solution using calorimetric techniques. Different techniques and associated experimental protocols have been proposed and will be presented in next paragraph. Mathonat et al. [14, 15], Arcis et al. [16, 17] use a flow calorimetric technique where the heat of mixing is detected by thermopiles surrounded a mixing cell. Kim and Svendsen [18], Carson et al. [19] and Kierzkowska-Pawlak [20] use a reaction calorimeter based on power compensation of the heat of mixing. Oscarson et al. [21], Merkley et al. [22] and Helton et al. [23] use a flow technique where the heat of mixing is determined from power compensation. In another hand enthalpies of solution can be derived from solubility data or from thermodynamic model fitted to solubility data. However as shown by Arcis et al. [24] and Kim et al. [25] such calculated values can show large divergence with direct experimental.

The representation of chemical reactions in solution in the thermodynamic models [21, 26, 27] necessitates the knowledge of the equilibrium constants of CO_2 dissociations, water dissociation, amine protonation and carbamate formation. For original amines the protonation or carbamate formation equilibrium constants are usually not available and must be measured. In order to derive enthalpy properties using Van't Hoff equations, these equilibrium constants must be determined as function of temperature. Such data can be obtained from a protonation constant determined at a

Table 14.1 Literature solubility data of CO_2 in aqueous solution

Source	m mol.kg^{-1}	T K	p_{CO_2} kPa	ASD %
(a). Monethanolamine (MEA)				
Mason and Dodge [28]	0.5–12.5$^{(b)}$	273–348	1.32–100	2
Jones et al. [29]	15.3	313–413	0.0027–930	0.5
Lee et al. [30]	2.5–5.0$^{(b)}$	313–373	1.15–6621	3
Lee et al. [31]	1.0–5.0$^{(b)}$	298–393	0.1–10000	4
Lawson and Garst [32]	15.2	313–413	1.32–2750	9
Isaacs et al. [33]	2.5$^{(a)}$	353–373	0.0066–1.75	15
Austgen and Rochelle [34]	2.5$^{(a)}$	313–353	0.0934–229	NC
Shen and Li [35]	15.3–30.0	313–373	1.1–2550	12
Dawodu and Meisen [36]	4.2$^{(a)}$	373	455–3863	13.5
Jou et al. [37]	30.0	273–423	0.0012–19954	3
Song et al. [38]	15.3	313	3.1–2359	12
Jane and Li [39]	2.5$^{(a)}$	353	3.57–121.8*	5
Mathonat et al. [14]	30.0	313–393	5000–20000	7
Ma'mun et al. [40]	30.0	393	7.354–191.9	2
Arcis et al. [41]		323; 373	500–5000*	7
(b). Methyldiethanolamine (MDEA)				
Jou et al. [42]	2.00–4.28$^{(a)}$	298–393	0.001–6630	NC
Merkley et al. [22]	20.0–23.5	298–393	1121	5
Chakma and Meisen. [43]	1.69–4.28$^{(a)}$	373–473	103–4930	14
Austgen and Rochelle [34]	2.00–4.28$^{(a)}$	313	0.0056–93.6	NC
Shen and Li [35]	30	313–373	1.1–1979	12
Jou et al. [44]	35	313–373	0.963–236	3
Dawodu and Meisen [36]	4.28$^{(a)}$	373–393	162–3832	14
Oscarson et al. [21]	20.0–60.0	289–422	22-6164	5
Kuranov et al. [27]	1.95–2.00$^{(b)}$	313–413	73.5–5036.7	3
Xu et al. [45]	1.72–6.85$^{(a)}$	328–363	137.5–808.5	NC
Mathonat et al. [15]	30	313–393	2000 – 10000*	7
Rho et al. [46]	5–75	323–373	0.775–268.3	5
Silkenbäumer et al. [47]	2.632$^{(b)}$	313	12 – 4080*	2
Baek and Yoon [48]	30	313	1.02–1916	3
Rogers et al. [49]	23–50	313	0.00007–1.0018	6
Xu et al. [45]	3.04–4.28$^{(a)}$	313–373	0.876–1013	NC
Lemoine et al. [50]	23.63	298	0.02–1.636	5
Pacheco et al. [51]	35–50	298–373	73–738	NC
Kamps et al. [52]	3.95–7.99$^{(b)}$	313–353	176.5–7565	4
Park and Sandall [53]	50	298–373	0.78–140.40	10
Kierzkowska-Pawlak [20]	10-30	293–333	100–300	3
Bishnoi and Rochelle [54]	4.28$^{(a)}$	313	0.108–0.730	NC
Sidi-Boumedine et al. [55]	25.73–46.88	298–348		2
Ali and Aroua [56]	2$^{(a)}$	313–353	0.06–95.61	2
Kundu et al. [57]	23.8–30.0	303–323	1	3
Benamor and Aroua [58]	2 – 4$^{(a)}$	303–323	0.1–98.2	5
Ma'mun et al. [40]	50	328–358	65.75–813.4	2

(continued)

Table 14.1 (continued)

Source	m mol.kg^{-1}	T K	p_{CO_2} kPa	ASD %
Jenab et al. [59]	$2.0 - 2.5^{(a)}$	298–343	101–2320	4
Huttenhuis et al. [60]	35–50	283–298	0.054–986.80	NC
(c). Diethanolamine (DEA)				
Lee et al. [30]	0.5–8.0$^{(a)}$	273–413	0.69–6900	5
Oyevaar et al. [61]	0.984–2.463$^{(a)}$	298	1.844–14.8	5
Kennard and Meisen [62]	10.0–30.0$^{(c)}$	373-478	73.1–3746.7	10
Dawodu and Meisen [36]	4.2$^{(a)}$	373	455–3863	13.5
Lal et al. [63]	2.0$^{(a)}$	313–373	0.0026–3.336	15
Lee et al. [64]	3.0–4.0$^{(a)}$	298–403	0.001- 5000	NC
Lawson and Garst [32]	25.0$^{(c)}$	311–394	1.974–4315.79	9
Mason and Dodge [28]	0.5–8.3$^{(a)}$	273–348	1.32–100	2
Sidi-Boumedine et al. [55]	41.78$^{(c)}$	298- 348	2.46–4662.7	2
Seo and Hong [65]	30.0$^{(c)}$	313–353	4.85–357.3	3
Benamor and Aroua [58]	2.0–4.0$^{(a)}$	303–323	0.1–104.7	2
Haji-Sulaiman et al. [66]	2.0–4.0$^{(a)}$	303–323	0.1–104.7	2
Haji-Sulaiman and Aroua [67]	2.0$^{(a)}$	301–353	5–100.3	12
Arcis et al. [68]	15; 30$^{(c)}$	323; 373	500 -5000	7

$^{(a)}$ *molarity* (mol.L^{-1}); $^{(b)}$massic fraction; $^{(c)}$wt%; $^*p_{\text{tot}}$ in kPa

Table 14.2 Literature references for experimental enthalpy of solution

Source	*wt* %	T K	p_{CO_2} kPa	ASD %
		MEA		
Mathonat et al. [14]	30.0	313–393	2000–10000	7
Kim and Svendsen [18]	30.0	313–393	0.001	3
Carson et al. [19]	10.0–30.0	298	265	2
Arcis et al. [41]	15.0; 30.0	323; 373	500–5000*	5
		DEA		
Helton et al. [23]	20–50	300–400	87–1121	5
Carson et al. [19]	10.0–30.0	298	265	2
Oscarson et al. [21]	20.6–49.8	300–450	90–1121	5
Arcis et al. [68]	15.0; 30.0	323; 373	500–5000*	5
		MDEA		
Merkley et al. [22]	20.0–23.5	298–393	1121	5
Oscarson et al. [21]	20.0–60.0	289–422	156–1466	5
Mathonat et al. [15]	30.0	313–393	2000–10000	7
Kierzkowska-Pawlak [20]	10.0–40.0	293–333	100–300	2
Carson et al. [19]	10.0–30.0	298	265	2
Arcis et al. [17]	15.0–30.0	323	500–2000	5
Arcis et al. [16]	15.0–30.0	373	500–2000	5

$^*p_{tot}$ in kPa

reference temperature and the enthalpy of amine protonation. Experimental data can be obtained by combining electrochemical and calorimetric techniques [21]. For primary and secondary amine it is also essential to determine the equilibrium constants for carbamate formation. These constants can be obtained by the determination of carbamate concentration when adding CO_2 to the amine solution using available speciation techniques. As for amine protonation the study can be completed by measurements of enthalpies of carbamate formation using mixing calorimetry. Then it appears that to develop thermodynamic models for CO_2 dissolution in aqueous solution of amine, there is a real need of accurate experimental data for enthalpies of solution, enthalpy of amine protonation as well as for enthalpy of carbamate formation.

14.5 Calorimetric Studies of CO_2 Dissolution in Amine Solutions

14.5.1 Calorimetric Techniques for Measuring Heat of Mixing

The calorimetric techniques for measuring heats of mixing two fluids can be classified into their mode of measurement and their principle of heat detection. The isothermal displacement calorimetry will refer to a "static" mode and flow calorimetry, to a "dynamic mode". The principles of heat detection in the following examples will be power compensation or heat flux determination.

Isothermal Displacement Calorimeter

The technique used by Carson et al. [19] at University of Canterbur (New Zealand) is a modification of the liquid–liquid isothermal displacement calorimeter originally developed by Stokes et al. [69]. This static technique has then been primarily used to study binary liquid mixtures. The technique consists in filling a calorimetric cell with a known amount of one of the liquids (solvent) and injecting the solute from a burette. The injection is performed after reaching thermal equilibrium in the calorimetric cell. As reaction occurs, the temperature change is compensated in order to maintain the calorimeter to its initial temperature. Heat of mixing can be determined at temperatures range from 298 to 313 K and pressures comprised between 0.1 and 0.3 MPa.

In order to make it possible the determination of enthalpies of solution of gases in organic liquid solvents, Battino and Marsh [70] set up a modified burette arrangement. This modified technique proved to be effective for the particular systems where small gas solubilities are observed. However the technique was impractical in the case of CO_2 dissolution in aqueous solution of amine. This was mainly due to high solubility of carbon dioxide and consequently, large volumes of injected gas. Then Carson

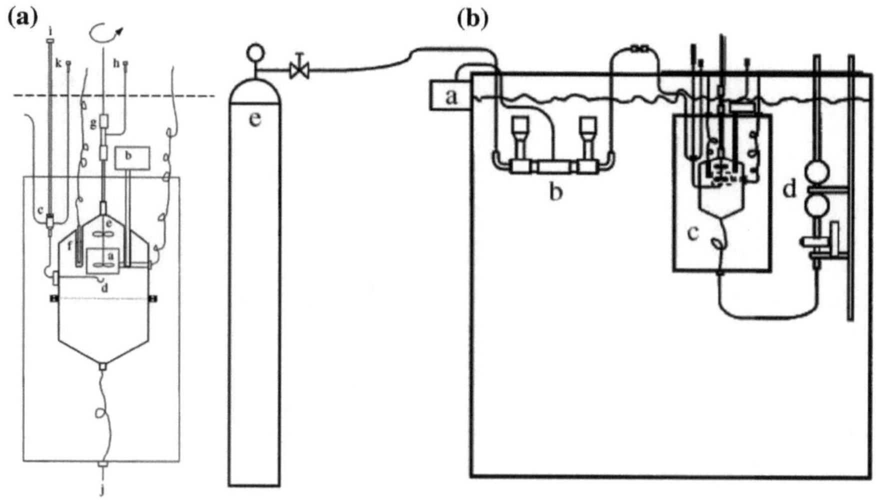

Fig. 14.2 Schematic representation of **a** classical displacement calorimeters [69]. (*a* annular heater; *b* Peltier cooling unit; *c* inlet valve; *d* injection tube; *e* stirrer; *f* thermistor; *g* stirrer gland; *h* solution outlet; *i* inlet valve control; *j* outlet to pipette; *k* by-pass outlet.) and of **b** modified calorimeter [19] (*a* pressure transducer; *b* gas injection device; *c* calorimeter; *d* pipette; *e* carbon dioxide reservoir tank)

et al. [19] carried out the last modifications of this technique. A special gas injection device constructed from stainless steel was used in place of the former burette. The mole number of gas introduced into the cell is calculated from the pressure changes after gas injection at known temperature. The exothermic effect when mixing CO_2 and amine solution is measured after successive injections of CO_2. The schematic representation of the experiment is shown Fig. a,b.

Kim and Svendsen [18] at Norwegian University of Science and Technology choose a similar technique for measuring heat of mixing of CO_2 in amine solutions. The main difference with Carson et al. modified calorimeter [19] results in a larger range of experimental temperatures (253–473 K) and pressures (up to 10 MPa) that can be investigated. Their experimental procedure (Fig. 14.3) consists as previously in multiple successive injections of CO_2 up to reach a CO_2 pressure in the gas reservoir tank close to pressure in the calorimeter vessel.

Isothermal Compensation Flow Calorimeter

This dynamic technique used by Oscarson and co-workers [21] at Brigham Young University (USA) to study CO_2 capture studies was developed in same University by Christensen and coworkers [71]. The principle of isothermal compensation calorimeter is based on the measurement of an external heat power used to maintain a reaction vessel at constant temperature during mixing. In flow mode the fluids flow to the

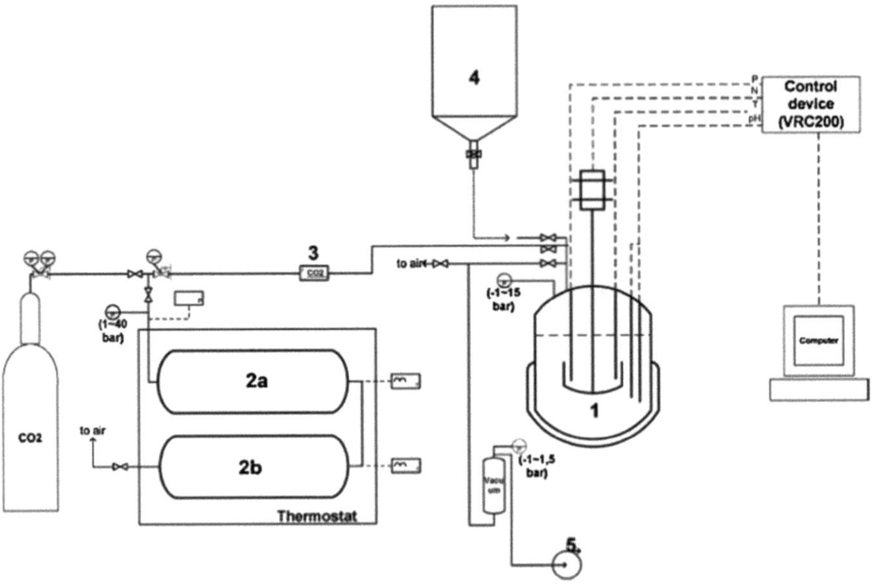

Fig. 14.3 Experimental setup of Kim and Svendsen [18] calorimeter. *1* reaction calorimeter ($2000 cm^3$); *2a, 2b* CO_2 reservoir tanks; *3* CO_2 mass flow controller; *4* amine solution feed bottle; *5* vacuum pump

reaction vessel where they mix. Different types of reaction vessel were proposed as for example that one described by Ott et al. [72] and represented in Fig. 14.4. In this example heaters and mixing tube are wound around and vacuum silver braised to a nickel plate cylinder. The mixing tube is made of hastelloy. Two concentric tubes drive the fluids to the mixing point located at the bottom of the cylinder. The energy generated by the reaction is balanced by adjusting the heater or the Peltier cooler device. Hence the change in heater or cooler power is proportional to the enthalpy change in the calorimeter. The two solutes are injected at defined flow rates by used of two syringe pumps. The pressure is maintained constant using a back pressure regulator. This kind of calorimetric technique can be be adapted to be used at temperature from ambient up to 473 K and at pressures up to 20 MPa.

Heat Conduction Differential Flow Calorimeter

This dynamic technique was initially developed by Mathonat and coworkers at University of Clermont-Ferrand (France). It is a flow technique where the heat of mixing is measured in a mixing cell located inside a Calvet type calorimeter (Fig. 14.5a). The Calvet sensors are a thermopiles constituted of thermocouples surrounding the mixing cell and measuring the heat power exchanged with a thermostated calorimetric block. The mixing cell represented in Fig. 14.5b consists of an hastelloy tube

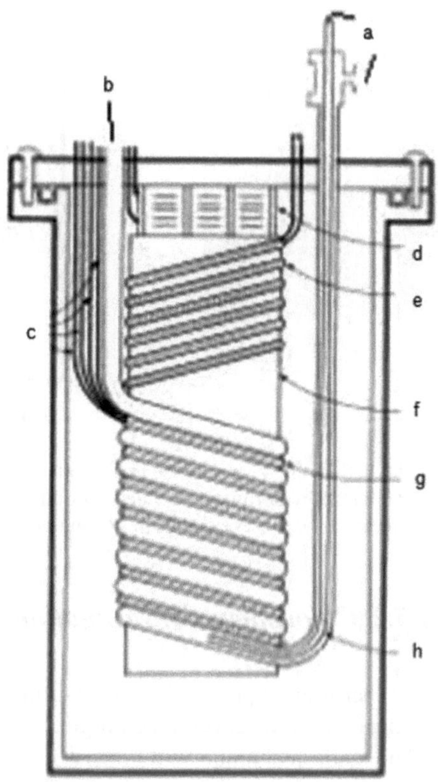

Fig. 14.4 Schematic representation of a reaction vessel [72] of isothermal compensation flow calorimeter. *a* inlets; *b* outlet; *c* electrical leads; *d* Peltier cooler; *e* controlled heater; *f* isothermal cylinder; *g* calibration heater; *h* inside tube

tightly coiled inside a metallic confinement cylinder which fits into the well surrounded by the thermopile detector. For measuring the heat of dissolution of carbon dioxide in an absorbent solution, the fluids are injected at constant flow rates by used of high pressure syringe pumps. The temperature of the fluids to be mixed is controlled by pre-heaters positioned on the incoming tubes before the mixing point.

The fluids encounter following two different mixing technologies schematically represented in Fig. 14.5-c. The first one developed by Mathonat et al. [15] consists in flowing the fluids to be mixed into two concentric tubes; the fluids meet where the thinner tube ends. The second one presented by Arcis et al. [16, 17, 73] consists in two parallel hastelloy tubes silver welded on a T type brass piece represented in Fig. 14.5-c. Experiments are carried out at constant pressure range from 0.1–40 MPa. The temperature of experiment will depend on the calorimeter. Using a commercial Calvet type calorimeters from Setaram, this technique permits measurements from 200 to 573 K.

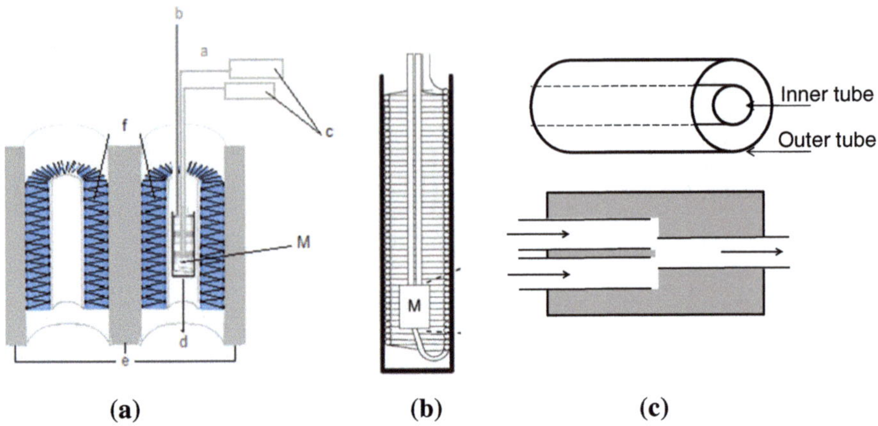

Fig. 14.5 Schematic view of the thermopiles **a**, mixing cell **b**, and mixing points **c** in the Setaram C80 or BT2.15 calorimeters *a* inlets; *b* outlet; *c* syringe pumps; *d* mixing cell; *e* thermostated calorimetric bloc; *f* thermopiles; *M* mixing point

14.5.2 Calorimetric Investigations

The experimental protocol for measuring heat of mixing of gas in liquid absorbent will depend on the chosen technique.

The static technique will consist in additions of known amount of gas in a cell containing the absorbent solution. The heat of mixing is obtained by integration of a peak corresponding to the heat power signal recorded as a function of time during gas dissolution. The end of dissolution is considered as reached when the heat power signal returns to baseline state. In the approximation of small gas addition, the heats of mixing expressed per mole of CO_2 can be assumed to be equal to differential enthalpies of solution H_{diff}. Carson et al [19] measured the enthalpy of solution of CO_2 in aqueous solutions of monoethanolamine (MEA) and methyl diethanolamine (MDEA) at 298.15 K. The experiments were carried out at atmospheric pressure for small CO_2 loading charges. In these conditions the gas is assumed to be totally dissolved in amine solution and no vapor phase was considerate. The method is particularly appropriate for the determination of enthalpy of solution at infinite dilution at atmospheric pressure. The uncertainty is estimated by the authors to $\pm 2\,\mathrm{kJ \cdot mol^{-1}}$. The results represented in Fig. 14.6 show that the enthalpy of solution is almost constant down to $\alpha = 0.05$ and then decreases.

Kim and Svendsen determined the enthalpy of solution of MEA for higher CO_2 loading charges α (Fig. 14.7). The maximum loading charge reported in this work is close to 0.8. The uncertainty on enthalpy of solution is estimated to 2.2 %. It is observed that for loading charge below 0.5 the enthalpy of solution is more or less constant. It is usually admitted that the enthalpy derived from the plateau (Fig. 14.7) corresponds to an enthalpy of solution at infinite dilution. However this behavior must be confirmed by measurements in the domain of small loading charges to con-

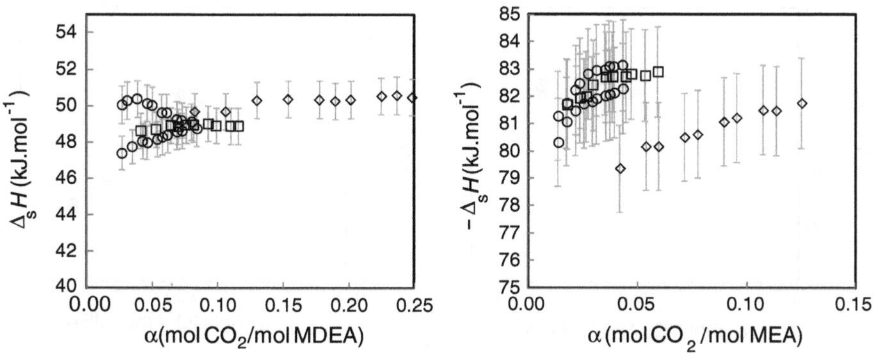

Fig. 14.6 Enthalpy of solution of CO_2 in aqueous solutions of monoethanolamine (MEA) and methyl di-ethanolamine (MDEA) at low loading charges using an isothermal displacement calorimeter [19]. Mass fractions of amine solutions ◇ 0.1, □ 0.2, ○ 0.3

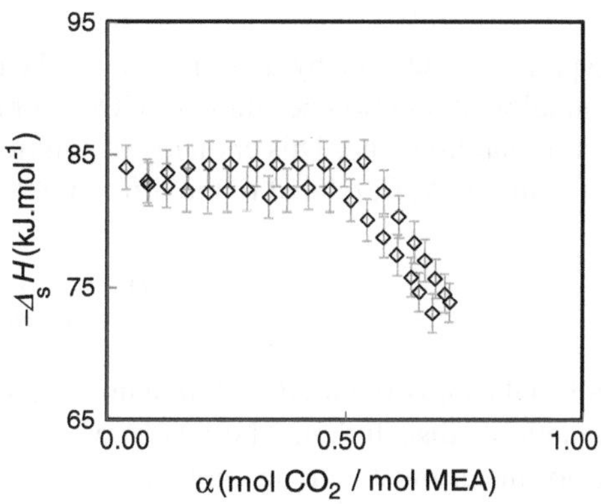

Fig. 14.7 Enthalpy of solution at 313.15 K and 0.3 MPa of CO_2 in aqueous solutions of monoethanolamine (MEA) at mass fraction 0.3, using static technique [18] on a wide range of loading charges α

firm the enthalpy decrease observed by Carson et al. [19]. This domain is particularly difficult to investigate due to problems of detection of small heat effects. The curve representing the enthalpy of solution in Fig. 14.7 shows a split at loading charge $\alpha = 0.5$. As the monoethanolamine (MEA) is a primary amine that forms carbamate with CO_2, the stoechiometric limit of gas absorption should be 0.5. However, when adding more gas, it was observed that the reaction of carbamate formation is displaced to release the amine. Then dissolution of CO_2 can continue with formation of carbonate as it is shown by the speciation curve in Fig. 14.9, representing the solution composition as function of loading charge. The change in the chemical mechanism of CO_2 absorption will modify the enthalpy of solution after loading charge 0.5 (Fig. 14.7).

Using flow techniques (or dynamic technique), the gas and the absorbent solution flow separately in narrow tubes to a mixing point. Then the mixture flows in a common mixing tube located in the sensible part of the calorimeter. The heat power due to gas dissolution is then detected along the mixing tube, by a thermopile in Calvet

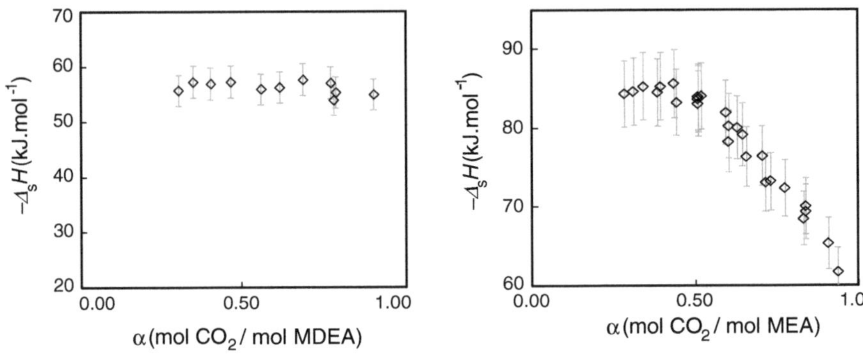

Fig. 14.8 Enthalpy of solution of CO_2 in aqueous solution of MDEA and MEA using dynamic techniques. Results for amine solutions at mass fraction of 0.30, at temperature 322.5 K and pressure 5.17 MPa [41, 74]

type calorimeter or by a compensation heat power in Picker type calorimeter. The enthalpy of solution per mole of CO_2 is obtained by dividing the heat power by the gas molar flow rate. This enthalpy of solution determined here is an integral enthalpy of solution $\Delta_s H$, related to the differential enthalpy following Eq. 14.5

$$\Delta_s H = \frac{1}{\alpha} \int_0^{\alpha} H_{diff} d\alpha. \tag{14.5}$$

Special care is required with dynamic technique to adjust the fluid flow rates to the kinetic of dissolution. The residence time of mixture in the sensible zone of the calorimeter must be long enough in order to reach total dissolution. The volumetric fluid flow rates are usually comprised between 0.05 and $1 \text{mL} \cdot \text{min}^{-1}$. The heats of mixing are measured at variable loading charges by changing fluid (gas/absorbent) flow rates ratios. The experiments are carried out increasing loading charge α up to the saturation of the absorbent solution. Investigation can be carried out at elevated pressure where physical dissolution can be observed.

The graphs reported in Fig. 14.8 represent enthalpy of solution as function of loading charge determined using fluxmetric detection; the graphs obtained with compensation detection will be similar. As initially mentioned, the enthalpies remain nearly constant in the domain of the low loading charges before decreasing when increasing loading charges. With monoethanolamine (MEA), the decrease of the enthalpy of solution starts at loading charge 0.5 corresponding to the change of chemical mechanism of absorption (Fig. 14.9). The speciation in the liquid can be estimated from thermodynamic model (Fig. 14.9). It indicates that significant physical dissolution starts for loading charge about 0.75. This is revealed by the presence of molecular CO_2 in the solution. Then enthalpy decrease is also physical dissolution of CO_2; the energy of dissolution associated to physical mechanisms is lower than the energy of chemical dissolution. On the graph representing the enthalpy of solution of CO_2 in

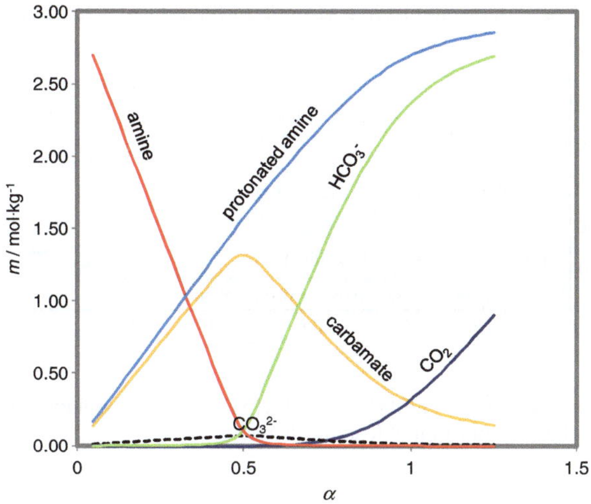

Fig. 14.9 Speciation in aqueous solution of MEA ($3.00\,\mathrm{mol \cdot kg^{-1}}$) as function of CO_2 loading charge. Value calculated at 313.15 K using thermodynamic model [75]

MDEA solution, the enthalpy decrease starts at loading charge about 0.9. The MDEA is a tertiary amine and then does not form carbonate. The chemical mechanism of CO_2 absorption corresponds only to formation of carbonate. However the enthalpy of solution decreases at loading charge about 0.9. This phenomena is observed because of one part of CO_2 is physically dissolved.

The thermodynamic models developed to represent rigorously CO_2 dissolution in aqueous solution of amine consider all possible chemical reactions. Then the system is complex and numerous interaction parameters, molecule-molecule, ion-ion and molecule-ion, have to be defined. These models correlate vapor liquid equilibria data and are used to predict carbon dioxide solubility as a function of its partial pressure in the gas phase, temperature and amine concentration of the absorbent solution. In addition the models can give a detailed composition of solution as function of loading charge α as presented in Fig. 14.9. The model then predicts if gas dissolution leads to formation of hydrogenocarbonates or carbamates but experimental data of the real speciation are very scarce to validate those results. However as the energy of carbamate or carbonate reaction are different, the enthalpy data can provide some indications on the nature of the compounds produced by reaction of CO_2 with the amine. For this purpose thermodynamic models are developed to decompose the total enthalpy of solution into contribution terms associated to the formation of all the different species in solution, such as protonated amine, carbonate, bicarbonate or carbamate. Example of calculation of enthalpy contribution terms is given in Fig. 14.10 for the dissolution at 313.15 K of CO_2 in aqueous solution of MEA of composition $3.00\,\mathrm{mol \cdot kg^{-1}}$. The comparison between the enthalpy of solution obtained by summing the contribution terms and experimental data can then be used to validate the model.

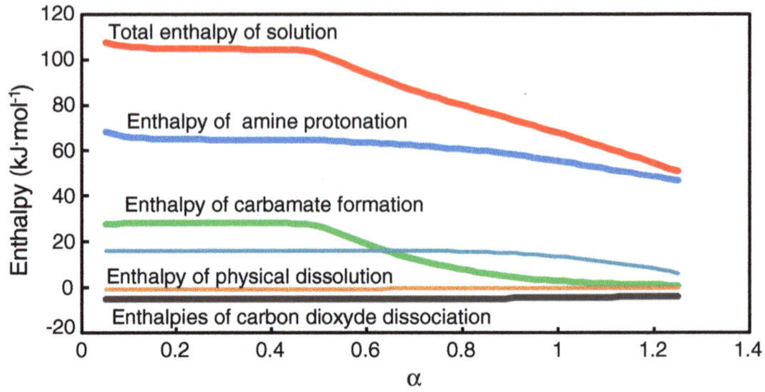

Fig. 14.10 calculated enthalpy contribution at 313.15 K for CO_2 dissolution in aqueous solution of MEA 3.00 mol · kg^{-1} as function of loading charge

Other interest of flow techniques is the short time required to obtained enthalpy of solution at given loading charge α, usually about 30 min. It is then not so much time consuming to investigate a large domain of loading charge, covering the domain of total and partial dissolutions of the CO_2 gas flux. In this way the flow techniques allow a simultaneous determination of the gas solubility at given temperature pressure and absorbent composition. For this purpose the enthalpies are expressed as energy per mole of amine and represented as function of loading charge. The graphic representation then makes it possible to identify the domains of total and partial CO_2 dissolutions. In the domain where the solution is unsaturated (all injected CO_2 is dissolved), the enthalpies expressed per mol of amine (Figs. 14.11 and 14.12) increase with gas loading charge. Over passing the loading charge corresponding to the limit of solubility, the additional CO_2 injected remains in vapor phase and the enthalpy per mole amine remains constant. It corresponds to the apparition of a plateau as showed in Fig. 14.11. The loading charge at the limit of gas solubility corresponds to the point where the plateau is reached. If the mechanism of dissolution is chemical reaction, the increase of enthalpy is almost linear and the intersection between the unsaturated and saturated domain is easily determined (Fig. 14.11-I). This determination is somewhere more difficult when physical mechanism is involved as the enthalpy slope decreases slowly before reaching the plateau (Fig. 14.11-II). In the case of primary amine as MEA (Fig. 14.12) a change in enthalpy slope is observed at loading charge $\alpha = 0.5$, due to the formation of a carbamate and only the points above $\alpha = 0.5$ must be considered to determine the linear part of the unsaturated domain that will cross the plateau. The uncertainties on the value of gas solubility are obviously more important than those obtained from direct pVT techniques. However it represents a simple and easy method for acquisition of solubility data at elevated pressures and temperatures. In addition the comparison of solubility obtained by mixing calorimetry with reference values obtained from pVT techniques will confirm that good mixing and total gas dissolution is achieved in the mixing cell.

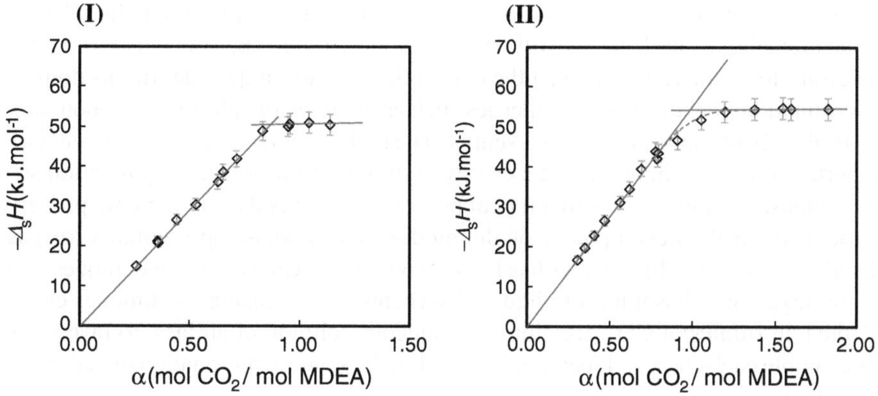

Fig. 14.11 Enthalpy of solution of CO_2 expressed as energy per mole of amine as function of loading charge. Case of gas dissolution at 322 K in aqueous solution of (I) MDEA 30 wt%, p = 0.5 MPa and (II) MDEA 30 wt%, p = 5.17 MPa

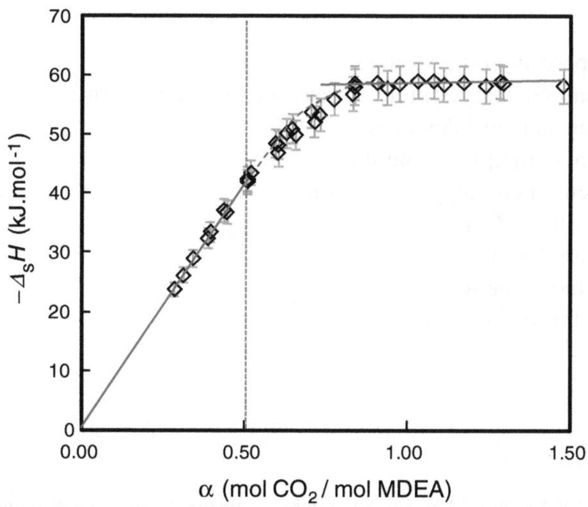

Fig. 14.12 Enthalpy of solution of CO_2 expressed as energy per mole of amine as function of loading charge. Case of gas dissolution in aqueous solution of MEA 30 wt% at 322 K and $p = 5$ MPa

14.6 Conclusion

The environmental impact of CO_2 must be considered and it is urgent to reduce its anthropogenic emission. Solutions have to be proposed for near future. One option is a decarbonation of post combustion effluents. Design and development of future industrial process for CO_2 separation from fumes will require studies of gas dissolution in various selective absorbent solutions.

Aqueous solutions of amine are considered as promising solvent but the selection of the best adapted molecule is still open. Then important experimental work will be necessary to characterize and test the dissolution of carbon dioxide. In this domain, determination of calorimetric properties such as enthalpy of solution of carbon dioxide in absorbent solutions will be essential. The enthalpies of solution are particularly important for the estimation of the energy required for the solvent regeneration step of an industrial process. For theoretical point of view, this thermodynamic property participates to the development of thermodynamic models representative of gas-absorbent systems. This chapter has focused on mixing calorimetric techniques used to investigate gas dissolution in liquid absorbents. The presented techniques can be adapted, in studies of CO_2 dissolution in aqueous solution of amines, to investigate for example each chemical reaction involved such as amine protonation or carbamate formation.

Symbols

wt %: mass percent composition
α: gas loading charge (mole of gas / moles of absorbent)
ASD: average standard deviation
$\Delta_s H$: integral enthalpy of solution
H_{diff}: differential enthalpy of solution
MEA: mono-ethanolamine
DEA: di-ethanolamine
AMP: 2-Amino-2-methyl-1-propanol
MDEA: methyl di-ethanolamine

References

1. M.E. Mann, R.S. Bradley, M.K. Hughes, Global-scale temperature patterns and climate forcing over the past six centuries. Nature **392**(6678), 779–787 (1998)
2. N. Thybaud, D. Lebain (2010) Panorama des voies de valorisation du CO_2. ADEME.
3. D.M. Ruthven, S. Farooq, K.S. Knaebel, *Pressure Swing Adsorption* (VCH, New York, 1994)
4. E.S. Kikkinides, R.T. Yang, S.H. Cho, Concentration and recovery of carbon dioxide from flue gas by pressure swing adsorption. Ind. Eng. Chem. Res. **32**(11), 2714–2720 (1993)
5. U. Kragl, Book review: basic principles of membrane technology. (Second edition) By M. Mulder. Angew. Chem. Int. Ed. Engl. **36**(19), 2129–2130 (1997)
6. R. Klaassen, P.H.M. Feron, R. van der Vaart, A.E. Jansen, B. Dibakar, D.A. Butterfield (2003) Chapter 6 Industrial applications and opportunities for membrane contactors. Membr. Sci. Technol. **8**, 125–145 (Elsevier)
7. D. Tondeur, F. Teng, M.L. Trevor, *Carbon Capture and Storage for Greenhouse Effect Mitigation Future Energy* (Elsevier, Oxford, 2008)
8. A. Fezoua, A. Bouchemoua-Benaissa, Y. Ouabbas, F. Chauvy, A. Cameirão, J-M Herri, CO_2 capture by gas hydrate crystallization. in *S4FE2009 (International Conference on Sustainable Fossil Fuels for Future Energy)*, Rome, Italy, 2009

9. N. Favre, M.L. Christ, A.C. Pierre, Biocatalytic capture of CO_2 with carbonic anhydrase and its transformation to solid carbonate. J. Mol. Catal. B Enzym. **60**(3–4), 163–170 (2009)
10. D. Wappel, G. Gronald, R. Kalb, J. Draxler, Ionic liquids for post-combustion CO_2 absorption. Int. J. Greenhouse Gas Control **4**(3), 486–494
11. B. Gwinner, D. Roizard, F. Lapicque, E. Favre, R. Cadours, P. Boucot, P.-L. Carrette, CO_2 capture in flue gas: semiempirical approach to select a potential physical solvent. Ind. Eng. Chem. Res. **45**(14), 5044–5049 (2006)
12. G.F. Versteeg, L.A.J. Van Dijck, W.P.M. Van Swaaij, On the kinetics between CO_2 and alkanolamines both in aqueous and non-aqueous solutions. An overview. Chem. Eng. Commun. **144**(1), 113–158 (1996)
13. P.-A. Bouillon, M. Jacquin, L. Raynal (2012) Method for deacidifying a gas by an absorbing solution with demixing control. FRANCE Patent EP 2 168 659 B1
14. C. Mathonat, V. Majer, A.E. Mather, J.-P.E. Grolier, Use of flow calorimetry for determining enthalpies of absorption and the solubility of CO_2 in aqueous monoethanolamine solutions. Ind. Eng. Chem. Res. **37**(10), 4136–4141 (1998)
15. C. Mathonat, V. Majer, A.E. Mather, J.-P.E. Grolier, Enthalpies of absorption and solubility of CO_2 in aqueous solutions of methyldiethanolamine. Fluid Phase Equilib. **140**(1–2), 171–182 (1997)
16. H. Arcis, L. Rodier, K. Ballerat-Busserolles, J.Y. Coxam, Enthalpy of solution of CO_2 in aqueous solutions of methyldiethanolamine at $T = 372.9$ K and pressures up to 5 MPa. J. Chem. Thermodyn. **41**(7), 836–841 (2009)
17. H. Arcis, L. Rodier, K. Ballerat-Busserolles, J.Y. Coxam, Enthalpy of solution of CO_2 in aqueous solutions of methyldiethanolamine at $T = 322.5$ K and pressure up to 5 MPa. J. Chem. Thermodyn. **40**(6), 1022–1029 (2008)
18. I. Kim, H.F. Svendsen, Heat of absorption of carbon dioxide (CO_2) in Monoethanolamine (MEA) and 2-(Aminoethyl)ethanolamine (AEEA) solutions. Ind. Eng. Chem. Res. **46**(17), 5803–5809 (2007)
19. J.K. Carson, K.N. Marsh, A.E. Mather, Enthalpy of solution of carbon dioxide in (water + monoethanolamine, or diethanolamine, orN-methyldiethanolamine) and (water + monoethanolamine + N-methyldiethanolamine) at $T = 298.15$ K. J. Chem. Thermodyn. **32**(9), 1285–1296 (2000)
20. H. Kierzkowska-Pawlak, R. Zarzycki, Calorimetric measurements of CO_2 absorption into aqueous N -methyldiethanolamine solutions. Chem. Pap. **56**, (2002)
21. J.L. Oscarson, X. Chen, R.M. Izatt (1995) A thermodynamically consistent model for the prediction of solubilities and enthalpies of solution of acid gases in aqueous alkanolamine solutions. Gas Processors Association, Research, Report, vol 130
22. K.E. Merkley, J.J. Christensen, R.M. Izatt (1986) Enthalpies of solution of CO_2 in aqueous methyldiethanolamine solutions. Gas Processors Association, Research, Report, vol 102
23. R. Helton, J.J. Christensen, R.M. Izatt (1987) Enthalpies of solution of CO_2 in aqueous diethanolamine solutions. Gas Processors Association, Research, Report, vol 108
24. H. Arcis, L. Rodier, K. Ballerat-Busserolles, J.Y. Coxam, Modeling of (vapor+ liquid) equilibrium and enthalpy of solution of carbon dioxide (CO_2) in aqueous methyldiethanolamine (MDEA) solutions. J. Chem. Thermodyn. **41**(6), 783–789 (2009)
25. I. Kim, K.A. Hoff, E.T. Hessen, T. Haug-Warberg, H.F. Svendsen, Enthalpy of absorption of CO_2 with alkanolamine solutions predicted from reaction equilibrium constants. Chem. Eng. Sci. **64**(9), 2027–2038 (2009)
26. J.I. Lee, F.D. Otto, A.E. Mather, Measurement and prediction of solubility of mixtures of carbon-dioxide and hydrogen-sulfide in a 2.5 N monoethanolamine solution. Can. J. Chem. Eng. **54**(3), 214–219 (1976)
27. G. Kuranov, B. Rumpf, N.A. Smirnova, G. Maurer, Solubility of single gases carbon dioxide and hydrogen sulfide in aqueous solutions of N-methyldiethanolamine in the temperature range 313 – 413 K at pressures up to 5 MPa. Ind. Eng. Chem. Res. **35**(6), 1959–1966 (1996)
28. J.W. Mason, B.F. Dodge, Equilibrium absorption of carbon dioxide by solutions of the ethanolamines. Trans AIChE **32**, 27–48 (1936)

29. J.H. Jones, H.R. Froning, E.E. Claytor, Solubility of acidic gases in aqueous monoethanolamine. J. Chem. Eng. Data **4**(1), 85–92 (1959)
30. J.I. Lee, F.D. Otto, A.E. Mather, Solubility of carbon dioxide in aqueous diethanolamine solutions at high pressures. J. Chem. Eng. Data **17**(4), 465–468 (1972)
31. J.I. Lee, F.D. Otto, A.E. Mather, Equilibrium between carbon dioxide and aqueous monoethanolamine solutions. J. Appl. Chem. Biotechnol. **26**(1), 541–549 (1976)
32. J.D. Lawson, A.W. Garst, Gas sweetening data: equilibrium solubility of hydrogen sulfide and carbon dioxide in aqueous monoethanolamine and aqueous diethanolamine solutions. J. Chem. Eng. Data **21**(1), 20–30 (1976)
33. E.E. Isaacs, F.D. Otto, A.E. Mather, Solubility of mixtures of hydrogen sulfide and carbon dioxide in a monoethanolamine solution at low partial pressures. J. Chem. Eng. Data **25**(2), 118–120 (1980)
34. D.M. Austgen, G.T. Rochelle, C.C. Chen, Model of vapor-liquid equilibria for aqueous acid gas-alkanolamine systems. 2. Representation of hydrogen sulfide and carbon dioxide solubility in aqueous MDEA and carbon dioxide solubility in aqueous mixtures of MDEA with MEA or DEA. Ind. Eng. Chem. Res. **30**(3), 543–555 (1991)
35. K.P. Shen, M.H. Li, Solubility of carbon dioxide in aqueous mixtures of monoethanolamine with methyldiethanolamine. J. Chem. Eng. Data **37**(1), 96–100 (1992)
36. O.F. Dawodu, A. Meisen, Solubility of carbon dioxide in aqueous mixtures of alkanolamines. J. Chem. Eng. Data **39**(3), 548–552 (1994)
37. F.Y. Jou, A.E. Mather, F.D. Otto, The solubility of CO_2 in A 30-mass-percent monoethanolamine solution. Can. J. Chem. Eng. **73**(1), 140–147 (1995)
38. J.H. Song, J.H. Yoon, H. Lee, K.H. Lee, Solubility of carbon dioxide in monoethanolamine + ethylene glycol + water and monoethanolamine + poly(ethylene glycol) + water. J. Chem. Eng. Data **41**(3), 497–499 (1996)
39. I.S. Jane, M.H. Li, Solubilities of mixtures of carbon dioxide and hydrogen sulfide in water + diethanolamine + 2-amino-2-methyl-1-propanol. J. Chem. Eng. Data **42**(1), 98–105 (1997)
40. S. Ma'mun, R. Nilsen, H.F. Svendsen, O. Juliussen, Solubility of carbon dioxide in 30 mass % monoethanolamine and 50 mass % methyldiethanolamine solutions. J. Chem. Eng. Data **50**(2), 630–634 (2005)
41. H. Arcis, K. Ballerat-Busserolles, L. Rodier, J.Y. Coxam, Enthalpy of solution of carbon dioxide in aqueous solutions of monoethanolamine at temperatures of 322.5 K and 372.9 K and pressures up to 5 MPa. J. Chem. Eng. Data **56**(8), 3351–3362 (2011)
42. F.Y. Jou, A.E. Mather, F.D. Otto, Solubility of hydrogen sulfide and carbon dioxide in aqueous methyldiethanolamine solutions. Ind. Eng. Chem. Process Des. Dev. **21**(4), 539–544 (1982)
43. A. Chakma, A. Meisen, Solubility of carbon dioxide in aqueous methyldiethanolamine and N, N-bis(hydroxyethyl)piperazine solutions. Ind. Eng. Chem. Res. **26**(12), 2461–2466 (1987)
44. F.Y. Jou, J.J. Carroll, A.E. Mather, F.D. Otto, The solubility of carbon-dioxide and hydrogen-sulfide in a 35 wt-percent aqueous-solution of methyldiethanolamine. Can. J. Chem. Eng. **71**(2), 264–268 (1993)
45. G.W. Xu, C.F. Zhang, S.J. Qin, W.H. Gao, H.B. Liu, Gas-liquid equilibrium in a CO_2-MDEA-H_2O system and the effect of piperazine on it. Ind. Eng. Chem. Res. **37**(4), 1473–1477 (1998)
46. S.W. Rho, K.P. Yoo, J.S. Lee, S.C. Nam, J.E. Son, B.M. Min, Solubility of CO_2 in aqueous methyldiethanolamine solutions. J. Chem. Eng. Data **42**(6), 1161–1164 (1997)
47. D. Silkenbaumer, B. Rumpf, R.N. Lichtenthaler, Solubility of carbon dioxide in aqueous solutions of 2-amino-2-methyl-1-propanol and N-methyldiethanolamine and their mixtures in the temperature range from 313 to 353 K and pressures up to 2.7 MPa. Ind. Eng. Chem. Res. **37**(8), 3133–3141 (1998)
48. J.I. Baek, J.H. Yoon, Solubility of carbon dioxide in aqueous solutions of 2-amino-2-methyl-1,3-propanediol. J. Chem. Eng. Data **43**(4), 635–637 (1998)
49. W.J. Rogers, J.A. Bullin, R.R. Davison, FTIR measurements of acid-gas-methyldiethanolamine systems. AIChE J. **44**(11), 2423–2430 (1998)
50. B. Lemoine, Y.G. Li, R. Cadours, C. Bouallou, D. Richon, Partial vapor pressure of CO_2 and H2S over aqueous methyldiethanolamine solutions. Fluid Phase Equilib. **172**(2), 261–277 (2000)

51. M.A. Pacheco, S. Kaganoi, G.T. Rochelle, CO_2 absorption into aqueous mixtures of diglycolamine and methyldiethanolamine. Chem. Eng. Sci. **55**(21), 5125–5140 (2000)
52. A.P.-S. Kamps, A. Balaban, M. Jodecke, G. Kuranov, N.A. Smirnova, G. Maurer, Solubility of single gases carbon dioxide and hydrogen sulfide in aqueous solutions of N-methyldiethanolamine at temperatures from 313 to 393 K and pressures up to 7.6 MPa: new experimental data and model extension. Ind. Eng. Chem. Res. **40**(2), 696–706 (2000)
53. M.K. Park, O.C. Sandall, Solubility of carbon dioxide and nitrous oxide in 50 mass methyldiethanolamine. J. Chem. Eng. Data **46**(1), 166–168 (2000)
54. S. Bishnoi, G.T. Rochelle, Thermodynamics of piperazine/methyldiethanolamine/water/carbon dioxide. Ind. Eng. Chem. Res. **41**(3), 604–612 (2002)
55. R. Sidi-Boumedine, S. Horstmann, K. Fischer, E. Provost, W. Fürst, J. Gmehling, Experimental determination of carbon dioxide solubility data in aqueous alkanolamine solutions. Fluid Phase Equilib. **218**(1), 85–94 (2004)
56. B.S. Ali, M.K. Aroua, Effect of piperazine on CO_2 loading in aqueous solutions of MDEA at low pressure. Int. J. Thermophys. **25**(6), 1863–1870 (2004)
57. M. Kundu, S.S. Bandyopadhyay, Modelling vapour—liquid equilibrium of CO_2 in aqueous N-methyldiethanolamine through the simulate annealing algorithm. Can. J. Chem. Eng. **83**(2), 344–353 (2005)
58. A. Benamor, M.K. Aroua, Modeling of CO_2 solubility and carbamate concentration in DEA, MDEA and their mixtures using the Deshmukh-Mather model. Fluid Phase Equilib. **231**(2), 150–162 (2005)
59. M.H. Jenab,M. bedinzadegan Abdi, S.H. Najibi, M. Vahidi, N.S. Matin, Solubility of carbon dioxide in aqueous mixtures of N-methyldiethanolamine + piperazine + sulfolane. J. Chem. Eng. Data **50**(2), 583–586 (2004)
60. P.J.G. Huttenhuis, N.J. Agrawal, J.A. Hogendoorn, G.F. Versteeg, Gas solubility of H2S and CO_2 in aqueous solutions of N-methyldiethanolamine. J. Pet. Sci. Eng. **55**(1–2), 122–134 (2007)
61. M.H. Oyevaar, H.J. Fontein, K.R. Westerterp, Equilibria of carbon dioxide in solutions of diethanolamine in aqueous ethylene glycol at 298 K. J. Chem. Eng. Data **34**(4), 405–408 (1989)
62. M.L. Kennard, A. Meisen, Solubility of carbon dioxide in aqueous diethanolamine solutions at elevated temperatures and pressures. J. Chem. Eng. Data **29**(3), 309–312 (1984)
63. D. Lal, F.D. Otto, A.E. Mather, The solubility of H2S and CO_2 in A diethanolamine solution at low partial pressures. Can. J. Chem. Eng. **63**(4), 681–685 (1985)
64. J.I. Lee, F.D. Otto, A.E. Mather, Solubility of mixtures of carbon-dioxide and hydrogen-sulfide in aqueous diethanolamine solutions. Can. J. Chem. Eng. **52**(1), 125–127 (1974)
65. D.J. Seo, W.H. Hong, Solubilities of carbon dioxide in aqueous mixtures of diethanolamine and 2-amino-2-methyl-1-propanol. J. Chem. Eng. Data **41**(2), 258–260 (1996)
66. M.Z. Haji-Sulaiman, M.K. Aroua, A. Benamor, Analysis of equilibrium data of CO_2 in aqueous solutions of diethanolamine (DEA), methyldiethanolamine (MDEA) and their mixtures using the modified Kent Eisenberg model. Chem. Eng. Res. Des. **76**(8), 961–968 (1998)
67. M.Z. Haji-Sulaiman, M.K. Aroua, Equilibrium of CO_2 in aqueous diethanolamine (DEA) and amino methyl propanol (AMP) solutions. Chem. Eng. Commun. **140**(1), 157–171 (1994)
68. H. Arcis, K. Ballerat-Busserolles, L. Rodier, J.-Y. Coxam, Measurement and modeling of enthalpy of solution of carbon dioxide in aqueous solutions of diethanolamine at temperatures of (322.5 and 372.9) K and pressures up to 3 MPa. J. Chem. Eng. Data **57**(3), 840–855 (2012)
69. R.H. Stokes, K.N. Marsh, R.P. Tomlins, An isothermal displacement calorimeter for endothermic enthalpies of mixing. J. Chem. Thermodyn. **1**(2), 211–221 (1969)
70. R. Battino, K. Marsh, An isothermal displacement calorimeter for the measurement of the enthalpy of solution of gases. Aust. J. Chem. **33**(9), 1997–2003 (1980)
71. J.J. Christensen, L.D. Hansen, D.J. Eatough, R.M. Izatt, R.M. Hart, Isothermal high pressure flow calorimeter. Rev. Sci. Instrum. **47**(6), 730–734 (1976)
72. J.B. Ott, C.E. Stouffer, G.V. Cornett, B.F. Woodfield, R.C. Wirthlin, J.J. Christensen, U.K. Deiters, Excess enthalpies for (ethanol + water) at 298.15 K and pressures of 0.4, 5, 10, and 15 MPa. J. Chem. Thermodyn. **18**(1), 1–12 (1986)

73. H. Arcis, L. Rodier, J.-Y. Coxam, Enthalpy of solution of CO_2 in aqueous solutions of 2-amino-2-methyl-1-propanol. J. Chem. Thermodyn. **39**(6), 878–887 (2007)
74. H. Arcis, L. Rodier, K. Ballerat-Busserolles, J.Y. Coxam, Enthalpy of solution of CO(2) in aqueous solutions of methyldiethanolamine at T=322.5 K and pressure up to 5 MPa. J. Chem. Thermodyn. **40**(6), 1022–1029 (2008)
75. H. Arcis, *Etude thermodynamique de la dissolution du dioxyde de carbone dans des solutions aqueuses d'alcanolamines* (Université Blaise Pascal, Clermont-ferrand, 2008)

Chapter 15
Adsorption Microcalorimetry, IR Spectroscopy and Molecular Modelling in Surface Studies

Vera Bolis

Abstract The advantage in coupling adsorption microcalorimetry with IR spectroscopy and/or *ab initio* modelling in surface studies is illustrated by a selection of examples dealing with metal oxides and silica-based (either non porous or microporous) materials. Correlations between thermodynamic and vibrational parameters are illustrated for the adsorption of CO and of NH_3, employed as probe molecules for studying the Lewis/Brønsted acidity of the investigated materials. Surface reconstruction processes, responsible for endothermic effects, are invoked to interpret the unexpectedly low heat measured in the calorimetric cell for the high-coverage adsorption of CO on transition aluminas, the adsorption of NH_3 on a defective all-silica zeolite and the adsorption of H_2O on an amorphous alumino-silicate.

15.1 Introduction

Heat evolved when (reactive) molecules contact the surface of a solid material is related to the molecule/surface site bonding energy. By processing the overall set of volumetric-calorimetric data at increasing equilibrium pressure, both magnitude of the enthalpy of adsorption and its evolution with increasing coverage are obtained. In such a way the possible heterogeneity of the surface (structural, chemical and/or induced) is quantitatively described.

The volumetric-calorimetric data obtained by a Tian-Calvet heat-flow microcalorimeter connected to a gas-volumetric apparatus (as the one described in Chap. 1) being intrinsically molar quantities, their molecular interpretation often requires a multi-techniques approach.

V. Bolis (✉)
Dipartimento di Chimica and NIS Centre of Excellence, Università di Torino,
Via Pietro Giuria 7, 10125 Torino, Italy
e-mail: vera.bolis@gmail.com

A. Auroux (ed.), *Calorimetry and Thermal Methods in Catalysis*,
Springer Series in Materials Science 154, DOI: 10.1007/978-3-642-11954-5_15,

Adsorption microcalorimetry is suitably and fruitfully coupled with complementary techniques, in particular with the ones which are intrinsically apt to detect the nature of both the pristine surface sites and the new terminations created by the chemical interaction of the (probe) molecules with the surface sites [1–3]. IR spectroscopy is an ideal technique for this purpose [4, 5]. On the other hand, the increasing amount of computational *ab initio* results on adspecies formed at simulated model sites is ideally suited to be compared with experimental data from microcalorimetry [6–10].

IR spectroscopy experiments and/or *ab initio* calculations being strategically planned and performed in parallel with the microcalorimetric measurements allow to describe the surface features of a solid material in an almost exhaustive way (nature/structure, population and strength of the surface sites) [11–13].

The combined use of adsorption microcalorimetry, IR spectroscopy and/or *ab initio* calculations will be illustrated for a selection of non-dissociative chemical adsorption cases. In the reported examples, no new surface terminations were originated during the adsorption: only a perturbation of the vibrational features of the adsorptive when adsorbed at the surface was so detected by IR spectroscopy. The experimental details of the microcalorimetric measurements, including the description of the investigated materials, were reported in Chap. 1.

15.2 CO Adsorbed on Coordinatively Unsaturated Metal Cations

The adsorption of carbon monoxide has been widely used over the years in studies aimed at characterizing the surface acidity of coordinatively unsaturated (*cus*) Me^{n+} cations. The *cus* metal cations, either exposed at the dehydrated surface of oxidic materials or located within the dehydrated zeolite nanocavities as charge-balancing cations, are known to act as Lewis acidic sites of variable strength [14].

On non d/d^0 *cus* metal cations, for which any d-electrons π-back-donation is prevented, CO is either simply polarized by the local electric field generated by the positive charge of the *cus* metal cation, or is σ-coordinated to the *cus* metal cation through the C – end lone pair [1–15]. In the former case weak electrostatic adducts, in the latter case reversible chemisorbed species of variable stability are formed. In both electrostatic and σ-coordinated adducts, the C–O stretching frequency of the adsorbed molecule is upwards shifted with respect to the stretching frequency of the free molecule in the gas phase ($\nu_{COads} > \nu_{COgas} = 2143\,\mathrm{cm}^{-1}$). It has been observed that the extent of the C–O stretching frequency shift, defined as $\Delta\nu_{CO} = \nu_{COads} - \nu_{COgas}$, is a measure of the extent of the lone pair donation to the *cus* cations and so of their electron-accepting properties [16–18]. In this respect, $\Delta\nu_{CO}$ can be taken as a measure of the acidic strength of the *cus* cations, and is correlated to the enthalpy of adsorption, which directly measures the strength of the $Me^{n+} \leftarrow CO$ bond [4, 15, 19–21].

In the case of d-block *cus* metal cations, the strength of the σ-coordinative bond in the carbonyl-like species is reinforced by the π-back donation of d electrons. The actual spectral position of the ν_{CO} band turns out to be a compromise between the upward shift due to the polarization/σ-coordination and the downwards shift due to the π-back donation [1, 22]. As a consequence, the stretching frequency of adsorbed CO can arrive to suffer a downwards shift with respect to the free molecule ($\nu_{CO ads} < \nu_{CO gas}$).

In Fig. 15.1 the possible *cus* metal cation/CO interactions are schematically illustrated. In Fig. 15.1a the plain electrostatic polarization/σ-coordination of the molecule on non d/d^0 *cus* metal cations, implying an upwards shift of the C–O stretching frequency with respect to the free molecule, is reported. In Fig. 15.1b the σ-coordination + π-back donation on d-block *cus* metal cations, for which the C−O stretching frequency can be either slightly upward shifted or downward shifted with respect to the free molecule, according to the weight of the π-back donation contribution with respect to the electrostatic/σ-coordination contribution.

In Fig. 15.2 the shift of the CO stretching frequency ($\Delta\nu_{co}$) is reported as a function of the negative enthalpy of adsorption ($-\Delta_a H$) for CO adsorbed on *cus* Me^{n+} cations (either non d, d^0 or d in nature). The reported plot summarizes a large number of experimental data. $\Delta\nu_{co}$ and $-\Delta_a H$ values were obtained through parallel experiments carried out in different laboratories (see ref. [1] for details on the samples and on the experimental conditions). The reported data will be discussed for CO adsorbed at vanishing coverage on: (i) non-d/d^0 metal cations (circle), (ii) copper and silver metal carbonyls (square and diamonds, respectively), and for CO adsorbed at high coverage on: (iii) transition catalytic alumina (star). Note that series (i) includes also a d^{10} metal cation (*cus* Zn^{2+}) which was found to behave similarly to the non-d metal cations, both when hosted in Y-zeolite cages or when exposed at the surface of dehydrated ZnO [23].

As far as non-d/d^0 metal cations are concerned, a linear correlation between the spectroscopic and thermodynamic quantities does exist (left-top side of the figure), resulting in the Eq. 15.1:

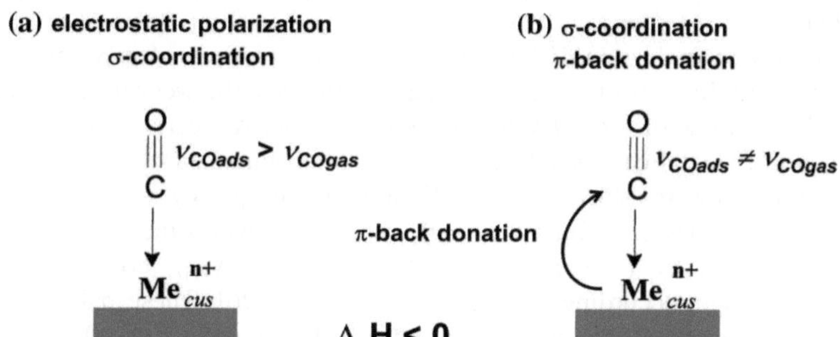

Fig. 15.1 The possible *cus* metal cation/CO interactions are schematically illustrated. **a** electrostatic polarization/σ-coordination (non d/d^0 metal cations); **b** σ-coordination + π-back donation of d electrons (d-block metal cations)

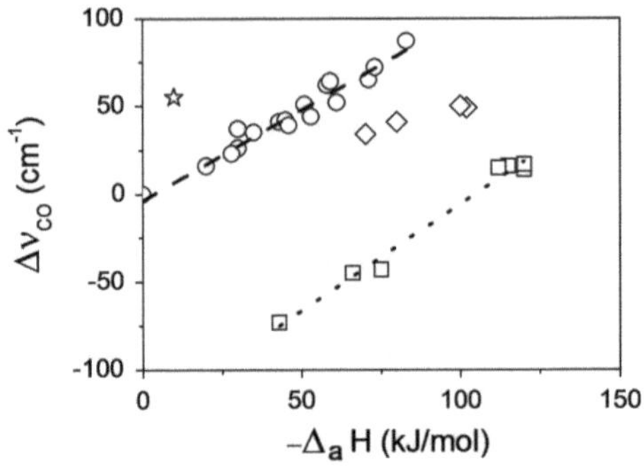

Fig. 15.2 $\Delta\nu_{co}$ *versus* $-\Delta_a H$ for CO adsorbed at vanishing coverage on *cus* metal cations exposed at the surface of either microporous or non-porous dehydrated systems (see Table 3 in ref. [1] for details). *Circle* non-d/d^0 metal cations; *square* Cu-carbonyls; *diamond* Ag-carbonyls; *star* high CO coverage on transition catalytic alumina. Adapted from ref. [1] Fig. 6

$$\Delta\nu_{CO} = \left[(1.03 \pm 0.05)\,\text{cm}^{-1}\,\text{mol}\,(\text{kJ})^{-1}\right]|\Delta_a H| + (-4 \pm 3)\,\text{cm}^{-1} \quad (15.1)$$

with $R = 0.981$ [1]. The observed linearity for non-d/d^0 metal cations in Fig. 15.2 arises from a statistically significant number of pairs of experimental data, so that Eq. 15.1 can be considered as an empirical rule of general validity for all non-d/d^0 CO complexes.

In the case of d-block metal cations, a lack of correlation with the Eq. 15.1 empirical rule is expected as a consequence of the π-back donation of d-electrons. It was already mentioned that the presence of a π-back-donation contribution does reinforce the strength of the carbonyl bond and does cause a downwards shift of the stretching frequency [22, 24, 25]. This was confirmed by plotting the pairs of IR spectroscopic and microcalorimetric data for Cu- and Ag-carbonyls formed either within the zeolite nanopores or at the surface of metal oxides. The reinforcement of the carbonyl bond was witnessed by: (i) the large heat of adsorption ($q_0 > 100\,\text{kJ mol}^{-1}$) measured for both for Cu- and Ag- carbonyls (see Chap. 1, Fig. 1.14) and (ii) the partial irreversibility of the adsorption upon evacuation of pressure (see Chap. 1, Figs. 1.10 and 1.11).

A deviation from the non-d/d^0 metal cations linear plot was clearly observed for both copper and silver species, with remarkable differences between the two *group 11* metals. The increase of $-\Delta_a H$ values for both Cu- and Ag- carbonyls with respect to the non-d/d^0 metal cations adspecies was not accompanied by a correspondent increase of the upwards shift of the C – O stretching frequency.

The formation of well-defined $[\text{Cu (CO)}]^+$ complexes within the zeolite nanopores was characterized by $(-\Delta_a H)_0 \approx 120\,\text{kJ mol}^{-1}$ and $(\Delta\nu_{co})$ comprised in the 16–14 cm^{-1} range, according to different authors [1, 26]. These pairs of values well agreed with those obtained for CO adsorbed on $Cu/SiO_2{\cdot}Al_2O_3$($-\Delta_a H = 115 - 112\,\text{kJ mol}^{-1}$ and $\Delta\nu_{co} = 16 - 15\,\text{cm}^{-1}$), as reported in ref. [27].

The low-enthalpy values for Cu-carbonyls reported in Fig. 15.2 refer to $\Delta\nu_{co}$ and $-\Delta_a H$ pairs obtained for partially reduced Cu- species. In one case CO was adsorbed on a Cu(I)-MFI specimen, in which Cu(I) sites were already engaged with NH_3 ligands, so yielding a mixed amino-carbonyl $[Cu\,(NH_3)\,CO]^+$ complex as illustrated in ref. [1]. The zero-coverage enthalpy change of $\approx$80 kJ mol^{-1} indicated a weakening of the CO/$\left[Cu\,(NH_3)_n\right]^+$ interaction with respect to the CO/Cu(I) one. The CO stretching frequency for the amino-carbonyl complex was negative ($\Delta\nu_{co} = -43\,cm^{-1}$), confirming the major change in the amino complex Cu/CO interaction with respect to that of the bare Cu(I) cations. In fact, the actual charge density of copper cations in $\left[Cu\,(NH_3)_n\right]^+$ was much lower than that of the pristine *cus* Cu(I) cations, owing to the presence of the charge-releasing NH_3 ligands. This was clearly demonstrated by XANES spectroscopy, as reported in ref. [1, 28]. In the other reported cases, CO was adsorbed on Cu-species grafted on a non-porous ZnO matrix. Cu- species were reduced by thermal treatments in H_2, either partially in the $Cu(\delta+)/ZnO$ sample or totally in the $Cu(0)/ZnO$ one, as reported in ref. [29]. For both partially and totally reduced Cu- species, the stretching frequency shift was negative ($\Delta\nu_{co} = -45$ and $-74\,cm^{-1}$, respectively) and the enthalpy change was significantly lower ($-\Delta_a H$ $\approx$66 and 43 kJ mol^{-1}, respectively) than that for copper sites located within the zeolite nanocavities, and characterized by a well-defined oxidation state of the metal.

By plotting $\Delta\nu_{co}$ against $-\Delta_a H$ values for the adsorption of CO on the various Cu-sites, as illustrated in Fig. 15.2, it is rather evident that the spectroscopic and thermodynamic parameters are once more linearly correlated, as witnessed by the Eq. 15.2:

$$\Delta\nu_{CO} = \left[(1.21 \pm 0.07)\,cm^{-1}mol\,(kJ)^{-1}\right]|\Delta_a H| + (-127 \pm 6)\,cm^{-1} \quad (15.2)$$

with $R = 0.993$ [1]. The linear decrease of $\Delta\nu_{co}$ with decreasing $-\Delta_a H$ by moving from $\left[Cu\,(CO)_2\right]^+$ to $\left[Cu\,(NH_3)_n\,(CO)\right]^+$ complexes, and down to both $Cu(\delta+)/ZnO$ and $Cu(0)/ZnO$ species was interpreted as due to the decrease of the actual charge of copper cations which brought about a progressive extinguishment of the carbonyl bond electrostatic component. This result strongly suggests the major role played by the electrostatic component in the carbonyl-like bonds.

By the inspection of Ag/CO data reported in Fig. 15.2, it is rather evident that the deviation from the Eq. 15.1 empirical line for Ag-carbonyls was not as large as for Cu- carbonyls. This datum indicates that the π-back-donation contribution was lower for the former than for the latter. In fact, the zero-coverage enthalpy of formation for Ag(I)-carbonyl ($\approx$100 kJ mol^{-1}) was lower than for Cu(I)-carbonyl ($\approx$120 kJ mol^{-1}), as reported in ref. [1] and pointed out in Chap. 1 Fig. 1.14. As for the stretching frequency shift, the upwards shift measured for Ag(I)-carbonyls was $\Delta\nu_{co} = 50\,cm^{-1}$, much larger than that measured for Cu(I)-carbonyls ($16 - 14\,cm^{-1}$), confirming the π-back donation minor contribution in the Ag/CO bond. Data for CO adsorbed on Ag(I) sites dispersed at the $SiO_2{\cdot}Al_2O_3$ surface were in fairly good agreement with those for Ag(I)-MFI, as it was reported in ref. [27]. Conversely, a lower enthalpy change was measured for Ag sites characterized by silver species in a not-well defined oxidation state, either hosted within a zeolite nanopores (≈80 kJ mol^{-1}) or dispersed at the SiO_2 surface (≈70 kJ mol^{-1}). The correspondent stretching frequency shift ($\Delta\nu_{CO} = 34\,cm^{-1}$ for the former and 41 cm^{-1} for the latter) was lower than for Ag(I)-MFI (50 cm^{-1}) [27].

In the silver case, owing to the limited set of partially reduced Ag-carbonyls data, it was not possible to draw any reasonable correlation. A clear trend was however observed: once more as far as the electrostatic component contribution decreases, as a consequence of the reduced actual charge of the metal cation, the strength of the interaction decreases.

By a further inspection of Fig. 15.2, it turns out that the pair of $-\Delta_a H$ and $\Delta\nu_{co}$ values for CO adsorbed on transition catalytic aluminas is definitely out of the Eq. 15.1 correlation line, if the high coverage values are considered.

In fact, the pairs of values for CO adsorbed at low coverage on both $\gamma-$ and $\delta,\theta-Al_2O_3$ were found to fit quite well with the Eq. 15.1 correlation plot: $\Delta\nu_{CO} = 60 - 70\,cm^{-1}$ and $-\Delta_a H$ ≈60–70 kJ mol^{-1}. Conversely, at high CO coverage the measured heat was surprisingly low ($-\Delta_a H$ ≈10 kJ mol^{-1}) with respect to the vibrational parameter ($\Delta\nu_{CO} = 55\ cm^{-1}$).

It is here recalled that in cation-deficient spinel transition aluminas, the Al(III) cations are located either in tetrahedral or octahedral cavities of the cubic close packed array of oxide ions [15, 30]. In the $\gamma-$ and $\delta,\theta-Al_2O_3$ here discussed, two CO adspecies were formed at the surface when outgassed at $T = 773$ K. The two adspecies were defined as $(CO)_A$ and $(CO)_B$ in refs. [15, 31], being their presence witnessed by the appearance of ν_{CO} bands at ≈2205 and 2215 cm^{-1}, respectively. A third CO adspecies defined as $(CO)_C$, witnessed by the appearance of a ν_{CO} band at ≈2230 cm^{-1}, was observed only when the surface was outgassed at $T =$ 1023 K [15]. The formation of more than one CO adspecies was ascribed to the presence of the tetrahedral *cus* Al(III) cations located in different crystallographic positions. The tetrahedral *cus* Al(III) cations are in fact known to be the only sites able to adsorb at room temperature, being the octahedral *cus* ones too weakly acidic to do so [4, 30, 31]. The vanishing coverage enthalpy of formation of the three species ($-\Delta_a H \approx 58$, 73 and 83 kJ mol^{-1} for $(CO)_A$, $(CO)_B$ and $(CO)_C$, respectively) was found correlated with the correspondent shift: 62, 72 and 87 cm^{-1} for $(CO)_A$, $(CO)_B$ and $(CO)_C$, respectively) as reported in refs. [1, 15].

In Fig. 15.3 the heats of adsorption for γ- and $\delta,\theta-Al_2O_3$, both activated at $T =$ 773 K, are reported as a function of CO uptake in comparison with the correspondent data for TiO_2-anatase (activated at $T = 673$ K). The zero-coverage heat of CO adsorption on γ- and $\delta,\theta-Al_2O_3$ (≈60 kJ mol^{-1}) was close to the value measured for CO adsorbed on *cus* Ti(IV) cations on TiO_2 [20] and was compatible with a plain electrostatic polarization/σ-coordination of CO at the Lewis acidic AI(III) sites [4, 32]. However, remarkable differences were observed between the titania and alumina plots as far as the surface coverage increased. The heat of adsorption on TiO_2 decreased only slightly upon increasing coverage, whereas on Al_2O_3 (both $\gamma-$ and δ,θ-phases) the heat values dropped abruptly down to very low values (q ≈10 kJ mol^{-1}, at the highest coverage reached at p_{CO} ≈80 *Torr*).

The assignment of this surprisingly low heat value to the adsorption of CO on Lewis acidic sites other than the tetrahedral Al(III) cations was discarded by the IR

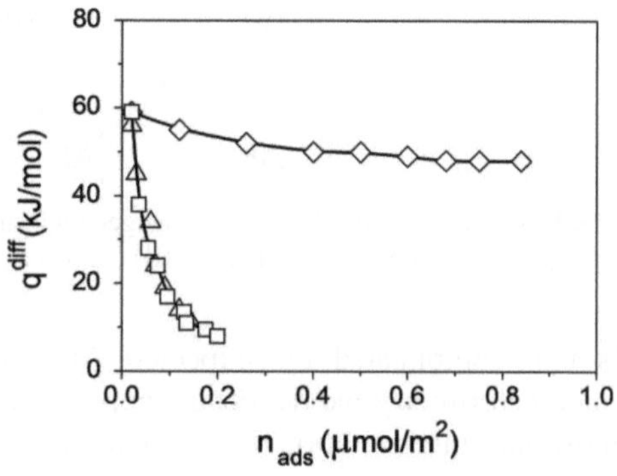

Fig. 15.3 Differential heat of adsorption (at $T = 303$ K) as a function of the increasing CO uptake on: TiO_2-anatase pre-outgassed at $T = 673$ K (*diamond*); $\gamma-Al_2O_3$ (*triangle*) and $\delta, \theta-Al_2O_3$ (*square*), both pre-outgassed at $T = 773$ K. Adapted from ref. [4] Fig. 3b

spectra which confirmed the presence of the sole σ-coordination of CO, through the C-end, on the Lewis acidic tetrahedral Al(III) sites [4, 15, 30, 31].

A process other than a plain σ-coordination taking place at the Al_2O_3/CO interface was invoked to justify the lack of correlation with the linear plot described by Eq. 15.1 for the high-coverage $-\Delta_a H$ value. It was demonstrated that the heat measured at high coverage comprised two different processes: (i) the exothermic σ-coordination of CO on tetrahedral *cus* Al(III) cations, and (ii) an endothermic surface reconstruction accompanying the adsorption process. In this respect, it is here recalled that the overall heat measured within the calorimetric cell is irrespective of how many and what kind of processes are actually occurring at the gas–solid interface.

In Fig. 15.4a , the presence of surface Al(III) cations engaged with strained Al–O bonds (and so not available as such for the interaction with molecules) is schematically illustrated [4, 30]. The adsorption of on this kind of Al(III) cations, which causes the O–Al groups interaction to be broken, is schematically illustrated in Fig. 15.4b. This effect was in fact evidenced at high coverage of CO by IR spectroscopy in the region of the Al–O modes (1100–1000 cm^{-1}), and was found to be entirely reversible [4, 30, 31, 33].

The modification of the surface structure produced by the rupture of the Al$\cdots$O–Al moieties interaction is intrinsically endothermic, and caused the measured heat for the Al_2O_3/CO interaction to be lower than what expected for a plain σ-coordination. This result does justify the dramatically low heat of interaction measured within the calorimetric cell, and does explain the lack of correlation between the energetic and vibrational parameters [1, 20, 34, 35].

In conclusion, in transition aluminas only the zero-coverage heat of adsorption, which is correlated to the shift of the stretching frequency, can be taken as a measure of the electron accepting properties of *cus* Al(III) cations and of the strength of the σ-dative bonds formed.

The lack of correlation between thermodynamic and spectroscopic data for CO adsorbed on transition aluminas revealed that the interaction at the gas-solid interface

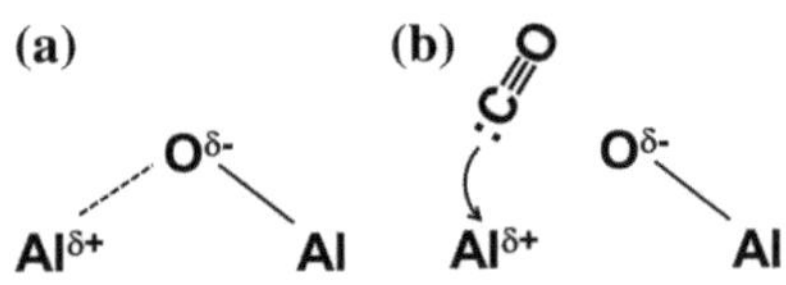

Fig. 15.4 **a** Surface Al(III) cations engaged with strained Al–O bonds in transition aluminas; **b** the adsorption of CO on the engaged Al(III) cations causes the Al(III)/O–Al interaction to be broken

is more complicated than expected. Also in this respect, the joint use of adsorption microcalorimetry and IR spectroscopy was proved very useful in arriving at a detailed molecular interpretation of the process.

15.3 NH_3 Adsorbed on All-Silica MFI Zeolites (Silicalite)

Defective Silicalite is an all-silica MFI zeolite, which is non-hydrophobic and weakly acidic as a consequence of the abundant polar defects (Si–OH nests) generated by the structure to compensate the Si atoms vacancies [7]. IR spectra in the ν_{OH} stretching frequency region give a clear evidence of the presence of Si–OH nests, which are characterized by a different geometrical arrangement according to the synthesis procedure and/or post-synthesis treatments, as illustrated in ref. [36].

The shift of the stretching frequency of the Si–OH terminations ($\Delta\nu_{OH}$) when the latter are engaged in H-bonding interaction with molecules is generally taken as a measure of the strength of the H-bonding interaction [6, 13, 37–39]. In the case of NH_3 adsorbed on a variety of silica–based materials (defective MFI-Silicalite and amorphous non-porous silica) $\Delta\nu_{OH}$ was found to be linearly correlated to the zero-coverage adsorption enthalpy $(-\Delta_a H)_0$, as illustrated in Fig. 15.5. See ref. [36] for details on the samples, and see also Chap. 1 Fig. 1.15b for the differential heat versus coverage plots of the MFI-Silicalite investigated specimens.

The pair of $\Delta\nu_{OH}$ and $(-\Delta_a H)_0$ values measured for one of the investigated MFI-Silicalite samples (Sil-B, see Chap. 1 Fig. 1.15b) was found to deviate from the linear correlation illustrated in the plot. The zero-coverage adsorption enthalpy for Sil-B

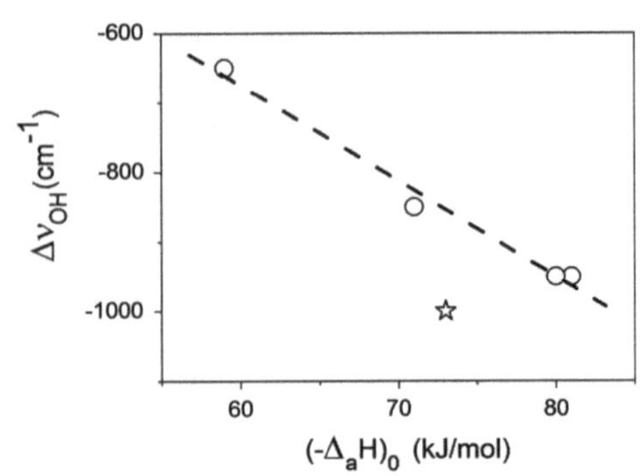

Fig. 15.5 Shift of the ν_{OH} stretching frequency ($\Delta\nu_{OH}$) versus the zero-coverage adsorption enthalpy $(-\Delta_a H)_0$ measured for NH_3 on a variety of silica–based materials (defective MFI-Silicalite and amorphous non-porous silica). Adapted from ref. [36] Fig. 6

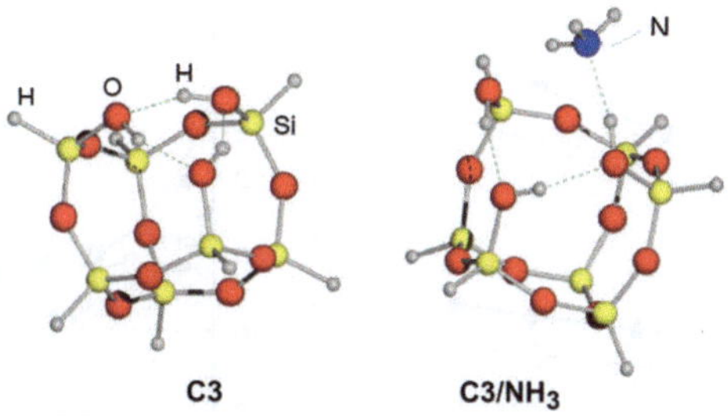

Fig. 15.6 Cluster model of a secondary building unit of MFI-Silicalite simulating a Si–OH nest arranged in a closed chain (ring), either free (C3) or in H-bonding interaction with one NH_3 molecule ($C3/NH_3$). Geometries were fully optimized without constraints at B3-LYP/6-31+G(d,p) level. Atomic symbols for cluster atoms are reported in C3 model; atomic symbol for ammonia N atom on $C3/NH_3$ model. Adapted from ref. [36] Fig. 8

($\approx$73 kJ mol^{-1}) was lower than that expected on the basis of the $\Delta_{\nu_{co}}$ value ($\approx$85 kJ mol^{-1}). It is here recalled that the heat released for the adsorption of NH_3 on Sil-B was lower than for the other investigated samples in the whole coverage examined (see curve 3 in Chap. 1 Fig. 1.15).

Also in this case, the apparently anomalous behavior of one sample was interpreted as an indication of the presence of phenomena others than a plain adsorption. In fact, the adsorption of NH_3 on Sil-B was demonstrated to imply an endothermic step which lowered the measured heat value.

The Sil-B sample was characterized by a structured spectroscopic response in the ν_{OH} region. The peculiar structure of the ν_{OH} band of this specimen was interpreted as due to the presence of small highly structured rings of mutually interacting silanol groups located within the nanopores of the zeolite. This was also suggested by neutron diffraction data [40, 41] and was confirmed by an *ab initio* modeling study. A nest arranged in a closed chain (ring) was simulated by a cluster model of a secondary building unit of MFI-Silicalite. The model is reported in Fig. 15.6, either free (C3) or in H-bonding interaction with one molecule ($C3/NH_3$). Geometries were fully optimized without constraints at B3-LYP/6-31+G(d,p) level. The *ab initio* modeling study demonstrated that rings must be first "opened" in order to become available for the interaction with NH_3. The insertion of one molecule in the ring required the breaking of the Si–OH$\cdots$OH–Si mutual H-bonding interaction, the energetic cost of which was computed to be as high as $\approx$12 kJ mol^{-1}. This value turned out to be in fairly good agreement with the experimental difference between the expected ($\approx$85 kJ mol^{-1}) and the measured ($\approx$73 kJ mol^{-1}) enthalpy of adsorption [36].

Also in this case, the accurate interpretation at molecular detail of the volumetric-calorimetric results was achieved thanks to the joint use of microcalorimetry, IR spectroscopy and *ab initio* calculations.

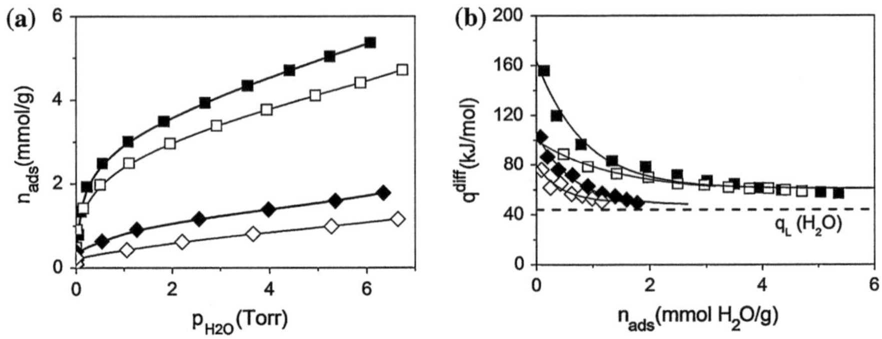

Fig. 15.7 **a** volumetric isotherms, **b** differential heats of adsorption versus coverage plots of H_2O_{vap} adsorbed at $T = 303$ K on H-BEA (*square*) and on $SiO_2{\cdot}Al_2O_3$ (*diamond*). H-BEA was pre-outgassed at $T = 873$ K, $SiO_2{\cdot}Al_2O_3$ at $T = 673$ K. *Solid symbols* first run; *open symbols* second run of adsorption. Adapted from ref. [7] Fig. 8

15.4 H_2O Vapor Adsorbed on Crystalline and on Amorphous Alumino-Silicates

The adsorption of water on a crystalline zeolite (H-BEA) and on an amorphous alumino-silicate specimen ($SiO_2{\cdot}Al_2O_3$) was studied in order to investigate the nature/structure of the acidic species originated by the presence of Al moieties the silica matrix [7]. It is generally accepted that proton-exchanged zeolites and alumino-silicate materials of similar composition exhibit different acidic features [42]. In particular, it was demonstrated that the Brønsted acidic strength, which is strongly dependent on the structural features of the material, is highest for H-zeolites [43].

In Fig. 15.7, volumetric isotherms (Sect. a) and differential heats (Sect. b) of water vapor (H_2O_{vap}) adsorption for the two investigated specimens are reported. The two samples,[1] were pre-outgassed for 2 h at either $T = 873$ K (H-BEA) or $T = 673$ K ($SiO_2{\cdot}Al_2O_3$), at a residual pressure $p \leq 10^{-5}$ *Torr* in order to ensure a maximum density of Lewis and Brønsted acidic sites.

The H – BEA specific adsorption capacity was much larger than that of $SiO_2{\cdot}Al_2O_3$, as witnessed by the volumetric isotherms. For instance at $p = 6$ *Torr*, uptake was 5.3 mmol g^{-1} for against only 1.7 mmol g^{-1} for $SiO_2{\cdot}Al_2O_3$. In both cases the adsorption was found to be partially irreversible (2nd run isotherms lie below 1st run isotherms) but the percent of the irreversible component was much larger for $SiO_2{\cdot}Al_2O_3$ than for H-BEA: ≈40 % *versus* ≈20 % of the total uptake, respectively. This datum suggested that H_2O_{vap} adsorption on the amorphous alumino-silicate caused an irreversible modification of the surface larger than on the crystalline specimen.

[1] H-BEA ($\frac{SiO_2}{Al_2O_3} = 4.9$) features are reported in Chap. 1 (Sect. 1.4.1). The amorphous alumino-silicate ($\frac{SiO_2}{Al_2O_3} = 5.8$) was purchased by Strem Chemicals, Inc.

$SiO_2{\cdot}Al_2O_3$ being more reactive towards water than H-BEA, a larger energy of interaction with H_2O_{vap} was expected for the former than for the latter. Surprisingly, the heat of adsorption curves reported in Fig. 15.7b indicate that the heat of adsorption was dramatically larger for the crystalline than for the amorphous alumino-silicate, in the whole range of coverage examined. The zero-coverage heat of adsorption was $\approx$160 kJ mol^{-1} for H-BEA and $\approx$110 kJ mol^{-1} for $SiO_2{\cdot}Al_2O_3$. At high coverage, the former sample heat values lie well above the latent heat of liquefaction of water, $q_L = 44$ kJ mol^{-1}, while the latter sample heat values approach q_L. The same trend was observed for the 2nd run reversible adsorption.

Also in this case, an endothermic step during the adsorption was invoked as a possible explanation for the large $SiO_2{\cdot}Al_2O_3$ surface reactivity surprisingly not accompanied by a high energy of interaction with water. The endothermic step of the overall process was due to the deformation/reconstruction of the surface, consequent to the interaction with water molecules. This process is expected to be facilitated by the flexible structure of the amorphous alumino-silicate and inhibited by the rigidity of the crystalline zeolite framework. In fact, the Lewis acidic *cus* Al(III) atoms, when making part of a rigid zeolite framework, acquire a close similarity with the Lewis acidic sites exposed at the ionic surface of transition aluminas [44]. This was confirmed by the closeness of the zero-coverage heats of adsorption of H_2O_{vap} on $\delta-Al_2O_3$ and H-BEA ($\approx$180 and $\approx$160 kJ mol^{-1}, respectively) [7]. Conversely, the covalent $SiO_2{\cdot}Al_2O_3$ *cus* Al(III) species being exposed at a pliable amorphous Si – O – Al surface were much less available for the interaction with molecules than those making part of the above mentioned rigid structures. As a consequence of the surface reconstruction allowing Al(III) species to interact with water molecules, the measured heat for $SiO_2{\cdot}Al_2O_3$ was lower than what expected for a plain adsorption.

This interpretation was supported by the *ab initio* modeling results described in ref. [7]. Owing to the large uncertainty from the experiments of the local structure around the Al atom, two topological different clusters (LS and LC structures) were designed to simulate the Lewis acidic site, as illustrated in Fig. 15.8, top side row. LS model was adopted to mimic highly strained moieties, typical of defects located in nanocavities. [32] The cluster LC was conversely adopted to simulate the surface species partially saturated by the coordination with an additional nearby framework O atom. This structure can be reasonably assumed as a model for $SiO_2{\cdot}Al_2O_3$,[7] in which the Al atom is allowed to expand its coordination thanks to the flexibility of the Si-O-Al amorphous structure. The structure of all clusters, either free or interacting with water (*vide infra*), were fully optimized at *ab initio* level using the B3-LYP/6-31+G(d,p) model chemistry [32, 45].

In the bottom side row of Fig. 15.8, the B3-LYP/6-31+G(d,p) optimized structures of LS and LC clusters interacting with one molecule (LSW and LCW, respectively) are reported. The computed water binding energies (BE) were corrected for the basis set superposition error, using the standard Boys-Bernardi counterpoise method [46]. It turned out that H_2O interacts much less strongly with LC (BE–LCW $\approx$109 kJ mol^{-1}) than with LS (BE–LSW $\approx$160 kJ mol^{-1}), in agreement with the lower local coordinative unsaturation of the LC-Al(III) atom with respect to that of the LS model. When H_2O is adsorbed at the LC site, a fraction of

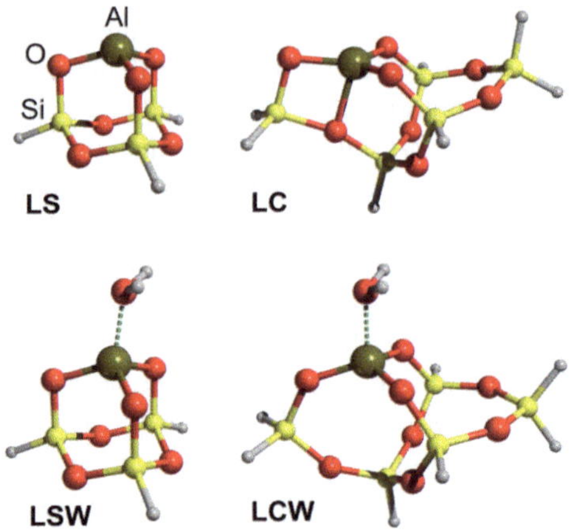

Fig. 15.8 *Top side row* B3-LYP/6-31+G(d,p) optimized clusters mimicking Lewis acidic Al(III) sites. LS model mimics highly strained Al(III)atoms, typical of defects present in H-BEA zeolites; LC cluster simulates the Al(III) atoms coordinating an additional nearby framework O atom. *Bottom side row* B3-LYP/6-31+G(d,p) optimized structures of the LS and LC clusters interacting with one H_2O molecule (LSW and LCW, respectively). Atomic symbols are reported in the model cluster LS. Adapted from ref. [7] Figs. 2 and 3

the adsorption energy is lost to pull out the Al atom from the amorphous framework. Conversely, in the crystalline material the acidic sites are already in place, as imposed by the rigidity of the structure. The "extraction" of the Al-containing site requires an energy cost which was computed to be $\approx 25\,\%$ of the total binding energy for the LCW case [7].

Once more the *ab initio* modeling results allowed to properly interpret the apparently anomalous results obtained by adsorption microcalorimetry, confirming the hypothesis formulated to justify the experimental data.

15.5 Conclusions

A few examples of adsorption processes accompanied by an endothermic step due to the deformation/reconstruction of the surface in interaction with molecules were illustrated. In the reported cases, the heat measured within the calorimetric cell was the combination of an exothermic (adsorption) and an endothermic (surface reconstruction) effect, which caused the calorimetrically measured heat to be lower than what expected on the basis of a plain adsorption. An extra-care in interpreting (at molecular level) the experimental calorimetric results should be addressed in several cases, and in this respect it is quite fruitful to complement the molar volumetric-calorimetric data with results from other approaches, typically the various spectroscopic methods and/or the *ab initio* molecular modeling.

References

1. V. Bolis, A. Barbaglia, S. Bordiga, C. Lamberti, A. Zecchina, Heterogeneous nonclassical carbonyls stabilized in Cu(I)- and Ag(I)-ZSM-5 zeolites: thermodynamic and spectroscopic features. J. Phys. Chem. B **108**(28), 9970–9983 (2004). doi:10.1021/Jp049613e
2. V. Bolis, S. Maggiorini, L. Meda, F. D'Acapito, G.T. Palomino, S. Bordiga, C. Lamberti, X-ray photoelectron spectroscopy and x-ray absorption near edge structure study of copper sites hosted at the internal surface of ZSM-5 zeolite: A comparison with quantitative and energetic data on the CO and NH_3 adsorption. J. Chem. Phys. **113**(20), 9248–9261 (2000). doi:10.1063/1.1319318
3. V. Bolis, S. Bordiga, C. Lamberti, A. Zecchina, A. Carati, F. Rivetti, G. Spano, G. Petrini, Heterogeneity of framework Ti(IV) in Ti-silicalite as revealed by the adsorption of NH_3 Combined calorimetric and spectroscopic study. Langmuir **15**(18), 5753–5764 (1999). doi:10.1021/la981420t
4. V. Bolis, G. Cerrato, G. Magnacca, C. Morterra, Surface acidity of metal oxides. Combined microcalorimetric and IR-spectroscopic studies of variously dehydrated systems. Thermochim Acta **312**, 63–77 (1998)
5. V. Aina, F. Bonino, C. Morterra, M. Miola, C.L. Bianchi, G. Malavasi, M. Marchetti, V. Bolis, Influence of the chemical composition on nature and activity of the surface layer of Zn-substituted Sol-Gel (bioactive) glasses. J. Phys. Chem. C **115**(5), 2196–2210 (2011). doi:10.1021/Jp1101708
6. J. Sauer, P. Ugliengo, E. Garrone, V.R. Saunders, Theoretical-study of Van-Der-Waals complexes at surface sites in comparison with the experiment. Chem. Rev. **94**(7), 2095–2160 (1994). doi:10.1021/cr00031a014
7. V. Bolis, C. Busco, P. Ugliengo, Thermodynamic study of water adsorption in high-silica zeolites. J. Phys. Chem. B **110**(30), 14849–14859 (2006). doi:10.1021/Jp061078q
8. C. Busco, V. Bolis, P. Ugliengo, Masked Lewis sites in proton-exchanged zeolites: a computational and microcalorimetric investigation. J. Phys. Chem. C **111**(15), 5561–5567 (2007). doi:10.1021/Jp0705471
9. M. Corno, C. Busco, V. Bolis, S. Tosoni, P. Ugliengo, Water adsorption on the stoichiometric (001) and (010) surfaces of hydroxyapatite: a periodic B3LYP study. Langmuir **25**(4), 2188–2198 (2009). doi:10.1021/La803253k
10. M. Corno, A. Rimola, V. Bolis, P. Ugliengo, Hydroxyapatite as a key biomaterial: quantum-mechanical simulation of its surfaces in interaction with biomolecules. Phys. Chem. Chem. Phys. **12**(24), 6309–6329 (2010). doi:10.1039/C002146f
11. V. Bolis, C. Busco, V. Aina, C. Morterra, P. Ugliengo, Surface properties of silica-based biomaterials: Ca species at the surface of amorphous silica as model sites. J. Phys. Chem. C **112**(43), 16879–16892 (2008). doi:10.1021/Jp805206z
12. V. Bolis, C. Busco, G. Martra, L. Bertinetti, Y. Sakhno, P. Ugliengo, F. Chiatti, M. Corno, N. Roveri, Coordination chemistry of Ca sites at the surface of nanosized hydroxyapatite: interaction with H_2O and CO. Phil. Trans. R. Soc. a-Math. Phys. Eng. Sci. **370**(1963), 1313–1336 (2012). doi:10.1098/rsta.2011.0273
13. M. Armandi, V. Bolis, B. Bonelli, C.O. Arean, P. Ugliengo, E. Garrone, Silanol-related and unspecific adsorption of molecular ammonia on highly dehydrated silica. J. Phys. Chem. C **115**(47), 23344–23353 (2011). doi:10.1021/Jp206301c
14. K.I. Hadjiivanov, G.N. Vayssilov, Characterization of oxide surfaces and zeolites by carbon monoxide as IR probe molecule. Adv. Catal. **47**, 307–511 (2002). doi:10.1016/S0360-0564(02)47008-3
15. V. Bolis, G. Magnacca, C. Morterra, Surface properties of catalytic aluminas modified by alkaline-earth metal cations: a microcalorimetric and IR-spectroscopic study. Res. Chem. Intermed. **25**(1), 25–56 (1999). doi:10.1163/156856799X00374
16. A. Zecchina, C. Lamberti, S. Bordiga, Surface acidity and basicity: general concepts. Catal. Today **41**(1–3), 169–177 (1998). doi:10.1016/S0920-5861(98)00047-9

17. A.M. Ferrari, P. Ugliengo, E. Garrone, Ab initio study of the adducts of carbon monoxide with alkaline cations. J. Chem. Phys. **105**(10), 4129–4139 (1996). doi:10.1063/1.472283
18. C. Lamberti, S. Bordiga, F. Geobaldo, A. Zecchina, C.O. Arean, Stretching frequencies of cation CO adducts in alkali-metal exchanged zeolites—an elementary electrostatic approach. J. Chem. Phys. **103**(8), 3158–3165 (1995). doi:10.1063/1.470249
19. C. Morterra, E. Garrone, V. Bolis, B. Fubini, An infrared spectroscopic characterization of the coordinative adsorption of carbon-monoxide on TiO_2. Spectrochim. Acta Part A-Mol. Biomol. Spectrosc. **43**(12), 1577–1581 (1987). doi:10.1016/S0584-8539(87)80051-X
20. V. Bolis, B. Fubini, E. Garrone, C. Morterra (1989) Thermodynamic and vibrational characterization of CO adsorption on variously pretreated anatase. J. Chem. Soc. Faraday Trans. I **85**, (1383–1395)
21. E. Garrone, V. Bolis, B. Fubini, C. Morterra, Thermodynamic and spectroscopic characterization of heterogeneity among adsorption sites—CO on anatase at ambient-temperature. Langmuir **5**(4), 892–899 (1989). doi:10.1021/la00088a002
22. A.J. Lupinetti, S.H. Strauss, G. Frenking, Nonclassical metal carbonyls. Prog. Inorg. Chem. **49**, 1–112 (2001). doi:10.1002/9780470166512.ch1
23. V. Bolis, B. Fubini, E. Garrone, E. Giamello, C. Morterra, in *Studies in Surface Science and Catalysis: Structure and Reactivity of Surfaces*, ed. by C. Morterra, A. Zecchina, G. Costa, vol 48 (Elsevier Sci. Publ. B.V., 1989) pp. 159–166
24. S.H. Strauss, Copper(I) and silver(I) carbonyls. To be or not to be nonclassical. J. Chem. Soc. Dalton Trans. **1**, 1–6 (2000). doi:10.1039/a908459b
25. Q. Xu, Metal carbonyl cations: generation, characterization and catalytic application. Coord. Chem. Rev. **231**(1–2), 83–108 (2002). doi:10.1016/S0010-8545(02)00115-7
26. Y. Kuroda, Y. Yoshikawa, R. Kumashiro, M. Nagao, Analysis of active sites on copper ion-exchanged ZSM-5 for CO adsorption through IR and adsorption-heat measurements. J. Phys. Chem. B **101**(33), 6497–6503 (1997). doi:10.1021/jp9710796
27. Y. Kuroda, H. Onishi, T. Mori, Y. Yoshikawa, R. Kumashiro, M. Nagao, H. Kobayashi, Characteristics of silver ions exchanged in ZSM-5-type zeolite, aluminosilicate, and SiO_2 samples: In comparison with the properties of copper ions exchanged in these materials. J. Phys. Chem. B **106**(35), 8976–8987 (2002). doi:10.1021/Jp020507r
28. V. Bolis, S. Bordiga, G.T. Palomino, A. Zecchina, C. Lamberti, Calorimetric and spectroscopic study of the coordinative unsaturation of copper(I) and silver(I) cations in ZSM-5 zeolite - Room temperature adsorption of NH_3. Thermochim. Acta **379**(1–2), 131–145 (2001). doi:10.1016/S0040-6031(01)00612-8
29. E. Giamello, B. Fubini, V. Bolis, Microcalorimetric investigation of the interaction of carbon-monoxide with coprecipitated cupric oxide zinc-oxide catalysts in well-defined oxidation-states. Appl. Catal. **36**(1–2), 287–298 (1988). doi:10.1016/S0166-9834(00)80122-0
30. C. Morterra, G. Magnacca, A case study: surface chemistry and surface structure of catalytic aluminas, as studied by vibrational spectroscopy of adsorbed species. Catal. Today **27**(3–4), 497–532 (1996). doi:10.1016/0920-5861(95)00163-8
31. C. Morterra, V. Bolis, G. Magnacca, IR spectroscopic and microcalorimetric characterization of Lewis-acid sites on (transition phase) Al_2O_3 using adsorbed CO. Langmuir **10**(6), 1812–1824 (1994). doi:10.1021/la00018a033
32. V. Bolis, M. Broyer, A. Barbaglia, C. Busco, G.M. Foddanu, P. Ugliengo, Van der Waals interactions on acidic centres localized in zeolites nanocavities: a calorimetric and computer modeling study. J. Mol. Catal. a Chem. **204**, 561–569 (2003). doi:10.1016/S1381-1169(03)00339-X
33. L. Marchese, S. Bordiga, S. Coluccia, G. Martra, A. Zecchina, Structure of the surface sites of delta-Al_2O_3 as determined by high-resolution transmission electron-microscopy, computer modeling and infrared-spectroscopy of adsorbed CO. J. Chem. Soc. Faraday Trans. 1 Phys. Chem. Condens. Phase **89**(18), 3483–3489 (1993). doi:10.1039/ft9938903483
34. V. Bolis, B. Fubini, E. Garrone, C. Morterra, P. Ugliengo, Induced heterogeneity at the surface of group-4 dioxides as revealed by CO adsorption at room-temperature. J. Chem. Soc. Faraday Trans. **88**(3), 391–398 (1992). doi:10.1039/ft9928800391

35. V. Bolis, C. Morterra, B. Fubini, P. Ugliengo, E. Garrone, Temkin-type model for the description of induced heterogeneity—CO adsorption on group-4 transition-metal dioxides. Langmuir **9**(6), 1521–1528 (1993). doi:10.1021/la00030a017
36. V. Bolis, C. Busco, S. Bordiga, P. Ugliengo, C. Lamberti, A. Zecchina, Calorimetric and IR spectroscopic study of the interaction of NH_3 with variously prepared defective silicalites—comparison with ab initio computational data. Appl. Surf. Sci. **196**(1–4), 56–70 (2002). doi:10.1016/S0169-4332(02)00046-6
37. A. Zecchina, S. Bordiga, G. Spoto, D. Scarano, G. Petrini, G. Leofanti, M. Padovan, C.O. Arean, Low-temperature fourier-transform infrared investigation of the interaction of CO with nanosized ZSM5 and silicalite. J. Chem. Soc. Faraday Trans. **88**(19), 2959–2969 (1992). doi:10.1039/ft9928802959
38. H. Knozinger, *Handbook of Heterogeneous Catalysis*, vol. 2 (Wiley/VCH, Weinheim, 1997)
39. M. Armandi, B. Bonelli, I. Bottero, C.O. Arean, E. Garrone, Thermodynamic features of the reaction of ammonia with the acidic proton of H-ZSM-5 as studied by variable-temperature IR spectroscopy. J. Phys. Chem. C **114**(14), 6658–6662 (2010). doi:10.1021/Jp100799k
40. G. Artioli, C. Lamberti, G.L. Marra, Neutron powder diffraction study of orthorhombic and monoclinic defective silicalite. Acta Crystallogr. Sect. B Struct. Sci. **56**, 2–10 (2000)
41. C. Lamberti, S. Bordiga, A. Zecchina, G. Artioli, G. Marra, G. Spano, Ti location in the MFI framework of Ti-silicalite-1: a neutron powder diffraction study. J. Am. Chem. Soc. **123**(10), 2204–2212 (2001). doi:10.1021/Ja003657t
42. M. Trombetta, G. Busca, S. Rossini, V. Piccoli, U. Cornaro, A. Guercio, R. Catani, R.J. Willey, FT-IR studies on light olefin skeletal isomerization catalysis III. Surface acidity and activity of amorphous and crystalline catalysts belonging to the SiO_2-Al_2O_3 system. J. Catal. **179**(2), 581–596 (1998). doi:10.1006/jcat.1998.2251
43. M.M. Huang, A. Auroux, S. Kaliaguine, Crystallinity dependence of acid site distribution in HA HX and HY zeolites. Microporous Mater. **5**(1–2), 17–27 (1995). doi:10.1016/0927-6513(95)00028-8
44. G. Della Gatta, B. Fubini, L. Stradella, Energies of different surface rehydration processes on "eta", "theta" and "alpha" aluminas. Journal of the Chemical Society. Faraday Trans 2 Mol. Chem. Phys. **73** (7):1040–1049 (1977) doi:10.1039/F29777301040
45. V. Bolis, A. Barbaglia, M. Broyer, C. Busco, B. Civalleri, P. Ugliengo, Entrapping molecules in zeolites nanocavities: a thermodynamic and ab-initio study. Orig. Life Evol. Biosph. **34**(1–2), 69–77 (2004). doi:10.1023/B:ORIG.0000009829.11244.d1
46. S.F. Boys, F. Bernardi, The calculation of small molecular interactions by the differences of separate total energies. Some procedures with reduced errors. Mol. Phys. **19**(4), 553–566 (1970). doi:10.1080/00268977000101561

Chapter 16
Characterisation of Catalysts and Adsorbents by Inverse Gas Chromatography

Eva Díaz and Salvador Ordóñez

Abstract Inverse Gas Chromatography (IGC), in contrast to analytical chromatography, consists on adsorption of a known solute on an adsorbent whose properties are to be determined. The shape and positions of the peaks supply information about the nature and reactivity of the solid surface. If different probe molecules are used (i.e. polar and apolar molecules, molecules with acid/base properties), it is possible to study the specificity of these interactions. Therefore, IGC can be used both as a tool for both characterizing the adsorption of a given compound on a given solid or for studying the nature (in terms of acid-base properties, polar or apolar interactions, etc.) of the active sites of a certain catalyst.

16.1 Introduction

Sorption measurements are a useful method in the characterization of solid materials. From these data, it is possible to obtain information about the capacity of adsorption, but also thermodynamic properties—enthalpies of adsorption, surface energy—as well as kinetic information, such as diffusion rates. Sorption measurements can be obtained either by static or dynamic methods. Static methods carried out the adsorption measurements under vacuum, after a pre-treatment at high temperature in order to clean the material surface. Dynamic methods use a flowing gas device. Inverse gas chromatography (IGC) is a dynamic method. In comparison to static adsorption systems, dynamic sorption techniques show shorter measurement time, and a wider range of experimental possibilities.

In contrast to analytical chromatography, the stationary phase is the sample under investigation, and the mobile phase acts as probe molecule. Thus, the roles of the

E. Díaz (✉) · S. Ordóñez
Department of Chemical Engineering and Environmental Technology, Faculty of Chemistry, University of Oviedo, C/ Julián Clavería s/n, 33006 Oviedo, Spain
e-mail: diazfeva@uniovi.es

A. Auroux (ed.), *Calorimetry and Thermal Methods in Catalysis*, Springer Series in Materials Science 154, DOI: 10.1007/978-3-642-11954-5_16, phases are inverted and hence, the name of "inverse" gas chromatography. This technique involves injecting a series of volatile probes and measuring their retention volumes. Retention volume is related to the interaction parameters between the probes and the solid and can be converted into a number of surface thermodynamic properties. Through an adequate choice of the probe molecules to be adsorbed, it is possible to obtain information about the surface structure and/or surface functionality of adsorbents. After the introduction of the technique and subsequent theoretical developments, the application of IGC in the materials sciences has developed fast [1]. IGC has been used for the characterization of polymers, copolymers, polymer bends, biopolymers, industrial fibers, wood and pulp fibers, composites, coatings, pigments, catalysts, glass beads, coal, chemicals, and steel tubing [2].

Inverse chromatography can be used in the gas phase as well as in the liquid phase. Although there is some interest in the research of the solid in liquid phase, a vast literature has done o gas phase (IGC), thus this chapter will be centred in IGC.

16.2 Experimental

IGC measurements can be carried out using a pulse or continuous technique. The pulse of probe molecule is introduced into the carrier gas stream. This pulse is transported by the carrier gas through the system to the column with the solid sample. On the stationary phase, adsorption and desorption occur and the result is a peak in the chromatogram. The ratio of adsorption/desorption is governed by the partition coefficient. At fixed conditions of temperature and flow rate, the time of retention of a compound is characteristic of the system. An alternative is the frontal technique. This is carried out by injection into the carrier gas stream of a continuous stream of the probe molecule. When the sample enters into the column, there is a distribution between phases, and the concentration profiles takes the shape of a plateau, preceded by a breakthrough curve. The shape of this curve is characteristic of each system [3]. The benefit of the frontal technique is that equilibrium can be always established due to its continuous nature while pulse chromatography requires the assumption of a fast equilibration of the probe molecule adsorption on the surface. Between both techniques, the main part of publications describes pulse experiences, since they are faster, easier to control and more accurate, especially if interactions between probe molecules and the adsorbent are weak.

The experimental set-up for the pulse chromatographic experiments consists of a column inside an oven, with an inlet of the carrier gas with the probe molecule, and the detector at the exit of the column. The pure carrier gas is introduced into the column (packed with the material under study). The injection of the sample takes place prior the oven of the chromatograph and it can be done by different methods [4]:

- by a syringe via the manual injector port of the chromatographic device, consisting of vapor or liquid;

- by a vapour headspace system, in this case a carrier gas is passed through a reservoir containing the prove molecule in its liquid form. The gas is saturated with the prove molecule and then flowing through the injection loop. Concentration can be controlled by the temperature in the reservoir and the loop volume. This saturated carrier gas is injected into another pure carrier gas stream.

Concerning the carrier gas, helium or nitrogen, the most important requirement is to ensure its purity and dryness, since some adsorption processes are highly sensitive to traces of impurities, in particular moisture. Add to the retention time of the probe molecule, it is necessary also to know the dead-time of the system—time that the probe molecule would require travelling along the column without any interaction. Obviously, this dead-time can not be measured using the probe molecule since interactions will be always present. Thus, another molecule, knows as tracer, with negligible interactions with both the adsorbents and the column walls is employed. Usual tracers molecules are methane, hydrogen, nitrogen, or even air. The retention time, as well as the area under the peak, is measured by flame ionisation (FID) or thermal conductivity (TCD) detector. The FID has the benefit of the sensitive, but it is limited to organic samples, while TCD is more versatile. Occasionally mass spectrometric detectors are also used. This is particularly interesting for experiments where two o more probe molecules are injected.

As far as columns are concerned, they are constructed from glass or metal tubes. Furthermore, in the case of metals, they are usually stainless steel column, with passivated inner walls to avoid interactions. In the literature, there is a wide variety about column lengths and diameters for different applications. The main criteria for selecting column dimensions are the following [4, 5]:

- Small column diameter, in order to keep gas-phase diffusion effects to minimum.
- Ratio between the column diameter and the particle diameter:

$$D_{in}/d_p > 10 \tag{16.1}$$

  This ratio between both diameters ensures minimization of the effect of channelling at the wall.
- Ratio between the column length and the particle diameter:

$$L/d_p > 50 \tag{16.2}$$

  This ratio minimizes the axial dispersion. The column length is not so crucial as the bed length of the packed stationary phase. Usually packed beds are supported on a porous filter or hold in place with glass wool plugs. For this reason, the column can be longer than the packing. To avoid additional peak broadening is recommended to pack the free place with inert material of the same particle size (glass), Fig. 16.1. The length of the packing depends on the uptake capacity of the sample and the amount of probe injected. It must be sure that the retention is strong enough (good separation between probe and tracer peak) to obtain reproducible and accurate results. This can be checked by repeating the column with different masses. If the

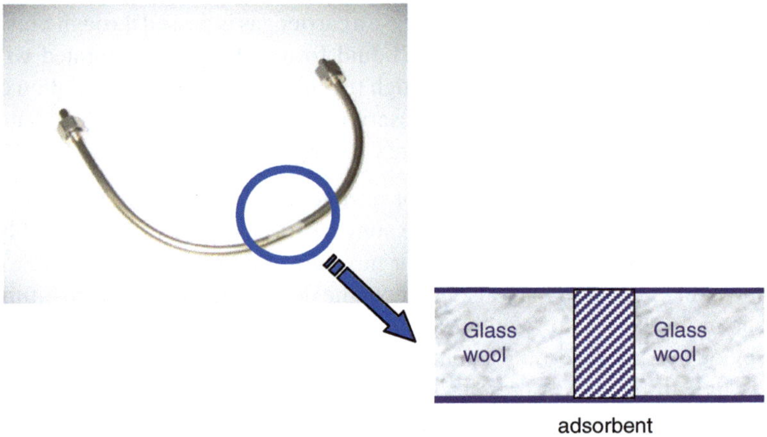

Fig. 16.1 Scheme of column packing and inert material disposition

measured parameters are independent of mass of the adsorbent, enough amount of packing is used.
- Particle size should be selected to minimize the effects of intraparticular diffusion.

Concerning the flow rate, it is supposed that the lower the flow rate the more likely the equilibrium of a system is reached. However, low flow rates mean longer experimental times and broader peaks, thus accuracy in the retention time determination is decreased. A chromatographic column can be considered as a sum of discrete but contiguous narrow layers, or plates. At each plate, equilibration of the solute between the mobile and stationary phase was assumed to take place. Movement of the solute down the column was then treated as a stepwise transfer of equilibrated mobile phase from one plate to the next. So, a chromatographic column is constituted by a number of steps, N, with a length, L. Efficiency studies of a chromatographic column have generally been carried out by determining H as a function of mobile-phase velocity u, according to the van Deemter equation (Eq. 16.3) [6]:

$$\mathrm{H} = \mathrm{A} + \frac{\mathrm{B}}{u} + \mathrm{C}u = \mathrm{A} + \frac{\mathrm{B}}{\mathrm{u}} + (\mathrm{C_S} + \mathrm{C_M})u \tag{16.3}$$

where H is the plate height in centimeters; u, the linear velocity of the mobile phase in centimeters per second; and the quantities A, B, and C are coefficients related to the phenomena of multiple flow paths, longitudinal diffusion, and mass transfer between phases, respectively. The C coefficient can be divided into two coefficients, one related to the stationary phase (C_s) and one related to the mobile phase (C_M). The van Deemter equation contains terms linearly and inversely proportional to, as well as independent of, the mobile phase velocity. Taking into account these considerations, it is recommended to repeat the experiment at different flow rates and determine the optimum (minimum H) via van Deemter equation, Fig. 16.2a.

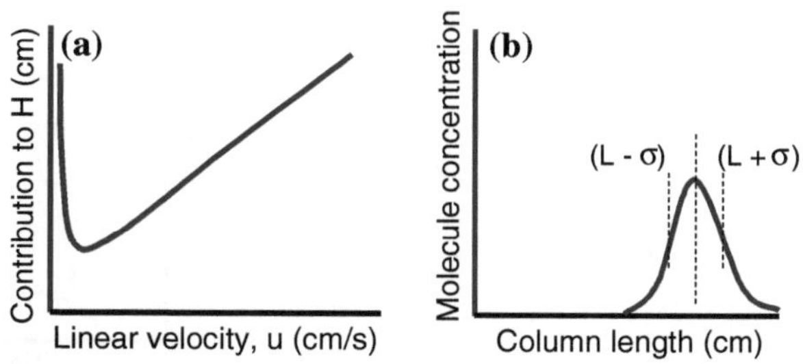

Fig. 16.2 **a** Plot of the plate height versus the mobile phase velocity; **b** Calculation of plate height

The plate height, H, is given by, Eq. (16.4):

$$H = \frac{\sigma^2}{L} \tag{16.4}$$

Thus, the plate height can be obtained from the length of column that contains a fraction of the probe molecule that lies between $L - \sigma$ and L (Fig. 16.2b). Because the area under a normal error curve bounded by σ is about 68 % of the total area, the plate height, as defined, contains approximately 34 % of the probe molecule.

Once the plate height is known, the number of plates of the column is obtained directly from the length of the chromatographic column ($N = L/H$).

Taking into consideration the aforementioned points related to both the column dimensions and packing and the gas flow rate, and working at very low concentrations of adsorbate, symmetric peaks are obtained for many materials. Under these conditions, the hypothesis of infinite dilution can be considered. Due to several both experimental conditions and instrumental problems, broadening of the peaks can be observed—avoiding the use of the Gaussian peak, explained later. Some of these causes are:

- Inadequate column dimensions, amount of stationary phase or gas flow rate, thus axial dispersion could be the responsible of the broadening.
- Large volume of adsorbate injections.
- Dead volumes in detector or injector.
- Imperfect column packing.

However, sometimes, even working at very low concentrations of the probe molecule and following the aforementioned recommendations, peaks with a large broadening are obtained. In these cases, it can be assumed that this asymmetry is not due to instrumental problems or experimental conditions. This effect is characteristic of the so-called "slow kinetic process" [7], Fig. 16.3, associated with markedly energetically heterogeneous surfaces containing preferential sites were desorption takes place in a slower way. Slow kinetic process depends on the concentration of the

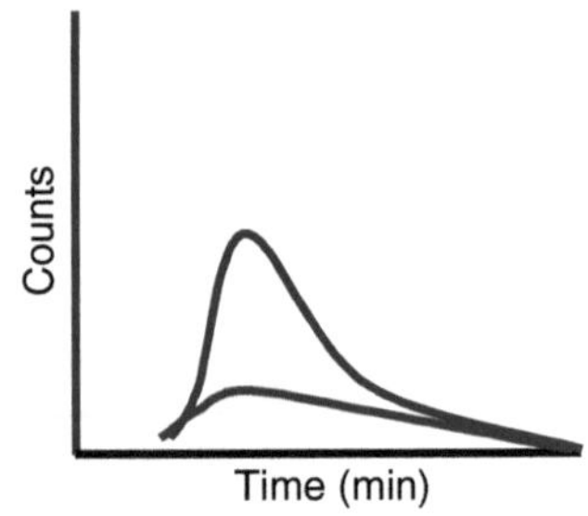

Fig. 16.3 Effect of slow kinetic process

solute, and is attenuated when the amount of adsorptive is decreased. Moreover, they correspond to non equilibrium situations, thus the gas flow rate have a significant effect on it.

16.3 Adsorption Isotherms

Adsorption isotherms of gases or vapors are the basis upon which the surface characteristics of adsorbents are defined. From this magnitude, specific surface area, porosity and other properties of the solid can be obtained. Even more, using adsorbates with diverse physical and chemical characteristics, it is possible to define the type of adsorbate-adsorbent interactions involved and the nature of the adsorption in the system tested. From the chromatographic peaks, the adsorption isotherms can be directly obtained. Figure 16.4 presents a general relation between the chromatographic peak and the adsorption isotherm shape. In the case of infinite dilution, a symmetrical (gaussian) peak is observed representing a linear Henry isotherm. At high concentration (finite dilution) tailing or leading will occur. In the case of a type I, II, or IV isotherm there is a tailing because adsorbent/adsorbate interactions are much stronger than adsorbate/adsorbate interactions.

Glückauf [8, 9] develop a method for obtaining the adsorption isotherms from chromatographic peaks in which a continuous stream of adsorbate is injected into the column until saturation, and the adsorbed material is then eluted by a pure carrier gas stream. The adsorption isotherm is calculated from the shape of the desorption curve. Gregg and Stock [10, 11] demonstrated that it was possible to obtained the adsorption isotherms from chromatographic data for all types of Brunauer isotherm. For this purpose, high concentrations of adsorbate are applied, but the effect of gradient pressure was neglected in most of the experimental work published before 1968 [12]. Taking into account these considerations, adsorption isotherms can be obtained either from the ideal GC, at conditions of infinite dilution, or by non-ideal and non-linear chromatography, at conditions of finite dilution. Ideal GC is described here, whereas finite chromatography isotherms determination can be found in the literature [3, 12].

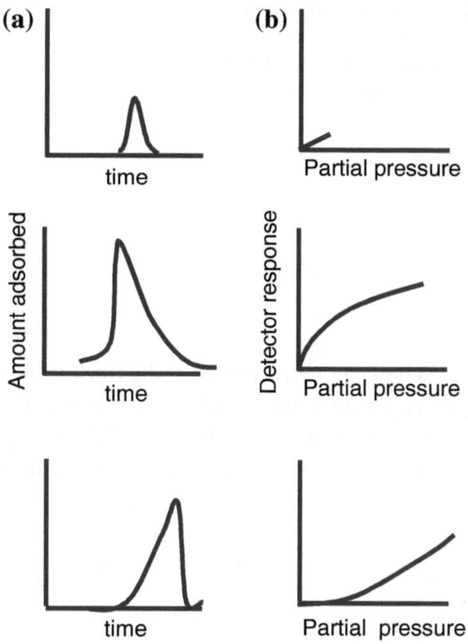

Fig. 16.4 Correlation of peak form **a** and adsorption isotherm **b** for finite and infinite dilution

At conditions in a chromatographic columns which approach the requirements of ideal IGC (minimum of van Deemter curve), the following hypothesis can be supposed:

- Flow through the column under isothermic and isobaric conditions
- Fast fluid-solid transfer
- Negligible intraparticular diffusion
- Axial transport due to convection
- No concentration or velocity gradients in radial direction
- Reversible adsorption and instantaneous equilibrium

In this case, diffusional and kinetical broadening of the chromatographic column is reduced, and any distortion of the chromatogram is due to the deviation of the adsorption isotherm from the Henry's law.

Therefore, it can be described the change in adsorbate concentration across an increment of the chromatographic column of length *dx* as:

$$-\,u_0 v\left(\frac{\partial c}{\partial x}\right)_t = v\left(\frac{\partial c}{\partial t}\right)_x + v_a\left(\frac{\partial c_a}{\partial t}\right)_x \qquad (16.5)$$

where u_0 is the lineal gas velocity; v, the gas phase volume in the column; v_a, the volume of adsorbate retained on the adsorbent; c, the adsorbate concentration in gas phase; c_a, the adsorbate concentration on the adsorbent; t, the time since the injection, and x, the length from the beginning of the column and the *dx*. The first part of the

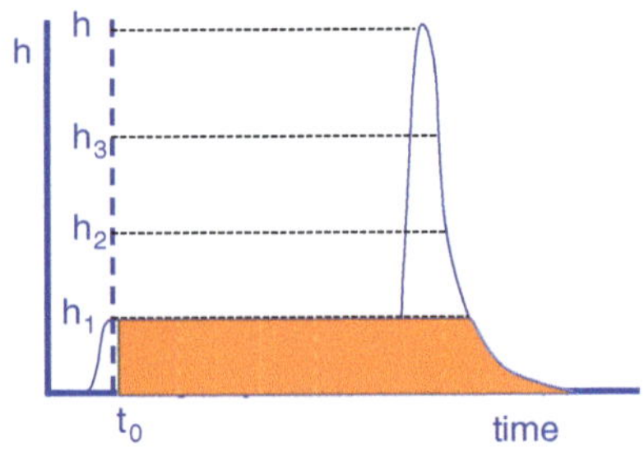

Fig. 16.5 Graphic integration of chromatogram and determination of the adsorption isotherm by the ECP method

mass balance corresponds to the mass of adsorbate accumulated, whereas the second part represents the rate of change of the amount of adsorbate in the layer *dx*.

By successive calculations [12], it is possible to determine the magnitude of adsorption, a, for an equilibrium concentration of adsorbate, c, in the mobile phase:

$$a = \frac{1}{m}\int_{0}^{c} V_R dc \tag{16.6}$$

where m is the mass of adsorbent in the column, and V_R, the retention volume.

The most common method of obtaining adsorption isotherms is the Elution of a Characteristic Point (ECP) [13], which consists of obtaining the isotherm from just one single injection. By introducing into Eq. (16.6) the magnitudes obtained from the chromatogram, the detector constant, k, and giving the detector deflections, h, the value of adsorption, a, is obtained:

$$a = \frac{m_a S_{ads}}{m S_{peak}} \tag{16.7}$$

where m_a is the mass of injected adsorbate; S_{ads}, is the area bounded by the height h between the tracer peak and the extender profile of the chromatogram (Fig. 16.5), and S_{peak}, the peak area.

The equilibrium concentration of adsorbate in the mobile phase can be expressed as:

$$c = \frac{m_a h}{F S_{peak}} \tag{16.8}$$

where F is the flow rate of the carrier gas. The equilibrium pressure is determined from the equation $p = cRT$. After substituting in Eq. (16.8), it is obtained the expression to calculate the equilibrium pressure:

$$p = \frac{m_a h RT}{F S_{peak}} \tag{16.9}$$

From a theoretical point of view, the method is accurate and a single elution profile allows the determination of a complete isotherm [14]. The applicability is restricted to very efficient columns allowing fast mass transfer, that it, columns possessing a high plate number. Low concentration measurements allow also to determine Henry constants, in this range the uptake is independent of the surface coverage. This regime is ideal also for the measurement of thermodynamic parameters since they can be obtained with the highest sensitivity. The span of the infinite dilution range depends on the probe molecules and the heterogeneity of the material. Especially for polar probe molecules adsorbing on very heterogeneous surface, non-symmetrical peaks are often observed even with the smallest injection size/concentration. This suggests that the values obtained under these conditions are not truly representing Henry conditions; however, they are still use for practical considerations.

16.4 Thermodynamic Parameters

16.4.1 Retention Volume

Application of IGC to study the properties of a solid is based on the assumption that the adsorbate equilibrium conditions are achieved between the mobile and stationary phases. Thus, chromatogram should be symmetric and the maximum of the peak must not depend on the amount of the injected adsorbate. Moreover, as the amount of adsorbate is very small, the concentration of the adsorbate in the gas phase is minimal and the adsorption process is conditioned by the real adsorbate-adsorbent interactions. Under these conditions, the retention volume—key parameter in IGC-of the solute depends on its partition between the stationary and mobile phase, and is an indication of the interaction strength between the solute molecule and the metal/adsorbent surface. The specific retention volume, V_g, in $\mathrm{cm^3/g}$, is given as:

$$V_g = Fj\frac{(t_R - t_d)}{m}\left(\frac{p_0 - p_w}{p_0}\right)\left(\frac{T}{T_{meter}}\right) \tag{16.10}$$

where F is the uncorrected flow rate detected by a bubble flow meter; t_R is the retention time in min; t_d, the hold-up time or time of a tracer compound in pass through the column; p_0, the outlet column pressure; p_w, the vapor pressure of water at the flowmeter temperature; T, the column temperature; T_{meter}, the ambient temperature, and j, the James-Martin compressibility factor. This parameter represents the volume of dry gas to elute the adsorbate, corrected at 273 K and per gram of stationary phase. Add to V_g, it is also very employed the net retention volume, V_N, defined as the volume of dry gas to elute the adsorbate, corrected at 273 K.

In both cases, the James-Martin factor for the correction of gas compressibility under pressure difference between column inlet, p_i, and column outlet, p_0, is introduced:

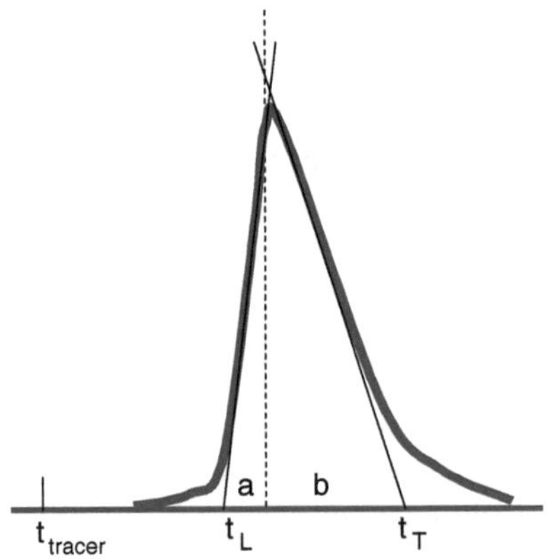

Fig. 16.6 Chromatogram presenting the Conder and Young method to obtain the skewness ratio (a/b) and retention time, t_R

$$j = \frac{3}{2}\left[\frac{(p_i/p_0)^2 - 1}{(p_i/p_0)^3 - 1}\right] \tag{16.11}$$

In the case of perfect symmetric peaks, the retention time can be determined directly from the peak maximum method, which is the simplest and most common. The peak maximum method is useful for determination of retention time if the skewness ratio is 0.7–1.3 [15]. The "skewness ratio" is defined as the ratio of tangent slope to the peak leading part and tangent slope to the peak tailing part whereas both tangents are drawn in the inflexion points. In such cases the skewness ratio is our of this interval, t_R is obtained from the first-order moment method or the Conder and Young method. Between these two methods, Conder and Young is recommended [16]:

$$t_R = (t_L + t_T)/2 \tag{16.12}$$

where t_L and t_T are the times at which the tangents drawn to the peak leading and tailing parts in their inflexion points intersect the zero line, Fig. 16.6.

The retention volume is related to the surface area and surface energy; that is, the higher the surface area and energy, the higher the retention time, and therefore, retention volume.

Moreover, the V_N and the slope of adsorption isotherm are related by Eq. (16.13) for small adsorbate injections, where conditions of "infinite dilution" are achieved:

$$V_N = K_S \cdot A = \frac{q}{c} \cdot A \tag{16.13}$$

where K_s is the inclination of the isotherm at infinite dilution, that is, the Henry's constant; A, the specific surface area of the adsorbent; q, concentration of the adsorbate in the stationary phase, and c, concentration of the adsorbate in the gas phase.

16.4.2 Free Energy of Adsorption

Thermodynamics information of the adsorption process at infinite dilution can be obtained from the retention volume. At infinite dilution, the standard free energy to transfer 1 mol of adsorbate from the gas phase to the surface at standard state, defined as the variation in the standard free energy of adsorption, ΔG^0_{ads} (J/mol), can be expressed as:

$$\Delta G^0_{ads} = -RT \ln \left[\frac{P V_g}{\pi_0 A} \right] \tag{16.14}$$

or its equivalent form:

$$\Delta G^0_{ads} = -RT \ln V_g + C \tag{16.15}$$

where P is the gas phase pressure, A is the specific surface area of the adsorbent, and π_0 is the spreading pressure of the adsorbed gas. Two different standard states can be considered: in the De Boer state, the spreading pressure has a value of 338 μN/m [17] at $p_0 = 1.01\,\text{kN/m}^2$, whereas in the Kemball and Rideal state, at $p_0 = 1.01\,\text{kN/m}^2$, the spreading pressure is 0.0608 μN/m [18]. The parameter, C, is a constant related to the standard reference states:

$$C = -RT \ln \left(\frac{A\pi}{P} \right) \tag{16.16}$$

16.4.3 Enthalpy and Entropy of Adsorption

When zero coverage (infinite dilution) conditions are fulfilled the standard differential heat of adsorption, q_0, is numerically equal to the opposite of the enthalpy of the process. This value can be obtained from the variation of ΔG^0_{ads} with temperature. For an equilibrium process this variation is given by Gibbs-Helmholtz equation:

$$\left[\frac{\partial \left(-\Delta G^0_{ads}/T \right)}{\partial (1/T)} \right]_P = \left[R \frac{\partial \left(\ln V_N \right)}{\partial \left(1/T \right)} \right]_P = q^0 \tag{16.17}$$

In the Fig. 16.7 is illustrated the dependence of $-\Delta G^0_{ads}/T$ as a function of $1/T$ for conventional carbon fibers and carbon fibers oxidized by electrochemical procedure. In the figure is illustrated the dependence of n-heptane. This behaviour, typical of hydrocarbons, implies that ΔH^0_{ads} is constant within the temperature range of characterisation. At this point, it is important to remark that comparison of differential heats of adsorption with the heats of liquefaction is recommended, in order to ensure the nature of the interaction. In the cases where the adsorbate-adsorbent interactions are stronger than adsorbate-adsorbate interactions, ΔH^0_{ads} is higher than the liquefaction heat.

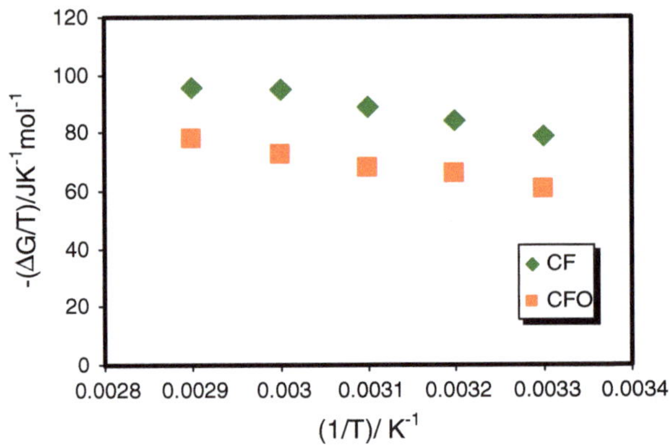

Fig. 16.7 Variation of $(-\Delta G^0_{ads}/T)$ with $(1/T)$ for the adsorption of n-heptane on carbon fibers (adapted from Ref. [19])

From the adsorption standard free energies and standard enthalpies, adsorption entropies can be calculated from:

$$\Delta S^0_{ads} = -\frac{\left(q^0 + \Delta G^0_{ads}\right)}{T} = \frac{\Delta H^0_{ads} - \Delta G^0_{ads}}{T} \tag{16.18}$$

In agreement with the linearity in the $-\Delta G^0_{ads}$/T versus $1/T$, the adsorption entropies are independent of the temperature.

Figure 16.8 shows the existence of so-called "thermodynamic compensation effect", i.e. a linear dependence of ΔH^0_{ads} on ΔS^0_{ads}. Thermodynamic compensation effect between the adsorption enthalpy and entropy was observed in different studies for n-alkanes. It indicates that the stronger adsorption of longer n-alkanes is accompanied by a higher loss of mobility of the molecules (it means higher interaction between the molecule and the surface). This type of plot is used currently to highlight the differences in adsorbate-adsorbent interactions. A good fit of the compensation effect data to a straight line indicates the non-specific nature of the adsorbate-adsorbent interactions.

In the case of Fig. 16.8, three straight lines are depicted, one corresponding to the data of HSAG-100 and HSAG-300, and two more fitting the CNTs and CNFs points. Although the line of CNFs is clearly shifted with respect to graphites, the slope of both lines is virtually the same and different from that of CNTs. Since the adsorbate type is the same for both sets of data (n-alkanes), the difference in slopes can be attributed to the existence of two different non-specific surfaces, one of the CNFs and graphites and the other represented by the CNTs [19]. Likewise, the shift between graphites and CNFs could be understood since HSAGs contain a large amount of structural defects, so the interactions could be modified. It is also evident the higher values of entropy for the nanotubes in comparison with the other materials, due to high entropy of adsorbate located inside the tubes [21].

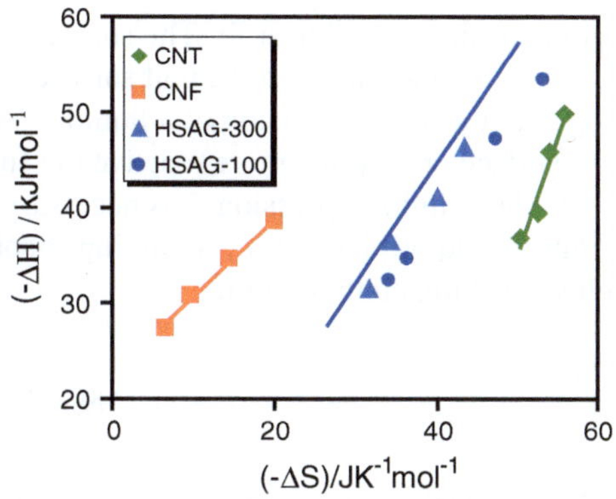

Fig. 16.8 Thermodynamic compensation effect of n-alkanes on carbon nanotubes (CNT), carbon nanofibers (CNF) and high-surface-area-graphites (HSAG) (adapted from Ref. [20])

16.4.4 Work of Adhesion: Dispersive and Specific Contribution

In the absence of chemisorption and interdiffusion, the work of adhesion is the sum of the different intermolecular forces involved and can be related to the surface free energies, where a is the compound and the superscripts D and S denote dispersive and specific interactions:

$$W_a = W_a^D + W_a^S \tag{16.19}$$

Generally, the work of adhesion is coupled to ΔG^0_{ads} according to:

$$\Delta G_{ads} = -NaW_a \tag{16.20}$$

where N is the Avogadro's number and a is the surface area of a single probe molecule.

For two materials interacting only via London dispersive forces across their interface, Fowkes [22–24] suggested that the work of adhesion, W_a, could be described as the geometric mean approach, where γ_L^D and γ_S^D are the dispersive component of the surface energy of the liquid (the probe) and the solid, respectively:

$$W_a = W_a^D = 2\sqrt{\gamma_L^D \gamma_S^D} \tag{16.21}$$

For probes interacting with the solid of interest via dispersive forces, a combination of Eqs. (16.15, 16.20, 16.21) will lead to:

$$RT \ln V_N = 2N \cdot a \cdot \sqrt{\gamma_L^D \gamma_S^D} + K \tag{16.22}$$

Thus, according to this approach, developed by Schultz et al. [25], by measuring the net retention volume for various n-alkane probes and plotting $RTlnV_N$ versus $a(\gamma_L^D)^{0.5}$, the dispersive component of the surface free energy can be determined from the slope of a linear fit. The surface free energy is the energy required to form (or increase the surface by) a unit surface under reversible conditions and is the analogue to the surface tension of a liquid. From the practical point of view, the higher the surface energy, the more reactive the surface.

In the Schultz expression, it is necessary the molecular area. It can be determined from the liquid density, ρ, assuming a spherical molecular shape in a hexagonal close-packing configuration:

$$a = 1.09 \cdot 10^{14} \left(\frac{M}{\rho N}\right)^{2/3} \tag{16.23}$$

where M is the molecular weight of the probe molecule.

However, some authors have stressed the difficulties associated with the determination of the molecular area, especially for non-spherical molecules such as straight alkanes. To avoid the problems derived from the molecular area, Dorris and Gray [26] considered the adsorption characteristics of single methylene groups in the n-alkane probes. By defining the increment in free energy of adsorption per – CH_2- unit:

$$\Delta G_{CH_2} = -RT \ln \frac{V_{N(n)}}{V_{N(n+1)}} \tag{16.24}$$

where V_N and $V_{N+1}$are the retention volumes of n-alkanes with (n) and $(n+1)$ carbon atoms, respectively. In this way, γ_S^D can be determined from:

$$\gamma_S^D = \frac{1}{4} \frac{\Delta G_{CH_2}^2}{\gamma_{CH_2} N^2 a_{CH_2}^2} \tag{16.25}$$

The benefit of this approach is the fact that despite the use of various n-alkanes probes, only methylene area, a_{CH_2}, and surface tension, γ_{CH_2}, have to be known. The CH_2 area is takes as 0.06 nm^2, based on C–C length of 0.127 nm and an average distance of 0.47 nm for two CH_2 [26]. Jacob and Berg [27] have found extraordinary an excellent agreement between n-alkane molecular areas as determined from the fitting of experimental adsorption isotherms to the BET model and as obtained by simply assuming the area of 0.06 nm^2 for each methylene group. The parameter $\gamma_{CH_2}$is estimated from the surface tension of a linear polyethylene melt as function of temperature:

$$\gamma_{CH_2}(\mathrm{mJ/m^2}) = 35.6 + 0.058\,(20 - T(^\circ C)) \tag{16.26}$$

The validity of this approach has established on the basis that IGC and wettability measurements lead to approximately the same γ_S^D value for poly(ethylene terephthalate) [28]. Furthermore, Dorris and Gray stated that the molecular area could be an adjustable parameter.

Once the dispersive interactions of a surface have been investigated, specific interactions can be studied by injecting polar probes. For these probes, W_a^S is usually

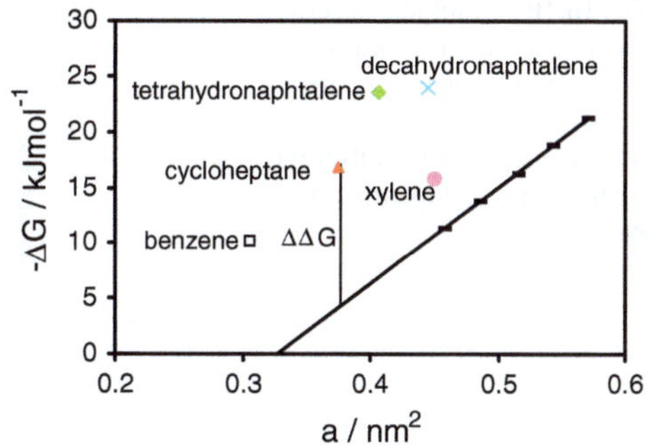

Fig. 16.9 Determination of the specific interaction parameter, I^{sp}, for polar probes on Al_2O_3 at 250 °C (adapted from Ref. [32])

larger than zero, which leads to increased net retention volumes as compared to n-alkanes. The adsorption of these molecules on the stationary phase is influenced, not only be dispersive interactions, but also by additional specific contributions. These specific contributions include dipole-dipole and acid-base interactions, the latter involving much higher energies than the former ones [35]. In fact, it is usually assumed that the specific contribution of the adsorption of polar probes are actually acid-base interaction only. In this way, Fowkes, by analogy with the dispersive work of adhesion, described the "extended Fowkes equation":

$$W_a^S = 2\sqrt{\gamma_L^S \gamma_S^S} \tag{16.27}$$

However, this expression can not predict accurately the magnitude of the non dispersive interactions, since it is wrong the assumption that the contribution to the work of adhesion of two polar compounds could be represented by the geometric mean value of their polar properties.

In order to quantify this specific contribution, several attempts were made. One is relate the specific component of the surface free energy to the parameter of specific interaction of polar solutes (I^{sp}). This parameter involves the surface properties in terms of potential and acid-base interactions and may be determined from the difference of free energy of adsorption, ($\Delta\Delta G$), between a polar solute and the real or hypothetical n-alkane with the same surface area, a, (Fig. 16.9):

$$I^{sp} = \frac{\Delta\,(\Delta G)}{Na} = \frac{\Delta G_{ads}^S}{Na} \tag{16.28}$$

Add to the surface area as parameter of comparison, the boiling point or the vapor pressure could also be used [29–31]. This treatment is essentially empirical, but it allows to compare the specific interaction between the surface and the solute molecules, based on a unified scale.

Donnet and coworkers [33, 34] obtained useless results with the above method when analyzing material with relative high dispersive component of the surface energy ($\gamma_S^D > 100\,\text{mJ/m}^2$), such as carbon nanofibers or graphite powders. For

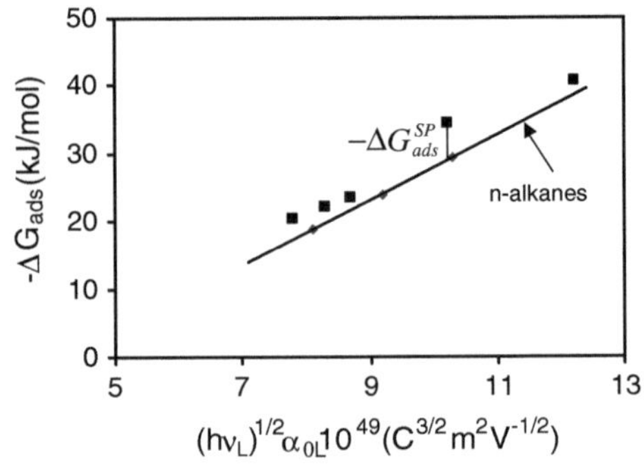

Fig. 16.10 Graphical description of the method followed to obtain the specific contribution of the adsorption free energy measured for different polar probes (adapted from Ref. [20])

these materials lower $(-\Delta G^0_{ads})$ values were found for the polar probes compared to reference n-alkanes. This problem can be encountered by plotting the free energy of adsorption as function of the molecular polarizability of the different polar adsorbates (Fig. 16.10):

$$-\Delta G^0_{ads} = \left(-\Delta G^D_0\right) + \left(-\Delta G^S_0\right)$$
$$= k_c\,(h\,\upsilon)^{1/2}\,\alpha_{0,s}\,(h\,\upsilon)^{1/2}\,\alpha_0 + \left(-\Delta G^S_0\right) \qquad (16.29)$$

where k_c, is the constant of the chromatographic process; h, the Planck constant (Js); υ, characteristic vibration frequency of the electron (s^{-1}); α_0, polarizability deformation (cm^2V^{-1}).

Since the free energy of ad Specific interactions $(-\Delta G^S_{ads})$ are determined from differences between $(-\Delta G^0_{ads})$ values of the polar probes and the reference line composed with data obtained from the elution of n-alkanes. In this way, $(-\Delta H^S_{ads})$ can be calculated from the variation of $(-\Delta G^S_{ads})$ versus $(1/T)$, as stated in Eq. (16.17). Therefore, the standard enthalpy of adsorption of polar probes is divided into two contributions, dispersive and specific:

$$\Delta H^0_{ads} = \Delta H^D_{ads} + \Delta H^S_{ads} \qquad (16.30)$$

The ability of the polar molecules to donate or accept electrons has been parameterized by means of the donor (DN) and acceptor (AN) number [35]. These parameters describe the basic and acidic nature, respectively. The DN values (kcal/mol) represent the enthalpy of formation for the adduct produced when the base in question reacts with the reference Lewis acid $SbCl_5$ in the 1,2-dichloromethane, as solvent. However, for the characterization of acids, no similar reference system was found. AN value (dimensionless) measuring the induced shift in the ^{31}P NMR spectra of the base Et_3PO_4 when this compound was dissolved in the acid under investigation. Riddle and Fowkes [36] corrected the AN scale to the enthalpy of reaction of Et_3PO_4 with $SbCl_5$. This new parameter, AN*, presents the same units as DN.

Thus, the measured $(-\Delta H^{S}_{ads})$ can be correlated to the acid and base indices AN* and DN of the solute probes [35], obtaining information about the surface acidity-basicity:

$$-\Delta H^{S}_{ads} = K_a \cdot DN + K_b \cdot AN^* \tag{16.31}$$

where K_a and K_b are indices reflecting the acidity (electron acceptor) and basicity (electron donor) of the solid surface. According to Eq. (16.31), a plot of versus (DN/AN*) should yield a straight line from which K_a can be obtained from the slope and K_b from the intercept. This equation is of empirical nature; other relationships have been also proposed in the literature [37].

16.4.5 Surface Heterogeneity

There is an important point that must be taken into account in the IGC determined parameters: for heterogeneous high energy surfaces, molecules will preferentially adsorb on the highest energy sites [38]. The distribution of energetic sites is usually called as surface heterogeneity. There exist two types of heterogeneity: structural and energetic. A typical example of a structural heterogeneity is a wide pore size distribution, where the geometrical effects determine the adsorption of the probe molecules. Energetic heterogeneity occurs with a wide distribution of various surface sites of different energetic levels. The energy heterogeneity can be described either by the adsorption energy distribution or the adsorption potential distribution.

For the determination of the adsorption energy distribution, $F(\varepsilon)$, from Eq. (16.32), and assumption on the shape of the local isotherms has to be made and usually complex numerical analysis is required [39, 40]:

$$\theta_t(p, T) = \int_{\varepsilon\,\min}^{\varepsilon\,\max} \theta_{1t}(\varepsilon, p, T) \cdot \mathrm{F}(\varepsilon) \cdot d\varepsilon \tag{16.32}$$

where ε_{min} and ε_{max} indicate the range in potential energy of adsorption, and θ_t refers to the local adsorption isotherm.

Concerning the adsorption potential distribution, it can be obtained easily form the adsorption isotherm. Furthermore, is less affected by experimental noise and produce reliable results. Once the adsorption isotherm is derived from IGC, the adsorption potential, *AP*, is calculated according to:

$$AP = RT \ln\left(\frac{p_s}{p}\right) \tag{16.33}$$

where p is the partial pressure, and p_s, the saturation pressure.

The distribution parameter, ϕ, represents the first derivation of the adsorbed amount of molecules, n, with the adsorption potential:

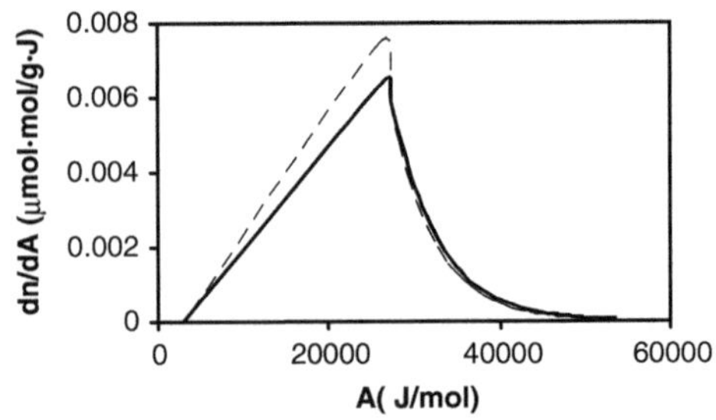

Fig. 16.11 Heterogeneity profiles of CNF (—) and CNF-oxi (--) at 250 °C for benzene (adapted from Ref. [41])

$$\phi = -\frac{\mathrm{d}n}{\mathrm{d}AP} \qquad (16.34)$$

Figure 16.11 shows, as example, the adsorption potential distribution of benzene over two carbon nanofibers, a parent and an oxidized one. It can be seen that there is a good coincidence of the profiles; therefore, the adsorbates interact with the same energy sites. It means that the treatment of the nanofiber does not create new adsorption sites, just varying slightly the adsorption capacities (area under the curve).

16.5 Applications and Comparison to Other Techniques

Adsorption techniques have been widely applied both to know the adsorption capacity of the adsorbents and to obtain thermodynamic parameter for increasing the knowledge of the surface of the material. Static methods are, probably, the most used for these purposes, since they are considered the most accurate. Thus, comparison of the adsorption parameters obtained by other techniques with static methods is a usual way to ensure the reliability of a technique. Thielmann and Baumgarten [42] investigated the adsorption properties of four aluminas with different microporosities by both IGC and a static method. Sorption measurements obtained form IGC gave similar results whereas the static experiments showed differences until 37 %. This difference can be explained by a different micropore structure of the aluminas. This is so because one of the hypotheses assumed for obtaining adsorption data from the IGC eluted peak is to consider instantaneous adsorption equilibrium. In the case of microporous materials, transport of the solutes through the porous structure could delay the equilibrium, becoming more difficult to satisfy this hypothesis. Thus, the application of the IGC to microporous materials has been discouraged in the literature for microporous materials [4]. With the same aim of comparison between IGC and static techniques, different ion-exchanged zeolites (microporous materials) have been studied by gas calorimetry coupled to a volumetric line and IGC [43]. *n*-Hexane was used as adsorbate and isotherms of adsorption as well as the isosteric heats of adsorption and enthalpies of adsorption obtained, respectively, by the earlier mentioned methods were determined. The comparison between the volumetric adsorption isotherms of n-hexane at 250 °C obtained both by IGC and calorimetry reveals that, at low pressures (when both techniques are applicable), the results obtained by both techniques are very similar. Moreover, the strength of adsorption is also quantify by the heat of adsorption (IGC) and the isosteric heat (calorimetry), obtaining deviations among them in the range of 4–20 %, depending on the material. Thus, at very low partial pressures, IGC could be a reliable technique even for microporous materials.

The wettability of powders is a valuable parameter in different fields such as the pharmacy [44], the ceramic [45], polymers [46] and nanomaterials [47]. Contact angle measurements are the most used in order to obtain information about the surface, however, IGC, by the dispersive component of the surface free energy could also offer a sensitive approach to surface characterization. In this way, comparison of the surface components using n-alkanes over theophylline and caffeine showed a good agreement [44]. It is important to mention that the method of calculating the γ_S^D value is applied to solid surfaces, flat at the molecular scales, presenting no chemical, structural or energetic heterogeneities. Thus, application of the method to other materials could give larger discordances. In this way, in a study of activated carbons characterization by IGC, Herry et al. [48] have obtained values of γ_S^D 10 times higher than the ones obtained previously by capillary wetting. This difference was attributed to the increase in the interaction potential in micropores. The discrepancy in the values obtained by the two techniques could be explained by assuming that the treatment of the column, prior to the IGC experiments, leads to carbon surfaces free from adsorbed species and other possible contaminants. Besides that, the contribution of high-energy sites will be significantly outweighed in IGC measurements under infinite dilution conditions. In fact, contact angle measurements are known to provide an average of the surface energy [49]. It is important to remark that in these situations, the ΔG_{CH_2} is more reliable, since it keeps a precise physical meaning [50].

Inverse gas chromatography parameters can also be applied in the field of catalysis. In this way, as example, parent NaX and CaA zeolites, as well as transition metal (Co^{2+}, Mn^{2+}, Fe^{3+})-exchanged zeolites, were evaluated for the catalytic oxidation of n-hexane. It was observed [51, 52], that although there was linear correlation between the acidity and the adsorption enthalpy of the n-hexane, there was no relationship between the acidity and the activity for n-hexane oxidation. However, if a reactivity parameter (such as T_{50}, temperature at which 50 % of conversion is attained) is plotted versus the adsorption heat, a so-called Volcano plot is obtained (Fig. 16.12), an optimum value of $(-\Delta H^0_{ads})$ being observed, higher and lower values yielding to worst catalytic performance.

IGC was also use to elucidate the nature of the interaction of both reactives and products with the surface of heterogeneous catalysts in order to obtain information for understanding the mechanism of the reaction. Xie et al. [31] proposed a mechanism for the partial oxidation of propylene to acrylic acid based on the adsorption parameters of reactive and products on styrene divinylbenzene copolymer (SDB) and Pd supported on SDB. Likewise, Díaz et al. [53] studied the performance of Fe-ZSM-5 catalysts for benzylation of benzene with benzyl chloride in terms of their chemical and adsorption properties.

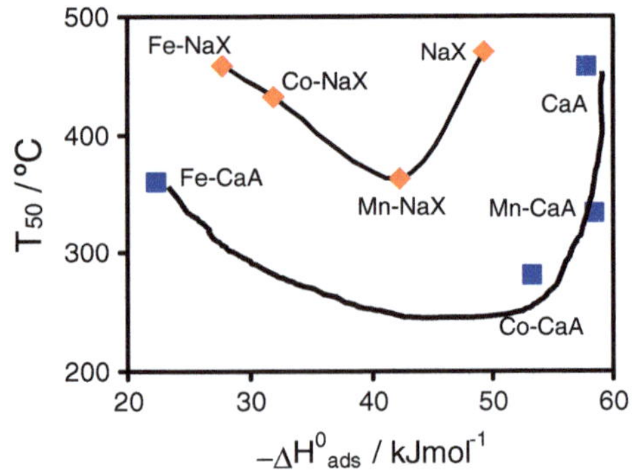

Fig. 16.12 Influence of the adsorption enthalpy of *n*-hexane of metal transition-exchanged zeolites on the T_{50} parameter (adapted from Ref. [52])

Finally, a mention to the reverse-flow (RF) GC technique, which is a variation of IGC where the probe molecule is injected at a middle point of the column containing the adsorbent as stationary phase, and the direction of carrier gas flow is reversed from time to time. This creates extra chromatographic peaks on the continuous signal. This technique allows to measured different physicochemical quantities, including rate constant of surface and gaseous reactions, and experimental isotherms [54, 55].

References

1. A.V. Kiselev, Adsorbents in gas chromatography, in *Advances in Chromatography*, ed. by J.C. Giddings, R.A. Keller (Marcel Dekker, New York, 1967)
2. A. Voelkel, Inverse gas-chromatography—characterization of polymers, fibers, modified silicas, and surfactants. Crit. Rev. Anal. Chem. **22**, 411–439 (1991). doi:10.1080/10408349108051641
3. V.R. Choudhary, L.K. Doraiswamy, Applications of gas chromatography in catalysis. Ind. Eng. Chem. Prod. Res. Develop. **10**, 218–237 (1971). doi:10.1021/i360039a002
4. F. Thielmann, Introduction into the characterisation of porous materials by inverse gas chromatography. J. Chromatogr. A **1037**, 115–123 (2004). doi:10.1016/j.chroma.2004.03.060
5. R.E. Hayes, S.T. Kolaczkowski, *Introduction to Catalytic Combustion* (Gordon and Breach Science Publisher, Amsterdam, 1997)
6. J. van Deemter, F.J. Zuiderweg, A. Klinkerberg, Longitudinal diffusion and resistance to mass transfer as causes of nonideality in chromatography. Chem. Eng. Sci. **5**, 271–289 (1956). doi:10.1016/0009-2509(56)80003-1
7. M. Montes-Morán, J.I. Paredes, A. Martínez-Alonso, J.M.D. Tascón, Adsorption of n-alkanes on plasma-oxidized high-strength carbon fibers. J. Colloid Interface Sci. **247**, 290–302 (2002). doi:10.1006/jcis.2001.8134
8. E. Glückauf, Adsorption isotherms from chromatographic measurements. Nature **156**, 748 (1945). doi:10.1038/156748c0
9. E. Glückauf, Theory of chromatography. Part II. Chromatograms of a single solute. J. Chem. Soc. 1302–1308 (1947). doi:10.1039/JR9470001302
10. S.J. Gregg, R. Stock, *Gas Chromatography* (Desty D.H, London, 1958)
11. S.J. Gregg, *The Surface Chemistry of Solids*, 2nd edn. (Chapman and Hall, London, 1961)
12. T. Paryjczak, *Gas Chromatography in Adsorption and Catalysis* (J. Wiley & Sons, New York, 1987)

13. E. Cremer, H. Huber, in *Gas Chromatogr: Instr Soc Amer Symp*, vol. 3, ed. by N. Brenner, et al. (Academic Press, New York, 1962), p. 169
14. A. Seidel-Morgenstern, Experimental determination of single solute and competitive adsorption isotherms. J. Chromatogr. A **1037**, 255–272 (2004). doi:10.1016/j.chroma.2003.11.108
15. J.R. Conder, S. McHale, M.A. Jones, Evaluation of methods of measuring gas-solid chromatographic retention on skewed peaks. Anal. Chem. **58**, 2663–2668 (1986). doi:10.1021/ac00126a019
16. B. Charmas, R. Leboda, Effect of surface heterogeneity on adsorption on solid surfaces: application of inverse gas chromatography in the studies of energetic heterogeneity of adsorbents. J. Chromatogr. A **886**, 133–152 (2000). doi:10.1016/S0021-9673(00)00432-5
17. J.H. De Boer, *The Dynamical Character of Adsorption* (Clarendon Press, Oxford, 1953)
18. C. Kemball, E.K. Rideal, The adsorption of vapours on mercury. I. Non-polar substances. Proc. R. Soc. A **187**, 53–73 (1946). doi:10.1098/rspa.1946.0065
19. M.A. Montes-Morán, A. Martínez-Alonso, J.M.D. Tascón, Effect of sizing on the surface properties of carbon fibres. J. Mater. Chem. **12**, 3843–3850 (2002). doi:10.1039/B202902B
20. E. Díaz, S. Ordóñez, A. Vega, Adsorption of volatile organic compounds onto carbon nanotubes, carbon nanofibers, and high-surface-area graphites. J. Colloid Interface Sci. **305**, 7–16 (2007). doi:10.1016/j.jcis.2006.09.036
21. S.Y. Bhide, S. Yashonath, Structure and dynamics of benzene in one-dimensional channels. J. Phys. Chem. B **104**, 11977–11986 (2000). doi:10.1021/jp002626h
22. F.M. Fowkes, Attractive forces at interface. Ind. Eng. Chem. **56**, 40–52 (1964). doi:10.1021/ie50660a008
23. F.M. Fowkes, Donor-acceptor interactions at interfaces. J. Adhesion **4**, 155–159 (1972). doi:10.1080/00218467208072219
24. F.M. Fowkes, M.A. Mostafa, Acid-base interactions in polymer adsorption. Ind. Eng. Chem. Prod. Res. Dev. **17**, 3–7 (1978). doi:10.1021/i360065a002
25. J. Schultz, L. Lavielle, C. Martin, The role of interface in carbon fiber-epoxy composites. J. Adhesion **23**, 45–60 (1987). doi:10.1080/00218468708080469
26. G.M. Dorris, D.G. Gray, Adsorption of n-alkanes at zero surface coverage on cellulose paper and wood fibers. J. Colloid Interface Sci. **77**, 353–362 (1980). doi:10.1016/0021-9797(80)90304-5
27. P.N. Jacob, J.C. Berg, Acid-base surface energy characterization of microcrystalline cellulose and two wood pulp fiber types using inverse gas chromatography. Langmuir **10**, 3086–3093 (1994). doi:10.1021/la00021a036
28. A. Pizzi, K.L. Mittal, *Handbook of Adhesive Technology* (Marcel Dekker, New York, 2003)
29. U. Panzer, H.P. Schreiber, On the evaluation of surface interactions by inverse gas chromatography. Macromolecules **25**, 3633–3637 (1992). doi:10.1021/ma00040a005
30. A. van Asten, N. van Veenendaal, S. Koster, Surface characterization of industrial fibers with inverse gas chromatography. J. Chromatogr. A **888**, 175–196 (2000). doi:10.1016/S0021-9673(00)00487-8
31. J. Xie, Q. Zhang, K.T. Chiang, An IGC study of Pd/SDB catalysts for partial oxidation of propylene to acrylic acid. J. Catal. **191**, 86–92 (2000). doi:10.1006/jcat.1999.2796
32. E. Díaz, S. Ordóñez, A. Vega, J. Coca, Adsorption properties of a Pd/γ-Al_2O_3 catalyst using inverse gas chromatography. Micropor. Mesopor. Mater. **70**, 109–118 (2004). doi:10.1016/j.micromeso.2004.03.005
33. S. Dong, M. Breadle, J.B. Donnet, Study of solid-surface polarity by inverse gas-chromatography at infinite dilution. Chromatographia **28**, 469–472 (1989). doi:10.1007/BF02261062
34. J.B. Donnet, S.J. Park, H. Balard, Evaluation of specific interactions of solid-surfaces by inverse gas-chromatography—a new approach based on polarizability of the probes. Chromatographia **31**, 434–440 (1991). doi:10.1007/BF02262385
35. V. Gutmann, *The Donor-Acceptor Approach to Molecular Interactions* (Plenum Press, New Cork, 1979)

36. F.L. Riddle, F.M. Fowkes, Spectral shifts in acid-base chemistry. 1. van der Waals contributions to acceptor numbers. J. Am. Chem. Soc. **112**, 3258–3264 (1990). doi:10.1021/ja00165a001
37. T. Hamieh, M. Nardin, M. Rageui-Lescourt, H. Haidara, J. Schultz, Study of acid-base interactions between some metallic oxides and model organic molecules. Colloids Surf. A **125**, 155–161 (1997). doi:10.1016/S0927-7757(96)03855-1
38. H. Ishida, *Characterization of Composite Materials* (Butterworth-Heinemann, London, 1994)
39. M. Pyda, G. Guiochon, Surface properties of silica-based adsorbents measured by inverse gas-solid chromatography at finite concentration. Langmuir **13**, 1020–1025 (1997). doi:10.1021/la950541f
40. H. Balard, A. Saada, E. Papirer, B. Siffert, Energetic surface heterogeneity of illites and kaolinites. Langmuir **13**, 1256–1259 (1997). doi:10.1021/la9515276
41. M.R. Cuervo, E. Asedegbega-Nieto, E. Díaz, A. Vega, S. Ordóñez, E. Castillejos-López, I. Rodríguez-Ramos, Effect of carbon nanofiber functionalization on the adsorption properties of volatile organic compounds. J. Chromatogr. A **1188**, 264–273 (2008). doi:10.1016/j.chroma.2008.02.061
42. F. Thielmann, E. Baumgarten, Characterization of microporous aluminas by inverse gas chromatography. J. Colloid Interface Sci. **229**, 418–422 (2000). doi:10.1006/jcis.2000.6958
43. E. Díaz, S. Ordóñez, A. Auroux, Comparative study on the gas-phase adsorption of hexane over zeolites by calorimetry and inverse gas chromatography. J. Chromatogr. A **1095**, 131–137 (2005). doi:10.1016/j.chroma.2005.07.117
44. J.W. Dove, G. Buckton, C. Doherty, A comparison of two contact angle measurement methods and inverse gas chromatography to assess the surface energies of theophylline and caffeine. Int. J. Pharm. **138**, 199–206 (1996). doi:10.1016/0378-5173(96)04535-8
45. C.-W. Won, B. Siffert, Preparation by sol-gel method of SiO_2 and mullite ($3Al_2O_3$, $2SiO_2$) powders and study of their surface characteristics by inverse gas chromatography and zetametry. Colloids Surf. A **131**, 161–172 (1998). doi:10.1016/S0927-7757(97)00149-0
46. R.A. Bailey, K.C. Persaud, Application of inverse gas chromatography to characterisation of a polypyrrole surface. Anal. Chim. Acta **363**, 147–156 (1998). doi:10.1016/S0003-2670(98)00084-1
47. K. Batko, A. Voelkel, Inverse gas chromatography as a tool for investigation of nanomaterials. J. Colloid Interface Sci. **315**, 768–771 (2007). doi:10.1016/j.jcis.2007.07.028
48. C. Herry, M. Baudu, D. Raveau, Estimation of the influence of structural elements of activated carbons on the energetic components of adsorption. Carbon **39**, 1879–1889 (2001). doi:10.1016/S0008-6223(00)00310-9
49. M.A. Montes-Morán, J.I. Paredes, A. Martínez-Alonso, J.M.D. Tascón, Surface characterization of PPTA fibers using inverse gas chromatography. Macromolecules **35**, 5085–5096 (2002). doi:10.1021/ma020069m
50. E. Papirer, E. Brendle, F. Ozil, H. Balard, Comparison of the surface properties of graphite, carbon black and fullerene samples, measured by inverse gas chromatography. Carbon **37**, 1265–1274 (1999). doi:10.1016/S0008-6223(98)00323-6
51. E. Díaz, S. Ordóñez, A. Vega, J. Coca, Characterization of Co, Fe and Mn-exchanged zeolites by inverse gas chromatography. J. Chromatogr. A **1049**, 161–169 (2004). doi:10.1016/j.chroma.2004.07.065
52. E. Díaz, S. Ordóñez, A. Vega, J. Coca, Catalytic combustion of hexane over transition metal modified zeolites NaX and CaA. Appl. Catal. B **56**, 313–322 (2005). doi:10.1016/j.apcatb.2004.09.016
53. E. Díaz, S. Ordóñez, A. Vega, A. Auroux, J. Coca, Benzylation of benzene over Fe-modified ZSM-5 zeolites: correlation between activity and adsorption properties. Appl. Catal. A **295**, 106–115 (2005). doi:10.1016/j.apcata.2005.07.059
54. N.A. Katsanos, N. Rakintzis, F. Roubani-Kalantzopoulou, E. Arvanitopoulou, A. Kalantzopoulos, Measurement of adsorption energies on heterogeneous surfaces by inverse gas chromatography. J. Chromatogr. A **845**, 103–111 (1999). doi:10.1016/S0021-9673(99)00262-9
55. N.A. Katsanos, Physicochemical measurements by the reversed-flow version of inverse gas chromatography. J. Chromatogr. A **969**, 3–8 (2002). doi:10.1016/S0021-9673(02)00992-5

Chapter 17
Liquid-Solid Adsorption Properties: Measurement of the *Effective* Surface Acidity of Solid Catalysts

Paolo Carniti and Antonella Gervasini

Abstract Attention should be devoted to the measurements of the adsorption properties of catalytic surfaces when they have to work in liquid-solid heterogeneous conditions. The mutual characteristics of the surface and the liquid affect the reagent interactions with the surface sites which could be engaged with the liquid interaction and then not-available for the reagent coordination. This leads to observe *effective* adsorption properties that could be different from the *intrinsic* properties of the surface. The possibility to quantitatively determine the *effective* acid properties of catalytic surfaces by base adsorption is here showed. The adsorption can proceed in any type of liquid of various characteristics (apolar, polar, aprotic, protic) with dynamic (pulse liquid chromatographic method) or equilibrium (liquid recirculation chromatographic method) methods. The measurements of *effective* acidity allows finding more sound relations with the catalytic activity for a better comprehension of the catalyst work and for a more correct determination of the turnover numbers in liquid-solid catalysis.

17.1 Introduction

Heterogeneous catalysts have different ability in the chemical transformation of reagents to form the desired reaction products due to various functionalities present at their surface. When reactions run in liquid by a liquid-solid heterogeneous catalysis, not only the characteristics of the surface (hydrophobic or hydrophilic surface) but the liquid properties (polarity, proticity, solvating ability) govern the process reactivity.

P. Carniti (✉) · A. Gervasini
Dipartimento di Chimica, Università degli Studi di Milano,
Via Camillo Golgi, 19, 20133 Milano, Italy
e-mail: paolo.carniti@unimi.it

A. Gervasini
e-mail: antonella.gervasini@unimi.it

A. Auroux (ed.), *Calorimetry and Thermal Methods in Catalysis*,
Springer Series in Materials Science 154, DOI: 10.1007/978-3-642-11954-5_17,

The liquid-solid interactions of physical (essentially van der Waals) and chemical (essentially acid-base) type can quench or put out some surface sites, so modifying their reactivity towards reagents. Therefore, in order to understand and predict the catalytic ability of given surfaces in different liquids, it is very important to measure the *effective* surface properties, besides the knowledge of the *intrinsic* properties.

Acid catalysts are among the most numerous and best studied solids finding use in oil refining processes, base industrial chemistry, and, at present, biomass conversion, and biorefinery processes [1, 2]. Acid surfaces possess either Brönsted or Lewis acid sites, or both Brönsted and Lewis sites simultaneously, with the possibility to operate different chemical transformations of given reagents. Depending on the properties of the liquid solvent, some acid sites can be silent; for example the Lewis acid sites of most surfaces are not lively any longer in water due to the solvating ability and weak basic character of water. The acidity measurement of the solid surface performed in the gas-solid phase (i.e., by using gaseous/vapor probes) could not be significant when the catalyst is then employed in liquid-solid phase. Analogous considerations apply if acidity characterization is performed in the liquid-solid phase with a solvent having a proticity and polarity very different from those of the reaction solvent. In this case, the influence of the solvent on the surface acidity cannot be correctly taken into account. From this viewpoint, the acidity of sulphonic resins in water [3, 4] and the influence of different solvents on the acidity of supported sulphonic acid catalysts [5] as well as of niobium containing catalysts [6] have recently been studied by acid-base titration calorimetry.

On a given acid catalytic surface, the amount and strength of the acid sites with lively characteristics in defined liquids (*effectivity* acidity) is of primary importance for its activity and they merit being determined. A technique widely used in the study of acidity of solids in aqueous media is the measurement of the isoelectric point (IEP) which expresses the overall ability of an acidic surface to attract protons [7], without providing any quantitative information on the surface sites.

Recently, some techniques have been presented in the literature making use of volumetric titrations of surface sites in liquids, of different polar and protic characteristics, to determine the amount of acid sites and relevant acid strength (*effective* acidity). Two different methods will be here discussed: the first one is based on a pulse liquid chromatographic method (a dynamic method) [8] and the second one on a liquid recirculation chromatographic method (a *quasi*-static method) [9]. When surface acidity studies are concerned, the measurements may be performed in apolar, aprotic liquid (like cyclohexane), for the determination of the *intrinsic* acidity, or in several other liquids with polar/protic characteristics, for the determination of the *effective* acidity. Basic probes of different basicity (following the pK_a scale) may be used (e.g., 2-phenylethylamine, PEA, aniline, AN, pyridine, PY, etc.). Titration temperatures may be varied from room temperature (r.t.) up to the normal boiling point of the liquid used in order to calculate, from the collected isotherms of adsorption, the isosteric heats of adsorption which can be related to the acid strength of the surface sites [10, 11].

17.2 Pulse Liquid Chromatographic Method

This method enables the characterization of the acidity, or more general of surface properties, of solids in liquids of different properties making use of an HPLC apparatus. The pulse gas chromatographic technique has been widely employed in the characterization of solid catalysts, particularly for the estimation of the dispersion of metal deposited on supports [12, 13] while little has been reported on the use of pulse liquid chromatographic technique for surface property determinations.

Experiments can be performed in a liquid chromatograph (HPLC), complete with pump, injector, and detector. The catalyst sample, in a sample holder surrounded by a thermostatted jacket, is placed in place of the chromatographic column and the chosen liquid is continuously eluted on the solid. In the adsorption test, pulses of a given amine solution of known concentration are sent at fixed time intervals onto the catalyst sample at a constant liquid flow rate. The non-adsorbed amounts of amine after each pulse are detected and measured up to sample saturation. The amount of amine adsorbed (mequiv/g) on the sample after the *i-th* pulse is calculated by the following equation:

$$(\text{amine adsorbed})_{\mathrm{i}} = (\mathrm{VC/w})(\bar{\mathrm{A}} - \mathrm{A_i})/\bar{\mathrm{A}} \qquad (17.1)$$

where V (ml) corresponds to the volume of a single pulse, C (M) is the concentration of the amine solution, w (g) is the mass of the catalyst sample, $\bar{\mathrm{A}}$ is the mean chromatographic area at saturation and A_i is the area corresponding to the *i-th* pulse. When the adsorption stoichiometry is known, the total number of acidic sites is related to the total amount of amine adsorbed on the sample. A typical example of a chromatogram for an amine adsorption onto a catalyst sample is depicted in Fig. 17.1.

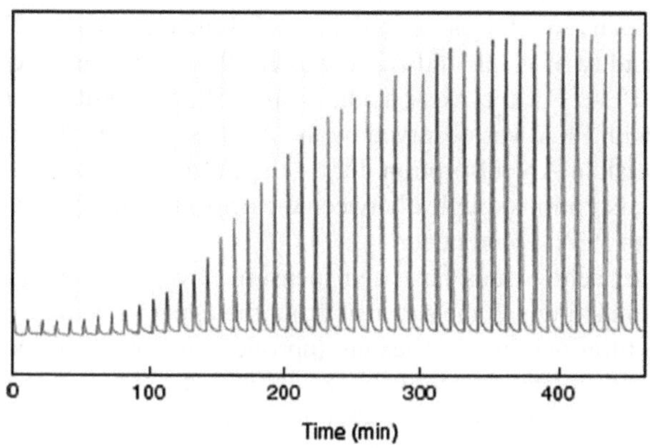

Fig. 17.1 Example of pulse chromatogram for the evaluation of the surface acidity of the IR-200 Amberlite in water employing PEA (from Ref. [8])

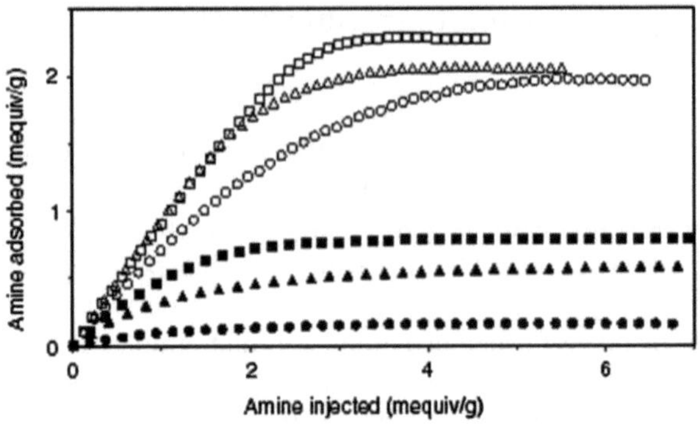

Fig. 17.2 Evaluation of the surface acidity for some Amberlites measured by the pulse chromatographic technique in water (from Ref. [8]): A120 (*squares*), A200 (*triangles*) and A50 (*circles*); probes: PEA (*open symbols*) and AN (*filled symbols*)

The amount of amine adsorbed up to a given pulse was obtained by the summation of the single adsorbed quantities via the equation above reported. Plot of the total amount of amine adsorbed versus the total amine injected provides clear evidence of the progressive coverage of the sample surface. At saturation, the curve attained a plateau corresponding to the total amount of amine adsorbed.

Figure 17.2 reports, as an example, the titration in water of some Amberlite samples by two different basic probes, PEA (strong base, $pK_a = 9.84$) and AN (weak base, $pK_a = 4.58$). The Amberlite IR-120 (A120) and IR-200 (A200) resins are strongly acidic gel resins (styrene-divinylbenzene copolymers) with SO_3^- functional groups, while Amberlite IRC-50 (A50) is a weakly acidic porous resin (metacrylic acid-divinylbenzene copolymer) with COO^- functional groups.

The A120 acidity measured by PEA is around 2.3–2.4 mequiv/g, and that of the A200 and A50 resins are of the same order (2.0–2.1 mequiv/g). By AN as base probe, the measured acidity of A120 and A200 was *ca.* three- or four-times less than that measured with PEA. For the weak acidic resin A50, an acidity 13-times less than that measured with PEA was observed by the AN base probe. These results are due to partially reversible AN adsorption during the pulse flow titration; the formation of a less stable adsorption complex between the acid sites and AN (in comparison to the situation with PEA) could occur.

The acidity of acid oxides has been measured by this technique, in this case the possibility to distinguish the *intrinsic* from the *effective* acidity is presented, by performing the titrations in cyclohexane (aprotic and very weak polar) and water (highly protic and polar), respectively [8]. A mixed oxide, silica-zirconia (SiZr) and the same base probes (PEA and AN) used for the titration of Amberlites have been chosen for this example. The acidity of SiZr determined by pulses of PEA in cyclohexane is 0.49 mequiv/g, i.e. ca. four-times higher than that in water (Fig. 17.3). This result can be compared with that obtained by titration with Hammett indicators [14] in the same solvent employing n-butylamine ($pK_a = 10.77$) as base probe.

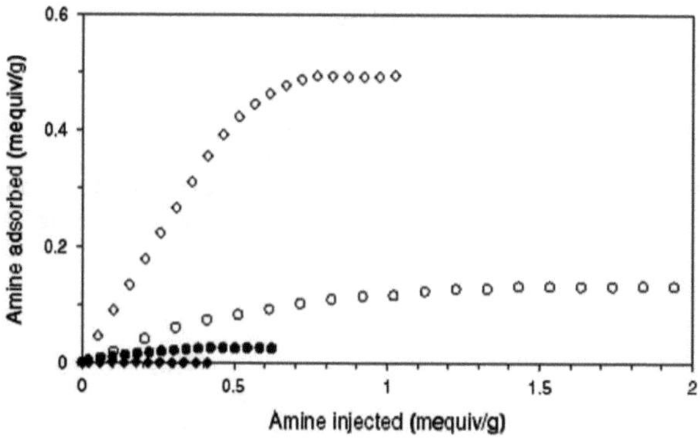

Fig. 17.3 Evaluation of the surface acidity in cyclohexane (*diamonds*) and in water (*circles*) for SiZr measured by the pulse chromatographic technique; probes: PEA (*open symbols*) and AN (*filled symbols*) (from Ref. [8])

In particular, the amount of acid sites titrated with 4-phenylazo-1-naphthylamine indicator was 0.48 mequiv/g, comparable to the value of 0.49 mequiv/g obtained in the chromatographic titration measurement. The acid sites titrated with bromothymol blue amounted to 0.15 mequiv/g. This value is very similar to that obtained in water with PEA (0.13 mequiv/g by pulse). PEA in water should only be able of titrating the strong sites of SiZr, due to the water presence which interacts with part of the surface acid sites. In contrast with the behavior observed with PEA, very little evidence for adsorption was observed when pulses of AN in cyclohexane were sent onto SiZr. This unexpected behavior with AN suggests that when the acidity of a solid in a liquid is measured, attention should be taken not only of the solvent–surface acid site interactions but also of the solvent–base interactions. Likely, the low basicity of the AN probe and the high AN affinity for the solvent are the cause of the weak adsorption on AN on the acid sites and of its desorption during interval between the successive pulses.

17.3 Liquid Recirculation Chromatographic Method

This method, which also makes use of an HPLC apparatus, enables the characterization of the acidity, or more general of surface properties, of solids in liquids of different properties by collecting solid-liquid adsorption isotherms [9, 15].

Concerning the acid measurements of solid samples, the experiments of acid-base titrations by base probes can be performed in various liquids, as above reported. In this case, the HPLC line, comprising the pump and detector coupled to a personal computer for the collection, storage, and processing of the data, works according to a recirculating method (Fig. 17.4).

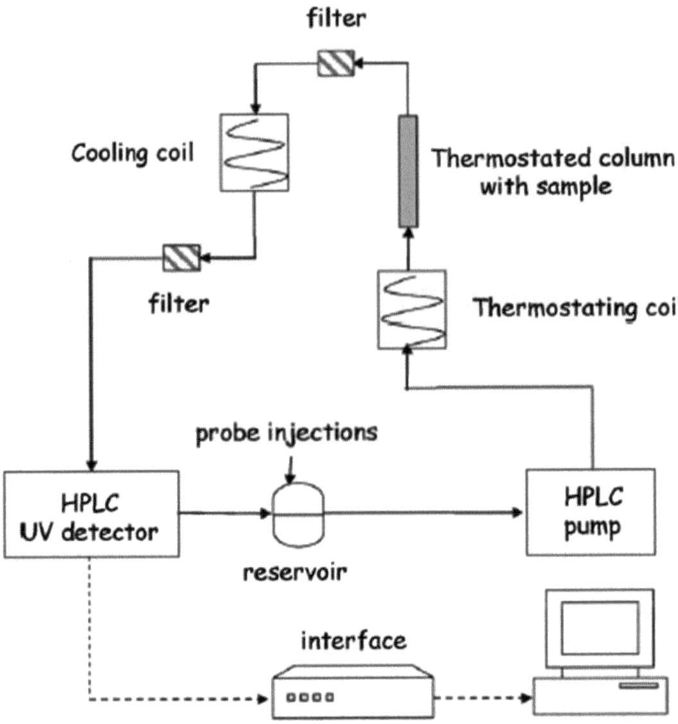

Fig. 17.4 Scheme of the apparatus used for the acidity determination in liquid by recirculation method (from Ref [9])

The sample holder, containing the sample and inserted in a glass jacket connected with a thermostat to maintain a constant temperature of adsorption, is mounted in place of the chromatographic column, as in the pulse method. During the titration experiment, successive dosed amounts of base probe in the chosen liquid solution are injected into a reservoir inserted into the line in which the solution continuously circulates. At the beginning of the experiment, the reservoir contains pure solvent. For each injection, the base solution recirculates onto the sample until adsorption equilibrium is achieved, revealed by the attainment of the detector signal stability. Figure 17.5 shows a typical example of raw data obtained from a titration experiment, consisting on a series of increasing steps, each one representing the attainment of the adsorption equilibrium. When the stoichiometry of the base adsorption on the acid site is known, the amount of acid sites per sample mass ($\mathrm{mequiv} \cdot \mathrm{g}^{-1}$) or sample area ($\mathrm{mequiv} \cdot \mathrm{m}^{-2}$) titrated at each equilibrium base concentration can be evaluated, obtaining the relevant adsorption isotherm on the sample.

The *intrinsic* (measured in cyclohexane, apolar and aprotic solvent) and *effective* (measured in water or other liquids) acidity of series of mixed silica-niobia oxides have been determined by this method with the principal aim of disclosing relationships between the surface acid properties and catalytic activity [9]. Dispersed niobia

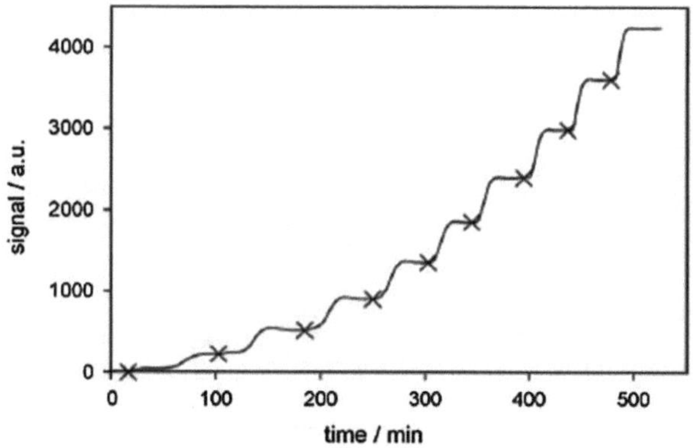

Fig. 17.5 Example of raw data obtained from the acid–base titration in liquid by recirculation method. The points representing the base probe injection are indicated by a cross (Ref. [9])

centers in silica (mixed $SiO_2 - Nb_2O_5$ oxides) constitute a good acid catalyst system used in various reactions [15–18]. In particular, It has been proved to be suitable for the dehydration of monosaccharides (like fructose) to obtain important chemical intermediates (like 5-hydroxymethyl-2-furaldehyde, HMF). The development of such reactive processes in agreement with a green chemistry approach imposes the use of friendly solvents, as water. Like Nb_2O_5[19], $SiO_2 - Nb_2O_5$ catalysts are able to maintain good activity in polar and protic liquids, and then to catalyze the conversion of fructose to HMF in water. They also show better stability than Nb_2O_5 in the reaction due to the dilution of the niobia centers into the silica matrix [9]. The results of the acidity measurements of $SiO_2 - Nb_2O_5$ at low (Si-Nb/15org) and high (Si-Nb/45org) niobia content in water by the recirculation chromatographic method are shown in Fig. 17.6. As expected, both the Nb-samples have higher *effective* acidity than the silica support and as higher the amount of niobia as more acid is the sample. Comparing series of $SiO_2 - Nb_2O_5$ samples with different niobia concentration, good relationship between the surface *effective* acidity of the samples and their activity in the fructose dehydration reaction has been found [9]. This result suggest some practical importance of the determination of *effective* surface properties of solids to understand and predict their activity in given condition, that is, in the presence of liquids which may interact with part of the surface sites.

Finally, by the liquid recirculation chromatographic method it is possible to determine the solid-liquid adsorption isotherms at different temperatures. Information about the energetic of the adsorption process can be obtained through the calculation of the isosteric heats of adsorption, once converting the isotherms into the corresponding isosteres for given surface coverages of the adsorbed species. Example of this procedure is given in Fig. 17.7 which shows the adsorption isotherms and relevant isosteres of PEA adsorption in cyclohexane on a typical cracking catalysis, a $SiO_2 - Al_2O_3$ sample (14.7 wt. % of Al_2O_3) with highly acidity property. The amount of acid sites titrated decreases with increasing temperature (from 17 to 57 °C) and

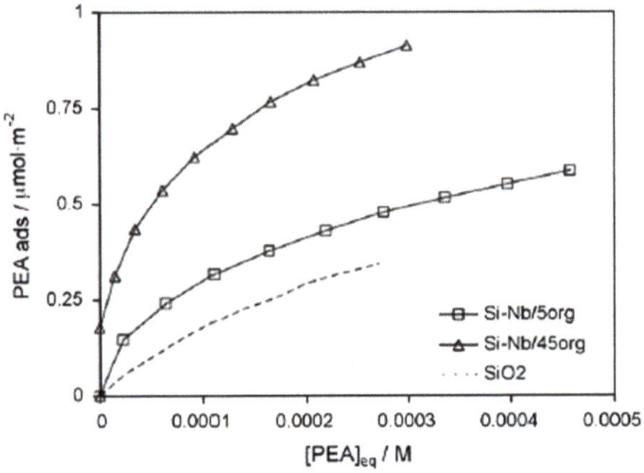

Fig. 17.6 Example of isotherms of adsorption of PEA on two silica-niobia samples at low (Si-Nb/5org) and high (Si-Nb/45org) niobia content collected in water at 17 °C by the recirculation method (from Ref. [9])

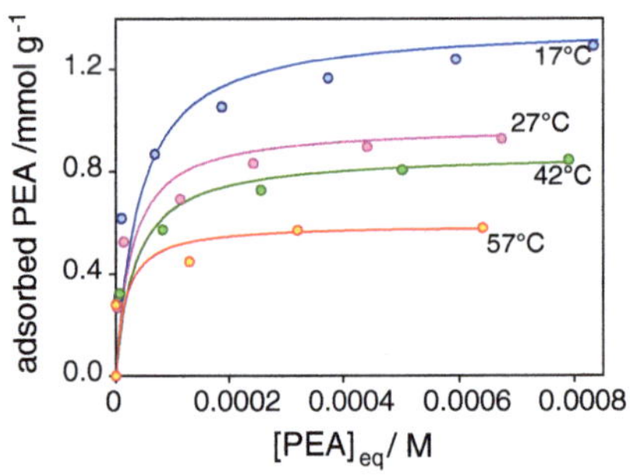

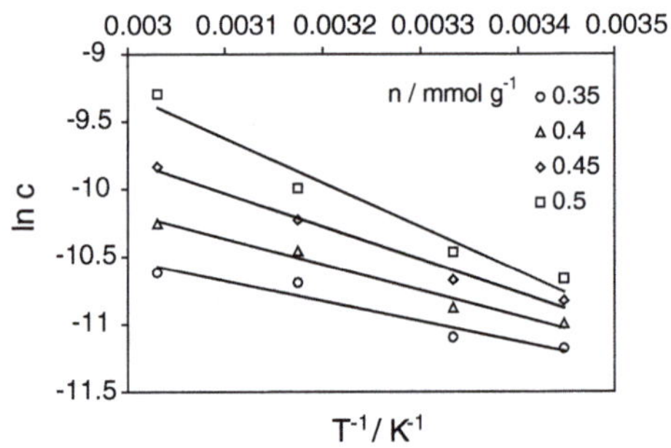

Fig. 17.7 Example of isotherms of adsorption of PEA on a SiO_2-Al_2O_3 sample collected in cyclohexane at different adsorption temperatures (left) with corresponding isosteres of adsorption at given surface PEA coverage (in $\mathrm{mmol}_{PEA} \cdot g^{-1}$) for the computation of the isosteric enthalpy of adsorption (*right*)

the isosteric adsorption enthalpy values are around 20 kJ mol^{-1} for PEA coverage from 0.35 to 0.50 meq g^{-1}.

17.4 Concluding Remarks

The adsorption properties of solids with their various surface functionalities are governed in liquids by both the hydrophobic or hydrophilic properties of the surfaces and the properties of liquids (polarity, proticity, solvating ability). Experiments based on pulse liquid or recirculation chromatographic methods, here illustrated, allow easily determining the adsorption properties of solids in various liquids which are similar to or the same as those in which the solids operate. The knowledge of the *effective* surface properties is useful to predict the solid activity in practical conditions of work.

References

1. T. Okuhara, Chem. Rev. **102**, 3641 (2002)
2. A. Corma, H. Garcia, Chem. Rev. **103**, 4307 (2003)
3. M. Hart, G. Fuller, D.R. Brown, J.A. Dale, S. Plant, J. Mol. Catal. A **182–183**, 439 (2002)
4. S. Koujout, B.M. Kierman, D.R. Brown, H.G.M. Edwards, J.A. Dale, S. Plant, Catal. Lett. **85**, 33 (2003)
5. S. Koujout, D.R. Brown, Catal. Lett. **98**, 195 (2004)
6. P. Carniti, A. Gervasini, S. Biella, A. Auroux, Chem. Mater. **17**, 6128 (2005)
7. C. Sun, J.C. Berg, Adv. Colloid Interface Sci. **105**, 151 (2003)
8. P. Carniti, A. Gervasini, S. Biella, Ads. Sci. Technol. **23**, 739 (2005)
9. P. Carniti, A. Gervasini, M. Marzo, Catal. Today **152**, 42 (2010)
10. J.A. Dunne, R. Mariwala, M. Rao, S. Sincar, R.J. Gorte, A.L. Myers, Langmuir **12**, 5888 (1996)
11. B. Dragoi, A. Gervasini, E. Dumitriu, A. Auroux, Thermochim. Acta **420**, 127 (2004)
12. T. Paryjczak, *Gas Chromatography in Adsorption and Catalysis*. (Ellis Horwood, Chichester, 1986), p 175
13. A. Gervasini, C. Flego, Appl. Catal. **72**, 153 (1991)
14. H.A. Benesi, B.H.C. Winquist, Adv. Catal. **27**, 97 (1978)
15. P. Carniti, A. Gervasini, M. Marzo, J. Phys. Chem. C **112**, 14064 (2008)
16. I.E. Wachs, J.M. Jehng, G. Deo, H. Hu, N. Arora, Catal. Today **28**, 19 (1996)
17. H. Yoshida, T. Tanaka, T. Yoshida, T. Funabiki, S. Yoshida, Catal. Today **28**, 79 (1996)
18. T. Tanaka, H. Nojima, H. Yoshida, H. Nakagawa, T. Funabiki, S. Yoshida, Catal. Today **16**, 297 (1993)
19. P. Carniti, A. Gervasini, S. Biella, A. Auroux, Catal. Today **118**, 373 (2006)

Index

A. Auroux (ed.), *Calorimetry and Thermal Methods in Catalysis*,
Springer Series in Materials Science 154, DOI: 10.1007/978-3-642-11954-5,

MIX
Papier aus verantwortungsvollen Quellen
Paper from responsible sources
FSC® C105338

If you have any concerns about our products,
you can contact us on
ProductSafety@springernature.com

In case Publisher is established outside the EU,
the EU authorized representative is:
Springer Nature Customer Service Center GmbH
Europaplatz 3, 69115 Heidelberg, Germany

Printed by Libri Plureos GmbH
in Hamburg, Germany